BIOCHEMISTRY
A CASE-ORIENTED APPROACH

REX MONTGOMERY, Ph.D., D.Sc.

ROBERT L. DRYER, Ph.D.

THOMAS W. CONWAY, Ph.D.

ARTHUR A. SPECTOR, M.D.

Department of Biochemistry
The University of Iowa
College of Medicine
Iowa City, Iowa

BIOCHEMISTRY

A CASE-ORIENTED APPROACH

SECOND EDITION

WITH 267 ILLUSTRATIONS

THE C. V. MOSBY COMPANY

Saint Louis 1977

SECOND EDITION

Copyright © 1977 by The C. V. Mosby Company

Printed in the United States of America

The C. V. Mosby Company
11830 Westline Industrial Drive, St. Louis, Missouri 63141

Library of Congress Cataloging in Publication Data

Main entry under title:

Biochemistry.

    Includes bibliographies and index.
    1. Biological chemistry. 2. Physiology, Patholog-
ical. I. Montgomery, Rex. [DNLM: 1. Biochemistry.
QU4 B6155]
QP514.2.B56 1977   574.1′92   76-41415
ISBN 0-8016-3469-5

TS/VH/VH 9 8 7 6 5 4 3 2

PREFACE

This book differs significantly from other biochemistry texts in that it uses the problem–solving case approach, a familiar method for teaching clinical concepts, but one that is less commonly used to teach the basic health sciences. The cases have been carefully selected to illustrate how biochemistry is used to help solve health-related problems. The text was written to serve a dual function: first, to make students aware of certain basic biochemical principles and, second, to provide opportunities for them to use these principles in analyzing commonly occurring health-related problems. We believe that the second goal is the ultimate objective of biochemical education for the health science student.

At first glance the chapters of this book are much like those of other biochemistry texts. Indeed, the first part of each chapter contains material appropriate for a one-semester course in general biochemistry. This "language" statement sets forth the biochemical principles needed to understand the chemical and molecular aspects of the health science problems. Although this section of the book is self-standing, it is limited in size and is not meant to be a detailed exposition of the entire panorama of modern biochemistry. The second part of each chapter applies these facts and concepts to the stated solutions of health-related problems. While the purpose of the case presentations is to illustrate the application of *biochemistry* to health problems, often students are curious to know more about the physiologic or pathologic aspects of the cases. This curiosity can be satisfied by reading the references given after each case. Generally, students realize that the purpose of the text is to teach biochemistry, to demonstrate *why* biochemistry is important in the health sciences, and to show how biochemical principles are involved in day-to-day professional practice. Experience has shown us that students grasp concepts more clearly when they see a relation between those concepts and their own professional goals. Learning also seems to be more enjoyable when the applications are made evident.

Selected case descriptions and biochemical questions about them are provided with and without written analysis. Those without analysis as well as a series of shorter additional questions at the end of each chapter are meant to be solved by the students. Many of the questions can be answered with a knowledge of the basic biochemistry presented in the first portion of the chapter. Sufficient references are given to solve the questions not answered in the main body of the text. These questions are included because it is imperative that students become accustomed to the method and value of bibliographic research. The knowledge gained by study of the basic principles of the first part of the chapter can be extended by independent study of other source materials.

The format of this book allows considerable flexibility. For some students the language statements plus the solved case problems may be sufficient. Others may wish to use the unsolved case problems or the questions for additional study.

The last chapter, which has been retitled Comprehensive Case Analysis, represents more of a challenge than the earlier ones and requires that the student have a total grasp of the preceding chapters. This chapter also requires reference to the original literature to answer certain questions.

As in the first edition we have selected nutrition as the first topic in order to introduce the case-oriented method of teaching from the very beginning, since many students have been exposed to this subject in their everyday living and in previous schooling. Knowledge of the properties of proteins (Chapter 2) and of enzymes (Chapter 3) is also required as background material for later chapters. Fluid and electrolyte control, which is commonly a problem in many diseases, is introduced in Chapter 4. The early introduction to these concepts allows the widest possible selection of case problems in later chapters. The remaining chapters may be studied in various orders, but we have found the sequence presented here to be satisfactory. Each chapter has been kept reasonably self-contained.

In our one-semester courses Chapters 1 through 13 are covered in a 14-week period, for the most part, one chapter per week. The chapters that often require a slightly longer time for completion are Acid-base, Fluid, and Electrolyte Control (Chapter 4), Carbohydrate Metabolism (Chapter 7), and Hormonal Regulation of Metabolism (Chapter 13). Weekly contact involves 4 lecture hours for the class as a whole and $2^1/_2$ hours of discussion in small groups. All 4 lecture hours are usually devoted to the basic biochemistry section of the text. In some instances, however, one of these lectures is used to present a topic taken from the clinical examples section; for example, phenylketonuria in Chapter 6, diabetes mellitus in Chapter 7, and hyperlipidemia in Chapter 10. One and one-half hour of the weekly small–group discussion sections is devoted to the clinical examples segment of the chapter. This session begins with a general discussion of the clinical cases that have been worked out in the text, in an attempt to correlate this material with the basic biochemistry contained in the chapter. Following this, students are asked to present one or more of the cases and additional problems that are not worked out in the text, either orally or in writing. The final hour of weekly discussion is a review session, devoted mostly to the basic biochemistry. Chapter 14 is covered during a final week of independent study, with daily 2–hour tutoring sessions scheduled on a voluntary attendance basis.

Examinations are based entirely on the analysis of cases. Questions include short answers, multiple choice, and brief calculations, but in all instances they are related to problems, real or not, designed to gauge the students' capacity to deal with applications of what has been learned.

The second edition benefits from continued refinement of this teaching program for students of medicine, which is now in its seventh year at the University of Iowa. More recently the program was extended to dentistry and physician's assistant students. Elsewhere, the book has been used successfully in colleges of pharmacy, nursing, allied health, osteopathy, and veterinary medicine.

The text of this edition has been revised and extended. Perhaps the most notable change is the significant increase in the number and variety of case problems. A larger number of cases have been worked out. Some of the cases used in the first edition have been retained with appropriate revisions and the addition of biochemical questions. In the basic biochemistry sections more extensive discussions are presented of cell membranes, enzyme catalysis, the theories of oxidative phosphorylation, renal regulation of electrolyte balance, glycoproteins, amino acid transport, and DNA synthesis. Emphasis on metabolic control and regulation has been strengthened throughout the text. New information has been incorporated concerning hormones and prostaglandins. The section on lipoproteins has been revised in accordance with the new nomenclature.

It is a pleasure to acknowledge advice and criticism from many colleagues and friends. To them we express our deep appreciation, especially to Drs. G. N. Bedell, R. L. Blakley, A. R. Boutros, J. D. Brown, G. F. DiBona, J. L. Filer, S. J. Fomon, E. M. Gal, N. Halmi, R. E. Hodges, H. P. C. Hogenkamp, K. Hubel, C. Lara-Braud, E. Mason, K. C. Moore, V. A. Pedrini, R. Roskoski, R. F. Sheets, L. D. Stegink, and H. Zellweger.

R. Montgomery
R. L. Dryer
T. W. Conway
A. A. Spector

CONTENTS

3 ENZYMES AND BIOLOGIC CATALYSIS, 98

4 ACID-BASE, FLUID, AND ELECTROLYTE CONTROL, 157

7 CARBOHYDRATE METABOLISM, 293

8 LIPID METABOLISM, 363

9 AMINO ACID METABOLISM, 418

12 NUCLEIC ACID AND PROTEIN BIOSYNTHESIS, 568

13   HORMONAL REGULATION OF METABOLISM, 615

14   COMPREHENSIVE CASE ANALYSIS, 664

ABBREVIATIONS, 705

APPENDIXES

CHAPTER 1

NUTRITION

OBJECTIVES

1. To analyze the biochemical role of a proper diet in maintaining homeostasis
2. To interpret the different dietary demands that result from alterations in work load, age, and normal physiologic conditions
3. To interpret the metabolic basis of some nutritional diseases

During a period of 1 year an average adult consumes more than 1000 pounds of food without any change in body weight. In order that this equilibrium state of health be maintained the total body requirements must be supplied to satisfy the demands of energy and tissue maintenance. In other cases food must also provide for growth, as in children and in pregnancy. These demands may vary on a day-to-day basis, depending on the work load and environment. They will change with age and physiologic state. Thus the needs of a manual laborer will change over the weekend when he is not at work, and those of an athlete will be different when he is in training. The nutritional needs of a person nearing retirement will be different than those of an adolescent.

Imposed on these differences are those of biologic individuality. The so-called average 70-kg man is not represented by any one person. Everyone is different, and nutritionally this is expressed at all biochemical and physiologic levels. For example, there are differences in the digestion and absorption of food, the transport of food to the cells, and the rate of waste elimination. However, the physiologic and biochemical regulatory mechanisms, responding to all these individual factors, arrive at an equilibrium for each person that is recognized as health. In some diseases, either genetic or acquired, the resulting nutritional deficiencies of the cells cannot be overcome without external assistance.

Like all other living systems, man survives only by means of a continual energy flux. In the broadest sense nutrition is the provision of needed energy. Sound nutrition depends on a proper dietary regimen, or food intake. This must include the six major components of the diet, *carbohydrates, proteins, fats, vitamins, minerals,* and *water.* Foods frequently contain nonnutritive components that, together with intestinal bacteria and waste materials manufactured by our cells, comprise the excreta in the form of sweat, urine, and feces.

Homeostasis

An organism as complex as the human body is an ordered aggregation of cells. Each cell obtains the nutrients essential for its well-being from the circulating extracellular interstitial fluid in which it is bathed. This same fluid also serves

1

to remove waste products excreted from the cell. The composition of living cells is remarkably constant so long as the interstitial fluid is normal.

The interstitial fluid represents the end of a transport system through which the blood and lymph exchange materials in the external environment with those in the cells. Thus nutrients are brought to the cells from the gastrointestinal tract and oxygen is brought from the lungs, while waste products are excreted in urine, feces, and expired air. Such transport in higher animals is subject to variations in load, since surges of nutrients arise from intermittent food intake. The body reacts to restore the concentration of the extracellular fluid constituents to normal and thus maintains the environment of the cells relatively independent of external change. The internal cellular composition must be kept constant within narrow limits, for otherwise the animal dies. The maintenance of this stable internal environment is termed homeostasis.

The biochemical mechanisms by which the compositions of the extracellular and intracellular fluids are controlled represent a major segment of biochemical knowledge and research. The relationship of disease to the breakdown in these controls is a subject of continuing study.

Body water

The average adult is composed of 55% water, 19% protein, 19% fat, and 7% inorganic material. The body water is distributed between two main components, that within the cells (*intracellular fluid*) and that outside the cells (*extracellular fluid*). The extracellular fluid is subdivided into *interstitial fluid*, which bathes the cells, and the *blood plasma*. On the average the body fluid makes up 65% to 70% of the *lean* body mass, that is, the weight of a person after removal of the excess body fat. With such an extensive distribution of water, the impression might exist that transport and metabolism are slow and perhaps inefficient. However, this is not the case. It has been shown in animals injected with a solution of isotonic saline containing either radioactively labeled ^{14}C bicarbonate or ^{14}C glucose that the label is distributed and incorporated into lipid and carbohydrate polymer (glycogen) molecules within 30 sec. As will be seen later, many cellular and subcellular particle membranes have to be crossed and many chemical reactions must occur in order to achieve such results. The fact that all of this occurs in a very short time demonstrates exquisite organization in and between the cells of the multicellular animal.

Principal food components

Our normal food is a mixture of complex plant and animal materials composed largely of protein, lipid, and carbohydrate. These must be reduced to simpler components before they can be utilized by our tissues. The processes by which food is broken into simpler components are known collectively as *digestion* and involve biologic catalysts called enzymes. The products of digestion are then taken into the bloodstream by the selective processes of *absorption*. A simple introductory description of these food components is necessary in order to understand how they function in nutrition.

Proteins

Proteins are biopolymers composed of monomeric units called α-amino acids. Twenty or so are biologically important (Table 2-1). The α-amino acids have the following general formula:

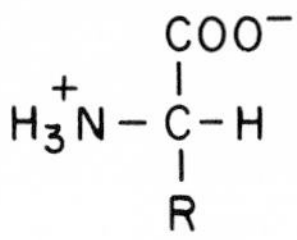

$$
\begin{array}{c}
\text{COO}^- \\
| \\
\overset{+}{\text{H}_3\text{N}} - \text{C} - \text{H} \\
| \\
\text{R}
\end{array}
$$

where R may be an alkyl, aryl, or heterocyclic group. The polymeric bond is formed through the interaction of the amino group of one amino acid with the carboxyl group of another amino acid to form a *peptide bond*. Thus two amino acids condense to form a dipeptide.

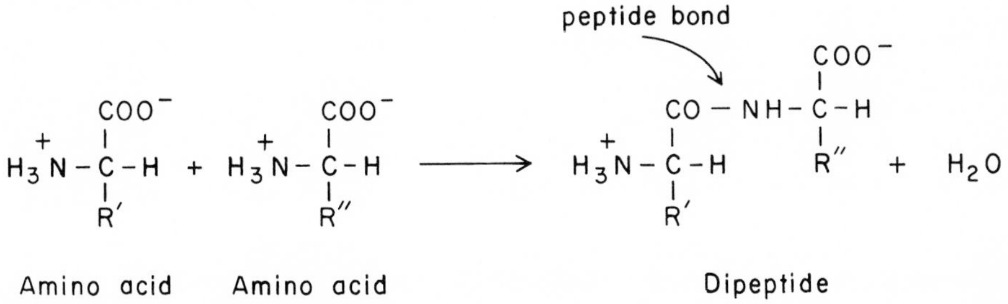

peptide bond

$$
\begin{array}{c}
\text{COO}^- \\
| \\
\overset{+}{\text{H}_3\text{N}} - \text{C} - \text{H} \\
| \\
\text{R}'
\end{array}
+
\begin{array}{c}
\text{COO}^- \\
| \\
\overset{+}{\text{H}_3\text{N}} - \text{C} - \text{H} \\
| \\
\text{R}''
\end{array}
\longrightarrow
\begin{array}{c}
\text{COO}^- \\
| \\
\text{CO} - \text{NH} - \text{C} - \text{H} \\
| \\
\overset{+}{\text{H}_3\text{N}} - \text{C} - \text{H} \quad\quad \text{R}''
\end{array}
+ \text{H}_2\text{O}
$$

Amino acid   Amino acid   Dipeptide

The further sequential addition of amino acid residues creates a polypeptide and finally a protein. Further details are presented in Chapters 2 and 12.

The sequence of the amino acids in each protein is unique and in some way determines the overall shape of the molecule. The particular shape or *conformation* in the living cell is called the *native state*. Proteins in the native state are not easily digested. For this reason the dietary proteins are *denatured* by such means as cooking (heat denaturation), beating (surface denaturation), or chemical transformation (for example, by acid in the stomach). The native structures are disordered in these processes so that the enzymes of the digestive system can hydrolyze the peptide bonds more rapidly to release the component amino acids. This reaction is initiated in the stomach by the enzymes pepsin and rennin at an acid condition of pH $\sim$ 1. It is completed in the small intestine, where the acidic stomach contents are made alkaline by the pancreatic secretions. The pancreatic juice also contains *proteolytic* (trypsin, chymotrypsin) and *peptidolytic* (carboxypeptidase) enzymes. *Aminopeptidases* and *dipeptidases* in the intestinal mucosal cells complete the hydrolysis of the resulting peptides to give amino acids, which are actively transported into the portal blood and thence to the cells of the body to be used as needed for homeostasis. The amino acids are quickly removed from the circulation; the liver removes most, but the kidney is also significantly involved. Other tissues take up less of this high postprandial amino acid flux in the bloodstream. After the high postprandial level subsides, 30 to 70 mg of amino acids/dl of plasma is continuously circulating through all tissues. The digestion of proteins is considered in greater detail in Chapter 9.

Lipids

Many different types of fat are present in the food that we eat. However, the bulk of the dietary fat that has nutritional value is *triglyceride*, or neutral fat. Triglyceride is composed of two parts, glycerol and fatty acids, the fatty acids being esterified to the three hydroxyl groups of the glycerol. The ester bond is the most common linkage in the lipids.

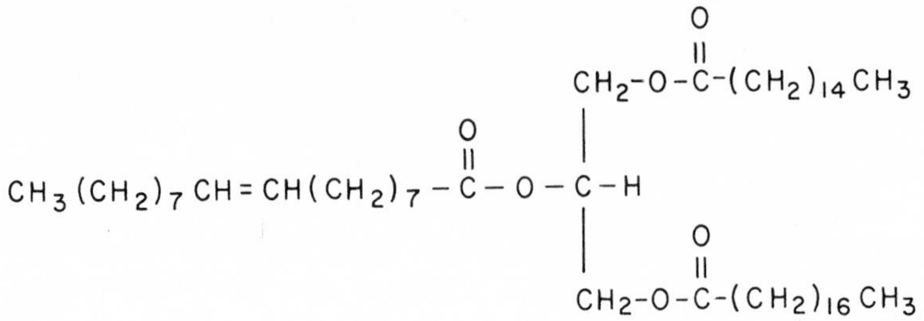

$$CH_3(CH_2)_7\ CH=CH(CH_2)_7-\overset{\overset{\displaystyle O}{\|}}{C}-O-\overset{\displaystyle CH_2-O-\overset{\overset{\displaystyle O}{\|}}{C}-(CH_2)_{14}CH_3}{\underset{\displaystyle CH_2-O-\overset{\overset{\displaystyle O}{\|}}{C}-(CH_2)_{16}CH_3}{C-H}}$$

Triglyceride

A triglyceride containing fatty acid residues of palmitic, oleic, and stearic acids, each in ester linkage to a hydroxyl group of glycerol, is illustrated above. It will be noted that the fatty acids may have a saturated or unsaturated alkyl chain. The more unsaturated the fatty acid, the lower the melting point for a given carbon chain length. Triglycerides that contain a high percentage of unsaturated fatty acid residues are more likely to be oils than solid fat. The unsaturated double bonds in the polyunsaturated fatty acids are two carbons apart. For example, linoleic acid, a biologically important fatty acid, contains two double bonds and has the following structure:

$$CH_3-(CH_2)_4-CH=CH-CH_2-CH=CH-(CH_2)_6-CH_2-COOH$$

Linoleic acid

Most lipids are poorly soluble in water, and they must therefore be emulsified in order to be digested and absorbed. The dietary triglyceride is emulsified by bile, a fluid that is secreted by the liver and stored in the gallbladder. Digestion occurs through hydrolysis catalyzed by a pancreatic enzyme, *lipase*. Not all of the ester bonds are hydrolyzed, and the mixture of fatty acids and partially hydrolyzed glycerides is absorbed. In the intestinal cells the mixture is converted again to triglycerides and then transported into the lymphatic system. The shorter, more water-soluble fatty acids are absorbed directly into the portal blood and delivered to the liver as fatty acid. Further discussion of the lipids will be found in Chapters 8 and 10.

Carbohydrates

The dietary carbohydrates are principally starch, lactose (milk sugar), sucrose (cane or beet sugar), and glucose (dextrose or blood sugar).

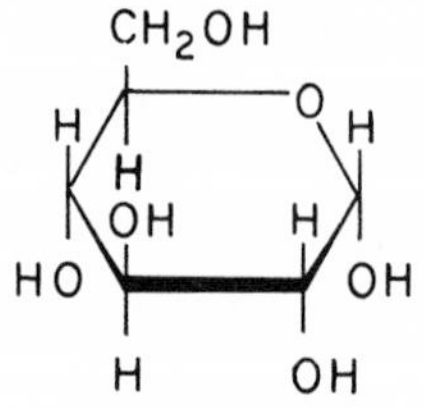

α-D-Glucose

α-D-Glucose is one of the simple building blocks of the carbohydrates. The structure is shown as a six-membered ring with substituents above and below the plane of the ring. The various simple sugars, called monosaccharides, differ in the arrangement of these substituents and sometimes in the size of the ring. For example, α-D-galactose, a constituent of the disaccharide lactose, differs from α-D-glucose at only one position of the ring, whereas β-D-fructose, present in sucrose, has a five-membered ring structure. The structures of these sugars are given below. It will be noted that the substituent hydrogen atoms on the ring have been deleted, as is the common practice, but it must be remembered that they are in fact present. The significance of α-D- or β-D- is a subject for discussion in Chapter 7.

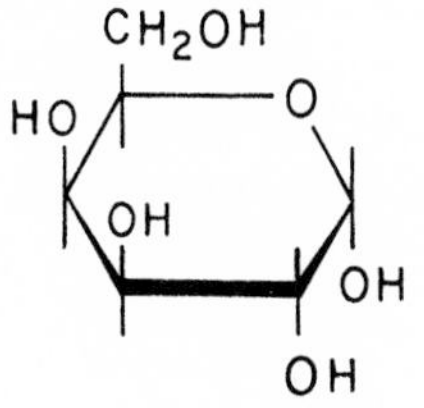

α-D-Galactose 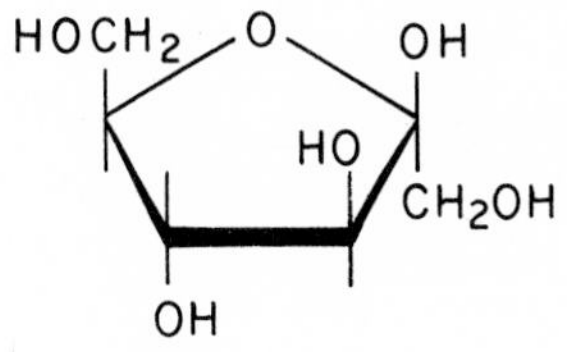   β-D-Fructose

The simple monosaccharides are linked one to the other by *glycosidic bonds*; two monosaccharides joined by such a bond form a *disaccharide*. The addition of a third monosaccharide forms a *trisaccharide* and so on until the high molecular weight polymers, called *polysaccharides*, are formed. The monosaccharides are usually abbreviated by their first three letters; for example, galactose is represented as Gal. Glucose is the exception and is indicated by Glc or G, the letter being used only when it cannot be confused with other conventions of abbreviation. The disaccharide sucrose might therefore be written as Glc—O—Fru, indicating that sucrose is a molecule of glucose linked glycosidically to fructose. The polysaccharide starch, in which many glucose units are linked glycosidically to form a polymer, might be represented as $Glc(—O—Glc—)_n O—Glc$.

The higher molecular weight carbohydrates can be hydrolyzed with acid to give their component simple sugars. The monosaccharides are characterized by having a reducing group, $>C=O$. They reduce Fehling's solution (Cu^{++}) to cuprous oxide (Cu_2O) and ammoniacal silver nitrate to metallic silver. They are therefore identified as *reducing sugars*.

Digestion of the carbohydrates involves splitting (hydrolyzing) all the glycosidic bonds with specific glycosidases down to the simple monosaccharide units. This action starts in the mouth with the enzyme salivary amylase, which is inactivated by the stomach acidity. The other glycosidases occur in pancreatic secretions (amylases) and in the intestinal mucosal cells so that only monosaccharides are absorbed into the portal system. These are commonly D-glucose, D-fructose, and D-galactose. The latter two sugars are converted in the liver to D-glucose, which is the sugar used by the cells to maintain normal metabolic function.

OVERALL VIEW OF METABOLISM OF PRINCIPAL DIETARY COMPONENTS

Following the digestion of the proteins and the carbohydrates, their simple building blocks, the amino acids and monosaccharides, are absorbed into the portal blood system. They are delivered for the most part to the liver, where some are transformed to other molecules. In addition, glucose and amino acids are released by the liver for delivery to other tissues. In this way homeostasis is maintained. The metabolic reactions that are concerned with the biosynthesis of large molecules from small molecules are commonly referred to as *anabolism*. Anabolism is mostly involved in the maintenance of tissue structures. The reverse metabolic process, the breakdown of molecules, is called *catabolism*. When these two processes balance, there is no change in body mass.

Lipids

The triglycerides are either stored in the adipose tissue or are hydrolyzed to glycerol and fatty acids. These products are transported to the liver and muscles, where they are catabolized as a source of energy. The glycerol is either converted to D-glucose or oxidized to a simple three-carbon keto acid, pyruvic acid. Pyruvic acid, usually present as the anion pyruvate, is then converted to acetyl coenzyme A (acetyl CoA), which is a thioester of acetic acid and the complex molecule coenzyme A (CoASH). By a series of oxidative reactions, known as β-oxidation, the fatty acids are also catabolized to acetyl CoA (see Chapter 8).

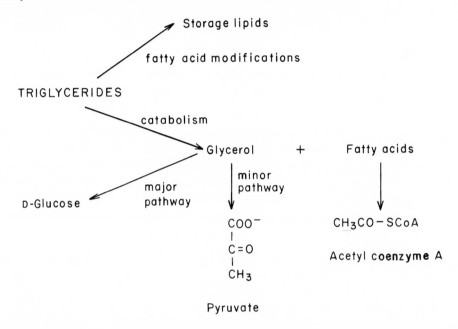

Finally, the acetyl residue passes into the oxidative reactions of the Krebs cycle (also known as the citric acid or tricarboxylic acid cycle), where the final products are carbon dioxide and water. These reactions are central to an understanding of metabolism and are discussed in detail in Chapter 6.

Carbohydrates and amino acids

D-Glucose is either anabolized to other carbohydrate-containing substances, including stores of glycogen, or it is catabolized to pyruvate. The pyruvate is converted to acetyl CoA and finally to carbon dioxide and water. The α-amino acids are either synthesized into tissue proteins or, if the body demands their use for other functions, they are converted to their corresponding α-keto acids. These keto acids are either converted to other amino acids or metabolites or are oxidized by way of the Krebs cycle to carbon dioxide and water. In general, the metabolism of the α-amino acids may be summarized as follows:

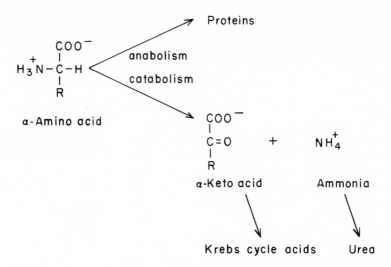

For example, L-alanine, the natural isomer of alanine in proteins, is converted to pyruvate, which is an intermediate in the catabolic pathway of D-glucose (p. 8).

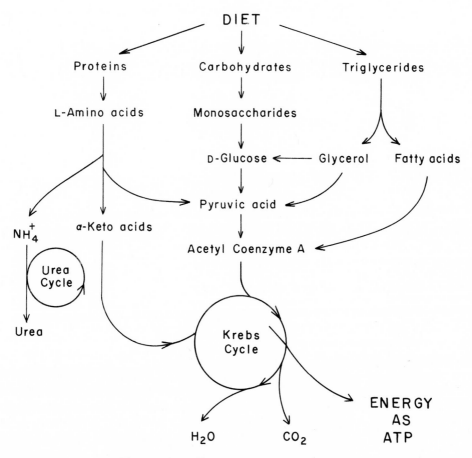

Fig. 1-1. Overall view of catabolic pathways.

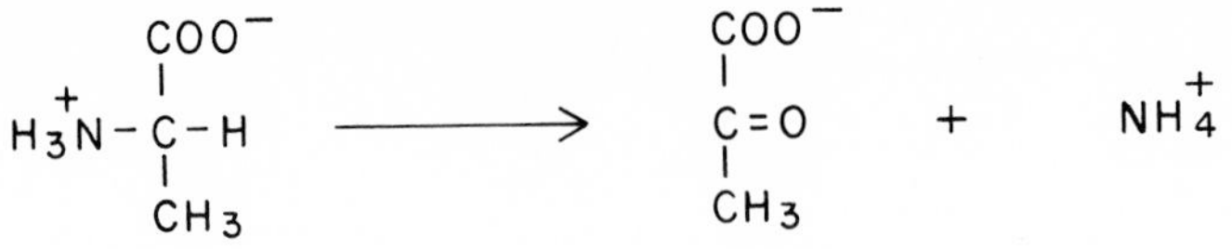

$$
\begin{array}{cc}
\underset{H_3\overset{+}{N}-\underset{\underset{CH_3}{|}}{\overset{\overset{COO^-}{|}}{C}}-H}{\text{L-Alanine}}
&
\longrightarrow
\end{array}
$$

L-Alanine → Pyruvate + NH_4^+

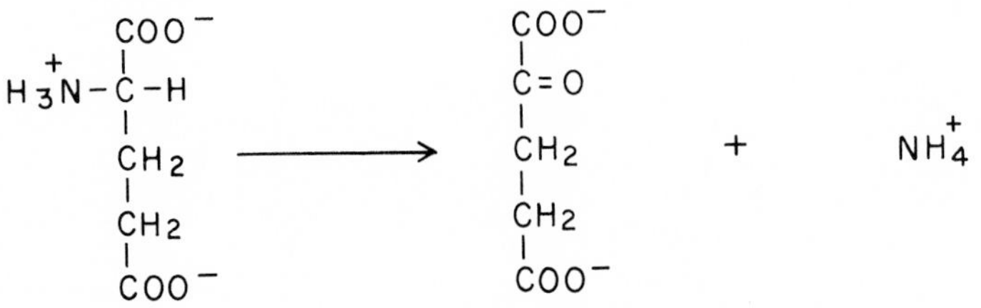

L-Glutamate → α-Ketoglutarate + NH_4^+

The deamination of L-glutamate, an amino acid present in large amounts in many proteins, produces α-ketoglutarate, which is an acid component of the Krebs cycle.

Krebs cycle and adenosine triphosphate

The overall reactions of catabolism are summarized in Fig. 1-1, where it is seen that the Krebs cycle is the focus of metabolism, being common to the interconversions of many metabolites. The Krebs cycle is also much concerned with the conversion of the chemical energy of the diet to the common currency of energy in the body, adenosine triphosphate (ATP).

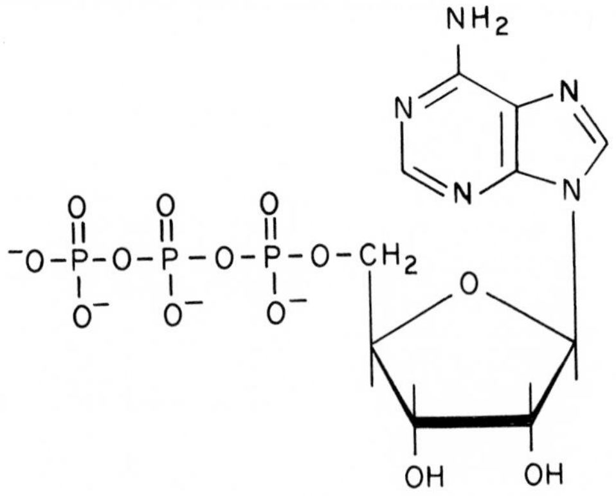

Adenosine−5′−triphosphate (ATP)

Modifications in response to stress

Many refinements of the simplified presentation in Fig. 1-1 will be developed as different areas of metabolism are considered in the later chapters. Modifications or extensions of these pathways occur in response to physiologic changes. For example, the acute need for energy in muscle contractions during intensive work outstrips the tissues' supply of oxygen for normal oxidative reactions. At such times pyruvate serves as an alternate oxidizing agent, thus relieving periods of relative tissue anoxia. The pyruvate is converted to lactate, which is released from the muscle cells into the blood for later conversion back to pyruvate in the liver when the supply of oxygen is adequate.

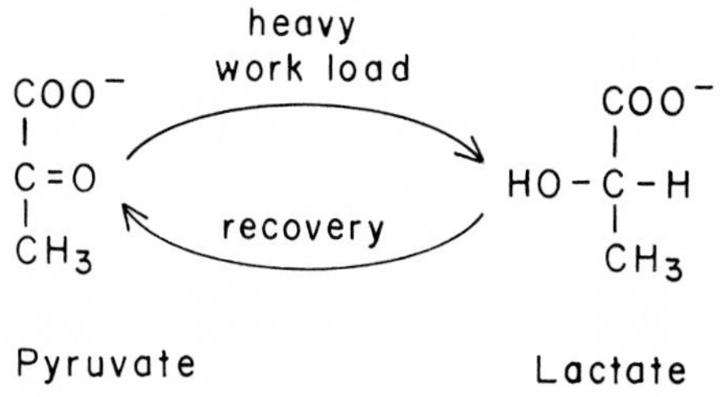

$$
\begin{array}{ccc}
\text{COO}^- & \xrightarrow{\substack{\text{heavy} \\ \text{work load}}} & \text{COO}^- \\
| & & | \\
\text{C}=\text{O} & \xleftarrow{\text{recovery}} & \text{HO}-\text{C}-\text{H} \\
| & & | \\
\text{CH}_3 & & \text{CH}_3 \\
\\
\text{Pyruvate} & & \text{Lactate}
\end{array}
$$

Thus, even during periods of oxygen deficiency, called *oxygen debt,* some energy can still be derived from the oxidation of glucose.

Disease can be related to the changes in the dynamic equilibrium of body constituents. For example, keto acids are overproduced from the metabolism of lipid when there is an inadequate supply of carbohydrates resulting from improper eating, dieting, or fasting, as well as diseases, such as uncontrolled diabetes mellitus. This results in elevated blood levels of acetoacetate, β-hydroxybutyrate, and acetone, these compounds being the so-called *ketone bodies.* This leads to *ketosis* and *acidosis,* metabolic imbalances that will produce very serious illness if they are not corrected quickly. At best ketosis represents a loss of calories since these compounds are lost in the urine. The formation of ketone bodies from the catabolism of fatty acids may be summarized as follows, the two arrows indicating that several metabolic steps are involved:

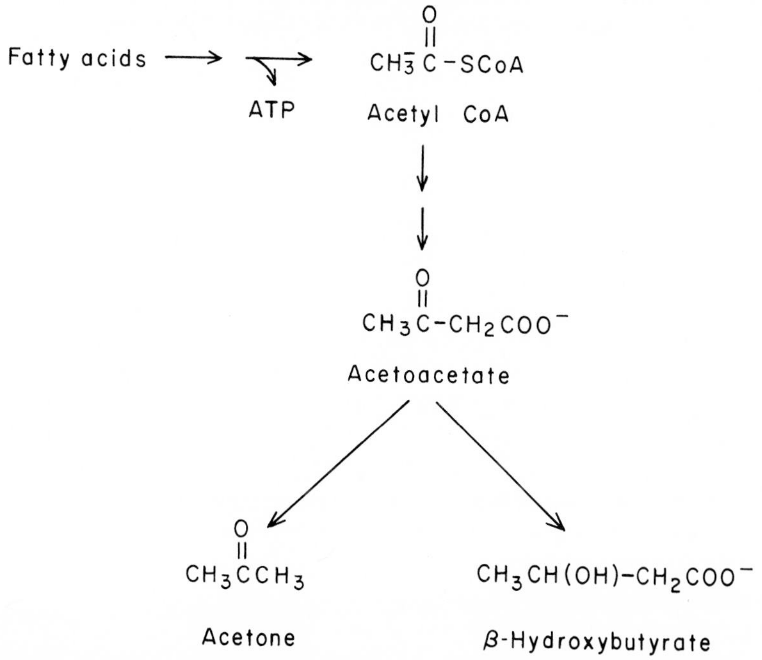

$$
\text{Fatty acids} \longrightarrow \underset{\text{ATP}}{\searrow} \longrightarrow \underset{\text{Acetyl CoA}}{\overset{\overset{\text{O}}{\overset{||}{}}}{\text{CH}_3\text{C}-\text{SCoA}}}
$$

$$
\downarrow
$$

$$
\underset{\text{Acetoacetate}}{\overset{\overset{\text{O}}{\overset{||}{}}}{\text{CH}_3\text{C}-\text{CH}_2\text{COO}^-}}
$$

$$
\underset{\text{Acetone}}{\overset{\overset{\text{O}}{\overset{||}{}}}{\text{CH}_3\text{CCH}_3}} \qquad\qquad \underset{\text{β-Hydroxybutyrate}}{\text{CH}_3\text{CH(OH)}-\text{CH}_2\text{COO}^-}
$$

GENERAL NUTRITIONAL REQUIREMENTS

Control of body functions is exercised in the whole animal at organ, cellular, and subcellular levels. The only voluntary step for man is that of selecting his diet, from which the six major food components provide both energy for the body and the basic materials for maintaining the body components. The molecules that make up the body structure or maintain homeostasis largely result from the metabolism of molecules in the food. In some instances, however, the synthesis of a metabolite might occur at a rate too slow to maintain health or it may not occur at all. In such cases the diet must include these molecules. They comprise the *essential amino acids*, the *essential fatty acid(s)*, and the necessary *vitamins* and *minerals*. From a nutritional standpoint the vitamins differ from the essential fatty acids and essential amino acids only in that much smaller amounts are required. The essentiality of these substances in the diet does not imply that the other metabolites are unimportant, but rather that the human body can synthesize sufficient amounts of the nonessential metabolites from the components of the diet.

Practical daily food plan

A practical and convenient food plan is based upon four basic food groups, each of which is chosen so as to contribute nutrients essential for a complete and balanced diet.

1. Milk group—provides high quality protein, calcium, and vitamin D
2. Meat group—supplies protein of high biologic value, thiamine, vitamin B_{12}, and minerals
3. Vegetable-fruit group—supplies vitamin C, carotene (vitamin A precursor), other water-soluble vitamins, minerals, and fiber ("roughage"); no one vegetable protein has a high biologic value, but when properly mixed it is possible for one protein to complement another
4. Bread-cereal group—high in carbohydrate for energy needs, also includes vitamins, fiber, and iron if the cereals and wheat are not refined; the proteins of this group do not have as high biologic value as animal protein

Substitutions within these groups are desirable in order to provide all the factors known to be essential for man and to exclude excessive components, such as cholesterol, that are injurious.

Recommended daily dietary allowances (RDA)

Any statement of a minimum daily nutritional requirement will be subject to some debate because, at present, the points of reference are ill-defined. For example, it is possible to omit vitamins from the diet for a few weeks without any apparent ill effects. However, prolonged omission of vitamins results in serious disease. For this discussion the minimum daily requirement (MDR) for the various food components is defined as that amount of a substance needed to maintain continuous health and freedom from any symptoms of nutritional deficiency. The MDR is now being replaced by the U.S. Recommended Daily Allowance (U.S. RDA). These values are appearing on food containers such as cereal boxes. The U.S. RDA is derived from the highest value for each nutrient given by the recommended daily dietary allowances, which serve as a guide for decisions affecting food intake for large groups. Thus the optimal nutritional needs of healthy individuals will be met by diets containing the correct RDA. The allowances are not determined for individuals, each of whom will have particular dietary requirements with the statistical probability that these will be met by the RDA. In the United States the Food and Nutrition Board of the National Academy of Sciences, National Research Council publishes the recommended daily allowances designed for the *average* American diet. This is defined as a diet in which 15% of the calories come from protein, 40% come from lipid, and 45% come from carbohydrate. It is clear from the table of recommended dietary allowances (Appendix A) that the requirements for food differ with age, body size, sex, work load, and physiologic condition. For example, during pregnancy the mother provides all the nutrients for the growing fetus, and her dietary intake must necessarily reflect this extra demand.

The RDA in no way reflects the special needs of diseased states or other traumatic situations, such as the severely burned patient or the premature infant.

The RDA makes no attempt to describe the source of the nutrients, so that carbohydrate intake may be derived from either refined foods or from those containing more cellulosic fiber, which will be excreted in the feces. Aside from the other nutrients that may be lost in such refining processes, for example, vitamins and minerals, it is a matter of concern and research to evaluate the need for roughage in the diet. Food should contain some fiber, which plays a part in the peristaltic stimulation of the bowel. Excessive intake of the nonutilizable, ingestible fiber, however, can lead to irritation of the intestinal mucosa. Even normal amounts of roughage may be undesirable in inflammatory diseases of the colon (see case 5, p. 38). Some studies suggest that the absence of cellulosic roughage may be related to colonic carcinoma and other diseases.

Energy requirements

Nearly all the dietary components that can be digested and absorbed are eventually catabolized to provide energy. In these processes the food is oxidized, as summarized in Fig. 1-1, and the energy is transformed to the intermediate chemical form—ATP. ATP is thus the common currency of energy in the body and is involved in driving various metabolic and physiologic functions that involve chemical reactions, electrical work, osmotic work, and mechanical work.

The maximum available energy contained in a food can be measured by burning it in an atmosphere of oxygen. This is done in a calorimeter. The resulting increase in temperature in the calorimeter is determined and the heat energy produced can be calculated as *kilocalories* (kcal or Cal). One kilocalorie is the amount of energy required to raise the temperature of 1 kg of water from 15° to 16° C.

Living systems oxidize the nutrients in more nearly reversible steps, as opposed to the irreversible burning that occurs in the calorimeter. Consequently, the energy available (metabolic energy) is less by a few percent. For example, D-glucose has a maximum metabolic energy, which will be referred to later as *free* energy for performing useful work. The maximum free energy is about 3% less than the calorimetric or *enthalpic* energy. Since the calorimetric measurements are relatively simple, and the purposes to which the caloric contents of foods are put do not demand the greatest precision, the calories listed in the dietary tables are those derived by calorimetry, with some adjustments for digestibility. They are expressed in kilocalories per gram, or kcal/100 g, or kcal per portion of food commonly used such as kcal per slice of bread.

The metabolic energy of food is converted principally to heat and ATP, the overall efficiency for the conversion to ATP being approximately 40%. In practice the proper diet for an adult is that which maintains health and a constant body weight. This requires approximately 14 to 20 kcal/pound of desired body weight, or about 2200 to 3000 kcal for an average, 154-pound (70-kg) adult man. These calories provide for (1) the basal metabolic rate, (2) the specific dynamic action of food, and (3) the extra energy expenditure for activity.

Basal metabolic rate

The basal metabolic rate (BMR) is the energy required by an awake individual during physical, digestive, and emotional rest. It is measured directly by the heat evolved or indirectly by the volumes of oxygen consumed and carbon dioxide evolved per unit time. It is not the minimum metabolism necessary for maintaining life, since the metabolic rate during sleep may be less than the BMR.

The BMR is influenced by many factors. It is changed in some hormone diseases, hyperthyroidism increasing and hypothyroidism lowering the BMR. The BMR changes with growth, reaching a maximum at about 5 years of age; in the adult it varies with body size. Exposure to cold or regular exercise routines causes increases; starvation produces a lowering – some small attempt at natural compensation for a reduced caloric intake. Increased cellular activity produced by disease is reflected in an increased BMR. With fever, there is approximately a 12% increase in BMR for each degree centigrade rise in temperature.

Many measurements of BMR have been made on humans, but the variations noted above have made it difficult to predict a value for any individual. For the purposes of this discussion the energy demand for basal metabolism in an adult will be taken as 24 kcal/day · kg body weight.

$$BMR = weight\ (kg) \times 24\ kcal/day$$

Specific dynamic action

Although the metabolic processes continuously produce heat, it is noted that heat production increases during digestion of food. This stimulatory effect of food on heat production is termed the specific dynamic action (SDA). The value is different for proteins, carbohydrates, and fat, being greatest for protein. These calories are wasted and on the average amount to about 10% of the total caloric intake.

The average caloric content, as measured by a calorimeter, is 4.1 kcal/g for carbohydrate, 9.4 kcal/g for fat, and 5.6 kcal/g for protein. All of this gross enthalpic energy value is not available to the body for metabolism. Some of the food is not absorbed and is lost in the feces. In addition, there are the losses resulting from SDA and through the excretion of urea, creatinine, and other organic materials in the urine. Since we are concerned with the amount of energy in food that is available for metabolism, the calorimetric values for each major foodstuff can be corrected for these losses and the *metabolic caloric contents* obtained. These are given in Table 1-1. These average values are used to calculate the metabolic content of foods as they appear in published diet tables and therefore include the corrections for SDA.

Activity

The effect of various types of muscular work on the energy requirements of the body can be measured. In very general terms a sedentary level of activity requires

Table 1-1. Metabolic caloric content of food components after correction for SDA and other losses

Carbohydrate	4 kcal/g
Lipid	9 kcal/g
Protein	4 kcal/g

additional calories equal to 30% of those for BMR. Moderate and heavy activity levels required 40% and 50% of the BMR, respectively. For example, a 90-kg man engaged in heavy activity requires approximately:

$$BMR = 90 \times 24 \text{ kcal/day} = 2160 \text{ kcal/day}$$

$$\text{Heavy activity} = 50\% \text{ BMR} = 1080 \text{ kcal/day}$$

$$\text{Total metabolic caloric requirement} = (2160 + 1080) \text{ kcal/day} = 3240 \text{ kcal/day}$$

During the weekend this same man, doing little work, would have a sedentary level of activity, and his daily need would fall to:

$$(2160 + 30\% \times 2160) \text{ kcal/day, or } 2708 \text{ kcal/day}$$

All calculations of the caloric requirement of a person are inexact, and the daily requirements are usually rounded off to the nearest 50 kcal. The 90-kg man would therefore be said to need 3250 kcal/day when at work and 2700 kcal/day during sedentary periods.

These calculations can be made more reliable if the individual activities are analyzed. Particular examples of additional caloric expenditure over and above the BMR for various activities are listed in Table 1-2 for a 70-kg adult. The corresponding caloric needs of other adults are proportional to their body weight. Thus a 50-kg woman would expend 395 kcal in addition to her BMR requirements when swimming for 1 hr at 2 mph. A 70-kg man driving a car for 8 hr/day, eating 2 hr/day, writing 6 hr/day, and sleeping 8 hr/day requires approximately:

$$BMR = (70 \times 24) \text{ kcal/day} = 1680 \text{ kcal/day}$$

$$\text{Activity} = (8 \times 63 + 2 \times 28 + 6 \times 28) \text{ kcal/day} = 728 \text{ kcal/day}$$

No calories are added for sleeping.

$$\text{Metabolic caloric requirement} = BMR + \text{activity} = (1680 + 728) \text{ kcal/day} = 2408 \text{ kcal/day}$$

Such a daily activity would be considered moderate and on this basis the caloric need could be estimated as 1680 + 40% of 1680 or 2350 kcal/day. This method of estimation provides an answer that is close to the 2408 kcal/day given by the preferred approach of individual analysis of the various activities. This falls into the range of 2200 to 3000 kcal/day, a figure that was initially suggested based on 14 to 20 kcal/day · pound body weight.

Table 1-2. Additional energy expenditure for 70-kg adult during different activities

ACTIVITY	EXTRA KCAL/HR
Bicycling (~8 mph)	175
Dishwashing	70
Driving car	63
Eating	28
Lying still, awake	7
Running	490
Swimming 2 mph	553
Typist (good typing speed, electric typewriter)	70
Walking 3 mph	140
Walking 4 mph	238
Writing	28

Lipid requirements

Lipids are principally a source of calories. They represent a concentrated energy supply in the diet as well as an efficient form of energy storage in the body. Lipids are found primarily in the adipose tissue. There seems to be no limit to the amount of fat that can be stored in adipose tissue from excessive caloric intake.

Since the 1950's it has become increasingly clear that for many people large proportions of vegetable fats in the diet produce a major and sustained fall in serum cholesterol, while the same proportions of animal fats produce opposite effects. As will be noted from Table 1-3, the vegetable oils have higher contents of polyunsaturated fatty acids in their composition than the animal fats or the

Table 1-3. Lipid composition of foods

FOOD GROUP	PERCENT LINOLEIC ACID	P/S RATIO*
Milk		
Butterfat	2	1:34
Meat		
Lard	10	1:34
Beef fat	2	1:24
Vegetables and fruits		
Corn oil margarine	32	1:0.45
Corn oil	57	1:0.2
Safflower oil	77	1:0.1
Soybean oil	52	1:0.3
Peanut oil	28	1:0.5
Bread and cereals†		
Wheat flower, white		1:0.3
Rolled oats		1:0.5

*P/S ratio = total polyunsaturated fatty acids/total saturated fatty acids.
†Composition of extracted lipid.

Table 1-4. Cholesterol content of foods

FOOD GROUP	CHOLESTEROL (mg/100 g)
Cow's milk	
Skimmed	1
Whole	11
Butter	250
Cheese	100
Meat	
Beef, pork, chicken, bacon, fish	60-70
Brain	2000
Kidney	375
Liver	300
Sweetbreads	250
Egg yolk	1500
Egg whole	550
Oysters	200
Vegetables, fruits	
All	0
Bread, cereal	
All	0

hydrogenated oils. The relationship that exists between serum cholesterol levels and the risks of cardiovascular disease thus indicate the advisability to control the intake of foods containing high percentages of cholesterol (see Table 1-4). It is also desirable to replace saturated animal fat with vegetable oils, which have high polyunsaturates and no cholesterol, or hydrogenated vegetable oils (margarines), which have no cholesterol and moderate amounts of essential fatty acids.

Essential fatty acids

All but one fatty acid, linoleic acid, can be synthesized from other fatty acids or non-fatty acid precursors, mainly carbohydrates. Linoleic acid must be provided in the diet. It is essential to health, being a component of cell membranes and the precursor of one family of the prostaglandins, a ubiquitous group of cyclopentane derivatives with hormonal properties. Another family of prostaglandins is derived from the unsaturated fatty acid arachidonic acid, which can be synthesized from linoleic acid. Arachidonic acid is therefore not an essential fatty acid, but it is sometimes classified as such.

In general, the dietary supply of essential fatty acids should be approximately equivalent to 1% of the caloric intake. Thus a diet of 2400 kcal/day should include at least 24 kcal, that is, (24/9) g or 2.7 g, of linoleic acid.

Carbohydrate requirements

In theory there is no need for carbohydrate in the diet since all the carbohydrate components of the body can be synthesized from proteins. However, ketosis and acidosis develop when carbohydrate is unavailable to cells. This occurs in the disease diabetes mellitus; although there is plenty of glucose in the extracellular fluid, the tissues cannot properly absorb it. For optimum nutrition and to avoid ketosis it is necessary to provide 5 g carbohydrate for each 100 kcal of diet. This means that 20% of the total calories should be provided in the form of carbohydrates.

Protein requirements

Amino acids derived from proteins in the diet serve as the source of the nitrogen for the biosynthesis of nonprotein compounds such as the porphyrins of hemoglobin and the nucleic acids. Amino acids also provide for the biosynthesis of tissue proteins. As will be described later, all twenty amino acids must be available together in order for protein synthesis to occur.

Essential amino acids

While half of the twenty amino acids can be synthesized by the cells, the other ten, the essential amino acids, cannot always be synthesized at a sufficient rate and must be supplied in the diet. Of the ten essential amino acids, eight are essential at all times during life. The other two are required in the diet during periods of rapid growth, for example, in childhood. These essential amino acids* are listed in Table 1-5, and their chemical formulae are given in Table 2-1. It will be noted that the infant requires all the amino acids that are basic, branched, or aromatic except tryosine. Tyrosine is essential only when phenylalanine is absent from the diet. A diet that is very low in phenylalanine is prescribed for children

---

*"Any *help* in *learning* these *little* molecules *proves* truly *valuable*" is a mnemonic device that may prove helpful.

Table 1-5. Amino acids essential for man

Arginine Histidine	} Additional amino acids essential for infants
Isoleucine Leucine Lysine Methionine Phenylalanine Threonine Tyrptophan Valine	} Essential for adults

suffering from phenylketonuria, a genetic disease in which the normal metabolism of phenylalanine is disturbed. A premature infant may also require tyrosine and cysteine, depending on whether the appropriate enzyme systems for the synthesis of these amino acids have developed at the time of birth.

The inclusion of adequate quantities of nonessential amino acids in the diet will spare the essential amino acids from being catabolized and the fragments used in nonessential ways. Carbohydrates also have a sparing effect on amino acids. If the diet is deficient in carbohydrates, amino acids will be deaminated and the carbon skeletons will be used for the maintenance of glucose levels, the generation of ATP, and the formation of Krebs cycle intermediates. Consequently, the diet must supply adequate amounts of both total nitrogen and essential amino acids.

Nitrogen balance

An adult is said to be in nitrogen equilibrium when the amount of nitrogen consumed equals the amount of nitrogen excreted in the urine and feces. *Positive* nitrogen balance occurs when the nitrogen intake exceeds the nitrogen excreted. A positive nitrogen balance is associated with growth, as in children and pregnancy or during convalescence when injured tissues are being repaired. *Negative* nitrogen balance occurs when the intake of nitrogen is less than the nitrogen excreted. This happens during starvation and in malnutrition. It also occurs in a variety of illnesses. Burns, trauma, or surgery will produce a period of negative nitrogen balance.

The amino acids essential for humans were determined by studies of how withdrawal of a single amino acid from the diet affected the nitrogen balance. Withdrawal of a single essential amino acid produces an abrupt change to a negative nitrogen balance.

Biologic value of proteins

Generally, the most nutritious diet with respect to amino acid content is one that provides a good balance of all of the amino acids, both essential and nonessential. In this regard some protein foods are better than others. For example, the proteins of human milk would score higher than beef steak, which is better than rice.

The biologic value (BV) of a protein is a measure of the degree to which its nitrogen can be used for growth or maintenance of total body function. The relative biologic value or quality of proteins may be determined in several ways, one of which is by nitrogen balance. The biologic value is expressed as:

$$BV = \frac{\text{N intake} - \text{N loss in feces, sweat, urine}}{\text{N intake} - \text{N loss in feces}} \times 100 = \frac{\text{N retained}}{\text{N absorbed}} \times 100$$

The nitrogen in the feces represents approximately the nitrogenous material in the diet that is not digested and absorbed into the body. However, the nitrogen in the sweat and urine arises in large measure from the catabolism of nitrogenous material. The biologic value is therefore the percentage ratio of the nitrogen balance to the nitrogen absorbed into the body. It is a measure of the extent to which the dietary protein satisfies the amino acid requirement.

In order for the dietary proteins to have any biologic value they must provide *all* of the essential amino acids. In general, animal proteins have a high biologic value. A major exception is gelatin, which lacks the essential amino acid tryptophan and therefore has no biologic value. Vegetable proteins have a low biologic value for man because each one has a low level of one or more essential amino acids. This is the main reason why plant geneticists try to develop plants with higher levels of essential amino acids in their protein, for example, the corn strains high in lysine. The vegetarian clearly demonstrates, however, that a proper mixture of plant foods can provide all the necessary protein requirements in the diet.

Biologic values have been placed on a relative scale, with whole-egg protein or egg albumin scored at 100. The recommended daily allowance for protein is calculated for a biologic value of 70, which is the biologic value for beef and the average biologic value of protein in an American diet. Corrections are necessary for dietary proteins of different average values. Thus an adult has a recommended daily allowance of 65 g of protein. If the dietary protein were of plant origin, with a biologic value of 40 instead of the 70 assumed in the recommended daily allowance, then the daily intake should be increased to (65 × 70/40) g or 114 g (see also case 2).

Some foods contain proteins that are toxic under special circumstances or may produce allergic reactions. Patients with celiac disease are sensitive to a certain fraction of wheat protein, called gluten. With elimination of wheat from the diet, improvement occurs in these patients (see case 7, p. 39).

Vitamin requirements

As noted earlier, vitamins are small organic molecules in the diet that either cannot be synthesized by humans or are synthesized at a rate less than that consistent with health. The vitamins are usually divided into those that are fat soluble and those that are soluble in water. The division may sound artificial, but it does relate to nutrition in the sense that a low-fat diet may be deficient only in the fat-soluble class of vitamins. The role of vitamins in metabolism will be considered at a molecular level in later sections. However, in some cases their functions are known only in a descriptive way.

Fat-soluble vitamins

Vitamin A. Vitamin A is an alcohol, retinol, that is present as the 11-*cis* or all-*trans* isomers. The structure of all-*trans*-retinol is shown below.

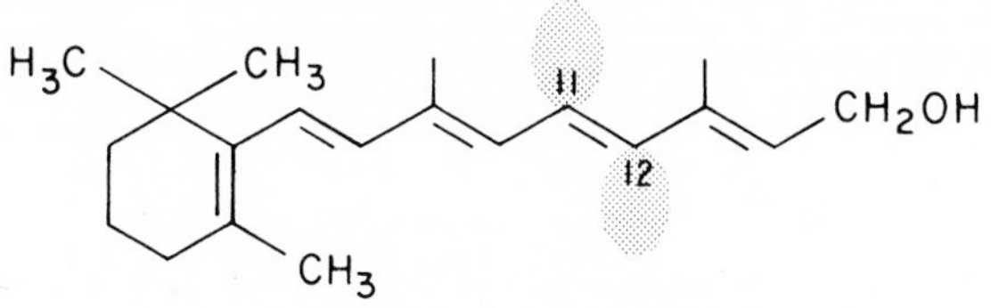

all − *trans* − Retinol

It is found primarily in liver, notably in the fish liver oils. Precursors of vitamin A occur in plants in the form of the α-, β-, and γ-carotenes. β-Carotene is physiologically equivalent to vitamin A because it is oxidatively cleaved in the intestinal mucosa at the central double bond to give two molecules of the corresponding aldehyde, retinal, which is reduced to retinol. These reactions are discussed at greater length in case 3 in this chapter, which deals with night blindness. Except for the visual cycle, the mode of action of vitamin A is largely unknown. Vitamin A deficiency leads to disturbances in the growth and remodeling of bone, skin lesions, keratinizing of many epithelial cells, and abnormal cellular function of the adrenal cortex. On the other hand, the ingestion of excessive amounts of vitamin A produces a toxicity that produces lethargy, abdominal pain, headache, excessive sweating, and brittle nails.

Vitamin D. Vitamin D, the antirachitic member of the vitamin family, is involved in the calcification of bone. In deficiency states the growing portions of the bones are affected, producing bowlegs, knock-knees, and enlarged joints. The beneficial effects of sunlight in the prevention of rickets have long been recognized and are known to be caused by the photo-oxidation in the skin of a derivative of cholesterol to a form of vitamin D, cholecalciferol.

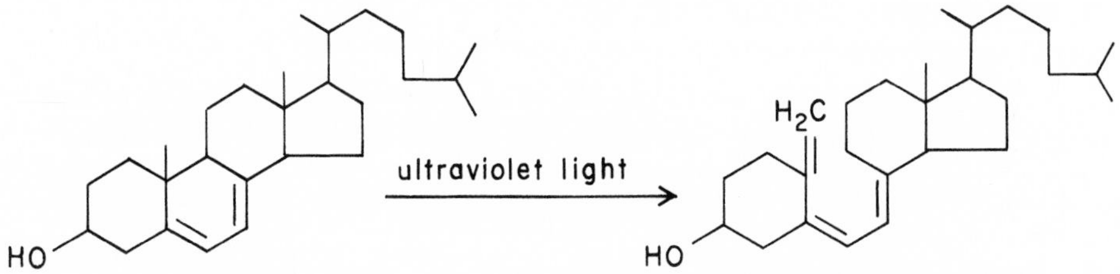

7-Dehydrocholesterol   ultraviolet light →   Cholecalciferol

This compound is also found in animal products, particularly fish liver oils. Another similar molecule with vitamin D activity is calciferol, which is found in plants. Both are converted to physiologically active forms as will be discussed later (Chapter 13).

Vitamins A and D are excreted from the body at a slow rate. Therefore, if excessive amounts of either of these vitamins are ingested, toxicity will result. Vitamin D toxicity produces elevated levels of circulating calcium that affect the blood pressure and cause deposition of calcium under the skin and in the renal tubules.

Vitamin E. Like vitamins A and D, vitamin E is found in animal or plant fats and oils. Wheat germ is a popular source. The most active vitamin E is α-tocopherol.

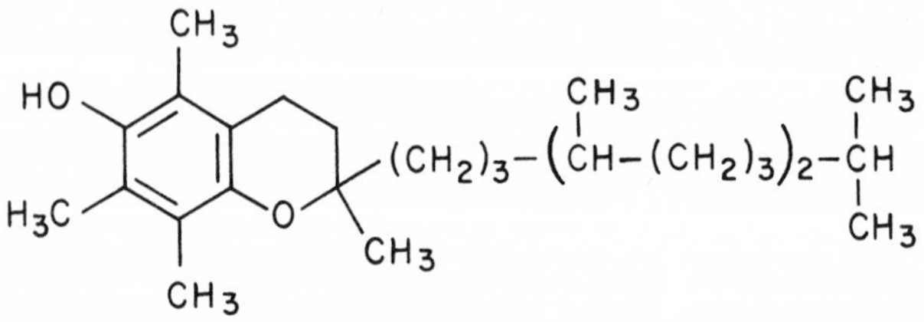

α-Tocopherol

β- and γ-Tocopherols, which differ from the α-form in their substituents on the aromatic ring, have 25% and 19% of the activity of α-tocopherol, respectively.

The biologic function of vitamin E in man is not clear. It probably functions in part as an inhibitor of the peroxidation of the lipid in cell membranes, and it may function to maintain the integrity of cellular membranes. There is no firm evidence to indicate that it acts in humans as an antisterility vitamin, as it does in rats. Fortunately, tocopherols are relatively nontoxic in man, since there is a popular misconception that vitamin E improves sexual performance.

Vitamin K. Vitamins K_1 and K_2 are required for the blood clotting process, which is discussed in greater detail in Chapter 14. These molecules are derivatives of menadione, a 1,4-napthoquinone that acts more rapidly when injected than vitamin K_1 and is thus commonly used as an antihemorrhagic pharmaceutic agent.

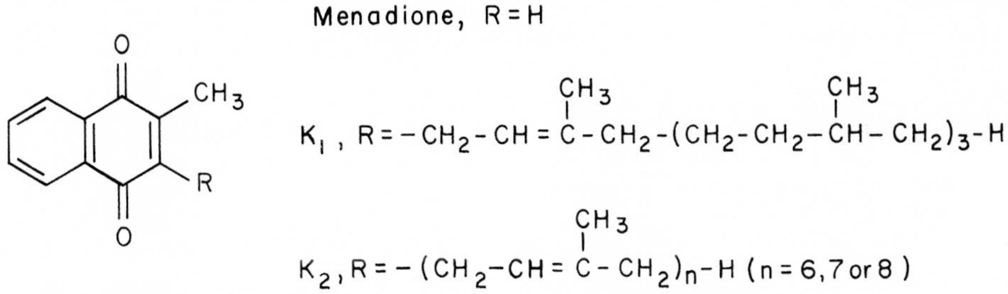

Menadione, R = H

$$K_1, R = -CH_2-CH=C(CH_3)-CH_2-(CH_2-CH_2-CH(CH_3)-CH_2)_3-H$$

$$K_2, R = -(CH_2-CH=C(CH_3)-CH_2)_n-H \quad (n = 6, 7 \text{ or } 8)$$

Vitamin K_1 is present in green leaves and other plant tissues that are eaten in the diet. Vitamin K_2 is a bacterial product and is produced by the bacterial flora of the intestine. Therefore, under normal circumstances the human is well provided with this vitamin. The situation may change when prolonged antibiotic therapy is instituted and some of the bacteria in the intestinal lumen are killed. A favorable symbiosis is thus lost.

Since vitamin K is absorbed with the other fat-soluble vitamins in the upper part of the small intestine, the process is dependent upon the presence of bile salts. As a result, any absence of bile formation or any biliary obstruction will reflect itself in a deficiency of the fat-soluble vitamins. Thus in obstructive jaundice, when bile cannot pass from the liver to the intestine, there may be prolongation of prothrombin or clotting time because of inability to absorb adequate amounts of vitamin K.

Water-soluble vitamins

Thiamine (vitamin B_1). Thiamine is present in meat, particularly pork, and in yeast, whole grains, and nuts. A deficiency of thiamine produces beriberi, a disease characterized by extensive damage to the nervous and circulatory systems, muscle wasting, and edema.

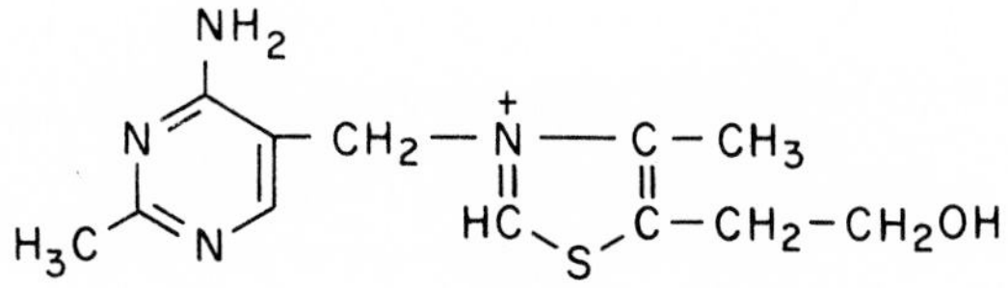

Thiamine

The vitamin is converted in the body to a pyrophosphate ester, which is the meta-bolically active form. It functions in the conversion of pyruvate to acetyl CoA and in many similar reactions that involve carbon dioxide removal coupled with oxidation, that is, oxidative decarboxylation.

Riboflavin (vitamin B_2) and niacin. Riboflavin is found in meat and plant products along with thiamine and niacin. Riboflavin is also produced by the bacteria that are always present in the intestine. As a result, little is known about the effects of riboflavin deficiency in man. Niacin, however, is known as the protective dietary component against pellagra. The symptoms of pellagra include a swollen tongue, dermatitis, and nervous and gastrointestinal distur-bances.

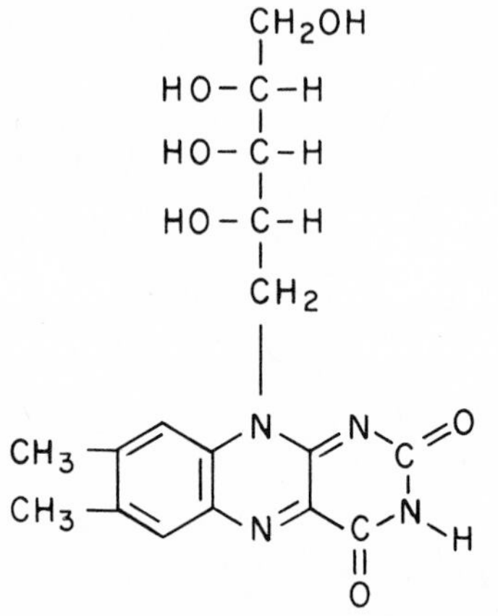

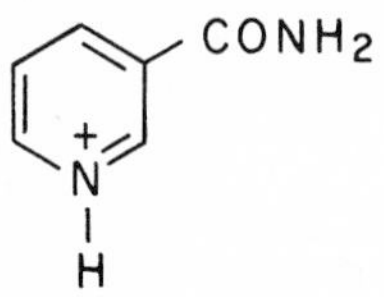

Niacin

Riboflavin

Niacin can be synthesized by humans from the amino acid tryptophan, but at a rate that is inadequate to maintain good health. Both vitamins are building blocks for the biosynthesis of compounds that act in oxidation-reduction reactions, namely, nicotinamide adenine dinucleotide (NAD) and flavin adenine dinucleo-tide (FAD). These vitally important compounds are discussed in later chapters.

Pyridoxine (vitamin B_6). Pyridoxine is the common form of vitamin B_6 in most commercial preparations. The other forms are pyridoxal and pyridoxamine.

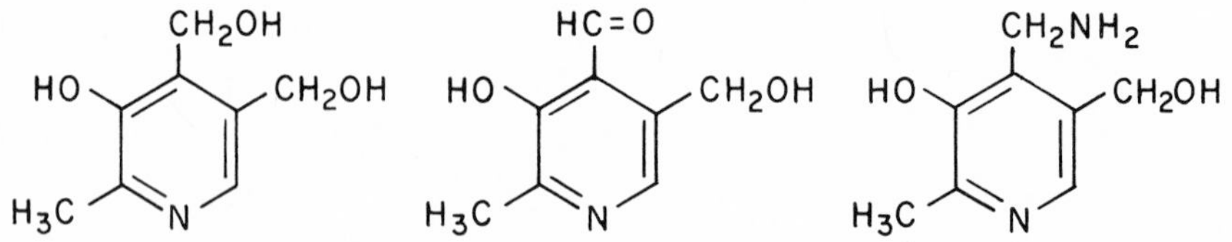

Pyridoxine   Pyridoxal   Pyridoxamine

The phosphate ester derivatives of these forms are involved in many reactions in metabolism, including the conversion of amino acids to the corresponding keto acids.

Pantothenic acid. Pantothenic acid is present in many tissues and was first isolated from liver and yeast. This vitamin is necessary for the biosynthesis of coenzyme A, which, as noted earlier in the overall view of biochemical processes, is a most important molecule in the metabolism of many nutrients.

Coenzyme A

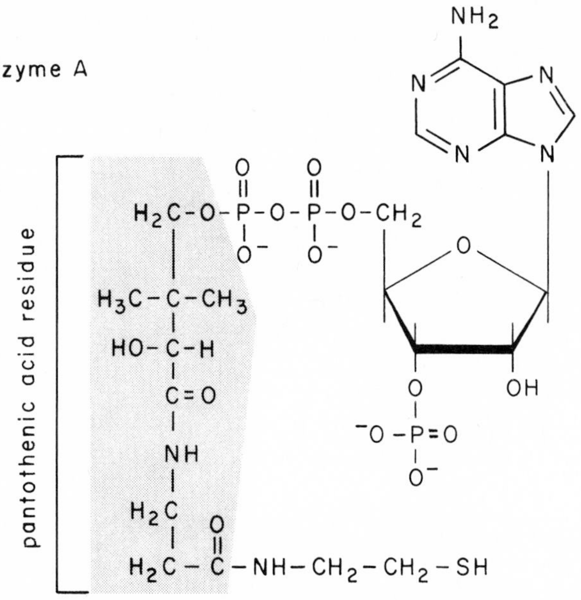

The pantothenic acid moiety is shaded in the above structure of coenzyme A, which is central to the anabolic and catabolic processes. Therefore, it is understandable that pantothenic acid deficiency would have profound consequences. Fortunately, there is little evidence of pantothenic acid deficiency in humans.

Biotin.   Like pantothenic acid, biotin is widely distributed in foods, so that a deficiency state rarely exists. Egg white contains a protein, *avidin*, that binds tightly with biotin. Therefore, biotin deficiency can occur in people who eat large quantities of raw eggs. The relatively simple, bicyclic molecule is involved in biochemical reactions where carbon dioxide is added to a molecule to produce a carboxyl (—COOH) group. This is in contrast to decarboxylations, reactions in which carbon dioxide is removed and in which thiamine is usually involved.

Biotin

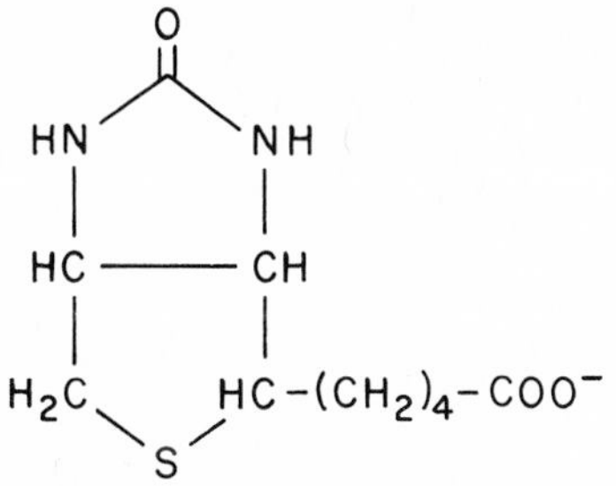

Folic acid.   Folic acid is composed of residues of glutamic acid, *p*-aminobenzoic acid, and a heterocyclic ring system, a pteridine. It is found in meats and vegetables, particularly green leaves. In most food sources folic acid is combined with one to five additional glutamic acid residues.

Folic acid, (R=H)

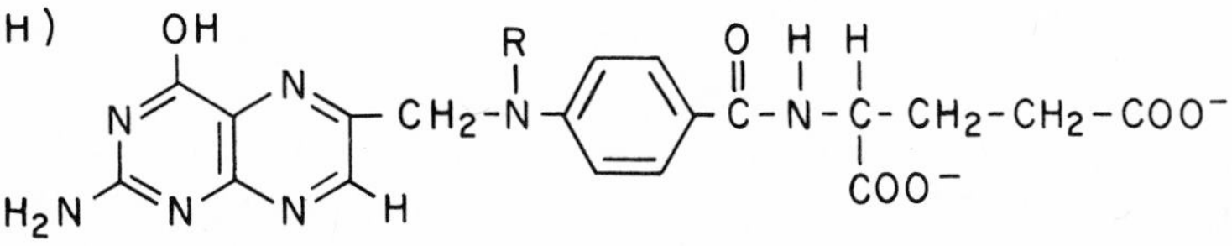

Additional variations in the molecule at substituent R also exist. Some of these derivatives are destroyed by food preparation such as boiling. Folate deficiencies are more widespread than is generally appreciated; this may be a result of low dietary intake, but the situation is accentuated by the destruction of folate in food preparation and by poor intestinal absorption.

Folate is involved in many biochemical reactions, particularly those related to the biosynthesis of deoxyribonucleic acid. Folate deficiency anemia is preceded by a reduction in serum folate levels, which is followed in a few months by a reduced number of erythrocytes. Demands for folate are increased during periods of growth and in hemolytic diseases or parasitic invasions. Growth demands the laying down of more tissue (protein biosynthesis), and hemolytic diseases require the increased rate of replacement of the erythrocytes. Parasites may limit the intestinal absorption of folate.

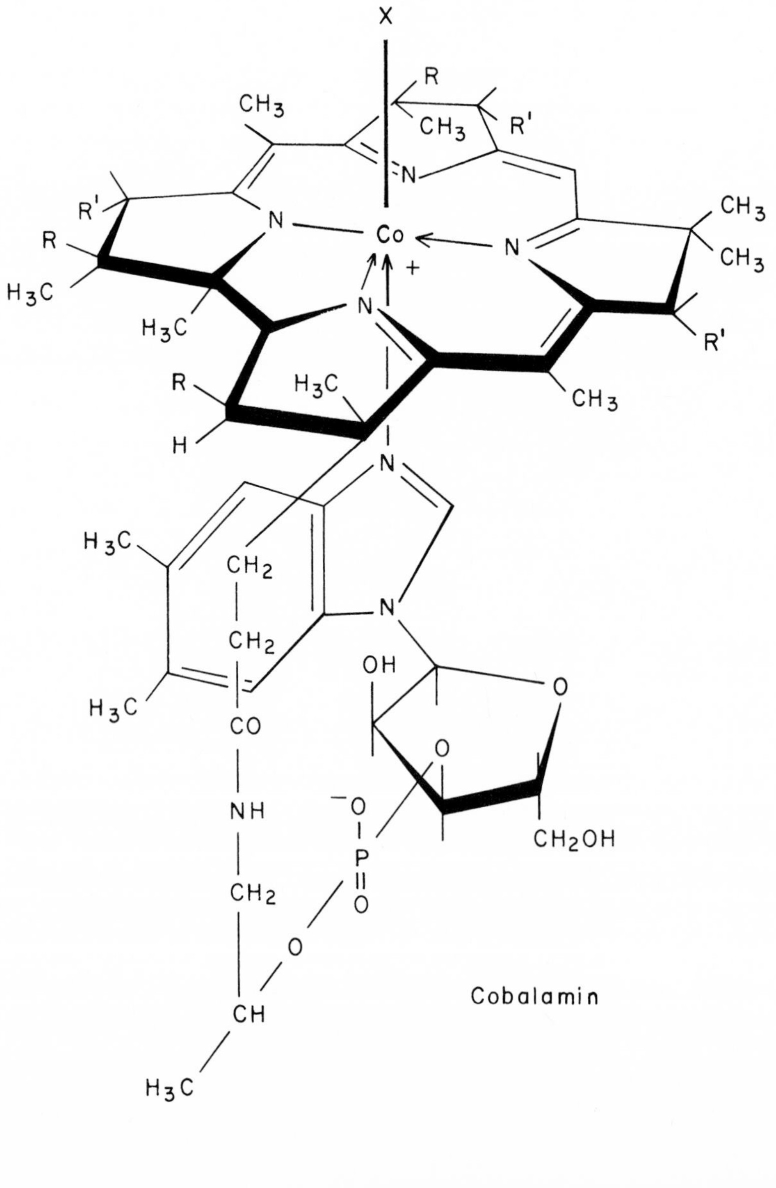

Cobalamin

Cobalamin (vitamin B$_{12}$).   Vitamin B$_{12}$ is found in foods of animal origin such as meat, especially liver and kidney; little, if any, is present in plants. In man there are negligible amounts provided by intestinal flora. It is particularly important therefore to supplement the vitamin B$_{12}$ supply in a strictly vegetarian diet. Vitamin B$_{12}$ is a complex molecule that is isolated as a cyanocobalamin (X = —CN). The active form in vivo is the methyl (X = —CH$_3$) or 5'-deoxyadenosyl derivative, as seen at the bottom of the opposite page. These substituents are attached directly to the cobalt, as shown in the partial structural formula. The total body content of vitamin B$_{12}$ is very low, about 2 μmoles. In man only two reactions have been shown to be dependent on it (pp. 262 and 534). Deficiency in the vitamin, however, results in anemic symptoms similar to those seen in folate deficiency. In addition, vitamin B$_{12}$ deficiency causes irreversible degeneration of the spinal cord by demyelinization. Because folate supplementation may mask B$_{12}$ deficiency, it is customary to delete folate from vitamin preparations.

Pernicious anemia is a disease state produced by a deficiency of vitamin B$_{12}$. It is caused by the absence of a protein called *intrinsic factor* that is normally synthesized in the gastric mucosa and is required for B$_{12}$ absorption in the intestine. It is therefore necessary to inject B$_{12}$ parenterally to treat pernicious anemia and its associated nervous system defects.

Ascorbic acid (vitamin C).   Vitamin C is found in fresh fruits and vegetables. Deficiency of vitamin C produces scurvy. The condition is corrected by feeding vitamin C, more correctly called L-*xylo*-ascorbic acid. Ascorbic acid is easily oxidized to dehydroascorbic acid, the ratio of these two forms varying somewhat in the tissues but always with the reduced form in large excess. Vitamin C is involved in the biosynthesis of collagen, the structural protein of connective tissue.

$$
\begin{array}{cc}
\begin{matrix}
\text{C=O} \\
| \\
\text{C-OH} \\
\| \\
\text{C-OH} \\
| \\
\text{H-C} \\
| \\
\text{HO-C-H} \\
| \\
\text{CH}_2\text{OH}
\end{matrix}
\quad
\xrightleftharpoons[+2H]{-2H}
\quad
\begin{matrix}
\text{C=O} \\
| \\
\text{C=O} \\
| \\
\text{C=O} \\
| \\
\text{H-C} \\
| \\
\text{HO-C-H} \\
| \\
\text{CH}_2\text{OH}
\end{matrix}
\end{array}
$$

Ascorbic acid   Dehydroascorbic acid

This accounts in part for the symptoms seen in scurvy of pathologic changes in teeth and gums, the tendency to hemorrhage resulting from weakened blood vessels and capillary beds, poor wound healing, and ulceration.

The level of vitamin C required to avoid scurvy is usually less than 50 mg/day. The exact level depends on many factors such as age, pregnancy, and individual variables, but a daily intake of 100 to 150 mg is usually recommended. An intake of several g/day has been recommended to maintain optimum health and maximum defense against infection. These recommendations are not found in the RDA (1974). Fortunately, insofar as is presently known, there are no toxic effects of ascorbic acid when the body is saturated with this vitamin, since any additional intake is either excreted intact or metabolized.

Mineral requirements

The inorganic components of the human body are principally sodium, potassium, calcium, magnesium, iron, phosphorus, chloride, and sulfur. To a large extent they represent the minerals of the skeleton and the buffer ions of the body fluids. They are essential parts of the diet. A number of elements that are present in much smaller amounts, called the trace elements, are also essential dietary components. These include copper, molybdenum, cobalt, manganese, zinc, chromium, iodine, and fluoride. Also found in the ash of the body are other elements such as barium, strontium, lead, nickel, and mercury. These are not known to be essential for metabolism and represent accumulations without controlled absorption or excretion—a body pollution.

The following comments are not intended to be extensive, but they do give some indication of the principal physiologic roles of the inorganic ions.

Sodium, potassium, and chloride. Sodium, potassium, and chloride represent the major ions of the body fluids. Sodium and chloride are concentrated mainly in the extracellular fluids, whereas potassium is found largely within the cells and is essential for many enzymic processes, for the transmission of nerve impulses, and for the functioning of muscle. The interplay of these ions, the control of body fluid volume and ionic concentration, and the energy involved in maintaining these ions in their respective fluid compartments are critical problems of homeostasis.

Sodium and potassium are present in most diets, and dietary allowances have not been established for these. Dietary deficiencies do not occur in a normal diet.

Calcium and phosphorus. A great variety of foods contain calcium and phosphorus. Calcium phosphate in the form of hydroxyapatite is the principal component of the hard structures of bones and teeth. There is a continuous exchange of these ions between the circulating fluids and the solid skeletal tissues and cell membrane structures. Calcium is involved in nerve and muscle excitability, blood coagulation, and some enzyme activities. Phosphate is an essential part, as organic esters, of the reactive form of most intermediary metabolites. It also plays an important role in the storage of chemical energy in the form of ATP.

Magnesium. Most foods, especially those of plant origin like potatoes, whole grain cereals, and fruits, contain magnesium, recommended daily dietary allowances for which are around 300 to 450 mg for adults (see Appendix A). Magnesium is largely bound with phosphates in the skeleton. It is also essential in metabolism, particularly in those reactions involving ATP. Magnesium is a depressant of the central nervous system and also elicits hypotension. High concentrations of magnesium reduce the heart rate and ultimately produce cardiac arrest. Some antagonism is noted between Ca^{++} and Mg^{++}. High levels of Ca^{++} tend to aggravate the effects of low Mg^{++}, probably because of interference with Mg^{++}-requiring enzymes.

Sulfur. Most of the required dietary sulfur is provided by the dietary protein. Sulfur is a component of the amino acids cysteine and methionine, the coenzyme CoASH, and the vitamins thiamine and biotin. It is present as an organic sulfate ester in some biopolymer components of connective tissue, in complex lipids known as glycolipids, and in conjugated bile acids that were referred to earlier as emulsifying agents for dietary fat in the intestine.

Iron. Iron is required for the synthesis of the heme portion of hemoglobin and myoglobin. It is also needed in nonheme proteins and other intracellular molecules called the cytochromes, which are involved in the oxidation of metabolites. Approximately 25% of the body iron is stored in the liver, spleen, and bone.

It is also present in the plasma protein transferrin. Iron is present in a variety of foods of animal and plant origin, particularly liver, egg yolk, and legumes; milk is a poor source. A deficiency of dietary iron gives rise to iron-deficiency anemia. Iron is conserved very efficiently by the body, but significant losses occur during hemorrhage.

Iodine. A high proportion, 70% to 80%, of the iodine in an adult is concentrated in the thyroid gland, where it is involved in the biosynthesis of the hormones thyroxine and triiodothyronine (Chapter 13). The natural sources of dietary iodine are vegetables, which vary in their iodine content according to the soil in which they are grown and the mineral content of their water supply. Deficiencies of iodine in the diet result in hyperplasia of the thyroid gland, a condition known as goiter. The natural occurrence of iodine in the water and soil can be correlated with the incidence of endemic goiter. Where the natural iodine supply is low, food additives such as iodized salt must be employed. Endemic goiter persists as a serious health problem in some parts of the world in spite of these very simple prophylactic measures that have been widely known for many years.

Fluoride. The classification of fluoride as an essential element depends to some extent on the criterion of essentiality. There is ample evidence that the consumption of fluoride confers increased resistance to dental caries, and there is indication that proper fluoride intake is related to normal skeletal maintenance. Thus it is reasonable to classify fluoride as an important ingredient of the diet but, like carbohydrate, not absolutely essential.

Few foods contain more than 1 to 2 parts per million (ppm) of fluoride, and the natural drinking water of many areas contains less than 0.1 ppm. In local areas where the fluoride content of water is very high, greater than 8 ppm, there is evidence of chalky white patches on the surface of the permanent teeth because the enamel is weak and easily pitted. These areas may also become stained. This mottled effect is only found in permanent teeth and develops during their formation. Severe mottling can also be accompanied by abnormal bone mineralization. The incidence of mottling reduces to near zero as the fluoride level of the drinking water falls to around 1 ppm.

Mottling of the enamel was noted with an accompanying increased resistance to dental caries. This resistance can be achieved without the mottling effect. The optimum protection against caries without the development of mottling occurs at around 1 ppm in the drinking water. Such concentrations are established in many communities by fluoridation of water with sodium fluoride or fluorosilicate. It appears that the fluoride is involved in a modification of the hydroxyapatite crystal in the enamel of the tooth, possibly by substituting for the hydroxyl group.

$$(Ca_3P_2O_8)_3 \cdot Ca(OH)_2 - 2F^- \longrightarrow (Ca_3P_2O_8)_3 \cdot CaF_2 + 2OH^-$$

Hydroxyapatite   Fluorapatite

Whether this is the mechanism by which it imparts resistance to dental caries is still a matter of debate.

Skeletal conditions of demineralization such as osteoporosis, which result in decreased radiologic bone density, have been related in some cases to low fluoride levels in drinking water. Reports of improvement in calcium balance and bone density as a result of fluoride therapy indicate a promising direction for research into such diseases.

Other trace elements. Copper, manganese, molybdenum, and zinc have each been shown to function as part of different enzyme systems or to be present in serum, where they are transported as metalloproteins. Cobalt must be provided in the form of vitamin B_{12}. There is demonstrable interplay between some pairs of metals; for example, high levels of copper reduce the intestinal absorption of zinc. Low serum levels of copper have been accompanied by anemia, but the role of copper in hemoglobin synthesis is not known.

Since most foods or water contain these trace metals, it is difficult to evaluate daily requirements. Excesses of most minerals result in symptoms of toxicity. In many instances the abnormalities resulting from intoxication are reversible; even heavy metal poisons such as lead tend to be excreted to some extent. More rapid elimination requires the administration of chelating agents such as the use of 2,3-dimercaptopropanol in the chelation of mercury and arsenic (Chapter 3).

SPECIAL PROBLEMS IN NUTRITIONAL MAINTENANCE
Parenteral nutrition

When a patient is unconscious or unable to take food by mouth, it is necessary to provide nutrients by special feeding. Parenteral (intravenous) feeding has routinely provided calories in the form of glucose, dissolved in isotonic sodium chloride (0.153 M), for periods of a few days. A total of 2000 to 3000 ml/day may be infused. This volume may be increased if other fluid losses are to be corrected. Such a fluid diet places the patient in (1) negative nitrogen balance since some nitrogen is always being excreted in the urine, (2) caloric insufficiency (5% aqueous glucose is equivalent to 20 kcal/100 ml), and (3) deficiencies in the bulk minerals such as potassium, magnesium, and calcium.

Hyperalimentation

Prolonged parenteral feeding procedures must provide a complete diet that is calorically adequate and nutritively balanced. This will allow the patient to maintain body weight or, if an infant, to grow. Provision of sufficient calories for an adult requires the infusion of large volumes, and this presents the danger of pulmonary edema or cardiac failure. This problem may be solved by feeding more concentrated solutions containing glucose, amino acids contained in a hydrolysate of a nutritionally complete protein such as casein, essential minerals (except iron), and vitamins. This solution could be infused directly into the superior vena cava, where the blood flow is great enough to dilute the hypertonic intravenous solution. The procedure is known as *parenteral* or *intravenous hyperalimentation*. However, the problem of administering water-insoluble lipids is only partially solved. Emulsification is not permanent enough, and the addition of lipids in this manner produces the danger of capillary occlusions. Therefore, fat is usually omitted from these intravenous feeding preparations, and fat-soluble vitamins are provided by injection. Essential fatty acid is partially provided by transfusing blood plasma once a week. Iron is also provided in the plasma; thus the problem of precipitation of iron hydroxides in the parenteral solutions is avoided.

Gavage feeding

A more natural feeding method is accomplished by stomach tube when the gastrointestinal system is functioning properly. A typical gastrostomy tube has an internal diameter of approximately 3 to 4 mm, a size that permits homogenates of most foods to be fed. Nutrition can be provided for weeks by this route without danger of venous thrombosis, pulmonary edema, or cardiac failure. The stomach contents can be emptied at any time that this should become necessary. However, there are occasions when gavage feeding requires precautions, for example, the unconscious patient who might aspirate if he or she vomits.

Weight changes

Weight loss. The body obeys all the classic laws of thermodynamics and requires a specific number of calories for homeostasis. Weight loss occurs when there is a deficit of calories. Balance studies indicate that at times of dietary restriction there is an initial loss of tissue protein. However, adaptation quickly takes place to conserve protein, and the fat of adipose tissue becomes the predominant source of the missing dietary calories. One pound (454 g) of adipose tissue contains 85% lipid and 15% water and therefore can be metabolized to produce

$$\left(454 \times \frac{85}{100} \times 9\right) \text{ kcal/pound}$$

or 3500 kcal/pound of body fat. From this figure one can predict the rate of weight loss within about 10% over an extended period of time. There are daily and weekly fluctuations that are associated largely with water losses.

Weight gain. More than 3500 kcal are required to add 1 pound of body fat, the difference in part being the added metabolic energy needed for biosynthetic as opposed to catabolic processes. It is more difficult, however, to predict the rate of weight gain in an adult for several reasons. The new tissue weight may be fat or protein, depending on the nutritional state of the individual. A well-nourished person will convert the additional calories into fat, but it is found experimentally that an excess of 3500 kcal adds only about 0.5 pounds of body weight. An emaciated individual on a balanced diet will gain weight by the development of both muscle protein and adipose tissue. Body protein, however, is approximately 80% water by weight and is equivalent to

$$\left(454 \times \frac{20}{100} \times 4\right) \text{ kcal/pound}$$

or 365 kcal/pound of muscle tissue. This is significantly less than adipose tissue.

In general, for initial estimates one may assume that twice as many calories are required to produce a weight gain as would be expected from the caloric content of the tissue. Thus to add 1 pound of adipose tissue requires 7000 kcal and 1 pound of muscle requires 900 kcal. Just how many calories can be fed will be controlled by the appetite, although forced feeding is possible by gavage if necessary. When it is desired to build up a patient, the protein content of the diet is usually increased. High-protein diets, however, require good kidney function. Also, the development of muscle tissue requires associated exercise.

REFERENCES

Davidson, S., and Passmore, R.: Human nutrition and dietetics, Baltimore, 1972, The Williams & Wilkins Co.

Dickie, R. S.: Diet in health and disease, Springfield, Ill., 1974, Charles C Thomas, Publisher.

Goodhart, R. S.: Modern nutrition in health and disease, Philadelphia, 1973, Lea & Febiger.

Guthrie, H. A.: Introductory nutrition, ed. 3, St. Louis, 1975, The C. V. Mosby Co.

Williams, S. R.: Nutrition and diet therapy, ed. 3, St. Louis, 1977, The C. V. Mosby Co.

The texts listed above present a detailed statement of the problems of human nutrition and its application to dietary planning.

Apfelbaum, M., Bostsarron, J., and Lacatis, D.: Effect of caloric restriction and excessive intake on energy expenditure, Am. J. Clin. Nutr. 24:1405, 1971.

Bury, K. D., Stephens, R. V., and Cha, C.: Chemically defined diets, Can. J. Surg. 17:124, 1974. *A review.*

Dudrick, S. J., and Rhoads, J. E.: Total intravenous feeding, Sci. Am 227:73, 1972. *This article describes, with good illustrations, the experiments conducted in one hospital to arrive at a complete parenteral feeding program for prolonged periods.*

Evaluation of protein quality, Publ. No. 1100, Washington, D.C., 1963, National Academy of Sciences, National Research Council. *This article deals with methods of determining biologic value.*

Freeman, J. B., et al.: Evaluation of amino acid infusions as protein sparing agents in normal adult subjects, Am. J. Clin. Nutr. 28:447, 1975.

Latham, M. D.: Nutrition and infection in national development, Science 188:561, 1975.

Plant, G. W. E., et al.: Biosynthesis of the water-soluble vitamins, Am. Rev. Biochem. 43:899, 1974.

Recommended dietary allowances, ed. 8, Washington, D.C., 1974, Food and Nutrition Board, National Academy of Sciences, National Research Council. *The tables of recommended dietary allowances, reproduced in part in Appendix A, are valuable sources of reference and illustrate various nutritional states in an American population. Equivalent tables are published for several other countries.*

Salans, L. B., and Wise, J. K.: Metabolic studies of human obesity, Med. Clin. North Am. 54:1533, 1970. *This volume is also concerned with many other aspects of human nutrition.*

Underwood, E. J.: Trace elements in human and animal nutrition, ed. 3, New York, 1971, Academic Press, Inc. *This is a complete statement on the present state of knowledge concerning all the trace elements and their toxicities.*

Clinical examples

CASE 1: VITAMIN C DEFICIENCY

A housewife and factory worker, 37 years of age, complained for 3 days of progressive pain and discoloration of both legs. She claimed to eat normally, including fresh fruits and vegetables, but on closer examination it appeared that she had been depressed for several months and in fact had eaten very little. She was unable to walk or stand unaided because of the pain. There were extensive areas of hemorrhage over the backs of the calves. There were a few follicular keratoses below the knees. No other hemorrhagic areas were found on the body, nor were there any changes in the gums. Laboratory analysis revealed 2.2 mg ascorbic acid/100 mg leukocytes. Treatment with vitamin C (700 mg/day) was started immediately. After 2 weeks the level of urinary excretion of vitamin C was 400 mg/24 hr.

Biochemical questions

1. What are the biochemical functions of vitamin C?
2. What is the change in the leukocyte levels of vitamin C as the vitamin is withheld from the diet?
3. What structural similarities are there between vitamin C and monosaccharides?
4. Since the catabolism of vitamin C varies greatly between species, yet recommendations for man are often extrapolated from other animals, how are excessive amounts of dietary vitamin C dealt with in man compared to animals? (See references.)
5. How does vitamin C function to prevent hemorrhage into various tissues?
6. What is the argument for high levels (g/day) of vitamin C?

Case discussion

Vitamin C is not excreted by the kidney until the body tissues are saturated and the concentration in the blood exceeds a certain value. Vitamin C is not stored, as are vitamins A and D, in the body fat, and deprivation results in a gradual decrease of the tissue levels.

The decrease in vitamin C levels is reflected in the blood, particularly in the leukocytes, as was seen in the patient under consideration. Replacement of vitamin C in the diet brings the patient back to health quite rapidly. Saturation, shown by excretion of the vitamin in the urine, occurs in less than 2 weeks.

The major function of vitamin C appears to be in the formation of normal collagen. After the "precollagen" has been synthesized, some of the proline residues are hydroxylated to hydroxyproline. Ascorbic acid is involved in this oxidative enzymatic reaction. Similar reactions convert lysine residues to hydroxylysine. These are two of the modifications of precollagen that are necessary for the protein to function properly in tissues. Considering therefore the ubiquitous nature of collagen in the body, it is not surprising that persons with scurvy have capillary hemorrhages (mechanically weakened vessels), swollen and painful joints, slow wound healing, and scar tissue formation, resulting in a decreased ability to combat infections.

Other biochemical roles are related to the oxidation-reduction property of ascorbic acid. This property is concerned with the hydroxylation of aromatic compounds, and thus the metabolism of tyrosine, and is possibly involved in hydroxylations of the steroid nucleus.

The involvement of ascorbic acid in biologic oxidations has been proposed for a number of years in light of the existence of both oxidized and reduced forms in blood and tissues. The molecule has many of the properties of the simple reducing sugar acids and was first referred to by Szent-Györgyi as a "hexuronic acid." (This name followed Szent-Györgyi's earlier attempts to publish the discovery of vitamin C as a sugar called "godnose" or "ignaose".) The unusual property of ascorbic acid as a sugar is the ease with which it is converted between the oxidized and reduced forms. This may be an advantage biochemically, but it also results in its destruction in foods by heating and alkaline hydrolysis to the biologically inactive 2,3-diketo-L-gulonic acid (p. 336).

The human dietary requirement for ascorbic acid has been a matter of much debate in the face of an extensive literature that extends back to the 1930's. It is established that the minimum amount of ascorbic acid required to prevent or cure scurvy in an otherwise healthy adult is about 10 mg daily or less. It has been argued, however, that the daily intake should be sufficient to saturate the body tissues. This amount is 60 to 100 mg daily depending upon age, size, and associated stressful conditions, such as wound healing or whether a woman is pregnant or lactating. Recent studies by Hodges and associates indicate that in a healthy, normal adult, 45 mg daily should maintain a full body pool of about 1500 mg of ascorbic acid, 3% of which is used daily for metabolic functions. These figures are consistent with many epidemiologic and nutritional studies, such as the absence of scurvy in England during World War II when the rationing of food permitted no more than about 20 mg of ascorbic acid daily. In those cases where supplementary provisions of the vitamin occurred there was little evidence of improvement in the general health.

Estimates of the need for large daily requirements of vitamin C are based in part on extrapolations from animal needs to those for human beings. Aside from the primates, the guinea pig is the only common animal that requires ascorbic acid in the diet. Not only is the guinea pig much smaller than human beings, but its BMR is much greater per kg body weight. Furthermore, the metabolic pathways for vitamin C are different in the various species. Unlike the guinea pig, man does not metabolize any of the ascorbic acid to carbon dioxide. Together with other differences in the metabolism of this vitamin, the assumptions made to support high dosage rates do not seem justified. Any pharmacologic benefits distinct from the nutritional effects of ascorbic acid are yet to be proved in man.

REFERENCES

Baker, E. M., et al.: Metabolism of [14]C- and ^{3}H-labeled L-ascorbic acid in human scurvy, Am. J. Clin. Nutr. 24:444, 1971.

Hodges, R. E.: What's new about scurvy, Am. J. Clin. Nutr. 24:383, 1971.

Hodges, R. E., et al.: Clinical manifestations of ascorbic acid deficiency in man, Am. J. Clin. Nutr. 24:432, 1971.

King, C. G., and Burns, J. J.: Second conference on vitamin C, Ann. N.Y. Acad. Sci. 258:1-552, 1975.

Miller, E. J.: Biosynthesis of collagen, Fed. Proc. 33:1197, 1974.

Pauling, L.: Vitamin C and the common cold, San Francisco, 1970, W. H. Freeman & Co. Publishers.

Walker, A.: Chronic scurvy, Br. J. Dermatol. 80:625, 1968.

CASE 2: OBESITY*

A young typist, Ms. O. B., had been plagued by obesity for a long time. She was heavy throughout childhood but gained more rapidly during her high school and college years. She had a family history of both obesity and diabetes. Her father and several aunts and uncles on her mother's side were diabetic. Her maternal grandmother had had a coronary occlusion. She herself had had glycosuria at one time.

At the age of 22 years she weighed 168 pounds and was 5′9″ tall. Her weight had

*Case courtesy Dr. R. E. Hodges, Department of Medicine, University of California, Davis, Calif.

been quite variable, ranging from a low of 150 pounds to a high of 210 pounds. On her first clinical visit at age 25 years she weighed 176 pounds and wished to reduce to 130 pounds, a good weight for her height. Her activity could be described as sedentary to moderate. Her dietary habits were poor in that she usually skipped breakfast, ate a light lunch, and then had a large evening meal followed by numerous snacks. Although prepared to diet, she did not wish to reduce her protein intake below the present level for reasons of palatability. Her dietary history is shown on the opposite page.

	Weight (g)			Calories (kcal)
	Protein	Carbohydrate	Fat	
Sample menu				
Breakfast:				
None				
Lunch:				
Luncheon meat sandwich				
2 bread	4	30		140
1 slice luncheon meat	4.6	0.5	6.8	81
1 tsp margarine			5	45
1 apple		10		40
1 cup whole milk	8	12	8	152
Total	16.6	52.5	19.8	458
Supper				
4 oz baked chicken	28		20	300
½ cup mashed potatoes	2	15		70
½ cup peas with mushrooms	2	7		35
1 tsp margarine			5	45
sliced tomatoes on				
a lettuce leaf				
1 tbsp Thousand Island		2.2	7.0	70
dressing				
1 hot roll	1	7		35
1 tsp margarine			5	45
⅙ apple pie	3.4	61	17.8	410
½ cup ice cream	2	15	10	158
1 cup whole milk	8	12	8	152
Total	46.4	119.2	72.8	1320
Snacks in the evening				
1 peanut butter and				
jelly sandwich				
2 bread	4	30		140
4 tbsp peanut butter	15.6	12.8	29.8	340
2 tsp jelly		10		40
1 tsp margarine			5	45
18 oz coke		48		192
1 oz Hershey chocolate	3.1	15.7	11.9	176
almond bar				
20 wheat thins spread with	4	28		140
6 tsp cream cheese		12		60
Total	26.7	156.5	46.7	1133

Total kcal — approximately 2900

	g	% w/w	Approximate percentage of kcal
Total protein	89.7	16%	12
Total carbohydrate	328.2	59%	45
Total fat	139.3	25%	43
	557.2		

Biochemical questions

1. Does the history of Ms. O. B. provide any clues as to the cause of her obesity?
2. What is the main chemical substance that accumulates in the body in obesity, and what interconversions of nutrients are responsible for this accumulation?
3. How should the present diet be changed in terms of caloric intake?
4. What rate of weight loss should be recommended, and how should the diet be modified periodically to achieve this?
5. What proportions of protein, carbohydrate, and fat should be incorporated into the new diets?
6. What kind of protein should be used in the new diet?
7. Should there be a change in physical activity? Explain.

Case discussion

Obesity. Obesity is most commonly defined as an accumulation of excess body fat. Obesity is not always the reason for a person being overweight, for example, the great musculature of an athlete. It is a disease that carries risks, as it is frequently associated with shortened life expectancy, coronary artery disease (Ms. O. B.'s grandmother), and other problems such as backache that arise because the person is carrying around too much weight. Obesity often has a familial component, as illustrated in this case. Several studies have shown that if one parent is obese then approximately 40% to 50% of the children will develop obesity. Furthermore, most diabetics are obese, and it is noted that Ms. O. B. came from a family with a history of diabetes. This is not to say, however, that all the obese are diabetic. The relationship between diabetes mellitus and obesity is clinically well recognized, but the common biochemical factors are poorly understood. Although the questions will be raised when diabetes is discussed in connection with carbohydrate metabolism, it may be noted here that the handling of dietary glucose by the obese patient is abnormal and improves after the excess adipose mass has been reduced. Similarly, the obese are more resistant to the action of the hormone insulin, which is involved in the metabolism of glucose and is reduced in the diabetic. An elevation of circulating plasma insulin is characteristic of the obese (p. 353). These aberrations are improved following weight reduction.

It has been calculated that in a normal adult the number of cells in the adipose tissue is relatively constant and that increase in adipose mass occurs through an increase in cell size. Obesity that is characterized by an excessive increase in cell size (hypertrophy) of a normal number of cells is the more common type. The other type of obesity is characterized by increased numbers of cells in the adipose tissue (hyperplasia or hypercellularity) (p. 385). This type of obesity commonly develops early in life and is thought to result from excessive feeding in infancy. Obesity that occurs later in life as a result of hyperplasia is the more difficult to control through dietary restrictions. However, everybody, normal or otherwise, loses weight as the result of dietary restriction of calories; therefore, such an approach was followed for Ms. O. B.

Caloric requirements. Ms. O. B.'s dietary habits were poor. The analysis indicates a caloric intake too high for her desired weight of 130 pounds. For example, taking the figure of 2900 kcal/day from her diet analysis and assuming that she could be considered sedentary to moderately active, the caloric requirement for her present 176-pound weight (176 pounds is equal to 176/2.2 kg) might be estimated.

Metabolic caloric requirement
$$= BMR + \text{average activity}$$
$$BMR = 24 \times \frac{176}{2.2} \text{ kcal/day}$$
$$= 1920 \text{ kcal/day}$$

Activity (average of sedentary and moderate)
$$= \frac{30 + 40}{2} \% \text{ of BMR kcal/day}$$
$$= \frac{35}{100} \times 1920 \text{ kcal/day}$$
$$= 670 \text{ kcal/day}$$
$$\approx 700 \text{ kcal/day approximately}$$

Metabolic caloric requirement
$$= (1920 + 670) \text{ kcal/day}$$
$$= 2590 \text{ kcal/day}$$
$$\approx 2600 \text{ kcal/day approximately}$$

It is no wonder that she is maintaining her overweight status.

Ms. O. B. is a typist and wishes to reduce to 130 pounds, which is a good average weight for her height and build. Her daily activities might be increased to those shown below, corrected for a 60 kg final weight:

	Caloric equivalent/day
Typing, 8 hr	$8 \times 60 = 480$
Cycling (8 mph), 1 hr	$1 \times 150 = 150$
Writing (or equivalent), 3½ hr	$3\frac{1}{2} \times 24 = 84$
Swimming (2 mph), ½ hr	$\frac{1}{2} \times 474 = 237$
Dishwashing (or equivalent), 1 hr	$1 \times 60 = 60$
Eating, 2 hr	$2 \times 24 = 48$
Sleeping, 8 hr	$8 \times 0 = 0$
	1059

Note that this activity is significantly increased over the 670 kcal assumed for her present activity. Also, since her present weight is 80 kg, this activity level would require 1410 kcal/day at the beginning of the dieting.

Rate of weight loss. Assuming that the period of dieting will be accompanied by a constant level of activity, the approach to the problem can follow three principal regimens.

1. The diet plan can be designed to effect a constant amount of weight loss each week, for example, 2 pounds/week, which corresponds to an intake of 1000 kcal/day less than required to maintain the weight and activity for the day in question. Such a regimen would lead to a very low caloric intake at the end of the dieting period. For example, Ms. O. B. will require about 2500 kcal/day when she reaches her ideal weight. Using plan 1 she would be eating 1500 kcal/day as she neared the end of the diet period. Therefore, the diet would have to be terminated by raising the caloric intake in order to maintain the desired weight. There is a danger that she would continue on the low calorie diet too long and become underweight.

2. Calculate the caloric intake that will be required once desired weight is achieved at the appropriate level of activity. This amount of food is taken constantly over the whole diet period with the result that the weight loss will be greatest initially and then taper off to zero. The time span required to achieve the desired weight will be extended, but no adjustment in diet is necessary during the diet period or once the ideal weight is reached. For Ms. O. B. the diet that would be prescribed is 2500 kcal/day, the amount needed to maintain her at 130 pounds (70 kg) if the schedule of activities is maintained. This figure is calculated as follows:

$$\text{BMR} = 24 \times \frac{130}{2.2} \text{ kcal/day}$$
$$= 1420 \text{ kcal/day}$$
$$\text{Activity} = 1050 \text{ kcal/day}$$
$$\text{Total} = 2470 \text{ kcal/day}$$

3. A compromise of the two preceding regimens. Although many physicians might recommend the 2500-kcal diet it was decided to begin with a 1600-kcal diet. This would produce more initial weight loss and would give Ms. O. B. encouragement that her efforts were worthwhile and were going to succeed.

It was decided to reduce the present caloric intake of about 2900 kcal/day to 1600 kcal/day. This is still significantly above the level of approximately 900 kcal/day that is estimated as being minimal for an adult to receive the necessary nutrients to maintain health. Since the tissue lost will be principally adipose, a reduction of 1000 kcal/day or 7000 kcal/week is equivalent to the loss of 2 pounds of body fat/week. Further loss will be achieved by the increased level of activity, which was originally estimated to be 700 kcal/day and is 1400 kcal/day at the beginning of the dieting. This *increase* of 700 kcal/day or 4900 kcal/week is equivalent to 4900/3500 pounds weight loss or about 1½ pounds/week. The rate of weight loss should be initially 3½ pounds/week. This rate will become less as the dieting proceeds. As the desired weight of 130 pounds is approached, so the caloric intake will need to be increased to about 2500 kcal/day if the schedule of activities is maintained. This figure was calculated for regimen 2.

Distribution of calories. Normal adequate diets in the United States provide 15% of the calories from proteins, 45% from carbohydrates, and 40% from fats. If this were the case for a 1600-kcal diet, then:

1. Protein provides $1600 \times \dfrac{15}{100}$ kcal/day
$$= 240 \text{ kcal/day}$$

and since 1 g protein is equivalent to 4 kcal energy, then:

$$240 \text{ kcal protein} = \frac{240}{4} \text{ g} = 60 \text{ g}$$

Similarly:

2. Carbohydrate provides $1600 \times \dfrac{45}{100}$ kcal/day
$$= 720 \text{ kcal/day}$$
$$= \frac{720}{4} \text{ g/day}$$
$$= 180 \text{ g/day}$$

3. Fat provides $1600 \times \dfrac{40}{100}$ kcal/day
$$= 640 \text{ kcal/day}$$
$$= \frac{640}{9} \text{ g/day}$$
$$= \text{approximately } 70 \text{ g/day}$$

If such a distribution of foods is modified in a diet designed for reducing, the following points need to be considered.

Nature of protein. In a low-calorie diet the intake of protein, mineral, and vitamins should not fall below the levels recommended by the National Research Council. For Ms. O. B. this requires approximately 55 g protein with a biologic value of 70 (based on the biologic value of egg protein as 100). The biologic value of a protein measures its nutritional value relative to a standard protein and takes into account its digestibility as well as its amino acid composition. Protein diets that completely lack any of the essential amino acids will have no biologic value as all the nitrogen will be excreted. On this relative scale, haddock protein is 89, milk protein is 79, vegetable protein is 59, and flour protein is 46. More

than 55 g of protein must be consumed to provide the minimum daily requirements if, for example, a vegetarian diet is requested by the patient. A protein intake of around 80 to 90 g, which Ms. O. B. desired for palatability, necessitates a caloric adjustment in the carbohydrates and fats.

Carbohydrate and ketosis. Enough carbohydrate should be provided to prevent ketosis and to ensure that the protein can be spared to provide amino acids for tissue maintenance. A minimum level of carbohydrate to prevent ketosis requires approximately 5 g/100 kcal in the diet. In a 1600 kcal diet this amounts to 80 g carbohydrate. However, the reducing diet is planned to allow the patient to initially lose $3\frac{1}{2}$ pounds/week or $3\frac{1}{2} \times 454$ g/week, or about 227 g/day, which is largely a loss of fat tissue. Since adipose tissue contains about 85% lipid, the body metabolism includes the oxidation of $227 \times \dfrac{85}{100}$ (or 193) g lipid/day to produce 193×9 (or 1737) kcal/day. This is added to the 1600 kcal/day in the diet, so that the total metabolic energy actually provided to Ms. O. B.'s tissues is $1600 + 1737$ or 3300 kcal/day. Based on the requirement of 3300 kcal, Ms. O. B. needs at least 5×33 or 165 g of carbohydrate.

Lipids and essential fatty acids. There is a nutritional requirement for essential fatty acids that is satisfied on the average by 1% of the caloric intake (16 kcal), equivalent in this case to about 2 g linoleic acid. Additionally, the fat intake could be reduced because Ms. O. B. will be oxidizing her own adipose tissue at a rate of approximately 1700 kcal/day, so that total lipid catabolism is quite high.

Vitamins and minerals. An adequate supply of vitamins and minerals should be ensured. Since most minerals are derived from vegetables, fruits, cereals, and proteins, it is likely that most average dietary programs will satisfy the mineral need. Vitamins, too, are usually adequately supplied, but these variables need to be checked as the diet program progresses.

Water. Water is required in the diet

	Weight (g)			Calories (kcal)
	Protein	Carbohydrate	Fat	
Sample menu				
Breakfast				
½ cup orange juice		10		40
1 egg poached on	7		5	73
1 slice toast	2	15		68
¾ cup cornflakes	2	15		68
1 tsp margarine			5	45
1 cup fortified skim milk	10	15	1	109
coffee, black				
Total	21	55	11	403
Lunch				
tuna and	14		10	146
dill pickle on rye bread	4	30		136
¾ cup shredded lettuce				
1 tbsp dressing			5	45
1 raw apple		10		40
1 cup fortified skim milk	10	15	1	109
Total	28	55	16	476
Supper				
1 cup hot bouillon				
3 oz broiled halibut steak	21		15	219
with lemon slice				
½ cup parsley potatoes	2	15		68
½ cup small whole beets	2	7		36
1 tsp margarine			5	45
½ cup tossed green salad with				
carrots and radishes				
3 saltine crackers	1	7		34
1 cup cantaloupe balls		10		40
1 macaroon cookie	1.3	14		107
coffee, black				
Total	27.3	53	20	549
Snacks in the evening				
¼ cup orange sherbert	2	15		68
1 cup milk, fortified skim	10	15	1	109
Total	12	30	1	177

Total kcal—approximately 1600 kcal

	g	Approximate percentage of kcal
Total protein	88.3	23
Total carbohydrate	193	49
Total fat	48	28

for many physiologic functions; an approximate amount would be 30 to 45 ml/kg body weight.

Proposed diet. Taking all of these factors into consideration, the diet shown above was initiated, with appropriate exchanges being made in the foods to provide variety. Ms. O. B. lost more than 50 pounds in the year following the commencement of the diet and has been able to maintain her weight within a normal range. It should be emphasized, however, that all initial calculations present a starting point in a program that must be carefully followed to check for unusual individual responses.

REFERENCES

Bortz, W. M.: Predictability of weight loss, J. A. M. A. 204:101, 1968.

Bowes, C. F., and Church, H. N.: Food values of portions commonly used, ed. 9, Philadelphia, 1963, J. B. Lippincott Co.

Mann, G. V.: The influence of obesity on health, N. Engl. J. Med. 291:178, 1974.

Meyer, J.: Overweight: causes, cost and control, Englewood Cliffs, N. J., 1968, Prentice-Hall, Inc.

Miller, D. S., and Mumford, P.: Gluttony, Am. J. Clin. Nutr. 20:1212, 1967.

CASE 3: VITAMIN A DEFICIENCY AND NIGHT BLINDNESS

A prisoner of war returned to his home country in a debilitated state. In the course of a physical examination he complained of dryness of the eyes, sticky eyelids, and temporary blindness when moving from areas of bright to dim lighting. Ophthalmologic examination disclosed no infection or foreign body causing the redness of the eyes, and a test for dark adaptation was given. The patient viewed a bright light for 5 min and then a dim light of varying brightness. The minimum brightness that could be seen was recorded each minute. The patient never could see as dim a light as he normally should have during the 45 min of the test. As part of his recovery diet, several times the normal daily requirement of vitamin A was provided over a period of several months. The patient's night vision gradually returned to normal.

Biochemical question

1. How is night blindness related to vitamin A deficiency?

Case discussion

Visual cycle. The biochemistry of vitamin A in the visual process involves both types of photoreceptor cells in the

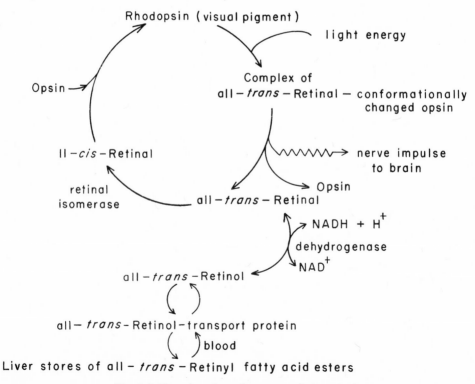

Fig. 1-2. Visual cycle and storage of vitamin A.

retina. The rods provide black and white vision and respond to dim light. The cones provide color vision and respond to bright light. Both of these cells require vitamin A, also called retinol, for the formation of the visual pigments. The molecular cycle is summarized in Fig. 1-2.

The protein opsin, when complexed with 11-*cis*-retinal by the formation of a Schiff base with a free amino group, produces the visual pigment rhodopsin. Under the influence of light, the retinal moiety is converted to all-*trans*-retinal. This isomerization is associated with a conformational change in the opsin. Such a change is transferred as a nervous impulse, possibly by alteration of the permeability to Na$^+$ or K$^+$ ions, from the photoreceptors to the optic nerve endings. Dissociation of the all-*trans*-retinal is followed by reduction of the

aldehyde group to the primary alcohol to give all-*trans*-retinol. Retinol returns to the circulating blood pool of vitamin A that is stored in the liver. The retinol is then converted back to all-*trans*-retinal in the retina, where it is also isomerized back to 11-*cis*-retinal. The cycle is completed when 11-*cis*-retinal reacts with opsin to produce rhodopsin.

Vitamin A replacement. For some reason, loss of vitamin A occurs during the photochemical event, so that a continuous dietary replacement of the vitamin is necessary. Deficiency results in lowered blood levels of vitamin A, and the time required for the re-formation of rhodopsin after bleaching by light is lengthened. Dark adaptation is thus slow, and treatment with vitamin A is needed to build up both the circulating and storage levels.

Vitamin A is stored in the liver in

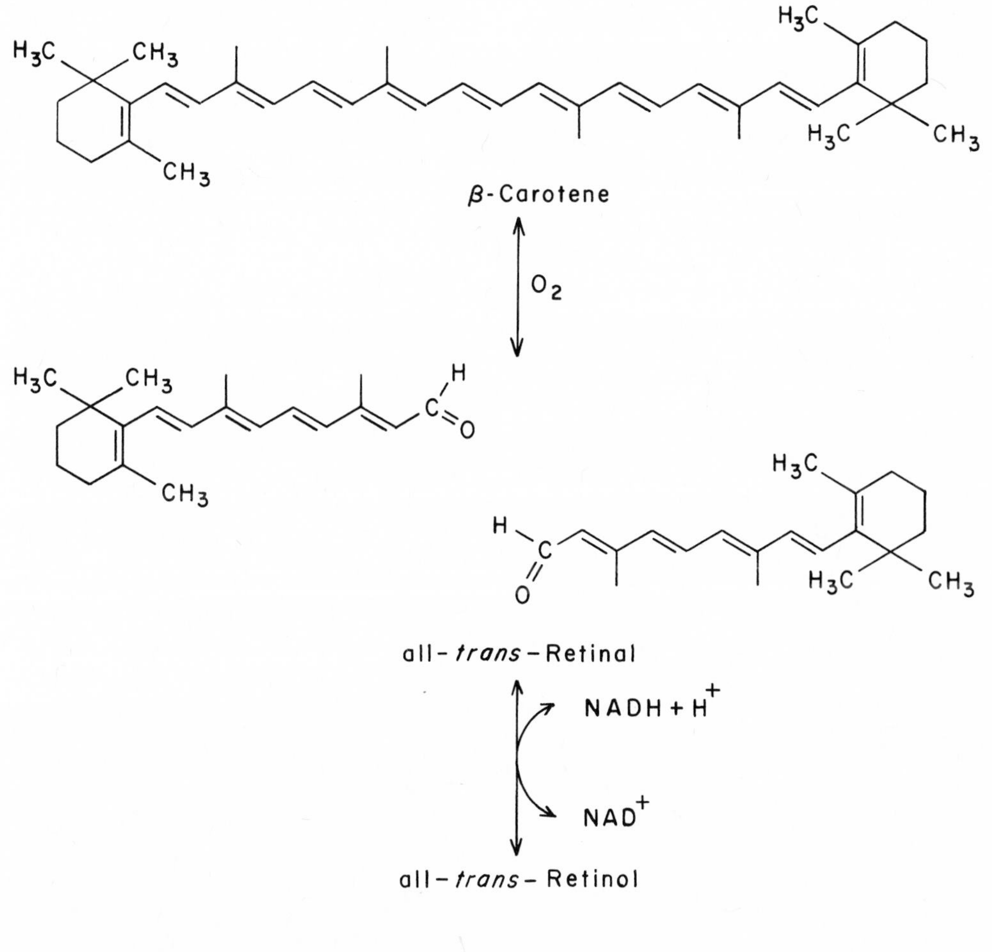

β-Carotene

all-*trans*-Retinal

all-*trans*-Retinol

such relatively large amounts (0.2 to 2.0 μmoles/g of liver) that deficiencies severe enough to cause clinical problems are not likely to develop unless deprivation is extreme and is continued for relatively long periods, at least many months.

Color vision. Similar photochemical events occur in both the rod and the cone cells. Each cone cell has one of three different color-sensitive pigments, blue (430 nm), green (540 nm), or red (575 nm). The three pigments contain the same 11-*cis*-retinal moiety, but their opsins are different. Each pigment has a different wavelength of maximum absorbance, noted above in parentheses, and its photochemical reaction responds to one of the three primary color components of the incident light. In each case the visual cycle is the same as that noted for rhodopsin.

Carotenes. β-Carotene is an orange pigment found in carrots and other plant tissues. It has the biologic potency of vitamin A. The β-carotene is oxidatively cleaved by the intestinal mucosa into two retinal molecules, each of which is then reduced to all-*trans*-retinol (see p. 37). Free retinol is absorbed into the intestinal mucosa, where it is esterified with a fatty acid such as palmitate and then transported in the lymph to the liver. Retinol is transported in the blood to the tissues by a retinol-binding plasma protein. When this protein contains retinol, it binds to prealbumin, another protein present in small amounts in blood plasma. This forms a protein-protein complex that is thought to protect the retinol in some way. The retinol is removed from the plasma-binding protein by the cells of the retina and the all-*trans*-retinol is oxidized to all-*trans*-retinal as noted in the visual cycle.

REFERENCES

Michael, C. R.: Color vision, N. Engl. J. Med. 288:724, 1973.
Rowan, R., et al.: Conformation of retinal isomers, Biochemistry 13:970, 1974.
Wald, G.: The molecular basis of visual excitation, Nature 219:800, 1968.
Wasserman, R. H., and Caradino, R. A.: Metabolic roles of vitamins A and D, Ann. Rev. Biochem. 40:501, 1971.

CASE 4: OBESITY

A graduate student, 6'2" tall and weighing a steady 290 pounds, damaged his knee while playing volleyball. Toward the end of his recovery he was convinced that he needed to make an effort to correct his obesity. He was referred to a dietitian by the student health service and his typical daily intake was found to be 140 g protein, 120 g fat, and 590 g carbohydrate. He expressed the desire to reduce to 190 pounds, a level close to his ideal weight, in about 8 months without significantly changing his level of activity.

Biochemical question

1. What diet would you recommend for weight reduction? Explain.

REFERENCE

Mayer, J.: Overweight: causes, cost and control, Englewood Cliffs, N.J., 1968, Prentice-Hall, Inc.

CASE 5: ULCERATIVE COLITIS

A 52-year-old man, 6 feet in height, had inflammation and ulceration of the colon. Careful examination showed a diffuse mucosal involvement, which was worse in the rectum. He had increasing episodes of diarrhea over the last several months and was now having an average of ten small watery stools a day, often mixed with blood. The cramping colonic pain was relieved by bowel activity. Before onset of the disease he had weighed 65 kg, but his weight had fallen to 56 kg during the last year and his appetite was very poor (anorexia). Remission was induced by administration of glucocorticosteroids.

Biochemical questions

1. What type of diet would be recommended for this patient?
2. With the possible associated problems in the intestinal digestion and absorp-

tion of nutrients, is there a possibility of any vitamin deficiencies and if so, which will be most acute?

3. How would the absence of some digestive enzymes result in changed composition of the stools?
4. If there was a higher than normal concentration of low molecular weight solutes in the stools, how might this contribute to the diarrhea?
5. The patient was observed to have a low protein concentration in the blood plasma. How would you explain this hypoproteinemia?

REFERENCES

Goligher, J. C., et al.: Ulcerative colitis, Baltimore, 1968, The Williams and Wilkins Co.

Truelove, S. C.: Medical management of ulcerative colitis, Br. Med. J. 539, June 1968.

Vogel, C. M., et al.: Intravenous hyperalimentation in the treatment of inflammatory diseases of the bowel, Arch. Surg. 108:460, 1974.

CASE 6: NIACIN DEFICIENCY AND PELLAGRA

A worker in the southern part of the United States reported to a physician with many of the symptoms of pellagra: swollen tongue, dermatitis, and nervous disturbances. The man's diet consisted principally of sweet corn with small amounts of other sources of protein. The identical twin of the man had no similar complaints, and although his diet was high in sweet corn, it was mixed with very significant amounts of beans.

Biochemical questions

1. Why is niacin considered to be a vitamin when it can be synthesized by man from tryptophan?

2. What are the tryptophan contents of the proteins of corn and beans?
3. What are the biologic values of the proteins of corn, beans, and a mixture of equal amounts of these vegetables?
4. How can the conditions of the two men be reconciled with their diets?

REFERENCES

Hundley, J. M.: Niacin. In Sebrell, W. H., Jr., and Harris, R. S., editors: The vitamins, New York, 1954, Academic Press, Inc., vol. 2.

Voiculescu, V., et al.: Eur. Neurol. 10:205, 1973. *Biochemical research in two cases of endogenous pellagra.*

CASE 7: ADULT CELIAC DISEASE

A thin 65-year-old woman was admitted to the hospital with a history of bouts of diarrhea since childhood. At 35 years of age she had experienced her first episode of anemia and had responded well to treatment with liver extract. Changing the therapy from liver extract to vitamin B_{12} did not cause remission of the anemia, suggesting that she was suffering from folic acid deficiency. Despite successful treatment of the anemia, the diarrhea persisted.

On admission, examination showed that the patient was otherwise well. The central nervous system was normal. A jejunal biopsy showed a subtotal villous atrophy. Xylose excretion was reduced, only 2.1 g being excreted in the urine (normal > 5 g) in 5 hr after a 25–g test dose had been given. Fecal fat excretion averaged 13.3 g/day during a 6-day balance period (normal < 6 g/day). Vitamin B_{12} absorption was also subnormal, even when it was administered with intrinsic factor. The hemoglobin level was 13.1 g/dl and the serum iron level was 95 μg/dl; blood clotting was slow but improved after treatment with 10 mg of vitamin K.

The patient was first given codeine phosphate as symptomatic treatment for the diarrhea and was then given a strict gluten-free diet. After 2 months she was much improved and her weight was increasing.

Biochemical questions

1. What is gluten?
2. What role does gluten play in celiac disease?

3. Would this patient's symptoms of diarrhea return if gluten were present in the diet?
4. What evidence is presented to indicate poor intestinal function?
5. Of what particular relevance to nutrition is the fact that the central nervous system was normal?
6. Would vitamin B_{12} injections have assisted the recovery?
7. How does this case differ biochemically from tropical sprue?

REFERENCES

Booth, C. C., et al.: A case of adult celiac disease with Addison's disease, Br. Med. J. 2:711, 1970.
Drummond, M. B., and Montgomery, R. D.: Acute sprue in Britain, Br. Med. J. 2:340, 1970.
Lamabadusuriya, S. P., et al.: Limitations of xylose tolerance test as screening procedure for coeliac disease, Arch. Dis. Child. 49:244, 1974.
McNeish, A. S., et al.: Coeliac disease: the disorder in childhood, Clin. Gastroenterol. 3:127, 1974.

CASE 8: VITAMIN D TOXICITY

J. B., 71 years of age, had a 3-week history of weakness, polyuria, intense thirst, difficulty in speaking and understanding commands, staggering walk, confusion, and a 30-pound weight loss. For 1 month he had been taking 200,000 units of vitamin D/day for the self-treatment of severe osteoarthritis. His blood was low in potassium and more alkaline than normal. Serum calcium was 13.5 mg/dl, and serum phosphorus was 3.5 mg/dl. Vitamin D intoxication was diagnosed.

Biochemical questions

1. Why did J. B. develop vitamin D intoxication while similar excesses of vitamin C have no apparent toxicity?
2. The metabolism of which minerals might be most affected by vitamin D intoxication?
3. Should any vitamin D therapy be needed with average exposure to the sun?
4. What tissue would be most affected by vitamin D toxicity?
5. Why are vitamins A, D, E, and K found in foods rich in fat rather than fruits and vegetables containing little fat?
6. How does this condition compare with vitamin A toxicity?

REFERENCES

David, N. J., Verner, J. V., and Engle, F. L.: The diagnostic spectrum of hypercalcemia, Am. J. Med. 33:88, 1962.
Elliot, R. A., and Dryer, R. L.: Hypervitaminosis A: a report of a case of an adult, J.A.M.A. 161:1157, 1956.
Frame, B., et al.: Hypercalcemia and skeletal effects in chronic hypervitaminosis A, Ann. Intern. Med. 80:44, 1974.
Verner, J. V., Engel, F. L., and McPherson, H. T.: Vitamin D intoxication, Ann. Intern. Med. 48:765, 1958.
Wieland, R. G., et al.: Hypervitaminosis A with hypercalcemia, Lancet 1:698, 1971.

CASE 9: NUTRITION FOR BURN PATIENTS

A 9-year-old female was admitted after sustaining a 50% total body surface and 50% full thickness hot gravy burn to her anterior and posterior trunk, legs, and arms. The past medical history was unremarkable, and there were no other abnormal physical findings. Her admission weight was 27 kg; 6 months postburn she weighed 28 kg. She was treated with topical sulfa drugs to control infection and was administered supportive fluid therapy. She became toxic from her wounds, ate poorly, and on the twenty-third day a subclavian catheter was inserted. Intravenous supplementation was begun and continued for 26 days. The rate of administration of fluid was pushed to the highest level tolerated without excessive urinary output or signs of congestive heart failure. This averaged 49.3 ml/day/kg body weight. During this time her intake averaged 1700 kcal/day, and her weight fluctuated between 26.8 and 30 kg. The patient was then able to tolerate a satisfactory oral intake, and she was discharged on the 150th postburn day.

The nutrient solutions contained all the amino acids, dextrose, water-soluble vitamins, and electrolytes. Plasma and whole blood was given to maintain serum protein levels of 3.0 g/dl and a hematocrit of 40%.

Biochemical questions

1. Assuming that the patient's normal weight was 28 kg and that her BMR was 25 kcal/day/kg body weight, what would be her normal resting caloric need?
2. With the severe tissue damage from the burn, what unusual losses of body constituents are occurring?
3. The IV parenteral fluids contained no fat. Is this of any consequence? Explain.

4. Why is not protein or sucrose used in IV feeding?
5. If the patient was getting adequate calories but was still in negative nitrogen balance and you could increase the concentration of amino acids in the IV fluid, which kind of amino acids would you select? Explain.
6. By what means did the patient receive some essential fatty acids?

REFERENCES

Pop, M. B., et al.: Parenteral nutrition in the burned child, Ann. Surg. 179:213, 1974.

Wilmore, D., et al.: Supranormal dietary intake in thermally injured hypermetabolic patients, Surg. Gynecol. Obstet. 132:881, 1971.

ADDITIONAL QUESTIONS AND PROBLEMS

1. Protein-calorie malnutrition: In the childhood disease called kwashiorkor the principal deficiency in the diet is of protein. Nutritional marasmus is a form of starvation. Compare the biochemical consequences of these two types of malnutrition.
2. Food faddism: "Eggs, bacon and grapefruit, all you can eat, but nothing else and you will loose weight." What are the possible biochemical consequences of this diet?
3. Bile: Discuss the effects of obstruction of the bile duct on the digestion and absorption of food.
4. Low-fat diets: What are the nutritional consequences of a low-fat diet?
5. Pancreatic secretions: Impairment of pancreatic function that reduces the flow of pancreatic secretions lowers the absorption of some nutrients from the diet. Discuss the nutritional consequences of reduced pancreatic secretion.
6. Thalassemia: Thalassemia major is a disease in which hemoglobin synthesis is impaired and frequent blood transfusions are required. This results in severe iron overload and deposition of iron in tissues (siderosis). Discuss the problems that might result from the administration of a drug that has strong chelating properties. (Seshadri et al.: Arch. Dis. Child. 49:621, 1974; Nienhuis et al.: Blood 46:905, 1975; O'Brien: Ann. N.Y. Acad. Sci. 232:221, 1974.)

CHAPTER 2

PROTEIN STRUCTURE

OBJECTIVES

1. To understand the structural elements of protein conformation
2. To analyze the interactions of proteins with small molecules
3. To relate the structural properties of particular proteins such as collagen, hemoglobin, and albumin to their biologic functions in health and disease

An understanding of the structural aspects of proteins is essential to an analysis of the biochemical components of living processes and to an appreciation of the biologic events that take place in the body at a molecular level. Very few biochemical reactions occur without catalysis, and all biocatalysts, called enzymes, are proteins. These catalytic events occur by the stereochemical fitting of the reactants to the highly ordered, three-dimensional structures of the enzyme molecules. Many of the nonenzymatic proteins also exhibit remarkable specificity in their interaction with other molecules. For example, antibodies are proteins that react with the antigens that originally stimulated their biosynthesis; in the visual cycle the protein opsin reacts specifically with retinal.

Life, with the cell as its unit, is continued by biochemical processes that occur within the confines of a cell membrane. Selective absorption and transport of all materials must occur across this lipid-proteinaceous envelope, which probably evolved to take part in such highly differentiated cellular functions as muscular contraction, electrical impulse transmission, and intestinal absorption. In all such cases there is a relationship between the structure of the protein and its function, details of which are slowly coming to light.

GENERAL PROPERTIES OF AMINO ACIDS AND PROTEINS
Amino acids

The α-carbon atoms of the amino acids, except glycine, are each linked to four different chemical groups. This is the characteristic of an asymmetric carbon atom. Considering the general formula for the α-amino acids (p. 3) and its relation in space to the tetrahedrally arranged valencies of the asymmetric carbon atom, isomers of the molecule can be represented by no more than two three-dimensional models (Fig. 2-1). If the R group is identical in each model and does not itself contain other asymmetric centers, then the two models are mirror images of each other and each isomer is optically active. The two isomers differ only in the direction in which they rotate the plane of polarized light.

Such pairs of isomers are called enantiomorphs. A compound that rotates the plane of polarized light in a clockwise direction is said to be dextrorotatory (+), as opposed to a compound that is levorotatory (−). Unfortunately this property does not relate in any simple way to the spatial arrangements of groups around

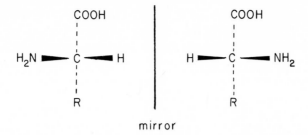

Fig. 2-1. General models of α-amino acid enantiomorphs.

the asymmetric carbon atom. Therefore, one cannot easily predict the optical rotation from the structural configuration, and conversely, the configuration cannot be predicted from the optical rotation.

The Fischer convention

The representation of stereochemical relationships of organic molecules with several asymmetric centers can become complicated. Emil Fischer first proposed a convention whereby these molecules can be represented in two dimensions. His directions are brief and simple. Literally translated, they are as follows: "One has to construct a model of the molecule (using balls for atoms and springs for the valency bonds) and put it into the plane of the paper in such a way that the carbon atoms are situated in a straight line with the groups of interest lying above the plane of the paper, then the formulae are obtained by projection."[*] Using this convention, the straight carbon chain can be at any direction in the paper. The stereomodels in Fig. 2-1 can be represented in many ways, one of which is shown below.

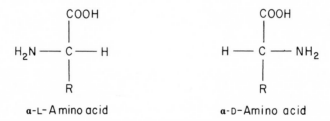

The "groups of interest" are the hydrogen atom and amino group attached to the α-carbon. Because of the way in which the carbon atoms are brought into the plane of the paper with the substituents on the asymmetric carbons above the plane, rearranging the projected formula by taking it out of the plane of the paper is not permitted. It cannot be flipped or partially turned out of the plane of the paper but must be rotated in the original plane. Additions have been made to the original convention of Fischer: one common convention implies that the vertical valency lines in the projected formula must refer to groups that lie behind the plane in the stereomodel and that the horizontal lines are restricted to those valencies that lie above the plane. This restriction was not used by Fischer but is now in fairly common usage. It is not appropriate to pursue the implications of this added convention here; it is mentioned only to emphasize the existence of more than one set of

[*]Fischer, E.: Über die konfiguration des traubenzuckers und seiner isomeren, Ber. Dtsch. Chem. Ges. 24:2683, 1891.

criteria. The projection formulae used in this text will be true for any set of conventions used to date. In another common representation the asymmetric carbon atoms in the chain may be indicated by the intersection of two lines, and an equation may be written thus:

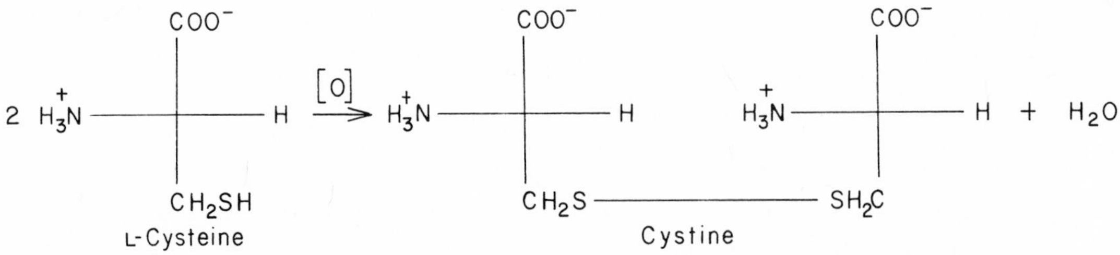

$$2 \ \overset{+}{H_3}N \underset{CH_2SH}{\overset{COO^-}{\rule{0pt}{0pt}}\!\!-\!\!\rule{0pt}{0pt}} H \ \xrightarrow{[O]} \ \overset{+}{H_3}N \underset{CH_2S}{\overset{COO^-}{\rule{0pt}{0pt}}\!\!-\!\!\rule{0pt}{0pt}} H \qquad \overset{+}{H_3}N \underset{SH_2C}{\overset{COO^-}{\rule{0pt}{0pt}}\!\!-\!\!\rule{0pt}{0pt}} H \ + \ H_2O$$

L-Cysteine   Cystine

As noted above, two optical isomers exist for those α-amino acids with one asymmetric carbon atom. The most commonly occurring amino acids in nature as represented by the Fischer formula are shown with the identification as L-; the mirror image is the enantiomorph designated as D-.

Like the amino acids, simple sugars also contain asymmetric (optically active) carbon atoms, and the Fischer convention has also been applied to these. The absolute configuration of D- and L-glyceraldehyde was established by x-ray crystallography, and the crystallographic findings showed that the dextrorotatory glyceraldehyde in fact had the D configuration. It is not possible to use these data to predict the L or D configuration of other molecules.

Classification of amino acids

When proteins are hydrolyzed to their constituent building units, sixteen α-L-amino acids and one L-imino acid are usually produced. The different structures are summarized in Table 2-1, where the R group is identified together with the conventional three-letter abbreviation of the amino acid and data on the ionization of the functional groups. All seventeen of these residues, together with tryptophan, L-asparagine, and L-glutamine, are identified by the genetic code for sequential inclusion into the polypeptide chain during protein biosynthesis. Thus, when one speaks of there being twenty amino acids in protein, it is the genetically coded amino acids that are usually meant. During chemical hydrolysis of the proteins, the L-asparagine and L-glutamine residues are converted to L-aspartic acid and L-glutamic acid, respectively. Other amino acid residues that may be identified, cystine, hydroxy-L-lysine, and hydroxy-L-proline, are residues produced by modifying one or more of the basic twenty amino acids in the parent structures of the protein. Such modification occurs after biosynthesis of the polypeptide chain. Cystine is the result of the oxidation of two residues of L-cysteine, an —S—S— linkage being formed from the two —SH groups shown above; a similar reaction occurs between two properly spaced cysteinyl residues in the polypeptide chain. The hydroxylation of L-lysyl or L-prolyl residues in the proteins collagen and elastin results in the introduction of hydroxyl groups in the corresponding R side chains.

The arrangement of amino acids in proteins is unique for each protein. This may be seen in the amino acid composition and, more precisely, in the amino acid sequence. Table 2-2 gives the amino acid composition of the proteins considered in some detail in this chapter. Note the similarity in composition of the subunits that make up hemoglobin and human immunoglobulin. Contrast these compositions with those of tropocollagen and tropoelastin, which lack several essential amino acids but contain unusually high amounts of glycine and hydroxyproline.

Table 2-1.   Classification of amino acids found in protein

$$H_3^+N-\underset{\underset{R}{|}}{\overset{\overset{COO^-}{|}}{C}}-H$$

Name	Abbreviation	R-	pK$_a$ Values -COO$^-$	-NH$_3^+$	Other
Aliphatic monoamino monocarboxylic acids					
Glycine	Gly	H-	2.34	9.60	
Alanine	Ala	CH_3-	2.35	9.69	
Valine	Val	CH_3-CH- $\quad\quad CH_3$	2.32	9.62	
Leucine	Leu	$CH_3-CH-CH_2-$ $\quad\quad CH_3$	2.36	9.60	
Isoleucine	Ile	CH_3-CH_2-CH- $\quad\quad\quad\quad CH_3$	2.36	9.68	
Serine	Ser	$HO-CH_2-$	2.21	9.15	
Threonine	Thr	$CH_3-\underset{\underset{OH}{|}}{\overset{\overset{H}{|}}{C}}-$	2.09	9.10	
Aromatic amino acids					
Phenylalanine	Phe	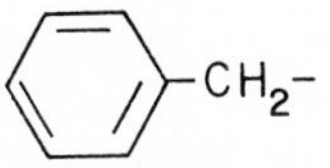 ⬡$-CH_2-$	1.83	9.13	
Tyrosine	Tyr	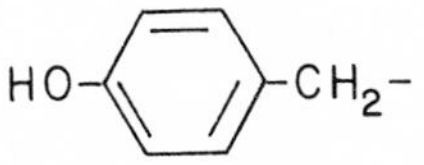 $HO-$⬡$-CH_2-$	2.20	9.11	10.07
Tryptophan	Trp	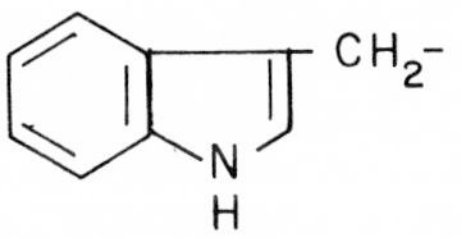 indole$-CH_2-$	2.38	9.38	
Acidic amino acids and their amides					
Aspartic acid	Asp	$^-OOC-CH_2-$	2.01	9.93	3.80
Asparagine	Asn	$H_2N-\overset{\overset{O}{\|}}{C}-CH_2-$	2.02	8.80	
Glutamic acid	Glu	$^-OOC-CH_2-CH_2-$	2.13	9.76	4.31

Continued.

Table 2-1.   Classification of amino acids found in protein — cont'd

Name	Abbreviation	R-	pKₐ Values		
			$-COO^-$	$-NH_3^+$	Other

Acidic amino acids and their amides — cont'd

Name	Abbreviation	R-	$-COO^-$	$-NH_3^+$	Other
Glutamine	Gln	$\overset{\overset{\textstyle O}{\|\|}}{H_2N-C}-CH_2-CH_2-$	2.17	9.13	

Basic amino acids

Lysine	Lys	$\underset{NH_3^+}{CH_2-CH_2-CH_2-CH_2-}$	2.18	8.95	10.53
Arginine	Arg	$\underset{NH}{H_3^+N-\overset{\|\|}{C}-NH-CH_2-CH_2-CH_2-}$	2.17	9.04	12.48
Histidine	His	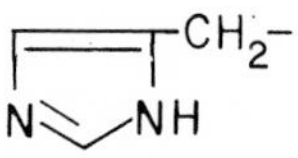 imidazole$-CH_2-$	1.82	9.17	6.0

Sulfur-containing amino acids

Cysteine	Cys	$HS-CH_2-$	1.91	10.36	8.24
Methionine	Met	$CH_3-S-CH_2-CH_2-$	2.28	9.21	

Imino acid

Proline*	Pro	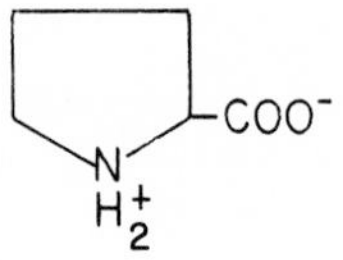 pyrrolidine ring $-COO^-$	1.95	10.64	

Other amino acids

Hydroxylysine†		$\underset{NH_3^+}{\overset{OH}{CH_2-CH-CH_2-CH_2-}}$	2.13	8.62	9.67
Cystine†		CH_2- ... $S-S-$ ... CH_2-	1.04 / 2.1	8.02 / 8.71	
Hydroxyproline*†		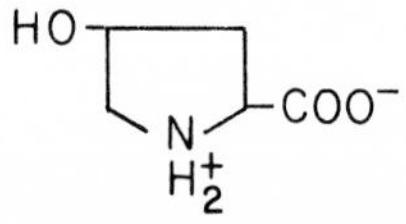 HO— pyrrolidine ring $-COO^-$	1.92	9.73	

---

*Complete amino acid structure shown; not R-group.
†Amino acids that are formed after the peptide chain has been synthesized.

Table 2-2. Amino acid composition of some proteins (residues/100 residues)*

AMINO ACIDS	HUMAN HEMOGLOBIN α	HUMAN HEMOGLOBIN β	HUMAN IMMUNOGLOBIN L	HUMAN IMMUNOGLOBIN H	CALFSKIN TROPO-COLLAGEN	PIG TROPO-ELASTIN	BOVINE MYOSIN	HUMAN SERUM ALBUMIN
Essential								
Arg	2.1	2.0	2.3	2.5	5.0	0.7	5.7	4.5
His	7.1	6.2	0.9	2.0	0.5	0	1.4	2.9
Ile	0	0	2.8	2.2	1.2	1.9	3.8	1.7
Leu	12.8	12.3	7.0	6.7	2.7	4.6	10.8	11.6
Lys	7.8	7.5	7.0	6.9	2.5	4.8	9.8	10.8
Met	1.4	0.7	1.4	1.3	0.5	0	2.7	1.1
Phe	5.0	5.5	3.7	3.4	1.3	2.8	3.3	6.1
Thr	6.4	4.8	8.4	7.4	1.5	1.4	4.7	5.4
Trp	0.7	1.4	1.4	1.6	0	0	0.4	0.1
Val	9.2	12.3	7.0	10.1	2.2	12.1	4.5	8.4
Nonessential								
Ala	14.9	10.3	6.1	4.9	10.7	21.8	8.6	–
Asn	2.8	4.1	3.3	3.6	–	–	–	–
Asp	5.7	4.8	4.7	3.4	4.8	0.3	10.4	10.0
Cys	0.7	1.4	2.3	2.5	0	0	1.1	6.7
Gln	0.7	2.0	7.0	4.3	–	–	–	–
Glu	2.8	5.5	4.7	5.1	7.4	1.8	21.3	1.5
Gly	5.0	8.9	6.1	7.4	33.5	33.4	3.9	2.7
Hyl	0	0	0	0	0.7	0	0	0
Hyp	0	0	0	0	9.1	1.1	0	0
Pro	5.0	4.8	4.7	8.3	12.4	10.9	1.8	5.7
Ser	7.8	3.4	14.9	11.9	3.4	0.9	4.3	4.5
Tyr	2.1	2.0	4.2	4.0	0.4	1.6	1.7	3.3
Amide N	–	–	–	–	–	–	–	8.1

*Data on hemoglobin and immunoglobin from Dayhoff, M. O.: Atlas of protein sequence and structure, Washington, D. C., 1972, National Biomedical Research Foundation, vol. 5; data on tropocollagen from Balazs, E. A., editor: Chemistry and molecular biology of the intercellular matrix, New York, 1972, Academic Press, Inc.; data on tropoelastin from Stevens, F. S., and Jackson, D. S.: Biochem. J. **104**:534, 1967; data on myosin from Horiaux, F., Hamoir, G., and Oppenheimer, H.: Arch. Biochem. Biophys. **120**:274, 1967; data on serum albumin from Tristram, G. R., and Smith, R. H.: In Neurath, H., editor: The proteins, New York, 1963, Academic Press, Inc.

Peptide bond

Nomenclature for peptides

A polypeptide is formed as the result of the condensation of amino acids, which produces a linear molecule. Usually one end of this string of amino acid residues has a free amino group and the other end has a free carboxyl group. Polypeptides are named as derivatives of the amino acid with the free carboxyl group; for example, the tripeptide alanyl-glycyl-serine would be:

Ala——Gly——Ser

Alanyl——glycyl——serine

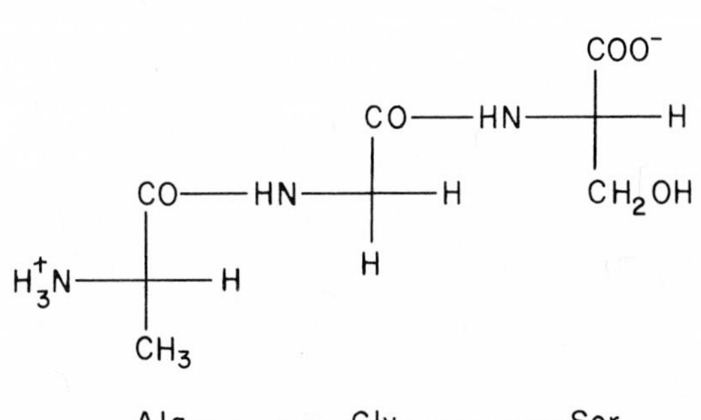

By convention the peptide sequences are written with the NH_3^+-terminus on the left. Each amino acid may be abbreviated by a three-letter symbol (Table 2-1), usually the first three letters of its name.

The differentiation between polypeptide and protein is unclear; the term "protein" usually refers to a biopolymer with a molecular weight greater than several thousand. A polypeptide is of lower molecular weight, and if the number of amino acid residues in the molecule is known, it may be indicated. For example, the hormone glucagon is a nonacosapeptide (twenty-nine amino acid residues), and oxytocin and vasopressin are nonapeptides (nine amino acid residues, cystine being counted as two residues).

The structure of vasopressin, the antidiuretic hormone, is presented below to illustrate several points. This nonapeptide becomes cyclized by the formation of an —S—S— bond between the two cysteine residues. The L-aspartyl and L-glutamyl residues exist in the cyclic peptide as amides and are therefore L-asparaginyl and L-glutaminyl residues. The COOH-terminus of glycine is also present as an amide, abbreviated —Gly (NH_2).

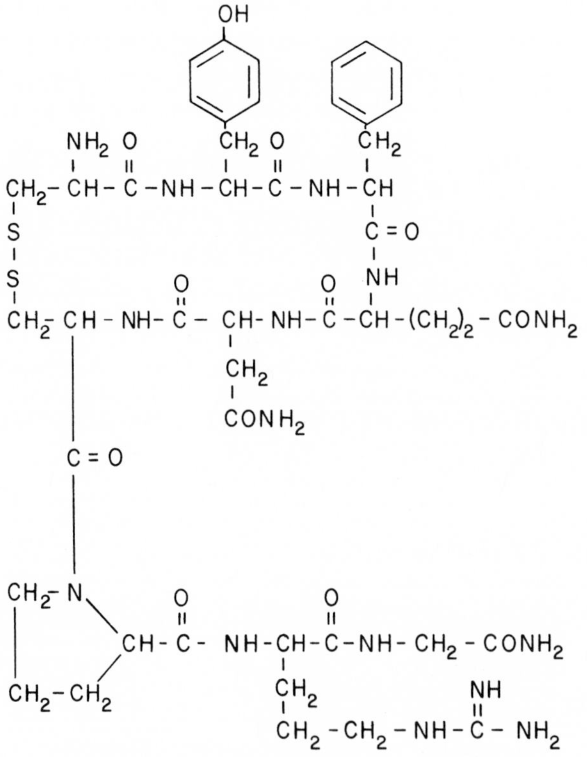

Vasopressin

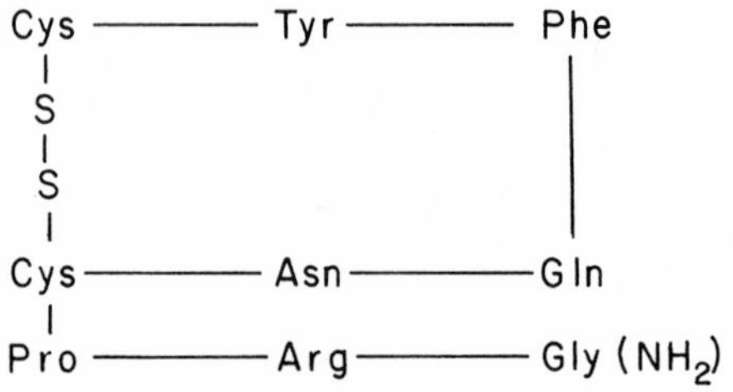

Cys ——— Tyr ——— Phe
|
S
|
S
|
Cys ——— Asn ——— Gln
|
Pro ——— Arg ——— Gly (NH_2)

Peptide conformations

The condensation of the amino group of one amino acid with the carboxyl group of another forms a peptide bond that has a resonance stabilization that results in a planar arrangement of the six atoms involved (Fig. 2-2). The partial double-bond character of the $C\cdots N$ bond places a distinct restriction on the shape of the molecule, free rotation being permitted only around the C—R, C—NH$_3^+$, and C—COO$^-$ bonds. When two or more peptide bonds are joined together, the most flexible movement would be somewhat equivalent to a chain in which each link, representing the peptide $C\cdots N$ bond, can move through a limited angle with respect to the next link. A polypeptide structure with such maximum flexibility is referred to as having a *random coil* conformation, the conformation being the three-dimensional shape of a molecule. Other properties of the polypeptide, however, will tend to stabilize the molecule into a more rigid form. Excluding any covalent cross-linkages, such as the —S—S— bonds in a cystinyl residue, these stabilizing elements of structure are hydrogen bonds, ionic interactions, and hydrophobic interactions.

1. *Hydrogen bonds.* Hydrogen bonds may be formed between —NH— or —OH groups and the C=O groups in the peptide bonds or —COO$^-$ in the R group. For example, two adjacent peptides may form hydrogen bonds, as indicated by the broken lines.

$$
\begin{array}{ccccc}
| & & & & | \\
C=O & ----- & H & — & N \\
| & & & & | \\
H—N & & & & C=O \\
| & & & & |
\end{array}
$$

2. *Ionic interactions.* Ionic interactions include ionic bonds or repulsions between charged groups such as —NH$_3^+$, —COO$^-$, guanidinium groups of L-arginyl residues, —S$^-$ from cysteinyl residues, or phenolic —O$^-$ from tyrosinyl residues. The interactions between oppositely charged groups such as —COO$^-$ and H$_3$N$^+$— give rise to salt bridges.

3. *Hydrophobic interactions.* Hydrophobic interactions represent affinities of groups for each other in much the same way that the aliphatic chain of the fatty acid is attracted to the oil droplet in a water-oil emulsion. The hydrophobic groups are driven out of the water. Thus there is an affinity between

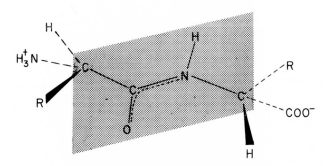

Fig. 2-2. Peptide bond conformation.

the R groups of the aliphatic amino acids or the aromatic groups of phenylalanine and tyrosine.

The stabilizing contribution of each ionic interaction to the conformation of
a molecule depends greatly on the distance between the groups and could be
greater than the stabilization resulting from a single hydrogen bond. In an average
protein the number of hydrogen bonds is great, and their total contribution is probably more significant. However, it is likely that the hydrophobic interactions are
the most important in determining the conformation. Although the contribution of
each interaction may be small, the total effect is a stabilization that is considerable.

It should be noted that these stabilizing factors are noncovalent. Stabilization
is also afforded by the covalent disulfide cross-linkages formed between L-cysteinyl
residues.

Conformational segments of polypeptide chains

The flexible random coil that is formed by a polypeptide can be arranged in
either a helix or a pleated sheet to give the maximum number of hydrogen bonds
between the peptide linkages. In the absence of any other stabilizing factors those
arrangements would give a more stable conformation than the random coil. A
large polypeptide or protein may contain regions of one or more of these conformational segments, identified here as random coil, pleated sheet, or α-helix.

Pleated sheets

1. *Parallel pleated sheets.* Polypeptide chains arranged with their carboxyl
   terminal ends in the same direction may be represented as shown in Fig. 2-3.
   Each peptide bond is involved in hydrogen bond formation with an adjacent
   peptide bond from another chain. Such a conformation places the R groups
   on both sides of the sheet, an important point when considering the layering
   of one polymolecular sheet on top of another. Hair or wool can be stretched
   in moist heat to form this type of conformation.
2. *Antiparallel pleated sheets.* A similar conformation is derived if the two

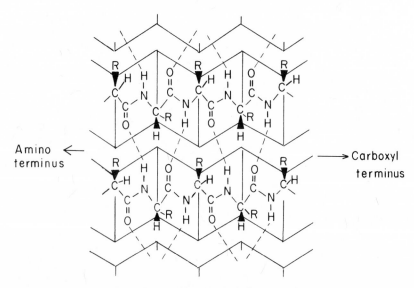

Fig. 2-3. Peptide parallel pleated sheet conformation.

adjacent polypeptide chains run in opposite directions. This is less common in human proteins; the best example in nature is silk fibroin.

Alpha helix

The polypeptide chain can twist to a limited extent to form a spiral or helix in which the planar elements of the peptide bond remain but are arranged around a central axis. A right-handed α-helix results when a polypeptide chain is arranged so that the maximum number of intrachain hydrogen bonds is formed between atoms of the peptide groups. Such a structure has an average of 3.6 amino acid residues per turn, with hydrogen bonds between every fourth amino acid. A right-handed α-helix is shown diagrammatically in Fig. 2-4. Other helical coils are possible but are less stable.

Up to this point the influence of the R groups on forming these conformational segments has not been considered. These groups have a potential to make their own contribution to stabilization and destabilization. This is particularly evident in the effects of different amino acid residues on the α-helix. Those residues that are uncharged permit the formation of stable α-helical units unless the groups are very bulky, as is the leucyl residue. A prolyl residue, having no substituent hydrogen atom on the peptide nitrogen, causes a break in the α-helix and produces a characteristic bend at the junction of the two helical portions of the chain on either side of it. Charged groups of like sign, for example, the carboxylate ions of the aspartyl or glutamyl residues or the $-NH_3^+$ groups of the lysyl and arginyl residues, cause destabilization as a result of charge repulsion. Ionic interactions of oppositely charged groups may modify the helix, as will the hydrogen bonds that may form between R groups or between an R group and the

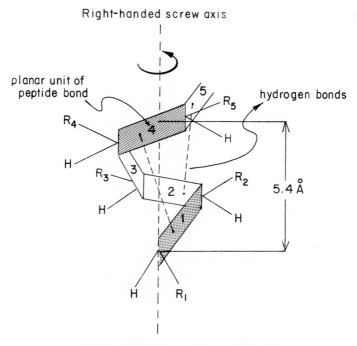

Fig. 2-4. Peptide α-helix conformation.

peptide bond. In like manner all of these factors can disrupt the pleated sheet conformations.

Folding of polypeptide chains

Having briefly examined possible conformational segments and stabilizing factors of a polypeptide chain, it is useful to consider the possible steps involved in arriving at a more stable structure in aqueous solution. A polypeptide chain might be considered as simply collapsing into a conformation in which hydrophobic interactions dominate. These interactions are most numerous and are not dependent on forces in any particular direction. Thus the aliphatic R groups would cluster together as much as possible, while the hydrophilic residues from glutamate, serine, and the like would be attracted to the solvent water molecules. Such an initial state most probably has many residues in a less than optimal environment for maximum stability. During the constantly occurring natural gyrations of the parts of the polypeptide chain, there are opportunities for a relatively weak hydrophobic interaction to be overcome by the formation of a more stable hydrogen bond or ionic interaction, until eventually the polypeptide chain is either in the most stable conformation or in one of several nearly equivalent stable conformations. In such a state it is likely that every group that can form a hydrogen bond has done so in one of the several pairings discussed above or with the aqueous solvent molecules. The stable conformation will remain until perturbed by a change in a condition such as temperature, acidity, solvent, or interacting ions that may chelate or form salt bridges. Likewise, a change in the oxidation state may break $-S-S-$ bridges between proximal cysteinyl residues in the stable conformation.

Protein conformation

Each protein or polypeptide has a unique sequence of amino acid residues in its chain, which in large measure determines its biologic and physicochemical properties. Methods of determining the composition and sequence of the amino acid units in proteins are well developed. With this knowledge, it is possible to chemically synthesize the simpler proteins and polypeptides such as the hormone insulin (51 amino acids) and the enzyme ribonuclease (124 amino acids). The biologic activity is also related to the way in which these polypeptide chains (polypeptide is used here in an adjectival sense to indicate any biopolymer chain that has a peptide polymer linkage) are arranged in space, the conformation of the molecule.

The sequence of amino acids in a protein chain is referred to as its *primary* structure. The protein molecules assume some three-dimensional shape called the *tertiary* structure, which is stabilized by covalent and noncovalent forces. The component of the tertiary structure that is stabilized by hydrogen bonds is called the *secondary* structure and is recognized in such segments as α-helices or pleated sheets, but the subdivision is somewhat arbitrary. Finally, some proteins are composed of aggregates of simple protein units; such a level of organization is referred to as the *quaternary* structure. The preceding levels of organization and structure are sometimes difficult to separate, for example, secondary and tertiary structure, and the student should not be discouraged by the lack of exact definition of these terms.

The particular three-dimensional shape of a molecule can be indicated by various physical methods, the most recent being that of x-ray crystallography. The conformation in the living cell, or that of the isolated protein where it shows maximum biologic activity, is called the *native state*. A protein must be in the native state to best perform its functions, be they structural (for example, collagen), catalytic (all *enzymes* are proteinaceous biocatalysts), or transport (for example, hemoglobin).

Denaturation

There are many conformations between the native state of a protein molecule and that in which the protein chain is spatially completely random. In the latter conformation the protein is said to be completely denatured. However, complete randomization of the protein chain is not necessary for loss of biologic function to occur. The biologic function has maximum activity in the native state, and any change in structure that leads to the loss of natural function is called denaturation.

Denaturation can be produced by many agents. These include heat and chemicals that destroy hydrogen bonds in the protein such as 6M urea, detergents like sodium dodecyl sulfate, and sulfhydryl reagents like mercaptoethanol. Denaturation does not involve cleavage of the primary protein structure, so that it is sometimes possible to reverse the denaturation by removing the causative agent.

Properties of proteins in solution

Proteins are polypeptides with an approximate range of molecular weight from 10,000 to 1×10^6. Some proteins are soluble in water; others require a dilute salt solution for the solvent; and still others such as the keratin of hair and skin are insoluble in any aqueous system. Many proteins have been fractionated and purified on the basis of their molecular size and solubility.

Gel permeation chromatography

Gel permeation chromatography is one technique for separation by molecular weight. A protein mixture in solution is slowly percolated through a long tube packed with a synthetic gel containing many pores of a known and controlled size. If the molecular weight of a solute protein is such that the molecular size is larger than the pores of the gel, the solute will pass through the column with the bulk flow of solution relatively unimpeded by the presence of the gel. Solute proteins of a smaller molecular weight and size that can penetrate into the gel pores will spend part of their time within the gel and part outside the gel. The smaller the molecule, the higher the probability that it will be in the gel. Since the gel is stationary, trapped particles will be relatively retarded or impeded by the presence of the gel, and their bulk flow through the column will be slower. By collecting the fluid emerging from the bottom of the tube in small fractions, it is possible to separate the solute mixture into individual components. Numerous gels have been prepared and marketed, and this type of chromatography is now an important tool of the protein biochemist.

Since separation by gel permeation chromatography is dependent on the size and shape of the molecule, and since in similar types of proteins these factors are related to molecular weight, it is possible to estimate rapidly the molecular weight of a protein by the rate at which it moves through a gel column relative to proteins of known size.

Ultracentrifugation

A more precise method of separating protein molecules according to molecular weight is ultracentrifugation. Under high gravitational forces, molecules sediment at a rate that is proportional to their molecular weight and shape. This principle has been used to measure the molecular weight of many kinds of molecules, particularly proteins. Ultracentrifuges have been built to operate at force levels as high as 500,000 times the force of gravity. The rotors of these devices are fitted with cuvettes through which light of a suitable wavelength may be passed during rotation, so that the light absorbed by a dissolved protein may be measured. At a given level of centrifugal force the dissolved protein molecules will move through the solution until their buoyant force is just equal and opposite to the force developed by the rotational field; thus molecules of different molecular weight can be separated. While they are extremely valuable, ultracentrifugal techniques are generally slow and require a very elaborate and expensive apparatus. It is common to describe the behavior of molecules in a centrifugal field in terms of their *sedimentation rate*.

$$S = \frac{v}{\omega^2 r} \times 10^{13} \text{ Svedberg units}$$

The sedimentation rate (S), expressed in Svedberg units, is equal to the velocity (v) in centimeters per second that the molecule moves at a distance (r) in centimeters from the center of rotation when rotating at ω radians per second. (Recall that there are 2π radians per revolution.) The factor of 10^{13} brings the S values to reasonable numbers (Table 2-3).

Some molecules, particularly those containing residues of fatty acids, will move upward in the centrifugal field. In such cases the rate of movement is expressed in flotation units (S_F).

It can be seen from Table 2-3 that the molecular weights are not directly proportional to the S values. This is because of the different overall shapes of each protein in solution, the two extremes being the *globular* proteins, which have a nearly spherical shape, and the *fibrous* proteins, which are more nearly rodlike.

Classification of proteins

Proteins have been classified in several ways, for example, as globular or fibrous types. One of the earliest classification systems was based on solubility; more recently, classifications have been drawn according to the components in the complex or conjugated protein molecule that are not amino acids. These systems are summarized in Tables 2-4 and 2-5 as they apply to animal proteins. Although the classification by solubility is not commonly used now, it gave rise to a number of protein class names such as albumin and histone that are still part of protein nomenclature.

The classification shown in Table 2-4 is based on the same properties of the proteins that are used on occasion for their preparation and purification. A mixture of proteins such as might be extracted from a tissue with water or dilute salt solution can be fractionated by the gradual addition of ammonium sulfate. The globulins will be precipitated first and can be removed by centrifugation or filtration. The albumins are precipitated when the ammonium sulfate is saturating the solution. These salt fractionations, combined with changes in the acidity of the solutions, can fractionate many mixtures of proteins reasonably well. Further separations and purifications may require the more delicate chromatographic procedures discussed previously.

The proteins may be divided into two groups, those composed only of amino acids, or *simple* proteins, and those that are *conjugated* proteins. The conjugated proteins have residues other than amino acids linked to the structure by covalent

Table 2-3. Molecular weights of selected proteins

PROTEIN	MOLECULAR WEIGHT	SVEDBERG UNITS (S)
α_1-Acid glycoprotein	28,000	2.8
Human serum albumin	69,000	4.6
Fibrinogen	330,000	7.9

Table 2-4. Protein classification by solubility

PROTEIN CLASS	EXAMPLES	SOLUBILITY
Albumin	Serum albumin, ovalbumin	Soluble in water in dilute solution; precipitated by saturation with $(NH_4)_2SO_4$
Globulin	Immunoglobulins, fibrinogen, ovomucoid	Insoluble in pure water; soluble in dilute neutral salt solutions; precipitated from solution by half-saturation with $(NH_4)_2SO_4$
Albuminoid or scleroproteins	Keratin, elastin, collagen	Insoluble in water, dilute neutral salt solutions, and alcohol; some soluble in dilute acid
Histone	Thymus histone	Soluble in water; insoluble in dilute ammonia; soluble in dilute acids and alkalies
Protamine	Sperm protamine	Soluble in water, dilute acids, alkalies, and ammonia

Table 2-5. Protein classification by composition

PROTEIN CLASS	EXAMPLES	NONPROTEIN COMPONENTS
Simple proteins	Serum albumin	None
Glycoproteins	Immunoglobulins, mucins, connective tissue, proteoglycans	Carbohydrate
Nucleoproteins	Viruses, chromosomes	Nucleic acids
Metalloproteins	Ferritin, carboxypeptidase	Metal ions
Lipoproteins	Chylomicrons	Lipids of various types
Chromoproteins	Hemoglobin, rhodopsin	Colored prosthetic group, for example, heme, riboflavin, and retinal

bonds. They are classified according to the principal nonprotein component, as indicated in Table 2-5. Several of the conjugated proteins will be described in greater detail later, as will the more elaborate scheme used for the classification of enzyme proteins.

Ionic properties of amino acids, peptides, and polypeptides

Because of the effect of ionic interactions on the conformation of polypeptides and because of the fact that biologic systems and their metabolic processes are critically dependent on the ionization of the acidic and basic groups in the biochemical components, it is important to consider the factors that determine the charges on these molecules. These charges will vary according to the H^+ concentration in the solution, which alters the extent to which a carboxyl group is undissociated or carries a negative charge as a —COO$^-$ ion. Similarly, the amino and other basic groups will be changed to positively charged ions or neutral groups. Therefore, it is necessary to review the problem of weak acids and weak bases and the reversible equilibria involved in such systems.

Reversible reactions

The majority of reactions in biochemical systems are reversible. For example:

$$H_2O + CO_2 \rightleftharpoons H_2CO_3$$

Such reversible systems are in fact the summation of two reactions, the forward and the reverse. The hydration of CO_2 yields undissociated carbonic acid, while the reverse reaction dissociates carbonic acid into H_2O and CO_2. These reactions are continuous, but at equilibrium the rates are equal and the ratios of the concentrations of H_2O, CO_2, and H_2CO_3 are constant, such that:

$$K_{eq} = \frac{[H_2CO_3]}{[H_2O][CO_2]}$$

where [ ] in each case should be read "the concentration of . . . ." The above expression follows from the general reaction:

$$aA + bB + ---- \rightleftharpoons xX + yY ----$$

where, at equilibrium:

$$K_{eq} = \frac{[X]^x[Y]^y ----}{[A]^a[B]^b ----}$$

In calculations involving equilibria it is necessary to have all the concentrations expressed in the same units, which are best converted to moles per liter since, by convention, tables of equilibrium constants are derived using these units.

For example, an important equilibrium reaction in the catabolism of D-glucose involves the isomerization of D-glyceraldehyde 3-phosphate and dihydroxyacetone phosphate. In a solution of 10 ml volume, 1.7 mg of D-glyceraldehyde 3-phosphate is in equilibrium with 42.2 mg of dihydroxyacetone phosphate. This equilibrium may be represented:

$$\text{D-Glyceraldehyde 3-P} \rightleftharpoons \text{Dihydroxyacetone P}$$
$$(1.7 \text{mg}) \qquad\qquad\qquad (42.2 \text{mg})$$

The molecular weight of the two isomers is 169. Therefore, the molar concentration of D-glyceraldehyde 3-phosphate is:

$$\frac{1.7 \times 10^{-3}}{169} \times \frac{1000}{10} = 1.0 \times 10^{-3} M$$

Similarly, for dihydroxyacetone phosphate:

$$\frac{42.2 \times 10^{-3}}{169} \times \frac{1000}{10} = 25.0 \times 10^{-3} M$$

For the above equilibrium equation, therefore:

$$K_{eq} = \frac{25.0 \times 10^{-3}}{1.0 \times 10^{-3}} = 25$$

Again by convention, the equilibrium constant is calculated by dividing the product of the components on the right side of the reversible reaction *as written* by the product of the concentrations on the left side. The unqualified statement for an equilibrium constant does not indicate the direction to which the value refers. The K_{eq} for reaction $A \rightleftharpoons B$ is the reciprocal of the K_{eq} for the reaction $B \rightleftharpoons A$. Thus if the above equilibrium had been written:

$$\text{Dihydroxyacetone P} \rightleftharpoons \text{D-Glyceraldehyde 3-P}$$

the same concentrations of reactants would have applied to the system described in the example, but:

$$K_{eq} = \frac{1.0 \times 10^{-3}}{25.0 \times 10^{-3}}$$

$$= \frac{1}{25}$$

$$= 4.0 \times 10^{-2}$$

Thus the larger the equilibrium constant, the greater the tendency of the reaction to proceed as written from left to right.

Since all biochemical reactions to be considered are assumed to occur in dilute aqueous solutions, the concentration of water does not change significantly and is set to unity, rather than to 55.5 M. For the H_2CO_3 system above:

$$K_{eq}' = \frac{[H_2CO_3]}{[CO_2]}$$

where:

$$K_{eq}' = K_{eq} \times [H_2O]$$

or

$$K_{eq}' = 55.5 \, K_{eq}$$

This approximation is valid for the dilute solutions of extracellular fluids or cytoplasm. A different situation may exist for those reactions occurring in membranes and similar cellular structures, where the free water may be limited.

Weak acids and weak bases

One of the largest groups of reversible reactions is the dissociation of weak acids and bases, which has been defined by Brønsted as follows:

$$\underset{\text{acid}}{HA} \rightleftharpoons H^+ + \underset{\text{conjugate base}}{A^-}$$

$$\underset{\text{base}}{B} + H^+ \rightleftharpoons \underset{\text{conjugate acid}}{H^+B}$$

By this definition, an acid ionizes to produce a proton and conjugate base. A base accepts a proton to form a conjugate acid. Thus the ammonium ion NH_4^+ is the conjugate acid of the base ammonia, NH_3. Similarly, the acetate ion CH_3COO^- is the conjugate base of acetic acid, CH_3COOH. From the Brønsted definitions it is also true that the acetate ion is a base and acetic acid is the conjugate acid, so that the CH_3COOH and CH_3COO^- species are called a conjugate acid-base pair. This relationship is also seen in the dissociation of the α-amino acids or the polyprotonic acids, for example, phosphoric acid. Taking glycine, the simplest of α-amino acids, the reactions may be written:

$$NH_3^+ - CH_2 - COOH \rightleftharpoons NH_3^+ - CH_2 - COO^- + H^+ \qquad (2\text{-}1)$$

$$(H_2A^+) \qquad\qquad\qquad (HA)$$

$$NH_3^+ - CH_2 - COO^- \rightleftharpoons NH_2 - CH_2 - COO^- + H^+ \qquad (2\text{-}2)$$

$$(HA) \qquad\qquad\qquad (A^-)$$

HA is the conjugate base of the fully protonated glycine H_2A^+ and the conjugate acid of the glycinate ion A^-. A^- is also the conjugate base of HA.

The equilibria of the glycine species expressed in equations 2-1 and 2-2 are written as stepwise dissociations of the diprotic acid H_2A^+. The equilibrium constants are 4.47×10^{-3} for equation 2-1 and 1.70×10^{-10} for equation 2-2, as would be expected from the general knowledge that the $-COOH$ group is a stronger acid (more dissociated) than the $-NH_3^+$ group. Equilibrium constants for such ionizing molecules are called dissociation constants and are represented as K_a for weak acids.

$$K_a = \frac{[A^-][H^+]}{[HA]} \qquad (2\text{-}3)$$

Weak bases are best considered in the context of dissociation reactions in their conjugate acid form, for which K_a refers to the dissociation constant for the loss of a proton. This is illustrated in equation 2-4. This unifying concept avoids the introduction of K_b, a second type of dissociation constant.

Definition of pX

Weak acids are usually considered to have dissociation constants smaller than 10^{-2}; that is, the acid is less than 10% ionized in solution. It is advantageous to manipulate very small or very large numbers in their logarithmic form. In general, for a value X:

$$-\log_{10} X = pX$$

so that if X is 10^{-2}, then pX is 2. The convention used for pH is:

$$pH = -\log_{10}[H^+]$$

The convention used for the dissociation constant K_a is:

$$-\log_{10} K_a = pK_a$$

Note that the *weaker* the acid, the *larger* the pK_a in a *logarithmic* proportion. An acid with pK_a 4 is not twice as strong as one of pK_a 8; rather it is 10^4 times stronger, a logarithmic change. Similarly, a change of pH 7.25 to pH 7.55 represents a *halving* of the hydrogen ion concentration.

As examples of the principles introduced to this point, the following situations may be considered.

1. The normal pH of arterial blood is 7.40. That is:

$$-\log_{10}\left[H^+\right] = pH$$

$$\text{or} \quad \log_{10}\left[H^+\right] = -pH$$

$$\text{In blood} \quad \log_{10}\left[H^+\right] = -7.4$$

Since the mantissa in logarithm tables is always positive, then -7.4 must be written $-8 + 0.6$, the arithmetic equivalent. This is usually written $\bar{8}.6$ to indicate that the characteristic, -8, is negative. Therefore, the $[H^+]$ for blood is:

$$\left[H^+\right] = \text{antilog} -8 \times \text{antilog } 0.6$$

$$= 10^{-8} \times 3.98 \text{ moles per liter}$$

Similarly,   pH 7.25 corresponds to 5.62×10^{-8} moles per liter of H^+

and   pH 7.55 corresponds to 2.82×10^{-8} moles per liter of H^+

2. The K_a and pK_a of common weak acids are as follows.

	K_a	pK_a
Acetic acid	1.82×10^{-5}	4.74
Ammonium ion	5.50×10^{-10}	9.26
Glycine	(1) 4.47×10^{-3}	2.35
	(2) 1.70×10^{-10}	9.78

In the case of the ammonium ion:

$$NH_4^+ \rightleftharpoons NH_3 + H^+ \tag{2-4}$$

$$K_a = \frac{\left[H^+\right]\left[NH_3\right]}{\left[NH_4^+\right]}$$

The value of K_a is 5.50×10^{-10}

Therefore, $pK_a = -\log_{10} 5.50 \times 10^{-10}$

$$= -(\log_{10} 5.50 + \log_{10} 10^{-10})$$

From log tables $\log_{10} 5.50 = +0.7404$

and by inspection $\log_{10} 10^{-10} = -10.0000$

$$pK_a = -(+0.7404 - 10.0000)$$

$$= -(-9.26)$$

$$= 9.26$$

This is the value in the tables.

The Henderson-Hasselbalch equation

The dissociation of a weak acid is represented as:

$$HA \rightleftharpoons H^+ + A^-$$

$$K_a = \frac{[H^+][A^-]}{[HA]}$$

$$= [H^+] \cdot \frac{[A^-]}{[HA]}$$

Taking logarithms

$$\log_{10} K_a = \log_{10} H^+ + \log_{10} \frac{[A^-]}{[HA]}$$

and changing the signs throughout, the equation gives

$$-\log_{10} K_a = -\log_{10} H^+ - \log_{10} \frac{[A^-]}{[HA]}$$

Since $-\log_{10} K_a = pK_a$ and $-\log_{10} H^+ = pH$, we may write

$$pK_a = pH - \log_{10} \frac{[A^-]}{[HA]}$$

or by transposing terms:

$$pH = pK_a + \log_{10} \frac{[A^-]}{[HA]} \qquad (2\text{-}5)$$

This form is known as the Henderson-Hasselbalch equation.

One application of the Henderson-Hasselbalch equation is in the consideration of the charges on the amphoteric molecules such as α-L-amino acids, peptides, and proteins as they relate to pH. The various ionizable groups on the peptide chain and its R groups each have their pK_a values, which as a first approximation may be taken to be the same as the corresponding amino acid. These are summarized in Table 2-1. Considering the dissociation of the proton from any one of these groups, the Henderson-Hasselbalch relationship states that as more of the conjugate acid is converted to the conjugate base, the log of their ratio will increase; the corresponding pH will become larger; and the solution will be more alkaline. At the point at which the concentrations of the conjugate acid and its conjugate base are equal, the pH is that given by pK_a. In terms of a titration curve, when half of a weak acid has been neutralized by a strong base such as NaOH, the pH of the solution is equal to the value of pK_a. This matter will be considered again in the discussion of buffers in Chapter 4.

Isoelectric point

In polyprotonic and amphoteric molecules it is assumed in the present discussion that the dissociation of each proton is discrete and is not influenced by the other ionizable groups. This is essentially true if the pK_a values are separated by more than two units. This means that in titrating a molecule such as glycine the —COOH group (pK_a 2.34) is completely ionized before the —NH$_3^+$ group (pK_a 9.60) loses its proton. At the point at which there are an equal number of negatively and positively charged groups, the net charge is zero and the pH of this solution is known as the isoelectric point (pI). In simple molecules it is calculated by taking the average of the pK_a values of the two groups that ionize either side of the molecular form with zero net charge. For a diprotonic amphoteric molecule such as glycine it is the average of the two pK_a values. The pI of glycine is $\frac{2.35 + 9.60}{2}$ or 5.97. For glutamic acid (Table 2-1, pK_a values 2.13, 4.31, and 9.76) and similar polyprotic ampholytes it is necessary to determine the pair of values to be averaged. The dissociation of glutamic acid may be abbreviated.

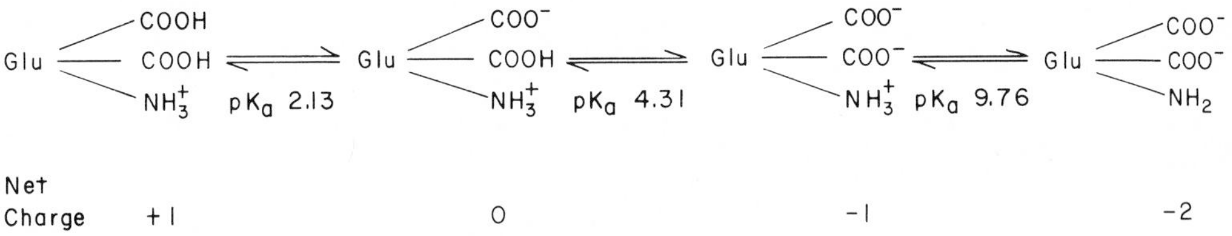

$$\text{Glu} \underset{\diagdown \text{NH}_3^+}{\overset{\diagup \text{COOH}}{\underset{}{\overline{}}} \text{COOH}} \quad \underset{pK_a\ 2.13}{\rightleftharpoons} \quad \text{Glu} \underset{\diagdown \text{NH}_3^+}{\overset{\diagup \text{COO}^-}{\text{COOH}}} \quad \underset{pK_a\ 4.31}{\rightleftharpoons} \quad \text{Glu} \underset{\diagdown \text{NH}_3^+}{\overset{\diagup \text{COO}^-}{\text{COO}^-}} \quad \underset{pK_a\ 9.76}{\rightleftharpoons} \quad \text{Glu} \underset{\diagdown \text{NH}_2}{\overset{\diagup \text{COO}^-}{\text{COO}^-}}$$

Net
Charge   +1   0   −1   −2

$$pI = \frac{2.13 + 4.31}{2}$$

$$= 3.22$$

The ionized form of the molecule that has a zero net charge is associated with the pK_a values of 2.13 and 4.31, which average to 3.22, the pH at which the molecule of glutamic acid is electrically neutral. Approximate pI values can be calculated for proteins and polypeptides from the numbers of each type of ionizable groups, but the assumptions are many. The procedure, however, can be illustrated by the following examples.

The completely protonated form of the antidiuretic hormone vasopressin (p. 48) is represented below, with the corresponding pK_a values taken from Table 2-1.

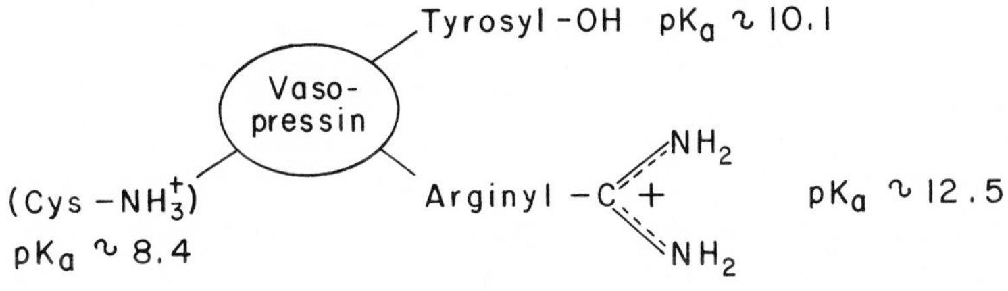

Tyrosyl −OH $pK_a \sim 10.1$

(Vaso- pressin)

(Cys −NH$_3^+$)
$pK_a \sim 8.4$

Arginyl −C$\begin{smallmatrix} \diagup \text{NH}_2 \\ + \\ \diagdown \text{NH}_2 \end{smallmatrix}$ $pK_a \sim 12.5$

$$pI = \frac{10.1 + 12.5}{2}$$

$$= 11.3$$

At about pH 10 the —NH$_3^+$ group of the cystinyl residue will be almost completely ionized to an uncharged —NH$_2$; the charge on the vasopressin is then approximately +1. Above pH 12.5 the molecule has a −1 charge. Thus the ionization of the tyrosyl residue, pK_a 10.1, is the one to be used in the pI estimation together with the final ionization of the arginyl residue, pK_a 12.5, since the molecule will be essentially uncharged between pH 10.1 and 12.5.

Glutathione, γ-L-glutamyl-L-cysteinylglycine, is an important tripeptide found in erythrocytes and other cells. It is an atypical tripeptide in the sense that the γ-carboxyl group, not the α-carboxyl group, of L-glutamic acid is in the peptide bond with the amino group of the cysteinyl residue. Of the five ionic forms of glutathione, the three of importance in calculating the pI may be represented:

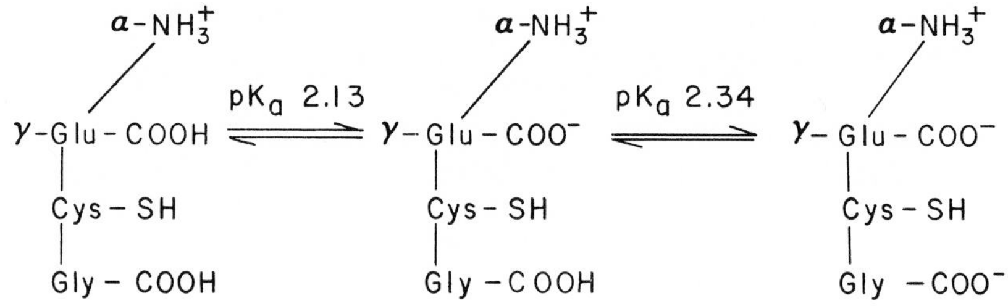

$$a-NH_3^+ \qquad\qquad a-NH_3^+ \qquad\qquad a-NH_3^+$$

$$\gamma-Glu-COOH \underset{}{\overset{pK_a\ 2.13}{\rightleftharpoons}} \gamma-Glu-COO^- \underset{}{\overset{pK_a\ 2.34}{\rightleftharpoons}} \gamma-Glu-COO^-$$

$$Cys-SH \qquad\qquad Cys-SH \qquad\qquad Cys-SH$$

$$Gly-COOH \qquad\qquad Gly-COOH \qquad\qquad Gly-COO^-$$

Net Charge $+1$ 0 -1

$$pI = \frac{2.13 + 2.34}{2}$$

$$= 2.23$$

The principle with regard to proteins is the same as that illustrated above, except that the number of ionizable groups is larger and the approximations in the calculation are greater.

Ion exchange chromatography

Ion exchange resins are water-insoluble polymers to which charged groups such as $-COO^-$, $-NH_3^+$, $-OSO_3^-$, and $-NH^+R_2$ have been covalently linked. These groups will interact with oppositely charged groups on molecules that are dissolved in an aqueous solution. The reaction is an equilibrium, the molecule (M) being distributed between the solid and the soluble aqueous phases.

$$\text{Resin} - \bar{X} \ + \ M-H^+ \ \rightleftharpoons \ R-\bar{X} \left\{ M-H^+ \right.$$

insoluble   soluble   insoluble complex

The relative distribution depends on the nature of M, the pK_a values of the groups, and the pH of the aqueous system.

These properties are used to separate ionic and amphoteric molecules such as amino acids, peptides, and proteins. The resin is placed into a chromatographic column and equilibrated with the solution used for elution; the mixture of materials is then filtered through the column. The molecules are bound to the resin and then move slowly down the column at rates that depend on the equilibrium position of each compound in relation to the resin. Separation is achieved if the relative rates are sufficiently different. Fractionation may require changes in salt concentration or pH, depending on the particular mixture.

Many kinds of ion exchange resins are available, and a very large number of variable conditions can be applied to achieve the needed separations. As a result, ion exchange chromatography is a very common technique in the analysis and preparation of biochemicals.

Electrophoresis

If a solution of a mixture of proteins is placed between two electrodes, the charged molecules will migrate to one electrode or the other at a rate that depends on the net charge and, depending on the supporting medium used, on the molecular weight. If the supporting medium is a solid substance such as an inert gel, a strip of paper, or cellulose acetate, sections may be cut out to provide purified or concentrated samples of the individual proteins making up the mixture. Alternatively, the entire gel or strip may be stained with reagents that react with proteins, thereby allowing a visual display of the separation. Techniques of this sort, which depend on the movement of charged molecules, are known collectively as electrophoresis. They are sometimes employed as diagnostic aids. Electrophoretic methods are usually simple, rapid, and require equipment of quite modest cost.

STRUCTURAL ASPECTS OF SPECIFIC PROTEINS
Plasma proteins

Normal plasma contains 6 to 8 g/dl of protein. Of this, 4.5 g is albumin and 2 to 3 g is composed of a mixture of globulins. The plasma proteins usually are classified on the basis of solubility and electrophoretic separation into the six categories listed in Table 2-6. The plasma proteins are separated and measured chemically by a procedure involving electrophoresis, staining, and densitometric scanning. The electrophoretic mobilities of the main plasma protein fractions are shown in Fig. 2-5, which illustrates the manner in which this information is usually presented clinically. The peaks correspond to those similarly identified in Table 2-6, with ϕ being an abbreviation for fibrinogen.

Most clinical electrophoretic assays such as this are run on a filter paper or cellulose acetate support medium using a barbital buffer of pH 8.6. Migration of the γ-globulins are an exception; the isoelectric points of the other plasma proteins are less than 8.16. Those plasma proteins with low isoelectric points are negatively charged at pH 8.6 and migrate toward the anode of the electrophoresis cell. In contrast, the positively charged γ-globulins migrate toward the cathode at pH 8.6.

γ-Globulins are synthesized by a class of lymphocytes known as plasma cells, while the other major plasma proteins are synthesized in the liver. Therefore, one of the hallmarks of severe liver disease is an abnormality, usually a decrease, in one or more of the plasma proteins.

When blood is allowed to clot, several plasma proteins contribute in forming

Table 2-6. Major classes of plasma proteins

TYPE	PERCENTAGE OF TOTAL PLASMA PROTEIN	SPECIAL PROPERTIES AND FUNCTIONS
Albumin	55	Transport of organic anions, osmotic pressure, binds Ca^{++}
α_1-Globulins	5	Glycoproteins, high-density lipoproteins
α_2-Globulins	9	Haptoglobin (hemoglobin transport), ceruloplasmin (Cu^{++} transport), lipoproteins of very low density
β-Globulins	13	Transferrin (Fe^{+++} transport), low-density lipoproteins
Fibrinogen	7	Blood clotting
γ-Globulins	11	Immunoglobulins

the matrix of the clot. The resulting solution, lacking fibrinogen, fibrin, and several clotting factors, is known as *serum*. Blood *plasma* is the supernatant fluid obtained by centrifuging out the blood's cells in the absence of clotting. Unlike serum, plasma contains fibrinogen and other clotting factors. Many clinical chemical determinations are made on serum rather than plasma.

Albumin

The most abundant protein in plasma is albumin. This protein has an isoelectric point of 4.8 and thus has a considerable negative charge at a physiologic pH. This explains the relatively high anodal mobility of albumin on the plasma protein electrophoretogram (Fig. 2-5). Albumin is a globular protein consisting of a single polypeptide chain and with a molecular weight of about 66,000. The two main functions of albumin are to *transport small molecules* through the plasma and extracellular fluid and to *provide osmotic pressure* within the capillary. Many metabolites such as free fatty acid and bilirubin are poorly soluble in water. Yet they must be shuttled through the blood from one organ to another so that they can be metabolized or excreted. A carrier is required to enhance the solubility of these substances in plasma so that they can be transported through this aqueous medium. Albumin fulfills this function and serves as a *nonspecific transport protein*. Moreover, albumin binds those poorly soluble drugs such as aspirin, digitalis, the coumarin anticoagulants, and barbiturates so that they also are efficiently carried through the bloodstream. In addition to carrying these large organic molecules, albumin binds small anions and cations. Indeed, about 50% of the calcium in the plasma exists as a complex with albumin. All of the small molecule–albumin interactions occur through physical bonds; that is, covalent linkages are *not* formed. The binding of a small molecule to a protein may be formulated by the general equation:

$$[P] + [A] \rightleftharpoons [PA]$$

where [P] is the protein concentration that does not contain any complexed small molecule, [A] is the concentration of the unbound small molecule, and [PA] is the concentration of the protein–small molecule complex. In such a system the effective concentration that determines the biologic activity of the small molecule depends on [A], the unbound concentration. For example, the physiologic effec-

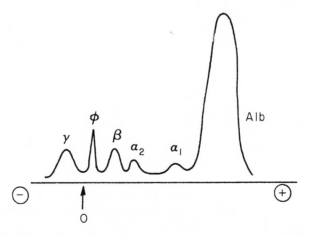

Fig. 2-5. Densitometric tracing of normal plasma protein electrophoretogram.

Table 2-7. Illustration of effect of albumin binding on "effective" plasma calcium concentration

PATIENT	PLASMA TOTAL Ca^{++} (mg/dl)	PLASMA ALBUMIN (g/dl)	BOUND Ca/ ALBUMIN RATIO (mg/g)	BOUND Ca^{++} (mg/dl)	FREE Ca^{++} (mg/dl)
Normal	9	4	1.5	6	3
Nephrotic	9	2	1.5	3	6

tiveness of calcium in the blood depends on the unbound calcium concentration, which in turn is in equilibrium with the calcium that is bound to albumin. This concept is crucial to an understanding of physiologic and pharmacologic function. The following hypothetical example illustrates the point. Suppose the clinical data indicate hypercalcemia, but repeated laboratory examinations reveal that the plasma total calcium concentration is only 9 mg/dl, the *lower* limit of normal. What one must realize is that the so-called normal calcium value is based on the assumption that the plasma albumin concentration is also normal. If the patient had nephrosis, a disease in which the plasma albumin concentration is low, the apparently normal total calcium concentration of the plasma would actually be in the hypercalcemic range as far as the unbound concentration is concerned. Under these conditions the unbound (ionized) calcium concentration would be high because there is less albumin available to bind calcium. This is illustrated in Table 2-7 by typical data that might be obtained for such a hypothetical example.

A second vital function of albumin is that it provides 80% of the osmotic pressure effect of the plasma proteins. Osmotic pressure is the main force that draws interstitial fluid back into the capillary at its venous end. Albumin provides much of the osmotic effect for two reasons: it is the most abundant protein in plasma on a weight basis, and it has a low molecular weight relative to the other major plasma proteins. Remember that colligative properties such as osmotic pressure depend on the number of particles in a solution. Moreover, the high negative charge that albumin exhibits at pH 7.4 causes water to cluster at its surface, producing a greater osmotic effect than would be predicted simply from the number of solute molecules present in the solution.

Alpha globulins

There are two classes of α-globulins in plasma, known as α_1 and α_2. The main α_1-globulins are *glycoproteins* and high-density *lipoproteins*. The main α_2-globulins are *haptoglobin*, the transport protein for any free hemoglobin that escapes into the plasma; *ceruloplasmin*, the copper transport protein; *prothrombin*, a proenzyme that is involved in blood coagulation; glycoproteins; and lipoproteins of very low density.

Beta globulins

The major β-globulins are *transferrin*, the iron-transport protein, and low-density lipoproteins.

Gamma globulins

The γ-globulin fraction is made up of *immunoglobulins*, or *antibodies*. There are 1.0 to 1.5 g/dl of γ-globulins in normal human plasma. The major plasma immunoglobulins that make up the γ-globulin fraction are listed in Table 2-8.

Table 2-8. Human plasma immunoglobulins

TYPE	PERCENTAGE OF TOTAL Ig	MOLECULAR WEIGHT × 10⁻³	SEDIMENTATION CONSTANT (S)
IgG	80	150	7
IgA	13	160, 320, 480	7, 9, 11
IgM	6	900	19
IgD	1	185	7
IgE	0.002	200	8

The commonly used abbreviation of immunoglobulin is Ig, and this is followed by the letter signifying the particular subclass, for example, IgG and IgA. Other immunoglobulin classes are present in smaller amounts (IgD, IgE). These five classes of immunoglobulins are heterogeneous. Indeed, each class is made up of hundreds or thousands of individual immunoglobulins.

The synthesis of a specific immunoglobulin is stimulated by an antigen, a protein or complex carbohydrate that is foreign to the species or individual. The newly synthesized immunoglobulin has the property of recognizing the antigen that stimulated its synthesis and combining very tightly with it. This antigen-antibody interaction is through specific noncovalent bonds. An analogous kind of reaction in terms of the bonds involved and their specificity is the reaction of an enzyme with its substrate. Antigen-antibody complexes are marked for elimination from the body. The mechanism for elimination may vary with the complex being considered, but a common mechanism of eliminating these complexes is by phagocytosis.

The immunoglobulins have a number of structural similarities even though at first glance at the molecular weights given in Table 2.8 one might think that each immunoglobulin type was totally different. All types are similar in that they have two small (light) and two large (heavy) polypeptide chains. These are called light and heavy to indicate how the isolated chains sediment in the ultracentrifuge. Most of our knowledge concerning immunoglobulin structure has been obtained from studies with IgG. The IgG fraction is the major immunoglobulin of the plasma, accounting for 80% of the total. The structure of a typical IgG molecule is shown below.

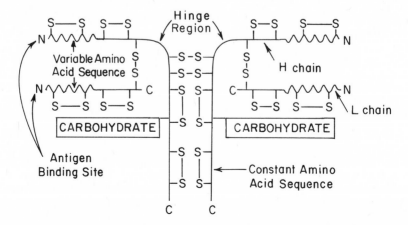

These molecules have a molecular weight of approximately 1.5×10^5 and a sedimentation coefficient of 7S. They are symmetric, the two halves being joined covalently by two disulfide bridges. Each half is composed of a heavy (H) and a light (L) chain. Thus an immunoglobulin can be abbreviated H_2L_2. This configuration of chains makes up the basic four-peptide unit that composes all immunoglobulins. The H chain contains 420 amino acids; the L chain contains 214 amino acids. Each H chain is joined to the corresponding L chain by a single disulfide bond; hence the structure of the IgG molecule can be represented as H_2L_2. In addition to the four disulfide bonds that maintain the quaternary structure of IgG, there are intrachain disulfide bonds that help to maintain the tertiary structure of each L and H chain.

It is important to note at this point that the amino acids in a peptide or protein chain are numbered starting at the N-terminal residue.

The primary sequence of 107 amino acids at the N-terminal of the L and H chain of each species of IgG is unique; in other words, the sequence of amino acids of the N-terminal end is *different* for each antibody. (Remember that there may be millions of copies of a single antibody or IgG species.) In contrast, the primary structure of the remaining amino acids to the C-terminal end is the same in all IgG molecules. Hence the sequence of amino acids 108 to 214 is *identical* in the L chains of every IgG molecule. Likewise, the sequence of amino acids 108 to 420 is identical in the H chains of every IgG molecule. The antigen binding site is located at the N-terminal end of each pair of H and L chains; that is, at the variable end, which is unique for each antibody. In this way the variable primary structure confers specificity to each antibody for a given antigen or group of closely related antigens, whereas the constant C-terminal structure provides the basic framework that all IgG molecules require to carry out their functions.

The gross structural features of IgA and IgM molecules are similar to IgG molecules, but there are also some important differences. There are two basic kinds of light chains, called kappa (κ)- and lambda (λ)- chains. Two κ- or two λ-chains may pair with two heavy chains to form all of the different immunoglobulin classes (Table 2-8). Kappa chains can be distinguished from λ-chains primarily by differences in the amino acid sequences of the constant regions. In spite of these differences, the molecular weights of κ- and λ-chains are about the same. The heavy chains can also be classified according to their primary structure into broad groups. The different types of heavy chains, unlike the κ- and λ-chains, are characteristic of the immunoglobulin types. For example, IgG has a heavy chain called γ; IgA has a chain called α; and the IgM heavy chain is called μ. The Greek letters δ and ϵ represent the heavy chains in IgD and IgE. Thus a particular IgG molecule might be described as $\gamma_2\kappa_2$ or $\gamma_2\lambda_2$, an IgA as $\alpha_2\kappa_2$ or $\alpha_2\lambda_2$, etc. The different types of heavy chains also can be distinguished by a characteristic amino acid sequence primarily in the constant region; however, unlike κ- and λ-chains, the heavy chain types vary considerably in molecular weight from approximately 50,000 for the γ-chain to about 70,000 for the μ-chain. The α-chain is intermediate in size. The three major immunoglobulin types, IgG, IgA, and IgM, also differ in the number of basic four-peptide units that make up their structure. The simplest, IgG, has only one such unit, that is, H_2L_2, whereas IgM has five such units, that is, $(H_2L_2)_5$ (see also case 2, Fig. 2-16). IgA may have one, two, or three such units. Obviously the number of basic four-peptide units in an immunoglobulin determines its molecular weight, and the polymeric nature of IgA and IgM is reflected in the molecular weights given in Table 2-8.

Fibrous proteins

Most of the proteins considered so far have been of the globular type. These have only one or a few polypeptide chains arranged so that the many polar groups on the surface make them quite soluble. Most of the enzymes and plasma proteins, with the exception of fibrinogen, are globular. Fibrous proteins, in contrast, have exceedingly extended structures consisting of several polypeptide chains often very tightly associated one to the other. These proteins are more difficult to work with in the laboratory because they are relatively insoluble in most physiologic fluids. Examples of fibrous proteins are the keratins, collagens, and elastins. The keratins are the proteins in hair, fingernails, and horny tissues. The collagens are constituents of skin, bones, teeth, blood vessels, tendons, cartilage, and connective tissue. Elastin is also an important structural element of skin, arteries, ligaments, and connective tissue. The primary role of the fibrous proteins is architectural: they provide the strong structure and support that the body requires to perform its many complex functions.

Collagen

More is known about the collagen class of fibrous proteins than the others, perhaps because of its wide occurrence and its association with a number of disease processes. More than 30% of the protein contained in the human body is collagen, making it by far the most abundant protein, and its concentration in some tissues is remarkably high.

Table 2-9 lists the collagen and elastin content of several tissues. Soft organs such as the liver contain only small amounts of collagen, while harder tissues such as the skin and tendons are almost entirely composed of it. This is consistent with the idea that the role of collagen is primarily structural (Table 2-10). Although most of the collagen in the adult is metabolically stable, it is far from being inert. Skin, for example, is continuously being broken down and resynthesized. Furthermore, the collagen content changes appreciably during growth, development, and

Table 2-9. Collagen and elastin content of some tissues (g/100 g of dry weight)

TISSUE	COLLAGEN	ELASTIN
Bone, mineral free	88.0	
Achilles tendon	86.0	4.4
Skin	71.9	0.6
Cornea	68.1	
Cartilage	46-63	
Ligament	17.0	74.8
Aorta	12-24	28-32
Liver	3.9	

Table 2-10. Physical form of collagen in different tissues

TISSUE	ARRANGEMENT OF FIBRILS AND MICROFIBRILS*
Tendon	Parallel bundles
Cartilage	Associated with mucopolysaccharides, no distinct microfibril arrangement
Skin	Planar sheets of microfibrils layered at many angles
Cornea	Planar sheets stacked crossways for strength

*Fibers can be seen by the naked eye, fibrils are visible by means of light microscopy, and microfibrils are seen only by means of the electron microscope.

morphogenesis. Collagen also is important in repair processes such as the formation of scar tissue in wound healing. Its exact role in the regulation of these
events is not understood.

Primary structure. Early studies showed that the collagen molecule was very
large and had a peculiar amino acid composition. Over one third of its amino
acid residues are glycine and about 20% are proline or hydroxyproline (Table 2-2).
Of all proteins studied, only elastin has such a high concentration of glycine. The
hydroxylated derivatives of lysine and proline are almost unique to collagen.
Elastin contains a smaller amount of hydroxyproline but no hydroxylysine. Tyrosine is present in low amounts, and the essential amino acid tryptophan is absent,
so collagen is a protein of zero biologic value. Cystine also is absent and so there
are no disulfide cross-links.

Collagen subunits. When collagen fibers are warmed slightly in dilute acid,
they dissociate and go into solution. If the solution is cooled, fibers will reform, but
if it is boiled, gelatin results and the denaturation is irreversible. Ultracentrifugation of the warm acid solution separates the mixture into three fractions designated α, β, and γ. The so-called β-chain has a molecular weight twice that of the
α-chain, and the γ-chain is three times as large as the α-chain. The α-chain is the
basic polypeptide unit that, when covalently cross-linked in a dimeric form, gives
the β-chain and, when cross-linked in a trimeric form, gives the γ-chain. Fig. 2-6
illustrates these relationships. There are several separable α-chains for most
mammalian collagens; these differ in amino acid composition but are about the
same size (see also case 4, Chapter 14). They have molecular weights of 97,000
and are abbreviated α_1, α_2. The common collagen of skin consists of two α_1-chains
and one α_2-chain twisted together to form a triple helix called *tropocollagen*. The
degree of cross-linking in tropocollagen may vary so that there will be differing
proportions of α-, β-, and γ-forms.

Tertiary and quaternary structure. None of the three protein units of tropocollagen have α-helical conformations because they contain a high proportion
of L-prolyl and hydroxy-L-prolyl residues. Also, the large proportion of glycyl residues, which by itself would destabilize the α-helix, contributes to a more extended
conformation, called the β-conformation, and makes possible a greater number
of hydrogen bonds between adjacent chains. These interchain hydrogen bonds
place a constraint on the stretching of the tropocollagen along the long axis. It
is also found that these polypeptide chains readily wrap around each other to
form a left-handed helix much like a three-stranded rope. Thus the unusual amino

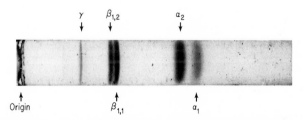

Fig. 2-6. Sodium dodecyl sulfate (SDS) polyacrylamide gel electrophoresis of guinea pig
collagen. The molecular weight of the α_1- and α_2-chains is approximately 100,000 with the
α_1-chain being slightly smaller. The proteins move in SDS polyacrylamide gels as an inverse
function of their molecular weights. The $\beta_{1,2}$-band consists of an α_1-chain cross-linked to
an α_2-chain; the $\beta_{1,1}$-band has two α_1-chains cross-linked. The $\beta_{1,1}$- and $\beta_{1,2}$-bands have
molecular weights of about 200,000. The γ-chain, or tropocollagen, has a molecular
weight of 300,000 and consists of $(\alpha_1)_2\alpha_2$, where the chains are cross-linked to one another.

acid composition of collagen is directly responsible for creating the triple helix. This point can be further emphasized. Sequencing data show that in the most polar regions of the α-chains the tripeptide Gly-X-Y is repeated in sequence over and over again. The symbols X and Y refer to amino acids other than glycine. The tripeptide Gly-X-Hyp is another common sequence, where Hyp is the abbreviation for hydroxy-L-proline. Considering these facts, it has been proposed that all of the small glycine residues are packed tightly by hydrophobic interactions against one another at the center of the triple helix. Each glycine —NH— group hydrogen bonds to a carbonyl group of an amino acid in the X position on an adjacent chain. The charged amino acids and those with bulky groups are on the outer part of the helix.

Tropocollagen is the building material of which the microfibrils of collagen are made. The triple helix is a rod about 3000 × 15 Å. These rods align themselves in the parallel overlapping manner illustrated in Fig. 2-7. The tropocollagen molecule has polar regions at either end and at intervals of about 680 Å along its length; these appear as dark vertical lines in the diagram. It is the interaction of these polar regions that contributes to the typical cross-striations seen in electron micrographs of collagen fibers. The electron micrographic appearance is consistent with an arrangement produced by aligning the polar groups of one tropocollagen molecule with those of a parallel neighbor displaced longitudinally by about one fourth of its length (Fig. 2-7). The tropocollagen molecules do not quite touch when placed end to end. This creates a so-called "hole" in the structure next to a region of slight overlap. The overlapped region next to the hole is thought to represent the most electron-dense striations. The hole is believed to be the nidus for the formation of the hydroxyapatite crystals that trigger bone formation.

Cross-linkages. When microfibrils are formed in the laboratory by reaggregating tropocollagen and are examined by means of the electron microscope, the structures formed appear identical to microfibrils isolated from tissue. However, they are lacking both in tensile strength and in interchain covalent cross-linkages. The native cross-connections of collagen can be specifically broken by treatment with proteolytic enzymes such as pepsin without cleaving the long tropocollagen subunits. Apparently the peptide bonds of the tightly twined triple helix are in-

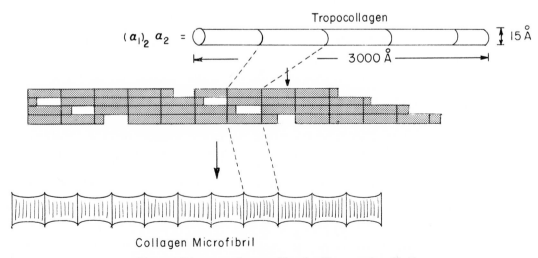

Fig. 2-7. Diagram of assembly of collagen microfibrils.

accessible to the proteolytic enzymes. Consequently, it is believed that the cross-links occur in those polar regions of tropocollagen that are less tightly compacted and therefore more accessible to the enzymes. This fits with what is known about the amino acids that participate in making the cross-links. They are lysine or hydroxylysine residues or derivatives of these residues. A basic amino acid such as lysine would be expected to be found in the polar regions.

An enzyme called peptidyl lysyl oxidase converts the ϵ-amino group of certain lysyl residues to an aldehyde group. Although this enzyme has been only partially purified, it has been shown to be quite distinct from the common monoamine oxidase that is important in drug metabolism. Peptidyl lysyl oxidase is specific for collagen-like molecules. It contains copper and pyridoxal phosphate, a coenzyme derived from vitamin B_6. The enzyme is drastically inhibited by nitriles, specifically β-aminopropionitrile, a compound that produces lathyrism. This disease is associated with deformation of the spine, demineralization of bones, dislocation of joints, and aortic aneurysms. A copper deficiency in pigs produces the same symptoms, undoubtedly because amine oxidase also is affected.

The semialdehydes of lysine form cross-links in a very slow nonenzymatic reaction. Two kinds of reactions of these aldehydes are possible. In one, the aldehyde reacts with an intact lysine residue on an adjacent chain to form a Schiff base, a rather unstable compound but one that can be reduced to a very stable secondary amine.

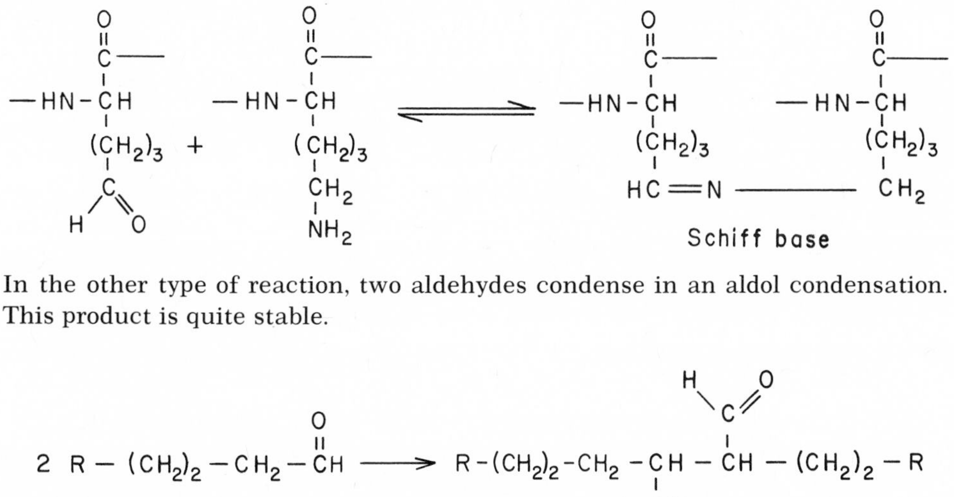

In the other type of reaction, two aldehydes condense in an aldol condensation. This product is quite stable.

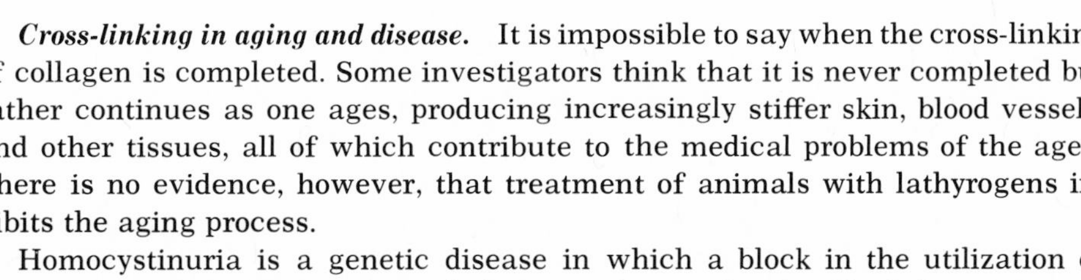

Cross-linking in aging and disease. It is impossible to say when the cross-linking of collagen is completed. Some investigators think that it is never completed but rather continues as one ages, producing increasingly stiffer skin, blood vessels, and other tissues, all of which contribute to the medical problems of the aged. There is no evidence, however, that treatment of animals with lathyrogens inhibits the aging process.

Homocystinuria is a genetic disease in which a block in the utilization of homocysteine results in increased plasma and urine levels of both homocysteine and its disulfide homocystine. Skeletal deformities and other symptoms similar to those seen in lathyrism are secondary to this defect. It is believed that the excessive amounts of homocysteine accumulated by these patients react chemically with the lysyl semialdehydes to block cross-linking. The similar sulfhydryl-containing compound D-penicillamine, used in the treatment of Wilson's disease and as an antidote for mercury or lead poisoning (case 1, Chapter 3; case 6, Chapter 14), produces lathyrism in animals and, after long administration to humans, extravasation of blood into the skin over the elbows and knees.

Marfan's syndrome is another genetic disease that is characterized by lathyritic symptoms. The defect here may be associated with an inability to form the proper substrate for the amine oxidase that makes the aldehydes necessary for cross-linking.

Elastin

Some of the similarities between elastin and collagen have been pointed out, but the differences are equally important. Elastin occurs with collagen in connective tissues. It is a yellow, fluorescent protein found mostly in ligaments and blood vessel walls (Table 2-9), but it also occurs in small amounts in skin, tendon, and loose connective tissue. Unlike collagen, fibers of elastin can be stretched to several times their length, and they snap back almost like a piece of rubber. Tropoelastin is the basic building unit for the elastic fibers. It contains components having molecular weights ranging from 30,000 to 100,000. It is not clear whether these represent subunits. Tropoelastin was first isolated from the aortas of copper-deficient pigs. Because of the lowered activity of the copper-requiring amine oxidase, this material lacked the stabilizing cross-links between interchain lysyl residues, and it was easily extracted from the insoluble fibers. The cross-links in elastin are considerably more complex than those in collagen. Condensations between lysyl and lysyl semialdehyde residues can cross-link two to four polypeptide chains. These lysine-derived structures, released by the hydrolysis of elastin, are called desmosine or isodesmosine, depending on the nature of the initial condensation.

Although elastin and collagen contain similarly high amounts of glycine and proline and both lack cysteine and tryptophan, elastin contains less hydroxyproline and no hydroxylysine (Table 2-2). The glycine content is about the same, but the Gly-X-Y repeating unit of collagen is not present. Consequently, elastin is resistant to hydrolysis by bacterial collagenases, which are proteolytic enzymes highly specific for collagen.

Keratin

The keratins are the most visible of all animal proteins. They are the highly insoluble fibrous proteins that constitute hair, feathers, fingernails, hoofs, and horns. They are good examples of proteins with α-helical structures. Other fibrous proteins with a high α-helix content are myosin, the major protein of muscle, and fibrinogen, a protein involved in blood clotting. Keratin is often called α-keratin to emphasize its α-helical structure. When α-keratin is heated, the strong intrachain hydrogen bonds of the α-helix are broken, and the molecules extend into a parallel pleated sheet (Fig. 2-3). This form is called β-keratin. Although the intrachain hydrogen bonds break, the stable interchain disulfide bonds remain intact. On cooling, β-keratin re-forms its intrachain hydrogen bonds and returns to the α-conformation.

Myoglobin and hemoglobin
Myoglobin

Myoglobin is a chromoprotein found in the muscle cells of mammals. Although appreciable quantities in man are found only in cardiac muscle, myoglobin is present in large quantities in the skeletal muscle of those mammals that dive deeply in the sea. Human myoglobin is a small globular protein with a molecular weight of 16,700 that contains 152 amino acid residues. It is a single polypeptide chain with L-valine at the NH_2-terminus. Coordinated to the protein is a heme residue that for purposes of the present discussion may be represented as in Fig. 2-8. Heme is an iron-containing *porphyrin* composed of Fe(II), four pyrrole rings linked by methene bridges, and eight side chains attached to the pyrrole rings (Fig. 2-8). The iron is inserted in the center, coordinately linked to the four nitrogen atoms of the pyrrole rings. In myoglobin the Fe(II) is also complexed with an imidazole nitrogen atom of a histidine residue in the protein chain, as shown in Fig. 2-9.

Myoglobin can react with O_2 to form oxymyoglobin (MbO_2), which is in equilibrium with the deoxymyoglobin (Mb).

$$Mb \ + \ O_2 \ \rightleftharpoons \ MbO_2$$

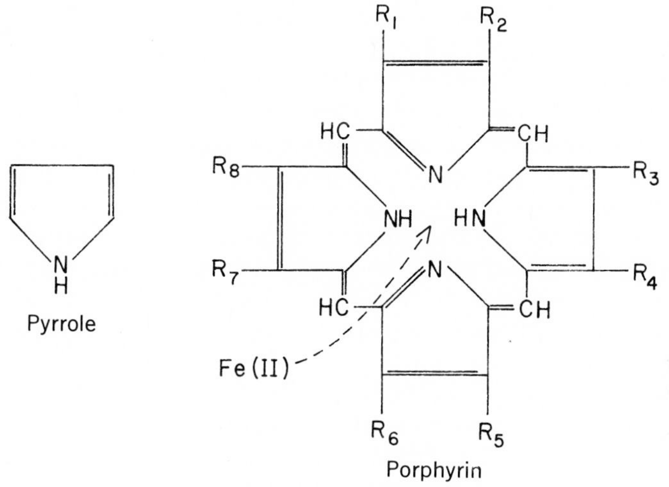

Fig. 2-8. Porphyrins.

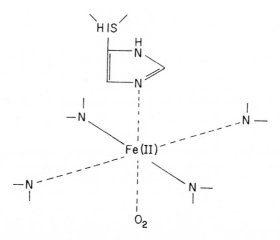

Fig. 2-9. Representation of oxymyoglobin structure around heme residue.

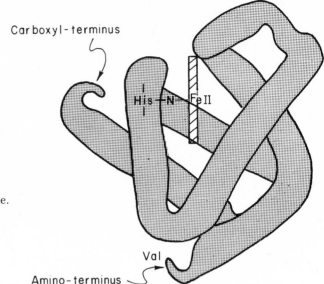

Carboxyl-terminus

His—N   Fe II

Val

Amino-terminus

Fig. 2-10. Myoglobin molecule.

The equilibrium position is dependent on the concentration of O_2 in the system. Therefore, myoglobin may be considered as a storage reserve for O_2, as it is largely in the oxymyoglobin form when the O_2 concentration in the cellular fluid is high but releases the bound O_2 for cellular use when the O_2 supply is reduced. Its function in deep-diving mammals is to provide a store of O_2, and its role in human cardiac muscle during periods of oxygen debt is similar (p. 9).

Myoglobin was one of the first proteins whose conformation was described through the application of x-ray analysis. Its three-dimensional structure consists of eight sections of relatively straight α-helical segments bent on each other to give a compact globular molecule. Comparing this conformation with the amino acid sequence of the primary structure, it is found that the hydrophobic interactions are maximum, as very few hydrophobic side chain residues are exposed to the aqueous solvent medium. The peptide chain is further stabilized by the significant formation of α-helical segments. Breaks in the α-helices are found when L-prolyl or other destabilizing amino acid residues occur in the chain. There is no stability introduced in myoglobin by —S—S— bonds, cysteine being absent from the molecule. The molecule is very compact; there is space for the heme residue in the crevice noted in Fig. 2-10, but otherwise space exists for only four water molecules in the interior portion of the molecule. The two negatively charged propionate side chains, R_6 and R_7 (Fig. 2-8), of the heme residue form ionic bonds with two L-lysyl $\epsilon\text{-NH}_3^+$ groups in the peptide chain, and this together with the coordination of the Fe(II) to the L-histidyl nitrogen binds the heme firmly into the globin.

Hemoglobin

The red color of blood is caused by the hemoglobin content of the erythrocytes. There are about 12 to 17 g hemoglobin/dl blood, depending on the age and sex of the individual. Under normal conditions all of the hemoglobin of the blood is present inside the erythrocytes. Hemoglobin is made in immature erythrocytes in the bone marrow and not during the 120– to 135–day life span of the erythrocyte in the circulation.

Centrifugation of a blood sample causes the cells to pack at the bottom of the centrifuge tube. These cells occupy about 40% to 47% of the blood volume; the percentage value of packed cells is called the *hematocrit*. Since the vast majority of these cells are erythrocytes, the hematocrit is a measure of the amount of erythrocyte hemoglobin to the first approximation.

Normal adult hemoglobin, Hb A, contains two globin subunits of one type identified as α-chains and two globin subunits of another type called β-chains. Hb A is therefore represented as $\alpha_2\beta_2$. In many ways these α- and β-chains are similar to myoglobin. They have L-valyl amino terminal residues and conformationally have a high proportion of α-helical segments. The α-chain has 141 amino acid residues with an L-arginine carboxyl-end terminus, and the β-chain has 146 residues with an L-histidine carboxyl-end terminus. The four subunits, with their identical heme residues, pack into a tetramer, molecular weight 65,000, which can be dissociated first into two dimeric $\alpha\beta$-units and finally into the mixture of α- and β-monomers. On removal of the dissociating agent, which may be an acid or an alkali, the monomers reassociate to the $\alpha_2\beta_2$-tetramer.

Hemoglobin variants

Normal. In the course of a lifetime most humans synthesize five different globin chains. All the hemoglobins assembled from these chains contain two α-chains. The variants are fetal hemoglobin, Hb F ($\alpha_2\gamma_2$); Hb A$_2$ ($\alpha_2\delta_2$), a variant that comprises about 2.5% of the hemoglobin in the adult; and embryonic hemoglobins (for example, $\alpha_2\epsilon_2$). The primary structures of these normally occurring globins are very similar, and each contains a protected pocket into which the heme residue fits.

Abnormal. Mutations that sometimes occur produce a substitution of a *single* amino acid residue in one type of globin chain. These mutations may be innocuous or fatal, depending on the nature of the substitution and on the point at which they occur in the chain. For example, sickle cell anemia, a serious and often fatal inherited disease, is caused by the presence of hemoglobin S (Hb S). In Hb S the sixth amino acid residue from the NH_2-terminal end of each β-chain is valine; in the normal β-chain it is glutamic acid. Thus in Hb S an amino acid containing an anionic side chain is replaced by one with an uncharged hydrocarbon chain. Because of this, the electrophoretic migration rate of Hb S is different from that of Hb A.

In many other abnormal hemoglobin variants that have been described only one amino acid residue is different in the α- or β-chains. Frequently, an amino acid residue with an ionic side chain is substituted so that the charge on the chain is changed and the variant is easily detected by electrophoresis.

Oxygen binding

Hemoglobin reacts reversibly with O_2 in the manner described earlier for myoglobin. The reaction equilibrium may be expressed in terms of the percentage of the oxygenated form, which will vary with the concentration of O_2. The relationship is represented graphically in Fig. 2-11, where the O_2 concentration is expressed as the partial pressure ($\bar{p}O_2$) in millimeters of mercury.

Oxygen concentration and $\bar{p}O_2$

As stated in Dalton's law, the total pressure of a gas mixture is the sum of the partial pressures of each of the components. Also, the partial pressures of the gases are proportional to the corresponding mole fractions of each gas in the

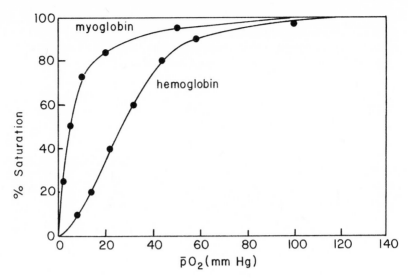

Fig. 2-11. Oxygen saturation curves for myoglobin and hemoglobin.

mixture. Finally, the concentration of the gas in the liquid phase of a gas-liquid system is proportional to its partial pressure in the gas phase, the other factor being the solubility of the gas in the liquid.

Since we are frequently concerned with air at 760 mm Hg, which contains O_2, N_2, CO_2, and water vapor, Dalton's law can be stated for air and body gases as:

$$P_{total} = \bar{p}O_2 + \bar{p}N_2 + \bar{p}CO_2 + \bar{p}H_2O$$

Then the mole fraction of any gas, say O_2, in the mixture is:

$$\frac{\bar{p}O_2}{P_{total}} \quad or \quad \frac{\bar{p}O_2}{760}$$

Now, the amount of O_2 dissolved in liquid at a given temperature is expressed in terms of a *Bunsen coefficient*, the volume (ml) of a gas at the specified temperature and 760 mm Hg pressure that will dissolve in 1 ml of liquid. The Bunsen coefficient, although corresponding to a particular temperature, is always corrected to the standard conditions of 760 mm Hg pressure and 273° K. Under such conditions the volume of a gas is quickly converted to a molar concentration, 22.40 ml of an ideal gas being equivalent to 1 millimole (mmole).

The use of partial pressures of gases is so common and important in problems of respiration, anesthesiology, and any quantitative considerations involving gases that it is well to illustrate these interrelationships of gas concentrations. Consider the concentration of O_2 in blood plasma. The O_2 is derived from the inspired air in which the partial pressure of O_2, $\bar{p}O_2$, is around 158 mm Hg. As a result of some mixing with air already present in the lung as well as a result of its saturation with water vapor, the air in the alveolae of the lungs has a $\bar{p}O_2$ of about 100 mm Hg. Diffusion of the O_2 into the capillaries of the lungs results in a further lowering of the O_2 tension ($\bar{p}O_2$) to about 90 mm Hg. Therefore the arterial blood plasma can be considered to be in contact with a gas that has a $\bar{p}O_2$ of 90 mm Hg. The temperature of the body is close to 38° C, at which temperature the Bunsen coefficient for O_2 in blood plasma is 0.024. This means that 0.024 ml of O_2, at

760 mm Hg and 273° K (0° C), will dissolve in 1 ml of plasma at 38° C when the partial pressure of the O_2 is 760 mm Hg. Since the amount of O_2 dissolved is proportional to its partial pressure and the partial pressure in alveolar air is 90 mm Hg, then the actual volume of O_2 in arterial plasma at the lungs is:

$$0.024 \times \frac{90}{760} \text{ ml } O_2 \text{ per ml plasma}$$

or

$$0.00284 \text{ ml } O_2 \text{ per ml plasma}$$

Since this volume of O_2 is derived from the Bunsen coefficient, which gives the volume of the dissolved gas corrected to standard conditions, then the volume is converted to millimoles by dividing by 22.40. That is:

$$0.00284 \text{ ml } O_2$$

$$= \frac{0.00284 \text{ mmoles } O_2}{22.4}$$

$$= 1.27 \times 10^{-4} \text{ mmoles } O_2$$

Since this is the amount of O_2 in 1 ml of plasma, the concentration of dissolved O_2 is 0.127 μmole/ml or 0.127 mM.

Similar calculations can be made for any of the other gases in plasma, given their Bunsen coefficients. If, like CO_2, they are not ideal gases, the volume of 1 mmole of gas at 760 mm Hg and 273° K also must be known.

Oxygen saturation curves of hemoglobin and myoglobin

Returning now to the oxygenation curves of myoglobin and hemoglobin, the saturation of myoglobin follows a rectangular hyperbola (Fig. 2-11), whereas hemoglobin describes a sigmoidal curve. The reaction of O_2 with myoglobin follows a simple equilibrium where the pK_{eq} is around 6:

$$K_{eq} = \frac{[MbO_2]}{[Mb][O_2]}$$

Throughout the range of $\bar{p}O_2$ values the O_2 is bound more readily to myoglobin than hemoglobin. Most important, the binding of O_2 to each molecule of myoglobin is independent of another molecule, because there is only one O_2 binding site on each. Each monomeric subunit of hemoglobin reacts with O_2 in a fashion similar to myoglobin, but in the tetrameric quaternary structure the binding of O_2 on each subunit is *not* independent of the binding at the other units. The binding of the first O_2 affects the binding of the second O_2, these two O_2 affect that of the third, and so on. This situation reflects a cooperativity in the binding processes. In effect, there are four different K_{eq} values for each of the binding sites, with K_1 being less than K_2 and K_3, and the final equilibrium constant demonstrates that the last O_2 molecule is bound more tightly than the first.

$$Hb_4 \quad + \quad O_2 \quad \underset{\longleftarrow}{\overset{K_1}{\longrightarrow}} \quad Hb_4(O_2)$$

$$Hb_4(O_2) \quad + \quad O_2 \quad \underset{\longleftarrow}{\overset{K_2}{\longrightarrow}} \quad Hb_4(O_2)_2$$

$$Hb_4(O_2)_2 + O_2 \xrightleftharpoons{K_3} Hb_4(O_2)_3$$

$$Hb_4(O_2)_3 + O_2 \xrightleftharpoons{K_4} Hb_4(O_2)_4$$

$$K_1 < K_2 \approx K_3 < K_4$$

Such a mathematical treatment can account for the sigmoidal saturation curve of hemoglobin.

Stereochemistry of hemoglobin oxygenation

Explanations for the cooperative effects in hemoglobin have been proposed on a molecular basis from evidence obtained by x-ray crystallography of oxy-hemoglobins and deoxyhemoglobins. It is known that the conformations of the α- and β-chains are altered when their respective heme residues combine with O_2. The stepwise process may be described using several models, two of which are presented here.

Induced conformation change. It is suggested that the conformational change in the oxygenated chain influences the conformation of a deoxy chain, making its binding with O_2 occur more readily in deoxyhemoglobin. Assume that the oxy chains, both α and β, are represented by a circle and that the deoxy chains are represented by a square. Changing the square to an octagon represents the con-formational modification brought about by the oxygenation of the previous chain, a sequence of events depicted in Fig. 2-12. Although the actual chains involved at each step are not identified, it is considered likely that one $\alpha\beta$-dimer reacts with

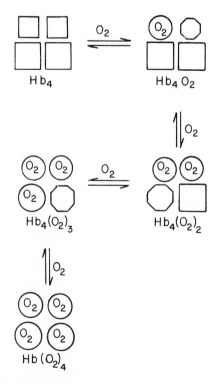

Fig. 2-12. One model for stepwise oxygenation of hemoglobin.

the first two O_2 molecules. The second dimer reacts similarly after some conformational modification, which enhances its binding with O_2.

Allosteric mechanism. The allosteric model relates more to the quaternary structure of hemoglobin. It assumes the existence of two structural forms, the relative position of the subunits in each form being different. In the deoxy form the α- and β-subunits are bound to each other by ionic or salt bridges in a more compact structure that is less reactive toward O_2 than is the oxy form. A number of the original salt bridges are broken in the oxy form, making it a more relaxed quaternary conformation that is more reactive to O_2. It is proposed that these two quaternary forms are in equilibrium at all stages of oxygenation of hemoglobin. The various equilibria may be summarized as shown in Fig. 2-13.

The quaternary structure $\left[\alpha_1 \quad \alpha_2 \quad \beta_1 \quad \beta_2 \right]^T$

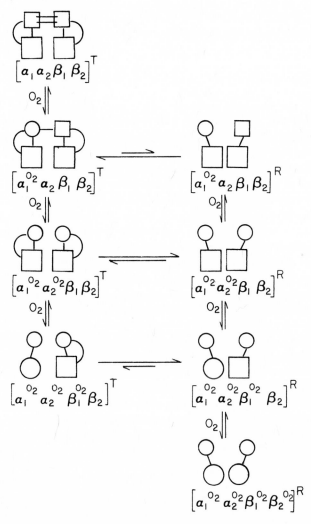

Fig. 2-13. Perutz allosteric model for oxygenation of hemoglobin.

which represents the tight (T) less reactive deoxyhemoglobin, reacts with one molecule of O_2 to give:

$$\left[a_1^{O_2} a_2 \ \beta_1 \ \beta_2 \right]^T$$

This is in equilibrium with the relaxed (R) quaternary form

$$\left[a_1^{O_2} a_2 \ \beta_1 \ \beta_2 \right]^R$$

which has a higher affinity for the next O_2, but it is still present in a smaller concentration than the T form. The second O_2 reacts with both R and T forms, but having caused the break in further salt bridges between the two α-units, the amount of

$$\left[a_1^{O_2} a_2^{O_2} \ \beta_1 \ \beta_2 \right]^R$$

is greater than the corresponding T form. The affinity for the third O_2 is therefore greater than that for the second O_2, and the following conformation predominates:

$$\left[a_1^{O_2} a_2^{O_2} \ \beta_1^{O_2} \beta_2 \right]^R$$

Therefore, the final O_2 molecule has a high affinity for the $Hb_4(O_2)_3$ molecule. Such a model accounts for the four different apparent equilibrium constants discussed earlier and the sigmoidal saturation curve of O_2 for Hb_4. The mechanism for this second model relates to the slight change in the Fe(II) position in the heme when it binds O_2. This shift is transmitted to the globin chain and produces displacement of an L-tyrosyl residue from a pocket in the globular polypeptide. With it is carried the NH_2-terminal residues that had been in salt linkage with nearby negatively charged groups. When two such subunits have been changed in this way, there is an increased tendency for the remaining L-tyrosyl residues to be displaced even without O_2 binding. The equilibrium has thus shifted toward the oxy (R) conformations, and these are more prone to bind O_2. The affinity for O_2 is greater in the R forms than the corresponding T forms, so that oxygenation tends to pull the equilibria in the direction of the oxy (R) forms.

Which of these two alternative models of cooperativity offers the more accurate explanation for O_2 binding by hemoglobin is not presently known. The model based on sequential change in tertiary structure induced by the binding of O_2 to another subunit was proposed by Koshland. The second model, proposed by Perutz, relates to the theory of allosterism developed earlier by Monod, Wyman, and Changeux for enzyme systems. The Perutz model is consistent with x-ray data of the deoxyhemoglobins and oxyhemoglobins. However, the crystalline intermediate oxygenated forms are not available for study at the present time, so that the proposed changes in conformation cannot be determined.

Allosterism

The models for the cooperative processes in the binding of O_2 to hemoglobin are applicable to other systems in biochemistry, particularly to those proteins that are designated as *regulatory* or *allosteric* enzymes. In such systems the binding of a low molecular weight compound, called an *effector*, influences the binding of the substrate to the enzyme. If the effector increases the binding of the substrate, it is termed a *positive* effector; if not, it is called a *negative* effector. For example, in the O_2-Hb_4 system, 2,3-diphospho-D-glycerate, which is described in the following section, acts as a negative effector.

Oxygen transport

Binding of 2,3-diphospho-D-glycerate to deoxyhemoglobin. 2,3-Diphospho-D-glycerate represents approximately 15% of the anionic content of the erythrocyte. It is produced by the catabolism of D-glucose, and its intracellular concentration is about 5 mM, nearly equimolar with hemoglobin.

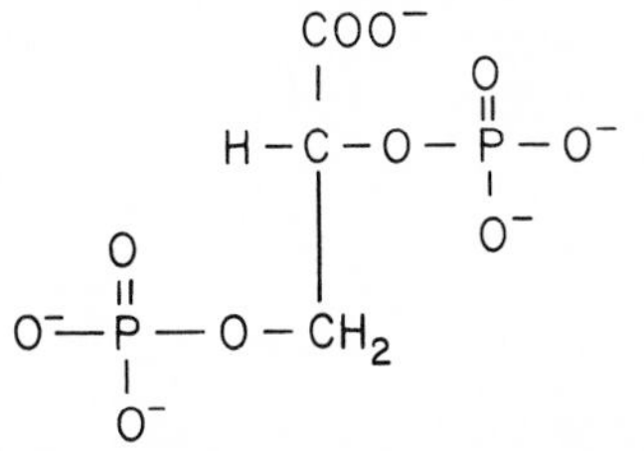

$$
\begin{array}{c}
COO^- \\
| \\
H-C-O-\overset{\displaystyle O}{\underset{\displaystyle O^-}{\overset{\|}{P}}}-O^- \\
| \\
O^- -\overset{\displaystyle O}{\overset{\|}{P}}-O-CH_2 \\
| \\
O^-
\end{array}
$$

2,3-Diphospho-D-glycerate

Its function at such high concentrations was not appreciated until its effect on the affinity of hemoglobin for O_2 was demonstrated.

One mole of 2,3-diphospho-D-glycerate is bound to one mole of deoxyhemoglobin. The 2,3-diphospho-D-glycerate is bound between the two β-subunits by ionic salt bridges. As a result, the equilibrium of the T and R forms is shifted to favor the T or deoxyquaternary structure. This reduces the tendency of the partially oxygenated intermediates to convert to the R or oxyquaternary structure. For example, the equilibrium between

$$
\left[\begin{array}{cccc} O_2 \\ \alpha_1^{} & \alpha_2 & \beta_1 & \beta_2 \end{array} \right]^T \text{and} \left[\begin{array}{cccc} O_2 \\ \alpha_1^{} & \alpha_2 & \beta_1 & \beta_2 \end{array} \right]^R
$$

would be more in favor of the former T form, which has less affinity for O_2 than the R form. This results in a saturation curve for hemoglobin that shows an overall lowering of the affinity for O_2 in the presence of 2,3-diphospho-D-glycerate (Fig. 2-14). However, it should be noted that at the $\bar{p}O_2$ in the alveolae, about 100 mm Hg, hemoglobin is nearly 100% saturated. In contrast, at the $\bar{p}O_2$ present in the capillary beds of peripheral tissues (40 mm Hg) there is less O_2 available to bind hemoglobin in the presence of 2,3-diphospho-D-glycerate. In other words, O_2 can be unloaded more easily and thus ensure an adequate supply to the tissues. It is interesting to note that fetal hemoglobin (Hb F) is not affected in this way by 2,3-diphospho-D-glycerate.

Carbon monoxide poisoning. CO binds in a competitive fashion at the same site on hemoglobin as does O_2 and shows similar cooperativity and the Bohr effect. CO binding is more than 200 times stronger than O_2 binding. The amount of O_2 that can be carried by the remaining unreacted Fe(II) of the hemoglobin molecule is reduced if CO is present at one of the heme residues. Moreover, the O_2 that *is* associated with hemoglobin molecules that contain some CO is bound more tightly, making it more difficult for the O_2 to be transferred from these hemoglobin molecules to the tissues. Therefore the toxic effect of CO is more severe than might be predicted simply from its concentration in the blood.

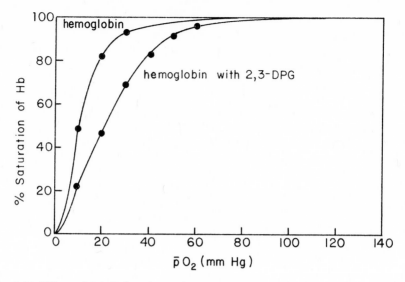

Fig. 2-14. Effect of 2,3-diphospho-D-glycerate on oxygen binding of hemoglobin.

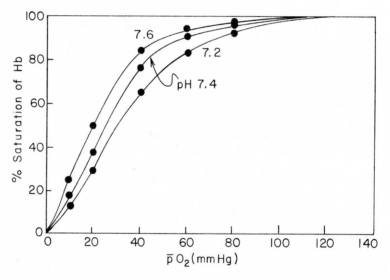

Fig. 2-15. Oxygen saturation curves at different pH values.

   The Bohr effect.   For each O_2 bound to Hb_4 there is movement of the penulti-
mate L-tyrosyl residues and the breaking of salt bonds. The proton that was held
in the salt linkage is now free to dissociate, which it does to a degree that varies
according to the pH. In the β-chain the COOH-terminal histidine residue releases
a proton. This is not the histidine residue coordinated to the heme Fe(II). In at
least one of the α-subunits the proton arises from the NH_2-terminal L-valine,
which, in the deoxy form, was forming a salt link with the COOH-terminal arginine in the other α-unit. Since these protons are arising from $-NH_3^+$ or imidazole nitrogen atoms, there will be a greater tendency for these groups to remain
protonated as the pH of the system becomes more acidic. Salt bridges are therefore favored, and these tend to stabilize the deoxy T form. This is reflected in the

O_2 binding curves for Hb_4 (Fig. 2-15). The result is a tendency to release O_2 at lower pH values, which is known as the Bohr effect. This effect depends on the quaternary structure of hemoglobin. It is not seen with myoglobin or in the α- or β-monomer subunits of hemoglobin. This and other changes of ionic character cause a change in the overall strength of hemoglobin acidity, such that the apparent pK_a of Hb_4 changes from 7.71 to 7.16 in $Hb_4(O_2)_4$. The oxygenation of hemoglobin is therefore associated with a release of protons. We can represent the reaction as:

$$HbH_x + O_2 \rightleftharpoons HbO_2 + xH^+$$

where x is approximately 0.7 mole/mole of hemoglobin. This property is extremely important in the physiology of O_2 transport and will be discussed further in Chapter 4.

Protein turnover

Why does an adult who is not growing require amino acids for the synthesis of protein? The reason is that protein is being continually broken down and re-synthesized—it is turning over. The use of radioisotopes makes it possible to measure the "turnover rate" of the proteins of a particular tissue. Most frequently the turnover rates are expressed in terms of the time required for half of the material to be changed. A half-life of 6 days, using a radioactive label as the indicator, means that half of the radioactive protein originally present has been degraded and replaced by newly synthesized protein from the unlabeled amino acids in the body pool within 6 days.

Most of the proteins of normal nondividing liver cells have half-lives of several days. Structural proteins of the muscle such as myosin have longer half-lives of about 180 days, and some of the collagens of connective tissue have very slow turnovers, with half-lives of approximately 1000 days. The average 70-kg man who is not gaining or losing weight synthesizes and degrades about 400 g protein/day.

Most of the nitrogen metabolized by the body is contained in the proteins. Those proteins that occur in the blood plasma have short half-lives of approximately 10 days, and the erythrocyte has a total life span of only 120 days. Changes in serum nitrogen and protein occur quickly and thus are useful indicators of a large number of disease states. Some common clinical tests measure serum proteins, including albumin and globulins. Blood urea nitrogen (BUN), nonprotein nitrogen (NPN), serum creatinine, serum uric acid, and a variety of other nitrogen-containing substances are also measured.

Other macromolecules besides proteins exhibit turnover. The template or messenger ribonucleic acids are good examples. The life spans of these substances are very important in regulating the amounts of proteins made in response to them. In all cases the metabolic reactions involved in making proteins or nucleic acids are different from those that break them down; therefore, the degradative reactions are not simply the reverse of the synthetic ones.

Genetic basis of protein structure

It is common knowledge that the biologic uniqueness of each individual has a genetic basis. The genetic information responsible for determining individual form and substance is stored in the nucleotide sequences of deoxyribonucleic acid (DNA). The expression of the information coded in these nucleotide sequences occurs because of the regulated synthesis of proteins whose amino acid sequences are directly determined by the nucleotide sequences in DNA. A few examples of the manner in which genetic variation determines the structure of proteins in man will be considered later, but to put these findings in their proper perspective, it is worthwhile to review the principles of mendelian inheritance.

Mendelism

The first experiment described in Mendel's classic paper, published in the obscure Czechoslovakian *Journal of the Brno Society of Natural Science* in 1865, was the following: Garden peas that produced round seeds were crossed with another type of peas that had wrinkled seeds. All of the seeds that resulted from this first filial generation (F_1) were *round.* The next season Mendel planted the round seeds produced by the previous mating. These plants were allowed to self-fertilize; this second filial generation (F_2) produced 5474 round seeds and 1850 wrinkled seeds, or 2.96 times more round than wrinkled seeds. Several other similar crosses were made, and other characteristics were observed. All of these experiments produced the same results. Of the two parental characteristics, only one was seen in the F_1 progeny, but in the second generation the lost characteristic reappeared in one fourth of the progeny. Based on this simple experiment, Mendel proposed that each characteristic was produced by a discrete entity that was passed from one generation to the next. We now call this discrete entity a *gene.*

Furthermore, Mendel realized that in the first generation one of the parental characteristics was *dominant;* the other, which failed to appear, was *recessive.* He reasoned that each character was controlled by two genes. For example, the parents with round seeds could be designated as RR, while those that always gave wrinkled seeds could be called rr. In the F_1 generation all of the progeny are of the Rr type, where R is dominant and r is recessive. A mating between the hybrids results in a random but *independent segregation* of the genes, so that the progeny of the two Rr parents will have an equal probability of receiving either an R or an r gene from either parent. Thus the progeny will be evenly distributed between the gene arrangements RR, rr, Rr, and rR. (Rr and rR are actually of identical genetic composition; they are designated in this order to emphasize the four different ways of pairing the two genes.) Only one fourth of the seeds, those with the *genotype* (genetic constitution) rr, would be wrinkled. On the other hand, seeds with the different genotypes RR and Rr, representing three fourths of the whole, would be round; that is, their *phenotype* is the same.

Mendel's interpretation of his data, based on the concept of paired genes, is applicable to the human set of diploid chromosomes. There are twenty-two sets of paired chromosomes and two sex chromosomes for a total of forty-six. The sex chromosomes in the female are also paired and are designated as XX, indicating that the somatic cells each contain a pair of X chromosomes. In male somatic cells the sex chromosomes are designated as XY. The other forty-four chromosomes are called *autosomal* chromosomes. Many human diseases can be analyzed from family pedigrees in terms of Mendel's rules. Often it can be deduced that a disease is inherited as an autosomal dominant or recessive trait or as an X-linked dominant or recessive. This kind of information is very useful in genetic counseling.

In one respect Mendel was very fortunate. He happened to study characteristics that were clearly either dominant or recessive. Many characteristics are of an intermediate type in which the heterozygote shows a characteristic intermediate between those of the homozygotes. As might be expected, the phenotypes of the human heterozygotes are not always pure dominant or recessive, and in these cases an intermediate severity of the disease is found.

One gene–one polypeptide concept

At first Mendel's ideas about the dominant factors were interpreted in terms of the presence of a dominant gene and the absence of a recessive gene in the F_1 generation. However, it subsequently became clear that the recessive gene is not actually absent but is present in either an inactive or a modified form. These changed genes occur as the result of mutations, those rare events caused by chemical changes in the nucleotide bases of the DNA.

The modified gene reflects this change in the phenotype. For many years investigators searched for the most immediate product of the gene's action, little realizing that in 1902 the English physician Garrod, aware of Mendel's ideas, had correctly concluded from his own observations on human genetic diseases that genes were responsible for producing enzymes. He proposed that a defect in a gene would produce an enzyme defect, and that this was the basis of inherited diseases. Some 30 years later the work of Beadle and Tatum conclusively proved this idea, and their names are generally associated with the one gene–one enzyme concept. Because of our more recent knowledge of the subunit structure of proteins and enzymes, we know that the one gene–one enzyme concept should be restated as the one gene–one polypeptide concept since the amino acid sequence of all proteins is determined by the nucleotide sequence of deoxyribonucleic acid, the chemical substance composing genes.

Thus we can think of one gene for the α-subunit of hemoglobin and another for the β-subunit. If, as in sickle cell anemia, the gene for the β-subunit is mutated, a person heterozygous for this autosomal disease would have all normal α-chains but half the β-chains would be normal and half would be abnormal. Consequently, one would predict that half of the hemoglobin in the carriers of this disease would be normal Hb A and half would be abnormal Hb S. It so happens that the carriers (heterozygotes), who are said to have the sickle cell trait rather than the disease, have somewhat more Hb A than Hb S. It is not completely clear why this is so. For most genetic diseases, however, a carrier will have approximately half the gene product of a normal individual. Because this is often enough of the protein to perform its function, most human genetic diseases are recessive.

Gene dosage and the X chromosome

Mutations in the X chromosome are usually not serious in females because they have a pair of X chromosomes; however, most X-linked diseases afflict males since they inherit only a single X chromosome. In respect to the genes on the X chromosome, the male is said to be hemizygous. The X chromosome is quite large compared to the Y chromosome and is expected to carry a considerable amount of genetic information. It is clear, however, that females make the same amount of the gene products of the X chromosomes as males, not twice as much. The reason for this is explained by the Lyon hypothesis, which states that very early in embryogenesis one of the X chromosomes of the cells in females becomes metabolically sequestered by forming a Barr body. The Barr bodies are seen in the interphase nuclei of these cells, one for each normal cell. Lyon further suggests that selection of the particular X chromosome to be inactivated is done *randomly*, so that approximately half of the cells contain the X chromosome of the female's father and the other half the X chromosome of the mother. Cloning experiments have borne out this prediction. The cells of the female are said to be mosaic in respect to the X-chromosome since some cells contain a functioning X-chromosome derived from the father, whereas other cells have the functional X-chromosome of the mother.

Because the commitment to sequester the X chromosome occurs early in embryogenesis when only a few cells exist for particular functions, the probability for an even distribution of the X chromosome is lower than it would be if many cells were involved. Consequently, a population of females will show a whole spectrum of activity, from that of the normal to that of the homozygote, for an X-linked product such as glucose 6-phosphate dehydrogenase. The number of individuals at each extreme is small; most will show the typical intermediate values of the usual heterozygote.

It is still possible for a female to have certain X-linked diseases. For example, she could have a disease that is inherited as a dominant trait, such as manic depression. She could also inherit an X-linked recessive disease if defective X-chromosomes came from both her father and her mother. In this situation the X-linked disease must be mild since the father, who would also have the disease, must be able to reproduce.

REFERENCES

Dickerson, R. E.: X-ray studies of protein mechanisms, Ann. Rev. Biochem. 41:817, 1972.

Dickerson, R. E., and Geis, I.: The structure and action of proteins, New York, 1969, Harper & Row, Publishers. *Describes in simple terms the relationship of the structure and function of those proteins that have been studied at an atomic level. The illustrations are excellent.*

Perutz, M. F., and Lehmann, H.: Molecular pathology of human haemoglobin, Nature 219:902, 1968. *A beautiful description of how the effects of amino acid substitutions in human hemoglobins can be explained in terms of protein structure.*

Traub, W., and Piez, K. A.: The chemistry and structure of collagen, Adv. Protein Chem. 25:243, 1971. *A review.*

Clinical examples

CASE 1: FALSE POSITIVE TEST FOR HEMOGLOBIN S

A 16-year-old black girl was admitted to the emergency room with complaints of lethargy, excessive menses, and heavy vaginal bleeding between menstrual periods. Results of the physical examination revealed only pale nail beds and conjunctivae. Values from the laboratory tests were as follows:

Hemoglobin	2.7 g/dl
Hematocrit	9.7%
Erythrocytes	$1.64 \times 10^6/mm^3$
White blood cells	$2000/mm^3$
Differential count	Normal
Serum iron	12 μg/dl
Sickledex test	Positive
Serum albumin	3.2 g/dl
Serum globulins	
α_1	0.2 g/dl
α_2	0.6 g/dl
β	1.2 g/dl
γ	1.9 g/dl

The clinical picture and the positive Sickledex test suggested a diagnosis of sickle cell anemia.

Biochemical questions

1. Discuss how the molecular structure of Hb S differs from that of Hb A.
2. How does the structural change affect oxygenation and deoxygenation? How does this influence the reliability of the Sickledex test?
3. What other diagnostic tests can you propose to confirm the type of anemia in this patient?
4. What information might be derived from the analysis of the serum proteins?
5. If sickle cell anemia were diagnosed, would treatment with urea be possible?
6. What is the molecular basis for the proposed urea treatment for sickle cell anemia?

Case discussion

Genetics of sickle cell anemia. In 1910 a Chicago physician first noted the crescent sicklelike shape of erythrocytes from a severely anemic black student. Several years later it was found that the sickling occurred only at low O_2 pressures and that if the cells were reoxygenated most took on the normal shape of a biconcave disk. Almost 40 years after the first report, genetic studies showed that sickle cell anemia was inherited as a defective gene on an autosomal chromosome and that the disease was of a recessive type. It is not typical of the usual recessive type of inherited disease, however. The full-blown disease of the homozygote, sickle cell disease, is very serious. The heterozygote, although usually healthy, can experience some of the symptoms, especially when stressed by anoxic conditions. Heterozygotes are said to have the sickle cell trait. While some individuals can withstand rather severe O_2 stress, other carriers will undergo a crisis simply by flying in commercial aircraft pressurized for an altitude of 5000 ft.

Half of the children resulting from the mating of a carrier and a normal person will have sickle cell trait; the other half will be normal. Unless there is a very rare, sporadic mutation, none of these offspring should have the disease. The risk of sickle cell anemia is very high in the children when both parents have the trait. While 50% of their children will have the trait and 25% will be normal, 25% will have sickle cell anemia.

The disease is very common in black populations all over the world, but it is quite rare in other populations. About 9% of black Americans are estimated to be carriers, and about one out of every seventy to 300 have the disease. (See Table 12-6 for comparison with other human genetic diseases.) The evidence suggests that sickle cell disease as well as thalassemia, another hemoglobin disease (case 3, Chapter 12), give the carriers an advantage in surviving a very dangerous malarial parasite, *Plasmodium falciparum*, which spends part of its life cycle in the red cell. This probably accounts for its high incidence in people

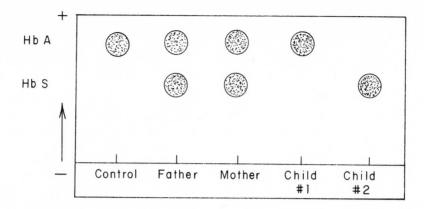

whose ancestors had to survive serious epidemics of malaria.

Structural differences between hemoglobin A and hemoglobin S. About the time that the genetics were worked out, Pauling and associates discovered that the hemoglobin from sickle cell patients was structurally different from the hemoglobin of normal controls in that the two were easily separated by electrophoresis. Hemoglobin from carriers showed approximately 35% Hb S and 65% Hb A. A typical electrophoretogram is shown above. The Hb A of the control migrates the fastest at pH 8.6, whereas both parents show the typical heterozygote pattern with both Hb A and Hb S. Child No. 1 has normal hemoglobin, but child No. 2, who shows only Hb S, would be expected to have the disease.

Amino acid substitution in hemoglobin S. At first it was believed by some that the secondary or tertiary structure of Hb S might account for its difference on electrophoresis. The conclusive proof that the hemoglobins differed in their primary structure by a single amino acid was provided by Ingram, who treated the aberrant β-chains with the enzyme trypsin. This proteolytic enzyme is very specific for hydrolyzing only lysyl and arginyl bonds; consequently, the β-chain was cleaved into twenty-eight shorter peptides. These were separated by applying the mixture to the corner of a large sheet of filter paper. Chromatography of the peptides separated several of them but left others clustered together and unresolved. The filter paper was then dried and rotated 90 degrees. Electrophoresis in the perpendicular direction completed

the separation. The resulting two-dimensional analysis clearly separated almost all of the peptides. Visualization was achieved by spraying the dried paper with ninhydrin, a reagent that reacts with amino groups to produce a violet color. This technique, called fingerprinting, is now in general use to analyze other proteins as well as many other variant hemoglobins. One of the peptides from Hb S was clearly different from Hb A. The amino acid sequences of the two peptides showed the following:

Hb A peptide   Val-His-Leu-Thr-Pro-*Glu*-Glu-Lys
Hb S peptide   Val-His-Leu-Thr-Pro-*Val*-Glu-Lys

This was the first example of precisely how a genetically determined variant protein differed from the normal wild type. Now dozens of similar examples are known, most of them being simple substitutions of one amino acid for another.

Solubility of deoxygenated hemoglobin S. Shortly after Pauling's experiment it was found that deoxygenated Hb S was twenty-five times less soluble than deoxygenated Hb A. Because position 6 of the β-chain occurs near the surface of the molecule, it is believed that in the deoxy structure the valine at this position in Hb S can interact hydrophobically with a similar residue on another Hb S molecule. Since there are two chains in each hemoglobin unit, each with a valine at position 6, long linear polymers of Hb S form. If the O_2 tension is low enough, the hemoglobin gels. Obviously this kind of hydrophobic interaction is not possible in Hb A, where position 6 contains the polar glutamyl residue. Similarly, in Hb

C, where the glutamyl residue is replaced by a polar lysyl group, gelling of the hemoglobin does not occur.

The gelled hemoglobin causes the cells to sickle. The abnormally shaped cells are phagocytized by the reticuloendothelial cells and removed from the circulation more rapidly than normal erythrocytes. This is the direct cause of the anemia seen in the disease. The sickled cells contribute to the crisis by blocking small blood vessels, causing the tissues to become painfully anoxic.

Sickledex test. The Sickledex test is the proprietary name for a test that is rapidly gaining popularity as a quick, simple, and convenient test for Hb S. It appears to depend on the lysing of the erythrocyte and the insolubility of reduced Hb S. However, serious errors in this test are possible unless one has a clear understanding of its molecular basis and the genetics of the disease.

There are always a few sickled cells in the blood of patients with the disease, but they are not present in the blood of those with the trait. The cells from the carrier can be made to sickle, however, as is done in the Sickledex test. The reagents used for this test consist of a phosphate buffer, saponin, which is used to lyse the erythrocytes, and dithionite, which is used to reduce the hemoglobin so that it will form visible aggregates if any Hb S is present.

As one might suspect, the test will pick up the trait as well as the disease. It also responds positively to several other abnormal hemoglobins, to elevated amounts of plasma proteins, and when too much blood is used in relation to the amount of reagents, as in the case presented. False negative reactions can occur if the reagents are old, if the turbid solution is viewed incorrectly, or if the hematocrit is below 30% and the hemoglobin is below 10 g/dl. The latter conditions applied to the patient described; consequently, the amount of blood was double that indicated in the directions supplied with the Sickledex kit. Since the serum protein concentration was normal in this patient, the excessive amount of protein in relation to reagents produced a false positive test result.

The signs of anemia in this patient were actually a result of an iron deficiency caused by chronic blood loss. A reliable Sickledex test can be performed in severely anemic patients by centrifuging down the red cells and removing enough plasma to artifically create a hematocrit of about 30%.

Screening programs for sickle cell anemia. Screening programs are unreliable, for they do not pick up the disease or trait in children under 6 months of age because of the persistence of fetal hemoglobin. The Sickledex test is often used, especially where large numbers of patients are screened. For the reasons just mentioned all positives should be confirmed by electrophoresis of the hemoglobin. By this means one can distinguish the homozygote from the heterozygote and perhaps pick up other hemoglobin variants that are not Hb S but that respond positively in the ordinary screening test.

Treatment of sickle cell disease with urea and cyanate. High concentrations of urea can disrupt the hydrophobic interactions that cause Hb S to form aggregates in vitro. In a few experimental cases a sterile 30% solution of urea in sucrose was administered intravenously to treat the crisis phase of sickle cell anemia. The results were encouraging despite the high concentrations of urea required. To denature proteins in vitro, 1 to 8 M urea solutions are commonly used. However, proponents of urea therapy estimate that a much lower concentration is effective in treating sickle cell disease. Most of the urea is excreted within 24 hr, but the benefits of the therapy may last as long as 4 months. From these observations it was predicted that it was not the urea that relieved the crisis but rather the small amount of isocyanic acid that is in chemical equilibrium with aqueous solutions of urea.

$$H_2O \ + \ \underset{\underset{\displaystyle NH_2}{|}}{\overset{\overset{\displaystyle NH_2}{|}}{C}} = O \ \rightleftharpoons \ HN = C = O \ + \ NH_4OH$$

Urea   Isocyanic
                        Acid

In a concentrated urea solution the isocyanic acid content can become appre-

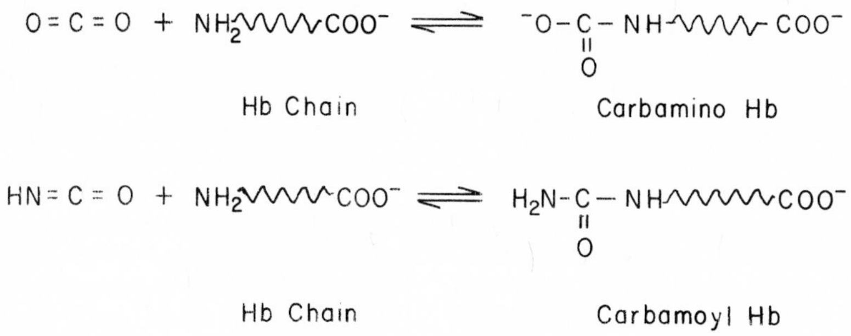

$$O = C = O \ + \ NH_2\text{-}\!\wedge\!\wedge\!\wedge\text{-}COO^- \ \rightleftharpoons \ {}^-O\text{-}C\text{-}NH\text{-}\!\wedge\!\wedge\!\wedge\text{-}COO^-$$
$$\underset{O}{\|}$$

Hb Chain   Carbamino Hb

$$HN = C = O \ + \ NH_2\!\wedge\!\wedge\!\wedge\text{-}COO^- \ \rightleftharpoons \ H_2N\text{-}C\text{-}NH\!\wedge\!\wedge\!\wedge\!\wedge\text{-}COO^-$$
$$\underset{O}{\|}$$

Hb Chain   Carbamoyl Hb

Fig. 2-16. Comparison of carboxylation of hemoglobin by CO_2 to its carbamoylation by isocyanic acid.

ciable. Isocyanic acid was found to irreversibly carbamoylate the amino terminal valine residues in both the α- and β-chains of hemoglobin (Fig. 2-16). No other amino groups were modified. Apparently the ϵ-amino groups of hemoglobin do not react under these conditions.

In Chapter 4 it is pointed out that about 10% of the CO_2 produced in the tissues is transported to the lungs as carbaminohemoglobin. Again it is the amino terminal valine residues that are carboxylated (Fig. 2-16). In a sense, CO_2 and isocyanic acid can be thought of as competing for the same site on the hemoglobin molecule. The difference is that the CO_2 is easily removed at the lungs, whereas the carbamino group is strongly bonded. The effect of the carbamoylation is to significantly increase the affinity of hemoglobin for O_2. Since the carbamoylated Hb S tends to remain oxygenated, it will not aggregate, and cell sickling is prevented.

A brief treatment with urea in vitro prevents the sickling of red cells taken from patients with the disease, but when the urea is dialyzed away, the cells revert to their sickled form. Cells treated with isocyanic acid are also prevented from sickling, but the effect is long lasting and is not reversed by dialysis.

The advocates of urea therapy insist that the urea effect is not caused by the small amounts of isocyanic acid contaminating their preparations. On the other hand, another group of scientists believes that only low levels of isocyanic acid need be used. They claim that isocyanate will form in water, since small amounts of potassium cyanate can be used in place of urea.

$$N \equiv C\text{-}O^- \ + \ K^+ \ \overset{H_2O}{\rightleftharpoons} \ {}^-N = C = O \ + \ K^+$$

Potassium cyanate   Isocyanate

Unfortunately, even at the low concentrations used, cyanate is toxic. Both the urea and the cyanate treatments were experimental and are not now used; nevertheless, the biochemical principles upon which they were based form the foundation for continued research.

REFERENCES

Arras, M. J., and Perry, R. E.: False-positive test for hemoglobin S, J.A.M.A. 220:126, 1972.

Cerami, A.: Cyanate as an inhibitor of red cell sickling, N. Engl. J. Med. 287:807, 1972.

Lee, C. K., and Manning, J. M.: Kinetics of the carbamylation of the amino groups of sickle cell hemoglobin by cyanate, J. Biol. Chem. 248: 5861, 1973.

Lehmann, H., and Huntsman, R. G.: The hemoglobinopathies. In Stanbury, J. B., Wyngaarden, J. B., and Fredrickson, D. S., editors: The metabolic basis of inherited diseases, ed. 3, New York, 1972, Academic Press, Inc.

Levene, R. D., editor: Sickle cell anemia and other hemoglobinopathies, New York, 1975, Academic Press, Inc.

Nalbandian, R. M., et al.: Sickledex test for hemoglobin S, a critique, J.A.M.A. 218:1679, 1971. *See also pp. 1680 and 1693 of the same issue.*

CASE 2: RHEUMATOID FACTORS IN RHEUMATOID ARTHRITIS

A 35-year-old female was diagnosed as having rheumatoid arthritis. She showed the typical symptoms of morning stiffness, painful and symmetrically swollen joints. Radiographs demonstrated the presence of bony decalcifications localized around the involved joints. There was a family history of similar illness. Serologic tests for rheumatoid factor were positive. In this test, dilutions of the patient's serum cause the agglutination of sheep red blood cells or polystyrene (Latex) particles that have been previously coated with human IgG.

Biochemical questions

1. Rheumatoid factor can be thought of as an antibody. What is the antigen that stimulates its production?
2. Rheumatoid factor is usually, but not always, IgM. IgM is thought to be composed of several monomeric units linked together. Considering the data in Table 2-8, how many monomers are necessary to make one molecule of IgM?
3. IgM cannot be made to yield monomers with dissociating substances such as urea, guanidine hydrochloride, or sodium dodecyl sulfate. How would you think that the monomers are held together in IgM? How would you prove your prediction?
4. Mixtures of IgM and IgG in human sera can be separated from one another by passing the serum proteins through a column of Sephadex-G200, a molecular sieve. Which component will emerge from the column first?
5. Treatment of IgG with pepsin, a proteolytic enzyme, produces a fragment called F_{ab_2}, which contains the N-terminal portions of both H and L chains. The rest of the molecule is broken into small fragments. Such treatment prevents the complexing with rheumatoid factor. What can you conclude about the location of the antigenic determinants of IgG?

Case discussion

Clinical features. Rheumatoid arthritis is a chronic inflammatory disease of the joints that affects mainly middle-aged people. The incidence in females is twice that of males. Its causes are unknown, although infection or trauma is thought to initiate or predispose to an immune response against the patient's own joint tissue.

Rheumatoid factors. Rheumatoid factors are present in the sera of 60% to 75% of patients with this disease, and the γ-globulin fraction is prominent on electrophoresis of the serum proteins (Fig. 2-5). False positive reactions are common, however, since the IgM factor can be produced experimentally by repeated injections of dead bacteria into animals. Apparently IgG antibodies formed in response to the microorganisms combine with the bacteria to produce a complex that is itself immunogenic. This complex causes the production of another set of antibodies (rheumatoid factors) against the IgG bound to the dead bacterial cells. This probably explains the high titers of rheumatoid factor in subacute bacterial endocarditis, a chronic infection of the heart valves. Thus the IgM rheumatoid factor is an antibody whose production can be stimulated by an individual's own IgG bound as an antibody-antigen complex to dead bacterial cells.

A similar triggering of an autoimmune response may be responsible for rheumatoid arthritis. Yet, what the initiating stimulus might be is far from clear. It need not be dead bacteria, but neither can they be ruled out. Mycoplasmas, slow viruses, and streptococci have all been suggested as initiating factors at one time or another. The inflammation of the joints may be caused by the presence of rheumatoid factor complexes and IgG within the leukocytes that cause them to rupture and spill out their lysosomal enzymes. These are a collection of very destructive hydrolytic enzymes that break down tissue, producing inflammation and perpetuating the autoimmune cycle.

Structure of IgM. Some investigators are interested in trying to chemically modify the structure of the rheumatoid factors so as to inhibit the inflammatory process. To understand the problems involved, it is necessary to describe the structure of the immunoglobins in more detail. When IgM is gently treated with the reducing agent 2-mercaptoethanol, the 19S molecule is broken into five equal parts that sediment at 7S, much like IgG. Unlike IgG, the IgM monomers are composed of $\mu_2\lambda_2$- or $\mu_2\kappa_2$-chains. Molecular sieves such as gels of Sephadex-G200 will separate any unreacted IgM from its IgG-like monomers. The smaller 7S proteins penetrate the gels while the larger 19S molecules, which are unable to penetrate into the retaining pores of the sievelike gels, rapidly flow around the gel particles and emerge from the column first. Because reduction is required to disaggregate IgM, it is thought that the polymeric molecule is held together by disulfide bridges. One model is given in Fig. 2-17. Rheumatoid factor that has been reduced to yield 7S monomers loses its activity to agglutinate IgG-coated Latex beads. The monomers, however, are inhibitors of IgM-promoted agglutination.

The IgM immunoglobin can be broken into several small identical fragments, F_{ab}, and one large fragment, F_c, by treatment with trypsin at 60° C. The cleavage points are indicated by the dashed lines in Fig 2-17, F_{ab} fragments still contain enough structure to bind their homologous antigen; thus only the variable portion of the molecule is required for antigen recognition. The subscript ab indicates that this F fragment retains the specificity of the intact antibody.

IgG molecules also yield F_{ab} and F_c pieces when treated with trypsin or papain, a proteolytic enzyme from papaya. Because there are many common amino acid sequences contained in the F_c fragments of IgG, F_c molecules can be crystallized, a property that is indicated by the subscript c.

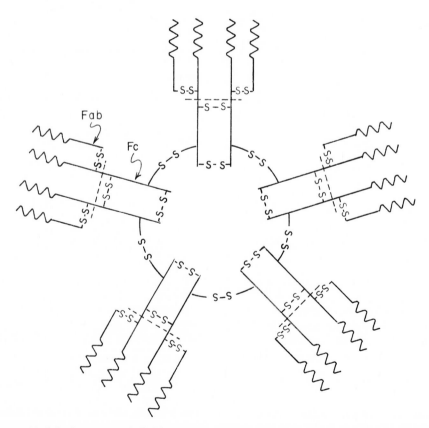

Fig. 2-17. Model of pentameric IgM molecule. Zigzag lines represent variable N-terminal regions of H and L chains. Dashed lines indicate sites cleaved by papain to form F_{ab} and F_c fragments.

Pepsin treatment of IgG produces a fragment similar to F_{ab} that contains the variable region of the molecule, and it breaks the F_c portion into several peptides. This treatment also prevents complexing of IgG fragments with rheumatoid factor. These experiments suggest that the antigenic portion of IgG, considering rheumatoid factor as the antibody, is located in the nonvariable F_c region of the molecule. The total sequence of one IgG is known, and since the F_c region is probably similar in all immunoglobulins, pharmacologic intervention using chemically synthesized portions of this region may become useful in the future for certain therapeutic situations. Remember, however, that the biochemical basis of rheumatoid arthritis is not understood, and this hypothesis should be considered in that light.

REFERENCES

Bennett, J. C.: The role and character of immunoglobins in rheumatoid inflammation, Fed. Proc. 32:138, 1973.

Eighteenth rheumatism review, Arthritis Rheum. 11:suppl., 1968.

Mason, M., and Currey, H. L. F., editors: An introduction to clinical rheumatology, London, 1970, Pitman Publishing Corp.

CASE 3: ORAL MANIFESTATION OF HEMOGLOBINOPATHY

A 23-year-old black woman came to the dental clinic requesting "gum treatment." A brief examination showed nothing remarkable, but the routine panoramic radiographs revealed atypical changes in the mandible. These defects were more obvious in the intraoral radiographs where the bone, in direct apposition, produced fine detail in sharp focus. The mandible had areas of decreased bone density, or osteoporosis. There was marked radiolucency in the molar-premolar area, transversed by prominent trabeculae (supporting structures) and sclerotic areas in the anterior mandible. Examination showed normal soft tissues, moderate periodontal disease, and few caries. Although the patient was apparently healthy, she had once been treated for "possible anemia." In view of these findings the patient was referred to the hospital for hematologic studies. The laboratory report showed hemoglobin, 11.8 g/dl; hematocrit, 31.5%; and hemoglobin electrophoresis, 100% Hb C. Electrophoresis of hemoglobin from the patient's three daughters, who were without symptoms of anemia, revealed that two had mixtures of Hb A and Hb C.

Biochemical questions

1. Is this woman homozygous or heterozygous for Hb C? Are her daughters? How can you tell?
2. Is this disease autosomal or X linked? Recessive or dominant?
3. What is the probability that the patient would have children with the disease? What is the probability of her children being carriers?
4. Did the patient inherit the disease from her mother or her father? Explain.
5. The defect in Hb C is in position 6 of the β-chain, where a lysine residue replaces a glutamic acid residue. Compared to Hb A, is Hb C more positively or negatively charged?
6. The erythrocytes of patients homozygous for Hb C show little evidence of sickling. How can this be explained from a knowledge of the amino acid substitutions?
7. Suppose the woman's husband had the sickle cell trait. What would be the probability of their children having both Hb S and Hb C? Can Hb S and Hb C be separated by electrophoresis? Explain.

Case discussion

The woman in this case was homozygous for hemoglobin C disease since all of her hemoglobin migrated on electrophoresis as hemoglobin C. If she had been heterozygous, like her daughters, approximately half of her hemoglobin would have been Hb A and half Hb C. The disease is recessive since the heterozygous daughters were without symptoms, and although there are not enough data given to prove the point, hemoglobin

C disease is inherited as an autosomal trait. If the woman had a son, one would predict that he would have also been a heterozygote with both Hb A and Hb C. If the disease were X-linked, all sons would have the disease and their hemoglobin would be 100% Hb C. Since the disease is autosomal and the patient is homozygous, she inherited the disease from both of her parents. The patients' children will not have the disease unless their father also has the disease or is a carrier of the disease; this is not the case in the present situation. The patient's two daughters, however, are carriers as would be any of her children, male or female.

The defect in hemoglobin C is in position 6 of the β-chain, the same amino acid modified in sickle cell disease. Ordinarily, this amino acid residue is glutamate, an acidic and negatively charged amino acid. The lysine replacement is positively charged. Recall that the defect in sickle cell disease is the replacement of glutamate with valine, which has a hydrophobic and uncharged side chain. The sickling is caused by the hydrophobic character of this amino acid. Since the amino acid replacement in the Hb C disease is one charged amino acid for another, this region of the β-chain will remain hydrophilic, and thus the cells will not sickle. Other deleterious effects do occur, however, and the red cell becomes more sensitive to lysis, resulting in anemia.

If one were to assume that this woman's husband had sickle cell trait, that is, was heterozygous for Hb S, one would expect that the CC genes of the mother and the AS genes of the father would on the average produce 50% SC children and 50% CA children. Thus all of the patient's children would be carriers of Hb C. In the mating proposed, 50% of the children would also be carriers of Hb S. None of the children would be homozygous for either hemoglobin C or sickle cell disease. The carrier states of these heterozygotes could be easily detected since hemoglobins A and C are widely separated at pH 8.6. Hb S and Hb C are closer together, but Hb S will migrate faster to the anode at this pH than the positively charged Hb C, which will be found close to the cathode, the negative pole.

REFERENCES

Halstead, C. L.: Oral manifestations of hemoglobinopathies, Oral Surg. 30:615, 1970.
Lehmann, H., and Huntsman, R. G.: The hemoglobinopathies. In Stanbury, J. B., Wyngaarden, J. B., and Fredrickson, D. S., editors: The metabolic basis of inherited diseases, ed. 3, New York, 1972, Academic Press, Inc.

CASE 4: SCAR FORMATION

A serious problem concerning the operative repair of lacerated tendons in the hand is the formation of scar tissue. This often will restrict the motion of the repaired or newly implanted tendon, seriously compromising motion of the finger or hand. A patient with a finger laceration involving the flexor tendon was referred to an orthopedic surgeon for operative repair. In the hope of reducing the chance of scar formation the patient agreed to an experimental protocol that involved administration of a lathyrogen. The informed consent of the patient was obtained for this procedure.

On the evening before the operation, β-aminopropionitrile fumarate was administered to the patient under conditions specified by the Food and Drug Administration. The next day flexor tendon grafts were inserted from midpalm to distal phalanx in order to correct laceration of both flexor tendons in the digital sheath. Junctions were planned to maintain the integrity of the repaired tendons without a major contribution from the healing process. There were no postoperative complications resulting from the surgical procedure. After the operation β-aminopropionitrile fumarate was administered orally every 6 hr, but the drug was discontinued after 20 days when a hypersensitivity reaction and a skin rash appeared. These symptoms promptly cleared after the drug was discontinued. To determine the effectiveness of the therapy, a 4 cm incision through skin and subcutaneous fat was made in the groin at the time treatment with the lathyrogen was started. At the conclusion of therapy the scar was excised, minced, and extracted with 0.45 M sodium chloride. Hydroxyproline determination on the extract showed 0.347 μg/mg of scar tissue. Two weeks after therapy was discontinued, a similar incision was made in the opposite groin. The patient then served as his own control. Samples of this scar tissue taken 12 days later were found to contain 0.103 μg of hydroxyproline/mg of tissue. The excretion of hydroxyproline in the urine was 35.8 mg/24 hr before treatment and 64.5, 61.6, and 74.5 mg/24 hr after 1, 2, and 3 weeks of therapy, respectively. Treatment appeared to speed the recovery of motion of the tendon and prevent serious stiffness of the joint when the cast was removed 3 weeks following the operation.

Biochemical questions

1. Explain how β-aminopropionitrile affects the structure of collagen.
2. Explain the significance of the threefold higher hydroxyproline content of the saline extract of the scar tissue after treatment with the drug.
3. Why are hydroxyproline levels higher in the urine during therapy when it is likely that the same amount of collagen is made in the absence as in the presence of the drug?
4. Why did the joints appear less stiff as a result of the treatment?

REFERENCES

Barrow, M. V., and Simpson, C. F.: Caution against the use of lathyrogens, Surgery 71:309, 1972.
Peacock, E. E., and Madden, J. W.: Some studies on the effects of β-aminopropionitrile in patients with injured flexor tendons, Surgery 66:215, 1969.

CASE 5: ASPIRIN-INDUCED ALTERATION OF HUMAN SERUM ALBUMIN

C. J. is a 56-year-old woman with rheumatoid arthritis. She had been taking 4 g of sodium salicylate daily for 4 months when a serum sample was obtained. This drug was stopped and she was given 4 g of aspirin daily. Another serum sample was obtained 5 weeks later.

The serum albumin was separated from the other serum proteins by sodium sulfate fractionation and was further purified by electrophoresis. The purified albumins, including a normal control sample, were reduced and carboxymethylated. The samples were dialyzed against water and lyophilized. Hydrolysis with trypsin was done in a 2 M urea solution containing 0.1 M ammonium bicarbonate, pH 9.0. The hydrolyzate was diluted and applied to a Dowex-50 column. The column was washed with water, the peptides eluted with 4 N NH$_4$OH, and dried. About 5 mg of the peptides was applied to paper and chromatographed in one dimension followed by electrophoresis at pH 3.55 in the other direction. The peptide maps of the patient's albumin showed a unique peptide, called "A," which was not present in tryptic digests from control albumin. As "A" was increased, two other peptides, "B" and "C," were reduced as compared to the control sample.

Biochemical questions

1. Why were the albumins reduced and carboxymethylated? Why was the trypsin treatment done in 2 M urea?
2. Considering the specificity of trypsin, what can be said about the primary structure of the peptides produced?
3. Explain how the peptides are bound to the Dowex-50 column and eluted with 4 N NH$_4$OH.
4. Peptide "C" was found to contain equimolar amounts of lysine and leucine. Is the structure of the dipeptide Lys-Leu or Leu-Lys? Explain.
5. About 2% of asthmatic patients are hypersensitive to aspirin but not to sodium salicylate. Try to explain this from the above results.

REFERENCES

Hawkins, D., Pinckard, R. N., Crawford, I. P., and Farr, R. S.: Structural changes in human serum albumin induced by ingestion of acetylsalicylic acid, J. Clin. Invest. 48:536, 1969.
Pinckard, R. N., Hawkins, D., and Farr, R. S.: In vitro acetylation of plasma proteins, enzymes and DNA by aspirin, Nature 219:68, 1968.

CASE 6: MONOCLONAL IMMUNOGLOBULINEMIA

A 40-year-old man, N. M., was suffering from recurrent epigastric pain. The physical examination was unrevealing. A mild hypochromic anemia (hemoglobin 10.6 g/dl) was present, and a diagnosis of thalassemia minor was proposed after the discovery of abnormally high proportions of Hb F and Hb A$_2$. Total serum protein was 7.6 g/dl. Electrophoresis of the serum proteins revealed the presence of an IgA monoclonal component. By a radial immunodiffusion procedure the concentrations of the major immunoglobulin classes were: IgG 2125 mg/dl, (normal range 600 to 1200 mg/dl), IgA 560 mg/dl (normal range 150 to 400 mg/dl), IgM 45 mg/dl (normal range 45 to 90 mg/dl). An M-component is any monoclonal immunoglobulin component; in this case it is IgA associated. There were no suggestive signs of amyloidosis, which is known to occur with increasing frequency in association with increased M-components, such as in multiple myeloma. The electrophoretic analysis of the serum of patient N. M. indicated that his M-component (concentration, 340 mg/dl) behaved as a β-globulin. The patient had five clinically healthy children. The family pedigree may be summarized:

	Age	Blood group	Ig abnormality	Thalassemia trait
Father	44	O	+	+
Mother	41	B	−	−
Son	16	B	+	−
Daughter	12	O	+	−
Daughter	9	B	+	−
Son	6	O	−	+
Son	2	B	−	+

Biochemical questions

1. What are the proportions of the three major immunoglobulins in the serum of N. M.? What is the Ig abnormality in N. M.? Explain.
2. What is the basis for the identification and determination of proteins by immunodiffusion? Describe a procedure whereby this might be combined with electrophoresis. Explain the theory of your procedure.
3. Are there any indications of the inheritance patterns for the blood group, for the Ig abnormality, and for the thalassemia?
4. What is the normal amount of Hb F in the hemoglobin of adults? How does this type of hemoglobin differ from Hb A in its primary structure and oxygen-binding character?
5. What is meant by the term "monoclonal"?

REFERENCES

Abramson, N., and Shattel, S. J.: M-Components, J.A.M.A. 223:156, 1973.
Cathcart, E. S., et al.: Immunoglobulins and amyloidosis, Am. J. Med. 52:93, 1972.
Michaux, J. L., and Heremans, J. F.: Thirty cases of monoclonal immunoglobulin disorders other than myeloma or macroglobulinemia, Am. J. Med. 46:562, 1969.
Ogawa, M., et al.: Clinical aspects of IgE myeloma, N. Engl. J. Med. 281:1217, 1969.

CASE 7: RED BLOOD CELLS AND SURGERY INVOLVING A HEART-LUNG BYPASS

Patient E. T. underwent open-heart surgery during which an oxygenator was used. The oxygenator extracorporeal circuit was primed with 1- to 3-day-old acid citrate dextrose (ACD) blood, diluted with bicarbonate-buffered saline to an 8% hemoglobin content. The volume of the priming blood was 40% to 50% of the patient's blood volume, and during the bypass the mixed blood averaged 10% to 11% hemoglobin. The determinations summarized in the following table were performed before, during, and after the bypass. The 2,3-diphospho-D-glycerate (2,3-DPG) level is expressed as micromoles of phosphorus per 100 ml of

	Priming blood	Patient control	Duration of bypass (min)				After bypass (hr)	
			10	20	40	60	1	24
2,3-DPG*	380	480		272	250	288	240	445
$\bar{p}O_{2_{50}}$*	14	26	18			20	23	27
pH_{RBC}*		7.2	7			6.9	7.2	7.2

*See case description.

packed red blood cells (RBC). The $\bar{p}O_{2_{50}}$ is the oxygen tension in mm Hg at which the hemoglobin was 50% saturated with O_2; during the determination of the $\bar{p}O_2$ it was shown that the 2,3-DPG and pH_{RBC} did not change. The pH of the red blood cells, pH_{RBC}, was measured in packed, hemolyzed cells. The pH of whole blood did not change from 7.4 during the whole procedure.

On the average, the priming blood contained 3% carbon monoxide, compared to 1% in the patient's blood.

Biochemical questions

1. Can the change in the 2,3-DPG content of the RBC be explained by admixture of the priming blood with the patient's blood?

2. What is the effect of 2,3-diphospho-D-glycerate and pH on the O_2-saturation curve of hemoglobin? Explain.

3. As a result of the combined decrease of 2,3-DPG and pH_{RBC} in the patient,

what will be the effect on the O_2 transport to tissues? Explain.

4. Would it be reasonable to assume that the donor of the priming blood was a smoker and that E. T. smoked less? What is the effect of carbon monoxide on O_2 transport?

5. It was noted that the exposure of RBC in vitro to elevated levels of O_2 accelerated the decline of 2,3-DPG in the cells compared to pumping the blood without the oxygenator in the circuit. How might this be explained?

REFERENCES

Bordink, J. M., et al.: Alterations in 2,3-diphospho-glycerate and O_2 hemoglobin affinity in patients undergoing open-heart surgery, Circulation, Suppl. I to vol. XLIII and XLIV, 1971, pp. 1-141.

Brewer, J. G., and Eaton, J. W.: Erythrocyte metabolism: interaction with oxygen transport, Science 171:1205, 1971.

Finch, C. A., and Lenfant, C.: Oxygen transport in man, N. Engl. J. Med. 286:407, 1972.

Proctor, H. J., et al.: Alterations in erythrocyte 2,3-DPG in postoperative patients, Ann. Surg. 173:357, 1971.

ADDITIONAL QUESTIONS AND PROBLEMS

1. Predict the direction of electrophoresis compared to Hb A of the following abnormal hemoglobins.

Hemoglobin	Amino acid substitution
α-Chains	
M Boston	58 His → Tyr
Bibba	136 Leu → Pro
J Capetown	92 Arg → Gln
β-Chains	
Zurich	63 His → Arg
M Saskatoon	63 His → Tyr
Sydney	67 Val → Ala
M Milwaukee	67 Val → Glu
Kempsey	99 Asp → Asn
Yakima	99 Asp → His

2. Calculate the isoelectric point of oxytocin (formula on p. 630).

3. A child with congenital agammaglobulinemia, the inability to synthesize appreciable amounts of γ-globulin, was constantly ill with bacterial diseases. Explain the underlying mechanism of this child's illness.

4. A patient with the nephrotic syndrome was losing large quantities of albumin in his urine. His plasma albumin concentration fell to 1.0 g/dl. He subsequently developed edema (swelling) caused by the collection of fluid in extracellular spaces. Explain.

CHAPTER 3

ENZYMES AND BIOLOGIC CATALYSIS

OBJECTIVES

1. To describe the nature of enzymes and the process of enzyme catalysis as the basis of biochemical transformations of cellular substances
2. To explain the relationships between the properties of enzymes and their physiologic or metabolic function
3. To demonstrate by appropriate examples that enzymes frequently operate in sequences or interrelated systems that form the basis of so-called metabolic pathways, which in turn are subject to various kinds of controls
4. To explain how quantitative assay of selected enzyme activities in blood and other body fluids can assist in the diagnosis and treatment of disease

Enzymes catalyze virtually all biologically important reactions. It is therefore essential for health scientists to understand the chemistry and function of enzymes if they are to be employed in diagnostic procedures. Important areas of medicine have substantially benefited from the application of enzyme analysis; such diseases as myocardial infarcts, hepatitis, cancer of the prostate, obstructive liver disease, and the muscular dystrophies may be cited as common clinical examples. Enzyme activity may be high in some diseases and low or lacking in others. Enzyme assays are also of growing importance in the area of genetic counseling, since they may be employed in the detection of heterozygous carriers of hereditary diseases.

It is clear that tissue enzymes are distributed in a highly organized fashion; that is, cells are not "loose sacks" of enzymes. Yet the products of an enzyme reaction in one tissue component may have significant effects on a separate enzyme process in another component of the given tissue or even in an entirely different tissue. Some detailed knowledge of enzyme distribution within cells and of the chemistry and function of enzymes is therefore essential for a detailed understanding of disease mechanisms and therapies.

What are enzymes?

Enzymes are *biocatalysts* produced by living tissue that increase the *rate* of reactions that may occur in the tissue. In the absence of enzymes the same or comparable reactions occur at too slow a rate to support life or require unphysiologic conditions. For example, carbon dioxide reacts with water to form carbonic acid, part of which immediately ionizes at a physiologic pH to form bicarbonate ions. The dissociation, of course, does not depend on enzyme catalysis but is a property of the acid structure. The reaction is shown on p. 99.

$$H_2O + CO_2 \rightleftharpoons H_2CO_3 \rightleftharpoons H^+ + HCO_3^-$$

The uncatalyzed decomposition of carbonic acid into water and carbon dioxide does not occur instantaneously. Because the reaction rate is in fact quite slow, true equilibrium may not be reached for 1 hr or more. If one takes a sample of carbonated water and adds the enzyme carbonic anhydrase to it, the true equilibrium is reached in minutes or less. Red blood cells in our bodies are especially rich in carbonic anhydrase, and its presence in the erythrocytes promotes the rapid interconversion of carbon dioxide and bicarbonate through the intermediate form of undissociated carbonic acid.

It should be noted that while this example tells us that some enzyme reactions are readily reversible, it does not tell us that all enzyme reactions have this property. Furthermore, while the definition speaks of increased rates, it tells us nothing of the precise equilibrium that exists between the reactants and products.

Enzyme structure

All known enzymes are proteins. The molecular weights of enzymes cover a wide range. For example, the enzyme ribonuclease, which hydrolyzes ribose-containing nucleic acids, is relatively small, having a molecular weight of approximately 13,700. It contains 124 amino acid residues in a single polypeptide chain, the exact sequence of which is known. In contrast, aldolase, an enzyme involved in glucose metabolism, has a molecular weight of approximately 156,500. It is composed of four subunits, each with a molecular weight of about 40,000. The individual polypeptides of liver or brain aldolase have a different primary structure than the polypeptides of the enzyme in muscle. The macromolecular complex known as pyruvate dehydrogenase is an even larger enzyme system. This enzyme complex catalyzes the conversion of pyruvate to acetyl CoA, a most important reaction. Pyruvate dehydrogenase is a multienzyme complex in which the components are so tightly organized that the entire system can be isolated as a discrete, particulate entity from many tissues. The complex from pig heart has a molecular weight of about 10×10^6; more detailed studies have shown that each complex contains no fewer than forty-two individual molecules, including several important and essential cofactors. The entire structure of the pyruvate dehydrogenase complex is required for catalysis.

Enzyme cofactors

In addition to the protein component, many enzymes require nonprotein constituents for their proper function. In some cases these components, broadly described as cofactors, may be metal ions and, in other cases, organic molecules of relatively low molecular weight. When the small organic molecules are tightly bound by either covalent or coordinate covalent bonds, they are called *prosthetic groups*. The heme group of the cytochromes is an example of a prosthetic group. Coenzymes are organic molecules, often derived from the B vitamins, that participate directly in enzymatic reactions. Some coenzymes are attached to their companion enzymes as tightly bound prosthetic groups, whereas others can be easily removed by dialysis. The complete functional complex of enzyme and cofactors is called a *holoenzyme;* the protein part, free of the cofactors, is called an *apoenzyme*. Remixing of the separated components is sometimes all that is needed to restore complete activity.

Carbonic anhydrase, mentioned above, is one of a class known as *metalloenzymes*, since at least one gram atom of zinc/mole of protein is an absolute requirement for its activity. Removal of the zinc by other reagents that bind it more tightly will completely inactivate carbonic anhydrase. In this instance, replacement of zinc by some other metal will also destroy or decrease the activity, but this is not true for all metalloenzymes. In some cases metal ions may be replaced by others without damage or significant loss of activity. For example, some enzymes that require magnesium ions will function equally or almost as well when the magnesium ions are replaced by manganese ions. The precise metal ion requirements need to be explored in each case. It is known that man requires numerous trace metals in his diet, and it appears that the requirement is a direct expression of the need for certain specific metals in the structures of various enzymes.

Vitamins and coenzymes

Pyridoxal phosphate, a slightly modified form of vitamin B_6 (pyridoxine), is an essential cofactor, or coenzyme, for an important group of enzymes known collectively as *transaminases*. As the name implies, these enzymes transfer α-amino groups from an amino acid to a keto acid. As we shall see, measurements of transaminase activity in serum are frequently useful in clinical medicine as a measure of tissue injury or death, since the enzymes frequently survive the breakdown of cells and then leak into the bloodstream. Details of the transaminase mechanism will be considered in Chapter 6.

Many dehydrogenases require the coenzyme nicotinamide adenine dinucleotide (NAD^+) or its phosphate derivative ($NADP^+$); others depend on flavin adenine dinucleotide (FAD). These coenzymes are derived from vitamins, as indicated in Table 3-1.

The enzymes can be divided into six major classes, as described below. Since many of the enzymes in a given class operate with the same coenzyme, it appears that the coenzyme provides a common mechanistic pathway, while the function of the apoenzyme is to select the proper reactants and to facilitate the chemical reaction.

Enzyme and cofactor turnover

Like all biologic materials, enzymes have a finite half-life; that is, they are subject to turnover and replacement. Therefore, a diet must include sufficient essential amino acids, metals, and vitamins to provide for enzyme replacement, among other needs. Human beings have only a limited ability to store most essential metal ions, and the biologic lability of most vitamins requires that the nutritional needs for the components of enzymes be met on a continuing, everyday basis.

Enzyme classification

To assist in the study of enzymes an international classification has been established that defines six major classes of enzyme function, each with several subclasses. Within each subclass formal names have been assigned to the known enzymes to describe the reactions they catalyze, in much the same way as the names of the Geneva IUPAC Conventions describe the structure of organic compounds. Since the trivial names for many enzymes, for example, pepsin, trypsin, and urease, have been deeply embedded in the literature, and since many other named enzymes are still easier to recognize by their trivial names, the newer nomenclature has been less widely adopted than the classification scheme itself. A summary of the international classification of enzymes follows; a more complete form is given in Appendix B.

1. *Oxidoreductases* catalyze a wide variety of oxidation-reduction reactions

Table 3-1. Common coenzyme structures

VITAMIN   COENZYME

Thiamine
(B$_1$)

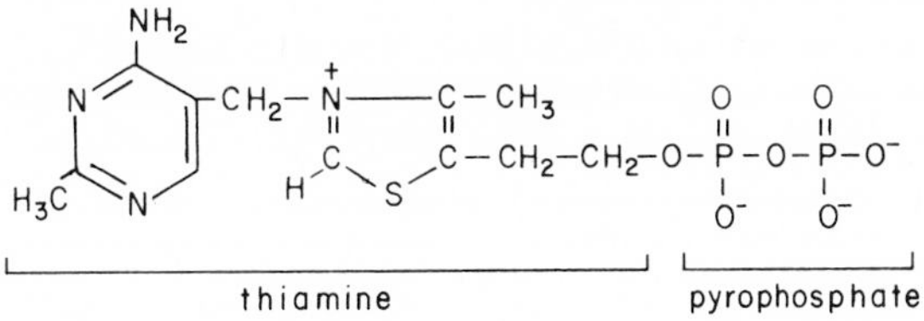

thiamine   pyrophosphate

Cocarboxylase, functions in decarboxylations, transketolase reaction

Riboflavin
(B$_2$)

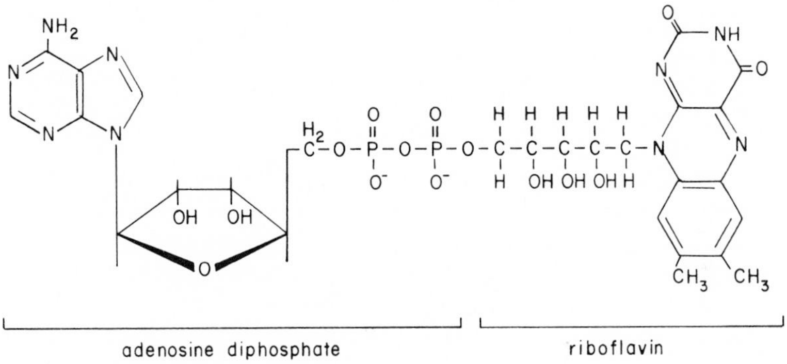

adenosine diphosphate   riboflavin

Flavin adenine dinucleotide (FAD), functions in dehydrogenations

Pyridoxine
(B$_6$)

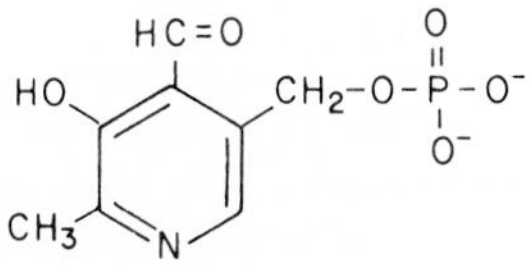

Pyridoxal phosphate, functions in transamination, deamination, decarboxylation and racemization reactions

Niacin[*]

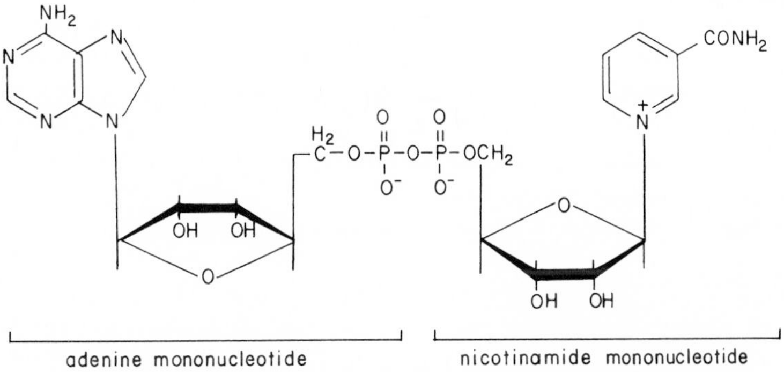

adenine mononucleotide   nicotinamide mononucleotide

Nicotinamide adenine dinucleotide (NAD$^+$), functions in dehydrogenations

Lipoic acid[†]

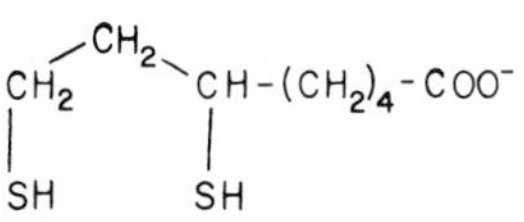

Dihydrolipoate   Dihydrolipoate-protein complex

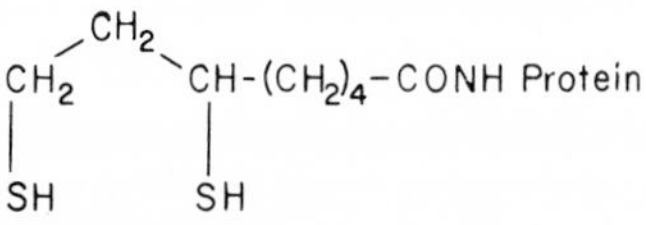

[*]Residue also in nicotinamide adenine dinucleotide phosphate (NADP$^+$) where the 2′-hydroxyl of the adenosyl moiety is also phosphorylated.
[†]Not a vitamin; included for reference.

and frequently employ coenzymes such as NAD⁺, NADP⁺, FAD, or lipoate as the hydrogen acceptor. Other acceptors include coenzyme Q or molecular oxygen. Common trivial names include dehydrogenase, oxidase, peroxidase, and reductase.

2. *Transferases* catalyze various kinds of group transfers. Many important steps in metabolism require transfer from one molecule to another of amino, carboxyl, carbonyl, methyl, acyl, glycosyl, or phosphoryl groups. Common trivial names include aminotransferase (transaminase), acyl carnitine transferase, and transcarboxylase.

3. *Hydrolases* catalyze cleavage of bonds between a carbon and some other atom by addition of water. Some common trivial names include esterase, peptidase, amylase, phosphatase, urease, pepsin, trypsin, and chymotrypsin.

4. *Lyases* catalyze breakage of carbon-carbon, carbon-sulfur, and certain carbon-nitrogen (excluding peptide) bonds. Common trivial names include decarboxylase, aldolase, citrate lyase, and dehydratase.

5. *Isomerases* catalyze racemization of optical or geometric isomers and certain intramolecular oxidation-reduction reactions. Trivial names include epimerase, racemase, and mutase.

6. *Ligases* catalyze the formation of bonds between carbon and oxygen, sulfur, nitrogen, and other atoms. The energy required for bond formation is frequently derived from the hydrolysis of ATP. Some trivial names include synthetase and carboxylase.

This scheme indicates how the enzymes discussed later in this book are classified in terms of reaction type and substrates employed.

INTRACELLULAR LOCATION OF ENZYMES

The exquisite organization of living cells first became apparent with the development of the light microscope, and our knowledge has proceeded to much greater sophistication with the development of the electron microscope. Fig. 3-1 is a reproduction of an electron micrograph of a typical mammalian liver cell, and Fig. 3-2 is a diagrammatic representation of a typical mammalian cell. The organized subcellular elements, or organelles, appear to have highly individualized complements of enzymes that are related to the functions that the organelles must perform.

Cytochemical methods can be employed to demonstrate the presence or absence of a specific enzyme in the following way. Suppose that a reagent reacts with the product of a given enzyme-catalyzed reaction to form either a colored material (for light microscopy) or an electron-dense material (for electron microscopy). If one then treats a suspension of intact cells or a tissue slice in the appropriate way, visualization of the product will indicate the presence or absence of the given enzyme. Using these techniques, it has been shown that individual enzymes have unique locations within a wide range of cell types. Localization of an enzyme within an organelle permits the assumption that it is related to the function of the organelle. More recently it has become possible to disrupt cells by careful use of mechanical or osmotic shock. The broken cells can then be separated into various subcellular fractions by differential centrifugation in media of graded density. Preparation of relatively pure samples of nuclei, mitochondria, microsomes, ribosomes, lysosomes, or other particulate structures from a given tissue is now a standard biochemical procedure. Careful examination of the enzymes in these isolated organelles supports and extends the argument that enzymes are arranged in a highly structured assemblage designed to further cell function.

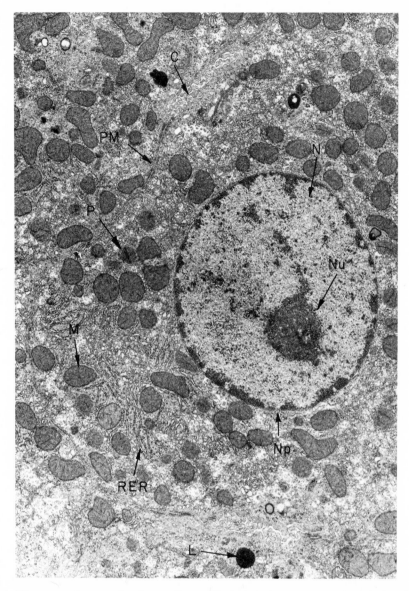

Fig. 3-1. Electron micrograph of a rat liver cell. (×9500.) Structures identified include **N,** nucleus; **Nu,** nucleolus; **Np,** nuclear pore; **M,** mitochondrion; **RER,** rough endoplasmic reticulum; **L,** lysosome; **PM,** plasma or cell membrane; **C,** bile canaliculus; and **P,** peroxisome, typical of liver cells.

Briefly, the subcellular location of enzyme systems may be summarized as follows. Many of the enzymes associated with the nucleus are involved with the maintenance, renewal, and utilization of the genetic apparatus. Most of the enzymes of mitochondria deal with reactions best described as energy-yielding or oxidative reactions that provide the driving force for many kinds of cellular work. Enzymes associated with the ribosomes promote the biosynthesis of proteins. Microsomal enzymes are responsible for a variety of hydroxylation reactions involved in steroid hormone biosynthesis and drug metabolism or inactivation. Lysosomes contain enzymes that catalyze the hydrolytic destruction of materials

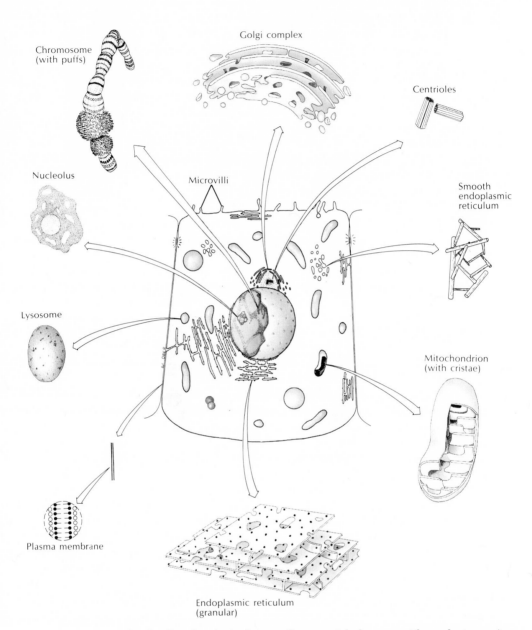

Fig. 3-2. Generalized cell with principal organelles, as might be seen with an electron microscope. Each of the major organelles is shown enlarged. Membranes of organelles are believed to be continuous with, or derived from, plasma membrane by an infolding process. Structures of other membranes (of nucleus, endoplasmic reticulum, mitochondria, etc.) are probably similar to that of plasma membrane, shown enlarged at lower left. Solid circles carrying short lines represent a lipid bilayer, and unfilled circles represent surface layers of proteins or peptides that constitute inner and outer surfaces of membrane. (From Hickman, C., and Hickman, C., Jr.: Biology of animals, St. Louis, 1972, The C. V. Mosby Co.)

no longer needed by the cell, an intracellular digestive process. The Golgi complex is a system of tubules and vesicles that apparently functions in secretory cells to collect and extrude protein or other cellular products from the cell. In cells that specialize in absorption, for example, the epithelial cells of the intestinal tract, the Golgi complex appears to work in the opposite manner; it serves to collect material taken up by the cells.

Certain other facts have emerged. It is clear that cytoplasmic enzymes catalyze the kind of carbohydrate metabolism known as glycolysis, which is discussed in Chapter 7. In fractionated extracts these enzymes appear to be simply dissolved, although in the intact cell they may be more highly organized. The enzymes responsible for fatty acid biosynthesis are also cytoplasmic, but unlike those involved in glycolysis, this system of enzymes is highly organized as a particulate complex containing all seven of the required enzymes.

Isolated organelles often exhibit a further detailed architecture of their own. For example, mitochondrial particles have a double membrane boundary. The outer membrane, the inner membrane, and the space between them constitute regions, each possessing specific groups of enzymes that are localized to perform their functions most efficiently. This is further described in Chapter 5. The same can be said of the limiting membrane of the cell, or plasma membrane, which is not merely an inert barrier; it contains its own characteristic array of enzymes, most of which facilitate or control the entrance of materials into the cell or the egress of substances from it. Some hormones, which circulate in the blood, exert their earliest metabolic effects on enzymes or receptor proteins that are part of the surface membranes of their target cells. This is explored in Chapter 13.

GENERAL ENZYME PROPERTIES

It has already been mentioned that enzymes, being proteins, have high molecular weights. This makes it very difficult to completely characterize them by classic physical or chemical means. The methods already described as being generally useful for proteins have also been specifically applied to the study of enzymes. These include ultracentrifugation techniques, optical light scattering and rotatory dispersion, x-ray diffraction analysis, and electrophoresis. By combining these techniques with suitable chemical determinations, it is possible to make quite precise statements about the structure of enzymes and about the forces that hold the protein and nonprotein portions in functional form. More general facts can also be learned about enzyme properties and behavior by very simple techniques. Enzymes isolated from their natural sources can be used in vitro to study in detail the reactions they catalyze. Reaction rates may be altered by varying such parameters as pH or temperature, by changing qualitatively or quantitatively the ionic composition of the medium, or by changing ligands other than the substrate or coenzymes. In most cases the effects produced by this kind of manipulation are consistent with the formal laws of nonenzymatic chemistry. In the remaining cases satisfactory theoretic explanations have been proposed that are in accord with accepted principles of thermodynamics and reaction mechanism theory.

Rates of enzyme catalysis may be significantly affected even by such modest changes of chemical environment as are within physiologic limits, and these changes undoubtedly participate in the *control* or *regulation* of interrelated enzyme systems necessary for living cells. More extreme changes (pathologic changes) may be the causes of certain diseases. In developing the details of this discussion there will be frequent opportunity to point out elements of enzymology that contribute to control.

Effect of temperature on enzymes

Since protein structure determines enzyme activity, anything that disturbs this structure may lead to a change in activity. The process of protein denaturation described in Chapter 2 applies also to enzymatic proteins, and the denaturants that bring it about are the same. For example, enzymes frequently show a marked thermal fragility. When heated to temperatures above approximately 50°C, most but not all enzymes are denatured. High temperature denaturation is usually irreversible because essential weak bonding forces are broken by increased thermal vibration of the component atoms, a phenomenon that damages the three-dimensional structure. Even under conditions where denaturation does not occur, most enzymes show an optimum temperature at which, other things being equal, activity is maximal. Fig. 3-3 shows that changes in activity above or below this temperature are not always symmetric. Some enzymes show a very steep decline in activity over a range of only a few degrees past the point where noticeable denaturation begins. This has been interpreted as a "melting" of the protein, with very rapid loss of essential weak bonding forces, analogous to the sharp melting point of simpler organic compounds. Some enzymes are also very sensitive to low temperatures. The mitochondrial ATPase, for example, is rapidly inactivated by cooling to 5°C, but is relatively stable at room temperature. One explanation is that at lower temperatures weak forces between different parts of a single subunit become greater than the forces between subunits. This causes disruption of the polymeric form, which is essential for activity. Finally, it must be noted that some enzymes are remarkably unaffected by heat. Some proteases and phospholipases can withstand boiling water temperature with little or no loss in activity. These facts indicate the difficulty in generalizing from one enzyme to another.

It is important to note that if an enzyme of a certain type is isolated from different tissues of the body, for example, lactate dehydrogenase from the heart, liver, lungs, and kidney, these dehydrogenases do not necessarily have an identical capacity to resist mild heat denaturation even though they all catalyze the same reaction. As will be shown later, this fact is useful in differential diagnosis of some diseases of specific organs. Other applications of heat denaturation are the sterilization of foods and surgical instruments or the pasteurization of

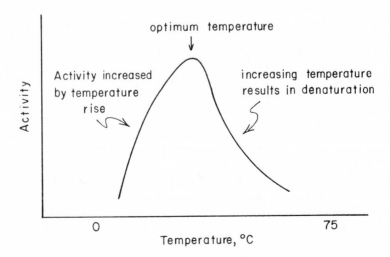

Fig. 3-3. Effect of temperature on enzyme activity.

milk, all of which depend on the rapid destruction by heat of enzymes essential to contaminant microorganisms.

pH and ionic dependence of enzymes

Enzyme activity is also related to the ionic state of the molecule and especially of the protein part, since the polypeptide chains contain groups that can ionize to a degree that depends on the prevailing pH. As is true of proteins generally, enzymes have an *isoelectric point* at which their net free charge is zero. The pH of the isoelectric point as a rule is not the same as the pH at which maximal activity is demonstrated. The pH optima shown by enzymes varies widely; pepsin, which exists in the acid environment of the stomach, has a pH optimum at about 1.5, while arginase, an enzyme that cleaves the amino acid arginine, has its optimum at 9.7. However, the large majority of enzymes have optima that fall between pH 4 to 8 (Fig. 3-4). Some enzymes show a wide tolerance for pH changes, but others work well only in a narrow range. If any enzyme is exposed to extreme values of pH, it is denatured. The sensitivity of enzymes to altered pH is one reason why regulation of body pH is so jealously controlled and why departures from normal may involve grave consequences.

Many enzymes exist in the body at a pH somewhat removed from their optimum value. In part this may be caused by a difference in the precise in vivo, as opposed to the in vitro, environment. It also emphasizes that control of pH may be a significant means of regulating enzyme activity. For example, the enzymes of the lysosomes, as a group, have fairly acid pH optima, not frequently reached outside the lysosome in a state of cellular health. When cells are damaged they may become acidotic, with consequent rupture of the lysosomal membranes and release of the contained enzymes. Under the acidotic conditions the enzymes are

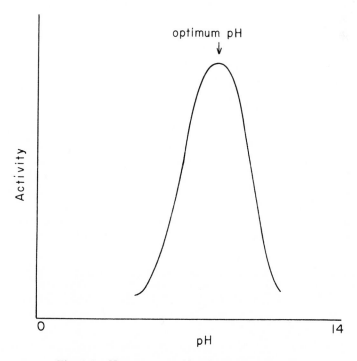

Fig. 3-4. pH activity profile of a typical enzyme.

in a better environment to function as scavengers of materials from severely diseased or dead cells.

Fructose diphosphatase and phosphofructokinase are enzymes with reciprocal effects on the conversion of fructose 1,6-diphosphate to fructose 6-phosphate (see p. 126). Other things being equal, phosphofructokinase is more severely inhibited when the pH drops below 7.5, so that mild acidosis favors formation of glucose and diminishes formation of pyruvate and lactate.

The pH is not the only factor that affects the ionic state of enzymes. For example, the concentration of water available to an enzyme bound in a subcellular membrane may be very different from the amount of water available to an enzyme in the cytoplasm. The precise requirements for the hydration of enzymes in their physiologic environment may be quite different from those observed in the isolated and purified state.

Active catalytic site

Enzymes differ from other proteins in that they possess what has been termed an active catalytic site; the nature of this site has long been a subject of study. The active site can be regarded as being composed of a relatively small number of amino acid residues, not necessarily in immediate sequence in terms of primary structure; however, these amino acids interact in a manner that allows catalysis to occur. Because of the peculiar and highly individualized ways in which peptide chains may be folded, amino acids some distance apart may contribute to the active site. At the same time, if some molecular change occurs, it is highly probable that the necessary interaction of the amino acids composing the active site may be weakened or lost. This accounts for the fact that relatively mild treatment can cause denaturation. Nature has established certain means of enforcing, at least to some degree, the folding patterns required for enzyme activity. One important means that has already been described involves the presence of amino acids that contain —SH groups. These groups can be oxidized to form disulfide (—S—S—) bonds that enforce propinquity between given residues, since the —SH groups from two cysteine residues form cross-links. There are instances in which enzyme activity depends on the oxidation of some specific —SH groups and on the reduction of others. Hence one may conclude that some enzymes are quite sensitive to the oxidation-reduction state of their environment.

By not requiring that all amino acids of the active center be immediately sequential in the peptide chain, a valuable advantage is gained in the control and regulation of enzyme catalysis. Two distinct conformations may exist, yet one may have the amino acids of the active center sufficiently close to cooperate in catalysis much more efficiently than another. A large number of enzymes are in fact controlled by conversion from one conformation to another. When these changes are affected by metabolites, the enzymes are known as *allosteric*.

Regrettably, too little is known about the intimate details of individual active sites; yet the concept is more than a theoretic comfort. Significant portions of some enzymes can be removed without loss of activity. Sometimes activity can even be increased or an inactive protein can be converted to an active enzyme by cutting off a portion of the peptide chain. Typical examples of the latter are the digestive proteinases, pepsin, trypsin, and chymotrypsin, each of which is produced and stored as an inactive *proenzyme* or a *zymogen*. When the zymogen pepsinogen is released into the gastric juice, it loses a peptide fragment in the acid gastric environment and is converted to active pepsin. The process has long been thought to be autocatalytic. Except for the pH of the environment and the locus at which they are released, the activation of trypsinogen and chymotrypsinogen seems to occur by a similar mechanism. It is possible but not yet proved that the fragment cut off from proenzymes interferes with the folding of the zymogen molecule that is required for enzyme function.

This hypothesis of autoactivation has been accepted for more than 30 years, but it begs the question of how the *first* molecule of active pepsin or trypsin was formed. More recent evidence indicates that at least some of the proteolytic zymogens, notable pepsinogen and trypsinogen, do have a detectable proteolytic activity under specific and appropriate conditions of pH. Acetyltrypsinogen cannot be activated by trypsin because the ε-amino group of the lysine residue at the bond normally split by trypsin is protected against cleavage by the acetyl group. Yet the acetyltrypsinogen was able to do two things that clearly demonstrate proteolytic capacity. It was able to activate unmodified chymotrypsinogen A, and it was able to hydrolyze tosylarginine methyl ester, a good synthetic substrate for trypsin. In view of this, the concept that zymogens are completely inert is probably no longer tenable. Activation of zymogens by scission of a peptide fragment is not limited to proteolytic enzymes. The conversion of proinsulin to insulin and of prophospholipase A to the active enzyme are similar cases.

Isomeric enzymes

In many species, including man, different molecular forms of the same enzyme have been isolated from the same as well as different tissues. The different molecular forms have been termed isoenzymes or isozymes. Isozymes were first detected and reported as a result of electrophoretic or column chromatographic attempts to purify certain enzymes, and the early reports were first discounted as technical or experimental artifacts. However, subsequent studies have clearly demonstrated that the measurable differences in physical properties are accompanied by differences in chemical properties, and thus the concept of isozymes has gained acceptance. In due course, several explanations for the existence of isozymes have been advanced and established.

Terminal deletion mechanism. In the case of the terminal deletion mechanism the isozymes are the result of cutting off a terminal segment of a single peptide chain. This might occur through a partial misreading of the single gene that codes for the protein or through a later cellular proteolytic cleavage of some of the polypeptide chains. This appears to be one mechanism for the formation of isozymes of carbonic anhydrase and hexokinase. There is no evidence for more than a single peptide chain in these enzymes nor for more than one gene.

Polymerization isozymes. According to the polymerization model, a functional enzyme is produced when several peptide chains form an aggregate that is held together by some type of weak bonding force or forces. Frequently three or more like chains must associate, but there may be some physiologic variation in the number of peptide chains that combine, so that more than one functional form of the enzyme results. Thus one isozyme may contain three chains and another four. Note carefully that the separate chains are identical in primary structure. This seems to be the principle behind the isozymes of phosphorylase and glucose 6-phosphate dehydrogenase.

Multiple subunit isozymes. Once again the enzymes are viewed as being composed of more than one peptide chain, or subunit, but in this instance the peptide chains are not all identical, since they appear to be the product of more than one gene. Lactate dehydrogenase and malate dehydrogenase have been thoroughly studied as examples of this type of isozyme. Lactate dehydrogenase is composed

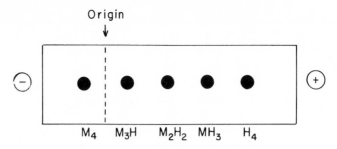

Fig. 3-5. Electrophoresis of lactate dehydrogenase isozymes at pH 8.6.

of four subunits. There are two different subunits that can be combined into tetramers in five ways. The possible combinations can be separated by electrophoresis, as shown in Fig. 3-5. If one subunit type is identified as "M" (the major form found in muscle or liver) and the second as "H" (the major form found in heart), then the tetramers could have the compositions M_4, M_3H, M_2H_2, MH_3, or H_4. These can be separated by electrophoresis. Note in Fig. 3-5 that the net charge on the isozymes differs; those with an increasing content of the H subunit have an increasingly larger negative charge, while the M_4 isozyme has a slightly positive net charge.

In man the content of the several isozymes differs in heart and in liver. Although the reason for this is not known, use is made of the fact in diagnostic differentiation of diseases of the liver and myocardium. In both types of disease states, lactate dehydrogenase leaks out of the damaged cells and increases the concentration of the enzyme in blood serum. Differentiation can be based in part on the pattern that appears on electrophoresis and in part on the fact that some of the isozymes from the myocardium are more resistant to heat denaturation than the corresponding hepatic isozymes. In the simpler differential test situation a serum sample may be analyzed twice, once before and once after heat denaturation under carefully controlled conditions.

Enzyme specificity

Many inorganic catalysts, for example, charcoal or finely divided platinum, show very little specificity toward the substances on which they exert a catalytic effect. Thus a mere handful of selected catalysts is quite sufficient for large areas of the synthetic organic chemistry practiced in industry. Enzymes, on the other hand, are more specific. Urease, an enzyme that hydrolyzes urea to carbon dioxide and ammonia, has nearly an absolute specificity toward urea, and only one other compound that can be split by urease is known. Similarly, catalase is almost completely specific toward hydrogen peroxide, which it converts into water and oxygen. Pepsin, a gastric proteinase, shows a somewhat lesser specificity toward its substrate. It prefers to cleave peptide bonds in which one participant amino acid has an aromatic ring; however, it is quite indifferent to the other amino acid in the peptide bond. It also preferentially attacks peptide bonds in the interior of a peptide chain, but even this requirement is relative. Other proteolytic enzymes show various types of specificity. Trypsin is one of the most discriminating proteolytic enzymes; it hydrolyzes only those peptide bonds to which arginine or lysine contribute the carboxyl group; chemists who are concerned with the determination of amino acid sequences of peptides or proteins can therefore make use of tryptic digestion. A few other proteolytic enzymes have specificity requirements similar to those of trypsin. It is worth noting that several of the factors involved in the process of blood clotting normally exist as proenzymes in the blood. Their activation by tissue injury promotes the formation of thrombin and the production of a clot. To guard against triggering the clotting process by inappropriate activation of the proenzymes, human plasma contains an antitryptic α-globulin that prevents blood from clotting within the vessels under normal circumstances.

Thus the degree of enzyme specificity is variable, ranging from the virtually complete requirement for a single substrate to other enzymes that may function with many different individual molecules, provided they contain some common structural feature, for example, a peptide bond composed of particular amino acids. Yet even where the specificity requirements are high they are not absolute, since it is possible to synthesize substrate analogues that can block or inhibit normal enzyme function.

MECHANISM OF ENZYME CATALYSIS
Catalytic mechanism of enzyme action

The sole driving force for chemical reactions is that the free energy of the products must be less than the free energy of the reactants. These matters are discussed in more detail in Chapter 5; for the moment it suffices to explain that, for a reaction to occur, the overall or net change ($\Delta G_{products} - \Delta G_{reactants}$) must be negative whether the reaction is catalyzed or not.

To be converted to product, a reactant species must somehow absorb enough energy to enter an *activated* or *transition* state. It is from this state, and this state only, that product species are formed. Enzymes lower the energy required for activation. These ideas are shown in Fig. 3-6, where the free energy of a reaction system is plotted against the time course of the reaction process. In the absence of enzyme, the activation energy, ΔG_A, of the forward reaction is so large that the rate is very slow or imperceptible under physiologic conditions. In the presence of enzyme, ΔG_A is reduced so that the reaction proceeds at a measurable rate. In either case since the reactant must absorb energy to enter the transition state, the sign of ΔG_A is positive, indicated by the direction of the arrows depicting the activation process. In an uncatalyzed reaction the activation energy may be supplied by heating the system. In a catalyzed reaction the much smaller activation energy is supplied by the random thermal energies of the reacting species at ambient temperatures. Once initiated, the reaction must have a negative free energy change, shown by the direction of the arrow to the right of the activation energy barrier. Because of this very high barrier, it is very unlikely that products will be spontaneously converted to reactants.

In analyzing the catalytic mechanism further, the following sequence of reactions is assumed to occur:

$$E + S \rightleftharpoons ES \rightleftharpoons ES^* \longrightarrow EP \longrightarrow E + P$$

where E represents the free enzyme, S the substrate, ES a complex between the enzyme and the substrate, ES* the activated or transition state of the comlex, EP a complex between the enzyme and products, and P the free product. (In less detailed formulations, ES* is omitted for brevity.) Generally, substrate is present in great excess compared to enzyme, so only a fraction of S is involved in ES at any given moment. That fraction of S that forms ES has a somewhat higher energy content than free S; because of the stabilizing effect of ES this statistically improbable condition persists through the time span required for catalysis. As ES is converted to EP there is a decrease in stability, so EP cleaves

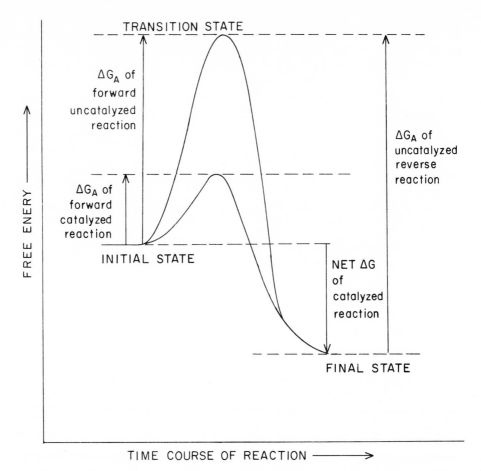

Fig. 3-6. Energy diagram for catalyzed and uncatalyzed reactions. ΔG_A represents the activation energy under the conditions specified. NET ΔG represents the change in available free energy as a result of the reaction; it is the difference between the free energy of products and the free energy of reactants. For a spontaneous reaction, NET ΔG is always negative.

to give free product plus enzyme. This cycle may be repeated hundreds or thousands of times per min.

The nature of enzyme catalysis

The more the process of enzyme catalysis has been studied, the more it becomes clear that it follows the chemical mechanisms described for noncatalyzed processes. The mechanisms of primary importance are these:

1. In organic chemistry, the concept of catalysis by generalized (Brønsted) acids and bases is well known. In enzymes certain amino acid residues can act as generalized acids or bases, that is, as proton donors or acceptors.

2. Enzyme-substrate "fit," resulting from the selectivity of an enzyme for its substrate, may raise the *effective* concentration of substrate at the reactive enzyme site a thousand times or more over the concentration in free solution. Similarly, one may argue that the effective concentration of catalytic centers is also increased, compared to what might be the case if they were free in solution.

3. Cooperativity exists between the enzyme and the substrate, or some other

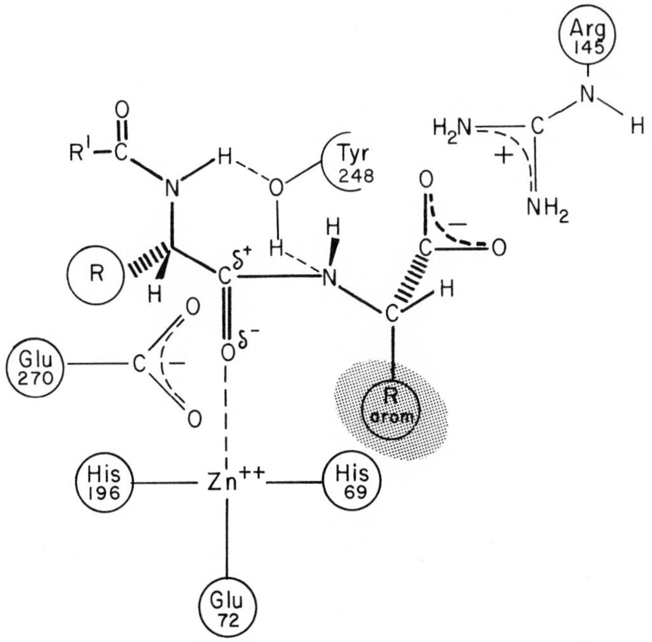

Fig. 3-7. Features of carboxypeptidase A that assist in forming the enzyme-substrate complex and the catalytic cleavage of a C-terminal amino acid from peptide or protein substrates. The substrate peptide is shown in heavy lines. The numbered residues refer to components of the enzyme protein.

ligand, whereby in the transition state the reactive part of the substrate is brought into very intimate contact with the active site of the enzyme. Such shifts may also be effective in increasing catalytic rates. Cooperativity may be positive, increasing the rates, or negative, decreasing rates. This can be an important feature in the control of enzyme reactions (p. 77).

These ideas can be illustrated by the case of the pancreatic exopeptidase, carboxypeptidase A. This is a metalloenzyme, with an absolute requirement for one gram atom of a divalent metal per mole of protein. In the natural state the metal is zinc, which is bound into the enzyme by three ligands, His 69, Glu 72, and His 196 (Fig. 3-7). In the free enzyme the fourth ligand is water, but in the

enzyme-substrate complex the fourth ligand is the $-\overset{O}{\overset{\|}{C}}-$ of the penultimate peptide bond from the C-terminal end of the substrate. The enzyme also has a hydrophobic region that forms a pocket into which the C-terminal residue of the substrate, preferably an aromatic residue, fits. Several other amino acid residues of the enzyme, including Arg 145 and Tyr 248, are involved in fitting and binding the substrate. Glu 270 and at least one other residue are involved in the catalytic process. The latter residue, not certainly identified, acts as a generalized acid. Most of these features are schematized in Fig. 3-7, where the substrate peptide is shown in heavy lines to distinguish it from essential parts of the enzyme. Note especially the partial double-bonded character of the Glu 270 carboxyl group and the Arg 145 guanidine group as well as the electronegativity of the carbonyl oxygen at the second peptide bond of the substrate. Finally, note the hydrogen bonding at Tyr 248, which assists in holding the substrate in the desired conformation.

A proposed mechanism for hydrolysis of the C-terminal amino acid of the substrate is shown, in very simplified form, in Fig. 3-8. The carboxyl group of Glu 270, known to be essential in catalysis, is shown at the left of each square. The zinc, also essential to catalysis, is shown at the right. In the center of square 1 is the substrate, with X representing the entirety of the C-terminal residue. At the bottom of each square is another residue that acts as a proton donor (generalized acid) in catalysis.

As the enzyme-substrate complex forms (square 1), attractive forces between the obligatory, ionized, C-terminal carboxyl of the substrate cause a spatial shift of some residues in the enzyme structure. This is indicated by a lack of "squareness" in square 2. The carbonyl of the peptide bond to be cleaved becomes liganded to zinc. The carboxyl group of Glu 270, acting as a nucleophile, forms an anhydride intermediate, with X now to be regarded as the leaving group. A further rearrangement occurs, as is shown in square 3, after which the anhydride intermediate is hydrolyzed by water. In square 4 it can be seen that the peptide substrate, now one residue shorter, and the split amino acid, HX, are set free. The enzyme returns to its free conformation and the process may be repeated. In effect, the enzyme contributes certain reactive centers, all part of its reactive site, which are not unlike those of laboratory reagents employed in noncatalyzed reactions. It obviously forms, at least temporarily, covalent bonds with the substrate. In this particular case it has not yet been possible to isolate an enzyme-substrate

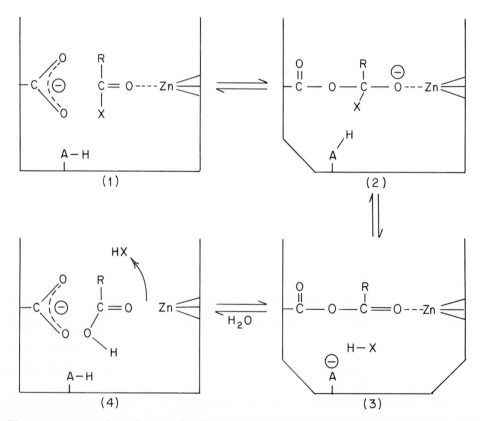

Fig. 3-8. A proposed mechanism for the hydrolysis of a C-terminal amino acid residue by carboxypeptidase. R is the peptide chain of the substrate, and X represents the C-terminal amino acid residue. They are joined by a peptide bond that carboxypeptidase hydrolyzes.

complex, but in other cases complexes have been prepared and separated from the reaction mixture.

The preceding presentation, it must be emphasized, deals only with the simplest case of a single-substrate system. Many enzymes, including those with loosely bound coenzymes, must be regarded as two-substrate systems, and the mechanistic and kinetic models are more complex. It is also highly probable that formation of the ES* complex is further modulated by some portion of the enzyme other than the active site. In allosteric enzymes, to be discussed later, formation of ES* is subject to modulation by effector molecules unrelated to the substrate. Such modulation constitutes an important aspect of enzyme regulation and control.

Quantitative analysis of single–substrate enzyme kinetics

The fundamental theory of enzyme catalysis is based on the classic studies of Michaelis and Menten and of Haldane. The quantitative analysis of enzyme action that has been developed depends largely on measured reaction rates.

If the *initial* reaction rate, defined as the rate observed for a given amount of enzyme when the concentration of the product formed is nearly zero, is plotted as a function of the substrate concentration, the results look very much like those shown in Fig. 3-9; the curve connecting the observed points would be hyperboloid and would asymptotically approach a maximum value, as shown by V_{max}. This is the maximum initial velocity that can be obtained without increasing the amount of enzyme.

The hyperbola described by a plot of reaction velocities as a function of substrate concentrations is difficult to use. If the reciprocals of the velocities are plotted as a function of reciprocal substrate concentrations, the hyperbola is converted to a straight line. Such linear double-reciprocal plots are far easier to construct and interpret. A transformation of the typical curve shown in Fig. 3-9 is presented in Fig. 3-10. The double-reciprocal plots are frequently called *Lineweaver-Burk* plots after the chemists who first described them.

If the usual convention is followed, representing concentrations by means of brackets, that is, by letting [S] stand for the molar concentration of the substrate,

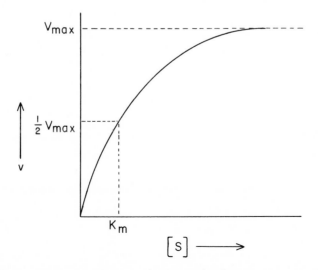

Fig. 3-9. Reaction velocity as function of substrate concentration.

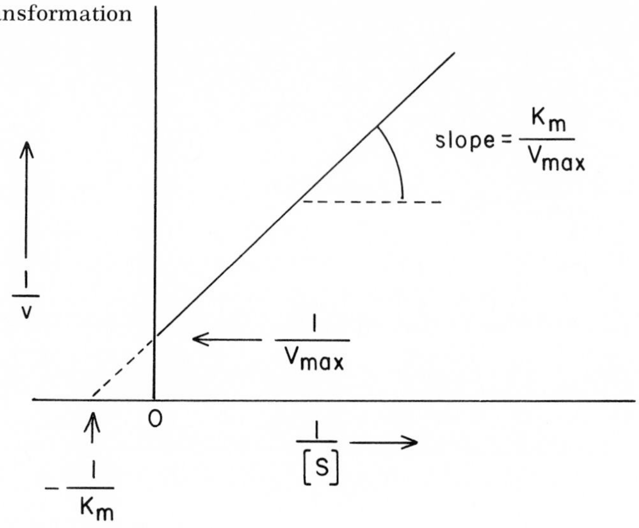

Fig. 3-10. Lineweaver-Burk transformation
of Michaelis-Menten curves.

and if a few assumptions are made regarding the experimental situation, it is possible to obtain a useful mathematical equation that describes the enzyme kinetics. The few assumptions are the following:

1. The system involves only a single substrate.
2. The system is at a steady state, i.e., $[ES^*]$ is a constant and the free enzyme E is in equilibrium with ES^*.
3. The system is established so that $[E] \ll [S]$ on a molar basis. Considering the high molecular weight of most enzymes, this is not a stringent limitation.
4. Since the analysis deals with initial reaction rates, $[S] \gg [P]$ and $[P]$ is negligible under these conditions.

In the case in which a single substrate goes to a single product, the reaction mechanism may be formulated as:

$$E + S \underset{k_2}{\overset{k_1}{\rightleftharpoons}} ES^* \underset{k_4}{\overset{k_3}{\rightleftharpoons}} E + P \qquad (3\text{-}1)$$

where k_1, k_2, k_3, and k_4 are the respective rate constants. At the steady state the concentration of ES is constant; that is, the rate at which it is being formed is the same as the rate at which it is being broken down. Under these conditions, equation 3-1 can be used to derive the rate equation 3-2:

$$k_1[E][S] + k_4[E][P] = k_2[ES^*] + k_3[ES^*] \qquad (3\text{-}2)$$

Since the analysis is restricted to initial reaction rates, $[P]$ is negligible and $[S]$ is virtually constant. Thus the term involving $[P]$ can be dropped and those involving $[ES^*]$ can be collected to give:

$$k_1[E][S] = (k_2 + k_3)[ES^*] \qquad (3\text{-}3)$$

$$\text{or} \quad \frac{[E][S]}{[ES^*]} = \frac{k_2 + k_3}{k_1} = K_m \qquad (3\text{-}4)$$

The ratio of rate constants, $(k_2 + k_3)/k_1$, can be replaced by a single constant, K_m.

Analysis from concentrations and reaction velocities

Instead of defining K_m in terms of rate constants, which are not easily determined by simple measurements, it is useful to have an expression cast in terms of chemical concentrations and reaction velocities, as these can easily be measured. It should now be clear that *maximal initial* velocity is achieved only when *all* of the enzyme (E_t) is in the form of the active complex (ES^*), from which it follows that:

$$V_{max} = k_3 [E_T]$$
(3-5)

Under any other conditions, the observed initial velocity will be:

$$v = k_3 [ES^*]$$
(3-6)

The free enzyme, which does not take part in catalysis, is related to the total quantity of enzyme and to the active complex by the equation:

$$[E] = [E_t] - [ES^*]$$
(3-7)

The information contained in equations 3-5 and 3-6, when applied to equation 3-7, gives:

$$[E] = \frac{V_{max}}{k_3} - \frac{v}{k_3} = \frac{V_{max} - v}{k_3}$$
(3-8)

This expression for [E] can now be substituted into equation 3-4.

$$K_m = \frac{(V_{max} - v)[S]}{k_3 [ES^*]}$$
(3-9)

Recalling again the facts contained in equation 3-6, equation 3-9 is transformed to read:

$$K_m = \frac{[S]}{v} (V_{max} - v)$$
(3-10)

and the relation between K_m, substrate concentration, and reaction velocities is established. Since it is customary to deal with enzymes in terms of reaction velocities, equation 3-10 is usually solved for v and written as:

$$v = \frac{[S] V_{max}}{[S] + K_m}$$
(3-11)

The constant K_m is generally known as the Michaelis constant. It can be seen from equation 3-10 that when the initial velocity is equal to half the maximum initial velocity (that is, $v = \frac{1}{2}V_{max}$), then [S] is equal to K_m. Both K_m and [S] are expressed in the same units, moles per liter.

It can be seen from equation 3-4 that K_m approximates the dissociation constant for the enzyme-substrate complex ES^*. As discussed earlier with regard to reversible reactions (p. 56), the reciprocal of K_m would therefore refer to the association constant of E for S. This is useful when considering a given enzyme in the presence of two alternate substrates; the substrate with the lowest K_m will be more effectively bound and will successfully compete for the binding site on the enzyme. Both K_m and V_{max} are constants of specific values for any enzyme, and they are the appropriate parameters for comparison of enzyme behavior. In designing an assay to measure the amount of an enzyme in blood or other material, it is important to know that sufficient substrate is present to completely saturate the enzyme, that is, to convert it completely to the enzyme-substrate complex.

Turnover number

In any experimental situation, V_{max} obviously depends on the amount of enzyme present. If an experiment could be performed containing *1 mole* of enzyme, the resulting activity would be termed the turnover number, which is expressed as moles of substrate converted to product per min (or per sec) per mole of enzyme. In those cases where the enzyme contains more than one active site per molecule, the turnover number is corrected accordingly. One would then use the turnover per mole of active site of the enzyme. Turnover numbers are of value in comparing the same enzyme from different tissues and in comparing different isozymes; they are of general theoretic interest as well.

Enzymatic activity

Another way of expressing the catalytic activity of an enzyme relates the micromoles of substrate reacted or product formed per min (or per sec) to the weight of protein in the sampled solution or body fluid. Because the amount of enzyme protein is not readily measurable, the amount of enzyme is frequently expressed in terms of the catalytic activity in the tissue or fluid. These and similar problems are discussed further in the section on clinical application later in this chapter.

Kinetic analysis of enzyme inhibition

Lineweaver-Burk plots can be used to assess the nature of enzyme inhibition. If catalysis is to occur, there must obviously be a certain structural correlation between the substrate on the one hand and the active site of the enzyme and its surroundings on the other. Anything that alters or interferes with this "fit" will inhibit or prevent catalysis. Several kinds of inhibition have been observed, but the present discussion will be limited to competitive and noncompetitive processes. Competitive inhibition is diagrammed in Fig. 3-11. Suppose that the active site of the enzyme, represented by the symbol E, contains three essential features, A, B, and C. Any substrate (S) must contain three corresponding features that are recognized by and acceptable to the enzyme, that is, A', B', and C'. An inhibitor (I) might contain only two of the three essential features and yet be sufficiently like the true substrate that it could still be bound by the enzyme. However, it could not undergo catalytic conversion to product. According to the Michaelis-Menten theory, equations such as the following may be written to describe the separate situations.

$$E + S \rightleftharpoons ES^* \text{ and } E + I \rightleftharpoons EI$$

In general, the forces binding most competitive inhibitors are of approximately the same magnitude as the forces binding natural substrates. Consequently, when mixtures of S and I are added to the enzyme system, the resulting degree of inhibition is proportional to the concentrations of either. Competitive inhibitors affect (increase) the apparent K_m but are without effect on V_{max}, the reason being that reaction velocity depends only on those uninhibited enzyme molecules where I is without effect, since it is not bound. The V_{max} occurs at high [S], where I cannot compete. These concepts are shown schematically in Fig. 3-12.

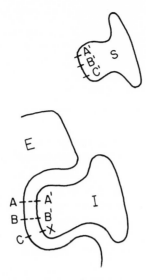

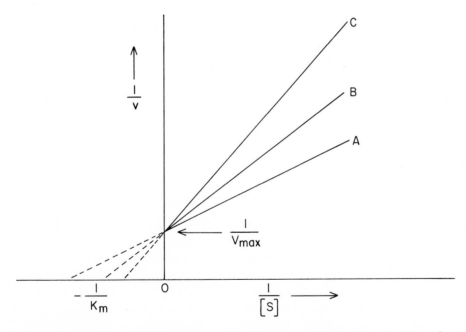

Fig. 3-11. Mechanism of competitive inhibition.

Fig. 3-12. Competitive inhibition depicted by Lineweaver-Burk plots. **A,** Normal uninhibited reaction; **B** and **C,** two different inhibitor concentrations (**B** < **C**).

Noncompetitive inhibition

Enzymes are also subject to noncompetitive inhibition by several mechanisms that are diagrammed in Fig. 3-13. Imagine an inhibitor, I_1, that does not attach to the active center at all but rather to some group in the near vicinity. If I_1 is sufficiently bulky or sufficiently charged, it may very well prevent a normal substrate from gaining access to the unaltered active site. A second possibility would be an inhibitor of the type indicated by I_2; it need not be very bulky, but it attaches irreversibly to one of the essential features of the active site and also prevents attachment of the unaltered substrate. Still another form of noncompetitive inhibition exists in which an analogue of the true substrate attaches to the normal active site. Here the analogue is attached so tightly (orders of magnitude more tightly) that even drastic increases in true substrate concentration are insufficient to displace it from the active site. In summary, noncompetitive inhibitors are those that (1) may not be bound to the active site but to some point in its vicinity, (2) may be bound at two or more points, one of which probably involves the active center, and (3) are bound to the active site but much more tightly than is normal. Noncompetitive inhibitors do not affect K_m, since with this type of inhibition the substrate cannot bind to the inhibited enzyme molecules. These concepts are shown schematically in Fig. 3-14.

Several examples may be cited to further describe these ideas. The enzyme succinate dehydrogenase in known to catalyze the following reaction:

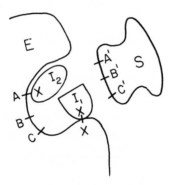

Fig. 3-13. Mechanism of noncompetitive inhibition.

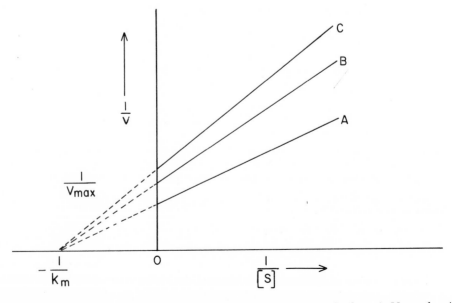

Fig. 3-14. Noncompetitive inhibition depicted by Lineweaver-Burk plots. A, Normal uninhibited reaction; B and C, two different inhibitor concentrations (B < C).

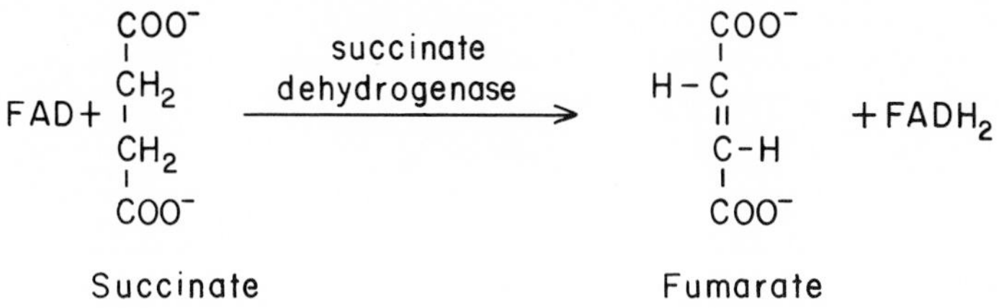

$$\text{FAD} + \begin{array}{c} COO^- \\ | \\ CH_2 \\ | \\ CH_2 \\ | \\ COO^- \end{array} \xrightarrow[\text{dehydrogenase}]{\text{succinate}} \begin{array}{c} COO^- \\ | \\ H-C \\ || \\ C-H \\ | \\ COO^- \end{array} + \text{FADH}_2$$

Succinate   Fumarate

Malonate, which has the structure $^-OOC \cdot CH_2 \cdot COO^-$, can be bound to the enzyme but cannot be dehydrogenated. Malonate meets the stipulations for a competitive inhibitor, because its effects can be offset or reversed by sufficiently increasing the concentration of succinate in the presence of the inhibitor.

Another important mitochondrial enzyme is malate dehydrogenase, which reversibly catalyzes the oxidation of L-malate to oxaloacetate. There is evidence that this enzyme first forms a complex with the coenzyme, and that the complex then reacts with the substrate. An inhibitor known as hydroxymalonate, $^-OOC \cdot CHOH \cdot COO^-$, displays inhibitor properties that vary according to the direction of the reaction, as seen in the following:

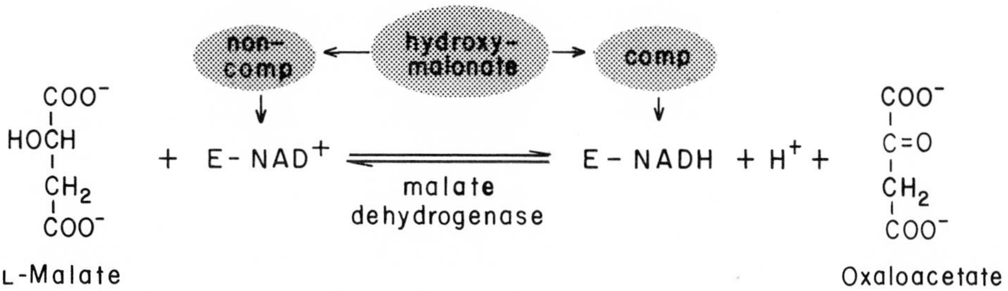

$$\begin{array}{c} COO^- \\ | \\ HOCH \\ | \\ CH_2 \\ | \\ COO^- \end{array} + \text{E-NAD}^+ \rightleftharpoons_{\substack{\text{malate}\\\text{dehydrogenase}}} \text{E-NADH} + \text{H}^+ + \begin{array}{c} COO^- \\ | \\ C=O \\ | \\ CH_2 \\ | \\ COO^- \end{array}$$

non-comp ← hydroxy-malonate → comp

L-Malate   Oxaloacetate

In other words, hydroxymalonate apparently reacts in a noncompetitive manner with the complex of the enzyme and oxidized coenzyme when malate is oxidized to form oxaloacetate. Conversely, it reacts in a competitive manner with the complex of enzyme and reduced coenzyme when oxaloacetate is reduced to form malate. This somewhat unusual case is described only to emphasize that the discrimination between natural substrate and inhibitors may be great enough to be different in each direction.

Two-substrate kinetics

The examples just cited are important for another reason. Michaelis-Menten kinetics strictly apply only to those enzymes that employ single substrates. A far larger number of enzymes actually require more than one substrate at a time. Enzymes of the first kind are known as *unireactant* while those of the second class are known as *multireactant*. For multireactant enzymes, illustrated by malate dehydrogenase, NAD^+ can be regarded not only as a coenzyme, but also as a cosubstrate. The kinetic analysis of multireactant enzyme systems requires an extension of the fundamental principles established by Michaelis and Menten. Since the enzyme and both substrates are simultaneously required for catalysis to occur, it is appropriate to describe the effective ternary catalytic complex by a symbol $(E \cdot S \cdot NAD^+)$. In forming this, two binary complexes are possible, represented by $(E \cdot S)$ and $(E \cdot NAD^+)$. Following catalytic transformation, $(E \cdot P \cdot NADH)$ could equally well produce $(E \cdot P)$ and $(E \cdot NADH)$; these, in turn, could decompose to liberate free enzyme and the second product.

Kinetic analysis sometimes permits an independent distinction between the two possible sequences of such a reaction mechanism. In the case of malate dehydrogenase, the mechanism is known as *ordered* because the addition of NAD^+ as the first, or leading, substrate is compulsory. Other examples are known in which the sequence of addition is not obligatory, but *random*.

Another pattern for multireactant enzyme systems is known as the "ping-pong" mechanism. Here one substrate must be bound and one product released before the second substrate is bound or the second product released. Typical examples of this pattern are found in the aminotransferases, also called transaminases.

Experimentally, one can proceed with kinetic analysis of a multireactant enzyme system in a manner quite similar to that already described. By adding a large excess of all but one component, and then varying that one, it is possible to obtain values of K_m and V_{max} for the substances varied. Repeating such experiments, varying in turn each cofactor or substrate, allows the successive calculation of each K_m and V_{max} as before. Thus, for malate dehydrogenase, one can determine $K_m^{NAD^+}$, K_m^{malate}, and so forth. It is also possible to examine the properties of various inhibitors as already outlined. Certain precautions are required, especially the determination that the large excess of the fixed components does not affect the properties of the enzyme.

There are some significant, practical applications of differential enzyme inhibition. Serum contains a group of enzymes known as acid phosphatases because they have an optimum activity under slightly acidic (pH 6.2) conditions. The members of this group are produced by many different tissues. From a clinical point of view, the most important one is produced by the prostate gland, since increases in its activity are very frequently associated with prostatic carcinoma. It has been shown that L-tartrate competitively inhibits about 95% of the prostatic activity but has a much lower inhibitory effect on acid phosphatases from erythrocytes, liver, kidney, or spleen. L-Phenylalanine exerts similar but not identical inhibition. Obviously samples from suspected carcinoma patients can be assayed in the presence and absence of L-tartrate or L-phenylalanine. Formaldehyde, on the other hand, inhibits erythrocyte acid phosphatase but not the prostatic enzyme. Calcium ions are noncompetitive inhibitors of many acid phosphatases, so it is usual to add a chelating agent to blood samples collected for this assay or to the sera derived from the blood. Ethanol and certain narcotic drugs are also noncompetitive inhibitors of acid phosphatase, a fact that should be kept in mind when interpreting the results of such tests.

Clinically important enzyme inhibitors

Enzymes may also be inactivated or denatured by a variety of chemical means, several of which have clinical importance. Many enzymes depend on essential sulfhydryl groups, which form tight covalent bonds with various heavy metals. For this reason mercury, lead, silver, etc. are extremely toxic. Even iron and copper, although they are classed as essential minerals, can produce intoxication when ingested in excess. A treatment of syphilis, popular not too long ago but no longer used, depended on the extreme sensitivity of *Treponema pallidum* to mercury or bismuth ions. Organic mercurials are still employed as disinfectants today.

Another class of inhibitors affects enzymes by introducing foreign alkyl groups into the structure. Most of these substances were initially developed as chemical warfare agents and are of high toxicity. The two best known are diisopropyl fluorophosphate (DFP) and the so-called nitrogen mustards, for example, methachloramine and 2,2'-dichloro-N-methyldiethylamine. They are now employed clinically as enzyme inhibitors in the treatment of certain types of neoplastic disease. DFP reacts readily with the OH group of serine residues, converting them to diisopropyl phosphate esters. The organic phosphates, of which DFP is only one example, are particularly good inhibitors of acetylcholinesterase, which breaks down the neurotransmitter substance known as acetylcholine. Inactivation of acetylcholinesterase produces violent spasms of the pulmonary system and interferes with normal neuromuscular and cardiac function. Similar agents employed as insecticides in agriculture may be severely or fatally toxic.

Enzymes can also be inhibited by affecting associated coenzymes or prosthetic groups. This is of considerable importance in designing chemotherapeutic agents and will be discussed in more detail later on as an aspect of enzyme kinetics.

ENZYME REGULATION AND CONTROL

In a system as complex as a living cell there must be some regulation or control of the multitude of simultaneous reactions that occur. These reactions are for the most part catalyzed by enzymes that are localized in a highly structured manner. It is entirely reasonable to inquire by what means these processes are controlled, other than by the types of inhibition that have already been discussed. In the simplest cases we must recognize the existence of enzyme *activators.* Many enzymes require the presence of free —SH groups and are inactivated or inhibited if these groups are oxidized. Glutathione, γ-glutamylcysteinylglycine, may be regarded as a natural intracellular activator for enzymes that contain sensitive and essential —SH groups. Note carefully that this is not a specific coenzyme effect; rather it depends on the fact that glutathione is a natural antioxidant that can be easily synthesized in the cell from readily available precursors. Glutathione *is* a specific coenzyme for a small number of enzymes, including maleyl acetoacetate isomerase, an enzyme involved in the oxidative degradation of aromatic amino acids. Thus, by controlling the intracellular concentration of glutathione, the activity of several enzymes is also manipulated.

Allosteric enzymes

Isocitrate dehydrogenase is made more active when the Mg^{++} that it may contain is replaced by Mn^{++}. Here the activation is relative, since both ions confer catalytic activity, but not to the same degree. Citrate also activates isocitrate dehydrogenase. It is hard to believe that effects brought to bear by such different materials as Mn^{++} and citrate could operate by identical mechanisms. Presumably the Mn^{++} activates by being more efficient at the catalytic site. Citrate, on the other hand, exerts what has been described as allosteric control. Enzymes that are subject to allosteric control have been designated as allosteric or regulatory enzymes (Chapter 2, p. 79). Some of these enzymes are composed of subunits containing a small number of identical or closely related peptide chains. The quaternary conformation is modified by the appropriate allosteric effectors. One or more of the functional sites on these enzymes may be *catalytic,* while one or more sities may be *regulatory* and not identical with the catalytic or active sites and may not even be on the same subunit. In other instances the allosteric and catalytic sites are located on the same subunit. When the reaction velocity of an allosteric enzyme is plotted as a function of substrate concentration, a sigmoid rather than a hyperboloid curve is obtained, as shown in Fig. 3-15, which provides an example of Michaelis-Menten kinetics for comparison. It can be seen in Fig. 3-15 that the shapes of the allosteric curves are changed considerably by altering the concentration of either positive or negative effectors, as indicated by the small arrows. In effect, decreasing the amount of negative effector or increasing the amount of positive effector produces a response equivalent to lowering the K_m of the substrate. In the most general case allosteric kinetics can be represented by the equation

$$v = \frac{[S]^n V_{max}}{[S]^n + K}$$

where n is a coefficient that represents the interaction of the binding sites, K represents a measure of the affinity of substrate for the enzyme, and the other symbols have their previously stated meanings. In the presence of sufficiently high concentrations of positive allosteric effectors, the binding of one substrate

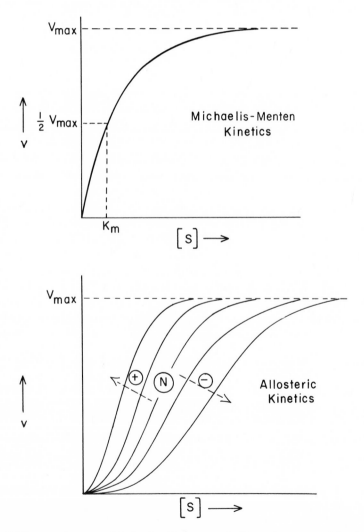

Fig. 3-15. Difference between Michaelis-Menten and allosteric enzyme kinetics.

molecule no longer affects the binding of a second or of a third, and n has a value of 1. Under these conditions the plot of initial reaction velocity versus substrate concentration reduces to the form indicated by Michaelis-Menten kinetics, and K is equal to K_m. The reverse is true, of course, for negative effectors, which make the plots become increasingly sigmoidal, as they lower the affinity of the enzyme for its substrate (Fig. 3-15).

It has been known for some time that oxygen binding by hemoglobin, when plotted as a function of oxygen tension, produces a sigmoid curve. For this reason the binding of oxygen by hemoglobin has been studied as a model analogous to allosteric enzymes, a topic discussed in Chapter 2. A second modifier of oxygen binding, 2,3-diphospho-D-glycerate, is analogous to the allosteric effectors of enzymes.

With the hemoglobin model in mind, let us return again to isocitrate dehydrogenase, which catalyzes the following reaction:

$$
\begin{array}{c}
\text{COO}^- \\
| \\
\text{H-C-OH} \\
| \\
{}^-\text{OOC-C-H} \\
| \\
\text{CH}_2 \\
| \\
\text{COO}^-
\end{array}
\; + \text{NAD}^+ \rightleftharpoons
\begin{array}{c}
\text{COO}^- \\
| \\
\text{O=C} \\
| \\
\text{CH}_2 \\
| \\
\text{CH}_2 \\
| \\
\text{COO}^-
\end{array}
\; + \text{CO}_2 + \text{NADH} + \text{H}^+
$$

Isocitrate   α-Ketoglutarate

Like hemoglobin, isocitrate dehydrogenase is a tetramer and all four of its peptide chains are required for activity. NAD^+ and ADP are positive allosteric effectors, while NADH and ATP are negative allosteric effectors. To be sure, NAD^+ is a required coenzyme (or cosubstrate, like oxygen in the case of hemoglobin), but it also markedly increases the effect of ADP in raising the reaction velocity. ADP increases the affinity of the enzyme for NAD^+ and vice versa. The corresponding but opposite effects are noted with the negative allosteric agents ATP and NADH. Thus this is an example of an enzyme that is controlled by at least three distinct allosteric effectors, one of which is related closely to isocitrate in the Krebs cycle, and one of which happens to be a required coenzyme. A schematic representation of these effectors and the ways in which they alter the activity of isocitrate dehydrogenase is shown in Fig. 3-16. Note particularly the opposite effects of NAD^+ and NADH, positive and negative effectors, respectively, and the opposite effects

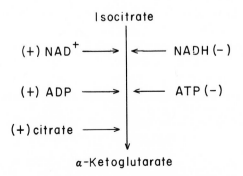

Fig. 3-16. Allosteric effectors of isocitrate dehydrogenase.

of ADP and ATP. Citrate is also a positive effector and, as will be seen later, there is a close relation between citrate and isocitrate in the Krebs cycle.

A second example is found in that portion of glucose metabolism known as glycolysis. Two enzymes are involved here: the first is known as phosphofructokinase, which catalyzes the following reaction:

$$
\begin{array}{ccc}
\begin{array}{l}
\text{CH}_2\text{OH} \\
| \\
\text{C}=\text{O} \\
| \\
\text{HOCH} \\
| \\
\text{HCOH} \\
| \\
\text{HCOH} \\
| \\
\text{CH}_2\text{OPO}_3^=
\end{array}
& + \text{ATP} \xrightarrow{\text{Mg}^{++}} \text{ADP} + &
\begin{array}{l}
\text{CH}_2\text{OPO}_3^= \\
| \\
\text{C}=\text{O} \\
| \\
\text{HOCH} \\
| \\
\text{HCOH} \\
| \\
\text{HCOH} \\
| \\
\text{CH}_2\text{OPO}_3^=
\end{array} \\
\text{D-fructose 6-phosphate} & & \text{D-fructose 1,6-diphosphate}
\end{array}
$$

A second enzyme, fructose 1,6-diphosphate 1-phosphatase, catalyzes the reverse reaction, as shown below.

$$
\begin{array}{ccc}
\begin{array}{l}
\text{CH}_2\text{OPO}_3^= \\
| \\
\text{C}=\text{O} \\
| \\
\text{HOCH} \\
| \\
\text{HCOH} \\
| \\
\text{HCOH} \\
| \\
\text{CH}_2\text{OPO}_3^=
\end{array}
& + \text{H}_2\text{O} \longrightarrow &
\begin{array}{l}
\text{CH}_2\text{OH} \\
| \\
\text{C}=\text{O} \\
| \\
\text{HOCH} \\
| \\
\text{HCOH} \\
| \\
\text{HCOH} \\
| \\
\text{CH}_2\text{OPO}_3^=
\end{array}
& + \text{H}_2\text{PO}_4^- \\
\text{D-fructose 1,6-diphosphate} & & \text{D-fructose 6-phosphate}
\end{array}
$$

These reactions are shown schematically in Fig. 3-17.

Phosphofructokinase is also an allosteric enzyme. Both citrate and ATP are strongly negative allosteric effectors, while inorganic phosphate and AMP are strongly positive effectors; these are shown at the left in Fig. 3-17. The allosteric effectors of the fructose 1,6-diphosphate 1-phosphatase are shown at the right and emphasize the opposite effects of AMP and ATP.

It should be clear that the opposing allosteric effects of the adenine nucleotides on this pair of enzymes represent a very functional control mechanism. At a given concentration of ATP, for example, 5 mM, phosphofructokinase would be significantly "turned off" and the phosphatase would be equally "turned on." Because the stated concentration of ATP represents a reasonably high value for most tissues, there would be little need for further breakdown of the primary fuel, D-glu-

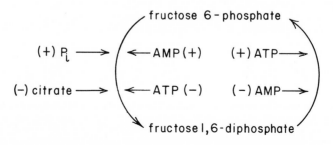

Fig. 3-17. Allosteric effectors of fructose 6-phosphate phosphorylation.

cose. If the ATP concentration were only 1 mM, the signals might be reversed, and increased oxidation of D-glucose would be facilitated by allosteric regulation of these enzymes. Note also that D-fructose 6-phosphate, the substrate of phospho-fructokinase, is also a positive effector in the sense that occupation of the catalytic site by the substrate makes it more difficult for ATP to occupy the regulatory site, and vice versa.

Data from recent studies of metabolism have indicated that numerous other enzymes may be added to the list of those subject to allosteric control, and further illustrations of the process will be provided in later chapters. The major features of allosteric control may be summasized as follows:

1. Allosteric enzymes are usually composed of more than one polypeptide chain.
2. Allosteric enzymes may contain two separate functional centers, one of which is catalytic and the other regulatory.
3. Allosteric enzymes may be subject to both positive and negative control by one or more factors.
4. Allosteric regulators may or may not be involved in the reaction as coenzymes, substrates, or products.
5. Allosteric enzymes undergo either a change in conformation or in the cooperativity of component polypeptides when influenced by appropriate effectors.

Product inhibition

The law of mass action states that as products of a reaction accumulate, the reaction will slow down. Enzyme-catalyzed reactions are like any others; they must also obey this law. Product accumulation is therefore a significant means of control in living tissue and in vitro. The reaction that converts D-glucose 6-phosphate to free glucose and inorganic phosphate is catalyzed by the enzyme glucose 6-phosphatase.

D-Glucose 6-phosphate   D-Glucose

If either free D-glucose or inorganic phosphate is added to a solution of the substrate and the reaction is started by addition of the enzyme, it is readily observed that the rate is slower in the presence of *initially* added products than in their absence. The initial addition of product shows that the decline in rate is not caused simply by a decrease in the concentration of substrate, as would be predicted by the law of mass action. Of course, when products accumulate at the expense of substrate, the effect is magnified. The hallmark of product inhibition is the reduction of the inhibitory effects of the product (or products) when the concentration of substrate is increased. Once again, this follows from the consideration of competitive inhibition.

A given enzyme may be subject to several kinds of control. The enzyme hexokinase catalyzes the conversion of D-glucose to D-glucose 6-phosphate.

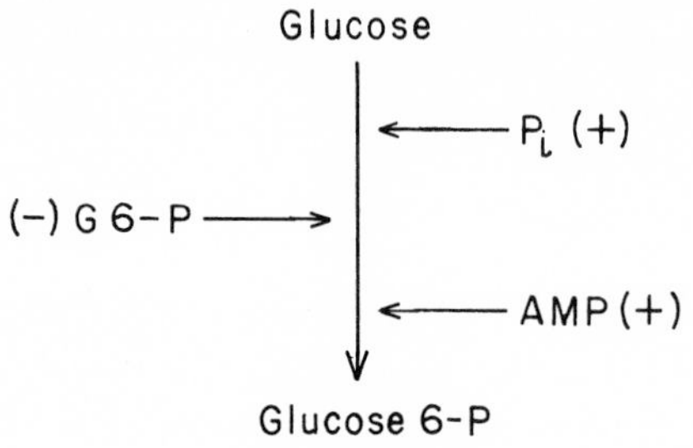
Glucose
Pi (+)
(−) G 6-P
AMP (+)
Glucose 6-P

Fig. 3-18. Effect of glucose 6-phosphate is most frequently described as noncompetitive inhibition of some mammalian hexokinase; some claim that glucose 6-phosphate is an allosteric effector.

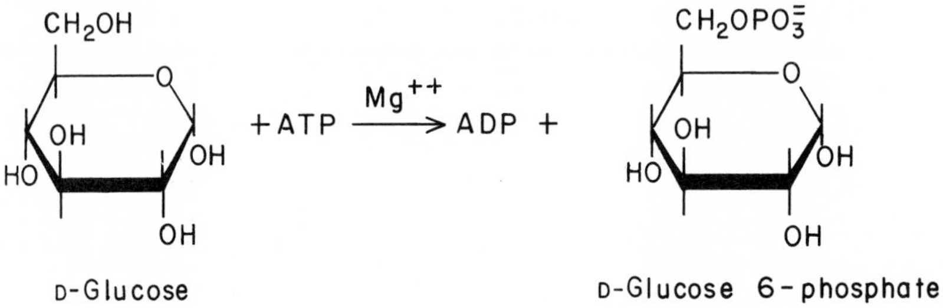
CH2OH
OH
HO
OH
OH
+ATP
Mg++
ADP +
CH2OPO3=
OH
HO
OH
OH
D-Glucose
D-Glucose 6-phosphate

Animal hexokinases are generally subject to product inhibition, as shown in Fig. 3-18. Indeed, in the case of animal hexokinases it may be that the product inhibition demonstrated by D-glucose 6-phosphate is also a type of mixed control, as its inhibitory effects are not so readily offset by proportionate increases in concentration of free glucose. The inhibitory effect of D-glucose 6-phosphate is countered by inorganic phosphate and AMP.

Feedback control

Certain substances of importance to cells are produced through lengthy sequences of reactions, which together constitute metabolic pathways. In some of these it has been found that a product exerts a negative effect either on the first reaction or on a very early reaction. A general representation of this kind of control is shown below.

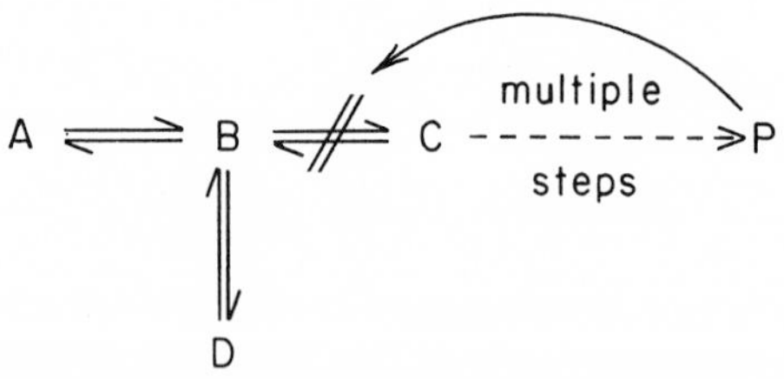
A
B
C
multiple
steps
P
D

As seen in the diagram, the product (P) acts to inhibit some early step in the pathway, but not necessarily the first. Frequently, substance B can be converted to more than one product; in this case the intermediate products are C and D. By feedback control it is possible not only to inhibit the production of P, but also to divert the flow of B from one pathway to another.

The term feedback control originated in electronics. Certain kinds of amplifiers can be stabilized (partly turned off) by feeding a portion of the output signal back into the input in a negative sense. Note carefully the essential difference between product inhibition and negative feedback. In product inhibition the substance exerting control does so on the enzyme immediately responsible for its formation. In negative feedback this is not the case. Note further that negative feedback can be regarded as a special case of allosteric control.

In bacteria, feedback control is common in pathways related to amino acid and pyrimidine synthesis (see Chapter 11, p. 539). In mammalian tissues, feedback control is less clearly demonstrable. A typical scheme involves pyruvate kinase, which converts phosphoenolpyruvate (PEP) to pyruvate. Major isozymes of this enzyme are found in muscle (M) and liver (L). The L isozyme is subject to negative feedback, as shown below.

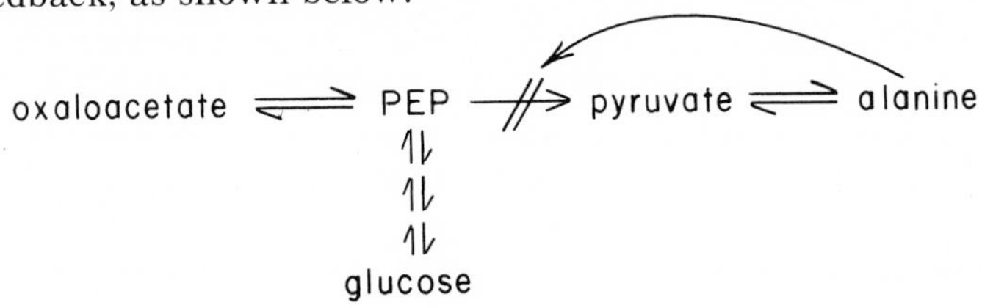

$$\text{oxaloacetate} \rightleftharpoons \text{PEP} \xrightarrow{\;/\!\!/\;} \text{pyruvate} \rightleftharpoons \text{alanine}$$
$$\text{PEP} \updownarrow\updownarrow\updownarrow \text{glucose}$$

Alanine is a feedback inhibitor of pyruvate kinase and diverts the flow of pyruvate from alanine to glucose. Although this pathway is a short one, it fits the requirements set forth above for feedback control.

Constitutive and inducible enzymes

Cellular enzymes can be divided into two primary classes in yet another way. The first class is known as constitutive enzymes, a term signifying that the enzymes in this class are present at virtually a constant concentration during the life of the cell. Presumably this is caused by a more or less constant relationship between the processes of enzyme synthesis and those of enzyme degradation. The second class is termed inducible enzymes, meaning that the amount of enzyme present in a cell at a given moment is variable, and that as the need for the particular enzyme increases, the rate of synthesis is increased or induced with respect to the rate of enzyme degradation. Thus some of the enzymes responsible for glucose metabolism may be induced by increasing the load of glucose that an animal is required to metabolize. Similarly, some of the enzymes involved in amino acid metabolism are inducible either by loading doses of the amino acid itself or by certain hormones. These subjects will be discussed in later chapters. For the moment it is sufficient to say that induction involves de novo synthesis of enzyme protein, since systems treated with specific reagents known to block the pathways for protein biosynthesis do not respond to inductive stimuli.

The functional activity of enzymes depends to a great extent on the physiologic status of the tissue in which they are found. As previously mentioned, starvation leads to decreased activity (presumably by decreased production) of digestive and certain other enzymes that are involved in nutrient absorption, since continued production of these would represent a waste of energy that a starving animal could ill afford. Similarly, an immobilized limb loses considerable mass; through the atrophy of disuse, many cellular components, including enzymes, are resorbed and are either excreted or converted to other substances. For example, removing a cast from a healed leg fracture reveals a weakened and wasted limb. Through graded exercise the function of the limb is restored, and with it cellular mass is regained. During the wasting process it is possible to measure increased serum concentrations of creatine phosphokinase, aldolase, and other enzymes not regarded as inductive in the normal sense of the term.

Regulation by covalent modification

It has become increasingly clear that an important means of enzyme regulation involves covalent modification not related to proteolysis. Most frequently the modification involves phosphorylation and dephosphorylation. These transformations are brought about by enzymes known as protein kinases and protein phosphatases, respectively. Protein kinases and phosphatases have other enzymes as their substrates. The interconversion of an enzyme from the phosphorylated to the dephosphorylated forms usually is associated with a marked change in the activity of the substrate enzyme, sometimes in one direction and sometimes in the other. Two amino acids, serine and threonine, are most frequently the sites to which phosphate groups are attached. The energy source to drive the phosphorylation is usually ATP, which also furnishes the phosphate moiety. Thus in glucose metabolism phosphorylation reciprocally activates glycogen breakdown and inhibits glycogen synthesis, so that the control mechanism of covalent modification acts to regulate energy flux through two otherwise competing systems. A growing list of enzymes in the areas of carbohydrate, lipid, amino acid, and protein metabolism emphasizes the importance of regulation by covalent modification.

Phosphorylation-dephosphorylation reactions are the most common means of covalent regulation, but they are not the only ones. In *Escherichia coli*, glutamine synthetase is regulated by adenylylation-deadenylylation, where a 5'-adenylyl group derived from ATP is attached to the OH group of a specific tyrosyl residue in each of the twelve peptides composing this enzyme. There is no known analogue of this system in mammalian tissues. Histones, basic proteins associated with chromosomal material in the nucleus, also can undergo phosphorylation-dephosphorylation reactions.

There are some unique advantages to enzyme regulation by covalent modification. It provides a means for modulating the extent or direction of energy flux by changing the proportion of an enzyme in the physiologically active form without the waste of removing or replacing the total peptide structure. Such a requirement would be needlessly costly in terms of amino acid requirements and the energy demanded for peptide assembly.

Summary

The preceding illustrations indicate that control or regulation of enzyme activity takes many forms. At the simplest level it is product inhibition; the effect on an enzyme is exerted largely by its own product and is essentially independent of other enzymes. Allosteric control is characteristic of more complex enzymes, and the regulation may be exerted by a substance or substances quite removed from the reaction catalyzed. The mechanism of control is largely a matter of shifts in equilibrium of at least two quaternary structures of a polymeric enzyme. The next level of control is feedback inhibition, in which an enzyme is affected by the product of another that is several steps removed in a multienzyme pathway. Feedback control may be used not only to control a single pathway but also to divert intermediates from one pathway to another. A very similar situation exists in control by covalent modification.

INTEGRATION OF ENZYMES INTO METABOLIC PATHWAYS

To this point we have considered only individual enzymes and the single reaction that each one catalyzes. In most cases, however, enzymatic reactions are arranged in sequences; through a series of transformations, a building block is converted into a complex product, or a complex material is degraded into building blocks or excretion products. Integrated series of enzymatic reactions are known as metabolic pathways. Two examples will be used to illustrate how enzymatic reactions are integrated into a pathway. One, the biosynthesis of the heme portion of hemoglobin, is a sequence of enzymatically catalyzed reactions that occurs in a single cell, the reticulocyte. The second metabolic pathway, hemoglobin degradation following death of the erythrocyte, takes place in several different organs. In this pathway, metabolic intermediates are passed from one organ to the next through the blood plasma and the bile.

Heme biosynthesis

All of the enzymatic reactions that are required for the complete synthesis of hemoglobin are contained in the reticulocyte, the immature form of the red blood cell. The synthesis of hemoglobin and other heme proteins requires the production of a complex precursor structure known as the porphyrin ring. An outline of the reactions by which this ring is made in mammalian tissues is shown in Fig. 3-19. The starting materials that condense to form δ-aminolevulinate (ALA) are succinyl CoA (SCoA) and glycine. The reaction is catalyzed by an ALA synthetase. Next, two moles of ALA condense to form one mole of a substituted pyrrole known as porphobilinogen (PBG). Again, the reaction is catalyzed by a specific synthetase.

Four moles of PBG are then tightly bound to an enzyme known as uroporphyrinogen synthetase that, by itself, is inactive. This enzyme requires a second protein, called a cosynthetase, in order to catalyze the synthesis of uroporphyrinogen III, the first of a series of modified porphyrin rings that ultimately result in the heme moiety of hemoglobin. Note the curious fact that here a coenzyme is a protein. Few such coenzymes are known to exist, and perhaps the cosynthetase is really a regulatory subunit that must interact with the catalytic subunit.

The remaining conversions shown in Fig. 3-19 involve decarboxylations and oxidations that are not completely understood but that result in protoporphyrin IX, which can react spontaneously but at a rather low rate with iron. The process is catalyzed by a specific iron-inserting enzyme known as ferrochelatase, and the reaction produces protoheme IX. Protoheme can react with oxygen to form protohemin IX, which inhibits the condensation of succinyl CoA and glycine to form ALA. In other words, protohemin IX is a feedback inhibitor of the entire reaction sequence. Normally, protohemin combines with globin to form hemoglobin, and so its concentration usually remains below that required for feedback inhibition. However, when the quantity of hemoglobin produced within the bone marrow is adequate for tissue needs, the process can be shut down at its initiation.

A word or two here will perhaps clarify some important but complex porphyrin nomenclature. The prefixes *uro-* and *copro-* were originally employed to indicate the source of material from urine and feces, respectively. It is now known that compounds of both types can be found in both sources, so from that point of reference they have lost much of their original significance. The terms are still useful, however, in understanding the chemistry of the compounds. Note that compounds of the uro- series are *octa*carboxylic, while compounds of the copro- series are *tetra*carboxylic. The Roman numerals, for example, III and IX, signify the possible order of substituents around the periphery of the prophyrin rings and were assigned arbitrarily on the basis of synthetic work done many years ago on model compounds. The only clinical significance that now attaches to the Roman numerals is related to the disease known as porphyria, which is a manifestation

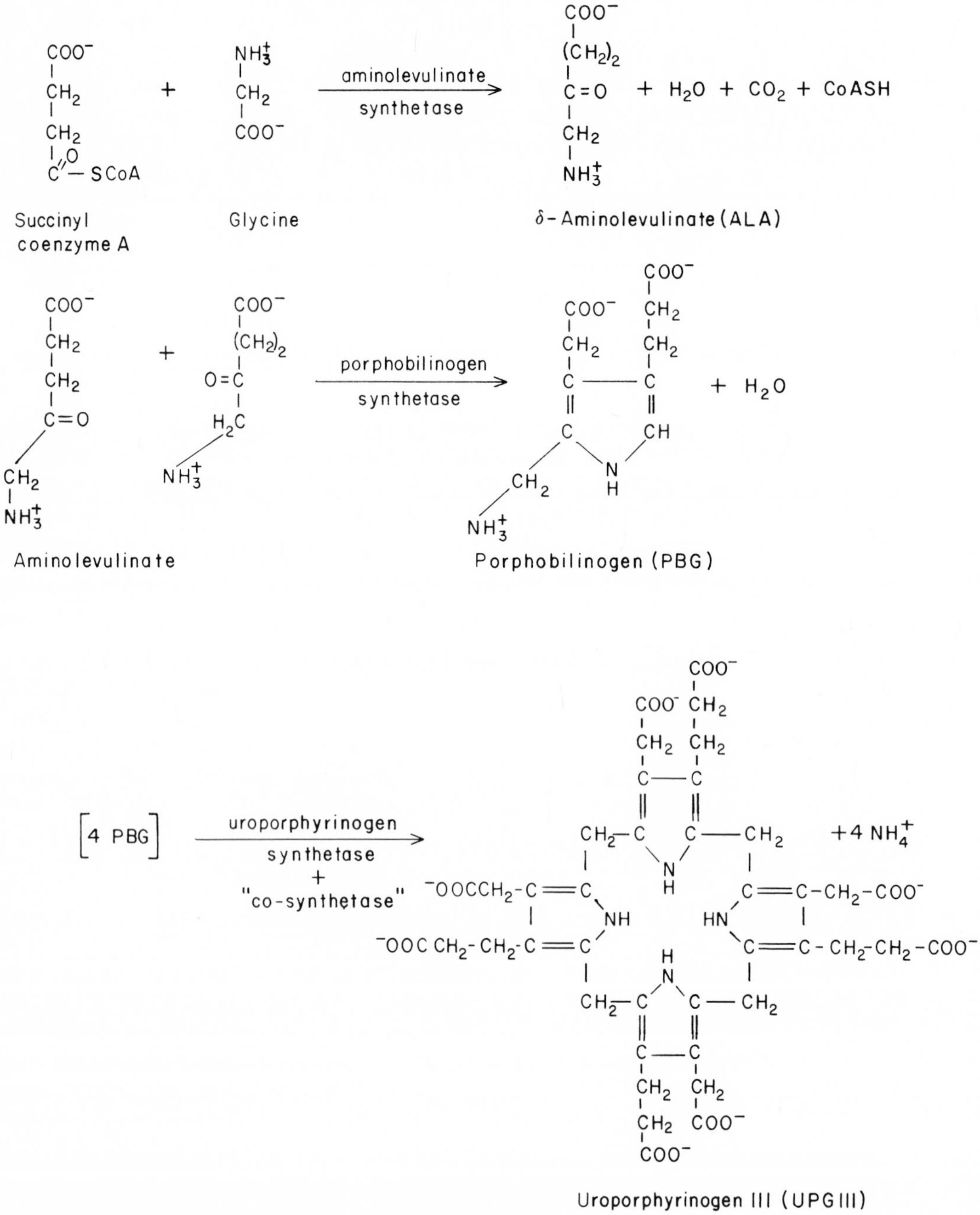

$$
\begin{array}{c}
\text{COO}^- \\
| \\
\text{CH}_2 \\
| \\
\text{CH}_2 \\
| \\
\text{C}^{\!\!\nearrow\!\!\,O} - \text{SCoA}
\end{array}
\quad + \quad
\begin{array}{c}
\text{NH}_3^+ \\
| \\
\text{CH}_2 \\
| \\
\text{COO}^-
\end{array}
\quad \xrightarrow[\text{synthetase}]{\text{aminolevulinate}} \quad
\begin{array}{c}
\text{COO}^- \\
| \\
(\text{CH}_2)_2 \\
| \\
\text{C} = \text{O} \\
| \\
\text{CH}_2 \\
| \\
\text{NH}_3^+
\end{array}
\; + \; \text{H}_2\text{O} + \text{CO}_2 + \text{CoASH}
$$

Succinyl   Glycine   δ-Aminolevulinate (ALA)
coenzyme A

$$
\begin{array}{c}
\text{COO}^- \\
| \\
\text{CH}_2 \\
| \\
\text{CH}_2 \\
| \\
\text{C} = \text{O} \\
\diagdown \\
\text{CH}_2 \\
| \\
\text{NH}_3^+
\end{array}
\quad + \quad
\begin{array}{c}
\text{COO}^- \\
| \\
(\text{CH}_2)_2 \\
| \\
\text{O} = \text{C} \\
| \\
\text{H}_2\text{C} \\
\diagup \\
\text{NH}_3^+
\end{array}
\quad \xrightarrow[\text{synthetase}]{\text{porphobilinogen}} \quad
\begin{array}{c}
\text{Porphobilinogen (PBG)}
\end{array}
\; + \; \text{H}_2\text{O}
$$

Aminolevulinate   Porphobilinogen (PBG)

$$
\left[\, 4 \text{ PBG} \,\right] \quad \xrightarrow[\substack{\text{synthetase} \\ + \\ \text{"co-synthetase"}}]{\text{uroporphyrinogen}} \quad \text{Uroporphyrinogen III (UPG III)} \; + 4\,\text{NH}_4^+
$$

Uroporphyrinogen III (UPG III)

Fig. 3-19. Diagrammatic outline of heme biosynthesis, shown here and on facing page.

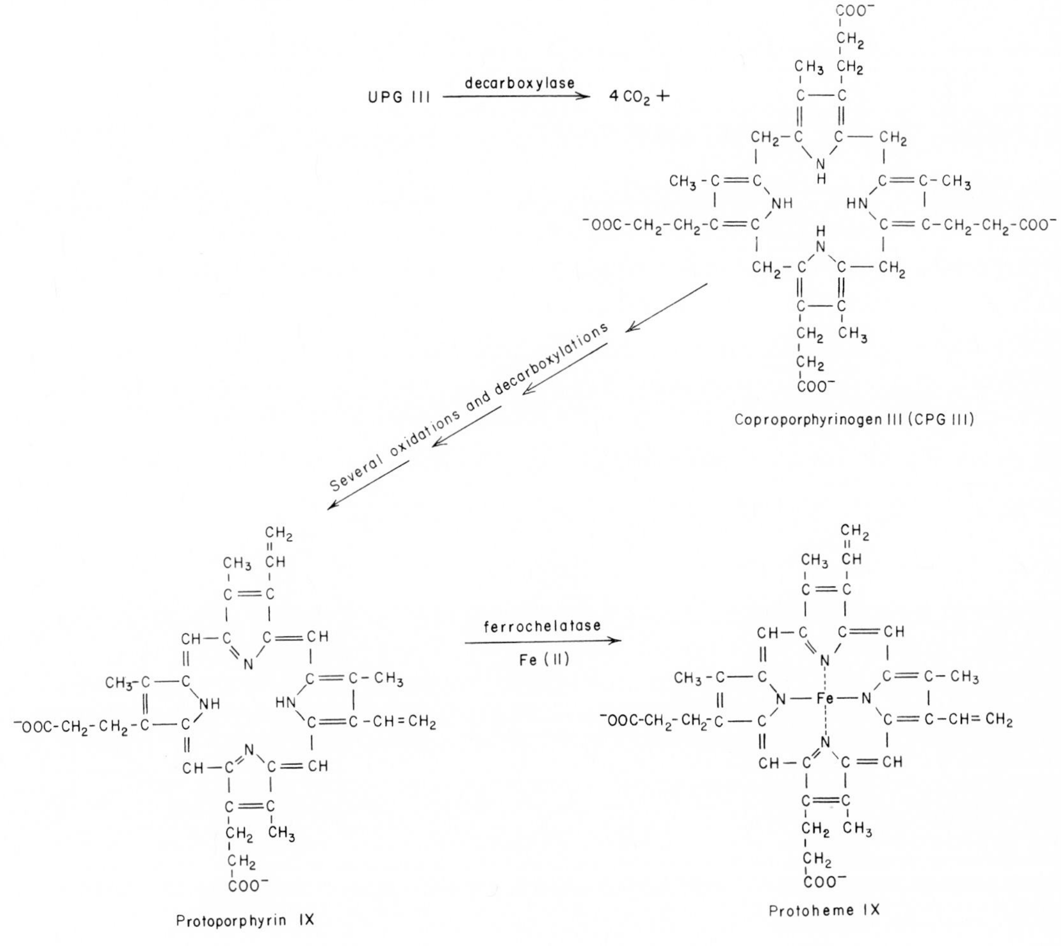
UPG III
decarboxylase
$4 CO_2$ +
Coproporphyrinogen III (CPG III)
Several oxidations and decarboxylations
ferrochelatase
Fe (II)
Protoporphyrin IX
Protoheme IX

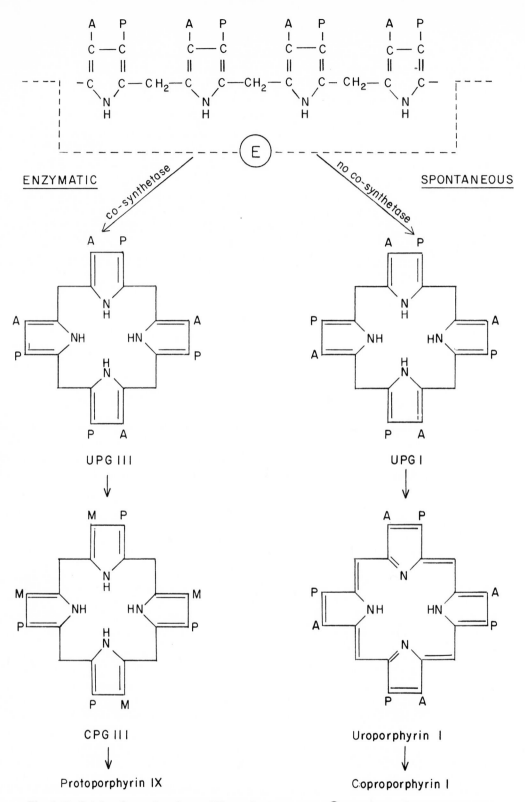

Fig. 3-20. Origin of type I and type III porphyrin isomers. Ⓔ represents the uroporphyrinogen synthetase. A and P represent the $^{-}OOC \cdot CH_2$— and $^{-}OOC \cdot CH_2 \cdot CH_2$— groups, respectively; M represents the CH_3— group. The normal path is on the left, the abnormal path is on the right. (See also *Enzymologia* 16:151-159, 1973.)

of the abnormal biosynthetic pathway that is shown schematically in Fig. 3-20. The cyclic structures are presented in a simplified shorthand notation.

Recall that four moles of PBG condense to form a tetrapyrrole derivative that is tightly bound to the inactive synthetase; this is shown in Fig. 3-20 by the dashed line. If there is adequate cosynthetase in the system, the tetrapyrrole rapidly condenses to form UPG III, as mentioned earlier. But examination of the A and P groups on the tetrapyrrole and comparison with the order found in UPG III clearly indicates that during the cyclization reaction the *ordering* of the A and P substituents must have changed in at least one location. It is this change of order that produces UPG *III*, CPG *III*, and ultimately protoporphyrin IX.

If the cosynthetase is nonfunctional, there is a spontaneous cyclization of the tetrapyrrole *without* a change in the order of A and P, which results in UPG *I* and its related compounds (Fig. 3-20). A genetically determined lack of adequate cosynthetase gives rise to *congenital porphyria*, a serious disease that is familial in display. Because the regulatory effects of protoporphyrin IX are reduced in degree in porphyria, large amounts of porphyrins are produced. Other types of porphyria are also known; these appear to reflect some insensitivity of ALA synthetase to regulation by protoporphyrin IX. These forms of porphyria are aggravated by certain drugs, especially barbiturates and other sedatives.

Fig. 3-21 shows the structure of a sedative drug (IPU) together with the structure of ALA. It is now known that, even in normal individuals, IPU in sufficient dosage will induce porphyria of a transient type. Study of the problem indicates that the drug induces increased concentrations of ALA synthetase in the liver. For this reason the drug no longer enjoys the popularity it once had. Similar results can easily be obtained in experimental animals. It can also be shown that injection of hemin can prevent the induction of ALA synthetase, as can the injection of ALA itself. The effect of hemin is probably related to the fact that it contains protoporphyrin IX. The drug-induced increase in ALA synthetase presumably produces so much tetrapyrrole precursor that some of it escapes from the enzymatic process directed toward heme biosynthesis and results in increased production of coproporphyrin instead.

Bilirubin glucuronide and bile pigment metabolism

Red blood cells have an average life span in the circulation of 120 days. Damaged and devitalized cells are removed from the circulation by various elements

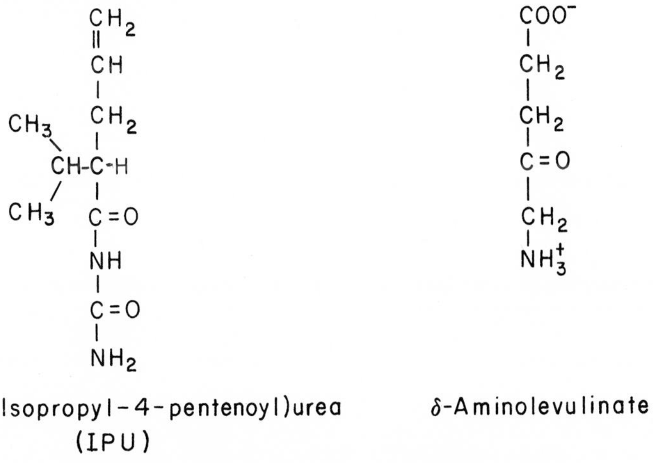

(2-Isopropyl-4-pentenoyl)urea   δ-Aminolevulinate
(IPU)

Fig. 3-21. Structural similarities of ALA and IPU.

of the reticuloendothelial system, especially by the spleen. Disposal of the erythrocyte debris involves many enzyme reactions. The initial reactions take place in the reticuloendothelial cells; metabolites produced there are transferred to the liver, where further enzymatic transformations occur. After exiting through the bile, the final metabolic transformations are accomplished by enzymes that are contained in intestinal bacteria.

The disposal of hemoglobin is important for several reasons. First, the body's stores of iron are limited. Second, the breakdown of the heme moiety involves the formation of bile pigments that are important in metabolism and that may also be

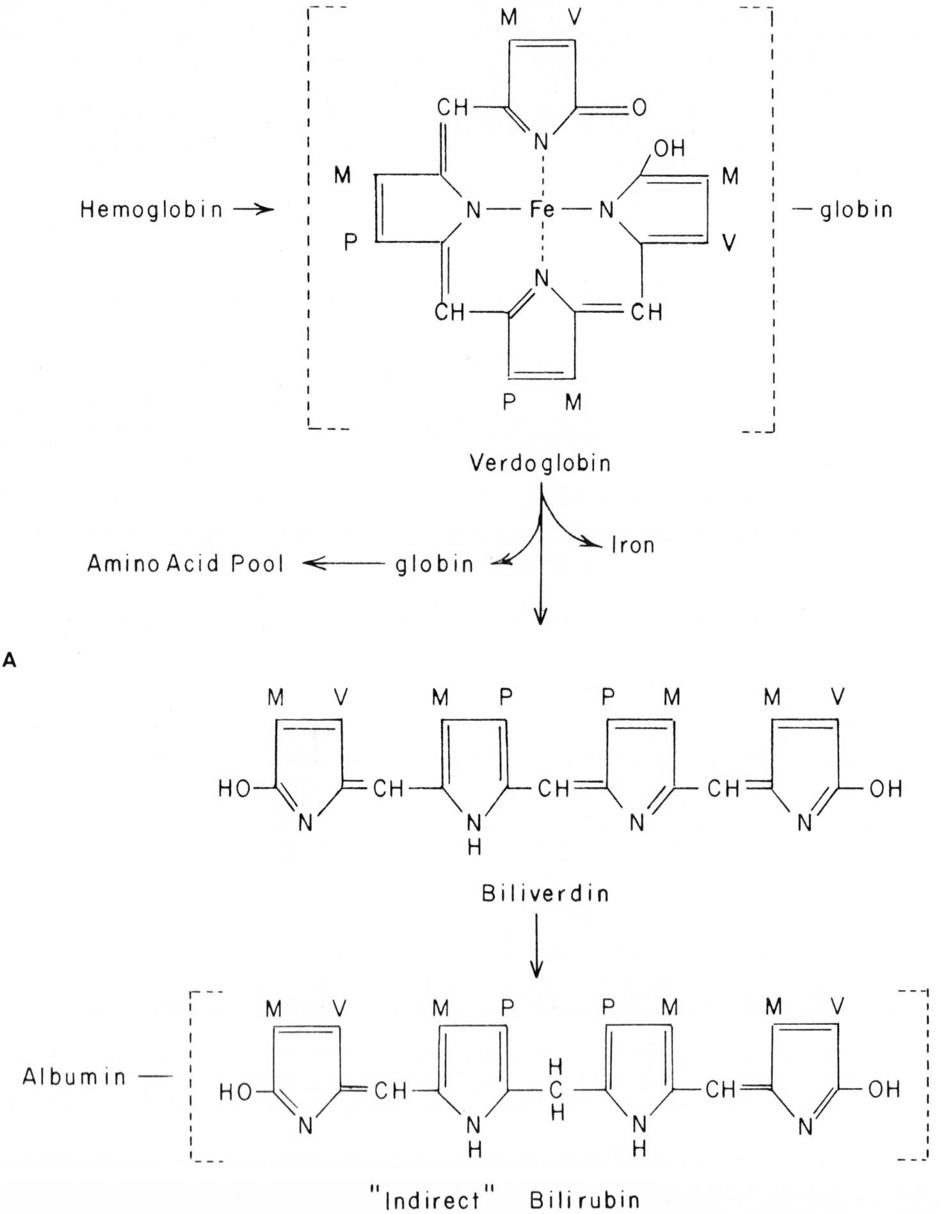

Fig. 3-22. Formation of bilirubin from hemoglobin. The differences in structure of "indirect" and "direct" bilirubin are shown. V represents the vinyl group, $CH_2=CH-$, M, and P represents CH_3- and $^-OOC \cdot CH_2 \cdot CH_2-$, respectively.

employed to determine the working state of the liver. Fig. 3-22 is a schematic outline of the pathway by which hemoglobin is catabolized. One of the methene bridges of the porphyrin ring in the heme is broken by an oxidation reaction, even though the central iron atom remains fixed in the ring and the ring is still complexed with the globin, as shown by the dashed lines surrounding the heme ring. The oxidized product is known as *verdoglobin*, and it has a dark, greenish color typical of a severe bruise or a "black eye." The ring system is destabilized by rupture of the methene bond, and the iron atom is removed and recycled by an iron transport protein that carries it to the erythropoietic centers of the bone marrow. At the same time the globin is removed, hydrolyzed by intracellular proteases, and the amino acids are returned to the amino acid pool. The residue is known as *biliverdin*, which as the name suggests is also dark green in color. Reduction of the central methene bond of biliverdin produces *bilirubin*, which has an orange-yellow color in solution. Bilirubin is removed from the reticuloendothelial cells to the circulation for transport to the liver, but because it is rather insoluble, proper transport can occur only after it has formed a complex with serum albumin, which is also shown by dashed lines in Fig. 3-22, A.

Bilirubin still contains two carboxyl groups that are not esterified. These are derived from the $-CH_2 \cdot CH_2 \cdot COO^-$ groups, indicated in Fig. 3-22, A, by P. Bilirubin in this form is known as "free" or "indirect" bilirubin, whether or not it has complexed with albumin. Protein-bound free bilirubin is sufficiently soluble to be transported through the blood, but it is not sufficiently soluble to be easily excreted from the body. On arrival in the liver, bilirubin is removed from the albumin complex and further modified by formation of the diglucuronide, shown in Fig. 3-22, B. This form of bilirubin, in which the carboxylate groups of the bilirubin have been esterified, is known as "direct" bilirubin.

The terms "direct" and "indirect" refer to the ease with which bilirubin can form a colored product when treated with diazotized sulfanilic acid, a procedure commonly employed to determine the quantity of bilirubin in serum. Direct bilirubin reacts rapidly because of its increased solubility, while the indirect form is less soluble and reacts much more slowly. Consequently, the efficiency of the enzymatic process by which the diglucuronide is formed can be of great value in the assessment of liver function or disease.

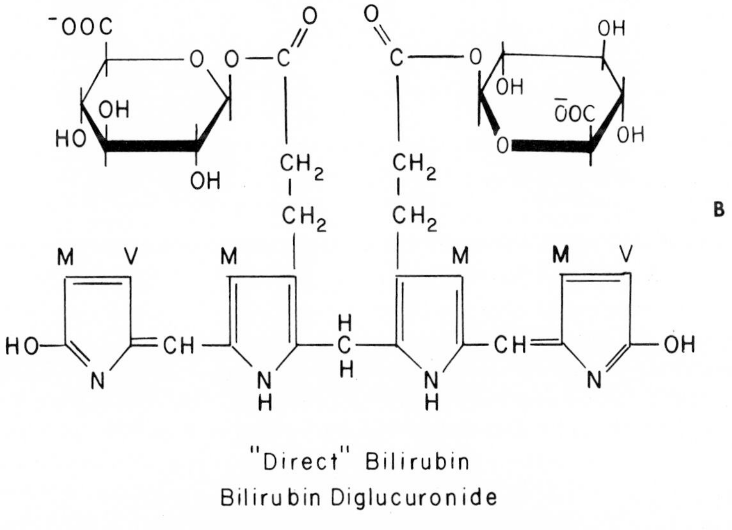

"Direct" Bilirubin

Bilirubin Diglucuronide

Fig. 3-22. For legend see opposite page.

Hemoglobin breakdown occurs constantly, and each day about 6 to 7 g of hemoglobin is converted to 250 mg of bilirubin, which circulates in blood serum at a normal concentration of about 1 mg/dl. The rate of glucuronide formation is such that the circulating ratio of direct to indirect bilirubin is approximately one fourth to one fifth. If the liver is healthy, most of the daily bilirubin production is collected in the bile (hence the eponym bile pigment) and is delivered to the intestinal tract. Action of intestinal microorganisms converts most of the bilirubin to a related compound, urobilinogen, some of which is reabsorbed along with other fatty products of digestion. Urobilinogen serves no direct metabolic purpose and is excreted in the urine, to which it imparts part of the typical amber color. The remainder of the urobilinogen, which was not reabsorbed into the intestine, passes to the lower portions of the bowel, where continued action of microorganisms converts it to stercobilin, a deeply pigmented material that gives much of their characteristic color to feces. When delivery of bile to the intestine is blocked, urobilinogen cannot be formed and stercobilin is not produced. In this condition the feces have a chalky gray color since they lack their natural pigment.

Bilirubin metabolism and jaundice

When the blood contains excessive amounts of bilirubin, which is a deeply colored substance, the sclerae (whites of the eyes) and the skin appear yellowish in color; this is the condition known as jaundice. Jaundice is clearly not a disease in and of itself, but it is an important symptom of underlying disease. Measurements of total serum bilirubin and of its major components are therefore valuable diagnostic and therapeutic aids. The means by which these measurements are made meaningful can be summarized as follows for prehepatic, hepatic, and posthepatic jaundice.

Prehepatic jaundice. Consider the case of a disease or intoxication that results in the breakdown of a greater than normal quantity of erythrocytes. The hemolyzed cells release their hemoglobin. The heme portion will be rapidly converted to indirect bilirubin and transported to the liver. Even if the liver is perfectly healthy, the increased flux of pigment cannot be metabolized rapidly enough to keep the plasma concentration within normal limits. The resultant increase in the total plasma bilirubin concentration is mostly indirect-reacting bilirubin. This condition is frequently seen in infants born with an Rh incompatibility. The total bilirubin may be ten to twenty times normal, and most of it will be indirect. These infants are frequently premature and, in addition to their hemolytic problem, they frequently lack the enzymes required to form the diglucuronide.

Hepatic jaundice. In diffuse damage or necrosis of the liver, as in hepatitis or cirrhosis, the hepatic cells lose some of their ability to extract bilirubin from the circulation and may also lose the ability to form the glucuronide. For these reasons the total bilirubin level is frequently elevated, and the indirect fraction is increased. Because the damaged cells allow some of the glucuronide to leak back into the circulation, the direct fraction may also be increased. Reabsorbed urobilinogen is filtered out of the blood plasma by the kidney circulation, and thus the quantity of this pigment in the urine also increases.

Posthepatic jaundice. Posthepatic jaundice is a term applied to disease that shuts off the delivery of bilirubin to the intestinal tract. The primary consequence is a greatly reduced formation of urobilinogen, so that less is found in the urine. The accompanying decrease in stercobilin formation causes the feces to appear chalky gray in color. Continued formation of bilirubin results in increased concentrations of total bilirubin in the serum, primarily resulting from increased direct bilirubin early in the disease process.

CLINICAL APPLICATIONS OF ENZYMES

The catalytic activity of enzymes can be measured by specific and sensitive methods, and these measurements of enzyme activity can be used as monitors of overt disease, of genetic tendencies toward a disease state, and of a patient's response to a particular type of therapy. In some instances it is not necessary to measure the enzyme activity itself; determination of whether a product of an enzyme reaction is present in normal amounts is clinically useful, as is the detection of abnormal isozymes.

Use of enzymes as reagents

A growing number of purified enzymes are becoming commercially available. These can be employed as reagents for the accurate determination of quite small amounts of such blood constituents as glucose, urea, uric acid, and triglycerides. Frequently these methods are more specific and faster than the chemical determinations available previously. Since many enzymes derived from plant or microbiologic sources can be used to assay human blood or tissue components, and since the enzymatic methods frequently lend themselves to automation, they have been a real boon to the health sciences and to the practice of medicine.

Table 3-2. Enzyme assays known or presumed useful in diagnosis or treatment of disease

ENZYME	ORGAN OR DISEASE OF INTEREST
Commonly assayed	
Acid phosphatase	Prostatic carcinoma
Alkaline phosphatase	Liver, bone disease
Amylase	Pancreatic disease
Glutamate transaminase	Liver, heart disease
Aspartate transaminase	Liver, heart disease
Alanine transaminase	Liver, heart disease
Lactate dehydrogenase	Liver, heart, red blood cells
Creatine phosphokinase	Heart, muscle, brain
Less commonly assayed	
Isocitrate dehydrogenase	Liver
Ceruloplasmin	Wilson's disease (liver)
Aldolase	Muscle, heart
Trypsin	Pancreas, intestine
Glucose 6-phosphate dehydrogenase	Red blood cells (genetic defect)
γ-Glutamyl transpeptidase	Liver disease
Guanase	Liver disease
Sorbitol dehydrogenase	Liver disease
Ornithine transcarbamylase	Liver disease
Pseudocholinesterase	Liver (poisonings, insecticides)
Leucine aminopeptidase	Liver, pancreas
5'-Nucleotidase	Liver disease
Pepsin	Stomach
Hexose isomerase	Liver disease
Hexose 1-phosphate-uridyl transferase	Galactosemia (genetic defect)
Malate dehydrogenase	Liver disease
Glutathione reductase	Anemia, cyanosis
Arginase	Liver disease
Lipoprotein lipase	Hyperlipoproteinemia
Elastase	Collagen diseases
Plasmin	Blood clotting disease

Measurement of enzyme activity in blood or tissue samples from patients can also be helpful in diagnosis and therapy, as has previously been described. Commercially available enzymes serve a purpose here also, since they are useful standards for evaluating the quality of laboratory performance.

For these reasons physicians and clinical chemists alike have been quick to take advantage of enzyme assays. Table 3-2 lists some of the enzymes that have already been employed on a clinical basis, together with the types of disease states in which the assays are used. It is apparent that there is considerable overlap in the use of enzyme assays for specific diseases. For the most part common clinical practice includes a much smaller list of routine tests; the remainder are employed only in special circumstances. In addition to measurement of the total enzyme activity, use is also made of the isozymal pattern of certain enzymes by electrophoresis and some of the other techniques discussed previously. The presence and quantitation of isozymes is of growing importance in differential diagnosis.

What is a valid enzyme assay?

This question can be answered by using the example of lactate dehydrogenase, and at the same time the mechanism of oxidation-reduction of the nicotinamide adenine dinucleotide (NAD^+) coenzymes can be explained. Lactate dehydrogenase (LDH) reversibly catalyzes the conversion of pyruvate to L-lactate, according to the following equation:

$$\begin{array}{c}
CH_3 \\
| \\
C=O \\
| \\
COO^-
\end{array} + NADH + H^+ \underset{pH\ 8.8-9.8}{\overset{pH\ 7.4-7.8}{\rightleftharpoons}} NAD^+ + \begin{array}{c}
COO^- \\
| \\
HO-C-H \\
| \\
CH_3
\end{array}$$

$$\text{LDH}$$

Note that although the reaction is reversible, the optimum pH value is different in each direction. Furthermore, the optimum pH varies somewhat with the temperature and with the substrate and buffer concentrations. LDH is rather sensitive to temperature changes, and if the temperature varies by as little as 3°C from the conventionally accepted assay temperature of 32°C, the observed value will vary from the correct amount by as much as 20%. Thus the assay must be performed under carefully controlled conditions.

The equilibrium strongly favors the reduction of pyruvate to lactate, although it is possible to drive it in the reverse direction. The majority of procedures are based on the reaction as written above, even though the required substrate, pyruvate, is less stable than lactate. As mentioned earlier, LDH protein does not undergo oxidation or reduction; that is the province of the coenzyme. The nicotinamide ring of NAD^+ is the point at which reduction takes place, as shown below.

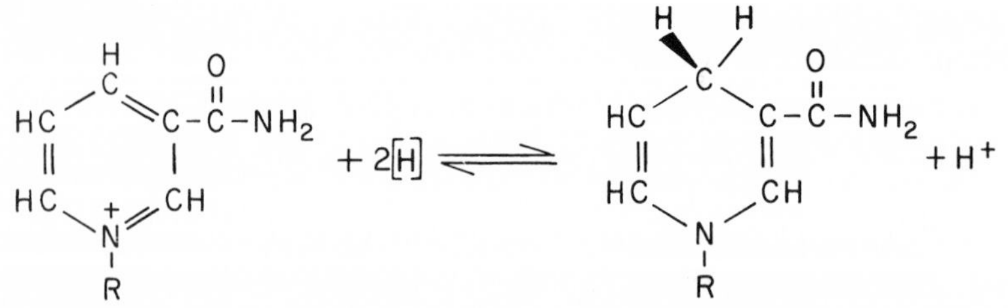

NAD$^+$ NADH
(oxidized coenzyme)   (reduced coenzyme)

Fig. 3-23. Block diagram of recording spectrophotometer.

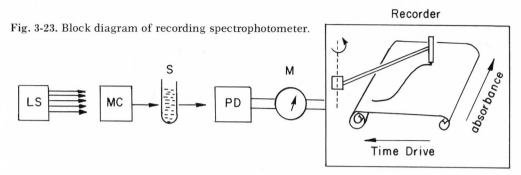

Note that the reduced form of the coenzyme accepts only one atom of hydrogen from the substrate and that the second atom remains free in solution as a proton. (The complete structure of the coenzyme is given in Table 3-2.)

There are several ways of measuring lactate oxidation and pyruvate reduction, but the most widely employed method depends on changes in the coenzyme. It is very practical because there is a change in the optical properties of the coenzyme as it goes from one oxidation state to the other. NADH readily absorbs light of wavelength 340 nanometers (nm), while NAD$^+$ has a much lower absorbance at this same wavelength.

Quantitative measurements are made with a spectrophotometer, a block diagram of which is shown in Fig. 3-23. The basic elements consist of a light source (LS) that emits "white" light, or light containing many wavelengths. This light passes through a monochromator (MC) that selects from the source a very narrow band of wavelengths, usually 2 to 20 nm in width, which it transmits through a sample cuvet (S). The amount of radiant energy transmitted through S depends on the concentration of absorbing material and on the length of the path through S. The radiant energy that passes through S then falls on a photodetector (PD), in which the signal is electrically amplified to affect a meter (M), drive a strip chart recorder, or both. In the strip chart recorder the paper chart moves at a fixed rate while the recording pen traces out an instantaneous record of the amplifier output.

The Beer-Lambert law describes the effect of an absorbing substance in a light path as follows:

$$A = \log I_o / I = \epsilon c L$$

where A is the absorbance, or extinction; I_o is the intensity of light incident on the solution; I is the intensity of light transmitted by the solution; c is the concentration of the absorbing substance; L is the length of the light path, usually expressed in centimeters; and ϵ is a characteristic of the absorbing substance, termed the extinction or absorption coefficient. When c is expressed in moles per liter and L has a value of 1 cm, ϵ is known as the molar absorption or extinction coefficient. The molar absorption coefficient of NADH is 6.22×10^3 liters/cm · mole. Consequently, with modern equipment it is possible to measure concentrations in the micromolar range.

Spectrophotometric assay makes it possible to show that for LDH the K_m for pyruvate is 9×10^{-5} M, while for lactate the K_m is 5×10^{-6} M; thus it is essential in any assay that the initial respective concentrations of the substrates be fixed well above these values. In the method of Wroblewski and LaDue there is at least a 100-fold excess of pyruvate in the final assay mixture.

A solution of NADH is added to a suitably buffered and diluted aliquot of serum. There is usually a short period in which the absorbance decreases as a result of the consumption of NADH by small amounts of pyruvate in the serum and by other NADH-linked enzymes as well, but the absorbance observed soon becomes constant. When this constant state is reached, the final reaction is initiated by the addition of sodium pyruvate solution and the time is noted. The decrease in absorbance is then monitored for the next few minutes; if the rate of change is essentially constant, the data can be used to calculate enzyme activity, since each mole of NADH oxidized is exactly equivalent to a mole of pyruvate reduced.

Two interesting variations on the LDH assay have been developed to assist in the differential diagnosis of heart and liver disease. It was mentioned earlier that the heart isozymes of LDH are more stable to mild heat denaturation than the corresponding liver isozymes. Samples of the serum can be assayed at 30° C as received, giving a value T. An aliquot of the sample can be heated to 57° C for 30 min and then assayed at 30° C to give a value L, while a third aliquot is assayed after being heated to 65° C for 30 min, giving a value H. The difference, $T - L$, represents the heat-labile fraction, which in normal serum amounts to 10% to 25% of the total and increasing to 33% to 80% in liver disease. The heat-stable fraction, designated by H, normally accounts for from 20% to 40% of the total. In patients with myocardial infarcts, H is 45% to 65% of the total value. The second modification of the LDH assay is based on the reactivity of the enzyme toward α-ketobutyrate instead of α-ketopropionate (pyruvate). Again, there is a difference in the rate at which the various isozymes can attack α-ketobutyrate. By measuring the activity toward both substrates, one can calculate a ratio, LDH/HBDH, in which HBDH refers to dehydrogenase activity using hydroxybutyrate as the substrate. In normal serum this ratio varies from 1.2 to 1.6; in liver disease it is raised and ranges from 1.6 to 2.5; while in myocardial infarcts it is decreased and ranges from 0.8 to 1.2.

The preceding discussion of an enzyme assay is presented as a typical case and is designed to show how many different properties of an enzyme can be employed to make maximum clinical use of laboratory determinations. Obviously each enzyme has peculiar properties of its own, and the nature of the reaction may permit different analytic procedures to be effectively employed. Correlating chemical findings with clinical or pathologic findings is an ongoing process that occupies a significant portion of the current literature.

Problems of enzyme assays

Enzyme assays have been an invaluable addition to clinical chemistry, but their use has been unnecessarily complicated by confusion resulting from arbitrary definitions of enzyme units. Each investigator who has developed an enzyme assay customarily has defined his own units in terms to suit his needs. This was permissible or necessary, since in most cases enzyme preparations pure enough for use as standards were not available at the time the tests were initially developed. This is still true of many enzymes. In fact, many enzymes now routinely assayed in clinical laboratories will work on different substrates, but the K_m for each of these may be different. The exact pH, time of incubation, and other variables contribute to the disparate conditions under which assays by different methods are performed. Some idea of the confusion engendered by proprietary units can be gained from a consideration of the data in Table 3-3, which deals solely with phosphatase assays. All of the assays mentioned measure the same enzyme activity, but it is clear that the substrates differ, as do the actual materials measured. The pH, the nature of the buffer system, and the time during which the reaction proceeds are not comparable. Similar confusion exists with regard to other enzyme assay methods. The hazards of this approach are all too obvious. In 1966 the International Union of Pure and Applied Chemistry and the Interna-

Table 3-3. Commonly used phosphatase procedures

AUTHORS	SUBSTRATE	AUTHORS'* UNITS	EQUIVALENT INTERNATIONAL UNITS† (IU)
Bodansky	β-Glycero- phosphate	$\frac{1.5\text{-}4.0}{100 \text{ ml}}$	$\frac{8\text{-}22 \text{ mIU}}{\text{ml}}$
Shinowara et al.	β-Glycero- phosphate	$\frac{2.2\text{-}6.5}{100 \text{ ml}}$	$\frac{15\text{-}35 \text{ mIU}}{\text{ml}}$
King and Armstrong	Phenyl phosphate	$\frac{3.7\text{-}13.0}{100 \text{ ml}}$	$\frac{25\text{-}92 \text{ mIU}}{\text{ml}}$
Bessey et al.	p-Nitrophenyl phosphate	$\frac{0.8\text{-}2.9}{1000 \text{ ml}}$	$\frac{13\text{-}38 \text{ mIU}}{\text{ml}}$
Babson	Phenolphthalein phosphate	$\frac{9\text{-}35 \text{ mU}}{\text{ml}}$	$\frac{9\text{-}35 \text{ mIU}}{\text{ml}}$

*The differences in proprietary units are caused in part by the different substrates, differences in time of incubation, and the pH and composition of the buffers.
†The equivalent international units are not directly calculable from the proprietary units. To be correct over the useful range, each set of equivalent data *must* be determined from the raw data of the analysis.

tional Federation of Clinical Chemistry joined to establish a program of international unit definitions that would put all enzyme assays on a comparable and rational basis. The fundamental unit proposed is the number of micromoles of substrate transformed per milligram of enzyme per min. For use in blood serum or other liquid samples an alternate proposal would express units in terms of micromoles of substrate transformed per milliliter per min (or per hr in some cases). Especially as more and more laboratories turn to computer control and machine printout of test results, it becomes imperative to accept some system that is uniform and compatible with machine systems of various types. The most recent proposal is the International System (SI) of units, proposed by the International Bureau of Weights and Measures. For the standardization of enzyme assays, the SI system proposed a new unit, the *katal*, defined as the moles of substrate transformed per sec. Like most similar efforts, this also seems to have a vast inertia built into it. Until all clinical chemistry laboratories join in an effort to standardize their reports, it is imperative that the health scientist become familiar with the local ground rules of archaic units.

REFERENCES

Bergmeyer, H. U., editor: Methods of enzymatic analysis, ed. 2, New York, 1975, Academic Press, Inc.

Holzer, H., and Duntze, W.: Metabolic regulation by chemical modification of enzymes, Ann. Rev. Biochem. 40:345, 1971.

Hyatt, M. A.: Electron microscopy of enzymes, New York, 1973, D. Van Nostrand Co. *A series of papers describing how electron microscopy is used to localize enzymes in organelles and membranes.*

Lehninger, A. L.: Biochemistry, ed. 2, New York, 1975, Worth Publishers, Inc. *Now regarded as the standard biochemistry textbook for general reading.*

Montgomery, R., and Swenson, C. A.: Quantitative problems in biochemical sciences, ed. 2, San Francisco, 1976, W. H. Freeman and Co. Publishers. *A useful selection of solved and unsolved problems.*

Segal, H. L.: Enzymatic interconversion of active and inactive forms of enzymes, Science 180:25, 1973.

Young, D. S.: Normal laboratory values (case records of the Massachusetts General Hospital) in SI units, N. Engl. J. Med. 292:795, 1975. *A discussion of what may become a future scheme for reporting clinical laboratory data in a uniform fashion.*

Zeffren, E., and Hall, P. L.: The study of enzyme mechanisms, New York, 1973, John Wiley & Sons, Inc. *A good, brief introduction to the details of enzyme mechanisms as examined by the kinetic approach.*

Clinical examples

CASE 1: LEAD POISONING

G. G., a $1\frac{1}{2}$-year-old child of migrant farm workers, was hospitalized because of weight loss, vomiting, and acute abdominal pain. It was noted that the child had mild muscular incoordination, weakness of the pedal dorsiflexor muscles, moderate hypertension, and questionable edema of the optic nerve ends.

Microscopic examination of a blood smear showed a moderate but distinct increase in the reticulocyte count. The red blood cell count was 4×10^6 cells/ mm^3, and the hematocrit was 37%. A 24 hr urine sample, collected in an acid-washed bottle, contained 840 μg of aminolevulinic acid and 1.2 mg of coproporphyrin III. Penicillamine therapy was started immediately and the laboratory was requested to perform a quantitative lead analysis on the urine. The report of that test indicated the sample contained 0.24 mg of lead in the 24 hr sample. X-ray examination of the patient's long bones showed electron dense deposits at the epiphyses.

Biochemical questions

1. How does lead exert a toxic effect on certain metabolic pathways?
2. Is the enzymatic inhibition produced by lead competitive or noncompetitive?
3. What is the mechanism of the increased urinary excretion of aminolevulinic acid and coproporphyrin III?
4. Why is lead deposited in the bones?
5. Besides the deposition in the bones, can you find evidence for any specific locus for the deposition of lead in human tissues?
6. What is the rationale for treatment of lead poisoning by penicillamine?

Case discussion

There is always a small amount of lead present in the body; normally the tissues contain lead in amounts of 4 to 10 ppm. Individuals with occupational exposures to lead may exhibit distinctly higher levels of lead content than are normal. Lead is absorbed in the gastro-intestinal tract, but it may also be absorbed through the lungs, as in chronic exposure to high concentrations of automobile exhaust fumes. Ingested lead is taken up most avidly by the sternum and long bones, with lesser amounts being deposited in the brain, kidneys, liver, and lungs. The remainder of the absorbed lead is excreted in the urine, which normally contains 4 to 15 μg/24 hr. Accidental lead poisoning results from ingestion of excessive amounts of lead, usually over a considerable period of time. In this case it is likely that the child was eating or chewing on something in his home that contained lead, such as flakes of paint from woodwork, furniture, toys, or radiators.

Noncompetitive inhibition by lead. Lead forms covalent bonds with sulfhydryl groups. Many proteins in the body, including a number of enzymes, contain cysteine residues that have free sulfhydryl groups. The proteins are almost always inactivated when lead combines with their sulfhydryl groups. The covalent complex between lead and the protein sulfhydryl groups is very tight, and the changes produced in the proteins are virtually irreversible. If the protein happens to be an enzyme, enzymatic activity is lost. A kinetic analysis of the particular enzyme would reveal noncompetitive inhibition. This is the type of inhibition that cannot be overcome simply by raising the substrate concentration. Those enzyme molecules that have formed a complex with lead no longer possess enzymatic activity. The remaining enzyme molecules, which are free of lead, function normally. The decrease in *functional* enzyme concentration results in a lowered V_{max}, but the K_m of the reaction would be unchanged.

Enzymes, being catalysts, are usually present in small amounts. Thus the presence of even a small excess of lead can exert pronounced metabolic effects. All that is required for toxic effects to be manifest is the inactivation of some fraction of one or more key enzymes.

Lead and heme biosynthesis. A metabolic pathway that is extremely sensitive to the presence of excess lead is heme biosynthesis; in particular, lead affects the enzyme ferrochelatase. Ferrochelatase catalyzes the insertion of ferrous iron into protoporphyrin IX, forming heme. In lead poisoning, aminolevulinate is excreted in increased amounts because this intermediate cannot be converted into heme rapidly enough when ferrochelatase activity is inhibited. Likewise, the protoporphyrin IX that is formed cannot be converted to heme because of ferrochelatase inhibition. The porphyrin builds up in the tissues and a portion of the excess is excreted as coproporphyrin III. The mild anemia noted in this patient resulted in part from inhibition of heme biosynthesis by toxic amounts of lead.

Lead and other biochemical processes. There is good reason to believe that the toxic effects of lead poisoning are not limited to the synthesis of hemoglobin; evidence of this are the neuropathy and muscle weakness mentioned in the case description. These conditions probably resulted from the inhibition of essential enzymes and proteins of the nervous system and muscles. In children, especially, lead intoxication produces severe cerebral edema, sometimes requiring trephining of the skull to prevent permanent brain damage caused by increased intracranial pressure. Most probably the affected proteins are those present in small amounts; clearly, they contain sulfhydryl groups. Moreover, it is well known that the heavy metal salts of fatty acids and of amino acids are relatively insoluble. This implies that transport and conversion to other products or to energy might be adversely affected by elevated lead concentrations in the tissues. It is possible, but not proved, that the weight loss sometimes seen in lead poisoning may be caused by this fact. Similarly, the fatty deposits in the livers of individuals with lead poisoning may represent an inability to properly mobilize triglycerides from the liver.

Experimental studies in lead-intoxicated rats have demonstrated that the proximal tubular epithelial cells of the kidneys are structurally and functionally abnormal. The basal mitochondria of these cells are swollen and have poorly defined cristae; as a consequence there is decreased reabsorption of amino acids from the tubular fluids. This is a significant nutritional loss. The membranes of isolated mitochondria are unusually fragile, and the respiratory control ratio is decreased, compared to normals. The phosphorylating ability is only partially restored by treatment with EDTA.

Lead and the skeletal system. Lead (group IV in the periodic table) has many properties similar to those of calcium (group II). The body does not clearly distinguish between them; thus lead is deposited in the mineral substance of bone, especially in growing children. As is true of calcium, the greatest amount of lead is laid down at the sites of active bone metabolism. Lead is more electron dense than calcium and appears on roentgenograms (x-ray photographs) as areas of greater density. The ability to deposit lead in the skeletal system may be a protective device, since the deposited lead is not available to react with protein sulfhydryl groups. However, it has become increasingly clear that mineral in bone exists in a dynamic equilibrium with other body minerals, so that any need to withdraw calcium from bony substance will bring with it deposited lead. Therefore treatment should be designed to solubilize deposited lead in a nonionic form that can be excreted without harm to proteins.

Chelation of lead. The child was treated with penicillamine, a nonantibiotic derivative of penicillin that forms very stable complexes with lead ions as well as with those of other heavy metals. Penicillamine is one example of a class of compounds known as chelating agents. These compounds have the capacity to combine with certain ligands (in this instance lead) to produce products that are only slightly dissociated. Other agents in this group include ethylenediamine tetra-acetic acid (EDTA) and 2,3-dimercaptopropanol (BAL). The structures of their lead complexes are shown in Fig. 3-24.

Lead in a stable nonionic complex is not toxic; therefore, chelating agents serve primarily to lower the toxicity

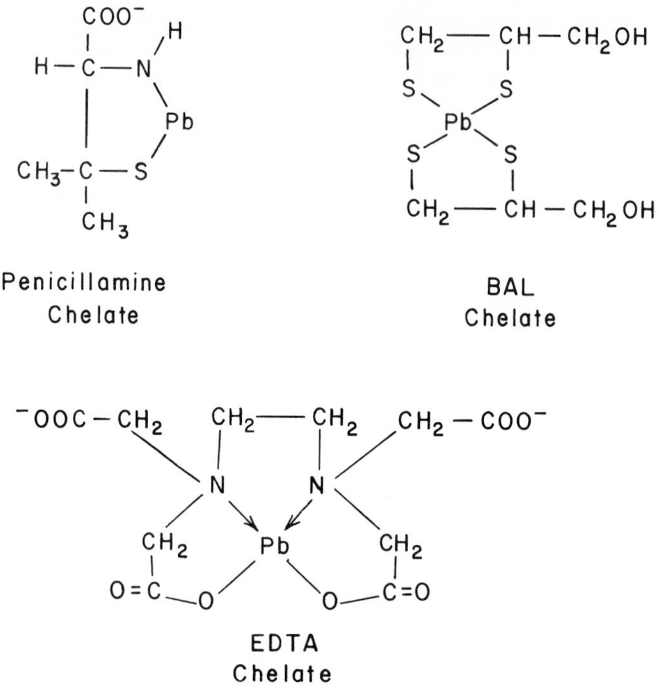

Penicillamine
Chelate

BAL
Chelate

EDTA
Chelate

Fig. 3-24. Structures of several lead chelates.

of circulating lead. Second, the chelated form of lead is more soluble than the nonchelated form; lead deposited in the bones will slowly be dissolved if it is bound as a chelate and will then be excreted by the kidneys. Unfortunately the reaction between the chelating agents and metal ions is not absolutely specific, and it is possible for various essential ions as well as those that are toxic to become involved in the reaction. Therapy with chelating agents requires considerable clinical skill and constant monitoring of the patient so as not to induce deficiencies of essential divalent cations such as Ca^{++}.

REFERENCES

Albahary, C.: Lead and hemopoiesis, Am. J. Med. 52:367, 1972.

Browder, A. A., et al.: The problem of lead poisoning, Medicine 52:121, 1973.

Chisholm, J. J.: Lead poisoning, Sci. Am. 224:15, 1971.

Lin-Fu, J. S.: Lead exposure and toxicity in children, N. Engl. J. Med. 289:1229, 1289, 1973.

Silbergeld, E. K., and Chisolm, J. J.: Lead poisoning: altered urinary catecholamine metabolites as indicators of intoxication in mice and children, Science 192:153, April 9, 1976.

Underwood, E. J.: Trace elements in human and animal nutrition, ed. 3, New York, 1971, Academic Press, Inc., pp. 437-443.

CASE 2: ACUTE PANCREATITIS

A 22-year-old woman was admitted to the hospital because of nausea, vomiting, fever, and severe abdominal pain. She stated that she had felt well until 3 days prior to hospitalization, when her symptoms came on rather acutely. The patient stated that she routinely consumed large quantities of alcohol and that she had drunk more than usual during the preceding 2 weeks. In addition, she reported that a man whom she had dated several months before had recently been hospitalized for hepatitis. A tentative diagnosis of acute infectious hepatitis was made. However, on physical examination the patient did not appear to be jaundiced, and a urine specimen was normal in color. Laboratory studies revealed the following data: plasma bilirubin concentration, 0.8 mg/dl; SGOT, 35 IU/ml; SGPT, 30 IU/ml; and lactate dehydrogenase, 110 IU/ml. On the other hand, serum amylase activity was 900 IU/dl, and serum lipase was 400 IU/dl.

Biochemical questions

1. Why do the values obtained for the serum enzyme activities *not* support the original diagnosis of acute infectious hepatitis?
2. Why does the elevation in serum amylase suggest the alternative diagnosis of acute pancreatitis?
3. What is the mechanism that explains the elevation of serum enzyme levels in illnesses such as hepatitis and pancreatitis?
4. How does the fact that the plasma bilirubin concentration is normal help in making a diagnosis in this case?

Case discussion

Infectious hepatitis, acute pancreatitis, and serum enzymes. Infectious hepatitis is an acute inflammation of the liver. The enzymes that are elevated in the serum in this disease are hepatocellular in origin. Lactate dehydrogenase and the transaminases, particularly SGPT, are present in large amounts in hepatocytes. Therefore, increases in serum lactate dehydrogenase, SGOT, and SGPT are associated with, but not diagnostic of, hepatitis. The absence of appreciable elevations in these three serum enzymes is very strong evidence against a diagnosis of acute hepatitis.

Alternatively, the finding of greatly increased serum amylase activities strongly suggests a diagnosis of acute pancreatitis. Amylase is produced in the pancreas and the salivary glands. Salivary amylase catalyzes the degradation of dietary starch to starch dextrins, maltose, and glucose. The digestion of starch is completed in the small intestine by pancreatic amylase, which converts the remaining starch and starch dextrins to maltose and glucose. Amylase is one of the main digestive enzymes biosynthesized by the pancreas, and pancreatic exocrine tissue is particularly rich in this enzyme. On the other hand, the pancreas is not particularly rich in lactate dehydrogenase or the transaminases. Because of the clinical picture of an acute abdominal illness, there is little reason to consider any abnormality of the salivary glands that might conceivably produce an elevation in serum amylase. However, the clinical picture is compatible with acute pancreatitis, especially in view of the history of excessive alcohol intake. In many cases it is difficult or impossible to make a definitive diagnosis based solely on clinical findings; additional information made available by the clinical laboratory, for example, the enzyme determinations done in this case, are invaluable diagnostic aids.

Serum enzymes and diagnosis. The serum enzyme elevations that accompany acute tissue inflammations such as pancreatitis or hepatitis are caused by a leakage of enzymes from damaged or dead cells. Consequently, the type of enzyme that is increased often pinpoints the tissue that is injured, for example, amylase and lipase in pancreatitis and creatine phosphokinase in muscle inflammation or injury. Usually the enzymes that escape into the blood are soluble rather than membrane-bound cellular proteins. The enzymes that leak into the blood from damaged tissues serve no physiologic function in the circulation. In most cases the elevated concentration of enzyme in the blood is not harmful; for example, since there is no starch in the blood, amylase has no substrate to degrade. None of the symptoms of pancreatitis are thought to result from the elevated amylase activity in the plasma. In this situation the plasma enzymatic elevation is merely a phenomenon associated with the disease process and happens to be fortuitously useful as a diagnostic aid.

Although a considerable amount of triglycerides may be present in the blood, they are not present in a form that can be hydrolyzed by the lipase that escapes from the injured pancreatic tissue. In this instance the enzyme and its substrate are present at the same time, but the physical state of the substrate makes

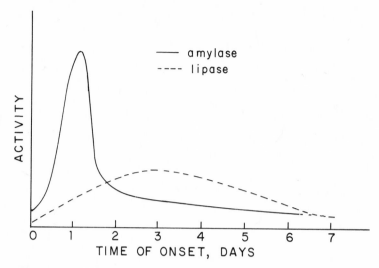

Fig. 3-25. Time course of serum amylase and lipase after onset of acute pancreatitis.

it unavailable for enzymatic hydrolysis. Again, the increased concentration of the enzyme may be quite useful as a diagnostic aid.

It should be noted that the time courses of the increased serum concentrations of enzymes in pancreatitis are not the same. The elevated amylase can usually be detected earlier, and the values return to normal more promptly than those of lipase. Typical graphs of the time courses of the two activites are shown in Fig. 3-25. Because of this difference, lipase assays may be useful in detecting those cases of acute pancreatitis that otherwise would not be evident until some time after the onset of the disease.

Bilirubin and liver disease. The finding that the plasma bilirubin was not elevated also helps to exclude a liver disease such as hepatitis. Bilirubin is conjugated in the liver to form bilirubin diglucuronide, a much more soluble form of bilirubin that can be readily excreted in the bile. Hepatitis is usually associated with increases of both direct and indirect plasma bilirubin that accompany jaundice. The inflammation, a swelling of

the liver, prevents bilirubin from gaining normal access to the hepatocytes. Less bilirubin is therefore conjugated. The level of bilirubin in the plasma rises even though the amount actually passing through the liver is not increased. The inflamed liver is simply incapable of metabolizing a normal bilirubin load. Moreover, because of the swelling and inflammation, the rate at which bilirubin diglucuronide moves from the hepatocytes into the biliary ducts is decreased. Because of stasis, increasing amounts of conjugated bilirubin escape into the hepatic sinusoids and hence into the blood; this accounts for the increase in direct reacting bilirubin in hepatitis. A normal plasma bilirubin concentration in hepatitis would be very unlikely, but such a finding would be entirely compatible with pancreatitis.

REFERENCES

Banks, P. A.: Acute pancreatitis, Gastroenterology 61:382, 1971.

Schmid, R.: Bilirubin metabolism in man, N. Engl. J. Med. 287:703, 1972.

Schmid, R.: The enzymatic formation of bilirubin, Ann. N.Y. Acad. Sci. 244:533, 1975.

CASE 3: SERUM AMYLASE IN THE DIAGNOSIS OF PANCREATITIS

An acutely ill patient was seen in the medicine clinic of a university hospital. The physical findings and much of the history were compatible with a diagnosis of acute pancreatitis, and a determination of serum amylase was requested. Shortly after the serum sample reached the laboratory, the physician was informed that the serum was turbid, with no evidence of hemolysis. The technologist agreed to dilute the sample in an effort to lower the turbidity and make the analysis more reliable.

Biochemical questions

1. Amylase acts on α-1,4 polymers of glucose. Are β-1,4 polymers also suitable as amylase substrates?
2. Are glycogen and cornstarch equally suitable as substrates?
3. On electrophoresis of an amylase specimen, several spots with approximately equal enzyme activity can be detected. What is the significance of this observation?
4. Assume you have a very lipemic serum sample. In order to assay it for amylase activity it must be diluted twenty-fold. What would you use as a diluent?
5. When dialyzed against EDTA, the activity of a certain amylase preparation fell to 20% of the control value. When dialyzed against EDTA *in the presence of starch*, the activity was 66% of the control value. Can you explain these observations?
6. Like most other proteins, amylase can be hydrolyzed to a mixture of amino acids and smaller peptides by proteolytic enzymes. Amylase is more resistant to proteolysis in the presence of Ca^{++} than in its absence. Explain this distinction.

Case discussion

Amylases are hydrolytic enzymes that hydrolyze polymers of glucose containing $\alpha(1 \rightarrow 4)$ glycosidic bonds. Starches and glycogen are typical of such polymers (see Chapter 7). Glucose polymers containing $\beta(1 \rightarrow 4)$ glycosidic bonds, such as cellulose, are not substrates because they do not meet the stereospecific requirements of amylase. The products of amylase digestion of starches and glycogen vary with the exact nature of the substrate, but formation of some reducing sugar is typical of amylase action. The form of starch known as amylopectin is not completely degraded; the relatively large residual polymer is known as a dextrin.

Human pancreatic amylase contains one or two major and one or more minor isozymes, and human salivary amylase contains one or two major and from three to six minor isozymes. The exact number of major and minor isozymes varies from individual to individual and appears to be genetically controlled. When amylase samples are subjected to electrophoresis, each of the isozymes present in the mixture can be detected as a separate spot or band by appropriate means.

Case 2 demonstrated that the measurement of serum amylase is of diagnostic value in acute pancreatitis. It has been found that some patients with pancreatitis have a high triglyceride concentration (500 to 6000 mg/dl) in their serum. The increased lipid causes a turbidity that makes it very difficult, because of light scattering, to detect the digestion of starch with precision. Dilution of the sample lowers the turbidity so that measurements can be made, but the results may be incorrect unless it is realized that amylase is activated by Cl^- ion. If the sample is diluted with solutions containing insufficient Cl^-, the results will still be incorrect. In a proper assay the final Cl^- concentration should be approximately 0.1 M.

It is also known that salivary and pancreatic amylases are metalloenzymes that contain a minimum of 1 gram atom of Ca per mole of protein. Amylases are inactivated by metal chelators such as EDTA or citrate; since these chelators are frequently employed in anticoagulant mixtures, amylase determinations should not be performed on plasma samples. The presence of Ca^{++} in amylase is not only a requirement for activity, but it also stabilizes the peptide structure of the enzyme against proteolytic digestion, probably by inducing a peptide conformation that makes certain bonds less accessible to proteolytic attack. This behavior is typical of numerous enzymes.

The stability of an enzyme is sometimes enhanced not only by cofactors, but also by the presence of substrates. Amylase provides a good example. It was noted before that chelation of the Ca^{++} by EDTA causes a loss in activity. If a fresh enzyme sample is dialyzed against EDTA in the presence of starch, the activity decrease is much less. One explanation is that the ternary complex between the enzyme, Ca^{++}, and starch has a much higher affinity for Ca^{++}

than the binary complex of Ca^{++} and the protein. Consequently, less Ca^{++} can be removed by the EDTA.

REFERENCES

Fallat, R. W., Vester, J. W., and Glueck, C. J.: Suppression of amylase activity by hypertriglyceridemia, J.A.M.A. 225:1331, 1973.

Spiekerman, A. M., Perry, P., Hightower, N. C., and Hall, F. F.: Chromogenic method for demonstrating multiple forms of α-amylase after electrophoresis, Clin. Chem. 20:324, 1974.

Street, H. V.: Inhibition of amylase activity of human body fluids by citrate, Biochem. J. 76:10, 1960.

Vallee, B. L., Stein, E. A., Summerwell, W. M., and Fischer, E. H.: Metal content of α-amylases of various origins, J. Biol. Chem. 234:2901, 1959.

CASE 4: CREATINE PHOSPHOKINASE AND MYOCARDIAL INFARCTION

An obese, middle-aged man was brought to the emergency room by police officers, who stated that the patient had been involved in an automobile accident. It appeared that he had "blacked out" and caused a collision. The patient stated that he had been short of breath and felt very dizzy just before the crash. Examination raised the suspicion of a cerebrovascular accident or a myocardial infarction. The patient was admitted for observation, and a blood sample was collected for creatine phosphokinase (CPK) assay.

Biochemical questions

1. In which tissues can CPK be found, and from which is it readily released into the bloodstream?
2. Would you expect an increased serum or plasma CPK activity following a stroke (cerebrovascular accident)?
3. Would an increased CPK activity help to distinguish between damage to the brain and damage to the heart?
4. Following electrophoresis on polyacrylamide gel, three major isozymes of CPK can be detected. The most anodic of these is known as the BB isozyme, the most cathodic is known as the MM isozyme, and the intermediate band is known as the MB isozyme (M stands for muscle and B for brain). Relate these findings to the charge on the M and B monomers.

5. It is necessary, in assays of CPK, to add mercaptoethanol (or dithiothreitol) to all of the buffers involved. What does this tell you about the structure of the enzyme?
6. Other than by electrophoresis, how might the individual CPK isozymes be separated?

Case discussion

Creatine phosphokinase is an enzyme that reversibly catalyzes the formation of ATP from ADP, at the expense of creatine phosphate (see Chapter 5, p. 233). It is probably present in all tissues but is released to the bloodstream, following injury, from only a few. The reason for this is not clear. It has been established that CPK is readily released from injured muscle and brain, but very little is released from injured liver or other tissues. Consequently, assays of CPK have been of particular use in assessment of myocardial damage in the presence of passive liver congestion resulting from chronic cardiac failure or other causes, where determinations of SGOT and SGPT might be equivocal.

As in all laboratory tests, there are pitfalls associated with the CPK assay. Any appreciable damage to major muscles will produce some increase in *total* serum CPK activity. Even repeated intramuscular injections will cause a transient serum CPK increase. Hypothyroidism is associated with an increased serum CPK; the reverse is true in hyperthyroidism. Chronic alcoholism, in the stages involving myopathy and neuropathy, may cause increased serum CPK activity.

There are some notable advantages of this assay compared to others. Following a myocardial infarct, CPK rises sooner than LDH or SGOT and decreases to normal earlier. The peak CPK value is often reached within 12 to 36 hr after injury, and the maximum value may be five to ten times normal. It is, therefore, especially useful as a sensitive test for early assessment of myocardial damage. The major problem with the assay is to relate an observed increase in serum activity with the tissue of origin.

At least three isozymes of CPK have been indentified by electrophoresis or column chromatography of serum or tissue extracts. Each of the isozymes is a dimer, which may involve an M (muscle) and B (brain) monomer. Serum from normal subjects contains largely the MM isozyme, with something less than 5% of the MB form and virtually none of the BB form. Myocardium contains a very large proportion of the MB isozyme and brain is correspondingly rich in the BB isozyme. Following surgery of muscle other than the myocardium, the increased serum activity is largely made up of the MM isozyme. After a myocardial infarct, the increase is largely in the MB isozyme and correlates well with the extent of injury.

Because of their relative tissue specificity, a simple and rapid assay for the individual isozymes of this enzyme would be very useful. Until recently, only electrophoretic or column chromatographic separations were available, and these were discouragingly slow. It has now been found that the MB isozyme, especially diagnostic of myocardial damage, can be rapidly and quantitatively absorbed from a serum sample by glass beads coated with a synthetic ion exchange resin. The desired isozyme can then be desorbed from the surface of the separated and washed beads by addition of 3.8 M sodium chloride. This novel separation, which depends on the difference in net charge of the isozymes, requires only a few minutes. It thus appears that a means has been developed to quantitate that isozyme specifically released from damaged myocardium. It is also known that all of the isozymes are absolutely dependent on the integrity of the —SH groups they contain. For this reason it is necessary to add antioxidants such as mercaptoethanol or dithiothreitol to all solutions employed in the separation procedure.

REFERENCES

Anido, V., et al.: Diagnostic efficacy of myocardial creatine phosphokinase using polyacrylamide gel electrophoresis, Am. J. Clin. Pathol. **61**:599, 1974.

Henry, P. D., et al.: Rapid separation of plasma creatine kinase isozymes by batch adsorption on glass beads, Clin. Chem. **21**:844, 1975.

Nevins, M. A., et al.: Pitfalls in interpreting serum creatine phosphokinase activity, J.A.M.A. **224**: 1382, 1973.

Roberts, R., et al.: Quantification of serum creatine phosphokinase isozyme activity, Am. J. Cardiol. **33**:650, 1974.

CASE 5: ANTIBIOTICS AS ENZYME INHIBITORS

A little girl was brought to the pediatric clinic with a badly infected wound on her knee. Her mother reported that the youngster had fallen from a skateboard several days before. The joint was red and swollen, and the child complained of pain when her leg was extended. The student in charge gave the mother a sample of penicillin G, instructed her as to dosage, and requested that the child return in several days. On the return visit the child was no better. The case was then presented to the resident physician, who gave the mother a prescription for oxacillin, and again suggested a return visit 5 days later. By the third visit the infection had apparently subsided, the swelling and redness had disappeared, and the joint was substantially free of pain on extension.

Biochemical questions

1. How is penicillin inactivated?
2. Bacteria sensitive to penicillin do not die if they are grown in an isotonic medium. What does this tell you about a possible mechanism of action of this antibiotic?
3. Significant quantities of some penicillins may be excreted in the urine of individuals given the drug. What does this signify in terms of human metabolism?
4. Penicillamine is sometimes employed in clinical medicine, even though it is not an antibiotic. Can you describe a possible use for penicillamine?
5. Some bacteria produce D–amino acids. What might be the fate of such compounds in the tissues of an individual suffering from an infection?

Case discussion

The penicillins constitute a valuable class of antibiotics, long used in the treatment of some infections. Typically, penicillins contain a thiazolidine ring fused to a β-lactam ring, with one of a series of organic acids attached to the α-amino group of the lactam ring through an amide linkage, as is shown below.

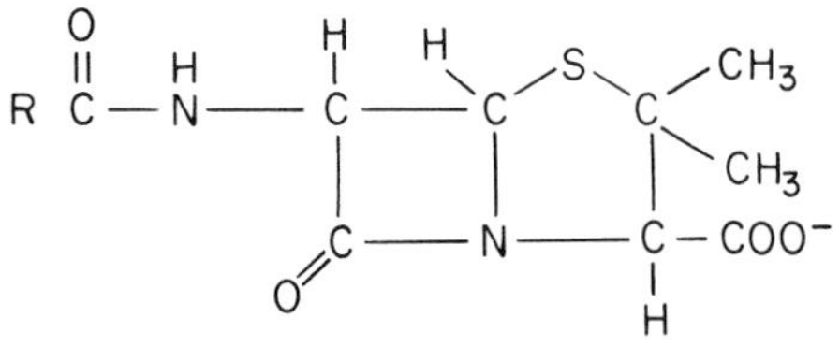

Three typical forms of the substituent group, R, are shown here and are identified with the penicillin that contains them.

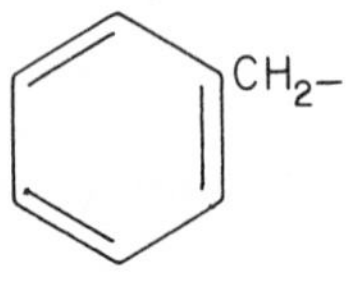

benzyl
group

Penicillin G

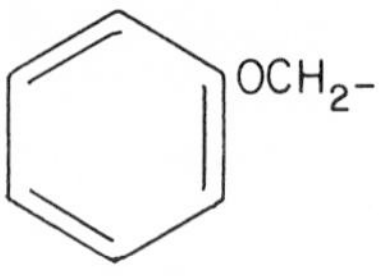

phenoxymethyl
group

Penicillin V

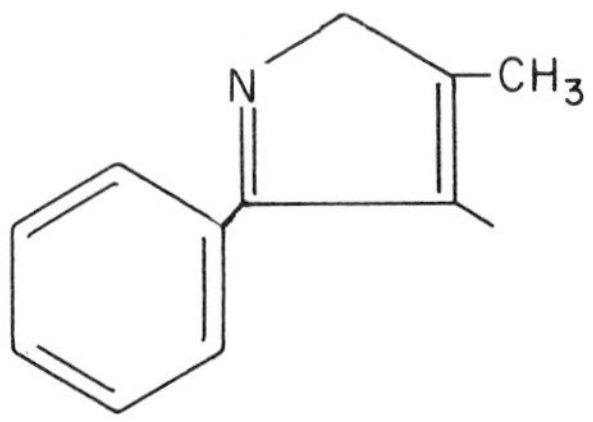

5-methyl-3-phenyl-
4-isooxazolyl group

Oxacillin

Penicillin G should not be used orally, and the first attempt at treatment was unsuccessful, because it is readily hydrolyzed by the acid gastric juice. Penicillin V and oxacillin can be given orally, since they resist exposure to the acid gastric environment. Both penicillin G and V are easily destroyed by the enzyme penicillinase, produced by some strains of microorganisms, through hydrolysis of the β-lactam ring. The semisynthetic oxacillin is more resistant to pencillinase, and so it can be used against organisms that produce the enzyme. Penicillinase production can be induced in some bacteria by treatment with concentrations of the drug insufficient to kill the invading organisms quickly or by previous exposure to penicillin in other hosts. In other words penicillinase is an inducible enzyme.

The action of penicillinase on penicillin produces penicillamine, which is without antibiotic action. Penicillamine is useful in clinical medicine because it has the ability to form chelates with metal ions, especially with heavy metal ions. It is frequently employed in the treatments of metal intoxications (see case 1, p. 144). Human tissues do not contain large amounts of penicillinase-like activity. A significant part of the administered drug passes into the urine in undegraded form because we lack the means to metabolize it.

Penicillin acts by inhibiting a transpeptidase that cross-links peptidoglycan polymers into a larger molecule, *murein,* which is an essential component of the cell wall in penicillin-sensitive bacteria. By blocking cell wall synthesis, the drug makes the infecting organisms very fragile, and they consequently burst and die. If the bacteria are grown in an isotonic medium and if care is taken to avoid mechanical shock, the cells survive surprisingly well, even though they are denuded of their walls. They will continue to grow in the presence or absence of penicillin. This is taken as evidence that the drug has little or no effect on the remainder of bacterial metabolism.

Further data are available on the mechanism of transpeptidase inhibition caused by penicillin. The cross-linking reaction mentioned above can be represented by the equation

$$NH_2 - Gly - R + R' - D - Ala - D - Ala - C \overset{O}{\underset{O^-}{<}}$$

$$\longrightarrow R' - D - Ala - Gly - R + D - Ala$$

where R and R′ represent the remainders of the two strands to be cross-linked. The R′ strand contains alanine as a C-terminal amino acid residue. Note that the amino acids indicated are in the D-configuration, as are the amino acids of many bacterial cell walls. The conformation of penicillin is believed to so closely resemble that of the polymer

$$R - D - Ala - D - Ala - C \overset{O}{\underset{O^-}{<}}$$ that it forms an

acyl-enzyme complex. However, since it contains an amide rather than a peptide bond, the complex cannot be deacylated by the transpeptidase, and murein biosynthesis is interrupted.

When invading microorganisms die within tissues of a human being, the D–amino acids they contain may be set free. Since D–amino acids cannot be employed for protein biosynthesis and related purposes, other means of disposal have been provided. The D-amino acid oxidase described earlier can be used to scavenge the energy resident in the carbon skeletons of D–amino acids.

REFERENCES

Lee, B.: Conformation of penicillin as a transition-state analog of the substrate of peptidoglycan transpeptidase, J. Mol. Biol. 61:464, 1971.

Strominger, J.: Actions of penicillin and other antibiotics on bacterial cell wall synthesis, Johns Hopkins Med. J. 133:63, 1973.

Strominger, J.: Penicillin-sensitive enzymes and penicillin-binding components in bacterial cell walls, Ann. N.Y. Acad. Sci. 235:210, 1974.

CASE 6: SERUM HEPATITIS

A 36-year-old man with chronic glomerulonephritis was first admitted to the hospital in September. He became progressively worse while being treated conservatively and developed severe azotemia. Peritoneal dialysis was instituted in early October, and hemodialysis was started shortly thereafter. The patient received two units of whole blood in December, and he received two additional units in January. In the middle of February he began to complain of epigastric fullness. His liver was enlarged and tender to palpation. Liver function tests were abnormal; plasma SGOT was 1200 IU/ml; SGPT was 375 IU/ml; and total plasma bilirubin was 4.5 mg/dl. The patient had received no hepatotoxic drugs and had not been exposed to hepatotoxic chemicals. A diagnosis of serum hepatitis was made.

Biochemical questions

1. What precautions should be taken in collecting blood for assays of enzymatic activities with regard to the temperature at which the sample is prepared and the anticoagulants or preservatives that might be added to the blood collecting tubes?

2. Why is it important to perform plasma enzymatic assays such as SGOT and SGPT at a standard temperature and pH?

3. Why does hepatocellular injury produce an increase in the plasma transaminase activities?
4. Is this patient's plasma bilirubin concentration consistent with hepatocellular damage?
5. How does "direct" bilirubin differ from "indirect" bilirubin?
6. What other enzyme assays, not reported in this case, might have been abnormal?
7. What data would you need to know in order to establish the best conditions for an assay to measure SGOT activity?
8. What vitamin is involved in the coenzyme of SGOT?
9. Urinary SGOT is not usually measured, but urinary amylase is frequently measured. How do you explain this distinction?

REFERENCES

Cossart, Y. E.: Serum hepatitis as a persistent infection, J. Clin. Pathol. (Suppl.) **6**:22, 1972.
Mackenzie, D. L., et al.: Advances in viral hepatitis, Va. Med. Mon. **102**:125, 1975.
Sherlock, S.: Chronic hepatitis, Gut **15**:581, 1974.

CASE 7: MUSCLE INJURY

A 29-year-old laborer developed chest pain while operating a pneumatic hammer on a construction project. The pain was of moderate intensity, but he felt well enough to continue work. As the day continued he became more apprehensive, particularly as he experienced sharp pain on respiration and a tight feeling across the anterior chest wall. He was rushed to the emergency room of a local hospital and was admitted following a brief examination. Electrocardiographic and chest x-ray examinations were negative, as were all of the laboratory tests that were performed except for an elevated level of plasma lactate dehydrogenase of 800 IU/ml. The elevation in plasma lactate dehydrogenase persisted during the next 4 days of hospitalization; no other laboratory or physical abnormalities appeared, and the chest pain gradually improved with bed rest.

Biochemical questions

1. What reaction does lactate dehydrogenase catalyze?
2. Is an elevation in plasma lactate dehydrogenase specific for damage to a given organ or tissue in the body?
3. Explain what isozymes are. How might a lactate dehydrogenase isozyme assay help to determine the etiology of this patient's illness?
4. In setting up a valid enzymatic assay for lactate dehydrogenase, the laboratory technician added lactate and NAD^+ to the plasma specimen. Why were these substances added? Should lactate and NAD^+ have been added in very small or in excessive amounts? Could NADH or FAD have been substituted for NAD^+ in this assay system?
5. Can you construct an LDH assay that depends on some measurement other than the spectral properties of NAD^+ or NADH? Describe it.
6. What other enzyme assay might be useful in diagnosing and following this case? Explain.
7. How many isozymes of LDH normally occur in serum?
8. Is there a similar proportional increase in the individual isozymes of LDH in all diseases? Explain.
9. Depending on the specific assay method employed, a suitable substrate could be pyruvate or lactate. Why? Explain.
10. Could you test for LDH equally well in plasma or in serum? Explain.

REFERENCE

Bergmeyer, H. V., editor: Methods of enzymatic analysis, ed. 2, New York, 1975, Academic Press, Inc.

CASE 8: OBSTRUCTIVE JAUNDICE

A 72-year-old man was admitted to the hospital because of weight loss and jaundice. Although he denied having abdominal pain, palpation of the abdomen indicated the presence of a firm, nodular, and irregular mass in the right upper quadrant of the abdomen. An upper gastrointestinal x-ray series confirmed the existence of an upper abdominal mass. An exploratory laparotomy was performed, and a carcinoma on the head of the pancreas obstructing the common bile duct was observed.

Biochemical questions

1. Explain the mechanism of this patient's jaundice.
2. As compared with the normal value, what result would you expect for the following laboratory studies: plasma indirect bilirubin, plasma direct bilirubin, urine bilirubin diglucuronide, urine urobilinogen, fecal stercobilin, liver glucuronyl transferase.
3. Would it surprise you if the patient reported that his urine or stools were extremely pale in color? Why?
4. Which one of the following plasma enzymes would most probably be elevated in this case: creatine phosphokinase, acid phosphatase, or alkaline phosphatase?
5. Where is urobilinogen produced?
6. How does urobilinogen enter the urine?
7. What is the source of the glucuronyl radical that is combined in direct bilirubin?
8. Is bilirubin a heme substance?
9. What other enzymatic abnormalities might be found in the serum of this patient?
10. Would you expect this patient to have a normal serum iron content? Why?

REFERENCES

Billing, B. H.: The enigma of bilirubin conjugation, Gastroenterology 60:258, 1971.
Lien, R. C., and Maddock, W. G.: Jaundice: differential clinical and laboratory diagnosis, Surg. Clin. North Am. 36:131, 1956.

CASE 9: UNUSUAL VARIANTS OF LACTATE DEHYDROGENASE

A patient entered the hospital with complaints diagnosed as *Pseudomonas* pneumonia. The patient died after 5 days in the hospital; autopsy revealed no evidence of tumors. During the hospital stay an unusual electrophoretogram of lactate dehydrogenase was obtained, so extracts of various organs were made from tissues obtained at postmortem examination. When these were examined in agarose media at pH 8.6, more than the typical five isozymic bands were observed. Samples were prepared from liver, lymph nodes, spleen, kidney, heart, skeletal muscle, and lung. Serum from a surviving daughter displayed a similar abnormal pattern.

Biochemical questions

1. How many monomeric units are involved in the formation of the typical, normal serum isozymes of lactate dehydrogenase?
2. Do all tissues contain the same isozymes?
3. Do all of the isozymes have the same physical stability?
4. How can the existence of H and M monomers be explained?
5. Is it possible to observe H and H' (unusual) monomers in the same individual? Why?
6. Would you expect the kinetics of the unusual variants to differ from the normal?
7. Would you expect the same variants to be present in all the tissues?
8. What is the source of the lactate dehydrogenase found in serum?
9. Where does lactate dehydrogenase occur in liver cells?
10. Would all of the isozymes function with the same coenzyme?

REFERENCES

Buchholz, D. H., and Donabedian, R. K.: Unusual variant of lactate dehydrogenase isozymes, Clin. Chem. 21:162, 1975.
Fujimoto, Y., et al.: Lactate dehydrogenase isozyme polymorphism in a patient with secondary carcinoma of the liver, Enzymol. Biol. Clin. 9:124, 1968.
Vesell, E. S.: Polymorphism of human lactate dehydrogenase isoenzymes, Science 148:1103, 1965.

CASE 10: ISONIAZID HEPATITIS

A 28-year-old nurse's aide had been exposed to a case of active tuberculosis. One year later she developed fever. Extensive examination did not disclose the cause. In 1973 she was seen in the surgery department for a right supraclavicular mass, which was biopsied. The mass was shown to be an abcess, from the drainage of which *Mycobacterium tuberculosis* was cultured. Therapy with isoniazid and ethambutol was started and her condition rapidly improved until 6 months later, which she ran out of drug supplies for a 3-week period. During this interval the fever, malaise, and adenopathy returned but were again rapidly controlled following resumption of drug therapy. Three months later she noted the onset of progressively severe jaundice and moderate anorexia. She also noted hypermenorrhea and easy bruising but did not consult her personal physician until frank scleral icterus was seen.

At that time the following laboratory data were obtained. The SGOT was 247 IU/dl, total bilirubin was 4.2 mg/dl, serum LDH was 150 IU/dl, serum alkaline phosphatase was 111 IU/dl, and the serum albumin was 2.9 g/dl. Her prothrombin time was 14 sec (control was 12 sec).

Treatment with isoniazid was discontinued, and her tuberculosis was treated with streptomycin plus ethambutol. Over a 6-week period her liver function abnormalities gradually subsided, except for modest elevation of alkaline phosphatase, which was still 130 IU/dl. The serum albumin had increased to 4.0 g/dl. She was discharged on the new drug program.

Biochemical questions

1. What is the structure of isoniazid? Of what vitamin can it be regarded as an analogue?
2. Do you think that isoniazid would inhibit serum LDH?
3. Could more than one enzyme be affected by ingestion of isoniazid?
4. Isoniazid is considered to be a bacteriostatic, rather than a bacteriocidal, drug. Explain this distinction in terms of enzyme properties.
5. The precise mechanism of ethambutol action is not established; it is known that its effects are additive with those of isoniazid. What possible explanations can you draw up for this synergism?

REFERENCES

Bacalao, J., et al.: Ethambutol-mediated alterations in ribonucleic acids of *Mycobacterium smegmatis*, J. Bacteriol. 112:1004, 1972.

Garibaldi, R. A., et al.: Isoniazid-associated hepatitis: report of an outbreak, Am. Rev. Respir. Dis. 106:357, 1972.

Rudoy, R., et al.: Isoniazid administration and liver injury, Am. J. Dis. Child. 125:733, 1973.

ADDITIONAL QUESTIONS AND PROBLEMS

1. What simple experiment could be designed to determine whether an enzyme exhibited classic (Michaelis-Menten) or allosteric kinetics?
2. Describe a simple analysis that would help to determine whether an enzyme inhibitor was competitive or noncompetitive.
3. How would you determine the K_m for lactate of plasma lactate dehydrogenase?
4. What are the possible ways in which pH changes could alter the rate of an enzymatic reaction?
5. Vitamin C is thought to be a coenzyme necessary for collagen synthesis. How would you set about investigating this possibility?
6. How might one employ the enzyme urease to measure blood urea nitrogen concentrations in a biologic sample? Could this be made a quantitative measurement?
7. The various isozymes of lactate dehydrogenase have distinctly different optimal temperatures. If they all catalyze the same reaction, how can the different temperature optima of the individual LDH isozymes be explained?
8. What methods could be employed to determine the intracellular location of the enzyme alcohol dehydrogenase in a suspension of freshly prepared mammalian liver cells?

CHAPTER 4

ACID-BASE, FLUID, AND ELECTROLYTE CONTROL

OBJECTIVES

1. To interpret from the body chemistry any variation of fluid pH and electrolyte composition from the normal
2. To relate information on pH and electrolytes to possible metabolic or respiratory imbalance
3. To understand the relation between body fluid solutes and osmolar regulation

Homeostasis implies a close control of the circulation and composition of fluids that contain both solids and gases. By means of such circulation, each cell in the body is bathed in a nutrient medium that is optimal for its function. The circulatory fluids also remove metabolic secretions and excretions from the cell. Only a very limited variation in circulating acid and base is consistent with life; therefore, even slight changes from the normal acid-base balance require proper clinical action to correct the cause as soon as possible. The control of the total concentration and volume of the body fluids and the control of specific ion concentrations occur in the normal individual principally through proper function of the lungs and kidneys. These organs interact with each other and all other parts of the body via the blood. In the first section of this chapter the normal function of the lungs, kidneys, and blood will be considered individually; the interaction of these three systems under conditions that vary from the normal will then be discussed.

BUFFERS

Acidic substances are constantly produced as a result of the normal metabolic processes. The acid produced in largest quantity is carbonic acid, but there is also a continuous formation of lactic acid and many ketoacids as intermediary metabolites. This continuous production of acid necessitates a means whereby the hydrogen ion can be removed without any extreme changes in pH. The lungs and kidneys, together with the circulating body fluids, constitute a system whereby this can be accomplished, holding the blood pH in the 6.8 to 7.6 range. Thus the erythrocytes produce approximately 0.3 to 0.4 moles of lactic acid per day from the anaerobic glycolysis of D-glucose, a nonionic substrate. This is released into the blood plasma where its effect on the pH is reduced by the buffering action of the blood components, such as bicarbonate and phosphate. The phosphates and bicarbonate are examples of weak acids, others being lactic acid, the amino acids, and proteins. It will be recalled from Chapter 2 that a weak acid represented by HA dissociates in an equilibrium to produce an anion A^- and a proton,

$$HA \rightleftharpoons H^+ + A^-$$

and that the neutralization of a weak acid by a strong base, such as NaOH, would be

$$HA + Na^+ + OH^- \rightleftharpoons Na^+ + A^- + H_2O$$

In such a titration (Fig. 4-1) the change in pH follows an S-shaped curve. The curve is flattest over that part of the titration in which between 9% and 91% of the weak acid is neutralized. Thus the change in pH with incremental additions of base is least over this range; that is, a solution containing appropriate proportions of weak acid and its salt is buffered.

Acid-base regulation can be considered in terms of the Henderson-Hasselbalch equation (equation 2-5, p. 60). If a weak acid is titrated with strong base, the concentration of the conjugate base A^- derived from the dissociation of HA may be considered negligible. It follows that the concentration of A^-, that is, $[A^-]$, increases in proportion to the amount of strong base added and therefore in proportion to the salt formed. This being the case, the following can be written:

$$pH = pK_a \ + \ log_{10} \frac{[salt]}{[acid]} \qquad (4\text{-}1)$$

where, at any point in the titration:

$$\text{Remaining acid} = \text{total acid added initially} - \text{salt formed} \qquad (4\text{-}2)$$

For example, consider the titration of 100 ml of 0.01 M lactic acid with 0.1 M NaOH. Since 100 ml of 0.01 M lactic acid contains a total of 0.001 equivalents of acid, the total titration will require $\frac{0.001}{0.1} \times 1000$ ml of 0.1 M NaOH, or 10 ml 0.1 M NaOH. When 5 ml 0.1 M NaOH has been added, 50% (or 50 ml) of the original 0.01 M lactic acid has been neutralized and converted to the equivalent of 50 ml 0.01 M salt. However, the volume of the total solution is changed from 100 ml to 105 ml by the alkali. Therefore, the actual concentration of acid remaining is:

$$\frac{50}{105} \times 0.01 = 0.00476 \text{ M}$$

and the salt formed also is 0.00476 M. Similarly, when 8 ml 0.1 M NaOH has been added, the concentration of the salt formed is:

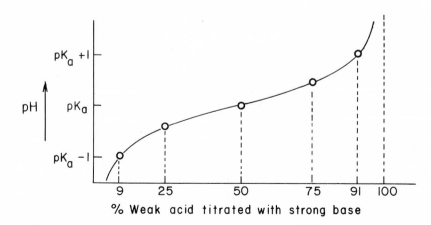

Fig. 4-1. Titration of weak acid with strong base.

$$\frac{80}{108} \times 0.01 = 0.00741 \, M$$

and the acid concentration remaining is:

$$\frac{20}{108} \times 0.01 = 0.00185 \, M$$

These values can be substituted in equation 4-1 to give the pH values at each point in the titration, given that the pK_a for lactic acid is 3.86. The relationship given in equation 4-1 between pH and the concentrations of the conjugate acid-base pair is most useful, particularly in calculations of titration curves.

Several properties of the Henderson-Hasselbalch equation should be noted:

1. When the weak acid and its conjugate base (salt) are equal in concentration, the pK_a of the weak acid gives the value of the pH of the solution. This condition occurs when half of the initial amount of a weak acid has been neutralized with a strong base, at which point the concentration of the residual acid is equal to that of the salt produced; that is, 50% of the initial acid is neutralized. Under these conditions, [acid] = [salt] in the above equation, and $\log_{10} \frac{[salt]}{[acid]}$ (that is, $\log_{10} 1$) is zero.

2. By similar reasoning, when the concentration of acid is ten times that of the salt, then:

A
$$\log_{10} \frac{1}{10} = -1.00$$

and

$$pH = pK_a - 1.00$$

In other words, the pH is one unit less than the pK_a when [acid] is ten times [salt]. By contrast, when [salt] is ten times [acid]:

B
$$\log_{10} \frac{10}{1} = 1.00$$

and

$$pH = pK_a + 1.00$$

Since, as in equation 4-2, the acid remaining equals the total acid initially minus the salt formed, and for condition A in relative terms:

$$10 \text{ units} = \text{total acid initially} - 1 \text{ unit}$$

then

$$\text{Total acid initially} = 11 \text{ units}$$

That is, the initial acid neutralized = $\frac{1}{11} \times 100\% = 9.09\%$. Similarly, for condition B the acid neutralized is $\frac{10}{11} \times 100\%$, or 90.9%. These titration points permit a rapid estimation of the titration process and may be represented graphically as in Fig. 4-1. More detail can be achieved by calculating other points in the titration from equation 4-1. For example, when 25% of the acid is titrated, the ratio $\frac{[salt]}{[acid]} = \frac{1}{3}$:

$$\log 0.33\underline{3} = \bar{1}.523$$
$$= -0.477$$

and

$$pH = pK_a - 0.48$$

Since the titration curve is symmetric around the pK_a value, at 75% titration:

$$pH = pK_a + \log_{10} \frac{3}{1}$$
$$= pK_a + 0.48$$

3. The approximations in this form (equation 4-1) of the Henderson-Hasselbalch equation do not permit calculations outside the range of $pH = pK_a \pm 1$. The contributions of A^- from the dissociation of the free acid HA cannot be ignored at the acid end of the titration curve. Also, the reaction of A^- with H_2O is significant when the salt concentration is near maximum.

4. Within the range $pH = pK_a \pm 1$ the conjugate acid-base pair has the largest buffering capacity, which is maximal at the hydrogen ion concentration value of the pK_a. The buffer capacity measures the *extent of change* of pH of a solution when acid or base is added and is represented in Fig. 4-1 by the slope of the titration curve. The smaller the slope, the smaller the pH change and the greater the buffer capacity. The buffer capacity is measured as the mole equivalents of H^+ or OH^- that are required to change 1 liter of 1 M buffer by 1.0 pH unit. The buffer capacity at a constant pH is directly proportional to the buffer concentration.

As noted earlier, a solution of any acid-base conjugate pair will act as a buffer with maximum buffering capacity at the pK_a value for the conjugate pair in question. In a polyprotonic molecule the appropriate conjugate pair must be selected; it is that which has a pK_a closest to the desired pH at which the buffer should act. If these values differ by more than one pH unit, another buffer system should be selected. The normal pH of blood and the tissue fluids is approximately 7.40. The pH of the intracellular compartments is less well established but is thought to be just below 7.0.

$$H_3PO_4 \underset{pK_{a_1} 2.1}{\overset{\overset{\displaystyle H^+}{\overset{\displaystyle +}{}}}{\rightleftharpoons}} H_2PO_4^- \underset{pK_{a_2} 6.7}{\overset{\overset{\displaystyle H^+}{\overset{\displaystyle +}{}}}{\rightleftharpoons}} HPO_4^= \underset{pK_{a_3} 12.3}{\overset{\overset{\displaystyle H^+}{\overset{\displaystyle +}{}}}{\rightleftharpoons}} PO_4^{\equiv}$$

In the phosphate buffer system the active conjugate pair in blood and tissues will be $H_2PO_4^-/HPO_4^=$, and the pH relationship can be expressed as:

$$pH = 6.7 + \log_{10} \frac{[HPO_4^=]}{[H_2PO_4^-]}$$

At pH 7.4 we have

$$7.4 - 6.7 = \log_{10} \frac{[HPO_4^=]}{[H_2PO_4^-]}$$

or the ratio $\dfrac{[HPO_4^=]}{[H_2PO_4^-]} = $ antilog$_{10}$ 0.7
$$= 5.01$$

In other words, the acid of the pair $[H_2PO_4^-]$ is approximately 20% as concentrated as the salt form, $[HPO_4^=]$. This provides a good buffer capacity, particularly in the direction of acidification, which is appropriate since the metabolism of the average diet produces about 70 meq more acid (excluding CO_2) than base per day. Proteins, phospholipids, and nucleoproteins yield sulfuric, phosphoric, and uric acids;

fruits and vegetables contain organic acids that are metabolized to CO_2, leaving the Ca^{++}, K^+, Mg^{++}, and Na^+ in excess to have an alkalinizing effect. The average mixed diet, however, produces an excess of nonvolatile acids. Catabolism therefore always tends toward acidification, and the ability to handle CO_2 and fixed (nonvolatile) acids is an important homeostatic control (see p. 166).

As an example of the application of these principles to a physiologic situation, consider a patient who excretes 1 liter of urine, pH 5.6, in 24 hr. The principal buffer system in urine is phosphate, the total concentration of which for this patient is 46.6 mM. This means that the concentration of each of the four ionic phosphate species added together amount to 46.6 mM. Each of these ionic species will be associated with its equivalent amount of counterion, which will be Na^+. The important point is that the kidney, by taking the phosphate from the blood at pH 7.4 and excreting it at pH 5.6, achieves a significant conservation of Na^+. How does this occur, and how much Na^+ is conserved for the patient in question?

Since we are dealing with the phosphate buffer system at pH 7.4 and pH 5.6, the equilibrium will involve $HPO_4^=$ and $H_2PO_4^-$, which, with Na^+ as the counterion, is as follows:

$$2Na^+ + HPO_4^= \rightleftharpoons Na^+ + H_2PO_4^-$$

(NOTE: The above equation is not stoichiometric but is balanced with respect to charge; see also case 2, p. 189.) For every millimole of $H_2PO_4^-$ excreted instead of $HPO_4^=$, 1 mmole of Na^+ is conserved, as explained below.

At pH 7.4 we have already calculated the following ratio:

A $\qquad\qquad [HPO_4^=] / [H_2PO_4^-]$ is 5.01

Since in this urine the sum of the ionic forms is 46.6 mM, the following can be written:

B $\qquad\qquad [HPO_4^=] + [H_2PO_4^-] = 46.6 mM$

But from A:

$$[HPO_4^=] = 5.01 [H_2PO_4^-]$$

which, when substituted in B, gives:

$$5.01 [H_2PO_4^-] + [H_2PO_4^-] = 46.6 mM$$

or

$$[H_2PO_4^-] = \frac{46.6}{6.01} mM$$
$$= 7.75 \ mM$$

therefore,

$$[HPO_4^=] = (46.6 - 7.75) mM$$
$$= 38.85 \ mM$$

The concentration of Na^+ needed to combine with these two phosphate ions is, therefore:

$$(7.75 + 2 \times 38.85) mM$$
$$= \qquad 85.45 mM$$

at pH 5.6, from (4.1)

$$5.6 = 6.7 + \log_{10} \frac{[HPO_4^=]}{[H_2PO_4^-]}$$

$$-1.1 = \log_{10} \frac{[HPO_4^=]}{[H_2PO_4^-]}$$

therefore,

$$\frac{[HPO_4^=]}{[H_2PO_4^-]} = 7.9 \times 10^{-2}$$

or

$$[HPO_4^=] = 7.9 \times 10^{-2} [H_2PO_4^-]$$

and substituting in (B) gives

$$7.9 \times 10^{-2}[H_2PO_4^-] + [H_2PO_4^-] = 46.6 \, mM$$

or

$$[H_2PO_4^-] = \frac{46.6}{1.079} \, mM$$

$$= 43.20 \, mM$$

and

$$[HPO_4^=] = 3.40 \, mM$$

In this relative proportion the associated $[Na^+]$ is $(43.20 + 2 \times 3.40)$ mM $= 50.00$ mM. Therefore, by excreting the urine at pH 5.6, the body has saved $(85.45 - 50.00)$ mmoles of Na^+/liter of urine or 35.45 mmoles of Na^+/liter of urine. The Na^+ was conserved in exchange for H^+.

Bicarbonate buffer

The bicarbonate and phosphate systems are the most important inorganic buffers in human physiology. An example of one role of the phosphate buffer has been given above. The bicarbonate buffer is now considered in detail because of its unique property of having a volatile acid component.

Carbonic acid, a dibasic acid with pK_a's of 3.88 and 10.22, would have little buffer capacity in humans if it were not for the fact that it dehydrates to CO_2, a gas that can be expired from the body. This reaction is relatively slow in the absence of the enzyme catalyst carbonic anhydrase.

$$CO_2 + H_2O \rightleftharpoons H_2CO_3$$

$$\frac{H_2CO_3 \rightleftharpoons H^+ + HCO_3^-}{CO_2 + H_2O \rightleftharpoons H^+ + HCO_3^-}$$

In this coupled system all the CO_2, whether it is physically dissolved as the gas CO_2 or in the hydrated form as H_2CO_3, is considered as the acid form. The apparent dissociation constant may be written:

$$K_a' = \frac{[H^+][HCO_3^-]}{[CO_2, \text{ both dissolved as } CO_2 \text{ and in the form of } H_2CO_3]}$$

The coupled system has a pK_a' of 6.1, and the equation relating pH to the concentration of bicarbonate, dissolved CO_2, and H_2CO_3 is:

$$pH = 6.1 + \log_{10} \frac{[HCO_3^-]}{[\text{dissolved } CO_2 + H_2CO_3]} \tag{4-3}$$

The denominator is in fact the *total* CO_2 in plasma in all forms *less* that which is present as bicarbonate. The following can be written:

$$pH = 6.1 + \log_{10} \frac{[HCO_3^-]}{[\text{Total } CO_2] - [HCO_3^-]} \tag{4-4}$$

The total CO_2 content of a sample of plasma is determined by measuring the volume of CO_2 liberated on acidification with strong acid. Total CO_2 may be expressed as milliliters of CO_2 corrected to standard temperature and pressure (STP) per 100 ml of plasma. In this form it is expressed as volumes percent (vol%).

For example, when a 5 ml sample of plasma was acidified, 0.11 ml of CO_2 was collected. The temperature in the clinical laboratory was 22° C, and the atmospheric pressure was 750 mm Hg. Correcting the gas volume to STP:

$$0.11 \times \frac{750}{760} \times \frac{273}{295} \text{ ml } CO_2 \text{ (STP)}$$

or

$$0.10 \text{ ml } CO_2 \text{ (STP)}$$

This volume was collected from 5 ml of plasma, so that the total CO_2 from 100 ml of plasma is:

$$0.10 \times \frac{100}{5} \text{ volumes \% } CO_2$$

$$= 2.00 \text{ volumes \% } CO_2$$

From the total CO_2 and the pH it is possible to calculate the $[HCO_3^-]$ from equation 4-4. By the mass balance of CO_2 given in equation 4-5:

$$[\text{Total } CO_2] = [HCO_3^-] + [\text{Dissolved } CO_2 + H_2CO_3] \tag{4-5}$$

A value for [dissolved $CO_2 + H_2CO_3$] is then obtained. This is now abbreviated [dissolved CO_2].

Throughout such calculations it is necessary to express all concentrations in the same unit, which is commonly millimolar (mM). Since CO_2 is a nonideal gas, 1 mmole occupies 22.26 ml at STP, so that:

$$\frac{\text{Volume of } CO_2 \text{ in ml}}{22.26 \text{ ml/mmole}} = \text{mmoles } CO_2$$

In the discussion of the O_2 binding of hemoglobin (pp. 75 to 80), it was shown that the concentration of a gas in a liquid was related to the partial pressure of the gas and the Bunsen coefficient (or solubility coefficient). Considering now the concentration of CO_2 in plasma, the solubility coefficient is 0.51 ml CO_2 (corrected to STP) in 1 ml of plasma at 38° C and 760 mm Hg of CO_2 pressure. Therefore, for each millimeter of mercury partial pressure of CO_2, there is 0.51/760 ml of physically dissolved CO_2/ml of plasma at 38° C. Now:

$$\frac{0.51}{760} \text{ ml } CO_2 = 6.71 \times 10^{-4} \text{ ml } CO_2 \text{ per ml of plasma per mm Hg}$$

To convert this volume to millimoles, it is divided by 22.26:

$$\frac{6.71 \times 10^{-4}}{22.26} \text{ mmoles of } CO_2 \text{ per ml plasma}$$

Since this is the amount of CO_2 dissolved in 1 ml of plasma/mm Hg partial pressure of CO_2, then the amount dissolved in 1000 ml plasma is:

$$\frac{6.71 \times 10^{-4}}{22.26} \times 1000 = 0.0301 \text{ mmoles per mm Hg partial pressure}$$

That is, such a solution is 0.0301 mM/1 mm Hg pressure of CO_2.

As an illustration of the use of these units, consider the equilibration of a 1 ml sample of plasma at 38° C with alveolar air, which normally has a partial pressure of 40 mm Hg of CO_2; that is, $\bar{p}CO_2$ is 40 mm Hg. (NOTE: The abbreviation of the partial pressure of a gas X as $\bar{p}X$ antedates the pX convention of pH or pK_a and bears no relation to it.)

$$\text{The volume of } CO_2 \text{ dissolved} = 40 \times 6.71 \times 10^{-4} \text{ ml}$$
$$\text{(measured at STP)/1 ml of plasma}$$
$$= 2.68 \times 10^{-2} \text{ ml/1 ml of plasma}$$
$$= 2.68 \times 10^{-2} \times 100 \text{ ml/100 ml of plasma}$$
or
$$2.68 \text{ volumes \%}$$

$$\textit{Also, the concentration of } CO_2 = 40 \times 0.0301 \text{ mM}$$
$$= 1.204 \text{ mM}$$

Equation 4-3 can be restated so that the acid term is expressed in terms of partial pressure of CO_2:

$$pH = 6.1 + \log_{10} \frac{[HCO_3^-]}{0.0301 \times \bar{p}CO_2}$$

Alternatively, from equation 4-5 the bicarbonate concentration can be restated in terms of total CO_2 and partial pressure of CO_2, as in equation 4-6.

$$pH = 6.1 + \log_{10} \frac{[\text{Total } CO_2] - 0.0301 \times \bar{p}CO_2}{0.0301 \times \bar{p}CO_2} \tag{4-6}$$

A normal, rested individual will have arterial blood plasma values of pH 7.4; total CO_2, 25 to 28 mM; $\bar{p}CO_2$, 40 to 43 mm Hg; and [HCO_3^-], 24 to 27 mM. These values are consistent with each other as calculated from the Henderson-Hasselbalch equation. For example, at pH 7.4 the ratio of the bicarbonate concentration to the acid concentration can be calculated from the pH and pK_a of this system, as in equation 4-7.

$$7.4 = 6.1 + \log_{10} \frac{[HCO_3^-]}{[0.0301\ \bar{p}CO_2]} \qquad (4\text{-}7)$$

or

$$1.3 = \log_{10} \frac{[HCO_3^-]}{0.0301\ \bar{p}CO_2}$$

or

$$\frac{[HCO_3^-]}{0.0301\ \bar{p}CO_2} = 20$$

Likewise, the average values of the salt and acid components listed above, when substituted in equation 4-7, give the same value, 20.

$$\frac{25}{0.0301 \times 41} = \frac{25}{1.23}$$

$$= 20$$

Alternatively, the same value of 20 will be calculated from the total CO_2 and $\bar{p}CO_2$ values, using the expression for the ratio of [salt]/[acid] given in equation 4-6.

$$\frac{\text{Total } CO_2 - 0.0301\ \bar{p}CO_2}{0.0301\ \bar{p}CO_2} = \frac{26 - 1.23}{1.23}$$

$$= 20$$

It is the *ratio* of [HCO$_3^-$]/[dissolved CO_2] that is 20 if the pH of the blood plasma is at the normal average value of 7.4. Any disproportionate change of either the [HCO$_3^-$] or the [dissolved CO_2] will change the above ratio from 20, and the blood will be either acidic or alkaline with reference to the normal. In such situations the compensatory mechanisms of the body will come into play in an attempt to correct the pH back to 7.4.

Venous blood plasma commonly shows about 1.1 mM more [HCO$_3^-$] and 0.14 mM more [dissolved CO_2] than the arterial blood plasma because venous blood transports CO_2 from the peripheral tissues to the lungs. Will the pH of venous blood be more acid or alkaline than arterial blood? The average values for bicarbonate and dissolved CO_2 in arterial blood are 25 and 1.23 mM, respectively. Therefore, in venous blood:

$$pH = 6.1 + \log_{10} \frac{(25 + 1.1)}{(1.23 + 0.14)}$$

$$= 6.1 + \log_{10} \frac{26.1}{1.37}$$

$$= 6.1 + \log_{10} 19.05$$

$$= 6.1 + 1.28$$

$$= 7.38$$

The ratio of 19.05 in venous blood indicates that it will be slightly more acid, a conclusion borne out by the calculated value of pH 7.38.

Control of pH in the body

The control of the pH of the body fluids centers largely around the functions of the lungs and the kidney, whereby excess H^+ is eliminated. The lungs function to reduce the $\bar{p}CO_2$ in the blood, thus increasing the $[HCO_3^-]/[H_2CO_3]$ ratio. The kidneys serve to retain as much HCO_3^- from the blood as necessary and to generate more by converting CO_2 to HCO_3^- and H^+. The H^+ is eliminated by the $HPO_4^{--}/H_2PO_4^-$ buffer system or as NH_4^+.

Since the H^+ is common to all buffers in the body fluids and is freely exchangeable with the intracellular constituents, all buffering reactions are coupled together.

$$H^+ \begin{cases} + \ NH_3 \ \longrightarrow \ NH_4^+ \\ + \ HCO_3^- \ \longrightarrow \ H_2O \ + \ CO_2 \\ + \ HPO_4^{=} \ \longrightarrow \ H_2PO_4^- \end{cases}$$

It is therefore not surprising to find that all of the above processes are interdependent.

RESPIRATORY CONTROL OF BLOOD pH

The $\bar{p}CO_2$ of blood can be changed quite rapidly by the rate and/or depth of breathing (pulmonary exchange). Slow, shallow breathing *(hypoventilation)* results in a buildup of alveolar $\bar{p}CO_2$ and a reduction in the diffusion of CO_2 from the blood into the pulmonary alveolar gas phase. This increases the $\bar{p}CO_2$ in the plasma and lowers the blood pH. *Hyperventilation* has the opposite effect. These two effects can be illustrated by the simple example of a normal, healthy person who develops hiccups. Knowing that such an attack can usually be stopped by increasing the levels of CO_2 in the blood, the subject holds his breath or breathes with a paper bag over his head (that is, inspires air with high $\bar{p}CO_2$) for a few min. The CO_2 content of the blood begins to increase. After the hiccups stop, the blood of this individual has a $\bar{p}CO_2$ of 60 mm Hg and a total CO_2 content of 30 meq/liter. At this time the blood pH can be calculated:

$$pH = 6.1 + \log_{10} \frac{30 - 0.0301 \times 60}{0.0301 \times 60}$$

$$= 6.1 + \log_{10} \frac{30 - 1.80}{1.80}$$

$$= 6.1 + \log_{10} 15.65$$

$$= 6.1 + 1.19$$

$$= \underline{7.29}$$

Wishing now to get rid of the excess CO_2, the individual breathes rapidly and deeply for a few min (hyperventilation). His blood plasma then contains 27.5 mM $[HCO_3^-]$, and the $\bar{p}CO_2$ is 30 mm Hg. By how much has the blood pH changed?

$$pH = 6.1 + \log_{10} \frac{27.5}{0.0301 \times 30}$$

$$= 6.1 + \log_{10} \frac{27.5}{0.91}$$

$$= 6.1 + 1.48$$

$$= \underline{7.58}$$

Thus the pH has moved from the acid side of normal to the alkaline side, changing 0.29 pH units by going from a state of hypoventilation to hyperventilation.

Transport of oxygen and carbon dioxide in blood

It was shown in Chapter 2 that O_2 is transported in blood as an oxyhemoglobin complex. The affinity of hemoglobin for O_2 is reduced by 2,3-diphospho-D-glycerate in the erythrocyte and by a decrease in pH, the latter effect being known as the *Bohr effect.*

Consider the circulation of blood through the lungs, where the $\bar{p}O_2$ is approximately 107 mm Hg and the $\bar{p}CO_2$ is about 36 mm Hg. It can be seen from Fig. 4-2 that in this condition the hemoglobin is nearly 100% saturated. As this blood moves to the peripheral capillary bed, the $\bar{p}O_2$ is reduced and the $\bar{p}CO_2$ is increased because the cells take up and utilize the O_2 to oxidize nutrients to CO_2, and they release the CO_2 that is formed back into the blood. This increases the blood $\bar{p}CO_2$, which in turn produces a lowered blood pH. The amount of O_2 combined with hemoglobin at a given $\bar{p}O_2$ is further reduced because of the decrease in pH, permitting more O_2 to be taken up by the tissues. Additional release of O_2 in the capillary bed is caused by the effect of 2,3-diphospho-D-glycerate in the erythrocyte (see Fig. 2-14). The overall function of the erythrocyte is therefore to bind the maximum amount of O_2 in the lungs and to release a portion of it to the tissues.

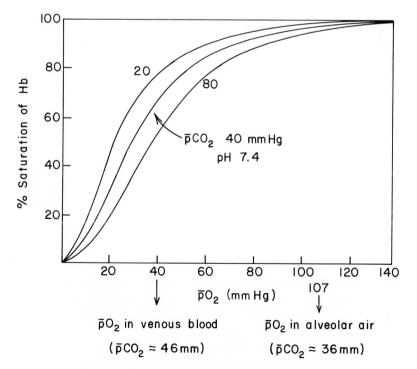

Fig. 4-2. Oxygen-hemoglobin dissociation curves.

Release of O_2 to the tissues is enhanced by the Bohr effect and the presence of 2,3-diphospho-D-glycerate.

Carbamino hemoglobin.   The CO_2 that is produced by catabolism must be transported to the lungs, where it is expired. About 10% is transported in the red blood cell as carbamino hemoglobin. In this form the CO_2 is linked covalently to the NH_2-terminal valine residues of the hemoglobin subunits.

$$CO_2 + HbNH_2 \rightleftharpoons HbNH.COO^- + H^+$$

The reaction is rapid, readily reversible, and probably not catalyzed by an enzyme.

Isohydric transport of CO_2.   The majority of the CO_2 is transported in the plasma as bicarbonate, which is produced in the erythrocyte from CO_2 (Fig. 4-3).

Several reactions occur in this transport process.

1. Since oxyhemoglobin is a stronger acid than deoxyhemoglobin (p. 83), the following must be written:

$$HbH_x^+ + O_2 \rightleftharpoons HbO_2 + xH^+$$

$$pK_a \ 7.71 \qquad\qquad pK_a \ 7.16$$

The pK_a values of these two weak acids are such that x = 0.7, approximately. This property of the hemoglobins demonstrates their buffering effect. It also explains the transport of an appreciable quantity of the CO_2 released from the tissues without change in pH – the so-called isohydric transport of CO_2. The oxyhemoglobin, arriving in the red blood cells at the capillary bed, dissociates to release O_2, a process that is facilitated by the state of the tissues relative to incoming arterial blood: low $\bar{p}O_2$, lower pH, and higher $\bar{p}CO_2$.

$$0.7H^+ + HbO_2 \rightleftharpoons HbH_{0.7}^+ + O_2$$

The H^+ in this reaction is derived from the dissociation of H_2CO_3, which is formed from the CO_2 diffusing from the tissue into the plasma and finally into the erythrocyte.

$$H_2CO_3 \rightleftharpoons H^+ + HCO_3^-$$

The uptake of H^+ by hemoglobin buffers the effect of the carbonic acid. The HCO_3^- formed in the erythrocyte diffuses into the plasma and is carried back by the venous blood to the lungs, where reduced hemoglobin is oxygenated. This results in the release of H^+, which reacts with HCO_3^- to give H_2CO_3. The concentration of HCO_3^- in the erythrocyte is therefore being reduced so that HCO_3^- from the plasma returns to the cell. This buffering effect reduces the pH change as a result of the oxygenation of Hb H^+. Also, the H_2CO_3 dehydrates (carbonic anhydrase) to form CO_2 for expiration by the lung. The reactions of hemoglobin with CO_2 and O_2 occur in support of each other and of the proper handling of the H^+ load.

2. Added to this H^+ load is that from the formation of carbamino hemoglobin.

3. The isohydric transport of CO_2 requires the counterdiffusion of Cl^- and HCO_3^- in the erythrocytes to maintain electroneutrality. Thus in the lung the movement of HCO_3^- into the erythrocyte for conversion to H_2CO_3 requires that a negatively charged ion move from the cell into the plasma to take its place. This part is played by Cl^-, so that the $[Cl^-]$ is higher and the $[HCO_3^-]$ lower in arterial than in venous blood. Conversely, in the capillary bed the CO_2 diffuses into the plasma and then into the erythrocyte; HCO_3^- is formed, as explained in the first reaction of the transport process, and moves to the plasma. Its place in the erythro-

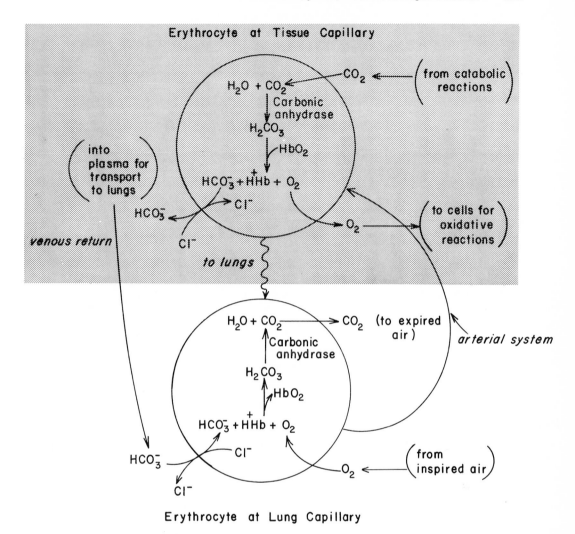

Fig. 4-3. Reactions of gases with erythrocytes.

cyte is taken by Cl^- from the plasma, so that venous blood has a lower $[Cl^-]$ and higher $[HCO_3^-]$ in the plasma than does arterial blood. This point can easily be demonstrated in vitro by blowing O_2 over a sample of venous blood.

Not all the H^+ produced by the first two reactions is accommodated by the protonation of deoxyhemoglobin. Therefore the venous blood plasma is more acidic than arterial blood, a state reflected by a higher $\bar{p}CO_2$.

RENAL REGULATION OF pH, WATER, AND ELECTROLYTES

Acid-base disturbances cannot all be corrected by respiratory function if the primary cause of the problem is a disturbance of nonvolatile anions and cations. In such cases appropriate correction is attempted through the selective excretion and reabsorptive properties of the kidney tubular cell membranes. Responses mediated through the kidney occur much more slowly than those mediated through respiration. The kidney plays an important role in acid-base balance as well as in maintaining the levels of electrolytes and water. In other words, it maintains homeostasis with respect to electrolyte composition and concentration.

Osmotic pressure: osmolarity

An explanation of the movement of water and materials of low molecular weight such as Na^+, glucose, urea, and Cl^- across membranes requires an understanding of osmosis. The osmotic pressure of a solution is proportional to the number of dissolved particles, charged or neutral. If the number of moles of these particles is N per liter, then the osmotic pressure (in atmospheres) is given by $P = NRT$, where R is the universal constant 0.082 (expressed in liter-atmospheres per mole per degree) and T is the absolute temperature. For nonionized solutes such as urea and glucose, 1 gram mole of solid gives 1 gram mole of solute particles. For sodium chloride the molecule dissociates into two charged particles when dissolved so that 1 gram molecule of NaCl behaves in osmosis as if there were two, Na^+ and Cl^-. The concentration of solute particles is therefore identified as the osmolarity of the solution. The following examples illustrate the relationship between molarity and osmolarity:

> One molar solution of glucose = 1 osmolar
> One millimolar solution of urea = 1 milliosmolar
> One milliequivalent of Na^+ per liter = 1 milliosmolar
> One millimole of NaCl per liter = 2 milliosmolar
> One millimole of Na_2SO_4 per liter = 3 milliosmolar
> One millimole of $MgSO_4$ per liter = 2 milliosmolar

In each case it is the number of particles produced in solution, not their charge, that governs the osmotic pressure and the osmolarity.

Tissue fluids are complex mixtures of materials of small and large molecular weights, some of which are partially ionized. Hence a method for determining the osmolarity of these fluids is necessary. The method used commonly is that of freezing point determination, which also reflects the number of particles in the solution. Thus the freezing point of a 1 M solution of glucose is theoretically $-1.86°C$; that of a 1 M solution of NaCl is $-3.72°C$. Urine may have a freezing point of $-0.40°C$, which corresponds to a concentration of $\frac{0.40}{1.86}$ osmolar, or 0.215 osmolar. Clinically, these values are expressed in milliosmolar units, so this value would be given as 215 milliosmolar (mOsm).

In more concentrated solutions the effective concentrations of solutes are lower than the molar concentration. The correction factor is called the *activity coefficient*. Thus a 10 mM solution of NaCl acts as if it were 18.5 mOsm (activity coefficient of 0.925). However, for this discussion it is assumed that this activity coefficient will always be 1.00. The basic concepts of the discussions that follow are not violated by this assumption.

The intracellular contents of a cell have a certain osmolarity. Solutions of equal osmolarity, or tonicity, are said to be *isotonic*. *Hypertonic* solutions will tend to draw water from a cell until the osmolarities inside and outside are equal. *Hypotonic* solutions cause water to move into the cell, resulting in an increase in cell size. The tonicity of the circulating extracellular fluids is clearly an important element of homeostasis.

Transport across membranes

The transport of a molecule across a membrane can proceed by a passive or active process.

Passive diffusion. Passive diffusion is the result of simple physical diffusion that is caused by the concentration gradient and the membrane potential; the sum of these two gradients is called the *electrochemical gradient*. Thus if the concentration of molecules or ions in aqueous solution is greater on one side of a membrane than another, water can diffuse from the side of its highest concentration to the other side. This is osmosis or osmotic diffusion. For charged molecules or ions, such as Na^+, there may be diffusion across a membrane toward the side with the most negative potential.

Active transport.   Active transport occurs in the transfer of materials against a concentration gradient. Active transport is one of the major energy-requiring cellular processes, with the energy provided by ATP. This energy is required to "pump" the molecules up the electrochemical gradient. One of the most important systems in this regard is that which maintains the ionic compositions of the intracellular fluid different from the extracellular fluid. The problem is summarized in Table 4-1.

K^+ is the principal cation in the cell; Na^+ diffuses into the cell and must be continuously removed. One of the mechanisms involves the Na^+/K^+ pump, whereby Na^+ is pumped from a low intracellular concentration (less that 10 mM Na^+) to a high extracellular fluid concentration (Na^+ around 140 mM). Coupled to this flux of Na^+ is the pumping of K^+ from a concentration of around 4 mM in the extracellular fluid to an intracellular concentration of about 140 mM. The system involves an enzyme, Na^+/K^+-ATPase, and is discussed in greater detail in Chapters 5 (p. 226) and 10 (p. 510). The Na^+/K^+-ATPase is widely distributed and in some cells, particularly brain and kidney, some 60% to 70% of the ATP produced is used to maintain the Na^+-K^+ distribution in the tissues.

The processes of diffusion and active transport may be represented in the cell as follows:

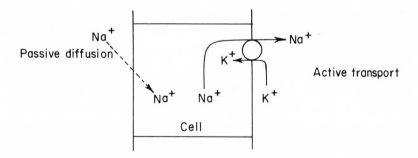

Mediated transport.   Mediated transport facilitates the movement of materials across a membrane by means of a "carrier," which reversibly binds specific materials to carry them. These carriers are sometimes called *translocases* (see

Table 4-1. Principal differences in composition of body fluid compartments and urine

	EXTRACELLULAR FLUID		INTRACELLULAR FLUID	URINE
	Plasma	Interstitial		
Water (percent lean body weight)	5	15	45 to 50	
Principal cation	Na^+	Na^+	K^+	Na^+
Principal anion	Cl^-	Cl^-	$HPO_4^=$	Cl^-
Other	Protein, glucose	Little protein, glucose	Protein, intermediary metabolites	Urea, creatinine, SO_4^-

Table 4-2. Systems for transport across membranes

TYPE	CHARACTERISTICS	EXAMPLES
Passive nonmediated	Movement down an electro-chemical gradient; osmotic diffusion	Na^+ into cells; H_2O exchange
Passive mediated (facilitated)	Movement down an electro-chemical gradient; substrate specificity; saturation kinetics; inhibited by specific compounds	Glucose in erythrocytes; ADP/ATP in mitochondria (p. 226)
Active mediated	Movement up an electrochemical gradient; requires metabolic energy; unidirectional	Na^+/K^+ pump; glucose in kidney; amino acids in intestine; H^+ in gastric secretion

Chapter 6). Mediated transport, also called *facilitated transport*, may be found in active or passive transport systems and is characterized by showing some specificity for the substances transported, by becoming saturated at high concentrations of substrate, such that the rate of flux becomes maximum at saturation, and by frequently being inhibited by some specific compound. The mediated transport system therefore behaves much like an enzyme.

Examples of all these types of transport systems will be found in the kidney and elsewhere. They are summarized in Table 4-2.

Kidney function

The kidney contains approximately 1.3×10^6 *nephrons* that operate in a parallel fashion. A nephron is represented diagramatically in Fig. 4-4. Each nephron consists of a *glomerulus,* which is supplied with blood in an arteriolar capillary system such that a high enough filtration pressure exists to effect ultrafiltration of the lower molecular weight materials in the plasma. The *glomerular filtrate* collects in *Bowman's capsule,* which leads successively to the *proximal convoluted tubule,* the *loop of Henle,* the *distal convoluted tubule,* and a branched *collecting duct* that is common to and drains a number of nephrons. The blood that leaves the glomerulus perfuses the tubules, collecting materials into this peritubular fluid that have been reabsorbed by the tubular cells from the glomerular filtrate.

Each part of the renal tubule is long with a narrow lumen, such that at no point is diffusion from the filtrate to the tubular cells rate limiting. Each day approximately 160 liters of fluid are filtered, yet the urinary volume is normally only between 500 to 2500 ml/day, depending on diet and activity. Thus approximately 99% of the filtered water is reabsorbed. Urine has a specific gravity of 1.003 to 1.030, depending upon the homeostatic control of total body fluid and solids.

Reabsorption of ions and water

Proximal tubule. The glomerular filtrate is about 300 mOsm, similar to the blood plasma from which it originates, and contains principally Na^+, K^+, Cl^-, HCO_3^-, $HPO_4^=$, $SO_4^=$, urea, glucose, and creatinine. As the filtrate flows along the *proximal convoluted tubules,* Na^+ diffuses down the electrochemical gradient at the luminal surface of the tubule cell and is activity transported from the cell into the peritubular fluid by a Na^+/K^+ pump in the basement membrane. The Cl^- follows the Na^+ flux passively. Water moves from the filtrate by passive osmosis

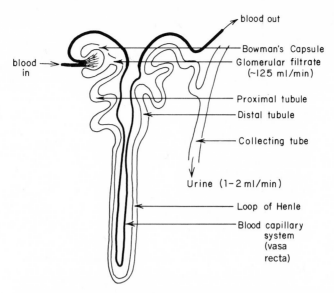

Fig. 4-4. Schematic representation of a nephron.

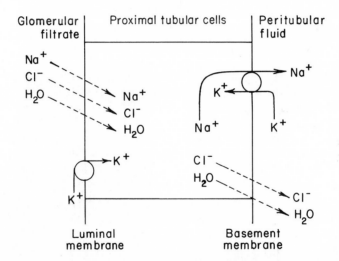

Fig. 4-5. Ion transport in proximal tubule.

so that the osmolarity of the filtrate remains approximately constant. Little concentration of urine occurs in the proximal tubule. The K^+ from the glomerular filtrate is activity pumped into the tubular cell and thence into the peritubular fluid. For 70% to 85% of the glomerular filtrate is reabsorbed in the proximal tubules. The process for Na^+, Cl^-, K^+, and H_2O resorption in the proximal tubule is schematically represented in Fig. 4-5.

Loop of Henle. The second part of the tubule, the *loop of Henle,* has a descending and ascending limb. As the luminar fluid coming from the proximal tubules moves down the thin descending limb, water diffuses out osmotically into the *vasa recta,* and Na^+ passively diffuses into the lumen. This continues until the volume of urine is much reduced at the bend of the loop. The concentration at the bend may be four times greater, about 1200 mOsm, than that at the entrance to the loop of Henle. The ascending limb is impermeable to water, but Cl^- is actively reabsorbed, bringing with it an equivalent amount of Na^+. The overall result is a removal of fluid from the loop of Henle, with further reabsorption of Na^+ and Cl^-. Therefore, the filtrate is approximately isotonic or hypotonic as it enters the distal tubule. Approximately 50% of the Na^+ entering the loop is reabsorbed by the time it reaches the distal convoluted tubule.

Distal segments of nephron. The volume of fluid entering the *distal convoluted tubule* is about 15% of the original glomerular filtrate. This segment of the nephron is sensitive to vasopressin, an antidiuretic hormone (ADH), in the absence of which the tubule is impermeable to water. Such a condition exists in diabetes insipidus, a disease resulting from the lack of ADH. With ADH present there is diffusion of water from the distal tubule and the *collecting ducts*, which are also sensitive to ADH. The sodium reabsorption continues as an active process in these distal segments of the nephron, the rate being influenced by hormones, in particular the mineralocorticoids such as aldosterone (see Chapter 13).

Chloride is reabsorbed by what may be an active process.

In normal circumstances approximately 70% of the K^+ is absorbed from the glomerular filtrate in the proximal tubule. Only about 5% of the filtered K^+ reaches the early segment of the distal tubule. If there is K^+ depletion in the body, K^+ may be absorbed further in the distal nephron. With normal or high K^+ intake in the diet, net K^+ secretion takes place, mainly in the distal segment of the distal tubule. The luminal surface of the tubule cell is permeable to K^+. Since intracellular K^+ is high, some K^+ will always leak out into the urine. The higher the urine flow rate the more K^+ will be lost, as is seen in diuretic therapy.

The renal excretion of K^+ is also influenced by changes in Na^+ reabsorption. As more Na^+ is reabsorbed so the electrical potential of the luminal membrane becomes more negative, which in turn increases the passive diffusion of K^+ into the luminar fluid. In contrast, when the fluid is nearly free from Na^+ (dietary Na^+ deprivation) there is drastic reduction in K^+ secretion. K^+ conservation is never complete, and K^+-depleted patients are usually not able to produce a K^+-free urine. However, if the dietary Na^+ is reduced to less than 10 meq/day, the Na^+ content of the urine will be zero in a few days.

K^+ secretion is also modified by acid-base changes, the excretion rates of K^+ and H^+ bearing a reciprocal relationship. Under conditions of acidosis, the K^+ secretion is depressed. Alkalosis increases K^+ secretion. In part these results can be reconciled by the movement of K^+ from cells during acidosis, thus reducing the intracellular concentration of K^+ upon which the passive diffusion of K^+ depends.

Whereas it was thought that K^+ and H^+ compete for secretion in exchange for Na^+, it has been found that the processes, although interrelated, are not coupled one to the other.

Renal threshold

Many solutes in the glomerular filtrate are reabsorbed by mediated transport processes. These systems can become saturated (p. 172). When more than saturation quantities or threshold concentrations are presented to the tubules, the reabsorption capacities are exceeded and the excess will be excreted in the urine. These concentrations, usually given as the blood plasma concentrations, are known as *threshold values*, for example, glucose, 140 to 170 mg/dl; total CO_2, 27 to 30 meq/liter; and K^+, about 11 to 12 mg/dl. The threshold value for urea is very low so that only a small portion is reabsorbed in the tubules. The threshold values vary with the physiologic condition, such as glomerular filtration rate, acid-base balance, and hormone concentrations (parathyroid hormone for phosphate).

Mediated transport in the reabsorption of solutes from the glomerular filtrate also exists for other sugars, vitamin C, phosphate, sulfate, some amino acids, some Krebs cycle organic acids, and uric acid. The transport of sugars and some amino acids have characteristics in common.

Glucose transport is coupled in some way to Na^+ transport in both the kidney and the intestinal mucosa. It is proposed that glucose enters the tubular cell by binding to a specific carrier molecule that also binds Na^+. The electrochemical gradient for Na^+ moves the complex across the membrane. The glucose can be transported against a glucose-gradient so long as Na^+ is cotransported. The Na^+ that enters the cell is "pumped" out by the Na^+/K^+-ATPase system. Glucose, fructose, galactose, and xylose have the same carrier, with glucose having by far the greatest affinity.

Specific transport systems in the intestine and the kidney have been described for five or so different groups of amino acids. These are the small neutral, large neutral, acidic, and basic amino acid groups and proline. Each group has its carrier and, like glucose, Na^+ is required for the transport of the amino acids into the cell. Other mechanisms of transport may also be present for the amino acids (see Chapter 9).

Creatinine clearance. There is a fairly constant daily production of *creatinine* from creatine (p. 234), the amount of which is determined chiefly by the size of the muscle mass. All of the creatinine that enters the glomerular filtrate is completely excreted. For all intents and purposes the determination of its daily excretion together with the urine volume is a measure of the glomerular filtration. Therefore, creatinine clearance is one measure of kidney function.

Renal control of acid-base balance

The kidneys regulate the concentration of HCO_3^- in blood by adjusting the amount of this anion that is reabsorbed. Under normal conditions little HCO_3^- is excreted because the renal threshold is 26 to 28 meq/liter; normal plasma contains 25 to 26 meq/liter of HCO_3^-.

Bicarbonate is reabsorbed in both the proximal and distal tubules, but about 90% of the filtered HCO_3^- is reabsorbed in the proximal segment. Hydrogen ions are actively secreted into the lumen and associate with HCO_3^- in the luminar

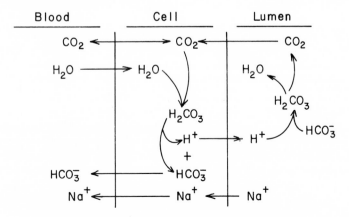

Fig. 4-6. Absorption of HCO_3^- from glomerular filtrate.

filtrate to form carbonic acid. In the presence of carbonic anhydrase at the luminar brush border of the tubular cell, the H_2CO_3 is decomposed to CO_2, which together with that already in the filtrate freely diffuses into the tubular cell. This CO_2 is partially hydrated to H_2CO_3, catalyzed by intracellular carbonic anhydrase. The newly formed H_2CO_3 then dissociates to HCO_3^- and H^+. These processes are illustrated in Fig. 4-6.

As noted in Fig. 4-5, Na^+ diffuses into the tubular cell down an electrochemical gradient and is pumped out to the peritubular fluid by a Na^+/K^+ pump in the basement membrane so that a low Na^+ concentration is maintained in the tubular cell. Although there may appear to be an exchange of Na^+ for H^+ in Fig. 4-6, there is no tight coupling of the two transport systems.

The mechanism for HCO_3^- reabsorption in the distal tubules and collecting ducts is qualitatively the same as in the proximal tubule. The distal segments, however, account for only about 10% of the total HCO_3^-.

Factors affecting bicarbonate concentration in blood

$\bar{p}CO_2$. An increase in $\bar{p}CO_2$ in the blood, and therefore in the glomerular filtrate, results in an increase in H_2CO_3 in the tubular cells (Fig. 4-6). Since some of the H_2CO_3 dissociates, this results in a corresponding increase in $[H^+]$ in the tubular cells. Any increase in the $[H^+]$ in the cell permits secretion of H^+ from the intracellular fluid into the luminar fluid so that more HCO_3^- can be reabsorbed. Alternatively the H^+ is excreted in the urine (see below) and the HCO_3^- generated within the cell adds to that already reabsorbed. Thus in hypoventilation (high $\bar{p}CO_2$, acidosis) there is a compensatory increase in $[HCO_3^-]$ as is required by the Henderson-Hasselbalch equation for readjustment of the blood pH. Contrariwise, hyperventilation (alkalosis) results in reduced HCO_3^- reabsorption.

Potassium. An inverse relationship was noted earlier for the secretion of K^+ and H^+ from the tubular cells. In high plasma $[K^+]$ (hyperkalemia), K^+ exchanges for intracellular H^+. This reduces the secretion of H^+, which is needed for the reabsorption of HCO_3^- from the filtrate in the lumen, as noted above. The urine then becomes alkaline as a result of this lack of H^+ secretion and contains $KHCO_3$. In hypokalemia the capacity to reabsorb HCO_3^- improves because of the increased intracellular $[H^+]$. The increased capacity for H^+ secretion thus leads to a total reabsorption of HCO_3^- from the filtrate and the potential for additional generation of HCO_3^-. Hypokalemia may therefore result in the plasma $[HCO_3^-]$ being maintained at a higher than normal value.

Chloride. When the body is depleted of Cl^-, hypochloremia results and plasma $[HCO_3^-]$ increases to maintain anion concentration. On the other hand, hyperchloremia causes decreased $[HCO_3^-]$ in the plasma. These processes are related to the extracellular volume. Intake of increased amounts of NaCl causes expansion of the extracellular fluid volume, reflected by a small gain in body weight. The higher intake of NaCl is matched in a few days by a progressively increased urinary excretion until a new balance of Na^+ intake and excretion is achieved. This higher flux of Na^+ through the distal segments of the nephron stimulates K^+ secretion but reduces H^+ secretion. The latter results in a reduction of HCO_3^- reabsorbed into the blood and thus a reduced $[HCO_3^-]$ in plasma.

Mechanisms for H^+ excretion

The H^+ secreted by the tubular cells is handled in three principal ways.

1. Reabsorption of HCO_3^-, which has been discussed earlier—the secreted H^+ is neutralized by HCO_3^- in the filtrate to form H_2CO_3 and thence H_2O and CO_2

2. Reaction with the $HPO_4^{--}/H_2PO_4^-$ buffer, which in effect exchanges one of the Na^+ in Na_2HPO_4 to give NaH_2PO_4 – this results in a conservation of Na^+, as has been illustrated in the problem worked out on p. 161.

3. Formation of ammonia in the tubular cells, partly from the catalytic hydrolysis of glutamine with glutaminase and partly from the oxidative deamination of α-amino acids

L-Glutamine
L-Glutamate → NH_3 — H^+ ← from H_2CO_3 or excess H^+ in filtrate
NH_4^+

The NH_3 diffuses from the cells to the lumen where it combines with the H^+ to form ammonium ion. Since the pK_a for NH_4^+ is 9.6 and the pH of urine when there are adequate amounts of acid is 6.0 or less, essentially all the ammonia in the urine is in the form of NH_4^+. Therefore, the H^+ is neutralized and excreted as an NH_4^+ anion. The control of deamination in the tubular cells therefore plays an important role in buffering the excess H^+ produced by body metabolism.

Summary of general movement of ions and water in nephron

Before considering the regulation of fluid concentration, it is well to summarize the flow of solutes and water in the nephron (Fig. 4-4).

Water. Passive reabsorption of H_2O occurs in the proximal tubule. Further reabsorption is controlled by vasopressin, also known as antidiuretic hormone (ADH), in the distal tubule, as ADH permits H_2O reabsorption.

Sodium. Na^+ is reabsorbed in the proximal and distal tubules and in the ascending limb of the loop of Henle. There is diffusion into the filtrate in the descending limb of the loop of Henle. In the collecting ducts the reabsorption of Na^+ by active transport is coupled more tightly to H^+ excretion than elsewhere.

Potassium. K^+ is normally totally reabsorbed by active transport in the proximal tubules. Depending upon the acid-base balance and the Na^+ excretion, various amounts of K^+ are secreted in the collecting ducts. K^+ secretion is depressed by acidosis and low Na^+ reabsorption in the collecting ducts.

Chloride. Reabsorption of Cl^- occurs principally by passive transport in the proximal tubule, with some evidence for active transport in the ascending limb of the loop of Henle. There is a reciprocal relationship between the plasma concentrations of Cl^- and HCO_3^- related to extracellular fluid volume.

Phosphate. Parathyroid hormone reduces the rate of inorganic phosphate (P_i) absorption from the filtrate.

Ammonium. NH_3 diffuses from the cells to the filtrate, where it reacts with H^+ to form NH_4^+.

REGULATION OF VOLUME AND CONCENTRATION OF BODY FLUIDS

The routes of intake and excretion of inorganic components in the body have been described, and it is necessary to turn to the regulation of extracellular volume before considering cellular homeostasis, the ultimate reason for it all.

The extracellular volume is directly related to its Na^+ concentration, which is exquisitely controlled by the kidney. Under normal conditions the intake of Na^+ is balanced by Na^+ excretion from the kidney, except for very modest losses in sweat and feces. In some ways the plasma volume is sensed by a volume-regulating mechanism, and this or some other mechanism also regulates the amount of Na^+ excreted. Thus hypervolemia and hyponatremia, produced experimentally by infusing water containing vasopressin to inhibit diuresis, result in the excretion of considerable amounts of Na^+ by the kidney in spite of the hyponatremia.

Water distribution in the body

Body fluids include *intracellular* fluid (that within the cells) and *extracellular* fluid; the latter includes *interstitial* fluid (between the cells) and circulatory fluids, principally blood plasma. On the average the total body fluid constitutes 70% of the *lean* body mass (LBM) (Chapter 1). This total is divided into intracellular fluid, which is 50% of LBM, and extracellular fluid, which is 20% of LBM. The latter is composed of interstitial fluid, 15% of LBM, and circulatory fluids, 5% of LBM.

For a normal 70-kg man the fluid balance may be broken down into daily losses and daily gains approximately as follows:

1. *Daily losses.* These estimates are based in part on body weight. For example, losses from the lungs are calculated as 0.5 ml/kg/hr. Values will also vary with the physiologic status; fever, hyperventilation, heavy urine loads, hot weather, or diarrhea causes increased water losses.

Typical normal values are as follows: urine, 1200 ml; stool, 200 ml; sweat, 360 ml; and respiration, 840 ml, for a total of 2600 ml.

2. *Daily gains.* Since the diet supplies energy by oxidative metabolic processes that produce CO_2 and water, the metabolic water serves partially to replace fluid losses. The amount of water produced will be proportional to caloric intake and the composition of the diet; it amounts to 0.6, 0.4, and 1.07 ml/g of carbohydrate, protein, and fat oxidized, or 10 to 14 ml/100 kcal derived from food. A value for a 70-kg man would be 300 ml/day for water of oxidation. This leaves (2600 − 300) ml or 2300 ml/day that must come from drinking and water present in the food.

It should be emphasized that the fluid balance in infants is more delicate. The infant body is about 80% water. A 7-kg child contains about 5.6 liters of water, of which 4.2 liters are intracellular and 1.4 liters are extracellular. The daily fluid losses are about 0.7 liter, or 50% of the extracellular fluid, compared with about 20% for the adult. Any excessive loss of water very quickly causes dehydration in the infant.

Since some water must be excreted in order to excrete the waste products of metabolism, there is a limit to the amount of water retention that the kidney can achieve through concentration of the urine. When the dietary intake of water is inadequate, body fluids become reduced (dehydration) until a limit is reached and kidney function ceases.

Aside from the consumption of water dictated by social customs and habits, water intake is regulated by a thirst reflex that is activated by the osmolarity of the extracellular fluid. The osmolarity of extracellular fluid fluctuates, largely as a result of changes in the concentration of NaCl. The greatest difference in composition between the plasma and interstitial fluids lies in the plasma proteins, which are restricted primarily to the plasma compartment and produce a small osmotic gradient that acts to draw water from the interstitial fluid (p. 65). It is partially countered by the blood pressure, so that abnormal variations in plasma protein concentration, capillary hydrostatic pressure, or capillary permeability will result in an abnormal distribution of fluids in these two compartments.

Electrolyte regulation

There is a continuous movement (leakage) of Na^+ into the cells that is opposed by its active transport back into the extracellular space (Na^+/K^+ pump, Chapter 5). This also serves to concentrate K^+ in the intracellular compartments. The distribution of these two important cations is thus maintained, with 90% to 92% of the cations being Na^+ and 3% K^+ in extracellular fluids. In contrast, 70% to 80% K^+ and about 6% Na^+ are contained in intracellular fluid. The active transport processes that maintain this distribution are essential to life. Chloride is largely excluded from the cell.

Having now considered the mobilities of the different solutes in the extracellular and intracellular spaces, the changes in volume that occur under various conditions may be illustrated. Bearing in mind that some solutes are essentially compartmentalized (for example, NaCl in extracellular spaces), that water is freely permeable, and that the volume of the extracellular space reflects its concentration of Na^+, then it is possible to predict the effect of variations in total body water and Na^+.

1. A transfusion of isotonic saline solution will increase the volume of the extracellular space without changing its tonicity. No change in cellular volume will result.
2. Removal of NaCl without removal of the equivalent amount of water (as would result from excessive sweating) results in a hypotonic extracellular fluid. Water will pass into the cells to restore osmotic balance, and the cell volume expands at the expense of the extracellular fluid volume.
3. Gain of NaCl without an equivalent amount of water (as by the infusion of hypertonic saline solution) produces an effect opposite to that described above; that is, the cell volume decreases and the extracellular fluid volume expands.
4. Ingestion of large volumes of water results in its distribution throughout all the body fluids, with a corresponding lowering of their osmolarity.

Because of the dynamic state of interacting equilibria and the homeostatic mechanisms, the actual changes produced by the above redistributions of water will be less than calculated. The kidney response will result in appropriate corrective measures, that is, water and/or Na^+ diuresis or reabsorption. Na^+ restrictions elicit the secretion of mineralocorticoids that result in appropriate conservation of this ion by reabsorption from the glomerular filtrate. If renal failure or a hormonal imbalance exists, these renal responses will be defective and homeostasis will be impaired.

A delicate regulation of extracellular volume is provided by the secretion of a proteolytic enzyme, *renin*, by the kidney. In response to decreases in blood pressure that are associated with a reduction in the plasma volume and Na^+ concentration, the kidney secretes renin. This enzyme acts on a circulating plasma α_2-globulin, angiotensinogen, to release a decapeptide, *angiotensin I*. As angiotensin I passes through the lungs, a second enzymic hydrolysis occurs, removing a dipeptide from the C-terminal end to produce *angiotensin II*, which is a powerful pressor agent. This causes an increase in blood pressure by constricting the arterioles and also stimulates the secretion of aldosterone. These events are depicted schematically in Fig. 4-7.

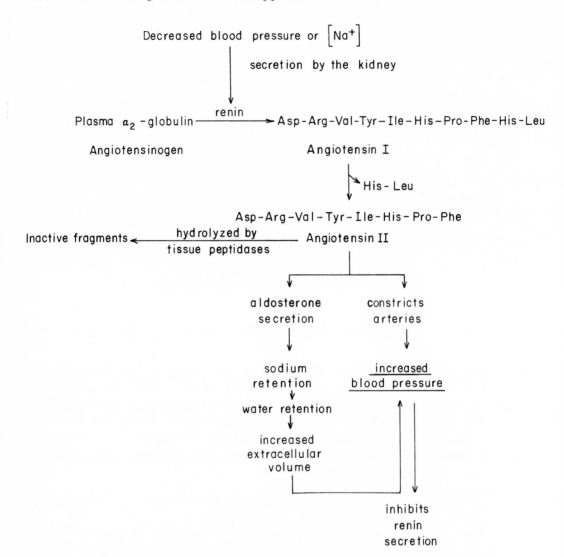

Fig. 4-7. Regulation of blood volume and composition by renin-angiotensin.

Diuretic agents

The removal of excess water from the body can be brought about by several processes:

1. The solute load on the kidney can be increased by a compound such as mannitol that cannot be reabsorbed by the kidney.
2. The glomerular filtration rate can be increased. This rate is determined by the hydrostatic pressure resulting from the cardiac output less the colloidal pressure (oncotic pressure). The oncotic pressure is proportional to the concentration of plasma proteins, principally albumin. The filtration pressure is normally around 75 mm Hg, but it can be increased by stimulation of the cardiac output.
3. Diuretics, which increase net renal excretion of solute and water, can be administered.

The most important action of diuretic drugs is to block tubular reabsorption of Na^+. The more important diuretics act at the following sites:

a. Carbonic anhydrase inhibitors. Sulfonamide was observed to lead to the development of acidosis, reduced plasma [total CO_2], reduced H^+ secretion into the urine, and loss of Na^+, K^+, and H_2O. Subsequently acetazolamide (Diamox) was developed, and more recently ethoxzolamide (Cardrase) and dichlorphenamide (Daranide). All these drugs have a free $—SO_2NH_2$.

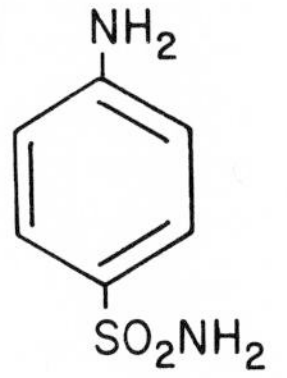

Sulfonamide

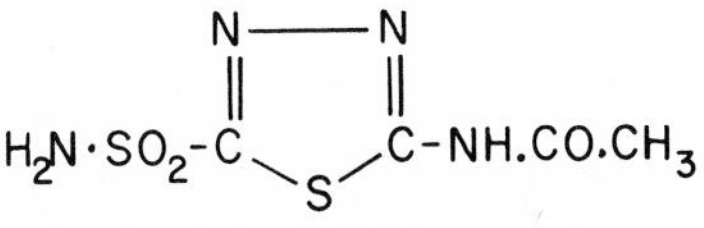

Acetazolamide
(Diamox)

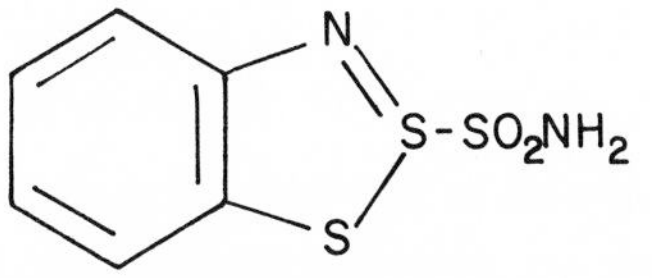

Ethoxzolamide
(Cardrase)

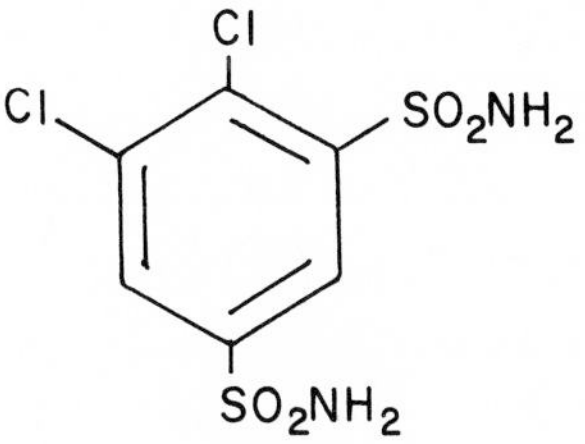

Dichlorphenamide
(Daranide)

The inhibition of carbonic anhydrase results in a decreased availability of H^+, a reduction in the reabsorption of HCO_3^-, and a reduced NH_4^+ excretion.

The reabsorption of Na^+ in the proximal tubule is reduced so that more Na^+ is delivered to the distal segments. This increases the secretion of K^+ in exchange for which some Na^+ is reabsorbed by active transport. Thus these types of diuretics produce a urine that is more alkaline, higher in Na^+, K^+, and HCO_3^- but lower in NH_4^+ than normal. Chloride is not significantly affected. The increased solute excretion carries with it more water, and a diuresis occurs.

b. Thiazides, such as chlorothiazide (Diuril), have some inhibitory effect upon carbonic anhydrase, but the principal site of action is thought to be the ascending limb of the loop of Henle. This action takes the form of reducing the amount of Na^+ and Cl^- reabsorption. The increased Na^+ concentration in the distal tubules will also increase K^+ secretion so that the diuresis is characterized by increased excretion of Na^+, K^+, Cl^-, and H_2O. Because of the increased K^+ excretion, one of the major side effects of the thiazides may be hypokalemia.

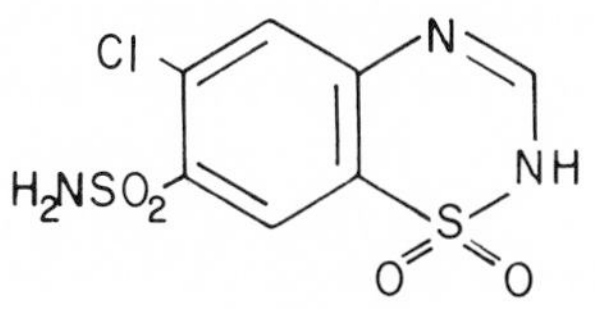

Chlorothiazide
(Diuril)

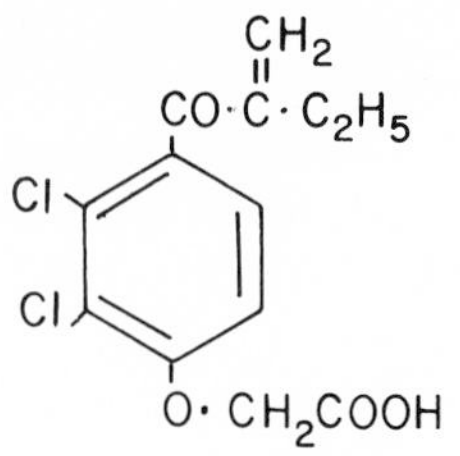

Ethacrynic acid

c. Furosemide and ethacrynic acid are both more active diuretics than the thiazides, acting also on the ascending limb of the loop of Henle. They cause excessive diuresis and Na^+ and K^+ loss. This can result in such a contraction of the extracellular fluid volume that the increase in $[HCO_3^-]$ may cause an alkalosis, a so-called *contraction alkalosis*.

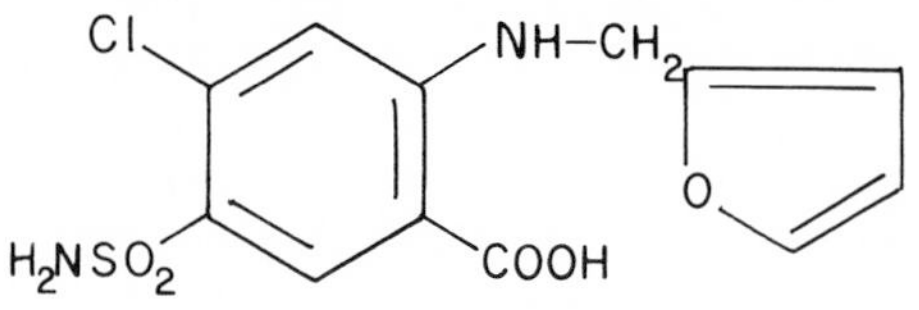

Furosemide

METABOLIC AND RESPIRATORY DISTURBANCES OF ACID-BASE BALANCE

The normal pH of blood plasma is approximately pH 7.4, usually 7.35 to 7.45. A lower pH identifies the condition of *acidosis*, as opposed to the state of *alkalosis* that exists at a higher pH. At pH 7.40 the ratio of $[HCO_3^-]/[$dissolved $CO_2]$ is 20, and the [total CO_2] is 26 to 27 meq/liter. Changes in these values may result from pulmonary or metabolic dysfunction, or both. The ratio can be changed, giving either acidosis or alkalosis resulting from the increase or decrease in either $[HCO_3^-]$ or $\bar{p}CO_2$. The [total CO_2] may also be increased or decreased. However, the imbalance may not be completely compensated. Uncompensated acidosis refers to the condition in which the $[HCO_3^-]/[$dissolved $CO_2]$ ratio is less than 20, and uncompensated alkalosis refers to a ratio greater than 20 (Table 4-3).

If the imbalance is caused by an alteration of $\bar{p}CO_2$, then it is respiratory in origin. Conversely, an alteration in HCO_3^- is considered to be of metabolic origin. These four conditions are therefore called *uncompensated respiratory* acidosis or alkalosis and *uncompensated metabolic* acidosis or alkalosis. The changes in the plasma buffer constituents are summarized in Table 4-3, where constituents are shown as either increased or decreased relative to the normal.

The body *attempts* to compensate for abnormal pH conditions by changing the bicarbonate buffer component that was normal. Thus in uncompensated metabolic acidosis the H_2CO_3 concentration is initially normal, but it is reduced by increasing the respiratory rate, so that the pH becomes as close to normal as possible. This produces an even more abnormal [total CO_2], and the condition is identified as *compensated metabolic* acidosis. The pH is compensated as much as possible, but the distribution of anions still resembles that in a metabolic acidosis. It will be noted from Table 4-3 that in certain cases the analysis of one constituent cannot identify the nature of the imbalance. For example, in both primary metabolic acidosis and primary respiratory alkalosis the [total CO_2] is reduced.

Table 4-3. Simple disturbances of acid-base balance*

	AVERAGE NORMAL VALUE	ACIDOSIS				ALKALOSIS			
		Metabolic		Respiratory		Metabolic		Respiratory	
		U*	C*	U	C	U	C	U	C
pH	7.4†	↓	7.4†	↓	7.4†	↑	7.4†	↑	7.4†
[HCO$_3^-$]/[dissolved CO$_2$]	20	↓	20	↓	20	↑	20	↑	20
[HCO$_3^-$] (meq/liter)	25-26	↓	↓	25-26	↑	↑	↑	25-26	↓
pCO$_2$ (mm Hg)	40	40	↓	↑	↑	40	↑	↓	↓
Total CO$_2$ (meq/liter)	26-27	↓	↓	↑	↑	↑	↑	↓	↓

*U = uncompensated (or primary); C = compensated.
†Approximate values.

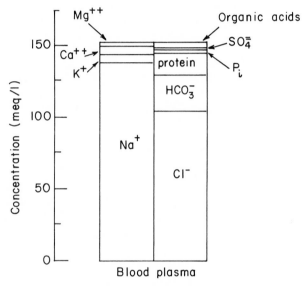

Fig. 4-8. Gamblegram of normal human blood plasma.

More definitive interpretations, therefore, require that at least two of the parameters be measured.

Gamblegrams

Before illustrating these respiratory and metabolic imbalances further, it is necessary to introduce the Gamblegram, a common pictorial representation of the composition of body fluids (Fig. 4-8). The Gamblegram shows the balance of anions and cations in normal blood plasma. The height of each block is proportional to the concentration, in milliequivalents per liter, of the constituent identified with it. It is strikingly clear that the principal cation is Na$^+$ and the principal anion is Cl$^-$. The other primary anions are HCO$_3^-$ and protein$^-$. Since the isoelectric point of most plasma proteins is less than 7.4, they will contain a net negative charge at physiologic pH values. The other ions, although important, are present in relatively small concentration and play little part in the maintenance of fluid osmolarity. The "organic acids" are important to the present discussion because metabolic acidosis is frequently associated with increases in the concentration of lactate or ketone bodies (Chapter 1). With the formation of keto acids, the cellular concentration of H$^+$ is increased, and this H$^+$ is exchanged for K$^+$ in the interstitial fluid. This process leads to a reduction in plasma pH (acidosis) with some buffering, however, by reaction of the HCO_3^- with the H^+. Thus the pH and the HCO_3^- concentration are reduced.

For simplicity the Gamblegram may show Na^+, K^+, Cl^-, HCO_3^-, and R, where R represents the remaining anions. As a first approximation it is found that normally $[Na^+] + [K^+]$ are 5 to 15 meq/liter greater than $[Cl^-] + [HCO_3^-]$. It is useful to bear this range in mind in those situations in which all the laboratory determinations are not available.

Respiratory acidosis

Chronic lung disease or depression of the respiratory rate by some disturbance of the nervous system will increase the $\bar{p}CO_2$, which produces a decrease in the $[HCO_3^-]/[\text{dissolved } CO_2]$ ratio. An uncompensated respiratory acidosis results, and the kidney responds by increased reabsorption of HCO_3^- and generation of HCO_3^- from CO_2. There is increased secretion of H^+ by the tubular cells in response the high $\bar{p}CO_2$. The patient may compensate to a pH that is close to normal, but as long as gas exchange is impaired, the bicarbonate will stabilize at an elevated concentration and the [total CO_2] will be high. If the onset of uncompensated respiratory acidosis is acute and kidney function is normal, the compensation may take 3 to 5 days. The urine would probably be more acidic, a reflection of the increased excretion of H^+. More Na^+ is reabsorbed in exchange for H^+ and K^+, the increased H^+ secretion also being reflected in greater formation and excretion of NH_4^+ in the urine. The plasma $[Cl^-]$ will be reduced in proportion to the increase in $[HCO_3^-]$, so that electrical neutrality is maintained.

Respiratory alkalosis

Hyperventilation may be seen in patients who have head injuries or who are under the influence of some drugs, such as salicylate poisoning. Hyperventilation for any reason results in a rapid decrease in $\bar{p}CO_2$, which decreases the availability of H^+ for secretion into the lumen of the kidney tubules. This leads to a corresponding reduction in the reabsorption of HCO_3^- and Na^+. More secretion of K^+ occurs because there is a greater concentration of Na^+ present in the distal segments of the nephron. The overall effect therefore is an increased excretion of Na^+, K^+, and HCO_3^- in the urine with an increased reabsorption of Cl^- to replace the HCO_3^- and so maintain the anion concentration. At such time as the hyperventilation is controlled, the increase in $\bar{p}CO_2$ will reverse the above responses of the kidney. HCO_3^- will be reabsorbed from the filtrate, and more will be generated by the tubular cells to return the $[HCO_3^-]/[H_2CO_3]$ ratio and [total CO_2] to normal.

Metabolic acidosis

A reduction in the concentration of plasma HCO_3^- leading to a metabolic acidosis may be caused by several factors:

1. Increased biosynthesis of metabolic acids such as ketone bodies or ingestion of acids such as salicylic acid or NH_4Cl (NH_4Cl is equivalent metabolically to $HCl + NH_3$, the latter being converted to neutral urea, leaving HCl to be excreted)
2. Excessive loss of HCO_3^- caused by diarrhea or other conditions resulting in a loss of pancreatic secretions, which are alkaline and contain higher $[HCO_3^-]$ than the blood plasma
3. Decreased excretion of H^+ by the kidneys resulting from acute kidney failure or an impaired ability to generate NH_3 for excretion of H^+ as NH_4^+

The biochemical mechanisms that result in these imbalances will be discussed in many of the following chapters. The resulting loss of HCO_3^- is compensated by increased pulmonary ventilation, which produces a decrease in $\bar{p}CO_2$.

In some types of metabolic acidosis the gap between the $Na^+ + K^+$ concentration and $HCO_3^- + Cl^-$ concentration is greater than normal because the other nonvolatile anions (fixed acids) are increased. These accumulated nonvolatile anions must be excreted. If the kidney is functioning properly, this excretion is accompanied by an increased loss of water and cations. This may produce dehydration and electrolyte imbalance, two problems that become interrelated.

Metabolic alkalosis

An increase in plasma $[HCO_3^-]$ occurs when abnormal amounts of alkali are retained. This can develop when salts of metabolic acids (sodium lactate or $NaHCO_3$) are administered or when ethacrynic acid is used to produce diuresis. It also results when acid is lost, as through vomiting of gastric HCl. The pulmonary compensation for metabolic alkalosis is hypoventilation, which increases the $\bar{p}CO_2$.

Mixed disturbances of acid-base balance

In most of these simple situations of acid-base imbalance, normal functioning of either the lungs or the kidneys has been assumed. Also, the simultaneous occurrence of two primary disturbances has not been considered; yet such situations are common. In these cases the simple scheme presented in Table 4-3 does not apply, although the compensatory mechanisms operate to their greatest capacity. The evaluation of the nature and extent of such conditions requires a careful

Table 4-4. Examples of disturbances in acid-base balance

CASE NUMBER	ARTERIAL $\bar{p}CO_2$ (mm Hg)	pH	BICARBONATE (meq/liter)	REMARKS
1	40	7.40	24.5	Normal
2	25	7.60	24.5	Severe respiratory alkalosis (patient being ventilated artificially)
3	31	7.51	24.5	Moderate respiratory alkalosis (mild hyperventilation)
4	60	7.22	24.5	Uncompensated respiratory acidosis, for example, hypoventilation resulting from narcotic overdose
5	60	7.37	35.0	Respiratory acidosis partially compensated by metabolic alkalosis (renal in origin), for example, patient with long-standing chronic obstructive lung disease
6	32	7.65	35.0	Combined respiratory and metabolic alkalosis; patient in case 5 on prolonged mechanical ventilation
7	22	7.35	11.0	Metabolic acidosis with secondary respiratory alkalosis, for example, patient with severe diabetic acidosis
8	50	7.07	15.0	Combined metabolic and respiratory acidosis, for example, patient in case 7 whose ventilation has been severely depressed by heavy sedation

history and physical examination in conjunction with an analysis of the laboratory data. Examples of such mixed disturbances are given in Table 4-4.

REFERENCES

Muntwyler, E.: Water and electrolyte metabolism and acid-base balance, St. Louis, 1968, The C. V. Mosby Co.

Pitts, R. F.: Physiology of the kidney and body fluids, Chicago, 1974, Year Book Medical Publishers, Inc.

Robinson, J. R.: Fundamentals of acid-base regulation, London, 1972, Blackwell Scientific Publications Ltd.

Weisberg, H. F.: Water, electrolyte and acid-base balance, Baltimore, 1953, The Williams & Wilkins Co.

Welt, L. G.: Clinical disorders of hydration and acid-base equilibrium, Boston, 1959, Little, Brown & Co.

The preceding references describe the basic regulatory mechanisms for acid-base, fluid, and electrolyte control. Analyses of several clinical cases are given in the works by Pitts, Weisberg, and Welt.

Gamble, J. L.: Companionship of water and electrolyte in organization of body fluids, Stanford, Calif., 1951, Stanford University Press.

Gamble, J. L.: Chemistry, anatomy, physiology and pathology of extracellular fluid, Cambridge, Mass., 1954, Harvard University Press.

The preceding works by Gamble are classic treatises with examples of clinical situations, including many Gamblegram illustrations of the electrolyte composition of body fluids and secretions.

Cannon, P. J., et al.: "Contraction" alkalosis after diuresis of edematous patients with ethacrynic acid, Ann. Intern. Med. 62:979, 1965.

Goldman, A. G., et al.: Salt wasting syndrome of obscure origin, Am. J. Med. 46:606, 1969. *The report, presented as a detective story, emphasizes the importance of obtaining the total medical history whenever possible to arrive at a correct diagnosis.*

Montgomery, R., and Swenson, C. A.: Quantitative problems in the biochemical sciences, San Francisco, 1976, W. H. Freeman & Co., Publishers. *Examples of calculations of acid-base balance illustrated. Also problems for practice.*

Clinical examples

CASE 1: DIABETES AND DIABETIC KETOACIDOSIS

A 21-year-old female with a 4-year history of juvenile onset diabetes was brought to the hospital in a coma. She had required ninety-two units of insulin daily in order to maintain her blood glucose concentration in an acceptable range and prevent excessive glucosuria. On admission she had a blood pressure of 92/20, pulse 122 per min, and deep respirations 32 per min. Laboratory data showed serum glucose 1200 mg/dl, hematocrit 49%, sodium 134 meq/liter, potassium 6.4 meq/liter, BUN 74 mg/dl, pH 6.80, and $\bar{p}CO_2$ 10 mm Hg. The plasma was strongly positive for ketones.

Eight units of regular insulin were given IV and eight units/hr were given by IV infusion pump. Her serum glucose fell at a rate of approximately 100 mg/dl/hr. In 7 hr her ventilation and blood pH were normal following IV injection of $NaHCO_3$ and vigorous fluid and electrolyte replacement.

Biochemical questions

1. What is the mechanism of the acid-base imbalance?
2. Calculate the total CO_2 concentration and from this explain the nature of the respiration.
3. Explain the purpose of the $NaHCO_3$ injection and the fluid replacement.
4. What electrolyte replacements should be given intravenously?

Case discussion

The patient arrived at the hospital in a serious comatose condition, with a low systolic and diastolic blood pressure. This resulted from a loss of body fluids, as further demonstrated by a high hematocrit. The blood pH of 6.80 indicates extreme acidosis caused by the production of ketone bodies, such as acetoacetic and β-hydroxybutyric acids (Chapters 8 and 10). These organic acids, the metabolic consequence of carbohydrate deprivation of the cells, produced an acidosis. Compensation was attempted by an increased respiration rate (normal 12 to 15 per min) where the body is trying to compensate for low blood serum pH by increasing the $[HCO_3^-]/[CO_2]$ ratio. However, this creates the further problem of reducing the total CO_2. Furthermore, the very low blood pressure is producing a reduced glomerular filtration rate with the result that the urea derived from amino acid metabolism is elevated above the normal value of 20 to 40 mg/dl in serum; the blood urea concentration is about twice that of the BUN (Why?) Very briefly therefore the acid-base imbalance derives from excessive ketone body production, hyperventilation, and an impaired kidney function. The metabolic analysis of the problem will be discussed more completely in Chapters 7 and 8, but the essentials were presented in Chapter 1, pp. 5 to 9.

The body attempts to buffer the incoming metabolic acids with the bicarbonate system. The excess CO_2 that is generated is blown off by hyperventilation, leading to decreased [total CO_2]. This is seen in the blood serum analyses.

$$\text{Metabolic acids} \longrightarrow H^+ + HCO_3^- \rightleftharpoons$$

$$H_2CO_3 \rightleftharpoons H_2O + CO_2 \longrightarrow \text{expired}$$

The relationship between these components of the bicarbonate buffer system is given by the Henderson-Hasselbalch equation.

$$pH = 6.1 + \log_{10} \frac{[\text{total } CO_2] - 0.0301\ \bar{p}CO_2}{0.0301\ \bar{p}CO_2} \quad (4\text{-}6)$$

In the present case the patient's blood has a pH of 6.80 and a $\bar{p}CO_2$ of 10 mm Hg. Therefore,

$$6.80 = 6.1 + \log_{10} \frac{[\text{Total } CO_2] - 10 \times 0.0301}{10 \times 0.0301}$$

that is,

$$\log_{10} \frac{[\text{Total } CO_2] - 0.301}{0.301} = 0.7$$

$$\frac{[\text{Total } CO_2] - 0.301}{0.301} = 5.01$$

$$\text{Total } CO_2 = (1.508 + 0.301)\,\text{meq/liter}$$

$$= 1.8\,\text{meq/liter}$$

Also

$$[HCO_3^-] = (1.8 - 0.3)\,\text{meq/liter}$$

$$= 1.5\,\text{meq/liter}$$

In other words the principal buffer system in the extracellular fluid was nearly exhausted because it was utilized to neutralize the incoming H^+ produced by the ketosis, and rapid replacement by $NaHCO_3$ injection was necessary.

An estimation of the solute concentration of the blood shows an osmolarity above normal. The $[Na^+ + K^+]$ is 140 meq/liter and there is about the same concentration of anions. In calculating osmolarity, however, the glucose (molecular weight, 180) and urea (molecular weight, 60) must be added. This calculates to $\frac{1200}{180} \times \frac{1000}{100} = 70$ meq/liter for glucose and $\frac{74}{60} \times \frac{1000}{100} = 12$ meq/liter for urea. From these estimates, excluding the unknown ions and other solutes that have not been taken into account, the osmolarity will be at least $(140 + 140 + 70 + 12)$ mOsm or 362 mOsm (normal 285 to 295 mOsm).

Therefore, fluid replacement needs to be hypotonic. The concentration of Na^+ in the blood serum is not too far removed from normal, and it might be expected from a consideration of the renal response to metabolic acidosis that the $[Cl^-]$ will be lower than normal, but not greatly so. The concentrations of the metabolic ketoacids in the blood will be elevated and will reduce the $[HCO_3^- + Cl^-]$, but principally the $[HCO_3^-]$ (see calculation on p. 187). All of these factors would argue therefore for an IV infusion of hypotonic saline.

For reasons that are explained by the severe acidosis, it is seen that there has been significant exchange of H^+ and K^+ between the intracellular and extracellular fluid compartments producing hyperkalemia. As the acidosis is corrected this elevated blood serum K^+ will be reduced. The blood glucose concentration is being reduced rapidly by the insulin therapy. As will be seen in Chapter 7, K^+ movement into the cell is further stimulated by glucose uptake. Both effects may lead to hypokalemia, which can be corrected by the appropriate addition of KCl to the IV fluids.

REFERENCES

Kitabchi, A. J., et al.: Comparison of therapy of diabetic ketoacidosis, Clin. Res. 22:650A, 1974.
Thompson, R.: The management of diabetes mellitus in the adolescent, Med. Clin. North Am. 59(6):1349, 1975.

CASE 2: POLIOMYELITIS

A 14-year-old boy who had never been immunized against poliomyelitis contracted the disease late in the summer. He was hospitalized and required the use of a respirator during the acute phase of his illness. When he appeared to be recovering, he was taken off the respirator with no apparent ill effects. Several days later, analysis of his blood revealed the following:

Na^+	136 meq/liter
K^+	4.5 meq/liter
Total CO_2	36 meq/liter
Cl^-	92 meq/liter
$\bar{p}CO_2$	70 mm Hg
Blood pH	7.32

Biochemical questions

1. What was the acid-base balance of the patient?
2. What anions probably were elevated in this patient's blood?
3. What was the probable cause of his condition?
4. The patient was put back on a respirator and the $\bar{p}CO_2$ reduced to 40 mm Hg. If this rate of ventilation was too rapid, what would be the result?

Case discussion

Analysis of laboratory values. Comparison of the composition of the patient's blood with normal values shows the following:

1. The concentrations of Na^+ and K^+

are within the normal limits of 136 to 145 and 3.5 to 5.0 meq/liter, respectively.

2. The concentration of Cl^- is slightly below the normal range (100 to 106 meq/liter).

3. Both the [total CO_2] and the $\bar{p}CO_2$ are high. From these values and the blood pH of 7.32, which is on the acid side of the normal range but approaching the normal, the concentration of HCO_3^- can be calculated.

$$CO_2 + H_2CO_3 = \bar{p}CO_2 \times 0.0301 \text{ meq/l}$$
$$= 70 \times 0.0301 \text{ meq/l}$$
$$= 2.11 \text{ meq/l}$$

$$[HCO_3^-] = [\text{Total } CO_2] - [CO_2 + H_2CO_3]$$
$$= (36.0 - 2.1) \text{ meq/l}$$
$$= 33.9 \text{ meq/l}$$

A quick calculation from equation 4-6 shows that the values of $[HCO_3^-]$ and $[CO_2 + H_2CO_3]$ are consistent with the experimentally determined blood pH of 7.32.

$$pH = 6.10 + \log_{10} \frac{33.9}{2.11}$$
$$= 6.10 + \log_{10} 16.06$$
$$= 6.10 + 1.21$$
$$= 7.31$$

Where possible, such checks for consistency in laboratory values are useful in order to exclude analytic errors.

4. The analysis of blood normally shows that the sum of the concentrations of the principal cations $K^+ + Na^+$ exceeds the sum of the concentrations of HCO_3^- and Cl^- by about 5 to 15 meq/liter. This difference represents the concentrations of the nonvolatile anions: $SO_4^=$, phosphate, lactate, and other organic acids of metabolism. In cases of metabolic imbalances such as uncontrolled diabetes mellitus, the serum concentrations of these organic acids may increase. This rise is associated with relative reductions in the HCO_3^- and Cl^-. In this patient the difference (D) is:

$$D = ([Na^+] + [K^+]) - ([HCO_3^-] + [Cl^-])$$
$$= (136 + 4.5) - (33.9 + 92) \text{ meq/l}$$
$$= 14.6 \text{ meq/l}$$

This suggests that no severe metabolic imbalance exists.

Renal compensation. Since both total CO_2 and $\bar{p}CO_2$ are high and there is a nearly compensated pH and no excess of metabolic organic acids, the patient had a compensated respiratory acidosis. A patient who required respirator support while recovering from poliomyelitis would have weakness of the chest muscles but no impairment of the efficiency of gaseous exchange across the alveolar membranes. However, there is impairment of pulmonary ventilation, and the $\bar{p}CO_2$ builds up. The patient cannot correct this until his chest muscle strength returns. Furthermore, high $\bar{p}CO_2$ acts as a narcotic and depresses the respiratory control center. In response to the acidosis and high $\bar{p}CO_2$ the proximal and distal tubular reabsorption of HCO_3^- is maximal. Bicarbonate is also generated in the tubular cells by the interplay of the bicarbonate-phosphate buffers (see below). There is thus a net loss of H^+ and

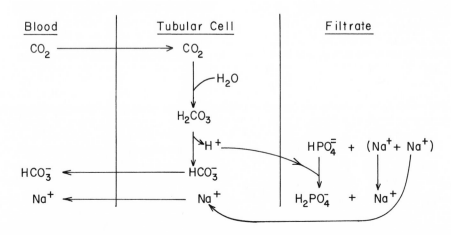

a conversion of CO_2 to HCO_3^-. This raises the $[HCO_3^-]$ so that the $[HCO_3^-]/[H_2CO_3]$ ratio approaches 20, returning the pH toward 7.4. While this is the correct direction for compensation of the respiratory acidosis, the result is an abnormally high [total CO_2]. As the patient's condition improves, this elevation in [total CO_2] is gradually reduced to normal.

Compensation by renal mechanisms is slower by several hr than the changes in $\bar{p}CO_2$ that can be effected by the lungs. If the lungs of a patient with a stable $\bar{p}CO_2$ elevation are ventilated to reduce $\bar{p}CO_2$ more rapidly than the renal compensation can follow, alkalosis would develop. The $\bar{p}CO_2$ would approach normal, but the $[HCO_3^-]$ and [total CO_2] would be high (as a residue from the compensated respiratory acidosis). Assuming that there has been no renal re-sponse to a reduction in $\bar{p}CO_2$ in this patient, the blood pH would be:

$$pH = 6.10 + log_{10} \frac{33.9}{(0.0301 \times 40)}$$

$$= 6.10 + log_{10} \frac{33.9}{1.20}$$

$$= 6.10 + 1.45$$

$$= 7.55$$

which is quite alkaline. This could be avoided by slowing the rate of reduction in $\bar{p}CO_2$.

REFERENCES

Brackett, N. C., Wingo, C. F., Nuren, O., and Solano, J. T.: Acid-base response to chronic hypercapnia in man, N. Engl. J. Med. 280:124, 1969.
Schwartz, W. B.: Defense of extracellular pH during acute and chronic hypercapnia, Ann. N.Y. Acad. Sci. 133:125, 1966.

CASE 3: ENCEPHALITIS

A 23-year-old male who had previously been in excellent health suddenly became acutely ill. His initial complaint was a feeling of general malaise. He then developed a headache of gradually increasing severity. His temperature was moderately elevated, and he became drowsy and lethargic. On admission to the hospital, stiffness of the patient's neck, general muscle weakness, and decreased reflexes were noted. He gradually lapsed into an unconscious state. A tentative diagnosis of encephalitis was made. Rapid breathing developed, and analysis of a blood specimen revealed the following gas and electrolyte picture.

Na^+	136 meq/liter
K^+	4 meq/liter
Total CO_2	12 meq/liter
Cl^-	114 meq/liter
$\bar{p}CO_2$	18 mm Hg

Catheterization showed an alkaline urine.

Biochemical questions

1. What are the pH and electrolyte balance of the patient?
2. If excessive levels of hyperventilation developed, the oxygen supply might be insufficient to support the patient's energy requirements. Oxygen debt would then result in high levels of lactate. What would the response of the kidneys be in this situation? What pH and electrolyte imbalances would exist in blood and urine?

Case discussion

Acid-base balance. The acid-base balance of the patient can be calculated from the total CO_2 and $\bar{p}CO_2$ data, as indicated in equation 4-6.

$$pH = 6.10 + log_{10} \frac{Total\ CO_2 - 0.0301\ \bar{p}CO_2}{0.0301\ \bar{p}CO_2}$$

$$= 6.10 + log_{10} \frac{12 - 0.0301 \times 18}{0.0301 \times 18}$$

$$= 6.10 + log_{10} \frac{11.46}{0.54}$$

$$= 7.43$$

The blood pH 7.43 shows that the patient has compensated for the imbalance. This compensation has resulted in a low [total CO_2] (hypocapnia). The difference in $[Na^+] + [K^+]$ and $[Cl^-] + [HCO_3^-]$ is 14 meq/liter, a value within the normal range. This indicates that there is minimal elevation of the levels of any organic acids. The condition is therefore one of compensated respiratory alkalosis, and the kidney has compensated for the low $\bar{p}CO_2$ by reducing the reabsorption of HCO_3^-, thus producing the alkaline urine.

Renal compensation. As noted earlier, the reabsorption of HCO_3^- is decreased by a reduction in the secretion of H^+ by the tubular cells as a result of the lowering of intracellular $[H^+]$ by the alkalosis. The decreased blood $\bar{p}CO_2$ also reduces the secretion of H^+ and thus HCO_3^- reabsorption.

The $[Na^+]$ and $[K^+]$ are normal, but the $[Cl^-]$ has increased to replace the reduced $[HCO_3^-]$. This is in accord with the common observation that $[Cl^-]$ fluctuates inversely as the $[HCO_3^-]$ with the sum of all anions providing the necessary counterions for the cations. Since H^+ is not being secreted at a normal rate by this patient, that part of the Na^+ reabsorption exchanging with H^+ is reduced. Therefore, more Cl^- is passively reabsorbed. The summated effect is one of a Cl^- increase and an HCO_3^- decrease in renal reabsorption.

The excretion of alkaline urine also results in an increased excretion of cations. There is a partial loss of the Na^+ conservation mechanism of the $HPO_4^=/H_2PO_4^-$ buffer, and the secretion of NH_3 is also reduced by the low $\bar{p}CO_2$. This means that Na^+ and K^+ will be lost at a greater rate during the compensation for respiratory alkalosis.

If the diagnosis of encephalitis were confirmed and there were no supporting devices available for treatment, the hyperventilation could overtax the capacity of oxidative metabolic processes to supply energy. Anaerobic metabolism would result, with release of lactic acid from the muscle cells. The extra metabolic (organic) acids would be buffered by the cellular fluids, so that more K^+ would be lost from the cells. An additional buffering effect would occur by reaction of the organic acids with bicarbonate in the extracellular fluids. It is estimated that the buffering contributions of intracellular cations and plasma HCO_3^- are approximately equal in

many instances, for example, metabolic acidosis without respiratory alkalosis. In the present case the buffering capacity of the bicarbonate system is severely impaired by the low total CO_2, which may become more severe. Since the respiratory function of this patient is uncontrolled, the compensatory response will reside in the kidney. The increasing blood $[H^+]$ produced by influx of lactate will cause an increase in HCO_3^- reabsorption from the glomerular filtrate. There will be an increase in lactate and a decrease in Cl^- in the urine. Therefore, the urine will become less alkaline.

The initial state of compensated respiratory alkalosis resulting from uncontrolled hyperventilation progressed into a mixed acid-base disorder. In other words, a metabolic acidosis was superimposed on the condition. The status may remain compensated, or the capacity of the kidney to compensate may be overcome unless there is clinical intervention.

The sequence of events in the plasma may be represented diagrammatically through the use of a Gamblegram, in which R^- represents the organic acids, $SO_4^=$, and phosphate. The change in the $[HCO_3^-]/[\text{dissolved } CO_2]$ ratio before severe metabolic acidosis developed may be summarized as shown on p. 192.

It should be stressed that throughout the discussions of acid-base balance and fluid and electrolyte control the arguments are based on analysis of the extracellular fluid, usually the blood. It is not convenient and sometimes impossible to study the tissues by taking biopsy samples. It is, for example, difficult to estimate the intracellular pH; in any case it probably varies in different subcellular compartments. It is surmised that the intracellular level of H^+ is greater than the extracellular level. In any event, the crucial value is that inside the cells. The extracellular fluids provide, to the greatest extent possible, a constant environment for them. This is homeostasis.

In the steady state of a normal or chronically ill individual, the distribution of the diffusible ions between the plasma and interstitial fluids is in equilibrium.

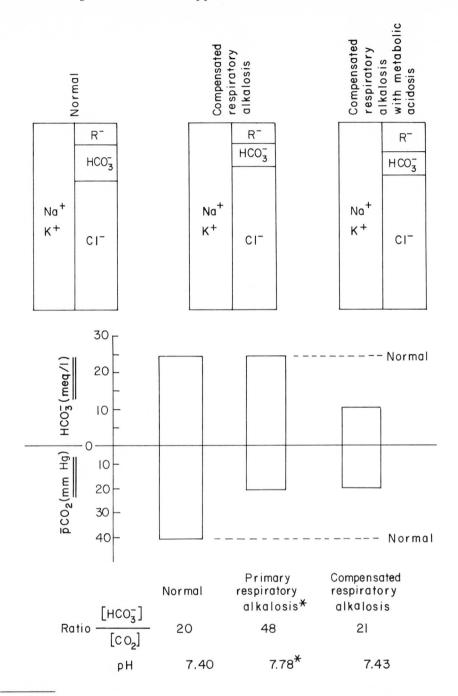

However, rapid changes, for example, in $[H^+]$ or $\bar{p}CO_2$, may result in a lag in the reestablishment of equilibrium. This is also true for particular compartments in the extracellular fluid, for example, the equilibration of the cerebrospinal fluid with plasma HCO_3^- is slower than with plasma CO_2.

CASE 4: DEHYDRATION

M. T., a lean, middle-aged man weighing 70 kg and with no medical history of disease, suffered a blow on the head. When he was admitted to the hospital 30 min after the accident, he was unconscious and perfectly still, and he remained in this condition until he died 16 days later. He was fed by stomach tube and received 160 g protein, 140 g carbohydrate, 60 g fat, 48 g sodium chloride, the necessary vitamins and minerals, and 2500 ml of water daily. After the first 5 days his salt intake was reduced to 5.0 g/day because of a hyperchloremia (120 meq Cl^-/liter). During the next 7 days his blood pressure fell gradually. An indwelling catheter was inserted on day 12; the 24 hr urine volumes were as follows:

DAY AFTER ADMISSION	URINE VOLUME (ml/24 hr)
13	2500
14	2100
15	2200
16	320 (morning only)

The urine specific gravity was 1.012 to 1.018 for days 13 to 15. On the morning of day 16, M. T. was in peripheral vascular collapse. Blood analysis showed the following:

	M. T.	NORMAL VALUES
BUN	100 mg/dl	10 to 20 mg/dl
Total protein	7 g/dl	6.8 g/dl
Na^+	175 meq/liter	136 to 145 meq/liter
K^+	3.7 meq/liter	3.5 to 5.0 meq/liter
Cl^-	134 meq/liter	100 to 106 meq/liter
Total CO_2	22 meq/liter	24 to 30 meq/liter
pH	7.4	7.4
Glucose	135 mg/dl	60 to 100 mg/dl
Urinary glucose	Negative	Negative

Biochemical questions

1. Was M. T.'s nutritional requirement satisfied?
2. Was fluid intake adequate?
3. What was the status of the acid-base and electrolyte balance by day 16?
4. What is one probable cause of the deterioration of M. T.'s condition?

Case discussion

Caloric requirements. It can be assumed that M. T. would have had normal blood and urine chemistries immediately prior to his accident. Since his unconscious state did not involve unusual muscular activity, his nutritional needs would be little more than those required to satisfy his BMR. That is:

Metabolic energy required by M. T. =
70 × 24 kcal/day = 1680 kcal/day

The metabolic energy provided by the gavage feeding amounted to:

Protein = 160 × 4 =		640
Carbohydrate = 140 × 4 =		560
Fat = 60 × 9 =		540
Total		1740 kcal/day

M. T. was therefore provided with adequate calories. However, it is a diet very high in protein, and if all the nitrogen of this protein were excreted as urea, it would amount to about 50 g/day. This is approximately twice the normal daily urea excretion rate, which is 25 to 30 g/day. This added solute load would demand an increase in urine volume.

Fluid balance. *Insensible* fluid loss in M. T. would be 1200 to 1300 ml/day. The fluid in the feeding (2500 ml) and that produced by metabolism (p. 178; $12 \times \frac{1700}{100}$ or 200 ml) provides for a urine flow of (2700 − 1250) ml or 1450 ml/day. An amount greater than this would cause dehydration. It is clear that during days 13, 14, and 15 there was a negative water balance of 1050, 650, and 750 ml/day, or an average of 800 ml/day. If this deficit had existed during the entire 16 days of M. T.'s hospitalization, his total body fluid loss would have been 15 × 800 ml or about 12 liters.

An indication that M. T. was becoming dehydrated is found in the gradual reduction in blood pressure after the first 5 days. The normal body response would have been the secretion of renin to produce angiotension II, which would have stimulated the secretion of aldosterone. However, the high-protein diet has placed about twice the normal solute load in the form of urea on the kidneys, requiring extra water for excretion. By day 16 there was a severe uremia, the blood urea nitrogen (BUN) elevation indicating that the kidney was not functioning to excrete the solutes. There was also a hypertonicity of the plasma (Na^+ 175 meq/liter), a compensated metabolic acidosis (pH 7.4, low total CO_2), and a moderate increase in R^- anions (23 meq/liter), probably organic acids. It is apparent that regardless of the necessity of conserving water, the continuous obligatory renal excretion of metabolic waste products required excessive water loss. This produced a progressive increase in the total solute concentration of body fluids. A conscious person would respond to this condition because of the sensation of thirst; the unconscious patient dehydrates unless extra fluid is provided.

Estimation of degree of dehydration. A measure of the degree of dehydration is provided by the increased concentration of Na^+, a valuable guide to the tonicity of body fluids. Since the patient appears to have dehydrated gradually, the buildup of electrolyte concentrations in the extracellular fluids would have been followed by the movement of water from the intracellular space to maintain an osmotic balance. In this situation the plasma $[Na^+]$ reflects the hypertonicity of the total body fluids.

The elevation of plasma Na^+ concentration will cause the osmotic transport of intracellular water to the extracellular space. Yet there will still be an obliga-tory loss of H_2O from the body in order to maintain kidney function as long as possible. A loss of total body fluid results, and this is inversely proportional to the plasma concentration of Na^+.

$$\frac{Normal\ [Na^+]}{Elevated\ [Na^+]} = \frac{Dehydrated\ total\ body\ fluid\ volume}{Normal\ total\ body\ fluid\ volume}$$

For the 70-kg man, M. T.:

$$\frac{140}{175} = \frac{Dehydrated\ total\ body\ fluid\ volume}{(0.70 \times 70)\ liters}$$

$$Dehydrated\ volume = \frac{49 \times 140}{175}\ liters = 39\ liters$$

$$Fluid\ deficit = 49 - 39\ liters = 10\ liters$$

Although the peripheral vascular collapse was related to the inadequate fluid intake, the problem probably stems from the decision to feed a diet unusually high in protein. Postmortem findings in this case argue against hypophyseal lesions that would impair the secretion of ADH and cause diabetes insipidus. The patient could have been better served nutritionally by an average diet containing 60 to 70 g of protein.

REFERENCES

Welt, L. G.: Clinical disorders of hydration and acid-base balance, Boston, 1959, Little, Brown & Co.

West, E. S., et al.: Textbook of biochemistry, ed. 4, New York, 1966, Macmillan Publishing Co., Inc.

CASE 5: EMPHYSEMA

A 57-year-old widow had a medical history of shortness of breath that had progressively worsened over the last 10 years. On examination, her blood chemistry findings were:

ARTERIAL BLOOD

$\bar{p}O_2$	45 mm Hg
$\bar{p}CO_2$	65 mm Hg
pH	7.42

VENOUS BLOOD

BUN	22.0 mg/dl
Total CO_2	42.7 meq/liter
Cl^-	87.0 meq/liter
K^+	3.6 meq/liter
Na^+	135.0 meq/liter

Biochemical questions

1. What is the nature of the acid-base balance?
2. What type of disease might be associated with this case?
3. Would unusual findings from the urine analysis be expected?
4. What conclusions can be drawn from the $\bar{p}CO_2$ value?

REFERENCES

Davenport, H. W.: The ABC of acid base chemistry, Chicago, 1969, University of Chicago Press.

Masoro, E. J., and Siegel, P. D.: Acid-base regulation, Philadelphia, 1971, W. B. Saunders Co.

Tuller, M. A.: Acid-base homeostasis and its disorder, New York, 1971, Medical Examination Publishing Co.

CASE 6: PULMONARY EMBOLISM

A 32-year-old man underwent surgery for a radical lymph node dissection on July 1. On July 3 a pulmonary embolism occurred. Shortness of breath developed rapidly and the patient experienced sweating and a tachycardia of 180/min. Blood chemistry analysis showed the following:

ARTERIAL BLOOD		VENOUS BLOOD	
$\bar{p}O_2$	65 mm Hg	BUN	10.0 mg/dl
$\bar{p}CO_2$	30 mm Hg	Total CO_2	23.0 meq/liter
pH	7.50	Cl^-	95.0 meq/liter
		K^+	4.1 meq/liter
		Na^+	143.0 meq/liter

Biochemical questions

1. What is the acid-base condition of this patient?
2. How can the $\bar{p}O_2$ and $\bar{p}CO_2$ values be reconciled?
3. Is there evidence as to the effectiveness of renal function?
4. What might be concluded from the $[Na^+ + K^+] - [Cl^- + HCO_3^-]$ gap?

REFERENCES

Refer to the references for case 5.

CASE 7: PNEUMONIA

P. T., a 65-year-old man with a history of pulmonary tuberculosis, was admitted to the hospital with a slight fever of 38.5° C and a severe shortness of breath. X-ray examination showed that pneumonia had developed in the left lung, the right lung having collapsed several years ago. P. T. had been on a low-salt diet and was taking diuretics and digitalis.

At the time of admission his blood chemistry is summarized in Table 4-5, column 1. Because of his respiratory distress, P. T. was placed on a respirator with the resulting changes in pH, $\bar{p}CO_2$, and $[HCO_3^-]$ shown in column 2 of Table 4-5. These values caused the clinician in charge to change the settings of the machine so that there was more retention of CO_2 by P. T., who was also given small doses of morphine. In 6 hr the blood values had changed (column 3).

Further adjustments of the respirator over the next few days resulted in a stabilized condition with the picture summarized in the final column of Table 4-5.

Biochemical questions

1. What might you gather from the creatinine, BUN, Cl^-, Na^+, and K^+ values on admission of P. T. to the hospital?
2. How do the blood chemistries relate (if at all) to the diuretic and low-salt therapy?
3. What different result might have been expected if the initial ventilation had not been so vigorous?
4. Explain how the patient went from one state of acid-base balance to another, as summarized in Table 4-5.

REFERENCES

Refer to the references for case 5.

Table 4-5. Blood chemistries of P. T.

	ON ADMISSION 1	ON RESPIRATOR 2	AFTER 6 HR 3	FINAL 4
pH	7.43	7.55	7.32	7.41
$\bar{p}CO_2$ (mm Hg)	64	40	79	50
$[HCO_3^-]$ (meq/liter)	42	36	41	31
Creatinine (mg/dl)*	1.1			
BUN (mg/dl)†	11			
Total CO_2 (meq/liter)	44			
Cl^- (meq/liter)	89			
Na^+ (meq/liter)	146			
K^+ (meq/liter)	3.6			

*Normal range is 0.7 to 1.5 mg/dl.
†Normal range is 10 to 20 mg/dl.

CASE 8: PULMONARY EDEMA

G. C., a 48-year-old man, had a myocardial infarct 3 years prior to and an ischemic episode 3 months before his arrival at the emergency room complaining of increasing dyspnea of 3 days' duration. Pulmonary edema developed immediately after entry, and his condition rapidly deteriorated despite treatment with aminophylline and oxygen. One hour after admission his blood pressure was 160/110 mm Hg and he was not responsive. An arterial blood sample was drawn and for the next $1\frac{1}{2}$ hr he was treated with morphine, aminophylline, rotating tourniquets, and a 500 ml phlebotomy. However, his condition deteriorated further: his blood pressure began to fall, and in another 15 min his pulse and respiration had ceased. Intubation was performed, and forced ventilation and external cardiac massage were initiated. A second arterial blood sample

Table 4-6. Analysis of arterial blood samples in case 7

SAMPLE NUMBER	pH	pCO_2 (mm Hg)	HCO_3^- (meq/ liter)	LACTATE (mg/ 100 ml)
1	6.99	97	18.5	67
2	7.09	68	18	63
3	7.30	58	26	44
4	7.41	43	25	—

was obtained. Shortly thereafter the pulse became palpable again. He was then given 224 meq sodium bicarbonate intravenously, and manual ventilation was continued for 2 hr. At this time he regained consciousness and a third blood sample was drawn. Ventilation was maintained by respirator. Within 12 hr this was disconnected and the endotracheal tube was withdrawn. Except for an episode of pulmonary infection, the remainder of his hospital course was uneventful. Chest films showed a regression of the pulmonary edema. At the time of discharge, analysis of a fourth blood sample showed that arterial blood values were normal. The analyses of the four blood samples are given in Table 4-6.

Biochemical questions

1. Explain the following aspects of treatment: oxygen therapy, withdrawal of 500 ml blood, administration of 225 meq $NaHCO_3$, rotation of tourniquets, and maintenance of ventilation by respirator.
2. How do the values obtained in the four blood samples reflect the course of treatment?

REFERENCE

Anthonisen, N. R., and Smith, J. H.: Respiratory acidosis as a consequence of pulmonary edema, Ann. Intern. Med. **62**:991, 1965.

CASE 9: HYPONATREMIA

B. F., a 74-year-old retired schoolteacher, was admitted to the hospital for surgery. Her medical problems included asthma of 10 years duration. She was currently asymptomatic and was receiving no therapy for this condition. She also had hypertension and was taking reserpine. On physical examination the blood pressure was 150/92 and her lungs were clear to percussion and auscultation. Her hypertensive therapy was changed to hydrochlorothiazide (50

mg twice daily), a derivative of chlorothiazide (Diuril). The following morning she became disoriented and began vomiting. The serum Na^+ was 119 meq/liter and the hydrochlorothiazide was discontinued. She was then given 3 liters of fluid containing 1.5 g/liter of NaCl. The following day she was given an additional 3 liters of fluid containing 7.5 g/ liter of NaCl. By the end of the day her confusion had resolved.

The laboratory data for the case were:

DAY	Na^{+*}	K^{+*}	Cl^{-*}	TOTAL CO$_2$*	BUN†	CREATININE†	WEIGHT‡	FLUID INTAKE§	URINE§
0 (Admission)	140	4.9	102	26	12	0.7	84.4		
1 Hydrochlorothiazide initiated									
2 Morning, hydrochlorothiazide discontinued									
2 1 PM	119	3.8	79	27	14	0.8		1640	?
3	120	4.4	82	26	9	0.7	85	1530	2675
4	134	4.3	96	26	11	0.7	84	3350	3000+
5	137	4.5	100	26	7	0.7	85.1	3795	?

*meq/liter.   †mg/dl.   ‡kg.   §ml.

Biochemical questions

1. What is the effect of hydrochlorothiazide on the fluid and electrolyte balance? Explain.
2. What information can be gained from the BUN and creatinine values?
3. How do fluid intake, urine volume, and body weight correlate?

REFERENCES

Fichman, M. P., et al.: Diuretic induced hyponatremia, Ann. Intern. Med. 75:853-863, 1971.

Kleeman, C. R.: Hypo-osmolar syndromes secondary to impaired water secretion, Annu. Rev. Med. 21:259-268, 1970.

Taclob, L. T., and Needle, M. A.: Hyponatremic syndromes, Med. Clin. North Am. 57:6, 1425, 1973.

ADDITIONAL QUESTIONS AND PROBLEMS

1. *Water imbalance:* Why is it inadvisable to change the water content of the body too rapidly?
2. *Diabetes insipidus:* What is the effect of this clinical condition on electrolyte and water balance? (Utiger, R. D.: Med. Clin. North Am. 52:381, 1968.)
3. Summarize the body's defense mechanisms to an introduction of a strong acid into the blood.
4. There are impairments in the production of ammonia in the malnourished person, particularly in patients with protein malnutrition. Discuss these effects upon the renal regulation of acid-base balance. Klahr, K., et al.: Am. J. Med. 48:325, 1970.)
5. Discuss the changes in the composition of extracellular fluid after diuresis of edematous patients using ethacrynic acid. (Cannon, P. J., et al.: Ann. Intern. Med. 62:979, 1965.)

CHAPTER 5

ENERGETICS AND COUPLED SYSTEMS

OBJECTIVES

1. To explain how energy changes associated with chemical reactions may be measured and compared
2. To demonstrate how a knowledge of free energy changes permits prediction of the direction of a reaction
3. To explain how the energy output from one reaction may be coupled to promote a second reaction
4. To define what is meant by the term "high-energy bond" relative to the large, negative change of free energy on hydrolysis of the bond
5. To explain the process of oxidative phosphorylation and its function as a source of high-energy bonds
6. To illustrate cellular uses of high-energy bonds by two examples, the transmembrane movement of ions and the mechanical work of muscle contraction

All material possesses a certain intrinsic energy. Living organisms convert accessible energy, in the form of food, for several purposes: repair and maintenance of cell substances, growth, reproduction, and the performance of various types of chemical and physical work. These functions must be performed in a controlled manner. The sum of the reactions by which an organism meets these needs is always associated with a *net decrease* in *free energy*. Free energy is defined as that portion of the total energy that is available for useful work. For any chemical reaction that proceeds spontaneously, the free energy of the products is always *less* than the free energy of the reactants. The enzymic reactions that catalyze cellular processes are no exception to this law.

The ultimate source of free energy used by living organisms lies in oxidative reactions. Animate matter has evolved distinctive patterns of oxidative reactions that are appropriately described as biologic oxidations. They are unique in that many of the individual reactions involve the removal of hydrogen from the substrate that is oxidized; ultimately, water is formed. Instead of being liberated as heat, part of the energy released by the oxidation is trapped, or conserved, in the form of compounds that have large, negative, free energies of hydrolysis. Such compounds are frequently termed high-energy compounds and are typified by adenosine triphosphate (ATP). The simultaneous oxidation of substrate and formation of ATP is therefore known as *oxidative phosphorylation*. The oxidative and phosphorylative processes are not only simultaneous but are also very tightly coupled; interference with one immediately affects the other. Many pathways lead to ATP formation, which unifies and simplifies metabolic systems, but when these pathways are deranged or when oxidative phosphorylation is damaged, the consequences may be disastrous. It is no exaggeration to state that deprivation of ATP would adversely affect so many vital processes that death would be inevitable.

It is, therefore, important for health scientists to understand the fundamental laws that govern energy changes in chemical reactions as they apply to living tissues, to know how these relate to the basic mechanisms of oxidative phosphorylation, and to appreciate the central role of ATP in the body's economy.

Fundamental laws of energy flux

To deal with problems of energetics, it is necessary to define what is meant by a system. An experimental system can be any part of the universe that one might wish to isolate for study. Systems may be divided into two classes, open or closed. *Closed systems* neither gain nor lose matter, but certain forms of energy may enter or leave. *Open systems* may gain or lose matter as well as certain forms of energy. Living systems are obviously open systems of considerable complexity, and it is very difficult to deal with the energetics of intact organisms. By dealing with simpler, closed systems it is possible to gain a great deal of useful information that is applicable to more complex situations.

It is not possible to measure the absolute value of the total energy in any system, but changes in the total energy, or in the specific components that comprise the total energy, can be determined. Provided a suitable reference point exists, comparison of energy changes can be made quantitatively in one case and easily related to another. Expressions describing the energetics of chemical reactions and related systems involve changes of heat content, free energy, or other thermodynamic quantities and are represented by the symbols ΔH or ΔG as the case requires.

First law

Certain fundamental laws govern the flux of energy in isolated, closed systems. The first law states that the total energy of such a system is constant. If some amount of heat (Q) is put into the system, it must either do work or increase the total energy of the system. Most biologic systems operate at nearly constant temperature, pressure, and volume; therefore, work represented by changes in volume and pressure is not regarded as biologically useful. If we represent the capacity to do biologically useful work by the symbol W and let H stand for *enthalpy*, or heat content, the first law can be expressed simply as:

$$\Delta H = Q - W \tag{5-1}$$

Note carefully that the first law applies to any ideal, reversible, closed system; it says nothing about the direction of any chemical reaction.

Second law

The second law states that all closed systems, left to themselves, will tend to approach a state of equilibrium, regardless of the path between the initial and final states. In simpler terms any system, no matter how highly organized it may be at any given instant, tends towards a state of greater disorder or randomization. Consider the letters on this page; in their present sequence they appear in a significant order and thus represent useful information. If they were merely scrambled on the page they would lose the order on which the communication of information is based. Restoration of the order, and with it the information, would require that work be done on the system. A measure of disorder is defined as *entropy*, designated by the symbol S. In chemical terms, entropy can be related to the random movements of molecules and can be measured by $T\Delta S$, where T is the absolute temperature of the system. When a chemical system is at equilibrium, there is no *net* reaction, the system has no capacity to do useful work, and:

$$Q = T\Delta S \qquad (5\text{-}2)$$

This is the condition of maximum entropy. The existence of entropy explains why heat energy cannot be completely converted into other energy forms; some of it is lost as entropy that can do no useful work.

Free energy

Work can be done by systems tending toward equilibrium and a measure of the work is given by the following equation.

$$W = -\Delta H + T\Delta S \qquad (5\text{-}3)$$

Willard Gibbs introduced the concept of free energy as another measure of the capacity to do useful work and defined it as:

$$\Delta G = \Delta H - T\Delta S \qquad (5\text{-}4)$$

Note carefully that $\Delta G = -W$, so that when the measure of W is *positive* (meaning that the system can do useful work), the measure of ΔG is *negative*, and vice versa.

Qualitative evaluation of reaction energetics

We are now in a position to make qualitative evaluations of the energy changes in any chemical reaction and to make judgments based on these evaluations. As an example, consider the reaction represented by the following equation:

$$\text{Glc 1-P} \xrightleftharpoons{\text{phosphoglucomutase}} \text{Glc 6-P}$$

where Glc 1-P and Glc 6-P represent glucose phosphorylated on the indicated carbon atoms, and phosphoglucomutase is the enzyme that catalyzes the interconversion. We can write a pair of auxiliary equations that also characterize the reactions. These are:

$$\Delta G = G_{\text{Glc 6-P}} - G_{\text{Glc 1-P}}$$

and

$$\Delta H = H_{\text{Glc 6-P}} - H_{\text{Glc 1-P}}$$

since the changes in free energy and in enthalpy are related only to the respective differences between the products and the reactants. From suitable measurements that reveal the algebraic signs of ΔG and ΔH, we could immediately draw the following conclusions.

1. If ΔG is negative ($\Delta G < 0$), the reaction is called *exergonic*. It will proceed spontaneously to the right; the reverse reaction will not be spontaneous.

2. If ΔG is positive ($\Delta G > 0$), the reaction is called *endergonic*. It will not proceed spontaneously to the right; the reverse reaction will be spontaneous.

3. If ΔG is zero, the reaction will be at *equilibrium*; there will be no net reaction in either direction.

4. If ΔH is negative ($\Delta H < 0$), the reaction is called *exothermic*; it will give off heat to the surroundings.

5. If ΔH is positive ($\Delta H > 0$), the reaction is called *endothermic;* it will take up heat from the surroundings.

6. If ΔH is zero, the reaction is called *isothermic;* there will be no net exchange of heat with the surroundings.

The predictive value of these expressions is obvious, and they also emphasize the significance of changes in free energy as the driving force behind all chemical reactions.

Standard state

Since it is necessary to deal with changes in free energy rather than with absolute values, some reference point must be established. The reference point is defined as the standard free energy, indicated by $\Delta G°$. Unless otherwise specified, standard free energies are calculated at 25°C and at 1 atmosphere (atm) pressure. For pure solids and liquids the standard state is taken to be the most stable form under these conditions. For gases the standard state is defined as that for the ideal gas. For solutions the standard state is defined as a concentration of one mole/liter. Changes in free energy are expressed in calories per mole.

Coupled reaction systems

We turn next to the concept of coupled reaction systems. Two or more reactions are said to be coupled if a product of one reaction is a starting material in the second, and so forth. The concept of coupling is a completely general one and applies to any properly related sequence of chemical reactions. Coupling is particularly important in biochemistry because of the many multi-enzyme, multi-reaction sequences that are known as metabolic pathways.

Consider an idealized sequence of reactions that can be represented by the following set of equations.

$$1. \quad A \rightleftharpoons B \quad (\Delta G_1)$$

$$2. \quad B \rightleftharpoons C \quad (\Delta G_2)$$

$$3. \quad C \rightleftharpoons D \quad (\Delta G_3)$$

$$\overline{Sum: A \rightleftharpoons D \quad (\Delta G\ )}$$

Provided the proper energy conditions prevail, material should pass from A to D. Proper energy conditions require only that $\Delta G_\Sigma < 0$. This conclusion is valid regardless of the individual values of ΔG_1, ΔG_2, or ΔG_3. Suppose for a moment that $\Delta G_2 = 0$; this would mean that B is in equilibrium with C, or that there is no net conversion of B to C. If at the same time $\Delta G_1 < 0$, there will be a significant conversion of A to B, and the increased concentration of B so produced will upset the equilibrium of reaction 2; by the law of mass action it will "push" the formation of C from B. Similarly, if we also assume that $\Delta G_3 < 0$, C will be converted to D, and this will "pull" the equilibrium of reaction 2 to the right. In other words, we can say that an exergonic reaction can be coupled to drive an endergonic reaction, provided the energy differences are of appropriate magnitudes.

Suppose instead that $\Delta G_2 > 0$; this would mean that *by itself* reaction 2 would proceed spontaneously to the left. If at the same time $\Delta G_1 < 0$ and were of lesser magnitude than ΔG_2, we could conclude that the effect of the first reaction would be offset in part by the leftward tendency of the second reaction. However, if ΔG_3 were also negative and of such a magnitude that $(\Delta G_1 + \Delta G_3) < \Delta G_2$, the system would again be driven to the right.

These arguments support the general stipulation that in coupled systems material will continue to flow from left to right, provided $\Delta G_\Sigma < 0$. Clearly, the number of reactions in the coupled sequence is not at all important, nor is the number of starting materials or products in an individual reaction. Only one common species of intermediate is needed between any two consecutive reactions in a coupled system. The nature of the common intermediate is not important; it may be an electron, a proton, or a more complex structure. There will be many examples of coupled systems in later discussions of metabolism and metabolic pathways. Frequently, the fact that two reactions are coupled is indicated by the notation:

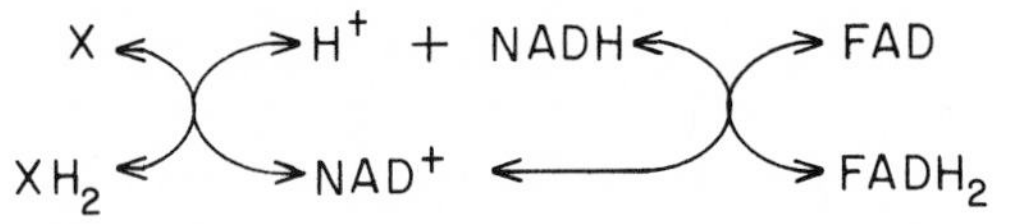

$$
\begin{array}{ccc}
X & \longrightarrow H^+ + NADH & \longrightarrow FAD \\
XH_2 & \longrightarrow NAD^+ \quad\longleftarrow & \longrightarrow FADH_2
\end{array}
$$

which is equivalent to the more formal expression:

1) $XH_2 + NAD^+ \rightleftharpoons X + NADH + H^+$

2) $NADH + H^+ + FAD \rightleftharpoons FADH_2 + NAD^+$

where XH_2 represents any theoretical substrate from which two hydrogens can be removed by an NAD^+-linked dehydrogenase, X represents an oxidized product, and FAD is the abbreviation for flavin adenine dinucleotide. Note that in the above equations there are two intermediates common to both reactions; these are the free proton and NADH.

Quantitative expressions of ΔG

The integration of individual reactions into complex metabolic pathways depends on the value of ΔG for each step in the sequence, since it is the sum of these values that determines the energetics of the pathway as a whole. Chemists have developed various ways of determining ΔG. A few of these that are important for our purposes will be described.

Equilibrium constants

For any reaction represented by:

$$aA + bB \rightleftharpoons cC + dD$$

it can be shown that:

$$\Delta G = \Delta G^\circ + RT \ln \frac{[C]^c [D]^d}{[A]^a [B]^b} \tag{5-5}$$

where R is the natural gas constant (1.987 cal/mole · degree), T is the absolute temperature ($^\circ C + 273$), ln stands for *natural* logarithm, and ΔG° is the standard free energy of the reaction. ΔG is equal to ΔG° if the concentration of each product and reactant is 1M. Note the formal resemblance of the above equation to the Henderson-Hasselbalch equation. ΔG and ΔG° bear a relation to one another that is analogous to the relation between pH and pK. Just as pH = pK only when the concentrations of the ionized and un-ionized forms of a weak acid are equal, so $\Delta G = \Delta G^\circ$ only when the product $[C]^c [D]^d$ is equal to the product $[A]^a [B]^b$.

At equilibrium, $\Delta G = 0$; it follows that:

$$\Delta G^\circ = -RT \ln \frac{[C_{eq}]^c \, [D_{eq}]^d}{[A_{eq}]^a \, [B_{eq}]^b} \tag{5-6}$$

The concentration ratio that exists at equilibrium is represented by K_{eq}, so the relation between ΔG° and K_{eq} is given by the equation:

$$\Delta G^\circ = -RT \ln K_{eq} \tag{5-7}$$

It may seem strange that ΔG° can be used to measure the equilibrium constant, or vice versa. The explanation is quite simple; the standard state is an arbitrary one, and in general ΔG is not zero if the reactants and products are at the standard state. Since ΔG is zero at equilibrium, the second term of equation 5-5 must be just equal in magnitude and opposite in sign to the value of ΔG°.

It must also be pointed out that standard thermodynamic conditions virtually never apply in living tissues; the typical concentration range of intracellular metabolites is from 1 to 5 mM, a far cry from the 1 M concentration that standard conditions require. Just as body pH is more dependent on the ionized and non-ionized concentrations than on the fixed pK of carbonic acid, so too is the energy available from a given reaction more dependent on the existing intracellular concentrations than on the fixed value of ΔG°. For example, assume that the intracellular concentration of ATP is 5×10^{-3} M; that of ADP is 4×10^{-3} M; that of P_i is 2.1×10^{-3} M; and that the cleavage of ATP to ADP + P_i can be represented by the reaction:

$$ATP + H_2O \rightleftharpoons ADP + P_i$$

According to equation 5-5, we can write:

$$\Delta G = \Delta G^\circ + RT \ln \left[ADP\right]\left[P_i\right] \Big/ \left[ATP\right]\left[H_2O\right]$$

Pure liquid water is the standard state; so $[H_2O]$ is 1 for dilute solutions. If the value for ΔG° at 30° C and pH 7.4 is assumed to be -7 kcal/mole, ΔG for the given nucleotide concentrations is approximately -10.2 kcal/mole. Considerably more free energy is, therefore, made available by the hydrolysis of ATP at this low concentration than at standard conditions.

Concentration differences

Free energy changes resulting from alterations in the concentration of a molecule or of an ion are also important in biochemistry and physiology. As was mentioned earlier, if molecules at a given concentration (C_1) are diluted and thus allowed to move farther apart to a new concentration (C_2), the disorder in the system is increased. As a consequence, the free energy decreases. This can be represented by the following equation:

$$\Delta G = RT \ln C_2 / C_1$$

Again, a simple problem can be used to illustrate the argument. What is the theoretical change in free energy required to produce gastric juice containing one mole of HCl from blood plasma? The pH of gastric juice is approximately 1.7, while that of plasma is approximately 7.4. The higher acidity of gastric juice is caused by the greater concentration of HCl, which is required to provide the acid environment in which pepsin can function as a digestive enzyme. Since the blood plasma is the source of the H^+ ion, the problem can be regarded as one of concentrating that ion. From the given pH data it follows that:

$$[H^+]_{(plasma)} = 4 \times 10^{-8}$$

and

$$[H^+]_{(gastric\ juice)} = 2 \times 10^{-2}\ M$$

Assuming that the process occurs at $37°$ C, and making use of the auxiliary equation that states that $\ln x = 2.303 \log_{10} x$, the equation becomes:

$$\Delta G = (1.987)(310)(2.303) \log (2 \times 10^{-2} / 4 \times 10^{-8})$$

or

$$\Delta G = 8084\ cal/mole$$

It must be emphasized that this value for ΔG assumes no change in the pH of either the plasma or the gastric juice; it is a measure of the energy required to move one mole of HCl from one fluid to the other under the stated conditions.

Note that the value for ΔG is positive, which means that work must be done on the system, or free energy must be expended, in order to move one mole of HCl from the plasma to the gastric juice. In fact the logarithmic form of the equation indicates that ΔG will always be greater than zero for concentration processes and will always be less than zero for dilution processes. Many cellular functions involve concentration processes similar to the formation of gastric acid. The "pumping" of Na^+ from intracellular to extracellular fluid and the formation of tears and urine are examples of concentration work processes. For these and other such processes, the change in free energy can be calculated as in the above example.

Oxidation-reduction reactions

Many biochemical systems involve oxidation-reduction reactions, which frequently involve the transfer of two electrons. Since oxidation is defined as the *loss of electrons* and reduction is defined as the *gain of electrons*, the two processes are tightly linked; that is, whenever something is oxidized, something else is simultaneously reduced. A typical oxidation-reduction system can be presented in the form of two half-reactions such as the following:

1) $NAD^+ + 2H^+ + 2e \rightleftharpoons NADH + H^+$

2) $XH_2 \rightleftharpoons X + 2H^+ + 2e$

$$\overline{NAD^+ + XH_2 \rightleftharpoons X + NADH + H^+}$$

where e stands for electrons. Describing an oxidation-reduction system in the form of its half-reactions emphasizes that it is a coupled system, since electrons are always, and hydrogens are frequently, products in one half-reaction and precursors in the other. Many biologic oxidations are catalyzed by dehydrogenases that employ coenzymes able to accept or donate hydrogens. NAD^+ is one important coenzyme and FAD is another. Note that NAD^+-linked oxidations always involve a proton that remains free in solution, but in FAD-linked oxidations both protons are tightly bound to the coenzyme, as shown in Fig. 5-1.

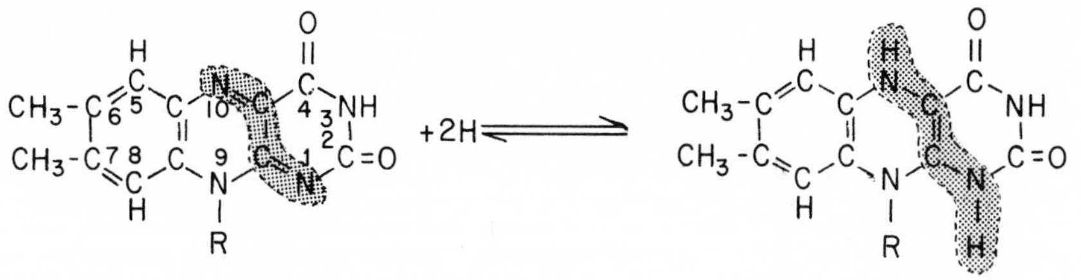

Fig. 5-1. Oxidation-reduction mechanisms of flavin nucleotides. Numbers in left structure indicate numbering sequence in isoalloxazine ring.

There is another advantage in describing oxidation-reduction systems as half-reactions. The explicit indication of electron transfer suggests that if it were possible to separate the half-reactions and to connect them by a wire, the movement of electrons through the wire would generate a potential difference (voltage) that could be detected by a potentiometer. The sign of the potential difference would reflect the direction in which the reaction proceeded, and the magnitude of the potential difference would be a measure of the energy that drove the reaction. Many systems have been studied in this manner, by joining a given half-reaction with another that serves as a standard of reference. The accepted reference standard is the *hydrogen electrode*, described by the half-reaction:

$$2H^+ + 2e \rightleftharpoons H_2(g) \qquad E_o = 0.00 \text{ volts}$$

The standard electrode potential of a hydrogen electrode obtained by bubbling hydrogen gas at $25°C$ and 1 atm pressure into a solution containing hydrogen ions at a concentration of 1 M is arbitrarily defined as zero volts. This is the *primary* international standard. Since few biochemically important reactions occur at the standard hydrogen ion concentration, a *secondary* reference electrode has been adopted by biologists. The secondary reference is made by bubbling hydrogen gas into a solution that contains hydrogen ions at a concentration of 10^{-7} M or at a pH of 7. The secondary standard is described at $30°C$ by the half-reaction:

$$2H^+ + 2e \rightleftharpoons H_2(g) \quad (pH=7.0) \quad E_o' - -0.42 \text{ volts}$$

where the negative potential is caused by the difference in hydrogen ion concentration with respect to the primary standard. Note the use of E_0' to indicate that the pH is 7. (By a corresponding convention, $\Delta G^{0'}$ represents energy values calculated or measured at pH 7.)

A given electrode or half-reaction can serve either as oxidant or reductant, depending on the nature of the other electrode with which it is coupled. For this reason, standard electrode potentials can be expressed either as oxidation or reduction potentials, but the algebraic signs in the two cases will be opposite. Table 5-1 lists a selection of standard reduction potentials. In using these data the following conventions apply.

1. Half-reactions with more positive potentials are written as the oxidizing agent for half-reactions with *less positive potentials*. In other words, molecular oxygen will oxidize the Fe^{+2} form of cytochrome *a*, acetoacetate will oxidize dihydrolipoate, etc., in any combination that might be made.

2. If the direction of a half-reaction is the reverse of that given in Table 5-1, the algebraic sign of the potential will also be reversed. By this convention, the oxidation of malate to oxaloacetate has a value for E_0' of $+0.17$ volts.

Table 5-1. Standard reduction potentials

Half - Reaction	E_o' (pH 7), volts [*]
$\frac{1}{2}O_2 + 2H^+ + 2e \longrightarrow H_2O$	0.815
cytochrome $a(Fe^{+3}) + e \longrightarrow$ cytochrome $a(Fe^{+2})$	0.29
cytochrome $c(Fe^{+3}) + e \longrightarrow$ cytochrome $c(Fe^{+2})$	0.22
coenzyme Q(ox) $\longrightarrow$ coenzyme Q(red)	0.10
fumarate $+ 2H^+ + 2e \longrightarrow$ succinate	0.031
$FAD + 2H^+ + 2e \longrightarrow FADH_2$	-0.06
oxaloacetate $+ 2H^+ + 2e \longrightarrow$ malate	-0.17
pyruvate $+ 2H^+ + 2e \longrightarrow$ lactate	-0.19
acetoacetate $+ 2H^+ + 2e \longrightarrow$ L-β-OH-butyrate	-0.27
lipoate $+ 2H^+ + 2e \longrightarrow$ dihydrolipoate	-0.29
$NAD^+ + 2H^+ + 2e \longrightarrow NADH + H^+$	-0.32
$NADP^+ + 2H^+ + 2e \longrightarrow NADPH + H^+$	-0.32
$2H^+ + 2e \longrightarrow H_2$	-0.42

[*] at 30°C

3. The free energy change $\Delta G'$ for an oxidation-reduction reaction is related to the electromotive potential by the equation:

$$\Delta G' = -nF(E_o') + RT \ln [oxidant] \Big/ [reductant]$$

In the special case where the ratio of the oxidant concentration to reductant concentration is unity, this reduces to:

$$\Delta G^{o'} = -nF(E_o')$$

where n is the number of electrons transferred, F is the caloric equivalent of the Faraday (96,500 coulombs/4.185 joules · calorie^{-1}, or 23,058 calories/volt), and E_o' is the standard electrode potential at pH 7. The other symbols are as defined previously. The value E_o' applies only to reactions at a pH of 7. In fact, intracellular reactions exist at pH values other than 7.4; because the potential of the hydrogen electrode and other electrode reactions involving protons are pH dependent, correction of the standard values is required. Such correction is accomplished by inserting an extra term in the equation for the electromotive potential, so that:

$$E_o'(pH\ 7.4) = E_o'(pH\ 7) + \frac{RT}{nF} \ln [oxidant] \Big/ [reductant] + \frac{RT}{nF} \ln [10^{-7.4}]$$

If the temperature is 30° C, the substitution of appropriate numerical values leads to the equation:

$$E_o'\ (at\ any\ pH,\ X) = -0.06X + E_o'\ (at\ pH\ 7)$$

Effect of concentrations on oxidation-reduction reactions

Biologic oxidations do not take place under standard conditions. It is, therefore, helpful to examine the effect of intracellular metabolite concentrations on electromotive potentials in oxidation-reduction systems. Consider the NAD^+-linked conversion of oxaloacetate to malate. From the values given in Table 5-1, and remembering that the half-reaction with the more positive potential will be the oxidizing agent, the half-reactions can be written as follows:

1) $2H^+ + 2e + $ oxaloacetate $\longrightarrow$ malate $\qquad E_o' = -0.17$ volts

2) $NADH + H^+ \longrightarrow NAD^+ + 2H^+ + 2e \qquad E_o' = +0.32$ volts

$$\overline{\qquad\qquad \Delta E_o' = +0.15 \text{ volts}}$$

Consequently, under standard conditions:

$$\Delta G^{o'} = -2(23,058)(0.15) = -6917.4 \text{ cal/mole}$$

Let us now assume more nearly intramitochondrial concentrations such as the following:

oxaloacetate	3×10^{-3} M
malate	1×10^{-3} M
NAD^+	5×10^{-4} M
NADH	2×10^{-4} M

It may be further assumed that the reaction is carried out at $37°$ C and a pH of 7. The values of E' for each half-reaction that would be observed under the above concentration conditions are given by the equations:

$$E'_{obs_{(1)}} = E_o' + \frac{(1.987)(310)(2.303)}{2(23,058)} \log_{10} 3 = -0.156 \text{ volts}$$

and

$$E'_{obs_{(2)}} = E_o' + \frac{(1.987)(310)(2.303)}{2(23,058)} \log_{10} 2/5 = +0.308 \text{ volts}$$

The difference in potential of the two half-reactions, or E'_{obs} for the system, is therefore 0.152 volts, so that:

$$\Delta G'_{obs} = -nF(E'_{obs}) = -7011 \text{ cal/mole}$$

Summary of energetics

1. Free energy is a parameter of chemical reactions that measures the amount of energy available to perform useful work.

2. A knowledge of reaction energetics permits the calculation of the change in free energy (ΔG) by which it is possible to predict the direction of a chemical reaction.

3. Any number of reactions may be linked in coupled systems, provided there is a common component between any two sequential reactions in the system.

4. Coupling allows an energetically favored reaction to drive a less favored

reaction, provided only that the net change in free energy for the system is less than zero.

5. Comparison of energy changes is made with reference to standard states, defined above, but the energy changes associated with more nearly physiologic concentrations of metabolites are frequently larger than those observed under standard conditions.

RESPIRATORY CHAIN

When NAD^+, $NADP^+$, FMN, and FAD are reduced by action of the dehydrogenases with which they are associated, it is essential that a means be provided whereby they may be reoxidized, since the supply of these coenzymes available to any cell is limited. We have already seen that the reoxidation of NADH is coupled with the reduction of FAD to $FADH_2$. This is merely the first step in a more complex system of coupled enzymes that exist within the mitochondria in very close physical as well as chemical association. Indeed, the physical association is so close that the entire assemblage of enzymes can be isolated as a submitochondrial particle by relatively simple means. These particles can perform all of the individual steps in a chemical chain of events, as summarized by the following equation.

$$2H^+ + 2e + \frac{1}{2}O_2 \rightleftharpoons H_2O$$

Collectively, the group of enzymes found in this assemblage is known by either of two names. The first term, "electron transport chain," refers to the fact that oxidation and reduction are defined as the loss or gain of electrons, so that any system designed to promote biologic oxidations must be so intimately related to electron flow, or transfer, that the name should properly represent that fact. Since the system involves a coupled sequence or "chain" of reactions, the name is highly descriptive and was long regarded as appropriate. More recently, for reasons shortly to be discussed, the dogma that electron flow is the only possible driving force has been challenged. Therefore, it is not necessary to completely accept the older point of view. To avoid difficulty with the electromotive hypothesis, a more neutral term, "respiratory chain," has been proposed. This newer name is also descriptive and appropriate, since it is clear that the reactions performed involve the uptake of oxygen, or respiration.

The fine structure of the respiratory chain is still a subject of debate, but there is general agreement concerning the basic elements of its overall structure. Fig. 5-2 diagrams the essential components, which are enclosed in the dashed line. Input to the system consists entirely of the reduced coenzymes NADH and $FADH_2$. There are two distinct means of entry, one shown to the left and the other toward the top of the dashed enclosure. The first entry is the most general, since it readily admits NADH from any of a large number of NAD^+-linked dehydrogenases. The second entry admits $FADH_2$, which is derived from a smaller number of FAD-linked dehydrogenases.

Major components of the respiratory chain

The major components of the respiratory chain fall into three distinct classes of molecules, of which two serve as hydrogen carriers and one serves as an electron carrier. Each of these classes deserves detailed comment.

Flavoprotein components

The prosthetic group of the first flavoprotein (FP_1) accepts hydrogens from NAD^+-linked dehydrogenases. It is not part of any individual dehydrogenase but

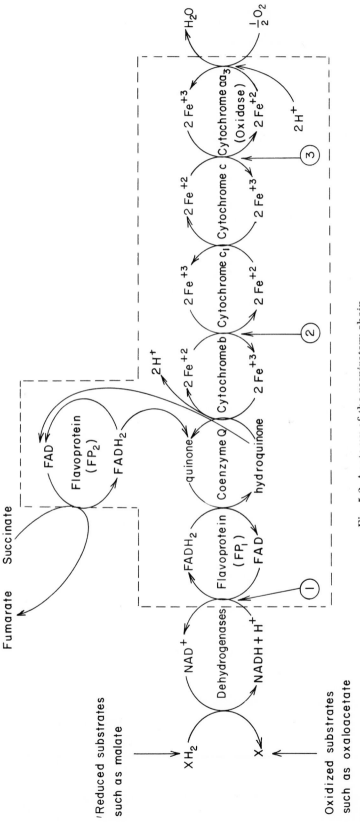

Fig. 5-2. Anatomy of the respiratory chain.

is contained in a separate enzyme known as *NADH–cytochrome b reductase.* FP_2 is also a hydrogen carrier and an integral part of the enzyme *succinate dehydrogenase,* which perhaps accounts for the specificity of the second entry. The mechanism by which flavoprotein prosthetic groups accept hydrogen has been shown schematically in Fig. 5-1, but the mechanism is probably more complex. It should be remembered that the chemistry of the flavin coenzymes and prosthetic groups is very subtle and not yet fully explained. The intrinsic reactivity of the isoalloxazine ring is heightened by the presence in flavoproteins of two additional moieties. These moieties include iron, as many as 8 gram atoms in some enzymes, which are known as nonheme iron since they are not incorporated into porphyrin ring systems. Also present are acid-labile sulfur atoms, so called because treatment with acid liberates H_2S. Still other flavoproteins may contain Mg^{++} or Cu^{++}, but their function in catalysis is uncertain.

Coenzyme Q, or ubiquinone

The compound known as coenzyme Q, or ubiquinone, connects the flavoproteins that function in the separate entries to the cytochromes (Fig. 5-2). This coenzyme exists in various forms, but the one found in mammalian tissues is frequently designated as coenzyme Q_{10}, which refers to the number of isoprenoid units in its side chain, as shown in Fig. 5-3. Bacterial and plant forms of the coenzyme usually contain fewer isoprenoid units.

Coenzyme Q appears to function as a collector of all the hydrogens that enter the respiratory chain. When it is reoxidized, electrons rather than hydrogen atoms are passed to the next chain component, with liberation of protons. These are transported to the chain terminus, where they react with oxygen to form water. An interesting antibiotic, piericidin, inhibits the action of coenzyme Q and blocks the reoxidation of NADH. The structure of piericidin is shown in Fig. 5-4.

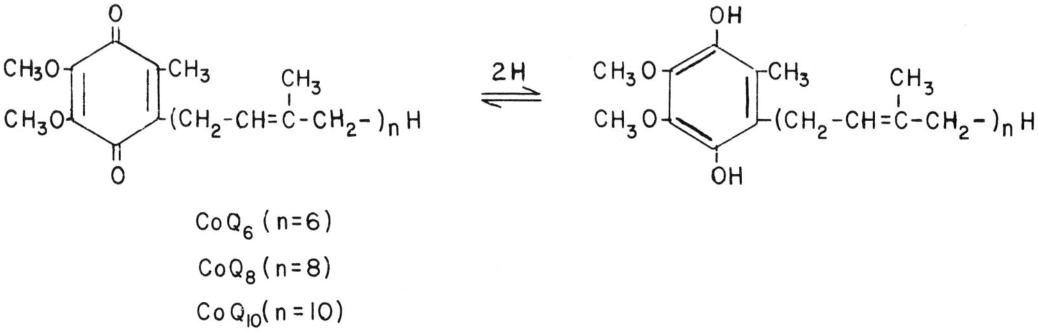

$$CoQ_6 \ (n=6)$$
$$CoQ_8 \ (n=8)$$
$$CoQ_{10} (n=10)$$

Fig. 5-3. Structure of coenzyme Q and its reduction product.

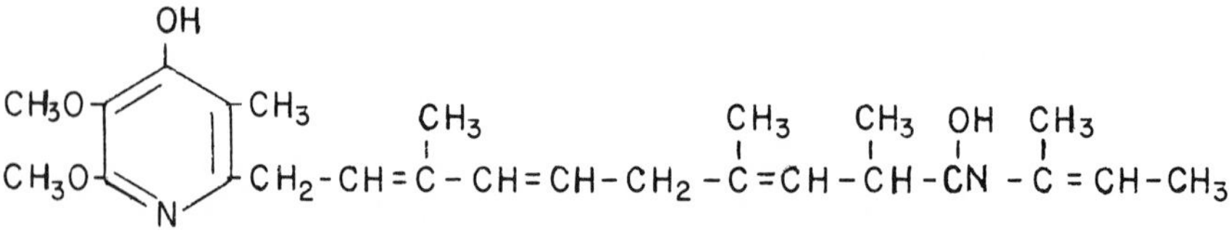

Fig. 5-4. Structure of piericidin.

Cytochrome components

There are at least four distinct cytochromes in the respiratory chain, as shown in Fig. 5-2. The cytochromes function as electron transferring agents, and all contain iron that is bound in a porphyrin ring. In the individual cytochromes the side chains attached to the periphery of the porphyrin ring are not the same as those in hemoglobin. The iron of the cytochromes can undergo reversible change from Fe^{+3} to Fe^{+2}. Except for cytochrome c, which is readily extracted from mitochondria, the other cytochromes are tightly bound to the membranes and are difficult to remove. For this reason it is doubtful that any except cytochrome c have been completely purified. Like hemoglobin, the cytochromes react with carbon monoxide and with cyanide ions. The complex between the cytochromes and carbon monoxide is also reversible, and its breakdown is hastened by exposure to light. However, considering the cellular locus of the cytochromes, this susceptibility to light is of little help in treating carbon monoxide poisoning. The best treatment is still exposure to high oxygen concentrations. The treatment of cyanide poisoning is based on the destruction of the cyanide; it is usually oxidized to thiocyanate by administration of thiosulfate ion. In addition, methylene blue, which can undergo reversible oxidation-reduction at a potential close to those of the cytochromes, is administered to serve as an artificial electron bypass around the poisoned respiratory chain.

Lipid matrix

While it probably does not play an active enzymatic role, some mention should be made of the matrix or supporting structure in which the enzymes are located. There is ample evidence that the matrix is composed of phospholipids that contain a large proportion of highly unsaturated fatty acids. Mammalian mitochondria contain as much as 26% by weight of phospholipid and are also characterized by the presence of a rather acidic phospholipid, cardiolipin (Chapter 8, p. 371), which accounts for nearly one sixth of the total phospholipid. This substance is almost specific for mitochondria, especially in the heart, and may well play a special role in fixing the enzymes into the respiratory chain and in facilitating the movement of protons through the matrix.

Respiratory chains of endoplasmic reticulum

Microsomes also contain a respiratory system. The microsomal chains act on $NADP^+$-linked enzymes preferentially, and the NADPH is oxidized by specific FAD-containing flavoproteins. The reduced flavoproteins are oxidized by a non-heme, iron-containing enzyme about which little is known. This enzyme takes the place of the cytochromes a, b, and c of mitochondria. The final enzyme in microsomal respiratory chains is also a cytochrome, analogous to cytochrome oxidase, which forms a very strong, light-absorbing complex with carbon monoxide. This complex has an absorption maximum for light at a wavelength of 450 nm and is known as cytochrome P_{450}. Since microsomes are involved in the desaturation of fatty acids, in hydroxylation of steroids, and in the detoxification of many drug substances, the need for an efficient oxidative system is clear.

Structural summary

A moment's reflection may well give rise to several questions. Why is this scheme so complex; why are so many steps needed; particularly, why are there at least four cytochrome carriers for electrons? In general, the questions can be answered as follows. Between any two chemical steps, for example, the oxidation of NADH and the combination of $2H^+ + \frac{1}{2}O_2 + 2e$ to form water, there is a finite difference in energy. It is easier to control the overall reaction if the total difference is broken into small, discrete, reversible units, which in turn increases the overall efficiency of the reaction. Furthermore, although the respiratory chain may seem complex, it is undoubtedly much simpler than other possible systems in which each dehydrogenase would have its own separate and distinct access to oxygen. Thus it should be regarded as an efficient means of unifying a very large number of reactions in a way that simplifies function and control. However, efficiency is not always an unqualified blessing. For example, if some accident impeded or interrupted the functions of the respiratory chain, the fact that so many reactions feed into it would produce repercussions in many other areas of metabolism. Cyanide, which blocks the action of cytochrome oxidase, is a notoriously quick-acting poison. Drugs and toxins may produce fever, disease, or death by interfering at some point in the respiratory chain. Vulnerability is thus the price paid for lack of biochemical redundancy.

HIGH-ENERGY COMPOUNDS

Having examined reaction energetics, it is necessary to explore how any part of the energy that passes through our cells is trapped or conserved for useful work. Such an exploration requires an understanding of what have been called high-energy compounds. Strictly speaking, there is no such thing as a high-energy compound or a high-energy bond, but these terms are useful and have been generally accepted. The term "high-energy compound" describes substances that contain one or more bonds of a special sort, sometimes represented by the symbol $\sim$ rather than by a straight line. The characteristics that differentiate this class of compounds are easily defined.

High-energy compounds are those that contain one or more $\sim$ bonds and that undergo a large, negative change in free energy when these bonds are hydrolyzed.

In quantitative terms, when hydrolysis of a bond yields a $\Delta G^{0\prime}$ of -1 to -5 kcal/mole, it is known as a low-energy or ordinary bond. When hydrolysis yields a $\Delta G^{0\prime}$ of -5 to -15 kcal/mole, it is known as a high-energy bond. Although this division seems arbitrary, experience indicates that it is useful for classifying most compounds. Typical data for representative high- and low-energy bond types are given in Table 5-2.

It is worth noting that among the high-energy compounds described in Table 5-2, all but two contain phosphate groups, and even one of the exceptions, acetyl CoA, can be derived from acyl adenylates without loss of energy. Yet it is equally clear that not all phosphate-containing compounds are high-energy compounds. Phosphorus-containing high-energy compounds include only those that can be classified as **enol phosphates, acyl phosphates,** or **pyrophosphates.** These types of compounds are examples of either simple- or mixed-acid anhydrides. A simple-acid anhydride is produced when the elements of water are abstracted from two molecules of the same acid; a mixed-acid anhydride results when the elements of water are abstracted from two different acid molecules.

The large amount of energy released on hydrolysis of a high-energy bond results from several factors. The primary cause appears to be the ionization of the charged groups generated on hydrolysis; contributions to the energy release are also generated by diminished resonance energy in the products of hydrolysis as compared to the intact compound.

The existence of high-energy compounds and the fact that they can be generated in our cells is of the highest importance to an understanding of metabolism.

Table 5-2. Typical high- and low-energy compounds

CHARACTERISTIC LINKAGE	GENERAL FORMULA	GENERAL DESIGNATION	BIOCHEMICAL EXAMPLE	$\Delta G^{0'}$ (kcal/mole)
High-energy compounds				
$-\overset{\text{O}}{\overset{\|}{\text{C}}}-\underset{\text{H}}{\text{N}}\sim$	$R\overset{\text{NH}}{\overset{\|}{\text{C}}}-\underset{\text{H}}{\text{N}}\sim PO_3^=$	Guanidinium phosphate	Creatine phosphate	-10.5
$-\overset{\text{O}}{\overset{\|}{\text{C}}}-\text{O}\sim$	$R\overset{\text{CH}_2}{\overset{\|}{\text{C}}}-\text{O}\sim PO_3^=$	Enol phosphate	Phosphoenol pyruvate	-14.8
$-\overset{\text{O}}{\overset{\|}{\text{C}}}-\text{O}\sim$	$R\overset{\text{O}}{\overset{\|}{\text{C}}}-\text{O}\sim PO_3^=$	Acyl phosphate	Acetyl phosphate	-10.1
$-\overset{\text{O}}{\overset{\|}{\underset{\text{O}^-}{\text{P}}}}-\text{O}\sim\overset{\text{O}}{\overset{\|}{\underset{\text{O}^-}{\text{P}}}}-$		Pyrophosphate	Adenosine triphosphate	-7.0
			Uridine diphosphoglucose	-7.3
			Cytidine diphosphocholine	-7.3
$-\overset{\text{O}}{\overset{\|}{\text{C}}}\sim\text{S}$	$R\overset{\text{O}}{\overset{\|}{\text{C}}}\sim SR$	Acyl thioester	Acetyl CoA	-7.5
	$R-\overset{+}{\underset{\|}{\text{S}}}-$		S-Adenosyl methionine	-7.0
Low-energy compounds				
$-\overset{\text{O}}{\overset{\|}{\text{C}}}-\underset{\text{H}}{\text{N}}-\text{R}$		Peptide bond	Tyrosylglycine	-0.5
$-\overset{\text{O}}{\overset{\|}{\text{C}}}-NH_2$		Amides	Glutamine	-3.4
$R\overset{\text{O}}{\overset{\|}{\text{C}}}-OR$		Esters	Ethyl acetate	-1.8
$RCH_2-O\overset{\text{O}}{\overset{\|}{\underset{\text{O}^-}{\text{P}}}}-O-$		Sugar phosphates	Glucose 6-phosphate	-2.9
			Glycerol-1-phosphate	-2.3

These compounds provide the means whereby approximately 40% of the total energy that passes through body tissues can be conserved. Conservation is accomplished by several specific reactions that couple the formation of such compounds with a biologic oxidation. The principles of coupling have already been discussed. In coupled reactions, as in others, the laws of conservation of energy must be obeyed. Table 5-1 indicates that the free energy of the hydrolysis of ATP to adenosine diphosphate (ADP) is approximately -7.0 kcal/mole. If we wished to reverse this reaction by some intracellular process, it would be necessary to put at least 7 kcal/mole back into the formation of ATP. Actually, the requirement would be even greater since biologic processes do not operate with absolute efficiency. According to the principles of energetics stated earlier, the synthesis of ATP from ADP would require coupling with an oxidation step having a standard reduction potential of at least 0.15 volts. This is a fairly large energy change, and there are not many sites at which potential differences of this magnitude can be found.

OXIDATIVE PHOSPHORYLATION

The terminal stages of biologic oxidation, which are catalyzed by the respiratory chain, provide three separate sites at which the energy change is of sufficient magnitude to permit coupling with the synthesis of ATP from ADP and inorganic phosphate (P_i). (The three sites are indicated in Fig. 5-2 by the encircled numbers.) Thus, provided supplies of ADP and P_i are sufficient, for each mole of NADH that enters the respiratory chain, three moles of ATP can be produced. Since this process also involves the consumption of one gram atom of oxygen, biochemists commonly speak of it as having a P/O ratio of 3, referring to the formation of three high-energy bonds. Fig. 5-2 shows that entry to the respiratory chain by way of succinate entails a distinct disadvantage, as this entry misses the first site of phosphorylation. Therefore, the oxidation of succinate provides a P/O ratio of only 2.

In normal mitochondria the coupling between oxidation and phosphorylation is usually so tight that one process does not occur without the other. Hence the name given to this process is oxidative phosphorylation, thus distinguishing it from other phosphorylating mechanisms that do not involve the respiratory chain. The second mechanism by which high-energy bonds can be synthesized is known as substrate-level phosphorylation; it will be discussed in Chapters 6 and 7.

Control of oxidative phosphorylation

Since there is such tight coupling between the oxidative and phosphorylating mechanisms of most mitochondria, it is possible to envisage several points at which high-energy bond synthesis could be controlled. One could go even further and define the respiratory or metabolic state of isolated mitochondria in terms of: (1) substrate concentration; (2) concentration of ADP, which serves as the acceptor for newly synthesized high-energy bonds; (3) concentration of ATP, the product of the system; or (4) the concentration of oxygen, which is required for operation of the respiratory chain. Experimentally, mitochondria can be suspended in suitable solutions and, by means of the so-called oxygen electrode, changes in oxygen uptake by the particles can be monitored following various additions to the system. Such an experiment is graphed in Fig. 5-5. At the upper left, mitochondria (*MITO*) are shown being added to the suspending solution, and consequently the chart indicates a slight decrease in the partial pressure of oxygen. The slow rate of oxygen uptake is caused by the relative lack of substrate and ADP. When substrate is added, the rate of oxygen uptake increases only slightly, because the ADP con-

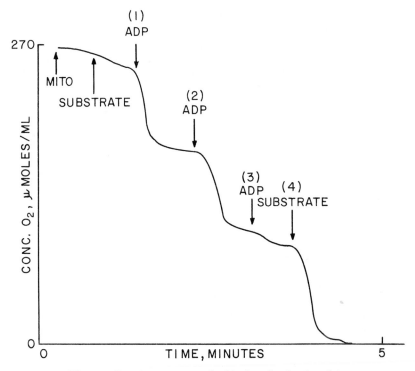

Fig. 5-5. Respiratory control of isolated mitochondria.

tent of the mitochondria is still rate limiting. When ADP is added, the rate of oxygen uptake increases significantly for a brief period but then levels off. Obviously, the addition of ADP in the presence of a suitable substrate stimulates respiration. The stimulation ceases when the added ADP has been completely converted to ATP. This is shown by the effect of a second addition of ADP, which once again has a stimulating effect. A third addition of ADP produces only a slight response; by this time the substrate has been consumed. When more substrate is added, respiration is again stimulated but soon ceases because the limited amount of oxygen that could be dissolved in the suspending solution is exhausted; that is, the solution has become anoxic. Although these data may be applied strictly only to isolated mitochondria, it is believed that similar situations can exist in vivo.

Concept of energy charge

The energy charge of the adenylate system can be defined as half the average number of $\sim$P groups per adenosine moiety; it has values that range from 0 to 1. The charge can be expressed in terms of the individual nucleotide concentrations as follows:

$$\text{Energy Charge} = \frac{1}{2} \frac{\left([ADP] + 2\,[ATP]\right)}{[AMP] + [ADP] + [ATP]}$$

The denominator of the fraction represents the total amount of adenine nucleotides in the system, while the numerator represents that fraction of the total that could, on hydrolysis, release the energy equivalent to one or more $\sim$P bonds. The numeric factor is introduced arbitrarily to make the values of the energy charge range from zero to one. The effect of ADP concentration in controlling the

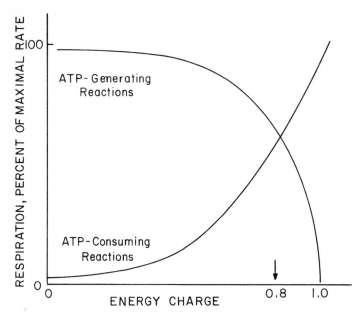

Fig. 5-6. Energy charge and mitochondrial respiratory rate.

mitochondrial respiratory rate is indicated in Fig. 5-6, where it is shown that if reactions that generate ATP predominate, the concentration of ADP will drop and respiration will diminish. If ATP-consuming reactions predominate, the concentration of ADP will rise, as will the rate of respiration. Under nomal conditions approximating the steady state, the curves intersect each other at an energy charge of approximately 0.8. An energy charge value of one would represent a state where all of the ADP, which acts as the charge acceptor, was phosphorylated; at this theoretic point oxidative phosphorylation would cease as a result of the lack of acceptor. A value of zero would represent that point where all oxidation had ceased (death), with a consequent cessation of phosphorylation. In fact, the physiologic range of value is quite narrow, ranging from about 0.6 to about 0.9. Even in as active a tissue as rat myocardium, the total adenine nucleotide concentration is roughly 16 μmole/g of wet tissue weight. This very small amount of adenine nucleotide not only acts as the major vehicle for capturing useful energy but is also involved in the primary control of oxygen consumption.

Inhibition and uncoupling of oxidative phosphorylation
Site-specific inhibitors

Although Fig. 5-2 shows three sites of oxidative phosphorylation in the respiratory chain, there is no indication as to whether the three sites are equivalent. Studies of isolated mitochondria provide ample evidence for the existence of site-specific inhibitors. How these inhibitors work is not clear, but at concentrations between 10^{-3} to 10^{-6} M they block the formation of high-energy bonds by blocking the oxidative process at specific sites on the respiratory chain. Some information about specific inhibitors is given in Table 5-3. Several of the inhibitors mentioned are sometimes employed as drugs, but this does not imply that the pharmacologic effect is necessarily related to their action as inhibitors. In many instances the effective concentrations for drug use differ significantly from the effective concentrations for purposes of inhibition.

Table 5-3. Site-specific inhibitors of oxidative phosphorylation

SITE	NATURE OF INHIBITORS	REMARKS
1	Alkyl guanidines*	Guanethidine, a hypotensive agent
	Rotenone	Insecticide
	Chlorpromazine	Tranquilizing drug
	Barbiturates	Seconal, Amytal, etc., used as sleep-inducing drugs
	Progesterone	Hormone, inhibits oxidation but to lesser degree than it inhibits phosphorylation
2	Antimycin a	
	BAL	British anti-Lewisite, an antidote to an old war gas, actually 1,2-dithioglycerol
	Phenethylbiguanides†	Phenformin, used as an oral insulin substitute
	Naphthoquinone	
	Hydroxyquinoline-8-oxide	
3	Carbon monoxide	
	Cyanide ion	
	Azide ion	

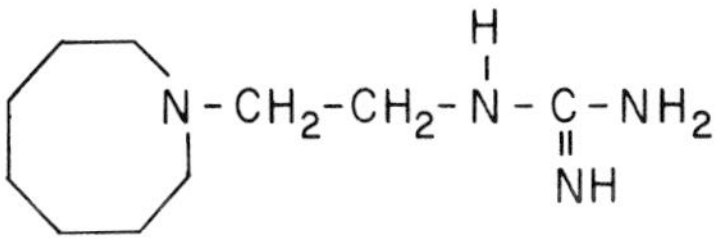

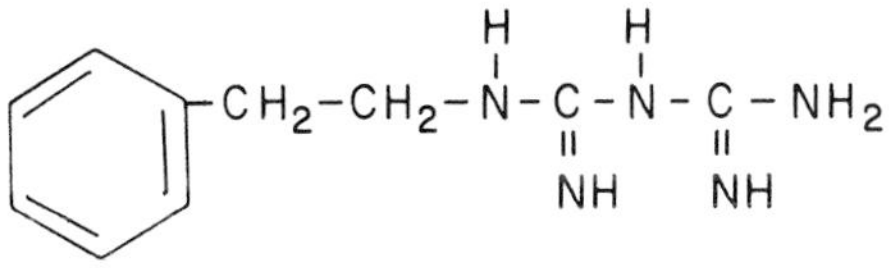

Non–site-specific inhibitors

The very fact that site-specific inhibitors exist might be regarded as an indication that the three sites are not equivalent. However, this indication is clouded by the existence of a second class of inhibitors that can act at any of the three sites and that appear to operate by different mechanisms than those of the first class. They function primarily by blocking phosphorylation rather than oxidation, but in tightly coupled systems the overall result is the same. Several but not all of the non–site-specific inhibitors are antibiotics, including oligomycin and aureovertin; others such as atractylate are toxic plant glycosides.

Uncouplers of oxidative phosphorylation

Oxidative phosphorylation can be interrupted by a third class of compounds known as uncouplers. Uncoupling agents either loosen or completely destroy the normally tight coupling between the oxidative and phosphorylating mechanisms; thus the P/O ratio is either reduced or becomes zero. Oxidation rates increase while phosphorylation decreases; the result is the production of extra heat. In intact animals extra heat production is manifested grossly as *fever*. Diminished formation of ATP may also produce clinical signs of other metabolic disturbances, since the relative lack of ATP can be associated with such important cellular processes as ion transport and membrane permeability.

Although synthetic uncoupling agents might be broadly described as lipophilic proton donors, a few are lipophilic proton acceptors. They apparently function by damaging the lipid-rich membranes that are essential for the integrity of the mitochondrial respiratory chain. Recent evidence indicates that these agents render the lipid-rich membranes abnormally permeable to protons. Common uncouplers include 2,4-dinitrophenol (once used as a weight-reducing drug), dicumarol, certain substituted salicylanilides, and even free salicylate, a metabolite of aspirin. Weight for weight, the salicylanilides are the most potent uncouplers yet devised. Natural uncouplers include the bile pigment, bilirubin, free fatty acids, and perhaps thyroxin, provided that they are present at sufficiently high concentrations within the mitochondria. Normally, these materials do not exist at uncoupling concentrations, but in some diseases it is possible. Infants born suffering from the result of Rh incompatibility (erythroblastosis fetalis) may have plasma free bilirubin concentrations of 20 mg/dl or more, levels that allow bilirubin to penetrate into brain cells and possibly cause permanent brain damage.

In order to prevent permanent damage from the very high circulating bilirubin concentrations, it is common clinical practice to perform exchange transfusions on erythroblastotic infants, totally replacing the blood volume one or more times until the anti-Rh agglutinins produced by the mother have been removed from the infant's body. As mentioned earlier, some pathogenic microorganisms may produce soluble toxins that can also act as uncouplers. Additional exposure to uncouplers may come from certain insecticides that are commonly used in food preparation establishments and even in the home. These agents have about the same toxicity for the human as for the housefly, but because of the larger body mass it is only by accident that human intoxication may occur. Nevertheless, it is well to be aware that such agents do pose a potential health hazard.

Structure of the mitochondrial inner membrane

By virtue of recent research involving chemical and electron microscopic study, a good deal is now known about the nature of mitochondrial inner membranes. Some of this is detailed in Figs. 5-7 and 5-8. In Fig. 5-7, *A*, a thin section of a typical mammalian mitochondrion is shown in schematic fashion. Because of the many infoldings of the inner membrane, its surface area is much larger than that of the outer membrane. These infoldings, or cristae, provide the surface on which enzymes of the respiratory chain are fixed, along with succinate dehydrogenase and other nonrespiratory chain components. The remaining enzymes of the Krebs cycle are located in the matrix space and are not shown in the figure. The matrix surface of the inner membrane is closely studded with knob-like projecting particles, which are known to contain the reversible ATPase involved in the process of oxidative phosphorylation. These are shown in Fig. 5-7, *B*. When the knob-like projections are removed from submitochondrial vesicles, which can be accomplished by simple chemical and physical means, the modified vesicles can adequately perform oxidations but not phosphorylation. When the knob-like particles are restored, which can also be accomplished by appropriate chemical and physical manipulation, the capacity to couple oxidation with phosphorylation is restored. These experiments, first performed in Ephraim Racker's laboratory, represent a biochemical milestone.

Some details of the arrangement of the respiratory chain components within the inner membrane are shown in Fig. 5-8. The matrix surface is shown at the top, with the protruding ATPase units. The major cytochrome components and succinate dehydrogenase are shown within the membrane thickness. Note that the respiratory chain is in the form of an open loop, with both ends facing the matrix side and with only cytochrome *c* facing toward the outer membrane. The symbols marked Ⓣ represent various translocases, or transferring systems, that are integral membrane components responsible for the controlled movement of materials into or out of the matrix compartment. These are discussed in greater

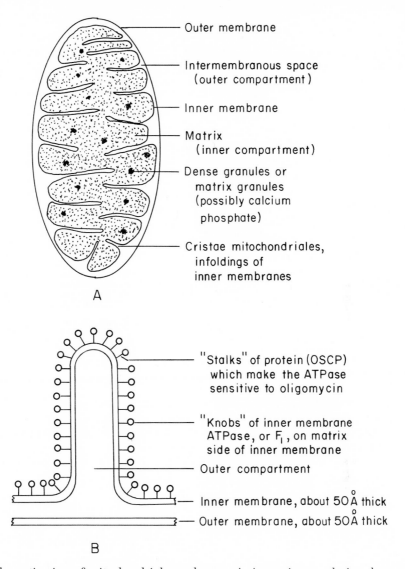

Fig. 5-7. Schematic view of mitochondrial membranes. **A,** Approximate relations between the outer and inner membranes, and **B,** relation of reversible ATPase to matrix surface of inner membranes. (OSCP = oligomycin sensitivity–conferring protein.)

detail in Chapter 6. The symbols marked with Ⓠ indicate the molecules of coenzyme Q, which are more numerous than the other chain components. Note also the circles marked ⊘, which represent the three coupling sites at which ATP synthesis is thought to occur. The energy captured at each of these three sites must ultimately be transferred to the reversible ATPase to drive the synthetic reaction.

Mechanism of oxidative phosphorylation

Although oxidative phosphorylation has been studied intensively for approximately 40 years, there are still unanswered questions regarding its mechanisms; the coupling process in particular is poorly understood. Two major theories, the chemical theory and the chemiosmotic theory, command our attention.

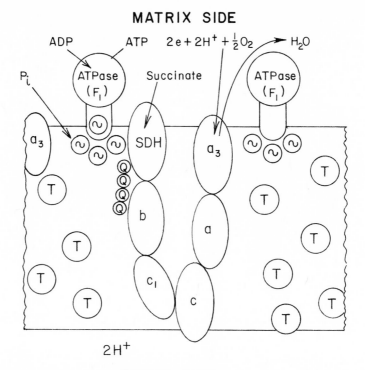

MATRIX SIDE

OUTER SIDE

Fig. 5-8. A schematic proposal for the localization of respiratory chain components within the inner mitochondrial membrane. Note that the chain is in the form of an open loop, with cytochrome c on the outer side, but with cytochrome a_3 and succinate dehydrogenase facing the matrix side. Note that some of the components are completely buried within the membrane thickness. The symbols ⓣ represent translocating systems, the symbols Ⓠ represent molecules of coenzyme Q, and the symbols ⌒ represent the three theoretical coupling sites. F_1 is Racker's designation of the reversible ATPase. These systems, together with other components not shown, are repeated to cover the entire inner membranes along the cristal surfaces.

Chemical theory

A broad outline of the chemical theory, the older of the two, is set forth in Fig. 5-9. This theory proposes that the coupling site is a discrete chemical entity that can be isolated. The evidence in its favor can be summarized as follows.

1. While both in vivo and in vitro experiments indicate that the normal function of the respiratory chain favors the uptake of oxygen and formation of ATP, there is significant evidence that the system is reversible. In the presence of excess ATP the system operates to form NADH from NAD^+.

2. The existence of site-specific inhibitors leads to the conclusion that within the coupling sites there are distinctive coupling factors (C_1, C_2, and C_3) that in some way are unique.

3. The coupling factors "trap" or "fix" oxidative energy by combining with a hypothetical intermediate (I) that is common to the three sites. $C_n \sim I$ is regarded as the first of a series of high-energy intermediates. Uncouplers such as 2,4-dinitrophenol operate by causing rapid hydrolysis of $C_n \sim I$, which clearly precludes the possibility of ATP formation.

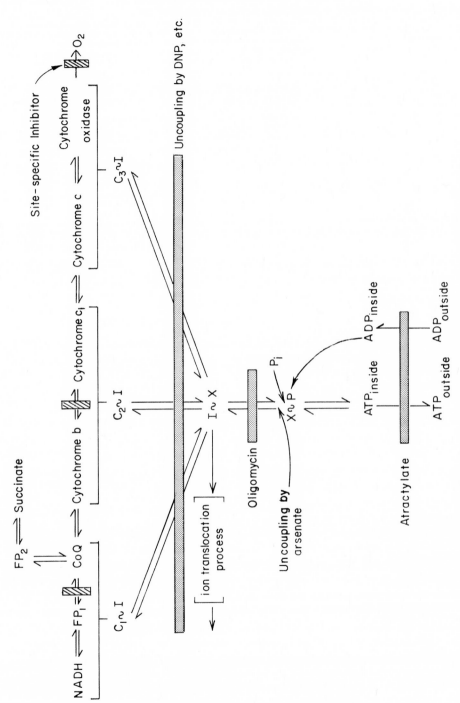

Fig. 5-9. Sites of inhibition or uncoupling of oxidative phosphorylation.

4. The individual complexes ($C_n \sim I$) funnel all of the $\sim I$ moieties to a second hypothetical intermediate (X), and the second high-energy intermediate, $I \sim X$, is formed. Normally, $I \sim X$ exchanges I for P_i to form $X \sim P$, but this exchange is blocked by the non–site-specific inhibitor oligomycin.

5. $X \sim P$ is the third high-energy intermediate to be formed and the first that contains phosphorus. Inorganic arsenate ion can also react with $I \sim X$ to form $X \sim As$, a complex that is unstable and that rapidly decomposes. This instability wastes energy. By aborting the formation of ATP, hydrolysis of the $X \sim As$ complex also serves as an uncoupling process, which, incidentally, partly explains the toxicity of arsenic compounds.

6. ADP can accept P from $X \sim P$ to form ATP, which is normally passed to the mitochondrial exterior by a specific system that exchanges ATP for ADP. If the mitochondria are treated with atractylate, the exchange of ATP for ADP is blocked. Accumulation of ATP and depletion of ADP will inhibit the respiratory chain, and oxygen uptake will ultimately cease. (The specific transport system for ATP and ADP is discussed in Chapter 6.) Thus atractylate blocks phosphorylation of *exogenous* but not *endogenous* ADP.

One may well ask why it is necessary to invoke a second hypothetical intermediate, X. The necessity arises because oligomycin-blocked mitochondria still contain a high-energy intermediate detectable by its ability to energize certain ion-translocating processes known to exist in mitochondria (p. 225). When mitochondria are treated with oligomycin and dinitrophenol, they can no longer translocate ions.

In spite of an enormous effort there is as yet no direct experimental evidence for the isolation of the hypothetical intermediates X or I, and scientists have begun to seek other explanations.

Chemiosmotic theory

Peter Mitchell and others have suggested that perhaps "X" and "I" do not exist in the form required by the chemical theory. It has also been argued that the only basis for describing oxidative phosphorylation in terms of electron transport is an ingrained prejudice that springs from an analogy with metallic conductors, where it is known that current flow depends on the existence of free electrons that are able to move through the crystal lattice of metallic substances. Mitchell has suggested that if a *protonmotive* force were substituted for an *electronmotive* force, new insight might be gained. Ordinary dry cell batteries are electronmotive devices. Protonmotive fuel cells have been designed and used in space projects by the National Aeronautics and Space Administration (NASA); more recently, fuel cells designed to continuously monitor blood glucose levels have been tested in animals.

Basis of Mitchell's hypothesis.   The hydrolysis of ATP can be written in the form of two half-reactions:

$$ATP + 2H^+ + O^= \rightleftharpoons ADP + P_i$$

$$H_2O \rightleftharpoons 2H^+ + O^=$$

which emphasizes not only the role of protons but also the fact that ATP formation depends on abstracting the elements of water from ADP and P_i. It is readily seen that a reversible ATPase, localized within a membrane of specialized properties, might create a protonmotive system and that it might very well be coupled with an oxidation-reduction system as well. The concept of a special membrane also provides the first clue as to why X and I have proved so elusive. There is now good evidence that the intermediates may be fixed elements of a membrane structure rather than of a bulk solution. Mitchell's hypothesis is schematically represented in Fig. 5-10 and entails the following characteristics:

1. There is a membrane-bound, proton-translocating, reversible ATPase system that may be composed of more than one protein molecule. It contains the hypothetical intermediates as X-H and I-OH, either as separate molecules or as functional groups.

2. There are proton-translocating, oxidation-reduction systems that couple with the reversible ATPase. In most details these systems are analogous to the oxidizing elements of the classic respiratory chain, except that in this theory they do not contain the coupling intermediates.

3. There exist exchange diffusion systems that link proton translocation to the exchange of selected ions (this topic is discussed in Chapter 6).

4. The membrane in which the entire system is assembled and in which the coupling of oxidation to phosphorylation takes place is ion-impermeable except through energy-linked processes. Free diffusion of ions is not possible.

The coupling complex extends through the entire thickness of the membrane and the C region protrudes on the matrix side in a small sphere (Fig. 5-10). X-H

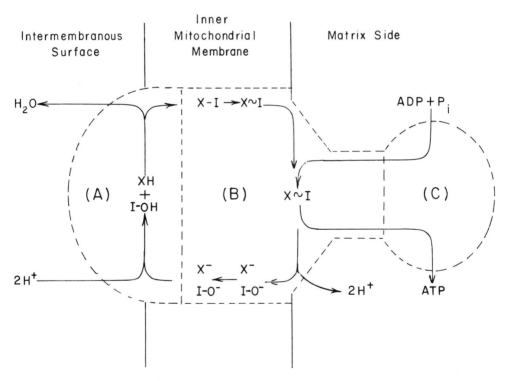

Fig. 5-10. Reversible proton-translocating ATPase according to Mitchell. Suggested functional regions are A, X-I hydrolase; B, X-I translocase; and C, X-I synthetase. In intact mitochondria the vectorial nature of the membrane is indicated by locations of each function given just above diagram. Note that in this proposal X and I are entirely in the membrane. They are distinct from the oxidation-reduction system. The dashed line outlines the limits of a single ATPase unit.

and I-OH are the agents that in some way collect the elements of water from ADP and P_i; in effect, each time one molecule of ATP is formed, two protons are extruded from the other side of the membrane, and in the process a proton gradient is built up across the membrane thickness.

In essence, Mitchell's argument states that the proton-impermeable membrane allows the generation of a proton concentration gradient, which is in itself a high-energy state, and that this energy can be utilized to drive the formation of ATP *or* the transport of ions. Accordingly, the search for coupling factors that could be isolated and purified is not likely to succeed, because the energy-transducing mechanism is in the nature of the membrane and not in free chemical intermediates.

The relation between the Mitchell hypothesis and the older chemical hypothesis, in terms of proton gradient generation, is shown in Fig. 5-11, in which the inner membrane separates two compartments with differing H^+ concentrations. Within the membrane are shown several components responsible for proton transport, the two major flavoproteins (FPI and FPII, coupling the oxidation of NADH and succinate, respectively), the nonheme iron protein (NHI), which is part of the reversible oxidation-reduction mechanism, coenzyme Q (CoQ), and the cytochromes. The electrons can move back and forth within the membrane but cannot escape. Protons generated by various biologic oxidations are passed from one side of the membrane (the inner side) to the other (the outer side). The final stoichiometry indicates that six H^+ are exported from the matrix during the production of three ATP.

Not all aspects of the chemiosmotic theory have won universal acceptance,

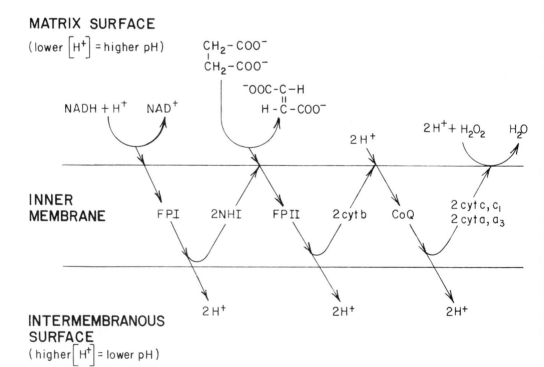

MATRIX SURFACE

(lower $\left[H^+\right]$ = higher pH)

Fig. 5-11. Mitchell's hypothesis in terms of inner mitochondrial membrane structure. NHI represents nonheme iron protein: FPI and FPII are flavoproteins. The electron-impermeable membrane serves as a means of transporting protons.

although the theory is consistent with known data on the composition of the respiratory chain and the associated coupling phenomena. The new concepts it introduces are entirely mechanistic, and although they are unproved, new hypotheses suggest new and meaningful experiments. This theory sheds light on why uncouplers are almost always lipophilic proton donors or acceptors and focuses attention on membrane-bound phenomena.

MITOCHONDRIAL ION TRANSLOCATION

It has been known for some time that actively respiring mitochondria accumulate Ca^{++} when it is provided in the medium in low concentrations (< 1 mM). Conversely, the addition of small amounts of Ca^{++} to resting mitochondria induces brief "pulses" of extra respiration. The respiration under these conditions is *not* associated with phosphorylation; it appears that either ATP may be formed or Ca^{++} may be accumulated (translocated). According to the chemical theory of oxidative phosphorylation, this phenomenon would best be explained as a diversion of energy at the level of $X \sim I$, since ATP is not involved. According to the chemiosmotic theory, Ca^{++} accumulation could be driven, or energy-linked, by a proton gradient. In either case, the phenomena are quite properly described as energy-linked translocations, inasmuch as ion uptake involves an obligatory dissipation equivalent to a high-energy bond. In the presence of either dinitrophenol or cyanide there is no Ca^{++} uptake.

When the concentration of ATP is increased through the oxidation of a suitable substrate, the uptake of Ca^{++} takes a different course. In this situation the Ca^{++} uptake is accompanied by P_i uptake, and a single proton is extruded from the mitochondria. A precipitate of calcium phosphate is formed within the mitochondria for as long as the oxidation of substrate continues; when oxidation is concluded, the precipitate apparently dissolves and the Ca^{++} is released back into the medium. Ca^{++} release is stimulated more when the mitochondria are suspended in K^+-containing buffers than when they are in Na^+-containing buffers. It is well known that highly irritable and contractile tissues are quite sensitive to Ca^{++} concentration, and it appears that ion translocation by mitochondria may be important in modulating muscle contraction through regulation of the Ca^{++} in the sarcoplasm.

The monovalent ions K^+ and Na^+ do not freely penetrate mitochondrial membranes, but K^+ can be taken up by an energy-linked process with the simultaneous ejection of a proton. In recent years a class of antibiotics termed *ionophorous* (ion-carrying) has been shown to markedly increase the penetrability of K^+, and to a much lesser extent that of Na^+, across mitochondrial membranes. A widely studied member of this group is valinomycin, which contains thirty-six atoms in its cyclic structure. It is composed of alternating hydroxy and amino acids, the sequence D-valyl-D-hydroxyisovaleryl-L-valyl-L-lactyl being repeated three times, as shown in Fig. 5-12. Nigericin is a similar cyclic antibiotic that also discriminates between K^+ and Na^+. Gramicidin is a linear polypeptide that is also ionophorous but less discriminating. All of the ionophorous antibiotics appear to function by increasing the ease with which alkali ions penetrate the mitochondria. When K^+ or Na^+ traverse the membrane in one direction, protons traverse it in the opposite direction, destroying the proton gradient that is required for oxidative phosphorylation according to the chemiosmotic theory. As a result, energy that would otherwise be used for ATP formation is dissipated. Although with nigericin this effect can be demonstrated only in the presence of K^+ (not Na^+), it is still that of uncoupling.

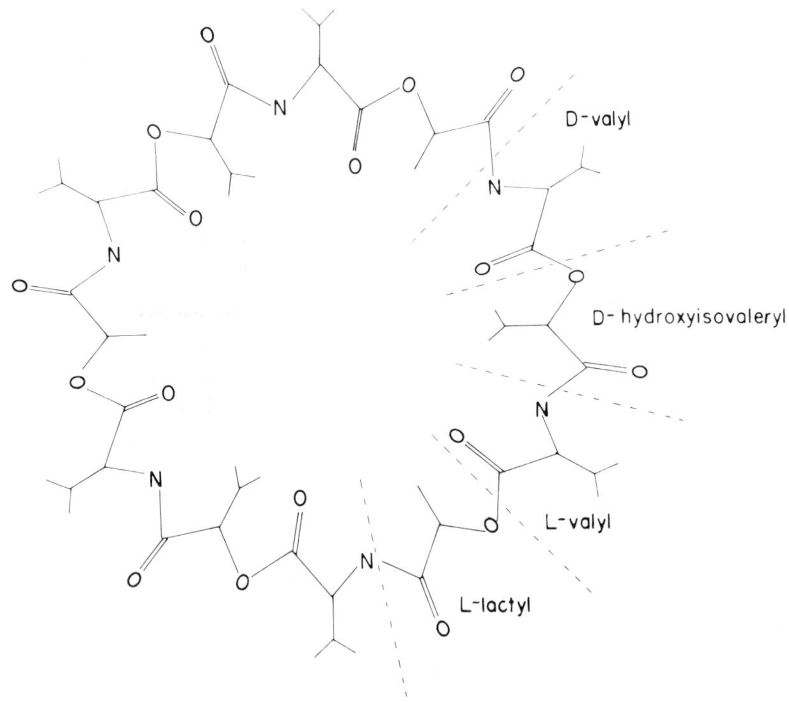

Fig. 5-12. Structure of valinomycin.

SODIUM-POTASSIUM PUMP

In Chapter 4 it was demonstrated that cells jealously guard their internal ionic composition. The major intracellular cation is K^+ and the major extracellular cation is Na^+. The clinical gravity of Cushing's disease and of primary aldosteronism (Chapter 13) results in large part from changes in the ionic composition of body fluids that may affect the intracellular ionic structure as well. Hypokalemia is dangerous to the heart, as is hyperkalemia, which is sometimes associated with renal disease. To maintain a homeostatic concentration of K^+ in the cells and to keep Na^+ out of the cells requires a constant expenditure of the energy derived from ATP. The enzyme system that performs this task is known as the sodium-potassium (Na^+/K^+) pump. It is bound to the plasma membrane of the cell and can also be described as a Na^+/K^+-dependent ATPase; that is, it is an enzyme system that hydrolyzes ATP at a rate that depends on the cation concentrations. The arrangement at the cell surface is such that the reaction is *vectorial,* or directed in space. In order for the reaction to proceed, both Na^+ and K^+ must be present, although in some instances NH_4^+ can be substituted for K^+. Note carefully that this system operates at the plasma membrane and is not identical with other ion-translocating systems that operate in the mitochondrial membranes.

Various explanations of the Na^+/K^+ pump have been proposed, but none of them has been completely accepted. One reasonable explanation has been offered by Opit and his colleagues. According to their theory the ATPase is a molecule that can be phosphorylated by ATP. When phosphorylation occurs the conformation of the ATPase undergoes a marked change, which is reversed by dephosphorylation. Phosphorylation and dephosphorylation are sensitive to Na^+ and K^+, respectively. A schematic representation of the mechanism is shown in Fig. 5-13.

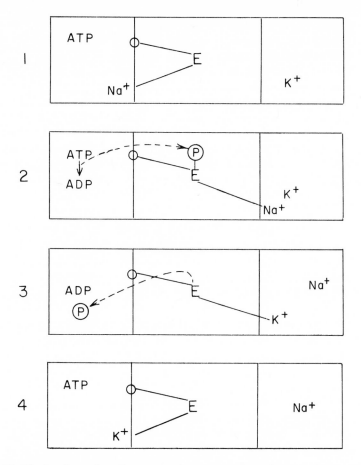

Fig. 5-13. Mechanism of sodium-potassium pump according to Opit and associates. The central square in each block represents the plasma membrane. To the left is the cellular interior; to the right is the external fluid. The pump is known to require Mg^{++} to function.

For simplicity the system has been divided into four stages, indicated in Fig. 5-13 by the numerals at the left.

1. Inside the cell there are ample supplies of ATP, and just inside the plasma membrane we find Na^+, which has penetrated to the interior of the cell, possibly in connection with the uptake of glucose or amino acids. The ATPase, represented by E, is shown as a nonlinear molecule. The circle represents one point on the enzyme that is supposed to be firmly fixed to the membrane. Another point, which can bear on alkali ion, is directed at this stage toward the inner plasma membrane surface, but it is able to move through the substance of the membrane. Outside the cell there are ample supplies of K^+.

2. After the concentration of Na^+ within the cell reaches some limiting value, it triggers the phosphorylation of the enzyme at the expense of ATP. The phosphorylation induces a change in the shape of the enzyme as well as a charge that allows it to bind Na^+, so that the bound ion is quite literally swept to the outer edge of the cell.

3. The Na^+ is exchanged for K^+ and the enzyme is dephosphorylated, causing the molecule to return to its original shape and sweeping the K^+ back across the thickness of the plasma membrane into the cell.

4. The bound K$^+$ is released into the interior of the cell as the charge on the dephosphorylated enzyme dissipates. The ADP has meanwhile been phosphorylated, and the cycle is completed.

The entire sequence of events depicted in Fig. 5-13 postulates the movement of ions *against a concentration gradient* by expenditure of energy from ATP. There are certain agents that inhibit this pump system; one of them is *ouabain*, a glycoside similar in structure to digitalis. In vitro a concentration of 10^{-4} M ouabain is sufficient to completely inhibit the Na$^+$/K$^+$-dependent ATPase. Ouabain was originally employed by South American Indians as a poison that they applied to arrowheads in order to kill their prey, demonstrating that interference with the Na$^+$/K$^+$ pump entails the gravest physiologic consequences.

MUSCLE CONTRACTION

In the examples just discussed, energy generated by the hydrolysis of ATP is coupled to the performance of osmotic work. Even though the Na$^+$/K$^+$ pump involves an enzyme that is supposed to undergo a change in shape, that change is not associated with any demonstrable change in the plasma membrane; that is, the system performs no mechanical work. In the case of muscle the situation is quite different, as the utilization of energy derived from ATP results in a gross change in the size and shape of the contractile cells. The length of a parallel-fibered muscle like the sartorius, with its insertion cut and with a small load, may decrease by as much as 55% of its resting length when subjected to tetanic stimulation with an inductorium. An understanding of the contractile process requires some knowledge of the structure of the three types of mammalian muscle cells.

Smooth muscle

Smooth muscle is composed of relatively small cells. Each contains a single nucleus, some mitochondria, and occasionally granules containing glycogen. The cytoplasm, or *sarcoplasm*, is relatively homogenous, but in suitably treated preparations numerous fine, threadlike structures, or *myofibrils*, can be seen embedded in it. The myofibrils constitute the contractile apparatus.

Striated or skeletal muscle

The second major type of muscle is known as striated or skeletal muscle, since the cells appear to be filled with distinct and regular transverse markings, or striations. When observed under polarized light, the fibers of striated muscle are seen to have alternating bands that are either optically iostropic (I bands) or anisotropic (A bands). The isotropic bands have uniform properties regardless of the direction in which they are observed, while the visible properties of the anisotropic bands depend on the direction in which they are observed or on the plane of the polarized light with which they are illuminated. A more detailed study of striated muscle fibers shows that A bands are thicker than I bands; I bands have a denser zone in their centers known as the Z line; A bands are bisected by a clearer zone called the H band; and the center of the H band contains another apparent feature known as the M line. These features are shown diagrammatically in Fig. 5-14.

Other faint striations can be seen running parallel to the length of the fiber; these mark the limits of individual fibrils, each of which is surrounded by a thin membrane known as the *sarcolemma*. In contrast to smooth muscle, striated muscle cells are multinucleated and are rather large, with diameters of from 10 to 100μ and lengths as great as 40 mm. Spacing between the I and A bands is vari-

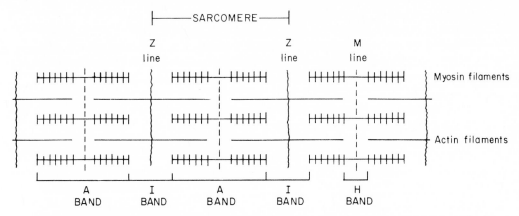

Fig. 5-14. Typical arrangement of myofilaments in striated muscle. Correlation of optical properties with chemical structure.

able; in general, the faster a muscle contracts the shorter the interval between bands and the greater the number of nuclei and mitochondria per fiber. Furthermore, very active muscles are darker in color than those that are less active because the former contain a greater number of mitochondria and more myoglobin.

Cardiac muscle

The third major muscle type is cardiac muscle, which shares some properties of both smooth and striated muscle. Its fibers do not exist as isolated muscle cells but exhibit numerous anastomoses between segments of fibers. In reality the myocardium is a syncytium, through which run the *Purkinje fibers*; these are highly specialized fibers that confer the property of rhythmic, independent contractility. Myocardial fibers exhibit transverse striations very similar to those of skeletal muscle.

Contractile proteins

Electron microscopic examination of striated fibers reveals a still more detailed structure than that already described. Fine filaments run parallel to the length of each fibril; some of these filaments are thicker than others. The thicker filaments are *myosin* filaments and the finer are *actin* filaments, terms derived from the major protein component of each. As is shown in Fig. 5-14, these filaments are interspersed in such a way that the ends of the actin filaments fall on either side of the H bands. When the fibers are viewed in cross section, it is evident that they are symmetrically arranged in a hexagonal form. The segment of a striated myofibril that falls between successive Z lines is known as a *sarcomere* and is regarded as the fundamental repeating unit of structure.

When a muscle is stimulated to maximal contraction, the sarcomere shortens more or less in proportion to the observed change in muscle length. From the change in position of the A and I bands, it appears that the myosin and actin filaments slide with respect to one another. We must therefore look to the chemistry of actin and myosin for a possible explanation of this phenomenon.

Striated muscle contains from 16.5% to 20.9% protein by weight; much of the protein can be extracted with dilute salt solutions. Such extracts contain three major proteins: myosin accounts for about 54%, actin for approximately 25%, and tropomyosin for approximately 11%. Myosin is precipitated on dilution of the salt solution and can be reextracted with 0.6 M KCl. Repeated application of this and other procedures yields a substance with a molecular weight of approximately 470,000. It is a rather elongated molecule (about 1600 Å), and physical evidence indicates that it contains two large polypeptide chains that are supercoiled; that is, each α-helical chain is wound about the other in helical fashion like the twisted strands of a rope, except at one end where the terminal portions of the peptide chains are folded into a relatively compact globular mass that is complexed with three smaller subunits. Each peptide chain contains about 1800 amino acid residues, making the myosin peptide one of the longest known. Myosin contains several proline residues in its peptide sequence. As was pointed out in Chapter 2, proline interferes with α-helix formation; therefore, it is interesting to note that all of the proline residues are contained in the globular portion of the myosin peptide chains. In the presence of Ca^{++}, myosin is a potent ATPase. When the globular portions of the peptides are removed from the helical portions by enzymatic digestion, the ATPase activity is retained by the globular fragment, and no ATPase activity is located in the helical fragment. The globular fragment, when freed, can even hydrolyze nucleotides other than ATP. Because of its asymmetric shape the intact myosin molecule shows a high birefringence of flow; this means that when viewed with polarized light through crossed Nicol prisms, the flowing solution alters the plane of polarized light. This phenomenon could occur only if the elongated molecules lined up in the fluid like matchsticks. Hence myosin shows a tendency toward ordered molecular stacking, not only in intact muscle but in simple solution as well. This tendency is a common property of highly asymmetric biopolymers. When intact muscle fibers are repeatedly extracted with KCl, examination shows that the A bands become much less distinct and much less anisotropic as the myosin is removed.

Actin occurs in two forms, G- and F-actin. G-actin is a globular protein consisting of a single polymeric chain with a molecular weight of approximately 46,000. A gram-molecular weight of actin tightly binds one gram atom of Ca^{++} and can also bind one mole of either ADP or ATP. Binding of ATP quickly converts the globular G-actin to a polymerized fibrous form known as F-actin. Each time a monomeric unit of G-actin is added to the highly polymerized F-actin, one molecule of ATP is hydrolyzed to ADP plus P_i, which remains bound to the polymer. F-Actin is also a supercoiled double helix, the diameter of which is very nearly the same as the diameter of the actin filaments of skeletal muscle. Actin has some other unusual structural features: it contains a relatively large number of L-proline residues, seven L-cysteine residues, and one residue of ε-N-methyl-L-lysine, the function of which is not clear. Obviously, these residues are not in a position to prevent formation of an α-helix.

Tropomyosin is the third major extractable muscle protein and exists as either tropomyosin A or B; both are similar to the helical portion of the myosin molecule. Tropomyosin B is found in nearly all muscles, but the A form is found only in some specialized "catch" or "locking" muscles that can remain in the contracted state for extended periods of time without continued energy expenditure. Tropomyosin B is an extended molecule that exists as a double-stranded, supercoiled helix with free —SH groups at one end. It can form a complex with F-actin.

Actin and myosin can form a complex known as actomyosin. Earlier preparations of myosin, made when extraction procedures were less advanced and products were less pure, were in all probability actomyosin preparations. Actomyosin is an even more potent ATPase than either actin or myosin alone. Again, Ca^{++} is required for the demonstration of ATPase activity. Of even greater importance is the fact that when actomyosin hydrolyzes ATP, its own physical properties undergo a great change. Solutions of actomyosin are very viscous; however, when ATP is added in the presence of Ca^{++}, the viscosity drops sharply, returning to the initial value when the ATP has been converted to ADP plus P_i. When solutions of acto-myosin are extruded through a fine orifice into dilute salt solution, fine threads will be precipitated. If these threads are connected to a suitable recording appara-tus while bathed in an appropriate buffer, addition of ATP to the buffer solution causes the actomyosin thread to shorten. It therefore appears that the energy released by hydrolysis of the ATP is employed to alter the structure of the acto-myosin complex, whether the complex is in solution or in the form of a thread. This is the basis of the mechanochemical coupling exhibited by muscle cells.

An approximation of the mechanism for mechanochemical coupling is shown in Fig. 5-15. Although for the sake of clarity only a single actin and a single myosin filament are shown, in actuality these surround each other in a hexagonal pattern. The globular heads of the individual myosin molecules extend from the surface of the myosin filament and are folded more or less parallel to the axis of the chain. They do not engage or contact the adjacent actin filament so long as the relaxed state is maintained by a sufficient supply of ATP.

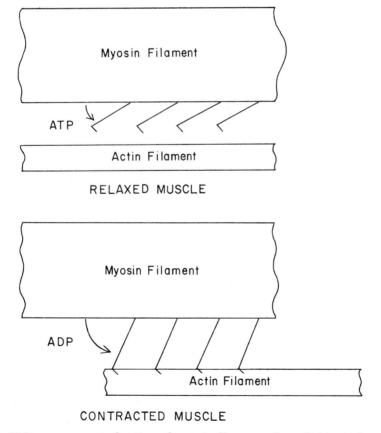

Fig. 5-15. Sliding movement of actin and myosin filaments. Cross-linking induced by acto-myosin ATPase causes conformational change of protruding myosin heads, which move actin chains by a ratchetlike action. ADP stabilizes cross-linking by weak bonds, and ATP destabilizes it.

On excitation, ATP is hydrolyzed by the actomyosin complex, and the conformation of the extended globular myosin head changes simultaneously. The increased angle between the myosin axis and the protruding head causes the latter to somehow engage the actin filament, pulling it in such a way that the sarcomere is shortened. This model accounts for the observed change in sarcomere dimensions and also for the fact that only the A zone shows no dimensional change. It does not explain how the two chains engage, or exactly how tropomyosin enters the final mechanism.

Excitatory mechanism

Any description of the excitation of muscle contraction must take into account that a single fiber contains many repeating units, or sarcomeres. These units must operate in a synchronous manner for maximum efficiency. This same stipulation also applies to the many fibers in a muscle bundle. It is now recognized that the sarcolemma conducts contraction impulses so efficiently that the electrical wave from a motor nerve end plate is spread over the entire surface of a muscle bundle. Furthermore, it has been established that the sarcolemma not only surrounds the muscle bundle but also penetrates it by means of numerous invaginations. These infoldings constitute the *sarcoplasmic reticulum*, which thus becomes an integral part of the excitatory or stimulatory apparatus. As the electrical impulse travels through it, the sarcoplasmic reticulum is depolarized and thus rendered distinctly more permeable to several ions, especially Ca^{++}. The ability of Ca^{++} to activate the actomyosin ATPase has already been mentioned; thus it is not surprising that the release of Ca^{++} into the sarcoplasm should be intimately involved in initiating the contractile process. The change in the potential of the sarcolemma as the excitatory wave passes over it is approximately 60 mv; from the earlier discussion of energetics it follows that the change in Ca^{++} concentration must be approximately tenfold.

Table 5-4 indicates that the total concentration of Ca^{++} in muscle is not very high; however, only the skeletal system, including the teeth, has a higher concentration. On the other hand, it has been estimated that the concentration of Ca^{++} required to trigger the contractile process need be only a small fraction of the observed value.

Table 5-4. Partial composition of mammalian skeletal muscle

COMPONENT	WET WEIGHT (g/100 g)
Total solids	20.0-28.0
Water	72.0-80.0
Total protein	16.5-20.9
Creatine + creatinine	0.27-0.58
Carnitine	0.19-0.30
Glycogen	1.0-2.0
Inorganic substances	1.0-1.5
Sodium*	0.07
Potassium*	0.39
Calcium*	0.006
Magnesium*	0.02
Phosphorus*	0.20
Sulfur*	0.21
Chloride*	0.05

*Mean values.

Relaxation mechanism

If Ca^{++} in free form is required to initiate the contractile process, then it is reasonable to assume that relaxation would require its removal. Mitochondria possess a mechanism for the uptake of Ca^{++} from the extramitochondrial (sarcoplasmic) space, as has already been discussed. It operates either at the expense of ATP or of a high-energy precursor of ATP. It has also been shown that the sarcoplasmic reticulum has a similar system. Together, these two systems can lower the concentration of ionic calcium to approximately 10^{-7} M, a level that is apparently below the threshold value required for contraction. The effects of Ca^{++} are antagonized by Mg^{++}; for this reason the intravenous injection of Mg^{++} causes a flaccid muscle paralysis. Although both ions are essential to life, they must be kept in carefully defined physiologic compartments. When this rule is violated, for example, by injections of Mg^{++}, the physiologic effects are profound.

Although the function of Ca^{++} in muscle contraction gives it a central role, intracellular K^+ and Na^+ are also important. Changes in K^+ cause serious alterations in membrane potentials. Since the skeletal mass provides a large reservoir of Ca^{++}, changes in the Ca^{++} concentrations that bathe body cells are rather difficult to induce. However, no such reservoir exists for K^+. Cardiac muscle is quite sensitive to changes in K^+ concentration, which occur on a detectable scale with each beat of the heart. Calcium concentrations appear to alter the threshold sensitivity to changes in K^+ concentration and to antagonize the rhythmic variations that occur. As the excitation wave passes over the myocardium, K^+ is expelled; it returns during the relaxation phase. The release of K^+ to the extracellular space is promoted by acetylcholine. Quinidine, which is used in the treatment of certain types of heart disease, strengthens the heartbeat by reducing the loss of K^+ during systole, the contractile phase of the heartbeat. Procaine and lidocaine, drugs also used in the treatment of cardiac irregularities, for example, fibrillation, apparently act in a similar way.

Sources of muscle energy

Life or death may frequently depend on intense bursts of muscle activity, and it is therefore essential that muscles have an abundant supply of energy at all times. The ready availability of energy has been ensured by a redundant system. Muscle is unique in containing significant quantities of creatine and creatinine, as shown in Table 5-4. Most of the creatine exists as phosphocreatine, a high-energy compound. Table 5-2 shows that phosphocreatine has a very large negative change in free energy when it is hydrolyzed; thus in time of need it can be used to replenish depleted stores of ATP, according to the equation:

$$\text{creatine} \sim P \ + \ ADP \ \rightleftharpoons \ \text{creatine} \ + \ ATP, \ \Delta G^{\circ'} = -3 \text{ kcal/mole}$$

The reaction is catalyzed by the enzyme ATP-creatine transphosphorylase, more commonly known as *creatine phosphokinase* (CPK). In muscle performing moderate work the concentration of phosphocreatine drops before there is any appreciable diminution of ATP, indicating that phosphocreatine is a primary means of maintaining the ATP level. Following exertion and the depletion of energy stores, creatine is once more phosphorylated to phosphocreatine. This is further substantiated by the treatment of intact isolated muscle with fluoro-2,4,-dinitrobenzene, which is known to inactivate CPK. Muscle poisoned in this manner shows a rapid drop in ATP when it is made to work.

Creatine is constantly lost from the body as the nonenzymatic dephosphorylation of phosphocreatine results in the formation of creatinine (Chapter 9). Creatinine is a normal constituent of urine, where it is found in levels that more or less depend on the active muscle mass of a given individual. The daily creatinine output in a healthy male is 20 to 26 mg/kg; in a healthy female it is 14 to 22 mg/kg. The daily urinary creatinine output is useful as a gauge of whether a complete sample has been collected for measurements of daily urine volume.

A second, important source of muscle energy depends both on the presence of the enzyme adenylate kinase, or *myokinase,* which is widely distributed in mammalian muscle, and on the fact that ADP contains one high-energy bond that is not normally used. In urgent situations, when other energy sources are inadequate, myokinase can catalyze the reaction:

$$2ADP \rightleftharpoons AMP + ATP, \quad \Delta G^{o'} = \text{approximately } 0$$

This system enables muscle to "wring out" the last measure of energy that resides in the adenylate system.

Oxygen toxicity, superoxide radical, and superoxide dismutases

The oxygen consumed in oxidative phosphorylation is the major component of cellular respiration. Through the reduction of oxygen, energy is trapped as ATP and water is formed. The complete reduction of a molecule of oxygen requires four electrons. If the reduction proceeded by single electron additions, one intermediate would be the superoxide radical O_2^-. More reduced products would include H_2O_2 (addition of two electrons per molecule) and the hydroxyl radical $OH\cdot$. Hydrogen peroxide is a toxic substance, but its concentration is kept very low by the action of catalases that decompose it to form water and free molecular oxygen.

It has recently been established that the superoxide radical is even more toxic and that its concentration must be kept low by a catalyzed dismutation reaction

$$2O_2^- + 2H^+ \longrightarrow H_2O_2 + O_2$$

The responsible enzymes are known as *superoxide dismutases.* These enzymes have been found in species ranging from bacteria to mammals. They may be isolated from the cytosol of liver, erythrocytes, and other tissues, as well as from mitochondria. Superoxide dismutases are metalloenzymes, but the metal requirements depend on the enzyme source. In liver the cytosol enzyme and the enzyme from the mitochondrial intermembranous space are dependent on copper and zinc, while the enzyme of the mitochondrial matrix is manganese-dependent, as is the enzyme from prokaryotic cells. At least four different isozymes have been detected in chicken liver. It has long been known that breathing high concentrations of oxygen over prolonged periods can be dangerous. An atmosphere of pure oxygen is lethal to rats, and use of high oxygen environments for extended periods has caused blindness in premature infants. It is now thought that O_2^- may be an important cause of oxygen toxicity. Premature infants may not have a fully developed capacity to produce superoxide dismutases, and, as a result, develop retrolental fibroplasia when they breathe an oxygen-enriched atmosphere.

Active phagocytosis by mature leukocytes involves a burst of oxygen uptake that has been associated with increased production of O_2^-. There is a genetic defect, known as chronic granulomatous disease, in which leukocytes do not demonstrate rapid oxygen uptake during phagocytosis. It is thought that diseased leukocytes have difficulty in killing organisms they engulf because their capacity to make the toxic superoxide radical is diminished or lacking.

Other sources of O_2^- have been found. Substantial quantities may be produced by xanthine oxidase, aldehyde oxidase, and dihydroorotate dehydrogenase. Some explanations for the toxicity of the radical have been established. O_2^- may be responsible for single-stranded breaks in DNA molecules, may cause depolymerization of acid mucopolysaccharides and may be capable of initiating the peroxidation of membrane lipids. The antibiotic streptonigrin is lethal to sensitive organisms such as *Escherichia coli* because the drug increases formation of O_2^-. It has also been suggested that part of the circulating oxyhemoglobin is actually a superoxoferriheme, and that one function of the large quantity of glutathione found in erythrocytes is to reduce O_2^-. Further studies are necessary to elucidate the role of superoxide radical and superoxide dismutases in disease.

REFERENCES

Carter, L. C.: Guide to cellular energetics, San Francisco, 1973, W. H. Freeman and Co. Publishers. *A programmed self-learning guide that is quite thorough and highly recommended for individual use.*

Dahl, J. L., and Hokin, L. E.: The sodium-potassium adenosine triphosphatase, Ann. Rev. Biochem. 43:327, 1974.

Fridovich, I.: Superoxide dismutases, Adv. Enzymol. 41:35, 1974.

Hoch, F. L.: Energy transformations in mammals: regulatory mechanisms, Philadelphia, 1971, W. B. Saunders Co. *An integrated review of many aspects of energetics with emphasis on control.*

Huxley, H. E.: The contraction of muscle, Sci. Am. 199:66, 1958. *An older but still fascinating discussion of the mechanical aspects of muscle.*

Lehninger, A. L.: Bioenergetics, Menlo Park, Calif., 1965, W. A. Benjamin, Inc. *A brief paperback review of classic energetics.*

Loewy, A. G., and Siekevitz, P.: Cell structure and function, ed. 2, New York, 1969, Holt, Rinehart and Winston, Inc. *An excellent review of muscle chemistry and mechanics.*

Luria, S. E.: Colicins and the energetics of cell membranes, Sci. Am. 233:30, December 1975.

Mitchell, P.: Chemiosmotic coupling and energy transduction, Bodmin, England, 1968, Glynn Research, Ltd.

Montgomery, R., and Swenson, C. A.: Quantitative problems in the biochemical sciences, ed. 2, San Francisco, 1976, W. H. Freeman and Co. Publishers. *Contains numerous illustrative problems useful in grasping fundamental energetics.*

Racker, E.: The two faces of the inner mitochondrial membrane. In Campbell, P. N., and Dickens, F., editors: Essays Biochem. 6:1, 1970. *An excellent review by a master of biochemistry and biochemical writing.*

Slater, E. C.: Coupling between energy-yielding and energy-utilizing reactions in mitochondria, Q. Rev. Biophys. 4:35, 1971. *A scholarly review of the theories of oxidative phosphorylation.*

Clinical examples

CASE 1: HYPOPHOSPHATEMIA

C. M., a 52-year-old woman, was hospitalized for progressive peripheral neuropathy and intractable steatorrhea. While extensive diagnostic evaluation did not produce data to explain the cause of her illness, intestinal biopsy findings were compatible with a malabsorption disorder. In an effort to provide nutritional support, intravenous feedings containing 12% glucose and 4% casein hydrolysate were administered through a venous catheter. Positive nitrogen balance was observed during the period of parenteral feeding. Five months later the patient died of infection and gastrointestinal bleeding. Autopsy showed extensive necrosis of the gastrointestinal tract from the esophagus to the colon.

During the extended periods of parenteral feeding, it was noted that the patient exhibited a marked hypophos-

phatemia. The data gathered from these observations are presented in Figs. 5-16 and 5-17. In addition, red blood cells taken from the patient and from a normal volunteer were collected, washed with saline, and resuspended at a hematocrit of 25% in plasma collected either from the patient or a normal volunteer. The resuspended cells were then incubated at 37° C for 5 hr and certain analyses were made; the results are summarized in Table 5-5.

Biochemical questions

1. Does normal plasma contain ATP? Explain.
2. Is the ATP found in the erythrocytes the product of oxidative phosphorylation? Explain.
3. Explain why P_i is required for ATP formation.

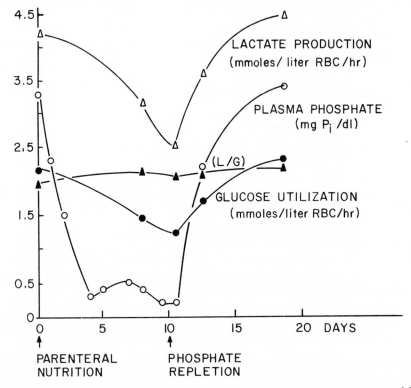

Fig. 5-16. Plasma phosphate, red cell glucose utilization and lactate production, and lactate/glucose ratio (L/G) during parenteral nutrition and phosphate repletion in vivo.

Table 5-5.   Red cell organic phosphates during incubation in hypophosphatemic and normophosphatemic plasma

	EXPERIMENT	ADENOSINE TRIPHOSPHATE (mmole/liter RBC)		2,3-DIPHOSPHO-GLYCERATE (mmole/liter RBC)		PLASMA P_i (mg/100 ml)
		At 0 time	At 5 hr	At 0 time	At 5 hr	At 0 time
Normal RBC and normal plasma	1	1.59	1.52	3.90	3.83	3.1
	2	1.58	1.52	4.18	3.79	3.0
Normal RBC and patient's plasma	1	1.59	1.38	4.00	3.30	1.2
	2	1.54	1.33	4.37	3.53	1.3
Patient's RBC and patient's plasma	1	1.10	0.98	2.85	2.69	0.4
	2	0.84	0.66	1.92	1.50	0.3
Patient's RBC and normal plasma	1	1.06	1.39	2.83	3.29	2.5
	2	0.87	1.25	1.96	3.42	1.9
Patient's RBC and patient's plasma + 1 mM P_i		0.75	1.54	2.11	4.02	3.7

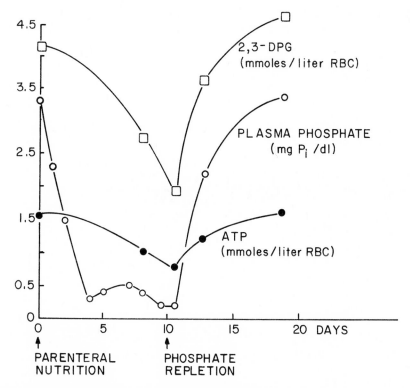

Fig. 5-17. Plasma phosphate, red cell adenosine triphosphate (ATP), and 2,3-diphosphoglycerate (DPG) during parenteral nutrition and phosphate repletion in vivo.

4. Would 2,4-dinitrophenol affect the ATP concentration of erythrocytes? Would atractylate?
5. What is the characteristic difference between oxidative and substrate-level phosphorylation?
6. Could a sodium pump exist in erythrocytes? Explain.

Case discussion

It is customary to think of ATP generation in terms of somatic cells that contain mitochondria and to emphasize the function of these organelles. The human erythrocyte lacks a nucleus and mitochondria and is entirely dependent on the glycolytic process (Chapter 7) for its ATP supply. Glycolysis yields much less ATP than does oxidative phosphorylation, but the yield is sufficient to meet the normal needs of the erythrocytes, since these atypical cells perform only limited biosynthesis. The case of C. M. demonstrates the effect of depressed ATP synthesis in erythrocytes, caused in large part by a profound lack of inorganic phosphate. Phosphate, of course, is an absolute requirement for synthesis of ATP by either substrate level or oxidative phosphorylations.

Fig. 5-16 shows the effects of parenteral hyperalimentation on certain blood chemistry data. At the onset of a feeding period the plasma inorganic phosphate and lactate concentrations were substantially normal, and the number of moles of lactate formed per mole of glucose utilized by C. M.'s erythrocytes was approximately two, indicating no significant impairment of the several intermediate steps of glycolysis. After 4 days the plasma phosphate concentration had dropped almost tenfold, and at the same time the consumption of glucose and the production of lactate had also declined. However, the molar ratio of lactate produced to glucose oxidized remained at a normal value, from which it is clear that the concerned enzyme mechanisms were normal, as before. Sometime during day 10 of the feeding period, inorganic phosphate was administered parenterally, with the result that the previously mentioned abnormalities were easily reversed. By day 19, values had returned to normal.

Certain other analytic data are presented separately in Fig. 5-17. (Note that the values of plasma inorganic phosphate are repeated for convenience.) The additional data describe the red cell concentration of ATP and a second important organic phosphate compound, 2,3-diphospho-D-glycerate (2,3-DPG) (Chapter 2). During the onset of phosphate depletion the concentration of ATP fell to approximately half its normal value, as did the concentration of 2,3-DPG. Since it has already been shown that the ratio of lactate production to glucose utilization was normal, the obvious conclusion is that failure to maintain a normal erythrocyte ATP concentration must be caused by a lack of inorganic phosphate required, along with ADP, for ATP synthesis. Direct experimental evidence in support of this conclusion is found in the data presented in Table 5-5. When normal cells were suspended in normal plasma and incubated for 5 hr, very little change was noted in either ATP or 2,3-DPG concentrations. The normal plasma contained about 3 mg P_i/dl. When normal cells were suspended in plasma from C. M., which contained a little less than half the usual amount of P_i, the cells were not able to maintain the concentration of ATP for 5 hr; note that the ATP concentration dropped approximately 12%. A similar drop was observed in the concentration of 2,3-DPG. When cells from C. M. were resuspended in normal plasma, the concentrations of ATP and 2,3-DPG actually rose during the 5-hr incubation period, demonstrating that the cells themselves were competent but starved for inorganic phosphate at the time of collection. The values observed at the end of the incubation were very nearly double those at the beginning.

The progressive course of hypophosphatemia was related to intensive intravenous alimentation with markedly hypertonic glucose and casein hydrolysate. Although casein is itself a phosphoprotein (it contains about 0.7% P), the plasma inorganic phosphate values declined. The explanation lies in the fact that the large glucose load administered caused very large quantities of P_i to enter the cells, since glucose metabolism requires formation of glucose 6-phosphate and other phosphorylated intermediates. For illustrative purposes, assume that C. M. was given 1500 ml of 12% glucose/24 hr, that is, one mole of glucose/day. Since the P_i concentration of blood plasma is normally only 1.0 to 1.5 meq/liter, it is not surprising that the pool of circulating P_i available to the erythrocytes was seriously depleted. We must conclude that the hypophosphatemia was induced by excessive administration of hypertonic glucose and that the available P_i had been transferred from the extracellular compartment to the intracellular compartment. Phosphate was obviously returned to the circulation at too slow a rate to permit normal erythrocyte metabolism, even though the P_i content of the red cells normally is about half that of plasma.

ATP and 2,3-DPG are both essential to the survival of circulating erythrocytes, which normally have a life span of approximately 120 days. Both phosphate compounds interact with hemoglobin, diminishing its affinity for oxygen and thereby promoting tissue oxygenation. Cells with depressed ATP concentrations show a lessened capacity to resist in vitro hemolysis, possibly because the Na^+/K^+-ATPase (sodium pump) cannot then operate efficiently. The cells lose their typical biconcave shape and become spheroidal as the Na^+ concentration increases. These swollen cells cannot readily traverse the finer capillaries of the circulatory system and are caught up in the reticuloendothelial cells, chiefly in the spleen, where they are destroyed. The swollen cells become more rigid, and the membranes do not regain plasticity when ATP is added. This may be caused by changes in an actomyosin-like protein that can be isolated from red cell membranes. This protein is apparently irreversibly dena-

tured when ATP concentrations fall to 15% to 30% of the normal values.

For the reasons already set forth, an adequate supply of ATP is as essential for the erythrocytes as it is for other cells. Being without mitochondria, erythrocytes are limited to substrate-level phosphorylation only. Since this process in the red blood cell is not restricted to compartmentalized organelles replete with controlling translocases but appears to be more generally distributed within the total cell volume, it is not surprising that translocase-specific reagents such as atractylate are without effect on erythrocyte phosphorylations. Similarly, dinitrophenol is without effect, since it is presumed to act at a specific site within mitochondrial membranes.

The erythrocytes are entirely dependent on the suspending plasma for supplies of glucose and phosphate. Furthermore, ATP cannot be obtained from the plasma, for two reasons. First, ATP is a highly charged molecule; as such it does not readily penentrate plasma membranes. Secondly, and even more importantly, the plasma is rich in phosphatases and pyrophosphatases that would very quickly destroy any ATP that did escape from other cells.

REFERENCES

Bunn, H. F., and Jandl, J. H.: Control of hemoglobin function within the red cell, N. Engl. J. Med. 282:1414, 1970.

Jacob, H. S., and Amsden, T.: Acute hemolytic anemia with rigid red cells in hypophosphatemia, N. Engl. J. Med. 285:1446, 1971.

Lichtman, M. A., et al.: Reduced red cell glycolysis, 2,3-diphosphoglycerate and ATP concentrations and increased hemoglobin-oxygen affinity caused by hypophosphatemia, Ann. Intern. Med. 74:562, 1971.

Travis, S. F., et al.: Alteration of red cell glycolytic intermediates and oxygen transport as a consequence of hypophosphatemia in a patient receiving intravenous hyperalimentation, N. Engl. J. Med. 285:763, 1971.

CASE 2: HALOTHANE INTOXICATION

A 21-year-old male was admitted to the hospital with a compound fracture of the tibia and fibula. He was less concerned about the fractures than about the fact that he was likely to need a general anesthetic, as several of his relatives

had died following the administration of ether. Because of this history a local anesthetic had been used when he had undergone an appendectomy in childhood. The patient's parents and general practitioner were interviewed and the history was confirmed; it was believed that in each case death had been caused by ether. It was decided to proceed cautiously with anesthesia, avoiding the use of ether, and to stop at any untoward sign.

The patient initially was given pethidine and atropine and was anesthetized with thiopentone followed by nitrous oxide, oxygen, and halothane. Halothane was administered to a maximum of 2% of inspired gas and then levels were reduced to 1.5%. After 10 min the halothane concentration was further reduced because the patient's blood pressure had fallen from 120 to 100 mm Hg. The total time during which halothane had been administered was 15 min. After another 10 min the patient appeared pale and cyanotic; his blood pressure had fallen to 80 mm Hg, and his pulse rate had risen from 100 to 160. Administration of anesthesia was stopped and surgery was concluded in the next 10 min. However, the patient remained deeply unconscious for an additional 30 min and his skin was hot and sweaty. He was rubbed down with ice-cold cloths and given a blood transfusion. He gradually recovered over the next $1\frac{1}{2}$ hr and the subsequent course was uneventful.

After his recovery, several investigations were made, but no abnormalities were detected. Included were chest x-ray films, an electrocardiogram, and estimations of urinary porphyrins, catecholamines, steroids, liver function, serum cholinesterase, proteins, calcium, phosphorus, and alkaline phosphatase.

Approximately 1 year later, this patient was admitted to the hospital for removal of an impacted ureteral calculus. He was given a spinal anesthetic and the stone was removed uneventfully.

An inquiry revealed that of thirty-eight relatives of the patient who had undergone general anesthesia, ten had died. In all of these cases it appeared that the agents used were ethyl chloride and ether. Three of the decreased had been cousins of the subject, and seven had been uncles or aunts. Where it was possible to obtain records of these cases, the course of events was seen to have been similar in each case. In all but one instance the operation had

been minor, successful, and unlikely to have been a direct cause of death. In the three best-documented cases, the patients had been returned to the ward after the operation in good condition, only to die about 30 min later. Two of the deceased had exhibited temperatures of 43° and 42° C. No abnormalities were found at postmortem examination.

The pattern of inheritance is compatible with that resulting from a dominant gene or genes, with one exception; the grandmother of the patient survived the administration of $CHCl_3$ for eclampsia.

Biochemical questions

1. What is the biochemical basis of this patient's fever? Explain.
2. How does the observed drop in blood pressure relate to possible ion concentrations in the myocardium?
3. What other tissues might have been affected by the halothane intoxication? Explain.
4. Would the P/O ratio have exceeded 3 in this patient? Explain.
5. Would you classify halothane as a site-specific inhibitor or as an uncoupler? What is the difference?

Case discussion

Halothane ($CF_3CHBrCl$) is an extremely potent anesthetic agent, the first of a series of fluorinated hydrocarbons to be used clinically as an anesthetic. It is generally regarded as considerably less toxic than $CHCl_3$ and much less toxic than CCl_4. Like all of the halogenated hydrocarbons, halothane is a mild hepatotoxin. Following administration to experimental animals there is moderate fat accumulation and some change in nuclear and mitochondrial morphology but no necrosis.

A saturated water solution of halothane at 37° C contains slightly more than 10^{-2} moles/liter. When aliquots of such a solution are added to suitably maintained preparations of rat liver mitochondria, both the oxygen uptake and P/O ratios show a precipitous fall when the levels of halothane reach a concentration of 10 mM, using succinate as the substrate. When glutamate is the substrate, even less halothane is required to produce similar changes. Loss of respiratory control precedes the uncoupling of oxidative phosphorylation and is evident even at halothane concentrations of 2 mM. When a suspension of liver mitochondria is exposed to a 14 mM concentration of halothane, they swell (indicated by a decrease in optical density) approximately four times as much as a control suspension. At lower concentrations of halothane, the swelling could be partially prevented by the addition of bovine serum albumin, ATP, or EDTA (a chelating agent for Ca^{++} and heavy metals) to the suspension. It is possible that serum albumin prevents swelling by binding fatty acids that have been liberated from mitochondrial membranes by low halothane concentrations, that ATP prevents swelling by offsetting decreased intrinsic ATP concentrations caused by uncoupling or loosening of coupling, and that EDTA binds Ca^{++} set free by the halothane (Lehninger has shown that Ca^{++} is itself a potent swelling agent). Reference to Fig. 5-9 will show that, at least according to the chemical theory of oxidative phosphorylation, energy can be diverted to ion translocation before the hypothetical step that results in the phosphorylation of ADP to ATP. Recent evidence indicates that the effects of halothane may include a relatively specific cyclic flux of calcium ions, which accounts for the loss of respiratory control. If mitochondria are treated with halothane in the absence of external calcium ions or if the external ions are chelated by some appropriate agent such as EDTA, then the respiratory control is properly maintained, and the apparent uncoupling does not occur. Further studies have shown that the presence of halothane specifically increases the activity of accumulated calcium ions in the matrix spaces of the mitochondria, thus inducing an enhanced efflux of the ions. It has been proposed that the effect is probably caused by an inhibition of calcium binding to matrix proteins. It thus appears that halothane produces many profound changes in the mitochondria.

To determine whether the loss of NADH-linked substrate oxidation is caused by a blockage of electron flow from NADH to cytochrome b, mitochondria were preincubated in halothane and NADH was added as a substrate. Normal mitochondria do not oxidize exogenous NADH unless cytochrome c is also added. The halothane-treated mitochondria were able to oxidize added NADH without the addition of cytochrome c, clear evidence that the membranes of the treated particles have an abnormal permeability. Since the addition of amytal completely stopped the oxidation of added NADH, it is evident that electrons could flow readily through the earliest segments of the respiratory chain. Electron transport was not inhibited and oxidation was not sensitive to a rate-limiting concentration of P_i; these are exactly the conditions that exist for such classic uncouplers as dinitrophenol. It must be remembered that these observations regarding halothane are not unique; other halogenated hydrocarbons show similar properties to varying degrees. In fact there is a linear relationship between the minimal required concentration for uncoupling and the oil-water solubility ratio for $CHCl_3$, CCl_4, and their related compounds. Even the nonhalogenated materials benzene and ether can be placed on the same straight line as halothane.

Isolated mitochondria constitute a specific, yet unphysiologic test system, but data congruent with the observations discussed above have also been obtained in experiments using isolated human liver cells in culture. When cultured cells are exposed to clinically useful concentrations of $CHCl_3$, $CF_3CHBrCl$, or $CH_3OCF_2CHCl_2$ (methoxyflurane, another fluorinated liquid anesthetic), phase-contrast photomicrography and time-lapse cinematography show varying degrees of cellular intoxication that are manifested by increased cytoplasmic vacuolization and formation of cytoplasmic granules. $CHCl_3$ caused reversible damage at commonly used concentrations, while halothane does not produce changes that could be readily visualized by the methods employed. In excessive concentrations, $CHCl_3$ produces permanent cell damage leading to disintegration in a significant fraction of the cells observed. Halothane causes less damage, and only a fraction of the cells die.

Oxidative phosphorylation traps energy in a useful form with an efficiency of approximately 40%. The remainder of the energy is released as heat, which is normally dissipated to the environment. If uncoupling occurs and the heat produced is greater than normal, the body temperature will rise above the typical 37° to 38°C and fever will result. While fever can result from causes other than uncoupling, the evidence here indicates that P/O ratios of isolated mitochondria exposed to halothane fell to values considerably less than 3. This change was associated with other signs of mitochondrial damage. Although the experiments described were performed on liver mitochondria, it is likely that other tissues were also affected, including the brain, kidney, heart, and muscles, all of which depend on oxidative phosphorylation for their energy requirements. This argument is supported in part by changes observed in the blood pressure, which fell precipitously during anesthesia. The fall in blood pressure is a direct reflection of altered myocardial contractility; from what is known of the contractile process, the myocardial concentration of calcium ion probably was affected by the intoxication.

The oxidation of $NADH + H^+$ occurs at three sites where ATP can be produced; consequently, the P/O ratio cannot be greater than 3. It can be less than 3, depending on the extent of uncoupling of oxidation and phosphorylation. Since oxygen consumption was not reduced, halothane did not affect electron flow, so it is not an inhibitor of any of the three coupling sites. It did appear to affect phosphorylation, so it can be described as an uncoupler. In this respect it is similar to dinitrophenol.

While normal individuals readily tolerate the insult of halogenated hydrocarbon anesthetics, the case just described demonstrates that in some individuals genetic predisposition creates a sensitivity that amounts to a clinically unacceptable intolerance to these potent uncouplers of oxidative phosphorylation.

REFERENCES

Belfrage, S., et al.: Halothane hepatitis in an anaesthetist, Lancet 2:1466, 1966.

Corssen, G., et al.: Effects of chloroform, halothane and methoxyflurane on human liver cells in vitro, Anaesthesia 27:155, 1966.

Grist, E. M., and Baum, H.: Evidence for a halothane-dependent cyclic flux of Ca^{++} in rat-liver mitochondria, Eur. J. Biochem. 57:617, 1975.

Grist, E. M., and Baum, H.: Mechanism of the halothane-dependent efflux of calcium from rat-liver mitochondria, Eur. J. Biochem. 57:621, 1975.

Kalow, W., et al.: Metabolic error of muscle metabolism after recovery from malignant hyperpyrexia, Lancet 2:895, 1970.

Snodgrass, P. J., and Piras, M. M.: Effects of halothane on rat liver mitochondria, Biochemistry 5:1140, 1966.

Trey, C., et al.: Fulminant hepatic failure, N. Engl. J. Med. 279:798, 1968.

Wilson, R. D., et al.: Malignant hyperpyrexia with anaesthesia, J.A.M.A. 202:183, 1967.

CASE 3: SNAKE VENOMS AS UNCOUPLING AGENTS

A young man was exploring a cave with a party of friends when he was bitten high on the left leg by a rattlesnake. No medical assistance was immediately available, except for a makeshift tourniquet applied centrally to the bite. The party then walked to the office of the nearest physician, which they reached 3 hours later. By this time the leg was swollen, hot, and very sore. The patient complained of numbness and "tingling all over" and showed signs of moderate shock. He was also slightly cyanotic.

The patient was immediately taken to the local hospital emergency room, where an intravenous infusion of glucose in water was begun. He was also given an injection of polyvalent antiserum (said to be effective against all toxic North American snakes), and the site of the bite was infiltrated with a 10% solution of magnesium sulfate. That night the nurses noted a trace of hematuria in a voided urine sample. The following day a second injection of antiserum was given, and the intravenous fluids were continued. No further hematuria was noted. The patient steadily improved and was discharged on the third day.

Snake venoms are very complex mixtures. In addition to various poorly defined protein components, they are known to contain proteases and phospholipases. Mitochondria isolated from animals deliberately injected with cobra venom show an uncoupling of oxidative phosphorylation. Careful analysis has indicated that phosphorylation at site I is more resistant to the venom than phosphorylation at sites II or III. Similar results have been obtained with other venoms, even when added to normal mitochondria.

Biochemical questions

1. What products of a phospholipase reaction might act as uncoupling agents?
2. Can the uncoupling effects of snake venom occur entirely at the outer mitochondrial membrane?
3. Does the arrangement of the components of the respiratory chain (Fig. 5-8) help explain the unequal effect of snake venom on sites I, II, and III? Propose a possible explanation.
4. Would leakage of NADH from damaged mitochondria affect the P/O ratio?

5. This patient was treated by infiltrating a solution of magnesium sulfate (injecting small amounts of the solution at many points) near the site of the snakebite. Is there a direct connection between ATP formation and Mg^{++} in the mitochondria? What would happen to intramitochondrial concentrations of Mg^{++} if cellular lipids were hydrolyzed by a phospholipase to liberate significant amounts of free fatty acids?
6. Is it likely that organelles other than the mitochondria might be damaged by snake venom? Explain your ideas.
7. Why would it be inadvisable to treat snakebite victims with antibiotics of the valinomycin type?
8. How is it possible to relate the fact that such disparate chemicals as bilirubin, palmitic acid and lysolecithin can all act as uncoupling agents when they are present in the tissues at high concentrations? Do factors other than the absolute concentration in any way explain this fact?

REFERENCES

Aravindakshan, I., and Braganca, M. B.: Preferential inhibition of phosphorylation in different parts of the mitochondrial respiratory chain in animals injected with cobra venom, Biochem. J. **79**:80, 1961.

Dryer, R. L., and Harris, R. R.: Metabolic fate of fatty acids in the carnitine cycle of brown adipose tissue mitochondria, Biochim. Biophys. Acta **380**:370, 1975.

CASE 4: CHRONIC ALCOHOLISM, DEATH INDUCED BY THE ACETALDEHYDE SYNDROME

A 43-year-old man with a history of chronic alcoholism sought medical assistance. His physician treated him with the drug known as disulfiram and at the same time cautioned the patient that the drug would produce severe symptoms, and possibily death, if he consumed alcoholic beverages while on this drug. He counseled the patient that such aversive therapy had previously been successful and that in the absence of alcohol the drug was innocuous.

The patient prospered for several

months but then attended an office party where, unbeknownst to him, the punch contained a significant amount of alcohol. Shortly after drinking rather freely, he felt unwell, was rushed to the hospital, and his physician was summoned. The patient complained of severe chest pain, vertigo, and blurred vision. He died later that evening.

The metabolism of ethanol proceeds in several steps; the first involves the NAD^+-linked alcohol dehydrogenase and produces acetaldehyde. Acetaldehyde is moderately toxic but is normally rapidly oxidized to acetate by an aldehyde dehydrogenase. Disulfiram induces the acetaldehyde syndrome by inhibition of the acetaldehyde dehydrogenase so that the blood concentration of acetaldehyde may rise to tenfold normal. The drug is also a potent inhibitor of dopamine β-hydroxylase, especially in the brain. Since dopamine is a major precursor of norepinephrine (see Chapters 9 and 13), the drug or its metabolites depress synthesis of this important hormone. These effects are presumed to be somehow responsible for the potentially serious effects of drinking while under therapy. (For further details regarding chronic alcoholism, see case 4, Chapter 14.)

Biochemical questions

1. Is the conversion of ethanol to acetate an example of a coupled reaction system?
2. Is the conversion of NAD^+ to $NADH + H^+$ involved in every coupled system? Why?
3. Assume that the intracellular concentration of NAD^+ is 5 mM, that the intracellular pH is 7.40, and that you have consumed sufficient amount of some alcoholic beverage to produce an intracellular concentration of 1 mM. If the standard oxidation potential for the half-reaction ethanol $\rightarrow$ acetaldehyde is known to be -0.197 volts, what is the free energy made available by oxidation of 0.5 moles of ethanol?
4. Would the energy yield in the above reaction be any different if you assumed the dehydrogenase was linked to $NADP^+$ instead of NAD^+? Explain why.
5. It has been claimed (Cedarbaum et al.: Arch. Biochem. Biophys. **161**:26, 1974) that acetaldehyde inhibits the respiratory control of isolated liver mitochondria. What does this mean in terms of cellular energetics?
6. Blood alcohol concentration declines as a linear function of time. It has been noted that when the blood alcohol falls from 30 to 20 mM, the blood acetaldehyde concentration remains virtually constant at approximately 30 mM. In a study of human liver aldehyde dehydrogenase, Kraemer and Dietrich (J. Biol. Chem. **243**:6402, 1968) determined the K_m for acetaldehyde was 7.5×10^{-7} M; for NAD^+ the corresponding value was 6×10^{-4} M. From other studies of alcohol dehydrogenase it is known that the K_m for ethanol is 1.8×10^{-3} M and the K_m for NAD^+ is 1.7×10^{-5} M. Using these data, can you explain the relative constancy of acetaldehyde concentration in the face of a falling ethanol concentration?

REFERENCES

Duritz, G., and Truitt, E. B., Jr.: Importance of acetaldehyde in the action of ethanol on brain norepinephrine, Biochem. Pharmacol. 15:711, 1966.

Goldstein, M., et al.: Inhibition of dopamine-β-hydroxylase by disulfiram, Life Sci. 3:763, 1964.

Grunnet, N.: Oxidation of acetaldehyde by rat-liver mitochondria in relation to ethanol oxidation and the transport of reducing equivalents across the mitochondrial membrane, Eur. J. Biochem. 35:236, 1973.

Korsten, N. A., et al.: High blood acetaldehyde levels after ethanol administration: differences between alcoholic and nonalcoholic subjects, N. Engl. J. Med. 292:386, 1975.

Musacchio, J., et al.: Effects of disulfiram on tissue norepinephrine content and subcellular distribution of dopamine, tyramine, and their β-hydroxylated products, Life Sci. 3:769, 1964.

Truitt, E. B., Jr.: Ethanol-induced release of acetaldehyde from blood and its effect on the determination of acetaldehyde, Q. J. Stud. Alcohol 31:1, 1970.

Wintrobe, M. M., et al., editors: Harrison's principles of internal medicine, ed. 7, New York, 1974, McGraw-Hill Book Co.

CASE 5: ZELLWEGER'S DISEASE (CEREBROHEPATORENAL SYNDROME)

A little girl was born with severe hypotonia, hepatomegaly, and acute cirrhosis. While she was in the pediatric intensive care unit, the resident noted apparent defects in the central nervous system with possible involvement of the kidneys as well. The child did poorly, in spite of intensive therapeutic measures. A tentative diagnosis of Zellweger's disease was reached, and the parents agreed to a skin biopsy for fibroblast culture as a means of further biochemical study. The child expired at 23 days of age.

The postmortem examination showed an enlarged liver and renal cortical cysts. Microscopic and electron microscopic studies showed that both peroxisomes and mitochondria were abnormal. Peroxisomes appeared to be missing from the brain and liver. The mitochondria were unusually dense, and the cristae were twisted and irregular.

Mitochondria were prepared from the cultured skin fibroblasts. When tested with either succinate or glutamate plus malate as substrates, the respiration was much less than normal. The addition of ADP failed to stimulate mitochondrial respiration. By addition of synthetic electron donors, known to be capable of furnishing electrons to the cytochrome portion of the respiratory chain, it was found that this portion of the respiratory apparatus was normal. It was concluded that the abnormality was located at some point prior to cytochrome *b*.

Biochemical questions

1. Succinate and glutamate are both energy-yielding substrates. Is the mechanism of energy conservation as ATP exactly the same for both of these substrates?
2. What features are common to the metabolism of succinate and malate? List as many as you can.
3. What sorts of enzymes are localized in the peroxisomes? Do these organelles contain a respiratory chain comparable to that found in the mitochondria?
4. Patients afflicted with Zellweger's disease are known to have abnormally increased deposits of iron in the liver, spleen, skin, and brain. Does this suggest any possibilities as to the location of the mitochondrial defect? Explain.
5. Hepatocytes from these patients appear to be swollen with glycogen. What might this signify?

REFERENCES

Bowen, P., et al.: A familial syndrome of multiple congenital defects, Johns Hopkins Med. J. 114: 402, 1965.

deDuve, C., and Baudhuin, P.: Peroxisomes (microbodies and related particles), Physiol. Rev. 46:323, 1966.

Goldfischer, S., et al.: Peroxisomal and mitochondrial defects in cerebrohepatorenal syndrome, Science 182:62, 1973.

Hruban, Z., and Recheigl, M.: Microbodies and related particles, New York, 1969, Academic Press, Inc.

CASE 6: HYPERMETABOLISM

A 36-year-old woman had suffered from an unknown metabolic disorder since childhood. Her BMR was approximately twice the normal. Clinical signs included profuse perspiration, polydipsia without polyuria, and thinness despite a large food intake. She was described by her personal physician as having been asthenic for several years. Her urine contained 3.4 g of creatinine/24 hr collection.

A muscle biopsy was performed, and mitochondria isolated from the tissue were tested in vitro. Using pyruvate plus malate as the substrate, P/O ratios between 1.5 and 2.6 were obtained.

Biochemical questions

1. Is it likely that this patient would have a normal body temperature? Explain.
2. What is the significance of a P/O ratio of 1.5?
3. Why is loss of creatinine in the urine a serious metabolic loss?

4. Would this patient be helped by a diet rich in phosphorus? Explain.
5. Is it possible to find a genetic defect in oxidative phosphorylation? Explain.
6. Fitch et al. (J. Biol. Chem. **249:**1060, 1974) have shown that γ-guanidino-propionic acid is an analogue of creatine that can be phosphorylated by muscle enzymes. At the same time, individuals fed this compound have a lower than normal concentration of ATP in their muscles. Explain these observations.
7. Patients fed the above compound demonstrate muscle wasting. As the muscles lose mass, what ions might be excessively excreted into the urine?
8. In an effort to offset the hypercreatinuria, what dietary adjustments might you recommend? Do you think the altered diet would be beneficial?

REFERENCES

Britt, B. A.: Malignant hyperthermia, a pharmaco-genetic disease of skeletal and cardiac muscle, N. Engl. J. Med. **290:**1140, 1974.
Ernster, L., and Luft, R.: Mitochondrial respiratory control: biochemical, physiological and pathological aspects, Adv. Metab. Disord. **1:**95, 1964.

CASE 7: ACUTE BACTEREMIA

A young veteran of military service in the Far East was hospitalized on his return to San Francisco. He complained of chills and fever and was moderately dehydrated. He was treated with intravenous fluids, a general diet, and antibiotics. When he failed to respond to treatment after several days and his temperature had increased to 40°C, he was placed in an oxygen tent. His condition did not improve; just before he died on day 11 of his hospitalization he was comatose and his temperature was 43°C.

Biochemical questions

1. How would you demonstrate that an unknown infectious agent had produced lethal quantities of an unknown uncoupling agent in this patient?
2. Was the use of oxygen therapy in this case based on sound biochemical reasoning? Explain.
3. What predictions could you make about the P/O ratio that would have been observed in this patient's liver mitochondria?
4. What are some of the general properties that might characterize the unknown uncoupling agent?
5. What ions might be involved in the uncoupling process?
6. How could you demonstrate that only the first of the three coupling sites was affected in this patient?
7. Would you expect the intracellular concentration of Na^+ to be decreased, increased, or unchanged? Why?
8. Would injections of ATP have helped this patient? Why? Explain.
9. If you were told that an animal treated with salicylanilide showed very similar symptoms, what conclusions could you draw regarding the nature of this patient's problem?
10. Assume you had to write the diet orders for this patient and you wanted him to receive 2000 kcal/day. What sort of diet would you prescribe? Why?

REFERENCES

Davis, B. D., et al.: Microbiology, ed. 2, New York, 1973, Harper & Row, Publishers.
Mela, L., et al.: Effect of *E. coli* endotoxin on mitochondrial energy-linked functions, Surgery **68:** 541, 1970.
Racker, E.: Mechanisms in bioenergetics, New York, 1965, Academic Press, Inc.

ADDITIONAL QUESTIONS AND PROBLEMS

1. Shivering to keep warm occurs on exposure to cold environments. Explain.
2. Some animals can produce heat without shivering (nonshivering thermogenesis). Explain how they might accomplish this.
3. Isolated muscle that is poisoned with cyanide can continue to contract, at least for a brief period. Explain.
4. What effects might severe acidosis have on the formation of high-energy bonds? Why?
5. Mitochondria contain GTP as well as ATP, but there is no known means of moving it into or out of the mitochondria. How does GTP get to its proper place?
6. Nigericin and valinomycin have been described as "super detergents" with hydrophobic exteriors and hydrophilic interiors. Explain how this quality might make them effective antibiotics. (Pressman, B.: Fed. Proc. 27:1283, 1968.)

CHAPTER 6

THE KREBS CYCLE

OBJECTIVES

1. To explain the terminal metabolic reactions that are common to the major biochemical pathways of energy catabolism
2. To illustrate that carbon dioxide is not only a metabolic end product but that it may also serve as an important intermediate in biochemical reactions
3. To recognize the central role of mitochondria in the catalysis and control of metabolic pathways

Man eats, when he can afford it, a large number of complex plant and animal products. The processes of digestion and absorption reduce food to a relatively small number of simpler substances, and further reduction by metabolic pathways that are specific for sugars, fats, and amino acids results in still simpler products. These simpler products frequently contain from two to six carbon atoms and are largely organic acids or acid derivatives.

Sir Hans Krebs first elucidated the *cyclic* and *catalytic* role of the dicarboxylic and tricarboxylic acids that, together with the associated enzymes, constitute a system termed the *common* or *final metabolic pathway.*

The Krebs cycle produces most of the carbon dioxide made in our tissues; it is the source of a significant part of the reduced coenzyme pool that drives the respiratory chain to produce ATP, and it is the means by which excess energy is made available for fatty acid biosynthesis prior to triglyceride formation for fat storage in the adipose tissues. It also provides important precursors for the subunits needed in the synthesis of various molecules, including hemoglobin and the nucleic acids. Some of its components provide direct (product-precursor) or indirect (allosteric) control of other enzyme systems.

In untreated diabetes mellitus the failure of normal glucose metabolism leads to reduced intracellular concentrations of glucose and pyruvic acid. The body responds by producing an excessive amount of ketone bodies from stored fat, but the excess levels of ketone bodies are toxic, leading to profound disturbances of acid-base and electrolyte metabolism that may result in coma or death. In treated but uncontrolled diabetes the disturbances may be less obvious, but they are no less malevolent, especially in such specialized tissues as the eye.

This chapter will focus on the common aspects of the metabolic pathways of carbohydrates, fatty acids, and amino acids. As the details are explored, it will be seen that, to a striking extent, these metabolic functions depend on a small number of coenzymes or prosthetic groups, including NAD^+, $NADP^+$, FMN, and FAD.

In the early days of modern biochemistry it was discovered that the addition of dicarboxylic or tricarboxylic acids to suitable tissue preparations causes the formation of CO_2, H_2O, and ATP, that they support the synthesis of certain amino and fatty acids, and they they can be incorporated into fixed tissue elements. In other words, they support many biochemical processes, for at least a brief period of time, as would more normal nourishment of the same tissues. Finally, Sir Hans Krebs conceived the brilliant idea that some of the dicarboxylic and tricarboxylic acids, together with a handful of associated enzymes, serve as a catalytic cycle for intermingling metabolites derived from food sources. This concept has been totally supported through the use of isotopically labeled compounds and by an enormous amount of research. The Krebs cycle can be described as a veritable biochemical traffic circle; material coming to it from carbohydrate sources might leave it to form fat, while material coming to it from amino acids might leave it to form carbohydrate. Only one "road" seems closed, that leading from fat to carbohydrate.

CELLULAR LOCATION OF THE KREBS CYCLE

In mammalian tissues all of the necessary components of the Krebs cycle are packaged in the mitochondria. A few individual acids and enzymes are also found elsewhere in the cell. Passage of materials from inside the mitochondria to the outside, and vice versa, is generally a controlled, active process. Mitochondria from different tissues vary slightly in the amount of enzymes or acids they contain and in the rate at which certain components can move across the membranes. However, in general, all mammalian mitochondria have very similar properties. This similarity is essential to the preservation of the catalytic concentration of dicarboxylic and tricarboxylic acids, on which the integrity and continued function of the cycle depends. The enzymes that function in the cycle are in close proximity to those of the respiratory chain, probably on the matrix side of the inner mitochondrial membrane or within the matrix space, loci that greatly facilitate the oxidative aspects of the cycle. Some of the enzymes are bound very tightly to the membrane structure, but others are more easily freed from the membrane and can be readily removed for purification and study.

NATURE OF CYCLE COMPONENTS

The individual acids that participate directly in the Krebs cycle are shown in Fig. 6-1; the enzymes are omitted for clarity. The overall reaction of the cycle can be expressed by the equation:

$$CH_3COOH + 2O_2 \longrightarrow 2CO_2 + 2H_2O$$

In other words, even though each of the component acids contains more than two carbon atoms, the *net effect* of the cycle is accurately described as the formation of CO_2 and H_2O from acetic acid. The two moles of CO_2 are equivalent to the carbon content of one mole of acetic acid, which in turn is equivalent to the oxidized moiety of acetyl CoA. It is important to realize that the carbons of the CO_2 are *not* identical to the carbon atoms of the acetyl group, at least in a single turn of the cycle. This will become more evident later on. As shown in Fig. 6-1, the system begins with pyruvate. Strictly speaking, pyruvate is not directly involved in the Krebs cycle, but there are several reasons for its inclusion here. First, pyruvate is a major source of acetyl CoA; energy derived from carbohydrate enters the Krebs cycle via pyruvate. Second, the enzyme complex that decarboxylates pyruvate to form acetyl CoA is very similar in composition and mechanism to the enzyme complex that decarboxylates α-ketoglutarate, and it is convenient to discuss them together. Fig. 6-2 lists the enzymes that are the catalysts for the cycle

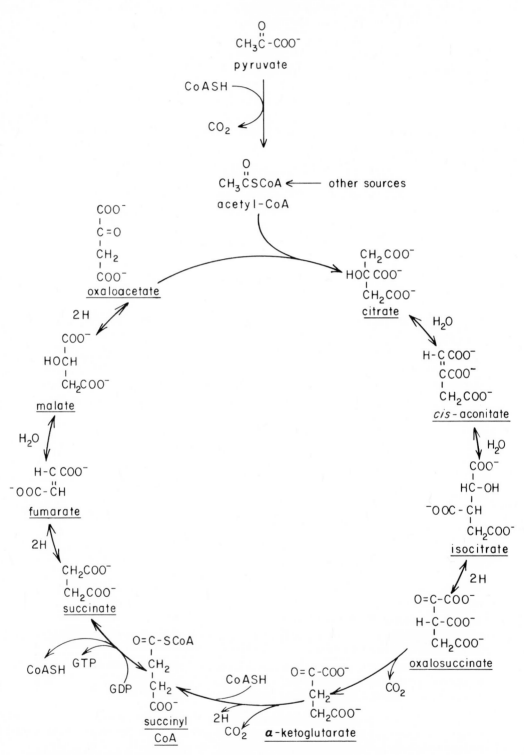

Fig. 6-1. Chemical components of the Krebs cycle. Enzymes are omitted for clarity.

and, together with the information given in Fig. 6-1, permits a detailed analysis of the whole system.

Pyruvate decarboxylation

Pyruvate is the anion of an α-keto acid; as a class, α-keto acids tend to decarboxylate even in the absence of enzymes. The decarboxylation rate is a good deal faster in the presence of the *pyruvate dehydrogenase complex*. The catalyst is

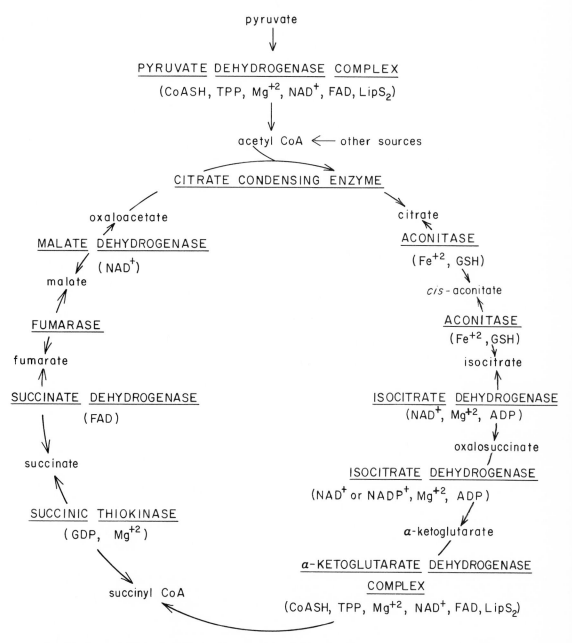

Fig. 6-2. Enzymes and coenzymes of the Krebs cycle. Cofactors are shown in parentheses below each enzyme.

called a complex because it is a multienzyme unit stable enough to be isolated as a large particle from many tissues. These particles have molecular weights in the neighborhood of 4×10^6 and are large enough to be visible under the electron microscope. Analysis of the particles from *Escherichia coli* reveals a number of details about the fine structure.

1. There are twenty-four molecules of pyruvate dehydrogenase, molecular weight 90,000, each containing one molecule of thiamine pyrophosphate.

2. There is one molecule of dihydrolipoyl transacetylase, molecular weight 864,000. This molecule is composed of twenty-four polypeptides, to each of which is attached one molecule of lipoate. This is a remarkable degree of complexity.

3. There are twelve molecules of dihydrolipoyl dehydrogenase, molecular weight 55,000, each containing one molecule of FAD.

The mechanism by which the complex operates is shown in Fig. 6-3, which presents a small portion of the entire unit. At the left, E_1 represents pyruvate dehydrogenase, to which is attached the α-hydroxyethyl group generated by decarboxylation of the pyruvate. Next to it is the oxidized lipoate, which presumably can swing in close proximity to either the E_1 or the E_3 side. As it approaches E_1, it can accept a proton on the sulfur attached to its terminal carbon (C_8) and accept the acetyl group on the sulfur attached to C_6, forming an acetyl thioester. By definition the acetyl thioester is a high-energy compound. As the acetylated lipoate swings away from E_1, it exchanges the acetyl group with free CoASH to form acetyl CoA, which is also a high-energy compound; this is the mechanism whereby the energy released in the decarboxylation process is trapped for later use. The lipoate is now in the reduced form, and in order to repeat the transacylation it must be oxidized again. This is accomplished by E_3, the dihydrolipoyl dehydrogenase that contains the FAD. In the final step, $FADH_2$ is reoxidized to FAD by NAD^+, producing NADH and H^+. NADH is ultimately oxidized in the respiratory chain to produce ATP, H_2O, and NAD^+.

Dihydrolipoyl transacetylase (E_2, Fig. 6-3) contains lipoate that is bound to an ϵ-amino group of a lysine residue of the peptide by means of an amide bond with

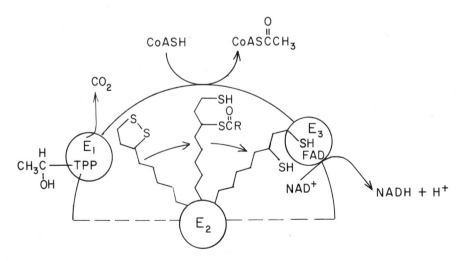

Fig. 6-3. Mechanism of pyruvate dehydrogenase complex. Three component enzymes are indicated by E_1, E_2, and E_3. Coenzymes are indicated by TPP for thiamine pyrophosphate, CoASH for coenzyme A, FAD for flavin adenine dinucleotide, and NAD^+ for nicotinamide adenine dinucleotide. Acetylated lipoate and lipoate are shown attached to E_2.

the carboxyl group of lipoate, as shown below. Lipoate is reduced in the process of accepting an acyl group, as shown below.

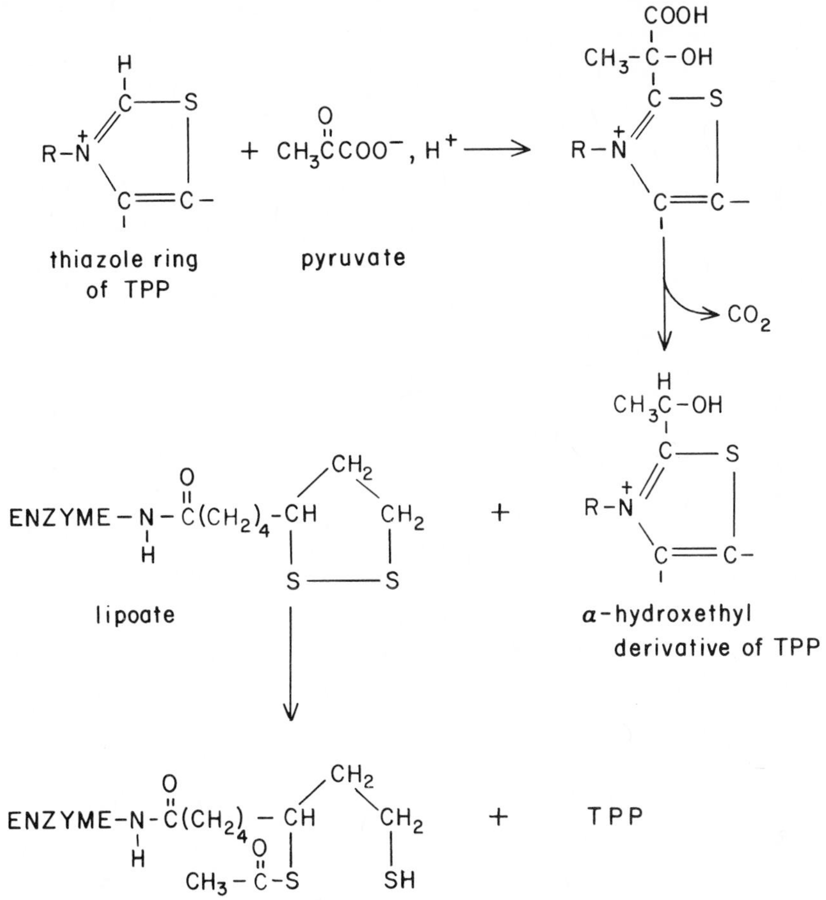

thiazole ring
of TPP

pyruvate

$+ CH_3\overset{O}{\overset{\|}{C}}COO^-, H^+ \longrightarrow$

CO_2

lipoate

a-hydroxethyl
derivative of TPP

+ TPP

acetylated form of lipoate

Regulation of the pyruvate dehydrogenase complex

Pyruvate lies at a major branch point in metabolism, so it is not surprising that this complex should be subject to regulation by several controls. It has been shown that both the bacterial and mammalian forms of the complex can be inactivated by phosphorylation and activated by dephosphorylation, as shown in Fig. 6-4. These covalent modifications are catalyzed by a *pyruvate dehydrogenase kinase* and a *pyruvate dehydrogenase phosphatase*, respectively. These regulating enzymes are an integral part of the dehydrogenase complex and appear to be attached to the pyruvate decarboxylase (E_1 of Fig. 6-3).

The major regulatory function is performed by the kinase; the phosphatase plays a smaller role. Cyclic AMP and a high concentration of ATP activate this kinase. Opposite effects depend on the concentration of ADP; when this is high a larger proportion of the complex is kept in the active state. Correspondingly, when the concentration of ATP is high it overcomes the ADP effect, more of the complex is converted to the inactive state, and glycolysis is slowed down. Ketosis or a high concentration of free fatty acids inhibits the phosphatase, and thus a greater proportion of inactive complex is maintained.

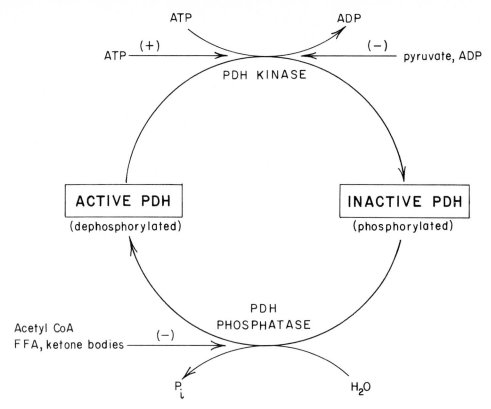

Fig. 6-4. Regulation of the pyruvate dehydrogenase complex. The kinase and phosphatase enzymes are part of the pyruvate dehydrogenase (PDH) complex. Phosphorylation and dephosphorylation are the major regulatory factors, with lesser contributions from the other allosteric effectors shown. FFA represents free fatty acids.

To summarize, the pyruvate dehydrogenase complex is regulated largely by a phosphorylation-dephosphorylation cycle that responds to the energy charge. To a lesser extent it is also regulated by the concentrations of pyruvate and acetyl CoA.

Condensing reaction

The condensing reaction entails the condensation of oxaloacetate and acetyl CoA to form citrate.

Citrate synthetase, or citrate-condensing enzyme, is exclusively a mitochondrial enzyme in mammalian tissues and is essentially a "one-way" enzyme, since the equilibrium strongly favors citrate formation. No coenzyme is required. This synthetase is remarkably specific regarding its substrates; fluoroacetyl CoA is one of the few compounds other than acetyl CoA that is an acceptable condensing partner for oxaloacetate. Fluoroacetate is sometimes sold as a rodenticide, and

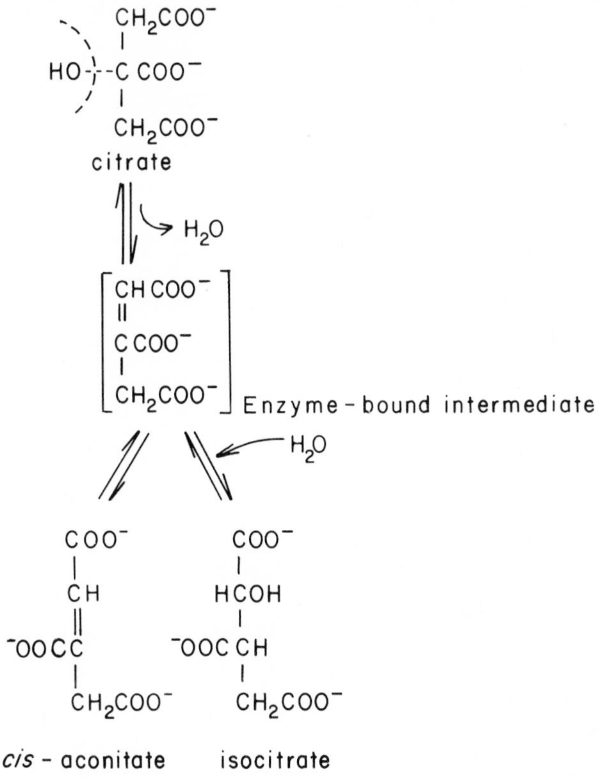

$$
\begin{array}{c}
\text{CH}_2\text{COO}^- \\
| \\
\text{HO--C COO}^- \\
| \\
\text{CH}_2\text{COO}^- \\
\text{citrate}
\end{array}
$$

$\Downarrow \searrow \text{H}_2\text{O}$

$$
\left[
\begin{array}{c}
\text{CH COO}^- \\
\| \\
\text{C COO}^- \\
| \\
\text{CH}_2\text{COO}^-
\end{array}
\right]
\text{Enzyme-bound intermediate}
$$

$\quad // \quad \searrow \longleftarrow \text{H}_2\text{O}$

$$
\begin{array}{cc}
\text{COO}^- & \text{COO}^- \\
| & | \\
\text{CH} & \text{HCOH} \\
\| & | \\
^-\text{OOCC} & ^-\text{OOC CH} \\
| & | \\
\text{CH}_2\text{COO}^- & \text{CH}_2\text{COO}^- \\
\\
\textit{cis} - \text{aconitate} & \text{isocitrate}
\end{array}
$$

Fig. 6-5. Isomerization of citrate by an enzyme-bound intermediate.

its deliberate or accidental ingestion by man is often fatal. The product of this reaction is fluorocitrate, which cannot be further metabolized by the enzymes of the Krebs cycle.

Isomerization of citrate

The conversion of citrate to isocitrate takes place in two steps, both of which are catalyzed by the enzyme *aconitase*. Aconitase is also virtually specific. It is known to require Fe^{+2} and probably *glutathione* (γ-L-glutamyl-L-cysteinylglycine, or GSH) for its activity, but the precise function of glutathione is not clearly understood. When aconitase acts on citrate, the mixture that results contains, at equilibrium, 89% citrate, 3% *cis*-aconitate, and 8% isocitrate. It is probable that a common intermediate accounts for the formation of the mixture, as is proposed by the scheme shown in Fig. 6-5.

Although citrate, the primary substrate for aconitase, indeed has a plane of symmetry such that the terminal carboxylate groups appear to be equivalent, this is not in fact the case from an enzymatic point of view. From in vitro experiments using variously labeled preparations of citrate, it has been determined that aconitase always operates on that part of the citrate structure that is derived from oxaloacetate. This puzzling observation can be explained by the hypothesis that the substrate attaches to the enzyme at three different binding sites. This hypothesis is depicted schematically in Fig. 6-6, where the citrate is represented in tetrahedral form, with the central carbon atom lying in the center. The idealized surface of aconitase is indicated below the tetrahedron, with the three binding sites identi-

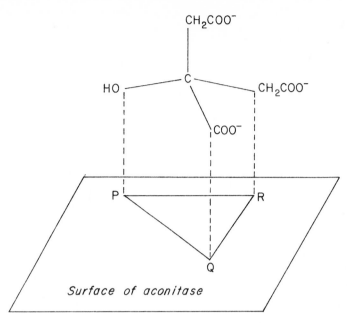

Fig. 6-6. Three-point attachment of citrate to aconitase, shown in tetrahedral form; this is an explanation of the biochemical asymmetry of citrate ion.

fied only as P, Q, and R. Assuming that the dashed lines represent specific affinities between the binding sites of the enzyme and the indicated functional groups of the substrate, citrate could attach to aconitase *only* if the citrate ion has the orientation shown. This can easily be verified with models. Therefore, if we have labeled acetyl CoA, and if we consider transformations brought about during the first turn of the cycle, the carbon atoms lost *must* originate from oxaloacetate carbons and not from those contributed by acetyl CoA.

First decarboxylation

Isocitrate is next dehydrogenated to oxalosuccinate, which remains in most mammalian tissues as a transient, enzyme-bound intermediate. Oxalosuccinate is also the anion of a α-keto acid and is decarboxylated to produce α-ketoglutarate.

Isocitrate dehydrogenase occurs in several forms. In most mammalian tissues the extramitochondrial form is $NADP^+$ linked and the intramitochondrial form is NAD^+ linked, but in heart mitochondria both forms occur. The NAD^+-linked enzyme exhibits complex and interesting allosteric behavior; for our purposes it is sufficient to state that ADP is a positive effector and that ATP is a negative effector. ADP causes the enzyme to associate as an active tetramer, which ATP then dissociates to an inactive dimer. When the enzyme is treated to remove all traces of metal ions, it completely loses activity. The NAD^+-linked form is also subject to strong inhibition by bilirubin, at concentrations of the latter sometimes found in severe jaundice.

Another difference between the NAD^+- and $NADP^+$-linked forms of isocitrate dehydrogenase involves the role of oxalosuccinate as an intermediate. There is no evidence for formation of this intermediate with the NAD^+-linked enzyme, and it appears in our version of the cycle only because numerous species of heart mitochondria appear to contain both enzymes. Nevertheless, *either* enzyme can readily decarboxylate added oxalosuccinate. As a consequence, isocitrate dehydrogenase is said to catalyze two separate reactions of the Krebs cycle, as does aconitase.

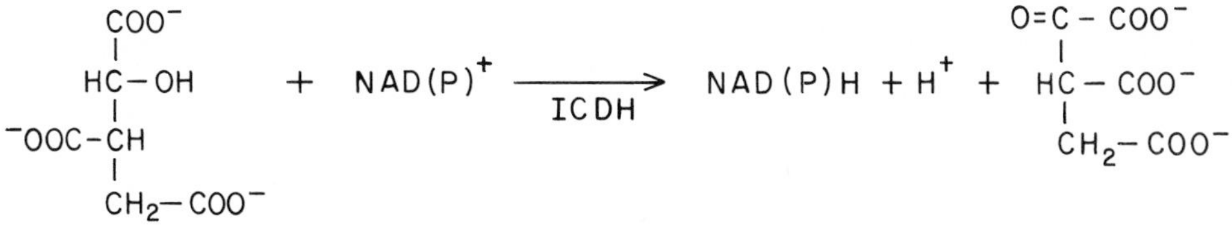

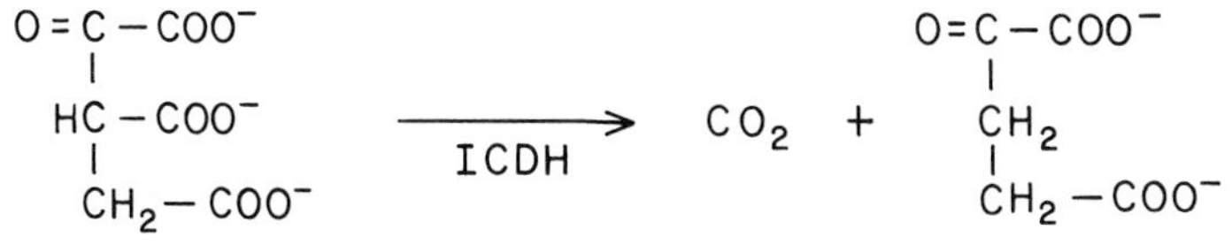

$$
\begin{array}{c}
COO^- \\
| \\
HC-OH \\
| \\
{}^-OOC-CH \\
| \\
CH_2-COO^-
\end{array}
\;+\; NAD(P)^+ \;\xrightarrow[ICDH]{}\; NAD(P)H + H^+ \;+\;
\begin{array}{c}
O=C-COO^- \\
| \\
HC-COO^- \\
| \\
CH_2-COO^-
\end{array}
$$

$$
\begin{array}{c}
O=C-COO^- \\
| \\
HC-COO^- \\
| \\
CH_2-COO^-
\end{array}
\;\xrightarrow[ICDH]{}\; CO_2 \;+\;
\begin{array}{c}
O=C-COO^- \\
| \\
CH_2 \\
| \\
CH_2-COO^-
\end{array}
$$

Second decarboxylation

The decarboxylation of α-ketoglutarate is quite different from that of oxalosuccinate; in fact, it is analogous to the decarboxylation of pyruvate that was discussed earlier. The first enzyme in this complex is specific for ketoglutarate rather than pyruvate, but otherwise the components of the complex and the mechanism by which it operates are as described for pyruvate. The product of this reaction is succinyl CoA, analogous to the acetyl CoA produced from pyruvate.

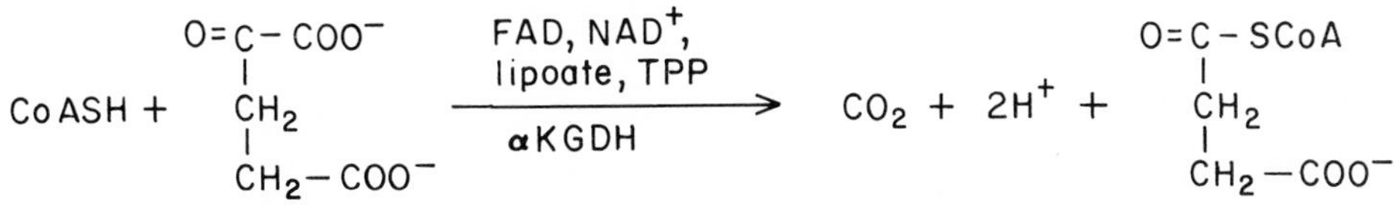

$$
CoASH \;+\;
\begin{array}{c}
O=C-COO^- \\
| \\
CH_2 \\
| \\
CH_2-COO^-
\end{array}
\;\xrightarrow[\alpha KGDH]{FAD,\; NAD^+,\; lipoate,\; TPP}\; CO_2 + 2H^+ \;+\;
\begin{array}{c}
O=C-SCoA \\
| \\
CH_2 \\
| \\
CH_2-COO^-
\end{array}
$$

Substrate-level phosphorylation

The succinyl CoA generated by decarboxylation of α-ketoglutarate is converted to free succinate and CoASH by *succinyl thiokinase*, and the reaction is tightly coupled to the phosphorylation of GDP to GTP.

$$
\begin{array}{c}
O=C-SCoA \\
| \\
CH_2 \\
| \\
CH_2-COO^-
\end{array}
\;+\; GDP \;\longrightarrow\; GTP + CoASH \;+\;
\begin{array}{c}
CH_2-COO^- \\
| \\
CH_2-COO^-
\end{array}
$$

Recall that this type of phosphorylation, in which the respiratory chain is not directly involved, is known as substrate-level phosphorylation. This is one of the few instances in which GDP is preferred over ADP as a high-energy acceptor. The formation of GTP traps significant energy that would be wasted as heat if succinyl CoA were simply hydrolyzed.

Final stages

Succinate is converted to oxaloacetate in a series of steps that employ mechanisms similar to those used in the conversion of citrate to isocitrate, that is, formation of a double bond, addition of water, and dehydrogenation.

In the first step, succinate is converted to fumarate, and the reaction is catalyzed by *succinate dehydrogenase*, a flavin-containing enzyme that is tightly bound to the inner mitochondrial membrane.

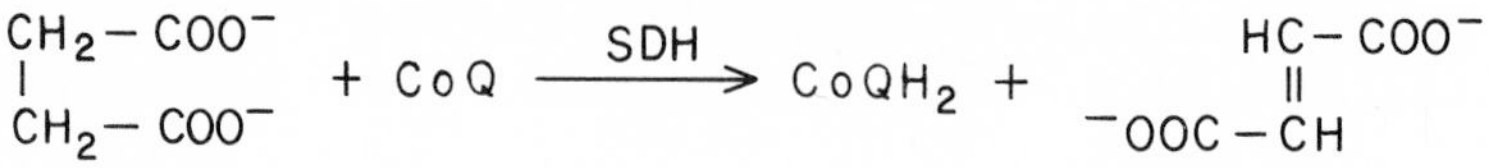

$$\begin{array}{c} CH_2-COO^- \\ | \\ CH_2-COO^- \end{array} + CoQ \xrightarrow{\text{SDH}} CoQH_2 + \begin{array}{c} HC-COO^- \\ || \\ ^-OOC-CH \end{array}$$

Until very recently it was not possible to prepare purified soluble forms of this enzyme; now that a soluble form has been prepared, it appears to contain not only FAD but also four gram atoms of nonheme iron/mole of enzyme. Succinate dehydrogenase is activated by reduced coenzyme Q and by ATP in intact mitochondria. Since the activity of the dehydrogenase in most mammalian tissues is very high when compared to the activity of the other Krebs cycle enzymes, it is advantageous to have this kind of control; if succinate dehydrogenase operated at maximum capacity, oxidation of NADH could be seriously impeded by competition for entry into the respiratory chain. Since the ATP equivalence of NADH is three and that of succinate is two, high-energy yields would suffer accordingly.

Fumarate is converted to L-malate by the enzyme *fumarase,* which is composed of four identical subunits.

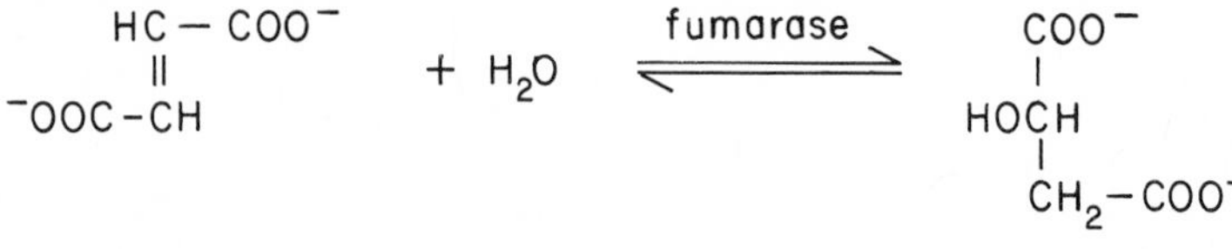

$$\begin{array}{c} HC-COO^- \\ || \\ ^-OOC-CH \end{array} + H_2O \xrightleftharpoons{\text{fumarase}} \begin{array}{c} COO^- \\ | \\ HOCH \\ | \\ CH_2-COO^- \end{array}$$

Each peptide has a molecular weight of 4.85×10^4 and contains three free —SH groups. It requires no coenzyme. The addition of water across the double bond is completely stereospecific, so that L-malate is always the product.

The last step is the conversion of malate to oxaloacetate by the enzyme *malate dehydrogenase* (MDH).

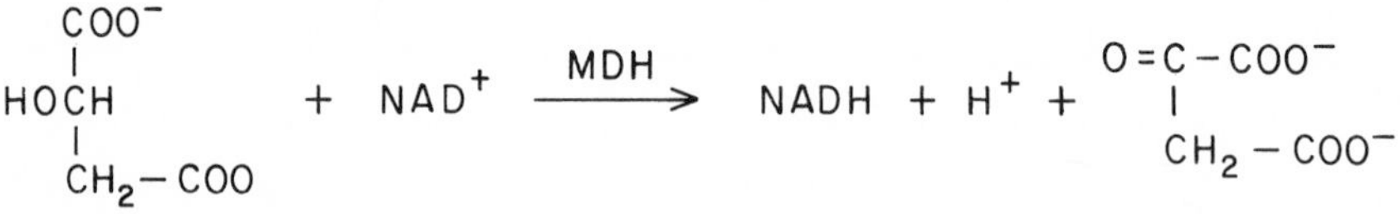

$$\begin{array}{c} COO^- \\ | \\ HOCH \\ | \\ CH_2-COO \end{array} + NAD^+ \xrightarrow{\text{MDH}} NADH + H^+ + \begin{array}{c} O=C-COO^- \\ | \\ CH_2-COO^- \end{array}$$

Six isozymes of MDH have been found in human tissues. Of the six, isozyme IV constitutes more than half of the total activity. As is true for lactate dehydrogenase (LDH), MDH's have electrophoretic properties comparable to the α_2-globulins of blood serum. There are probably covalent differences in the various MDH subunits, but these have not been established conclusively. MDH and LDH are normal components of blood serum. The detectable activity is sometimes increased as a result of disease, as will be explained later. Since it is technically easier to assay the isozymes of LDH than those of MDH, the latter assay has not yet become a part of routine clinical laboratory practice.

RECAPITULATION OF KREBS CYCLE ENERGETICS

When the free energy changes of the individual reactions considered thus far are summarized (Table 6-1), it is apparent that there is a large, negative net change, a clear indication that the Krebs cycle will operate spontaneously in a clockwise direction (Figs. 6-1 and 6-2). Even when the decarboxylation of pyruvate is omitted, the conclusion is the same. Strictly speaking, this conclusion implies that all of the components exist in tissues at an activity of 1 M, which is not the case. Nevertheless, there is ample experimental evidence that even at the prevailing tissue concentrations of 1 to 5 mM, the clockwise operation of the cycle is favored.

Table 6-1.   Energetics of pyruvate oxidation

Step	$\Delta G°'(-)$	$\Delta G°'(+)$
1. pyruvate $\longrightarrow$ acetyl CoA + CO_2	8.0	
2. acetyl CoA + oxaloacetate $\longrightarrow$ citrate	9.08	
3. citrate $\rightleftharpoons$ cis-aconitate		2.04
4. cis-aconitate $\rightleftharpoons$ isocitrate	0.45	
5. isocitrate $\longrightarrow$ a-ketoglutarate	1.70	
6. a-ketoglutarate $\rightleftharpoons$ succinyl CoA	8.82	
7. succinyl CoA $\rightleftharpoons$ succinate	2.12	
8. succinate $\rightleftharpoons$ fumarate		0
9. fumarate $\rightleftharpoons$ malate	0.88	
10. malate $\rightleftharpoons$ oxaloacetate		6.69
	-31.05	$+8.73$

TOTAL FOR ALL STEPS $\Delta G°' = -22.32 \, kcal/mole$

TOTAL FOR KREBS CYCLE STEPS $\Delta G°' = -14.32 \, kcal/mole$

A chemical summary of the Krebs cycle can now be given in a more exact form by the equation:

$$CH_3\overset{O}{\overset{\|}{C}}SCoA + 3NAD^+ + FAD + GDP + P_i + 2H_2O \longrightarrow$$

$$2CO_2 + CoASH + 3NADH + 3H^+ + FADH_2 + GTP$$

For each mole of acetyl radical that passes through the cycle, twelve moles of ATP can be generated by mechanisms already described. This is equivalent to 84 kcal (12×7 kcal). Three moles of NADH produce nine moles of ATP, and one mole of $FADH_2$ produces two moles of ATP; both of these processes involve the respiratory chain. The last mole of ATP is produced by an enzyme-catalyzed exchange between GTP and ADP. If the same mole of acetyl radical had been burned in a bomb calorimeter, the heat liberated would have been approximately 209 kcal/mole, from which it is evident that the efficiency of the biochemical trapping device is approximately 40%. By itself, the Krebs cycle is the largest single source of ATP.

ENTRY OF AMINO ACIDS INTO KREBS CYCLE
Transamination of amino acids

If the structures of the keto acids that are involved in the Krebs cycle are compared with the structures of certain amino acids, it is clear that they are *homologous* with respect to the carbon atoms they contain, that is, oxaloacetate and aspartate differ only by having a carbonyl oxygen in place of an amino group. The same is true for the pairs ketoglutarate-glutamate and pyruvate-alanine (Fig. 6-7). As we shall see, a means does exist whereby the carbonyl oxygen can be exchanged, reversibly, with the amino group. This device provides for conversion of amino acids into Krebs cycle intermediates or vice versa. When the supply of

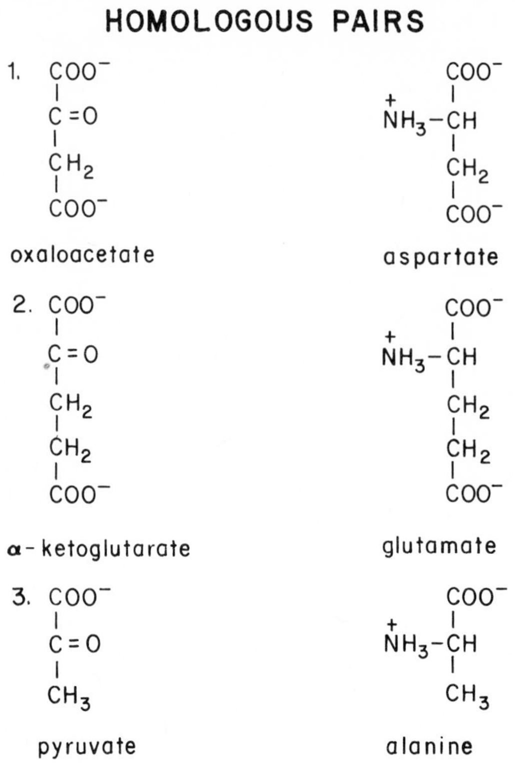

HOMOLOGOUS PAIRS

1.
$$
\begin{array}{c}
COO^- \\
| \\
C=O \\
| \\
CH_2 \\
| \\
COO^-
\end{array}
\qquad
\begin{array}{c}
COO^- \\
| \\
\overset{+}{N}H_3-CH \\
| \\
CH_2 \\
| \\
COO^-
\end{array}
$$

oxaloacetate   aspartate

2.
$$
\begin{array}{c}
COO^- \\
| \\
C=O \\
| \\
CH_2 \\
| \\
CH_2 \\
| \\
COO^-
\end{array}
\qquad
\begin{array}{c}
COO^- \\
| \\
\overset{+}{N}H_3-CH \\
| \\
CH_2 \\
| \\
CH_2 \\
| \\
COO^-
\end{array}
$$

α-ketoglutarate   glutamate

3.
$$
\begin{array}{c}
COO^- \\
| \\
C=O \\
| \\
CH_3
\end{array}
\qquad
\begin{array}{c}
COO^- \\
| \\
\overset{+}{N}H_3-CH \\
| \\
CH_3
\end{array}
$$

pyruvate   alanine

Fig. 6-7. Keto acids contained in the Krebs cycle and some homologous amino acids. These pairs are identical except for the groups attached to the α-carbon atom of each member.

NONHOMOLOGOUS PAIRS

1.
$$
\begin{array}{c}
COO^- \\
| \\
C=O \\
| \\
CH_2 \\
| \\
COO^-
\end{array}
\qquad
\begin{array}{c}
COO^- \\
| \\
\overset{+}{N}H_3-CH \\
| \\
CH_2 \\
| \\
CH_2 \\
| \\
COO^-
\end{array}
$$

oxaloacetate   glutamate

2.
$$
\begin{array}{c}
COO^- \\
| \\
C=O \\
| \\
CH_3
\end{array}
\qquad
\begin{array}{c}
COO^- \\
| \\
\overset{+}{N}H-CH \\
| \\
CH_2 \\
| \\
CH_2 \\
| \\
COO^-
\end{array}
$$

pyruvate   glutamate

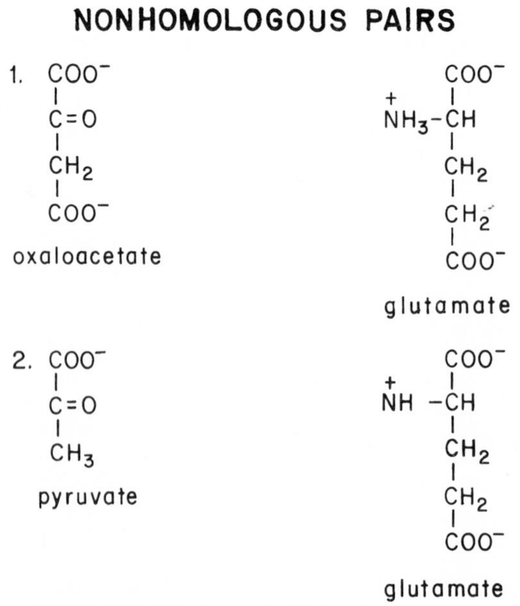

Fig. 6-8. Keto acids contained in the Krebs cycle and some nonhomologous amino acids. Note that these pairs differ in the groups attached at the α-carbon atom of each member and also in the number of carbons in either member of each pair.

these amino acids exceeds the requirements for protein biosynthesis, the excess can readily be converted to Krebs cycle intermediates so the energy content of their carbon skeletons is available for oxidation. When these amino acids are needed, they can be produced from the keto acid intermediates of the Krebs cycle.

Transaminases also exist that can catalyze the exchange of carbonyl oxygens and amino groups between *nonhomologous* pairs; in particular, two important transaminases catalyze such exchange between oxaloacetate-glutamate and pyruvate-glutamate (Fig. 6-8). A special significance of these two transaminases is the ease with which they can be measured in the blood serum and related to certain disease states. Numerous other transaminases involving glutamate are also known.

Mechanism of transamination

Pyridoxal phosphate contains an aldehyde group that in the mechanistic sense is functional group of the coenzyme. It can react with compounds containing amino groups to form *imines*, or Schiff bases.

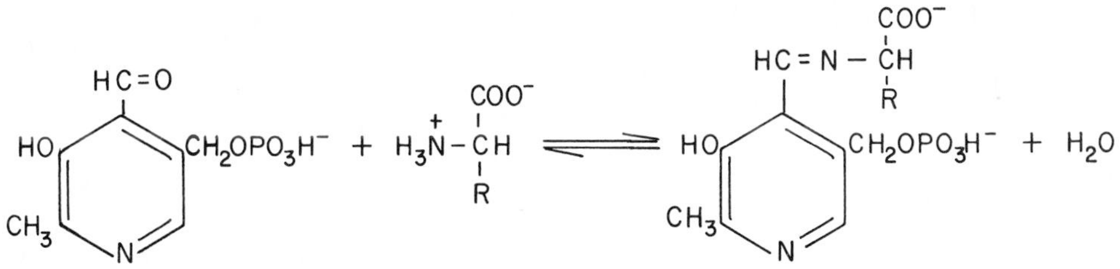

If the transaminase is now represented by $E\!-\!C\!=\!O$, the overall mechanism of transamination can be described in the four steps shown in Fig. 6-9. The first step is a shorthand statement of Schiff base formation. The second step shows that there are two distinct resonance forms of the imine, each of which is equally susceptible to hydrolysis or transfer. As a result, the products would be a keto acid and the pyridoxamine form of the enzyme, as shown in the third step. Finally, as shown in the fourth step, the amine form can react with a second keto acid to form a new amino acid and the aldehydic form of the enzyme. Thus the sum of these partial reactions amounts to the transfer of an α-amino group from one acid residue to another. Transaminase-catalyzed reactions have a value for $\Delta G^{o\prime} \approx 0$, so that they are freely reversible. The glutamate and aspartate synthesized by transaminase reactions pose no problem since they are frequently employed in other biosynthetic pathways to be discussed in later chapters.

Entry of other amino acids into the Krebs cycle

Direct transamination allows the entry of only a small number of amino acids into the Krebs cycle. Serine, threonine, glycine, and cysteine can also be converted to pyruvate by reactions that will be considered in Chapter 9. At this point it is sufficient to mention that these amino acids can be accommodated in the cycle by pathways that require no changes in their carbon skeletons. Interestingly, pyridoxal phosphate is also involved in some of the enzymes that convert the above-mentioned amino acids into Krebs cycle intermediates, but the reaction mechanisms frequently generate NH_4^+ as a product and therefore are not transaminations.

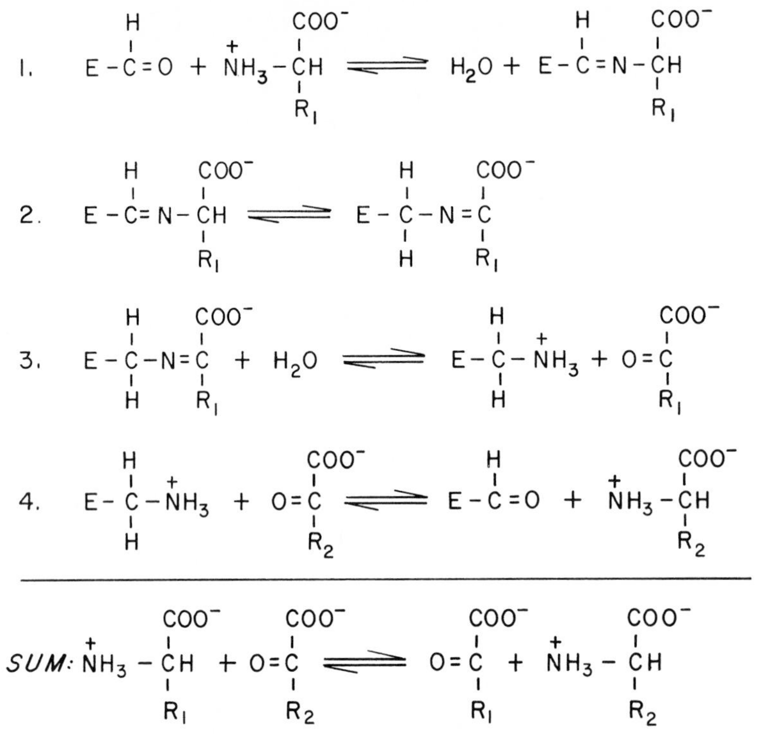

$$1.\quad \underset{}{E-\overset{H}{\underset{}{C}}=O} \;+\; \overset{+}{N}H_3-\underset{R_1}{\overset{COO^-}{CH}} \;\rightleftharpoons\; H_2O \;+\; E-\overset{H}{C}=N-\underset{R_1}{\overset{COO^-}{CH}}$$

$$2.\quad E-\overset{H}{C}=N-\underset{R_1}{\overset{COO^-}{CH}} \;\rightleftharpoons\; E-\underset{H}{\overset{H}{C}}-N=\underset{R_1}{\overset{COO^-}{C}}$$

$$3.\quad E-\underset{H}{\overset{H}{C}}-N=\underset{R_1}{\overset{COO^-}{C}} \;+\; H_2O \;\rightleftharpoons\; E-\underset{H}{\overset{H}{C}}-\overset{+}{N}H_3 \;+\; O=\underset{R_1}{\overset{COO^-}{C}}$$

$$4.\quad E-\underset{H}{\overset{H}{C}}-\overset{+}{N}H_3 \;+\; O=\underset{R_2}{\overset{COO^-}{C}} \;\rightleftharpoons\; E-\overset{H}{C}=O \;+\; \overset{+}{N}H_3-\underset{R_2}{\overset{COO^-}{CH}}$$

$$SUM:\quad \overset{+}{N}H_3-\underset{R_1}{\overset{COO^-}{CH}} \;+\; O=\underset{R_2}{\overset{COO^-}{C}} \;\rightleftharpoons\; O=\underset{R_1}{\overset{COO^-}{C}} \;+\; \overset{+}{N}H_3-\underset{R_2}{\overset{COO^-}{CH}}$$

Fig. 6-9. Mechanisms of pyridoxal phosphate–catalyzed transamination.

Still other amino acids can undergo *oxidative deamination*, a process that involves L-*amino acid oxidase*, an FAD-containing enzyme of broad specificity. Further details of these pathways will be discussed later, and the means by which the carbon skeletons are readily modified to produce Krebs cycle intermediates will be demonstrated.

Methionine and the branched-chain amino acids require more extensive and complicated modifications if any parts of their carbon skeletons are to be used in the Krebs cycle, since these acids may be converted to *propionyl CoA* or *methylmalonyl CoA*. These compounds will be discussed later in terms of fatty acid metabolism, but it should be kept in mind that they also arise from some amino acids.

Mechanism of methylmalonyl CoA mutase

The isomerization of methylmalonyl CoA to succinyl CoA is catalyzed by a mutase employing 5′-deoxyadenosylcobalamin, which is derived from vitamin B_{12}. It is one example of a larger class of reactions involving this coenzyme, which have in common the transfer of a hydrogen atom from one carbon atom to another. There is a simultaneous and opposite migration of another group, X.

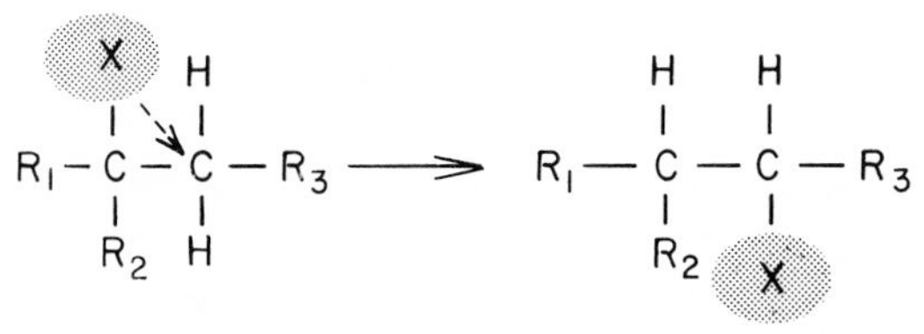

$$R_1-\underset{R_2}{\overset{X}{C}}-\underset{H}{\overset{H}{C}}-R_3 \;\longrightarrow\; R_1-\underset{R_2}{\overset{H}{C}}-\underset{X}{\overset{H}{C}}-R_3$$

In the specific instance the general equation becomes:

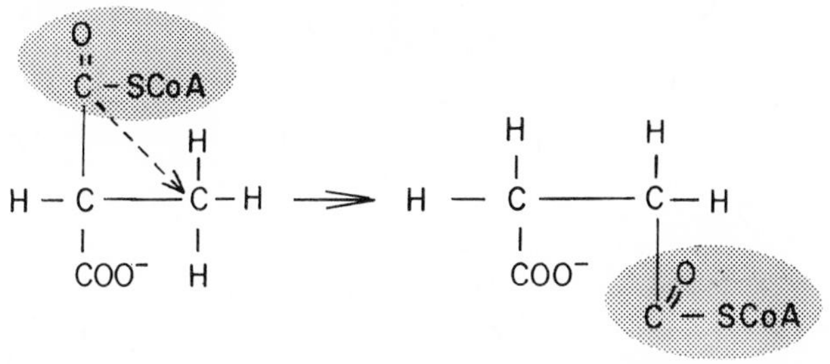

$$
\begin{array}{ccc}
\overset{\displaystyle O}{\underset{\displaystyle C-SCoA}{\|}} & & \\
\end{array}
$$

This intramolecular rearrangement takes place without incorporation of solvent hydrogen; the transfer occurs via the 5′-C position of the deoxyadenosyl moiety of the coenzyme by a mechanism that is not completely understood. The hydrogen attaches to the deoxyadenosyl group, and the cobalt serves as acceptor for the second group, X.

The conversion of methylmalonyl CoA to succinyl CoA permits the entry of residues derived from branched-chain amino acids (or fatty acids) into the Krebs cycle. Without this capacity, significant amounts of energy would be lost.

Summary

To recapitulate, amino groups may be removed from amino acids by transamination or by oxidative deamination. Transamination preserves the amino groups by collecting them primarily in the aspartate or glutamate pools, while the remainder may enter the alanine pool. Oxidative deamination converts the amino groups into ammonium ions or amide groups, many of which are finally converted to glutamine or asparagine. These reactions are of utmost clinical importance, since ammonium ion is quite toxic. Concentrations only a little above the normal may give rise to coma. All of this nitrogen is available for biosynthetic purposes that will be explored later. Transamination may also be employed for the biosynthesis of numerous nonessential amino acids.

The carbon skeletons of deaminated amino acids can be divided into several categories. Some of the keto acids can enter the Krebs cycle directly. A second category requires more or less extensive modification, frequently involving a cleavage into two fragments, one or both of which may enter the cycle. The last category produces odd-numbered carbon atom pieces that must be converted to even-numbered carbon units by a process of carbon dioxide fixation. Taken together, these systems permit a large percentage (but *not* the entirety) of the carbon skeletons of amino acids to enter the Krebs cycle. These reactions are made possible by the simultaneous existence of transaminating enzymes inside and outside the mitochondria and by the capacity of certain key intermediates to cross the mitochondrial membranes.

ANAPLEROTIC REACTIONS

It is conceivable that a sudden influx of pyruvate or of acetyl CoA to the Krebs cycle might seriously deplete the supplies of oxaloacetate required for the citrate synthetase reaction. Two reactions that are auxiliary to the Krebs cycle operate to prevent this situation. These are known as the anaplerotic ("filling up") reactions.

Fig. 6-10 shows an abbreviated version of the Krebs cycle, with the large influx of substrate indicated by the heavy arrows. Pyruvate from glucose and acetyl CoA from fatty acids might be produced in great quantity as a result of violent work or

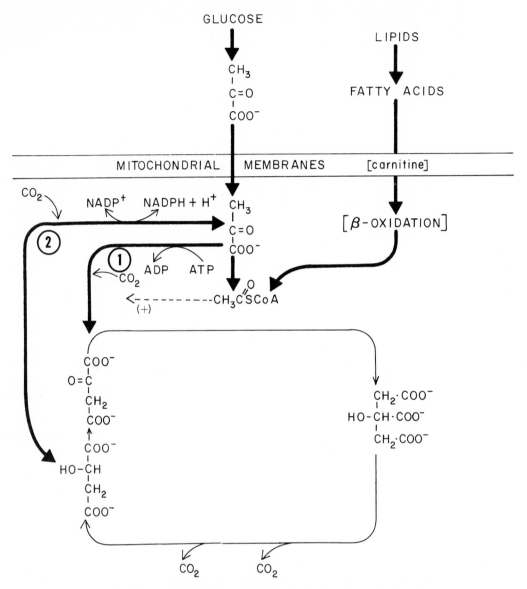

Fig. 6-10. Anaplerotic reactions of the Krebs cycle. Reaction 1 is catalyzed by pyruvate carboxylase, for which acetyl CoA is a required positive allosteric effector. Reaction 2 is catalyzed by the malic enzyme. These reactions produce the oxaloacetate required to condense with a large influx of acetyl CoA.

of the response to epinephrine release in emotional stress. Under these conditions, with a high intramitochondrial concentration of acetyl CoA, pyruvate may be converted to oxaloacetate by the mitochondrial enzyme *pyruvate carboxylase*, a reaction indicated in Fig. 6-10 by the circled number 1. Pyruvate carboxylase is an allosteric enzyme for which acetyl CoA is a required positive effector.

The second anaplerotic reaction is catalyzed by the so-called *malic enzyme* (not to be confused with MDH), the reaction shown by a circled number 2 in Fig. 6-10. This reaction provides extra stability by converting part of the pyruvate to malate, which adds to the oxaloacetate pool. Of the two anaplerotic pathways, the pyruvate carboxylase path is the most important. In diabetic subjects, malic enzyme activity is depressed, and its activity rises following insulin administration to either diabetic or normal subjects, but in vitro experiments indicate that it is not a primary pathway.

MITOCHONDRIAL COMPARTMENTALIZATION

Before considering the regulation of the Krebs cycle and of the respiratory chain, with which it is so closely linked, it is necessary to consider the matter of mitochondrial permeability. Mitochondrial membranes are designed to permit the entrance or exit of only a limited number of substances under conditions that are usually controlled. Control is required for several reasons: to maintain osmotic balance between the mitochondria and the cytoplasm, to maintain electroneutrality (ionic balance) between the mitochondrial interior and exterior, and to ensure minimal required catalytic concentrations of Krebs cycle intermediates within the mitochondrial matrix. From the standpoint of metabolic control we can recognize several intracellular compartments (see also Chapter 5). The extramitochondrial space is accessible to any cytoplasmic component; materials that are found in the cytoplasm may or may not be able to penetrate the mitochondrial membranes. Cytoplasmic enzymes, for example, are kept completely outside the mitochondria by virtue of their large molecular weights. Cytoplasmic coenzymes, for example, NAD^+, are also unable to penetrate the mitochondrial membranes even though they are of fairly low molecular weight. Other smaller molecules, for example, sucrose, can penetrate the outer mitochondrial membrane, but not the inner membrane. Thus sucrose can enter the intermembranous space, which is termed the *outer mitochondrial compartment*. The *inner mitochondrial compartment* is defined as the volume enclosed by the inner mitochondrial membrane. The outer and inner mitochondrial compartments each contain enzymes and coenzymes, which are not exchanged between compartments. Other materials can move between the compartments subject to certain limitations.

The laws that govern mitochondrial permeability have been experimentally derived from studies of intact mitochondria and of submitochondrial particles. When reference is made to the "inside," the space within the inner membrane is meant; the "outside" refers to the sucrose space. The rules can be summarized as follows:

1. Nicotinamide dinucleotides (NAD^+, $NADP^+$, NADH, and NADPH) do not cross the inner mitochondrial membrane. There is an intramitochondrial pool and an extramitochondrial pool; these do not mix. The same is probably true for other coenzymes, including CoASH, FAD, and FMN.

2. The Krebs cycle intermediates (with the exception of those discussed below) can move from the outside to the inside of the mitochondria; generally they move by means of specific carriers, or translocases, that require the simultaneous and oppositely directed movement of a counterion or ions. This follows from osmotic and ionic balance requirements. There are, however, exceptions to this rule. There is no good evidence for movement in either direction of fumarate. Recently evidence for the inward movement of oxaloacetate has been presented, but reliable evidence for movement in the opposite or outward direction is still lacking. For the present purposes we may take the position that oxaloacetate does not freely cross the mitochondrial membranes, and the explanations of mitochondrial regulation that follow are based on this premise.

3. Those amino acids that can be derived directly from Krebs cycle intermediates (or from pyruvate) by simple transamination can also penetrate to the inside of mitochondria.

4. A few compounds, chiefly phosphoenolpyruvate and dihydroxyacetone phosphate, can move outward from the mitochondria by means that are not clearly defined. In the case of dihydroxyacetone phosphate, α-glycerophosphate serves as a required counterion.

5. Of the purine and pyrimidine nucleotides, only ATP moves freely across the mitochondrial membranes. It does so by means of a specific translocase and requires ADP as a counterion. GTP is a possible exception to this exclusion law.

Nature of translocases

The translocases or transport systems just mentioned are part of the inner membrane structure of most mitochondria. They have some properites similar to those of enzymes that operate in bulk solution, but since most of them have not been isolated, they are not generally included in formal enzyme classifications. The properties of translocases include the following:

1. *Specificity.* The ATP translocase will not accept UTP, CTP, or ITP. (Where GTP has been reported to work, it is not yet known whether there is a specific translocase for it or whether GTP is an exception to the specificity rule.)

2. *Saturability.* A given translocase can be saturated with respect to the compound it transports; that is, it has what is equivalent to the K_m of a soluble enzyme.

3. *Inhibition properties.* There are specific inhibitors that block the transport activity of most of the known translocases, just as enzymes in bulk solution have specific inhibitors that depress or prevent their catalytic function. Indeed, it was by means of these inhibitors that the translocases were first observed. It is therefore entirely possible that additional translocases remain to be discovered.

4. *Vectorial nature.* Translocases operate in a spatially directed, or vectorial, sense. Thus whenever ATP moves out of the mitochondria, ADP must move in. This property has no counterpart in enzymology except for the action of the "pump" enzymes, which are components of plasma membranes. In the strictest sense, pumps may also be regarded as translocases.

Representation of translocases

Translocases are frequently represented by the symbols shown in Fig. 6-11, where the horizontal lines represent the membrane, the circle represents the translocase system that is bound into the structure of the membrane, and the arrows represent the vectorial movement of the substrate and the counterion.

The more important mitochondrial translocases are described in Table 6-2, where they are listed in terms of their substrates, counterions, and inhibitors

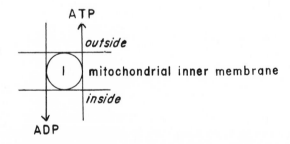

Fig. 6-11. Adenine dinucleotide translocase. This system is inhibited specifically by atractylate. Note that number refers to systems described in Table 6-2.

where these are known. As an arbitrary convention and to facilitate discussion, the individual translocase systems are identified in later figures by the numbers assigned to them in Table 6-2.

The first system in Table 6-2, already described, is the only means by which ATP produced within the mitochondria by the respiratory chain can be exported to other cellular sites where it is required. Blockage of this system poses a serious threat to the life of a cell. Recently, a protein said to be identical with the ATP-binding protein of the mitochondrial inner membrane has been isolated. Since it also strongly binds atractylate, the best known inhibitor of adenine nucleotide translocase, it is proposed that the isolated material is identical with the ATP translocating protein. It appears to be a dimer, with each of the two peptides having a molecular weight of approximately 29,000. It is estimated that this protein composes some 10% of the inner membrane mass. The second system illustrates one means by which ionic balance can be maintained. Two facts here are worth mention. First, note that arsenate, which acts as an uncoupler of

Table 6-2. Major mitochondrial translocases

	SUBSTRATES	COUNTERIONS	INHIBITORS	COMMENTS
1.	ADP	ATP	Atractylate	
2.	P_i, As_i, or $CH_3(CH_2)_nCOO^-$ ($n \leqslant 7$)	OH^-	p-Chloromercuribenzoate (PCMB) or mersalyl	Mersalyl is a mercurial anti-diuretic drug; PCMB is merely a reagent
3.	Succinate, malate, or malonate	P_i	Chlorosuccinate or 2-butylmalonate	
4.	Citrate, isocitrate, or *cis*-aconitate	Malate	2-Butylmalonate or benzene-1,2,3-tri-carboxylate	
5.	α-Ketoglutarate	Malate or malonate	2-Butylmalonate or MICA	MICA = 5-meth-oxyindole-2-carboxylate
6.	Glutamate	OH^-	4-Hydroxyglutamate, 2-aminoadipate, or N-ethylmaleimide	
7.	Aspartate	Glutamate, 2-aminoadipate, or 4-hydroxyglutarate		
8.	α-Glycerophosphate	Dihydroxy-acetone phosphate		This system seems to operate in one direction only (see text for details)
9.	Pyruvate	?	α-Cyano-4-hydroxycinnamate Phenylpyruvate α-Ketoisocaproate	Metabolite in phenylketonuria Metabolite in maple syrup urine disease

oxidative phosphorylation, can be transported into the mitochondria by the same translocase as inorganic phosphate. Second, note that ionic balance can also be accomplished by the translocation of short-chain fatty acid anions. The listed inhibitors can combine with SH groups, which must be essential in the translocase. The third system has been termed the *dicarboxylic acid* translocase, for which the common counterion is a divalent phosphate group. Similarly, the fourth system is known as the *tricarboxylic acid* translocase; the counterions are malate plus monovalent phosphate, since only the sum of these can provide an equivalent charge. The fifth and sixth systems are employed to transport α-ketoglutarate and glutamate, respectively. That the counterions are different emphasizes the amphoteric nature of the amino acid as compared to the keto acid. Systems seven and eight are somewhat peculiar in that no inhibitors for either have yet been found. System nine is the most recently described of the mitochondrial translocases. The inhibition exerted by α-cyano-4-hydroxycinnamate is quite specific for pyruvate; the reagent does not affect translocation of acetate or butyrate. It is noteworthy that this inhibitor also blocks translocation of pyruvate, and possibly lactate, in erythrocytes, even through they contain no mitochondria. It suggests that the translocase of erythrocyte plasma membranes may be similar in structure to the translocases of mitochondria in other cells.

The existence of these translocating systems underlines the concept that movement of most materials into or out of the mitochondria is highly organized and controlled. Of equal importance is the fact that although the translocases are independently constituted, they nonetheless work together in concert. This is depicted by Fig. 6-12, which shows how two separate systems can operate together to produce a net effect equivalent to that of the single hypothetical "system X."

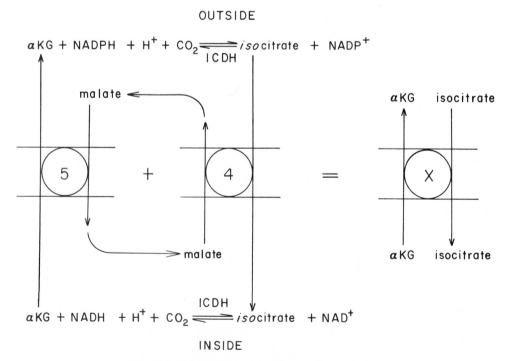

Fig. 6-12. Concerted action of translocases.

One can therefore speak of the *effective exchange of* α-ketoglutarate and isocitrate from one side of the mitochondrial membrane barrier to the other. Furthermore, because isocitrate dehydrogenase (ICDH) exists both inside and outside the mitochondria, system X constitutes a means for effectively moving 2H across the mitochondrial barrier thus providing an accessory means of oxidizing extramitochondrial NADPH + H⁺.

Fig. 6-13 shows another system that is somewhat more sophisticated in design but not in theory. It employs four individual translocating systems and functions with a net effect equivalent to moving oxaloacetate across the mitochondrial membrane barrier, even though in fact oxaloacetate cannot cross the barrier as a molecular species. The advantage of having transaminating enzymes on either side of the mitochondrial barrier is made apparent by this system. Clearly, without glutamate oxaloacetate transaminase (GOT) in both locations, the system would not work as shown. Similarly, one function of intramitochondrial and extramitochondrial MDH is explained by this system. The apparent accumulation of glutamate and the egress of P_i are handled by another coupled system.

Metabolic regulation of translocases

Some specific inhibitors of the mitochondrial translocases were cited in Table 6-2, but most of these are not materials normally encountered in cellular metabolism. It is important to appreciate that regulation by normal metabolites does occur, and that in all probability the regulation of certain translocases is a significant element in controlling the flux of material through the Krebs cycle.

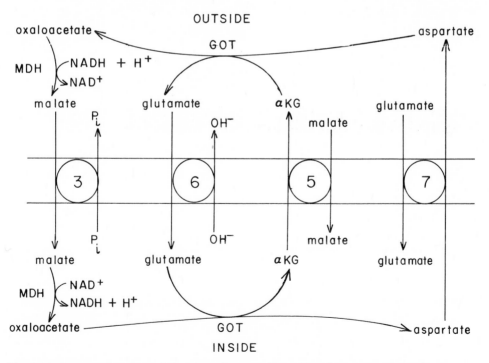

Fig. 6-13. Concerted action of four translocases. Note that this system depends on intramitochondrial and extramitochondrial forms of malate dehydrogenase (MDH) and glutamate oxaloacetate transaminase (GOT). Numbers identify systems defined in Table 6-2.

The central role of ATP in many cellular functions has already been pointed out, as has the fact that the major part of ATP biosynthesis depends on the Krebs cycle in conjunction with the respiratory chain. Adenine nucleotide translocation is therefore a logical site for a regulatory device. Translocation of ADP and ATP is reversibly inhibited by long-chain acyl CoA esters found in mammalian liver. It has been shown that liver mitochondria from diabetic rats respond very sluggishly to added ADP, as judged by measurements with ^{14}C-ADP or ^{32}P-ATP. These results could be reproduced in normal liver mitochondria to which oleyl CoA has been added. When the acyl CoA is depleted by the addition of carnitine, the sluggishness of respiratory response to added nucleotides is abolished. A very similar pattern is observed with mitochondria from hibernating animals, which also tend to have high hepatic acyl CoA concentrations. Thus it appears that the concentration of acyl CoA esters in the outer membrane or in the intermembranous space may very well act as a natural regulator of the adenine nucleotide translocase and, through it, of the entire process of oxidative phosphorylation.

Certain metal ions also have specific effects on the adenine nucleotide translocases. Low concentrations of calcium ion (200 μM) strongly enhance activity of the translocase; the K_m for ATP is reduced approximately fourfold, while the V_{max} is approximately doubled. Still further stimulation can be observed if the calcium is accompanied by potassium ion (20 mM). Two facts are worth emphasizing here. First, the required ion concentrations are well within the physiologic range. Second, the effects of calcium are quite specific and are not demonstrated by either magnesium or manganese ions, even though these form stronger chelates with ATP than does calcium. There is a good possibility that the effect of calcium results from some interaction between the calcium ion and certain phospholipids, probably phosphatidylserines, associated with the translocase itself.

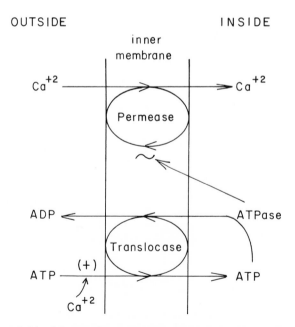

Fig. 6-14. Example of "feed-forward" mechanism, the Ca^{++}-stimulated uptake of ATP by adenine nucleotide translocase. Permease is a Ca^{++}-specific binding site, not part of the translocase itself.

An explanation of the calcium effects just described is diagrammed in Fig. 6-14 and can best be described as a type of "feed-forward" mechanism. As the calcium concentration of the outer mitochondrial compartment rises, it stimulates uptake of ATP by the translocase. The increased intramitochondrial concentration of ATP makes more high energy available to drive a calcium permease, which is also located in the inner mitochondrial membrane. Thus more calcium is taken into the inner compartment by a separate system. This system is called a feed-forward mechanism because the uptake of calcium is promoted, at least in part, by the calcium concentration itself. These facts illustrate the importance of the ionic environment in modifying or regulating key subcellular processes. Furthermore, they explain why the concentration of free (ionic) calcium is one of the more jealously guarded parameters of tissue fluids. For example, relatively small decreases in free calcium concentration in the plasma give rise to tetany. Fortunately the large reservoir of protein-bound calcium operates to assist in homeostatic control.

MITOCHONDRIAL FUNCTION IN LIPOGENESIS

Acetyl CoA is the primary intermediate from which long-chain fatty acids are biosynthesized. The enzymes that are involved in this process exist in the cytoplasm, and it is therefore important to examine the mechanism by which carbon atoms derived from amino acids and other noncarbohydrate precursors are made available as cytoplasmic acetyl CoA.

Fig. 6-15 shows a state of nutritional repletion in which the intake of glucose and amino acids is in excess of metabolic need. The surplus energy is therefore available for conversion to fat, which can be stored in adipose tissue. Fig. 6-15 shows that by direct transamination a large amino acid input to the Krebs cycle will lead to increased citrate concentration within the mitochondria. The citrate is exported from the mitochondria to the cytoplasm, where the enzyme *citrate lyase* cleaves it into oxaloacetate and acetyl CoA. The oxaloacetate can be converted to malate by cytoplasmic MDH, while the acetyl CoA is immediately available for fatty acid production. Through the formation of citrate, the mitochondria operate to promote fatty acid production. Citrate is not only the major vehicle for transport of acetyl groups from the mitochondria to the cytoplasm; it also acts as a positive allosteric effector for the very first step in the biosynthetic reaction sequence leading to fatty acids. These matters will be explored in detail in Chapter 8.

What of the amino acids that cannot enter the Krebs cycle by direct transamination? Here it need only be said that a very substantial amount of the carbon atoms derived from these amino acids can ultimately enter the Krebs cycle; however, the processes are quite complex, and they will be considered in detail in Chapter 9.

Fatty acid biosynthesis also requires NADPH. When the mitochondria are abundantly supplied, as in nutritional repletion of the type indicated in Fig. 6-15, the energy charge will also be high by virtue of the respiratory chain. The high concentration of ATP will lead to its export by the adenine nucleotide translocase, raising the cytoplasmic ATP concentration. In the cytoplasm, ATP tends to shift the pattern of glucose oxidation so that NADPH production is substantially increased and carbon dioxide is produced as well. Since carbon dioxide is also an intermediate in fatty acid biosynthesis, both of these factors tend to promote the production of fatty acids.

The mitochondrial function in lipogenesis can be summarized as follows:

1. Mitochondria collect carbon chains containing two to four carbon atoms from many different sources.

2. Citrate is the most important Krebs cycle intermediate involved in fatty acid biosynthesis, since at high intramitochondrial concentrations it can readily be exported into the cytoplasm.

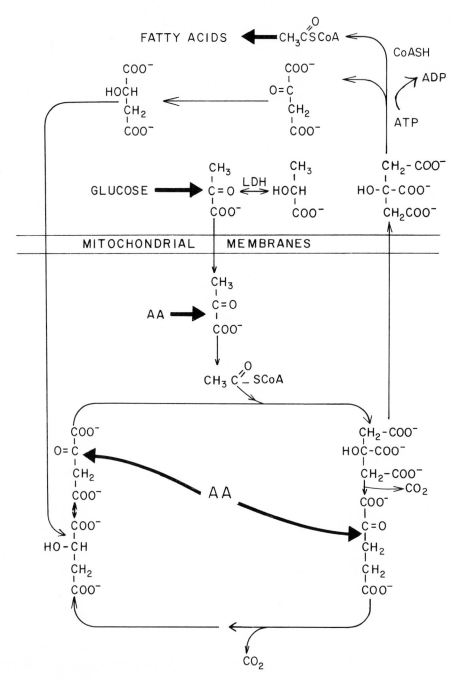

Fig. 6-15. Mitochondrial participation in lipogenesis. **AA** represents amino acids in excess of requirement for protein biosynthesis.

3. Citrate is the major normal source of cytoplasmic acetyl CoA, a primary precursor of fatty acid biosynthesis.

4. Citrate is also a required allosteric effector for the first step in fatty acid biosynthesis.

5. The higher concentration of ATP associated with repleted mitochondria is important in shifting the pattern of glucose oxidation so that production of NADPH, also required for production of fatty acids, is increased.

MITOCHONDRIAL FUNCTION IN GLUCONEOGENESIS

Just as the mitochondria play an important part in lipogenesis, so they play a role in the production of glucose from noncarbohydrate precursors, a process known as gluconeogenesis. Glucose produced in this way can directly enter the circulation or it can be stored in limited amounts as glycogen.

Fig. 6-16 is a schematic representation of gluconeogenesis as it involves the mitochondria. As indicated, the conversion of phosphoenolpyruvate (PEP) to glucose is a readily reversible process, but the conversion of pyruvate to PEP is not. PEP is a high-energy compound ($\Delta G^{0\prime} = -14.8$ kcal/mole) that donates a $\sim$P to ADP as it is converted to pyruvate. Furthermore, Fig. 6-16 shows that the enzymes that convert glucose to pyruvate are cytoplasmic, which raises a problem similar to that noted with fatty acid biosynthesis, that of transporting key intermediates out of the mitochondria. This problem is still far from settled, and various theories have been advanced to explain it; of these, the concept presented here is perhaps the most consistent with experimental observations.

There are three key enzymes that play a role in gluconeogenesis. They include pyruvate kinase, pyruvate carboxylase, and PEP carboxykinase, as shown in Fig. 6-16. The figure also shows that PEP carboxykinase is located both in the mitochondria and in the cytoplasm. However, it must be emphasized that this duality is not the same in all tissues, nor is it the same in a given tissue, for example, liver, from various species. It may therefore be true that gluconeogenesis takes place by various pathways, depending on the particular species and tissue. In spite of these vagaries, it has been shown that pyruvate carboxylase is absent from tissues that do not demonstrate gluconeogenesis, and that pyruvate kinase is quite low in tissues that exhibit significant pyruvate carboxylase activity. This nearly exclusive enzyme distribution prevents wasteful cycling between pyruvate and PEP, which would consume large amounts of ATP.

Pyruvate carboxylase is a tetrameric, biotin-containing enzyme that catalyzes the following two-step reaction.

1. $HCO_3^- + ATP + biotin\text{-}E \longrightarrow {}^-OOC\text{-}biotin\text{-}E + ADP$

2. ${}^-OOC\text{-}biotin\text{-}E + pyruvate \longrightarrow oxaloacetate + biotin\text{-}E$

It shows an absolute allosteric requirement for acetyl CoA in the first step. Presumably the allosteric effector acts by maintaining the tetrameric complex. The enzyme is also subject to end-product inhibition by ADP. Thus we see that this key enzyme is subject to regulation by the high-energy charge and that a product of fatty acid oxidation is an absolute requirement, as well. The allosteric requirement is indicated by the dashed arrow in Fig. 6-16.

Pyruvate carboxylase (reaction 1) converts intramitochondrial pyruvate to oxaloacetate, which is reduced to malate and exported as such to the cytoplasm. Once there, malate can be freely reconverted to oxaloacetate. Gluconeogenesis will occur during conditions when intramitochondrial citrate concentrations will

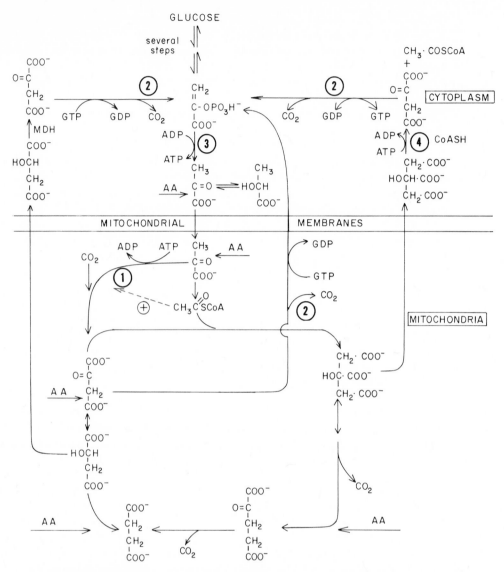

Fig. 6-16. Mitochondrial participation in gluconeogenesis. Note that glucose can be made from phosphoenolpyruvate by a series of reversible reactions. Note that conversion of phosphoenolpyruvate to pyruvate is an irreversible reaction. Reaction 1 is catalyzed by pyruvate carboxylase and reaction 2 by PEP carboxykinase. Reaction 3 is catalyzed by pyruvate kinase and reaction 4 by citrate lyase. Note that reaction 2 may occur inside and outside the mitochondria. AA represents amino acids in excess of amount required for protein biosynthesis.

be high enough for some citrate to be exported into the cytoplasm. There it is cleaved (reaction 4) to produce more oxaloacetate and acetyl CoA.

PEP carboxykinase is a monomeric enzyme that converts oxaloacetate to PEP at the expense of GTP (reaction 2). Most of the PEP is formed in the cytoplasm, but some exists in mitochondria. By means that are not at all clear, intramitochondrial PEP can escape to the cytoplasm. It also can be easily transformed to glucose or to glycogen. Conditions leading to rapid gluconeogenesis cause an increased synthesis of PEP carboxykinase. Fasting, diabetes, and particularly cortisol treatment induce production of the cytoplasmic enzyme. Note that when this reaction occurs in the mitochondria the source of GTP can readily be found, but when it occurs in the cytoplasm the source of GTP is obscure since, as already mentioned, it is not clear that guanine nucleotides cross the mitochondrial membranes. It may be that the cytoplasmic GTP is produced by a transphosphorylation, about which little is known.

In summary it can be said that, in gluconeogenesis as in lipogenesis, the mitochondria fulfill several functions.

1. They serve to collect carbon atoms from amino acids, either by simple transamination or by more detailed reaction schemes. The carbon chains enter the mitochondria and produce large amounts of citrate and oxaloacetate via the Krebs cycle.

2. Oxaloacetate is converted to malate or aspartate for export to the cytoplasm, where it is reconverted to oxaloacetate.

3. Citrate is exported to the cytoplasm, where it is cleaved to oxaloacetate and acetyl CoA.

4. Pyruvate is carboxylated to oxaloacetate in a reaction for which acetyl CoA is a required allosteric activator. This reaction may occur either in the cytoplasm or in the mitochondria, depending on the specific cellular distribution pattern of requisite enzymes.

5. Oxaloacetate is decarboxylated to form PEP, which may be converted to glucose or glycogen. Decarboxylation may be cytoplasmic, mitochondrial, or both, depending on the specific distribution of the requisite enzymes.

TRANSFER OF REDUCING EQUIVALENTS

The major site for oxidation of the pyridine adenine dinucleotides lies within the respiratory chain, and it has already been stated that these nucleotides cannot cross the mitochondrial membrane barrier. This raises the question of how extramitochondrial nucleotides can be oxidized. The solution to this problem depends on concerted action of translocase systems as indicated in Figs. 6-12 and 6-13. By employing such oxidation-reduction pairs as malate-oxaloacetate in one case and α-ketoglutarate-isocitrate in another, the net effect is to move 2H from one side of the membrane to the other; in both instances the hydrogens are associated with the substrate molecules and not with NADH or NADPH. Although the two illustratory systems can operate reversibly and in either direction, another important system for transfer of reducing equivalents appears to have a "one-way" character. This system depends on the oxidation-reduction pair composed of dihydroxyacetone phosphate and α-glycerophosphate. The extramitochondrial form of glycerophosphate dehydrogenase employs NAD^+ as coenzyme. The intramitochondrial form of the enzyme has a different structure and employs FAD as coenzyme. The reduced FAD is coupled directly to coenzyme Q in the respiratory chain. These properties, perhaps, give to this system its one-way behavior, as shown in Fig. 6-17. Thus this system can function to reduce intramitochondrial NAD^+ but not to accomplish the reverse. Systems that operate to transfer reducing equivalents are sometimes known as *shuttles*.

THE CLOSED ROAD: WHY FAT IS NOT CONVERTED TO GLUCOSE

Earlier in this discussion the Krebs cycle was compared to a traffic circle, with materials moving in and out. It was mentioned that one road, the conversion of fat to carbohydrate, was closed. It is therefore appropriate to end the discussion with an explanation of why there can be no *net synthesis* of glucose from fatty

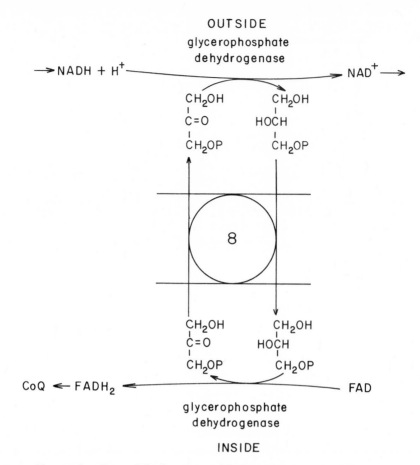

Fig. 6-17. Glycerophosphate dehydrogenase–dihydroxyacetone phosphate shuttle. This system is for transfer of reducing equivalents, or 2H, from one site to another. Number refers to translocases listed in Table 6-2.

acids, at least from the majority of fatty acids likely to be a part of the average diet.

As has already been described, fatty acids are broken down into acetyl CoA, which is converted by the Krebs cycle into citrate. In later stages of the cycle, two gram molecules of carbon dioxide are produced for each gram molecule of acetyl CoA taken up. Therefore, in the overall reaction there is no *net gain* of carbon atoms that could be converted into glucose.

Two features of the Krebs cycle are worth emphasis. The first is a simple matter of arithmetic. With two carbons coming in (via acetyl CoA) and two carbons leaving (via CO_2), the net gain is zero carbon atoms. The second feature is that the carbon atoms that leave as CO_2 are *not identical* with those taken up in the form of acetyl CoA. This is shown in Fig. 6-18, in which the carbon atoms of acetyl CoA are labeled. The decarboxylation steps involve carbon atoms initially derived from oxaloacetate. So, at the end of the first turn of the Krebs cycle, neither of the carbon atoms lost are labeled. Because succinate is symmetric with respect to attack by succinate dehydrogenase, all of the carbon atoms of the oxaloacetate will be labeled at the end of the first complete cycle. However, this does not alter the fact that no net surplus of carbon atoms can be generated by oxidation of acetyl CoA.

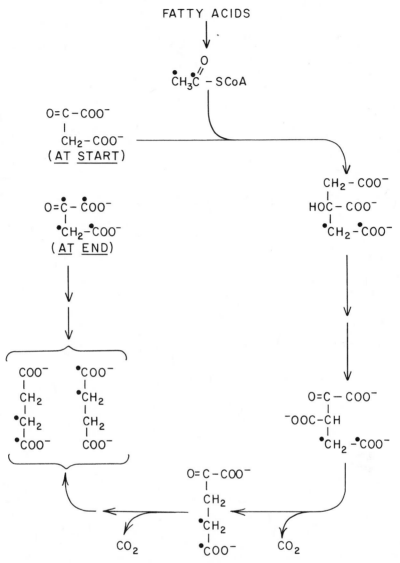

Fig. 6-18. Failure of fatty acids to provide for *net* synthesis of glucose. The two carbon atoms taken up from acetyl CoA are not identical with those lost but are equivalent to them. No label is lost as CO_2 in the first turn of the cycle. Because succinate is symmetric, the oxaloacetate at the end of the first turn will be labeled in all four carbon atoms.

It is important to understand that if a fatty acid contained an odd number of carbon atoms, the last three would be converted to propionyl CoA, as shown in Fig. 6-19. Propionyl CoA can be converted to methylmalonyl CoA; this in turn is converted to succinate. The succinate formed from propionyl CoA is in equilibrium with succinate formed by any other reactions. As was explained in Fig. 6-16, the Krebs cycle participates in gluconeogenesis by production of oxaloacetate, which is converted to PEP in the cytoplasm (Fig. 6-16). Succinate is a precursor of oxaloacetate, formed in this instance from a propionyl residue with no loss of carbon atoms. Therefore, unlike the input of acetyl groups, the input of propionyl groups does provide a net surplus of carbon atoms that can contribute

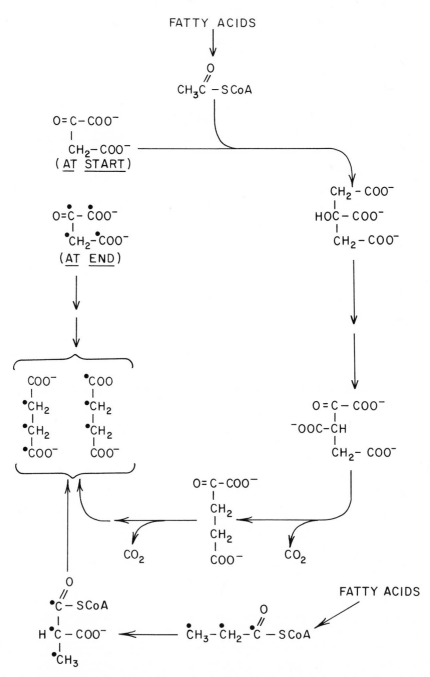

Fig. 6-19. Showing that propionate, derived from odd carbon fatty acids, can provide for net synthesis of glucose, since the three carbon atoms of propionate enter the Krebs cycle after the two steps involving decarboxylations.

to net synthesis of glucose. This will be equally true of any carbons that enter as methylmalonyl CoA.

Fatty acids with odd carbon numbers and branched-chain fatty acids are a very small portion of our total fatty acid intake. Therefore, the generalization that fatty acids do not contribute to the net synthesis of glucose is valid and is based on the equivalence of the two carbons that enter the Krebs cycle to the two carbons that are lost during each complete turn of the cycle.

SUMMARY

The original description of the Krebs cycle as a means of transforming materials of divers origin into a small number of intermediates common to a variety of catabolic and anabolic pathways should now be clear. The cycle also is able, by means of a few auxiliary enzymes, to maintain or replenish the concentrations of its own intermediate components. This function is served by what are known as the anaplerotic reactions. It should also be clear that movement of materials into or out of the mitochondria is carefully regulated and follows experimentally determined laws, so that osmotic and ionic balances can be maintained. Finally, it is important to remember that the cycle, together with the associated respiratory chain, provides several important regulatory substances that may act inside or outside of the mitochondria.

REFERENCES

Krebs, H. A.: Mitochondrial generation of reducing power. In Slater, E. C., Kaniuga, Z., and Wojtczak, L., editors: Biochemistry of the mitochondria, New York, 1966, Academic Press, Inc. *A brief statement of significant problem areas, some of them still unsolved.*

Krebs, H. A.: The history of the tricarboxylic acid cycle. In Perspect. Biol. Med. 14:154, 1970. *A look backward by the best authority.*

Larner, J.: Intermediary metabolism and its regulation, Englewood Cliffs, N.J., 1972, Prentice-Hall, Inc. *A somewhat condensed review of metabolism, with emphasis on regulatory mechanisms.*

Lowenstein, J. M.: The citric acid cycle, control and compartmentation, New York, 1969, Marcel Dekker, Inc. *A compilation of reviews by expert authors.*

Newsholme, E. A., and Start, C.: Regulation in metabolism, New York, 1973, John Wiley & Sons. Inc. *A recent review of control mechanisms.*

Clinical examples

CASE 1: HEPATIC COMA IN LATE CIRRHOSIS

A patient with advanced cirrhosis had been seen frequently in the Internal Medicine Clinic of a university hospital. His condition had progressively deteriorated; on his last visit he was brought to the hospital in a disoriented state, and in spite of vigorous treatment he sank into a coma from which he could not be roused. He was treated with intravenous fluids, and his electrolytes were carefully monitored. The BUN rose in spite of therapy, until on the fifth day the value was reported to be 120 mg/dl. At the same time the blood ammonia concentration also rose to a value of 765 μg/dl. In an effort to lower the blood ammonia concentration, α-ketoglutarate was added to the intravenous infusion, but before the infusion was complete the patient expired.

Biochemical questions

1. Is the conversion of α-ketoglutarate to glutamate readily reversible?
2. In which cellular compartment does transamination occur?
3. For α-ketoglutarate to be converted to glutamate, it must enter a particular organelle. How is that process controlled?
4. Lacking a sufficient source of α-ketoglutarate, could the deficit be made up by an excess of pyruvate?
5. Would it be helpful to attempt treatment of hepatic coma patients with a mixture of coenzyme A, thiamine, NAD$^+$, FAD, and lipoate?
6. Is it likely that the increased blood ammonia concentration is the sole explanation of hepatic coma?

Case discussion

Severe liver failure is accompanied by a comatose state, in which hyperammonemia is a prominent finding. It is argued by some that the increased ammonia content is causal in hepatic degeneration, while others contend it is an effect. Most treatments are designed to lower the blood ammonia because of the acutely toxic influence it exerts in the central nervous system. The deliberate injection of sufficient ammonium ions into the blood of an otherwise healthy animal will produce coma. It has been difficult to elucidate the mechanism of ammonia intoxication, since some patients with severe neurologic symptoms still show a normal or only a moderately increased ammonia concentration in the blood or cerebrospinal fluid.

Blood ammonia concentrations may be lowered by treatment with either glutamate or α-ketoglutarate; these normal metabolites can combine with ammonia to form glutamine, which is not neurotoxic and which can readily be excreted in the urine. Because formation of glutamine is an energy-requiring process (see Chapter 9), pyruvate is sometimes infused along with glutamate or α-ketoglutarate. A second argument advanced in favor of adding pyruvate is that some of its carbons can be converted to α-ketoglutarate. However, an identical argument can be advanced in favor of glucose, which is cheaper, is more stable than pyruvate, and produces two moles of pyruvate/mole of hexose.

It must be remembered that α-ketoglutarate does not, in all likelihood, react directly with NH$_4^+$; rather it must first undergo transamination with some other amino acid to form glutamate. Glutamate, in turn, can form glutamine. The transamination reaction takes place in the mitochondria, and entry of α-ketoglutarate is regulated by a dicarboxylic acid translocase for which malate is a required counterion. Large scale movements of these counterions may have undesirable effects in an already diseased liver.

Those who have proposed to employ a mixture of pyruvate and α-ketoglutarate have also proposed the addition of a mixture of coenzyme A, thiamine, NAD$^+$, FAD, and lipoate in an effort to promote conversion of pyruvate to acetyl CoA and thence to α-ketoglutarate. These additives are found in a protein-free yeast extract. There is little biochemical basis for this addition. It is true that thiamine and lipoate are absorbable as such, but coenzyme A and NAD$^+$ would almost certainly be destroyed by digestive processes prior to absorption. Moreover, NAD$^+$ cannot be transported across the mitochondrial membranes.

Recently, it has been proposed that ammonia may not be the only substance responsible for coma or that it may not be directly involved at all. It has been shown that α-ketoglutaramate can be formed as shown below.

It has been proposed that glutamine can undergo transamination to produce α-ketoglutaramate. This can be converted, by a nonenzymatic reaction, to a cyclic analogue of proline, or it can undergo hydrolysis by an amidase to produce free NH$_4^+$ and α-ketoglutarate. Cerebrospinal fluid from patients with hepatic coma contained roughly ten times the normal amount of α-ketoglutaramate; patients with severe hepatic disease in the absence of coma showed no such increase. In none of the patients examined could α-ketoglutaramate be detected in the blood.

It is possible, but has not been proved, that α-ketoglutaramate or its cyclic lactam, shown below, may compete with glutamate in those brain tracts where glutamate is reported to serve as a neurotransmitter.

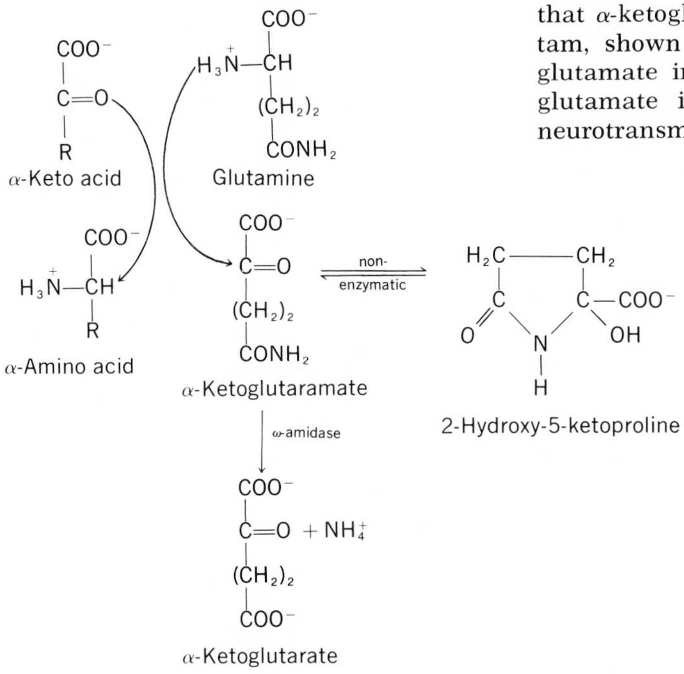

COO$^-$
|
C=O
|
R
α-Keto acid

COO$^-$
|
H$_3$$\overset{+}{N}$—CH
|
(CH$_2$)$_2$
|
CONH$_2$
Glutamine

COO$^-$
|
H$_3$$\overset{+}{N}$—CH
|
R
α-Amino acid

COO$^-$
|
C=O
|
(CH$_2$)$_2$
|
CONH$_2$
α-Ketoglutaramate

non-enzymatic

H$_2$C———CH$_2$
   C   C—COO$^-$
O   N   OH
       |
       H
2-Hydroxy-5-ketoproline

ω-amidase

COO$^-$
|
C=O + NH$_4^+$
|
(CH$_2$)$_2$
|
COO$^-$
α-Ketoglutarate

REFERENCES

Badley, B. W.: Some aspects of medical management of gastrointestinal disease, J. Can. Med. Assoc. 112:200, 1975.

Hindfelt, B.: Effect of acute ammonia intoxication upon the brain energy state in rats treated with L-methionine-D-L-sulfoxime, Scand. J. Lab. Clin. Invest. 31:289, 1973.

Vergara, F., et al.: α-Ketoglutaramate: increased concentrations in the cerebrospinal fluid of patients in hepatic coma, Science 183:81, 1974.

Zieve, L., et al.: Pathogenesis of hepatic coma, Ann. Rev. Med. 26:143, 1975.

CASE 2: PHENYLKETONURIA AND PYRUVATE METABOLISM

A fair-skinned, blue-eyed child of Scandinavian extraction was referred to the pediatric clinic of a university hospital. The mother complained that the child was extremely irritable and had a "musty" odor. The skin had a dry, eczematous appearance. Neurologic tests showed the child was mentally retarded for its chronologic age. A urine sample tested with ferric chloride solution gave a blue-green color. The blood concentration of phenylalanine was determined to be 0.6 mM (10 mg/dl). A diagnosis of phenylketonuria was made, and a special diet low in phenylalanine was prescribed.

Biochemical questions

1. What is the metabolic connection between phenylalanine and phenylpyruvate?
2. What is the metabolic connection between phenylpyruvate and pyruvate?
3. What compound is responsible for the blue-green color test with ferric chloride solution?
4. Do metabolites of other amino acids act as inhibitors of pyruvate uptake by mitochondria?
5. What are the consequences of inhibiting pyruvate uptake by mitochondria?

Case discussion

Phenylalanine is an essential amino acid that is metabolized by hydroxylation to form tyrosine (see Chapter 9, p. 430). Approximately one in 25,000 children is born with a defect in the hydroxylating system, leading to accumulation of the abnormal metabolite phenylpyruvate. There appear to be two forms of phenylalanine hydroxylase, one which is ubiquitous but of low activity, and a second, more active form restricted to the liver. It is the latter form that is reportedly lacking in children with this disease, which causes severe mental retardation.

It has recently been found that phenylpyruvate is a powerful inhibitor of the mitochondrial pyruvate translocase. It has also been found that the concentration of phenylpyruvate in the blood of untreated children may reach values of 0.1 to 0.5 mM, more than sufficient to severely reduce the supply of pyruvate to brain mitochondria, to correspondingly reduce the ATP that the mitochondria can generate, and so to account for the mental damage by interruption of myelination essential for normal brain development.

Chemical verification of phenylketonuria can be made in one of two ways. The simpler, less exact test depends on the formation of a colored complex between a ferric chloride test solution and phenylpyruvic acid in the urine. Normally, this test should be negative, but the presence of harmless complex-forming materials in the urine may give colors that mask the blue-green reaction that is diagnostic. A more precise test is a chemical determination of phenylalanine in the blood. The normal range of values is 0.7 to 4.0 mg/dl.

Phenylpyruvate is not the only natural inhibitor of the pyruvate translocase of brain mitochondria. The compound α-ketoisocaproate is formed and accumulates in the genetic error of leucine metabolism known as maple syrup urine disease (see Chapter 9). The compound gives a characteristic odor to the urine, hence the name. No simple test is available for its detection. α-Ketoisocaproate is a less powerful inhibitor than phenylpyruvate, but it may reach even greater concentrations (2 to 4 mM) in the blood, so it could have appreciable effects on pyruvate uptake.

A third inhibitor of pyruvate translocase is the synthetic compound α-cyano-4-hydroxycinnamate. It is noteworthy that this inhibitor also blocks translocation of pyruvate, and possibly lactate, in erythrocyte membranes, even though erythrocytes do not contain mitochondria. This suggests that the mitochondrial translocase and the corresponding system in erythrocyte membranes may have similar structures.

REFERENCES

Halestrap, A. P., Brand, M. D., and Denton, R. M.: Inhibition of mitochondrial transport of pyruvate by phenylpyruvate and α-ketoisocaproate, Biochim. Biophys. Acta **367:**102, 1974.

Halestrap, A. P., and Denton, R. M.: Specific inhibition of pyruvate transport in rat liver mitochondria and human erythrocytes by α-cyano-4-hydroxycinnamate, Biochem. J. 138: 313, 1974.

Papa, S., and Paradies, G.: On the mechanism of translocation of pyruvate and other monocarboxylic acids in rat liver mitochondria, Eur. J. Biochem. 49:265, 1974.

CASE 3: MYOCARDIAL INFARCTION

A 41-year-old faculty member, $5'7\frac{1}{2}''$ tall and weighing 165 pounds, was first seen after lunch with the complaint of severe "indigestion." He stated that he had felt unwell since midmorning, at which time he had left his office and gone home to rest. He was a moderate to heavy smoker and mentioned that smoking a cigarette made his discomfort worse. The indigestion was evidenced solely by mediastinal discomfort.

Following a physical examination the patient was advised to have chest x-ray films made and to have an EKG (electrocardiogram). The EKG tracing was abnormal and the patient was immediately admitted to the hospital. Blood was drawn for serum enzyme analyses on the morning following admission. The laboratory reported the following results: serum lactate dehydrogenase (LDH), 892 Wroblewski units; glutamate oxaloacetate transaminase, 540 Wroblewski units; glutamate pyruvate transaminase, 30 Wroblewski units; and serum cholesterol, 260 mg/dl.

The patient was treated conservatively and discharged with suitable advice regarding diet, exercise, and smoking.

Biochemical questions

1. Do the enzyme activities cited in the case description permit you to rule out disease of the abdominal organs?
2. Why is the serum activity of succinate dehydrogenase, which is an important enzyme of the Krebs cycle, not increased?
3. How does myocardial ischemia affect the operation of the Krebs cycle in the affected tissue?
4. Suitable measurements of anoxic mitochondria show an increased $NADH/NAD^+$ ratio. How does this affect the Krebs cycle?
5. What is the effect of myocardial anoxia on the production of ATP in the tissue? Will there be any effect on the production of GTP?

Case discussion
Chemical causes of myocardial infarcts

The "indigestion" experienced by this patient was actually an episode of myocardial ischemia (anoxia of tissue resulting from diminished blood flow to a portion of the myocardium). It resulted in a myocardial infarct (coagulation necrosis) of that part of the myocardium normally nourished by a plugged or occluded branch of the coronary circulation. Current opinion holds that this type of disease is frequently associated with elevated concentrations of serum cholesterol (Chapter 10). According to a large body of evidence, smoking makes matters worse because nicotine is a vasoconstricting agent.

Biochemical consequences of ischemia

As the capacity of a vessel to carry blood is reduced, the tissue that it normally nourishes may become more or less anoxic, and waste products of metabolism tend to accumulate in the tissues instead of being carried away. In addition, there will obviously be a reduced supply of nutrients, so that the affected cells are forced to depend on what they have previously stored. When ischemia is produced experimentally by ligation of a branch of the coronary artery, the pH of the ischemic tissue begins to drop within 1 min, and the contractions of the affected part of the myocardium become irregular or may cease within just a few min. Without an adequate supply of oxygen and with a substantially reduced blood flow, the affected tissue must rely on anaerobic pathways of metabolism, which brings about an accumulation of lactate that causes more localized acidosis. The tissue anoxia interferes with normal operation of the respiratory chain, which oxidizes the NADH produced by the Krebs cycle to NAD^+. Therefore, the $NADH/NAD^+$ ratio within the mitochondria will increase. It is important to remember that operation of the cycle is as dependent on reoxidation of NADH as it is on a supply of any other intermediate. This emphasizes the close metabolic, as well as the close physical relations (see Chapter 5) between the Krebs cycle and the respiratory chain.

Effects on cellular respiration. Anaerobic respiration of the deprived cells leads to a significant decrease in ATP and creatine phosphate within 30 min, and the mitochondria, where the majority of cellular oxidations occur, become swollen and lose their organization. Electron microscopic examination has shown that the cristae break down, and in their place dense granules of disorganized material appear. Within just a few hours, experimental infarcts lose much of the succinate dehydrogenase found in normal myocardial cells. The same is true for cytochrome oxidase. Within just a few hours the intracellular concentra-

tions of the Krebs cycle intermediates become too low to maintain normal metabolic pathways, and the tissue dies.

Other effects of tissue acidosis. The acidosis that accompanies tissue anoxia may also be responsible for the release of lysosomal enzymes, especially phosphatases, deoxyribonucleases, and other hydrolases, which begin the process of degradation necessary to remove debris from dead or dying cells. This process is a slow one and may require from several days to 1 week or more, depending on the extent and size of the lesion.

Significance of serum enzyme changes

Of greater diagnostic significance is the release, in an active form, of enzymes that are associated with the mitochondria and the Krebs cycle. These enzymes appear in the blood serum and can be used to substantiate other evidence that the lesion is indeed an infarct. Chief among these enzymes are serum glutamate oxaloacetate (SGOT) and serum glutamate pyruvate (SGPT) transaminases as well as creatine phosphokinase (CPK). While the release of these

Table 6-3. Brief summary of enzyme changes in myocardial infarct

SERUM ENZYME	EARLIEST INCREASE (hr)	MAXIMUM CONCENTRATION (hr)	RETURN TO NORMAL (days)	AMPLITUDE OF INCREASE (× normal)
Creatine phosphokinase	3-6	24-36	3	7
Malate dehydrogenase	4-6	24-48	5	4
Glutamate oxaloacetate transaminase	6-8	24-48	4-6	5
Lactate dehydrogenase*	10-12	48-72	11	3
α-Hydroxybutyrate dehydrogenase	10-12	48-72	13	3-4
Aldolase	6-8	24-48	4	4
Glutamate pyruvate transaminase	Usually normal, unless there are other complications			
Isocitrate dehydrogenase	Usually normal			

*Frequently done twice, once before and once after mild heat denaturation.

enzymes is not unique to damaged myocardium, the pattern of enzymes following myocardial damage is different from that found in damage to other viscera; thus enzyme assays for total activity, and more particularly for specific isozymes of LDH, have become important adjuncts to diagnosis. Several of the heart isozymes of LDH are more stable with regard to heat denaturation than are the corresponding liver enzymes. It is also important to note that the precise patterns of enzyme change in heart disease are characteristic. Thus in an infarct the ratio of total SGOT/SGPT is usually high; the reverse is true in liver disease. Whether this pattern represents some intrinsic differences in the enzymes as they occur in various tissues, or whether it represents some difference in cellular structures within the different tissues is not presently clear.

Not all of the enzymes known to be involved in the Krebs cycle are released to the serum in significant amounts following an infarct. For example, isocitrate dehydrogenase activity in the serum does not rise. However, in liver disease such a rise may be seen. Either the myocardial enzyme is inactivated before it is released or it may be released so slowly that the increase is not chemically significant. Whatever the case, there is no doubt of its presence in healthy cardiac muscle.

Numerous other enzyme assays have been proposed to aid in the diagnosis of myocardial infarcts, but they have not yet received as widespread acceptance as the ones mentioned here. The types of changes anticipated in myocardial infarcts as well as their time courses are given in Table 6-3.

REFERENCES

Braasch, W., et al.: Early changes in energy metabolism in myocardium following acute coronary artery occlusion in anesthetized dogs, Circ. Res. 23:429, 1968.

Jurkowitz, M., et al.: Ion transport by heart mitochondria, retention and loss of energy coupling in aged heart mitochondria, Arch. Biochem. Biophys. 165:98, 1974.

Kubler, W., et al.: Regulation of glycolysis in the ischemic and the anoxic myocardium, J. Mol. Cell. Cardiol. 1:351, 1970.

CASE 4: CONGENITAL DEFECT OF PYRUVATE DEHYDROGENASE

E. G., a 3-year-old girl, was the offspring of a consanguineous marriage, and her only sibling had died in infancy. Since birth she had been treated for a generalized neurologic disease. Her height and weight were normal, but her head circumference (45 cm) was in the microcephalic range.

On examination severe optic atrophy was noted; the patient was probably blind, but because of her very limited vocabulary (some twenty words) this was difficult to establish precisely. The child was hyperreflexive and extremely irritable; her movements were poorly coordinated, and muscular hypotonia was noted.

Blood lactate and pyruvate concentrations were consistently elevated (2 to 4 meq/liter and 2 to 10 meq/liter, respectively) despite thiamine hydrochloride or nicotinamide therapy. Citrate levels were also elevated (0.14 to 0.21 mmole/liter). The blood uric acid concentration ranged from 7.6 to 9.8 mg/dl during the period of observation. The serum CO_2 content remained below 20 meq/liter despite daily therapy with 4.4 g $NaHCO_3$. The serum creatine phosphokinase (CPK) level was elevated (127 U, normal = 6 to 50 U), as was the level of serum aldolase (11 U, normal = 3 to 8 U). SGOT and LDH levels were normal when measured.

Urinalysis revealed a slightly increased excretion of alanine (566 μmoles/24 hr, normal = 12 to 500 μmoles/24 hr). The urinary pH was 5.0, and the urine contained uric acid crystals.

Biochemical questions

1. What are the coenzyme requirements of the pyruvate dehydrogenase complex?
2. How do you account for the low concentration of serum bicarbonate?
3. Given that the pyruvate dehydrogenase complex contains several distinct enzymes, how could you deter-

mine if a defect existed in the first of these (pyruvate decarboxylase)?

4. If pyruvate cannot be adequately converted to acetyl CoA, what other sources of acetyl CoA might be found in the tissues?

5. How can you explain the increased urinary excretion of alanine?

6. How can you account for the increased serum lactate?

7. Why was it necessary to sonicate the cultured fibroblasts?

8. In studying the metabolism of the fibroblasts derived from skin biopsies of this patient, 1-^{14}C-pyruvate and 2-^{14}C-pyruvate were employed as substrates. If the enzyme defect is strictly limited to the pyruvate dehydrogenase complex, would the quantity of $^{14}CO_2$ produced differ with these substrates?

Case discussion

It is difficult to determine which tissues are most affected by certain genetic defects, and it is even more difficult to obtain ante mortem samples of a nature and size suitable for biochemical study. In this case it is clear that a large quantity of mitochondria would be required to elucidate the biochemical defects.

According to genetic theory, the nucleus of every cell contains all of the genetic information specified for that particular organism, but in highly differentiated cells, significant portions of of the genetic information remain repressed. Skin biopsies are quite safe and nearly painless, and the explants may be grown in tissue culture to provide fibroblasts. These proliferate rapidly and are sufficiently undifferentiated to be useful in examining a large number of genetic diseases. Fibroblasts can be employed as intact cells, as homogenates, or to provide samples of particular organelles. They therefore serve as an elegant tool for the study of metabolic processes, particularly when used with variously labeled substrates.

Studies with intact fibroblasts

When cultured fibroblasts derived from this patient were treated with either 1-^{14}C-pyruvate or 2-^{14}C-pyruvate and the rate of substrate oxidation estimated by collection of $^{14}CO_2$, it was found that the substrates were oxidized at approximately one-third the rate obtained with control fibroblasts from normal subjects. Results obtained when the substrate was either 1-^{14}C-palmitate complexed to defatted serum albumin or 1,5-^{14}C-citrate were similar. In contrast, the oxidation of 5,6-^{14}C-isocitrate or U-^{14}C-glutamate appeared to be comparable to the rates observed for normal fibroblasts.

Studies with disrupted fibroblasts

Because of permeability barriers, it is not always feasible to assay certain enzyme activities in intact cells; for this reason the cultured fibroblasts examined in this study were disrupted by sonication, a treatment that makes membranes permeable without denaturing the enzymes. Thus while blood analyses and examination of intact fibroblasts seemed to indicate that this patient suffered from a defect in the conversion of citrate to isocitrate, the activity of aconitase in sonicated preparations appeared normal. The same observations were made on several other enzymes. These results are summarized in Table 6-4.

Note that even in sonicated fibroblast preparations the activity of citrate synthetase from this patient was only about 75% of the control value. Direct measurements of the pyruvate dehydrogenase complex from this patient's disrupted fibroblasts were compared with values from her mother, H. G., and from controls. The observations are summarized in Table 6-5.

It is clear that pyruvate dehydrogenase activity in sonicated fibroblasts from this patient was low, just as it was in the intact cells. Activity from the mother's fibroblasts was also depressed but was at the lower limit of the range noted for the control subjects. Tissue from the father was not available for study. No increase in activity of the patient's sonicated fibroblasts was observed on addition of excess CoASH, NAD$^+$, pyruvate, or reduced and oxidized forms of lipoate. This indicated that the deficiency in pyruvate dehydrogenase activity was not caused by a deficiency of these coenzymes.

Table 6-4.   Enzyme activities in disrupted fibroblasts*

ENZYME	PATIENT (nmoles/min · mg)	CONTROLS (nmoles/min · mg)
Citrate synthetase	15.3 ± 0.26	24.9 ± 3.1
Aconitase		
Citrate to isocitrate	0.41 ± 0.06	0.51 ± 0.04
Isocitrate to aconitate	0.48 ± 0.06	0.67 ± 0.10
Isocitrate dehydrogenase		
NADP linked	1.56 ± 0.25	1.77 ± 0.33
NAD linked	0.12 ± 0.03	0.15 ± 0.33

*Values are the means of duplicate analyses from three cultures from the patient and four cultures from controls. Values represent means plus or minus the standard error of the means.

Table 6-5.   Pyruvate dehydrogenase activities in sonicated fibroblast preparations from E. G., H. G., and controls

SUBJECT	ACTIVITY (nmoles/min · mg)
Patient (E. G.)	0.059 ± 0.01
Mother (H. G.)	0.229 ± 0.06
Controls (nine individuals)	0.392 ± 0.03

Resolution of defect

The depressed activity of pyruvate dehydrogenase observed in this patient might have been the result of some soluble inhibitor abnormally synthesized by her cells. This was ruled out by mixing experiments in which varying proportions of sonicated material from the patient were mixed with material from the controls. In all instances the observed activity agreed with that calculated, leaving as the only tenable explanation a defect in one or more of the component enzymes of the complex.

Extensive studies have shown that the ferricyanide ion reacts with the hydroxyethylthiamine formed as an intermediate by the first, thiamine pyrophosphate-dependent enzyme of the complex (p. 252). Conversion of 1-^{14}C-pyruvate to $^{14}CO_2$ with ferricyanide as the electron acceptor was examined in sonicated fibroblasts from E. G.; the values observed were barely within the normal range. These results indicated that the activity of the first enzyme and its interaction with Mg^{++}, pyruvate, and thiamine pyrophosphate were all substantially normal, and it therefore appeared that this patient suffered from a defect in one of the later enzymes of the complex. E. G. differed in this respect from other patients in whom similar disease entities have been studied.

Metabolic repercussions of the defect

It is not surprising that a patient with a major defect in oxidative metabolism would suffer from profound neurologic disease. The brain is virtually dependent on the oxidation of glucose via glycolysis to pyruvate (see Chapter 7), which then enters the Krebs cycle as acetyl CoA. Interruption of this total pathway at any point leads to cerebral deprivation of ATP. Although the defect in this case was compatible with extrauterine life, it was incompatible with normal development and function.

Since production of ATP was impaired by a fault in conversion of pyruvate to acetyl CoA, glycolysis would continue, and pyruvate would accumulate. The only means by which the excess pyruvate could be removed are (1) conversion to lactate and (2) transamination to alanine. The serum of this patient did contain an increased concentration of lactate, which accounts for the acidosis and the low serum bicarbonate. Formation of alanine was probably increased also, since the urine contained a somewhat higher than normal quantity of this amino acid.

REFERENCES

Blass, J. P., Lonsdale, D., Uhlendorf, B. W., and Hom, E.: Intermittent ataxia with pyruvate decarboxylase deficiency, Lancet 1:1302, 1971.

Blass, J. P., Schulman, J. D., Young, D. S., and Hom, E.: An inherited defect affecting the tricarboxylic acid cycle in a patient with congenital lactic acidosis, J. Clin. Invest. 51:1845, 1972.

Schulman, J. D., and Blass, J. P.: Measurement of citrate synthase activity in human fibroblasts, Clin. Chim. Acta 33:467, 1971.

Skrede, S., et al.: Fatal congenital lactic acidosis in two siblings. II. Biochemical studies in vivo and in vitro, Acta Paediatr. Scand. 60:138, 1971.

CASE 5: DIAGNOSTIC PROBLEMS RESULTING FROM LABORATORY METHODOLOGIES

A newly admitted student appeared at the student health service bearing a letter from his family physician. The letter stated that the patient was in the late convalescent stages of infectious hepatitis but that he was, in the opinion of the attending physician, capable of undertaking graduate study. The letter further indicated that on several prior occasions the SGPT and bilirubin assays had returned to normal values. He suggested the patient might benefit from continued medical supervision and urged that this be arranged.

The health service physician agreed to follow the case and drew blood samples for check determinations of both the bilirubin and SGPT. The result of the SGPT test was abnormally high. When the family physician was informed of this, he insisted the finding did not accord with his data. Admission to the infirmary was arranged for further study. No overt problems were uncovered, so the patient was discharged, but the SGPT test remained high for several weeks longer. Subsequent discussions between the health service physician and a clinical chemist consultant to the laboratory explained the discordant results as probably caused by variations in laboratory methodology.

Biochemical questions

1. From what part of the liver cells does SGPT originate?
2. What is the source of the alanine that is the substrate for SGPT?
3. What is the metabolic fate of the pyruvate produced in the SGPT reaction?
4. How might ^{13}C-alanine, fed in the diet, end up in blood glucose?
5. What is the biochemical basis of the SGPT assay?
6. How is it possible to biochemically explain the discrepant results for the SGPT assay described above?

Case discussion

The transaminases are pyridoxal phosphate-dependent enzymes largely located in the mitochondria. They serve to interconvert certain amino acids into α-keto acids in a readily reversible manner, since the value of ΔG^0 for most transaminases is approximately zero. The conversion of amino acids into α-keto acids is frequently a first step in utilizing the carbon content of amino acids for oxidation to CO_2 and H_2O or for gluconeogenesis and lipogenesis. Because the transaminases show a broad specificity for substrates other than alanine or glutamate, they can operate on a large number of the dietary amino acids.

In conjunction with the transaminases, the Krebs cycle and a few ancillary enzymes provide a means for controlled transport of metabolites from the mitochondria to the cytoplasm. For example, citrate is the metabolite by which carbons derived from amino acids are delivered to the cytoplasm for lipogenesis (see Chapter 8), and malate is the metabolite by which carbons are exported to the cytoplasm for conversion to glucose (see Chapter 7). The Krebs cycle is therefore important in the operation of these major biosynthetic processes, as well as for the production of CO_2 and H_2O.

Transaminases are found in differing amounts in various tissues and show a differing pattern of release when a given tissue is injured. For example, the release of SGPT from injured liver is greater than from injured myocardium. It is not clear if this relates to the absolute quantity of the enzyme in liver as compared to myocardium, to a difference in cell fragility or "leakiness" between liver and heart, or to a difference in the rate of enzyme destruction before it can reach the circulatory system. Regardless of the cause, there is an incontrovertible relation between myocardial infarction and SGOT on one hand and hepatitis and SGPT on the other. These facts are frequently useful in differential diagnosis.

The alanine that is a substrate for the hepatic SGPT can arise from several sources. The most obvious is alanine resulting from a dietary intake beyond that needed for protein biosynthesis. A second source is from proteins degraded within the liver as a result of normal protein turnover. A third source is from amino acids resulting from breakdown of muscle protein during periods of nutritional deprivation. It has been established (Chapter 9) that muscle protein breakdown results in release of a mixture of amino acids of which alanine may comprise 10%, even though the actual content of alanine in muscle may be much lower. Since muscle does not liberate free glucose into the blood, the amino acids are a convenient means of transferring energy to the liver.

In the liver, alanine is transaminated to pyruvate. Decarboxylation of pyruvate provides acetyl CoA, while carboxylation of pyruvate provides oxaloacetate. Since these are the two compounds required for production of citrate, the large flux of alanine provides the input necessary for continued operation of the Krebs cycle even when glucose sources are minimal. For these reasons, it is clear that ^{13}-C-alanine could contribute its carbon atoms to the glucose pool.

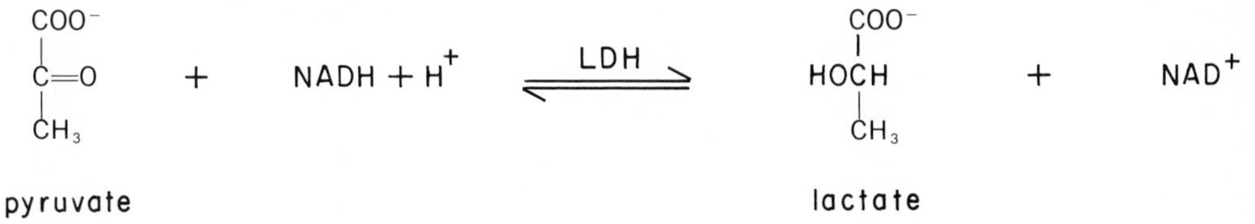

Because the assay of transaminase activity in serum has become an important adjunct to diagnosis, considerable effort has focused on development of precise methods; unfortunately, all laboratories do not conform to a common assay system. The most precise method is a spectrophotometric procedure based on the coupled reactions shown below. These reactions are coupled because the pyruvate that is formed in the first reaction is stoichiometrically consumed in the second. The second reaction is not directly involved in the transamination process, but it provides a means whereby transamination can be quantitatively measured, through observing a decrease in absorbance at 340 nm as NADH is converted to NAD^+. This is recorded as a function of time (see Chapter 3, p. 141). Even greater sensitivity can be obtained by measuring fluorescence instead of absorbance changes, but this is not yet common practice. The procedure is as follows: A sample of the serum to be analyzed is mixed with a considerable excess of NADH, L-alanine, and LDH in a suitable buffer. When this is observed in a spectrophotometer there may be a brief initial absorbance change resulting from the presence of extraneous NADH-consuming substances in the serum, but the absorbance soon reaches a steady value. The coupled reaction system is then initiated by addition of an excess of α-ketoglutarate, with proper mixing, and the absorbance change is measured as a function of time. This procedure gives normal values of 5 to 19 IU for males and 5 to 12 IU for females. All of the reagents are commerically available.

In careful studies of the above analytic system, it has been observed that problems may arise owing to the existence of differing proportions of the LDH isozymes in commercial preparations, since they are obtained from different organs as raw materials. The individual isozymes of LDH do not have equal K_m values for pyruvate. In fact, it requires about four times as much M_4-LDH as H_4-LDH to accurately measure an identical amount of SGPT activity. Mixed isozymes fall between these limits. It is therefore not surprising that discrepancies between laboratories may exist, depending on the nature of the LDH employed in the test system. Given the different kinetic properties of the LDH isozymes mentioned above, it is entirely possible that moderately elevated serum activities could be reported as normal, or the reverse. This case emphasizes that while enzyme assays are useful in diagnosis and management of disease, all conditions of the assay require complete and careful specification.

REFERENCES

Chang, M. M., and Chung, T. W.: Effect of lactate dehydrogenase isozymes on the coupled enzymatic assay for alanine aminotransferase activity, Clin. Chem. 21:330, 1975.

Gerhardt, W., et al.: Evaluation of methods for detection of inhibitors and specifications of NADH quality, Scand. J. Clin. Lab. Invest. 33(Suppl):139, 1974.

CASE 6: METHYLMALONIC ACIDURIA

An infant was admitted to the hospital with failure-to-thrive syndrome. The attending physician noted signs of mental deterioration and generalized physical disability. It was found that the urine collected from this child contained methylmalonate and methylcitrate.

Biochemical questions

1. Is methylmalonate a normal component of human urine?
2. What does the presence of methylcitrate in the urine signify?
3. What metabolic pathways other than the Krebs cycle itself might be affected by the presence of methylcitrate?
4. How might this condition affect the mitochondrial concentration of GTP?
5. What does this case indicate concerning the specificity requirements of citrate synthetase and of aconitase?

REFERENCES

Ando, T., et al.: Isolation and identification of methylcitrate, a major metabolic product of propionate in patients with propionic acidemia, J. Biol. Chem. 247:2200, 1971.

Ando, T., et al.: Propionic acidemia in patients with ketotic hyperglycinemia, J. Pediatr. 78: 827, 1971.

Oberholzer, G., et al.: Methylmalonic aciduria, inborn error of metabolism leading to chronic metabolic acidosis, Arch. Dis. Child. 42:492, 1967.

Rosenberg, L. E., et al.: Methylmalonic aciduria, an inborn error leading to metabolic acidosis, long-chain ketonuria and intermittent hyperglycinemia, N. Engl. J. Med. 278:1319, 1968.

Stokke, O., et al.: Methylmalonic acidemia, a new inborn error which may cause fatal acidosis in the neonatal period, Scand. J. Clin. Lab. Invest. 20:313, 1967.

CASE 7: CITRATE SYNTHESIS
BY ISOLATED LEUKEMIC
LYMPHOCYTES

Lymphocytes are important in cell-mediated immunity. Production of immunoglobulins requires a readily available energy source. It has been demonstrated that the Krebs cycle is adequate, along with the respiratory chain, to serve this purpose in lymphocytes collected from patients with chronic lymphocytic leukemia.

When 10^8 isolated lymphocytes were incubated with 0.5 μmoles of acetate or acetyl CoA and 10 μmoles of oxaloacetate, about 0.2 μmoles/hr of citrate was formed. Increasing the cell population over a fourfold range gave proportionate increases in citrate formation. When the incubation medium contained 20 μmoles of monofluoroacetate instead of acetyl CoA, an aliquot of 10^8 cells produced 0.8 μmoles/hr of citrate.

Biochemical questions

1. Could succinate have been used, instead of oxaloacetate, in the experiments described above?
2. In the above experiments, acetyl CoA and acetate appeared to produce citrate with the same efficiency. How is this possible?
3. What in vivo sources of acetyl CoA are available to lymphocytes?
4. Suppose you prepared mitochondria from collected lymphocytes, sonicated the mitochondria, and repeated the above experiments. How might the values measured differ from the ones reported above?
5. Could the experiments described have been performed equally well with leukemic erythrocytes?
6. In some of the experiments reported above, monofluoroacetate was used instead of acetate. Why is the observed citrate formation different when fluoroacetate was added?

REFERENCES

Aisenberg, A. C., and Bloch, K. J.: Immunoglobulins on the surface of neoplastic lymphocytes, N. Engl. J. Med. 287:272, 1972.

Dixit, P. K., and Cadwell, R.: Citrate synthesis by lymphocytes, Clin. Chem. 21:825, 1975.

Foster, J. M., and Terry, M. L.: Studies on the energy metabolism of human leukocytes, Blood 30:168, 1967.

Hedeskov, C. J., and Esmann, V.: Respiration and glycolysis of normal human lymphocytes, Blood 28:163, 1966.

CASE 8: MITOCHONDRIAL
METABOLISM IN POSTHEPATIC
JAUNDICE*

A 65-year-old man was seen in the clinic with severe jaundice. Following workup it was determined that the jaundice was posthepatic in origin, possibly caused by a malignancy of the head of the pancreas. On laparotomy his liver was noted to be enlarged and deeply stained with bile. A metastatic lesion was found in the right lobe of the liver. Biopsy of the lesion indicated an adenocarcinoma. The gallbladder was thickened and distended but was free of calculi.

The patient made a gradual recovery from surgery but shortly thereafter developed chills, fever, and diarrhea. Postoperatively the jaundice was slightly lessened, but then increased. As the jaundice worsened the patient became noticeably disoriented and lethargic, until he died in hepatic coma.

Biochemical questions

1. How would you determine that the jaundice was of posthepatic origin? Explain.
2. Bilirubin is known to be a potent inhibitor of NAD^+-linked isocitrate dehydrogenase. Would this affect mitochondrial CO_2 production? Explain.
3. Bilirubin is also known to inhibit NADH oxidation. Which steps in the

*We thank Dr. Brian Bross for his assistance with this case.

Krebs cycle would be inhibited for this reason?

4. How would severe jaundice affect the movement of metabolites into or out of the mitochondria? Explain.

5. Describe the possible effects of severe jaundice on the formation of ATP in the liver.

6. The toxicity of biliverdin is roughly one sixth that of bilirubin. Explain.

7. What changes might you expect in the blood lactate concentration? Explain.

8. If you could measure it, would you expect a normal value for citrate in the blood plasma? Explain.

REFERENCES

Cowger, M., et al.: Mechanism of bilirubin toxicity with purified respiratory enzymes and tissue culture systems, Biochemistry 4:2763, 1965.

Ozawa, K., et al.: Alteration in liver mitochondrial metabolism in a patient with biliary obstruction due to liver carcinoma, Am. J. Surg. 126: 653, 1973.

Ozawa, K., et al.: Role of portal blood in the enhancement of liver mitochondrial metabolism, Am. J. Surg. 124:16, 1972.

Ogasawara, N., et al.: Bilirubin, a potent inhibitor of NAD$^+$-linked isocitrate dehydrogenase, Biochim. Biophys. Acta 327:233, 1973.

Pelley, J. W., et al.: Lipoamide dehydrogenase in serum, a preliminary report, Clin. Chem. 22: 275, 1976.

ADDITIONAL QUESTIONS AND PROBLEMS

1. How might our patterns of metabolism be altered if all of the enzymes of the Krebs cycle were dissolved in the cytoplasm?

2. Erythrocytes contain no mitochondria. Would you expect them to contain large amounts of citrate?

3. Normal urine contains significant quantities of citrate, and in women the quantity measured is related to the menstrual cycle. Can you explain these observations?

4. Mammals do not convert stored fat to glucose, but many plant seedlings can easily do so. What enzymes contained in young plants allow this?

5. How would deficiencies of water-soluble vitamins affect the operation of the Krebs cycle? Are the effects the same for all of the water-soluble vitamins?

6. Hyperparathyroidism results in hypophosphatemia and hyperphosphaturia. How might these changes in phosphate metabolism affect the operation of the Krebs cycle?

7. A South African plant, *Dichapetalum cyanosum*, contains fluoroacetate in its leaves. Suppose these leaves were mixed with human food; what might be the effects on the person who ate such a mixture? How can it be explained in biochemical terms? How do you explain the ability of this plant to survive with a significant amount of fluoroacetate in its tissues? (Morrison, J. F., and Peters, R. A.: Biochem. J. 58:473, 1954 and Peters, R. A.: Adv. Enzymol. 18:1, 1957.)

8. While conducting an experiment on mitochondrial metabolism, a student accidentally swallowed a solution of atractylate. Explain the convulsions that followed.

CHAPTER 7

CARBOHYDRATE METABOLISM

OBJECTIVES

1. To interpret the role of carbohydrate metabolism in normal and selected common disease states
2. To relate carbohydrate metabolism to the functioning of the Krebs cycle and gluconeogenesis
3. To review the role of carbohydrates in nutrition and homeostasis
4. To discuss briefly the conjugated carbohydrate-containing biopolymers such as the glycoproteins

The largest and only replenishable store of carbohydrates is found in the plant kingdom, where it comprises most of the dry tissue weight. In contrast, animal tissue contains a comparatively small amount of carbohydrate, less than 1% in man. Yet in general, man's diet is high in carbohydrate, which represents approximately half the total caloric intake. Except for ascorbic acid (vitamin C), carbohydrates are not essential to the diet; through gluconeogenesis the body can synthesize necessary carbohydrates from other materials, principally from certain amino acids. However, low-carbohydrate diets usually result in metabolic imbalances. For example, a high-fat, low-carbohydrate diet results in a metabolic acidosis, whereas a high-protein, low-carbohydrate diet results in a protein imbalance with a high urinary nitrogen output; increasing carbohydrate in the diet prevents this loss of nitrogen, an illustration of the "protein-sparing" effect of carbohydrate.

The term "carbohydrate" originated with the idea that naturally occurring compounds of this class, for example, starch, glycogen, sucrose, and glucose, could be represented by the formula $C_x(H_2O)_y$, that is, a hydrate of carbon. It was soon found that this definition was too rigid, since it did not include the amino sugars such as glucosamine, the deoxy sugars such as 2-deoxyribose, and the sugar acids such as vitamin C.

NOMENCLATURE

The simplest sugars, the monosaccharides, are characterized by the reducing group $\diagdown C=O$. They are also polyhydroxy compounds and in general can be represented by the following structure:

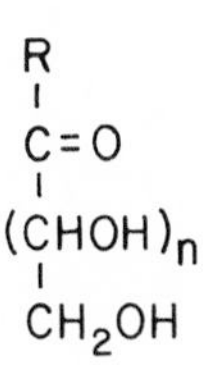

$$
\begin{array}{c}
R \\
| \\
C=O \\
| \\
(CHOH)_n \\
| \\
CH_2OH
\end{array}
$$

A reducing sugar

If R is hydrogen, the monosaccharide has an aldehyde group and is thus one of the *aldoses*. However, if R is —CH_2OH, the monosaccharide is the most common of the *ketoses*.

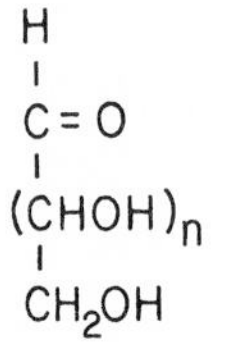

$$
\begin{array}{c}
H \\
| \\
C = O \\
| \\
(CHOH)_n \\
| \\
CH_2OH
\end{array}
\qquad\qquad
\begin{array}{c}
CH_2OH \\
| \\
C = O \\
| \\
(CHOH)_n \\
| \\
CH_2OH
\end{array}
$$

An aldose 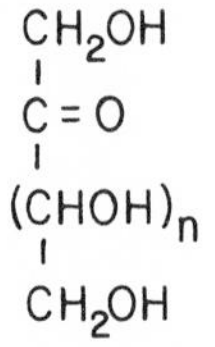 · · · · · · · · · · · · A ketose

We can further define the monosaccharides by the number of carbon atoms, for example, the *trioses* with three carbons and the *tetroses* with four carbons. These will be either *aldotrioses, aldotetroses, ketotrioses,* or *ketotetroses,* depending on whether they have an aldehyde or a keto group as the reducing function. Glucose is an aldohexose. Fructose is a ketohexose.

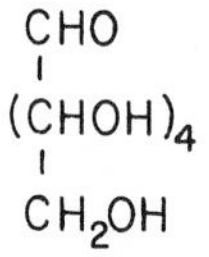

$$
\begin{array}{c}
CHO \\
| \\
(CHOH)_4 \\
| \\
CH_2OH
\end{array}
\qquad\qquad
\begin{array}{c}
CH_2OH \\
| \\
C = O \\
| \\
(CHOH)_3 \\
| \\
CH_2OH
\end{array}
$$

Aldohexose · · · · · · · · · · · · Ketohexose

(for example, glucose) · · · · · · (for example, fructose)

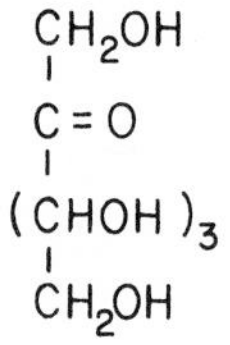

Galactose, mannose, glucose, and five other sugars are all aldohexoses represented by the same formula illustrated above. They are isomers that differ in the arrangement of the groups around the four asymmetric carbons in the molecule. When such asymmetry results in a *molecular* asymmetry, it gives rise to an *optically active molecule.*

Every optical isomer has a mirror image, the two isomers differing only in the direction in which the plane of polarized light rotates. A compound that rotates the plane of polarized light in a clockwise direction is said to be dextrorotatory (+), while those that rotate the plane of light in a counterclockwise direction are said to be levorotatory (−). Each of the carbohydrates is one or the other of the two possible optical isomers.

By convention, the family of sugars designated D, when represented by the planar Fischer formula, has the hydroxyl group on the asymmetric carbon farthest from the reducing group on the right-hand side of the carbon chain. The family of L-sugars has this hydroxyl group on the left.

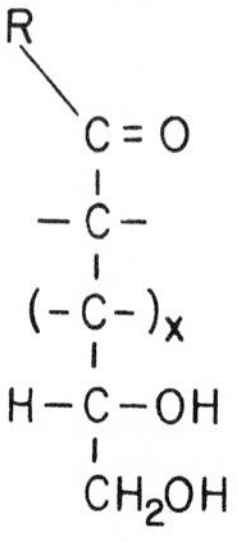

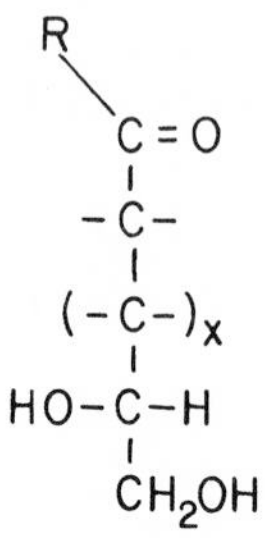

$$
\begin{array}{c}
R \\
\diagdown \\
C = O \\
| \\
- C - \\
| \\
(-C-)_x \\
| \\
H - C - OH \\
| \\
CH_2OH
\end{array}
\qquad\qquad
\begin{array}{c}
R \\
\diagdown \\
C = O \\
| \\
- C - \\
| \\
(-C-)_x \\
| \\
HO - C - H \\
| \\
CH_2OH
\end{array}
$$

D- · · · · · · Sugar · · · · · · L-

Note, however, that the D- and L-sugars are not necessarily dextrorotatory and levorotatory, respectively. In other words, the configuration around the asymmetric carbon farthest from the reducing group does not determine the rotation of the plane of polarized light.

The simplest of the aldose sugars is glyceraldehyde; the naturally occurring form is the D-isomer. The corresponding ketotriose is a symmetric molecule, dihydroxyacetone. The first asymmetric ketose is a ketotetrose, for example, D-erythrulose.

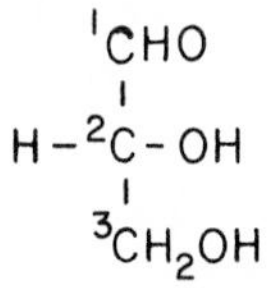

$$
\begin{array}{cc}
^1CHO & ^1CH_2OH \\
| & | \\
H - ^2C - OH & ^2C = O \\
| & | \\
^3CH_2OH & H - ^3C - OH \\
& | \\
& ^4CH_2OH
\end{array}
$$

D-Glyceraldehyde 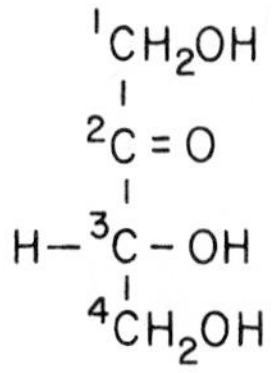   D-Erythrulose

As illustrated above, each carbon atom is identified by number; the reducing group carbon is C_1 for aldoses and C_2 for the common ketoses.

Other ketoses that will be discussed in relation to the metabolic paths of the carbohydrates are D-xylulose, D-ribulose, and D-fructose, the structures of which are shown below.

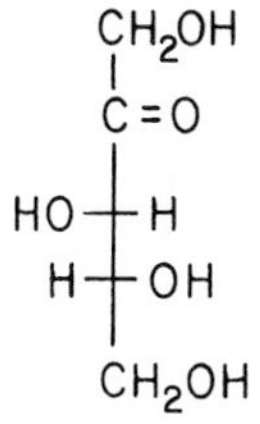

$$
\begin{array}{ccc}
CH_2OH & CH_2OH & CH_2OH \\
| & | & | \\
C = O & C = O & C = O \\
HO - H & H - OH & HO - H \\
H - OH & H - OH & H - OH \\
CH_2OH & CH_2OH & H - OH \\
& & CH_2OH
\end{array}
$$

D-Xylulose 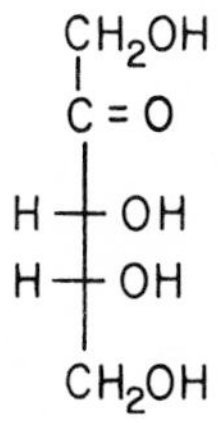   D-Ribulose 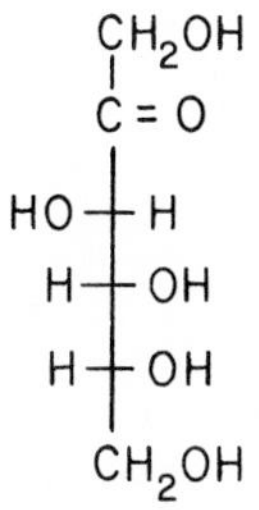   D-Fructose

RING STRUCTURES

It is important to note that the polyhydroxy aldoses and the ketoses, written above in the open-chain form, are reactive molecules. Although it is this particular form that reacts in some cases, the open-chain form is usually present in small amounts in aqueous solution. The $\diagdown C = O$ group reacts with a hydroxyl group in the same molecule to form a ring that may, for purposes of stability, be either five- or six-membered. These two different forms may be interconverted as shown for glucose in the reaction sequence presented in Fig. 7-1. The most immediate consequence of this ring formation is that the previously symmetric $\diagdown C = O$ group is now a new asymmetric carbon; thus there can be two forms of each ring sugar.

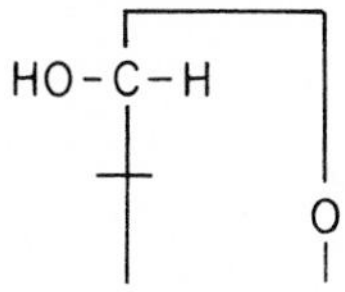

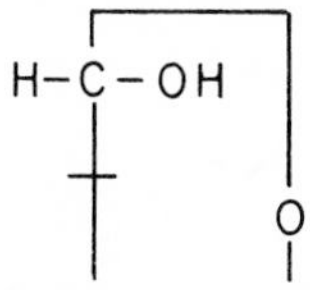

$$
\begin{array}{cc}
HO - C - H & H - C - OH \\
| & | \\
\quad O & \quad O \\
| & |
\end{array}
$$

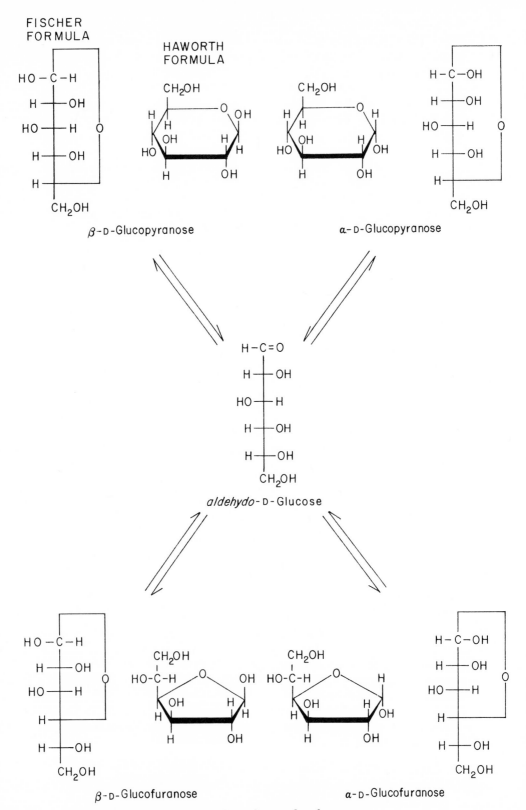

Fig. 7-1. Ring forms of D-glucose.

The new hydroxyl group, as represented above by the planar (Fischer) formula, can be on either side of the carbon chain. In the series of D-sugars the ring form with the hydroxyl group on the right side is designated the α-D-anomer and has the most positive optical rotation of the pair. The other ring form is called the β-D-anomer and has a more negative rotation.

The six-membered ring structures, composed of five carbons and one oxygen, may be related to a similar six-membered ring compound, pyran. It was suggested by Haworth that such sugars be identified as *pyranoses* to differentiate them from the five-membered ring forms, which are called *furanose* sugars. Although aqueous solutions of the reducing sugars may contain five different molecular forms (the open chain, the two pyranose rings, and the two furanose rings), it should be noted that only the pyranose sugars of the pentoses and hexoses are stable enough to exist to any large extent in aqueous solution. However, all forms are in equilibrium with each other; if a reaction occurs to remove one of them, then the principle of LeChetalier demands that the equilibrium be reestablished by the interconversion of all forms. This equilibrium can be demonstrated by observing the optical rotation when one dissolves crystalline dextrose (α-D-glycopyranose) in water. Such a change from a single form to an equilibrium mixture that includes its other forms is called *mutarotation*. Biologically, this change is catalyzed by the enzyme *mutarotase*.

The carbon atoms in the ring are identified by the same numbers as were given to them in the planar structures. Thus for α-D-galactopyranose and β-D-fructofuranose, the formulae are:

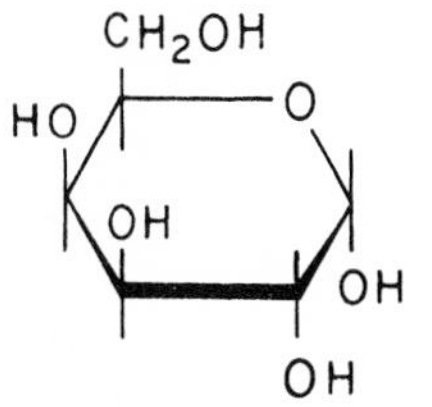

α-D-Galactose   β-D-Fructose

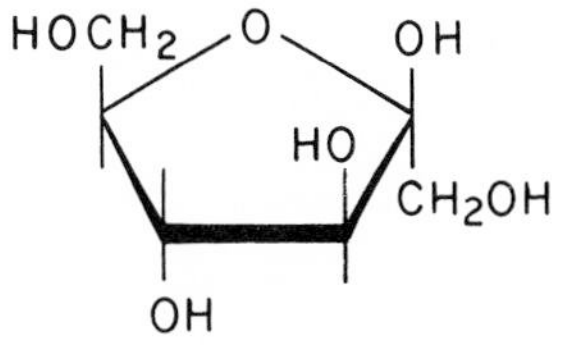

Note, as mentioned in Chapter 1, that the hydrogen atoms on the asymmetric carbons are usually deleted from the formula.

SUGAR DERIVATIVES
Glycosides

Probably the most important derivative of the reducing sugars is formed by their reaction with alcohols. Being a hemiacetal, the reducing sugar in the ring form will react with an alcohol to produce a glycoside, either the α-D- or the β-D-isomer. The hydroxyl group can be from a sugar, a sterol, an alkaloid, a protein, an inositol, or any similar compound, designated ROH. If the hydroxyl group comes from another sugar, the glycosidic bond will join the two monosaccharides to form a *disaccharide*. For example, the reaction of D-galactose with the ROH compound ethanol may be summarized in the manner shown at the top of p. 298. The D-galactose is in both α-D- and β-D-anomeric forms in solution during the reaction (indicated in the formula for the reducing sugar by not specifying the configuration around C_1). The arrangement is left as ⟩ H, OH. When referring to sugar derivatives in general, the generic prefix *glyc-* is used. This is replaced by

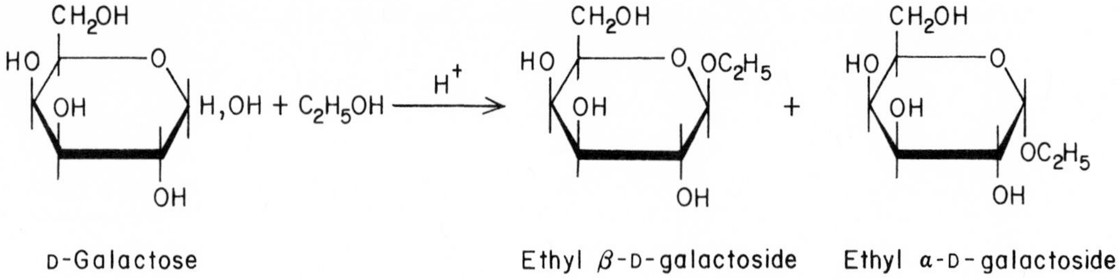

D-Galactose   Ethyl β-D-galactoside   Ethyl α-D-galactoside

the identifying name of the sugar when appropriate, for example, *gluc*oside or *galac*toside. The noncarbohydrate moiety of a glycoside is called the *aglycone*.

Disaccharides

While there are many possible combinations of pairs of monosaccharides and glycosidic linkages, four disaccharides are particularly important to our discussion; these are maltose, lactose, isomaltose, and sucrose.

Maltose

Maltose is a major product of the action of α-amylase on starch or glycogen. It is composed of two D-glucose residues, one of which is joined through its C_1 to the hydroxyl at C_4 of the other D-glucose. The glucosidic linkage is of the α-D-anomeric form. Maltose is a reducing sugar because the one D-glucose has a potentially free aldehyde group.

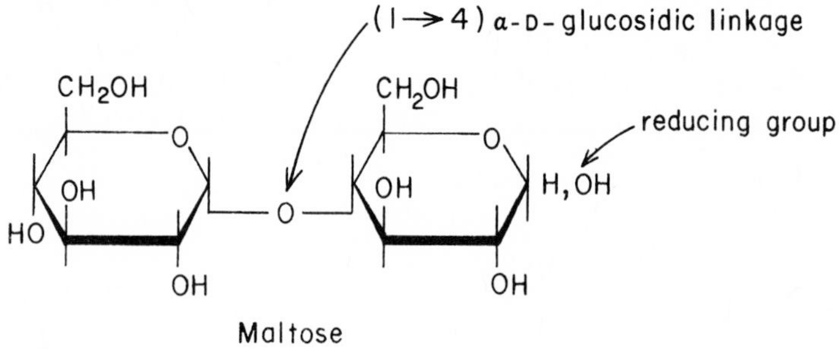

Maltose

Lactose

Lactose, the reducing disaccharide of milk, has a $(1 \rightarrow 4)$-β-D-galactosidic linkage to D-glucose.

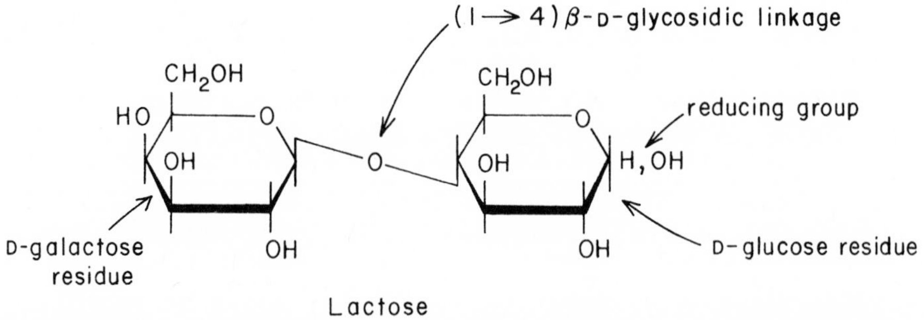

Lactose

Note that the structures of the D-galactose and D-glucose pyranose rings differ only in the configuration around C_4. These sugars are said to be epimeric at C_4, a point that is important in the metabolism of D-galactose and therefore of lactose.

Isomaltose

Isomaltose, a disaccharide derived from the branch point of starch, has a $(1 \rightarrow 6)$-α-D-glucosidic linkage to a second D-glucose residue.

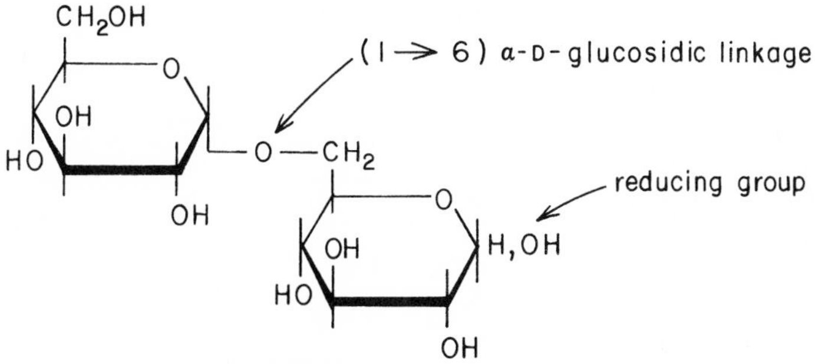

Isomaltose

Sucrose

Sucrose is produced from sugar cane or sugar beet. It is a nonreducing sugar since the reducing groups of α-D-glucopyranose and β-D-fructofuranose are glycosidically linked together.

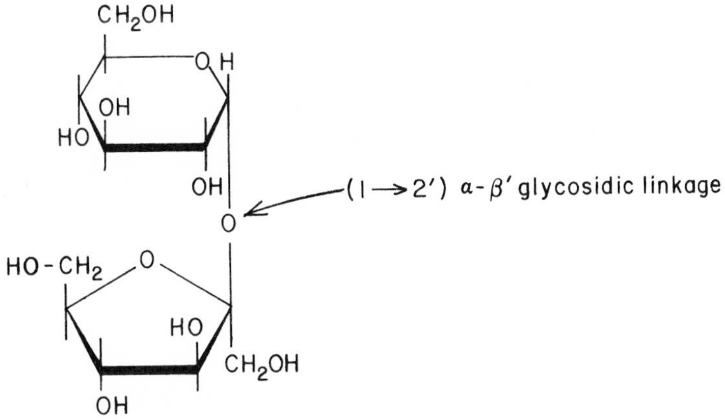

Sucrose

Oligosaccharides and polysaccharides

The glycosidic polymeric linkages continue to be formed by the sequential addition of monosaccharides; thus the trisaccharides, oligosaccharides, and eventually the high molecular weight polysaccharides are synthesized. Sugars are multifunctional compounds with four hydroxyl groups and a hemiacetal group (with α- or β-anomeric configuration) in the hexose ring structures. A large number of possible structures can be proposed theoretically for oligosaccharides and polysaccharides with the same sugar composition. Thus eleven disaccharides have been synthesized from D-glucose; 176 trisaccharides from D-glucose are possible.

The polysaccharides of greatest interest are starch and glycogen. Both are composed of D-glucose and are quite similar in general structure, but differences in their finer details show them to be two distinct materials. In both cases the principal glycosidic linkage is between the C_1 of one D-glucopyranosyl residue and the hydroxyl at C_4 of the adjacent residue, the anomeric form being α-D.

$$\alpha\text{-}D\text{-}G(1 \longrightarrow 4) \underbrace{\left[\alpha\text{-}D\text{-}G(1 \longrightarrow 4) \right]}_{n} \alpha\text{-}D\text{-}G$$

(Non-reducing end) (Reducing end)

Short chains of glucose residues linked in this manner are cross-linked through the hydroxyl groups at C_6 of some of the residues. Therefore, the structures resemble a branching tree. A small portion of the molecule is shown in Fig. 7-2.

Polysaccharides share a number of general structural features. All the polymeric linkages involve the original reducing carbon of the monosaccharide units and a hydroxyl group of an adjacent monosaccharide. Consequently, only one sugar residue in the polysaccharide can have a free reducing group. Branching in the polysaccharide is achieved by additional glycosidic linkages with a particular sugar residue in the existing chain. Since the biosynthesis of polysaccharides takes place through addition of monosaccharide units to the nonreducing ends of existing chains, every branch point represents an additional locus for polymer growth by chain elongation. In a highly branched polysaccharide such as glycogen the molecule is constructed of short chains whose average length can be calculated. In glycogen the average chain length is around twelve glucose units. Starch is actually a mixture of two polysaccharides. One, amylose, is essentially an unbranched chain of α-D-glucopyranosyl units. The other, a highly branched component called amylopectin, has an average chain length of approximately twenty-five glucose units.

Unlike proteins or nucleic acids, the polysaccharides are not biosynthesized on a template. This means that the size of the polysaccharide molecule is not fixed; rather, there is a range of molecular weights. Thus the molecular weight of normal glycogen may range from approximately 1×10^6 to more than 100×10^6, with a majority of molecules having molecular weights in the vicinity of 5 to 10×10^6. These values depend on the tissue of origin, the state of nutrition, and the presence of disease.

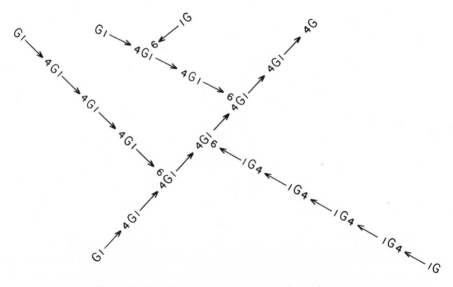

Fig. 7-2. Terminal segments of starch or glycogen.

Other naturally occurring sugar derivatives

Other important sugar derivatives that will be encountered in the course of this discussion are listed below.

Sugar phosphate esters

Sugar phosphate esters include α-D-glucopyranose 1-phosphate and D-fructose 1,6-diphosphate, the structures of which are shown below.

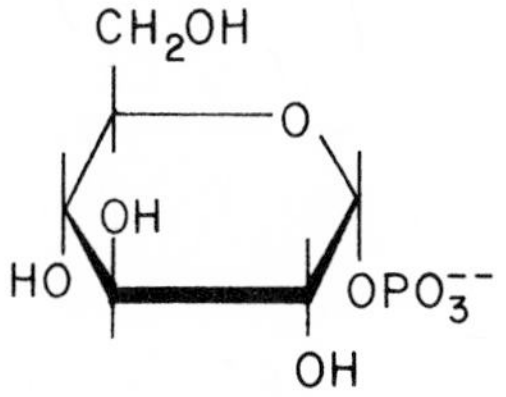

α-D-Glucopyranose 1-phosphate

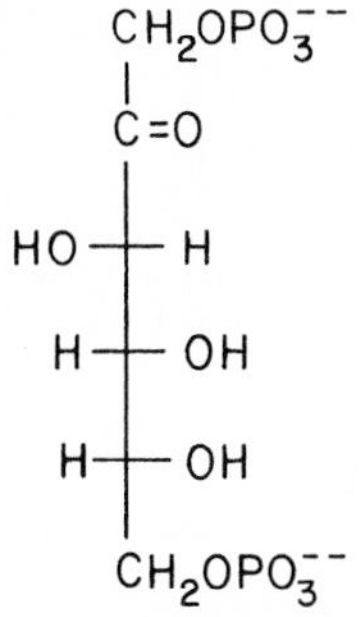

D-Fructose 1,6-diphosphate

The sugar phosphates are formed by reaction of the sugar with ATP in the presence of the appropriate sugar kinase. When the phosphate residue is attached to the hydroxy group of the reducing carbon, a glycoside is produced and the ring form is fixed. Esterification of any other hydroxyl groups in the sugar leaves the reducing function free, and the sugar phosphate may react in the open-chain form, as shown above for D-fructose 1,6-diphosphate.

Glyconic acids

The oxidation of the aldehyde group at C_1 in aldoses to the corresponding carboxyl function creates a class of sugar acids called glyconic acids. The name for the glyconic acid of any particular aldose is obtained by replacing -*ose* with -*onic acid*, for example, glucose and gluconic acid. Several derivatives of the glyconic acids are important in biochemistry; for example, vitamin C is an oxidation product of an L-glyconic acid. The metabolism of D-glucose by one pathway involves the 6-phosphate ester of D-gluconic acid, as shown below.

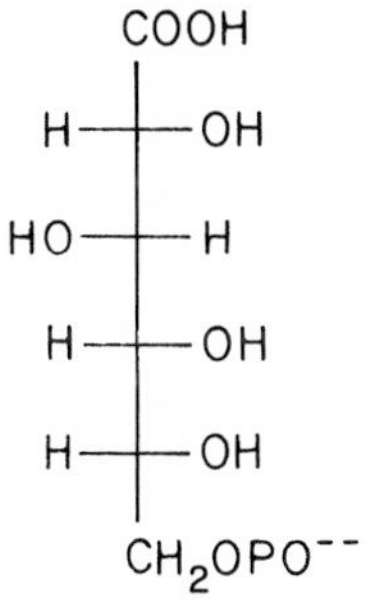

D-Gluconic acid 6-phosphate

Deoxypentoses

The most important member of the deoxypentoses is 2-deoxy-D-ribose. 2-Deoxy-D-ribofuranose is the ring form found in the deoxyribonucleosides and deoxyribonucleotides. These pentoses form the backbone of deoxyribonucleic acid (DNA).

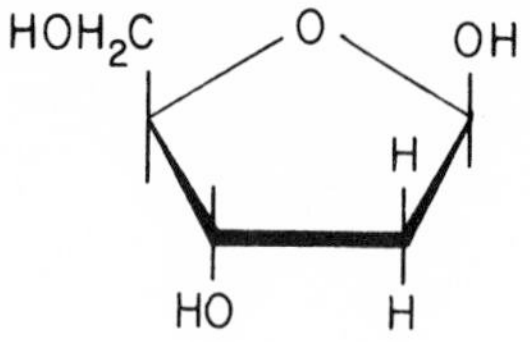

2-Deoxy-β-D-ribofuranose

Nucleosides and nucleotides

A sugar that is linked through its reducing carbon to a purine or pyrimidine base is called a nucleoside. In the older nomenclature this linkage was called an N-glycoside, and this designation is still appropriate. Phosphorylation of the sugar residue converts the nucleoside to a nucleotide. Common examples are adenosine and adenosine 5'-phosphate (AMP); 5' indicates that the phosphate is on carbon atom 5 of the sugar and not on the base, adenine.

Adenosine
(nucleoside)

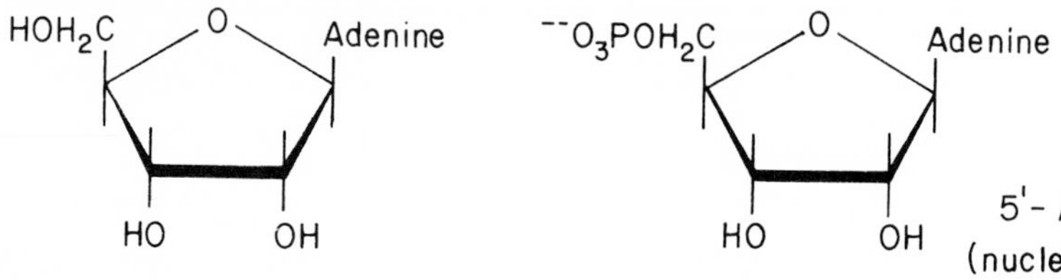

5'- AMP
(nucleotide)

Uronic acids

Uronic acids are produced through the oxidation of the primary hydroxy groups of sugars. One member of this group, D-glucuronic acid, is shown below.

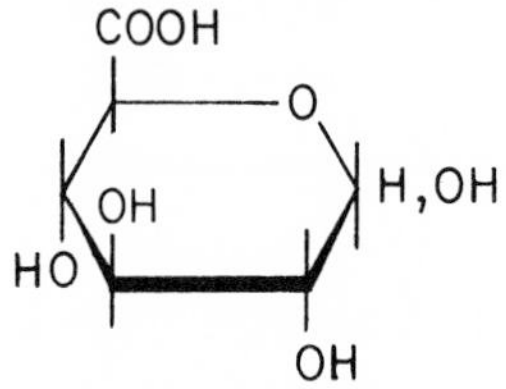

D-Glucuronic acid

The names of these compounds are formed by replacing the *-ose* with *-uronic acid.* These derivatives are particularly important in detoxification mechanisms, as seen earlier in discussions of bilirubin excretion (p. 137), where the bilirubin was linked glycosidically to D-glucuronic acid. The uronic acids D-glucuronic, D-galacturonic, D-mannuronic, and L-iduronic are important residues in polysaccharides. D-Mannuronic acid is the 2-epimer of D-glucuronic acid, and L-iduronic acid is the 5-epimer of D-glucuronic acid.

Amino sugars

Amino sugars are formed through the substitution of a hydroxyl group by an amino group on the sugar ring. The most common amino sugar is D-glucosamine, which occurs in many polysaccharides and is shown below.

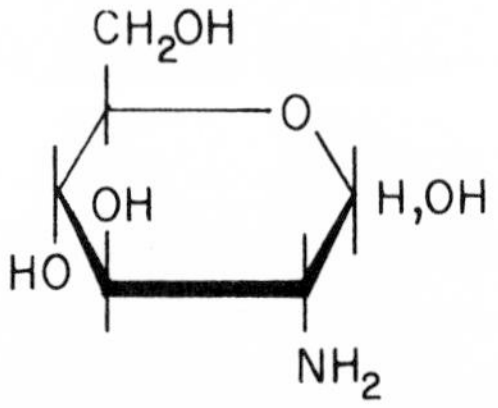

D-Glucosamine

The —NH$_2$ group is never free in nature. It is usually acetylated (—NH · COCH$_3$) but may also be sulfated (—NH—SO$_3^-$), as in the anticoagulant heparin.

A very large number of polysaccharides could be formed using only those sugars and sugar derivatives already mentioned. In man the main polysaccharide is the storage carbohydrate glycogen. The other carbohydrate polymers are found in covalent linkage with either proteins or lipids. They function in a structural capacity, as in chondroitin sulfate–protein complexes of extracellular ground substance, the hyaluronic acid–protein complex of synovial fluid, and the glyco-proteins and glycolipids of cell membranes. All these molecules are complex and play important biologic roles. Although their biosynthesis is becoming better understood, much remains to be learned of their relationships to pathologic conditions.

DIGESTION OF CARBOHYDRATES

The glycosidic bonds of oligosaccharides and polysaccharides are hydrolyzed by glycosidases to give the reducing sugar components. The glycosidases are usually specific for the glycosyl portion of the glycoside, including its anomeric form, but show little specificity for the aglycone. Thus lactase will hydrolyze lactose and many other β-D-galactopyranosides.

For the purpose of our discussion of the digestion of carbohydrates in man, the fate of starch, lactose, sucrose, and cellulose in the diet will be followed. The principal locations for digestion are the mouth, the lumen of the small intestine, and the brush border of the epithelial cells of the intestinal mucosa. The digestive process is diagrammed in Fig. 7-3.

As food is masticated in the mouth into a proper bolus for swallowing, the sali-vary α-amylase acts on the starch in a random manner. The $(1 \rightarrow 4)$-α-D-glucosidic bonds are split, producing maltose, some D-glucose, and smaller units of the starch molecule called starch dextrins, which contain all the original $(1 \rightarrow 6)$-α-D-glucosidic bonds. At this stage the starch has been reduced in size to an average chain length of less than eight glucose units, provided the food has been chewed thoroughly.

When the bolus of food meets a high level of acidity as it enters the stomach, the action of the α-amylase stops. Little hydrolysis of the carbo-hydrate occurs in the stomach; perhaps sucrose undergoes a slight hydro-lysis as a result of the greater sensitivity of its β-D-fructofuranoside linkage to acidity.

As the stomach empties, the pH of the material entering the small intestine is rendered alkaline by secretions from the bile and pancreatic ducts. The diges-tion of the starch dextrins is continued by the action of the pancreatic α-amylase, which is similar to the salivary enzyme but which has an absolute requirement for chloride. When the pancreatic α-amylase completes its hydrolysis of the starch, the intestinal lumen contains principally D-glucose, maltose, isomaltose, and the dietary lactose and sucrose. The ingested cellulose is a polysaccharide with $(1 \rightarrow 4)$-β-D-glucosidic linkages for which there is no hydrolytic enzyme in man; it is nondigestible.

The disaccharides are hydrolyzed at the brush border by specific disaccha-ridases that are contained in the cell membrane. The monosaccharides so formed and the glucose in the lumen pass into the portal blood system, through which they are transported first to the liver and then to the remainder of the body. Any hydroly-sis of the disaccharides that occurs in the lumen of the small intestine is caused by the sloughing of the brush border into this space rather than the secretion of the

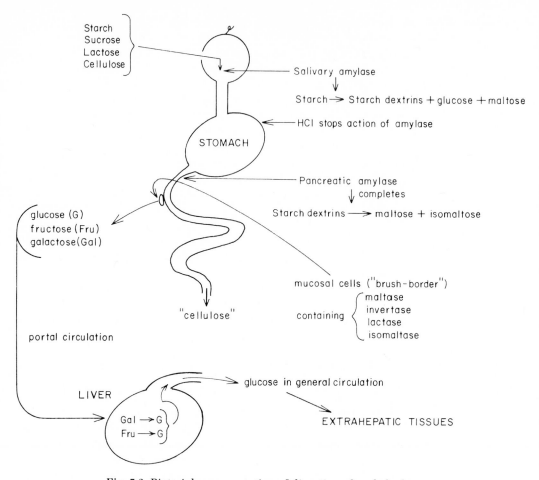

Starch
Sucrose
Lactose
Cellulose

Salivary amylase

Starch $\rightarrow$ Starch dextrins + glucose + maltose

HCl stops action of amylase

STOMACH

Pancreatic amylase
completes

Starch dextrins $\longrightarrow$ maltose + isomaltose

glucose (G)
fructose (Fru)
galactose (Gal)

mucosal cells ("brush-border")
containing
maltase
invertase
lactase
isomaltase

"cellulose"

portal circulation

glucose in general circulation

LIVER

Gal $\rightarrow$ G
Fru $\rightarrow$ G

EXTRAHEPATIC TISSUES

Fig. 7-3. Pictorial representation of digestion of carbohydrates.

enzymes into the intestinal fluid. However, depending on the nature of the ingested food, there could be high levels of D-glucose and D-fructose in the lumen.

ABSORPTION OF CARBOHYDRATES FROM INTESTINE

The mechanism by which sugars are absorbed from the intestine is complex and not completely understood. Some sugars, most commonly the pentoses, pass across the intestinal barrier by simple passive diffusion. The other sugars (notably D-glucose, D-fructose, D-galactose, and possibly D-mannose) can be transported against a concentration gradient; the last traces of these sugars will be absorbed from the intestine in spite of higher concentrations existing in the blood. However, under physiologic conditions the concentration of sugars is greater in the intestine than in the blood during much of the absorption process; thus this downhill concentration gradient does not require energy. It can be shown experimentally that the hexoses mentioned previously can be absorbed against a concentration gradient and that some active transport mechanism is available. It is likely that for 90% to 95% of the sugar absorption such a mechanism is not needed and may indeed function to prevent sugars from leaking back into the intestinal lumen.

The sugars absorbed by facilitated diffusion involve carrier proteins. Glucose is absorbed into the mucosa along with Na$^+$. The concentration of the Na$^+$ in these cells is kept at approximately 40 meq/liter by an active sodium pump that main-

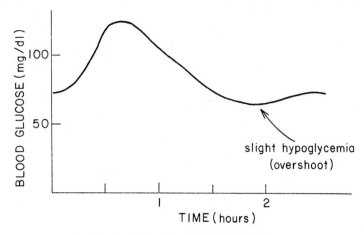

Fig. 7-4. Normal glucose tolerance curve.

tains transport into the blood, in which the concentration is approximately 140 meq/liter.

The digestion of the disaccharides and the absorption of the sugars so produced occur in the brush border, mostly in the upper jejunum. The two processes are largely supportive of each other, as little of the monosaccharide produced by the disaccharidases leaks back into the lumen.

Rate of glucose absorption

After a meal containing sugars the processes of digestion and absorption occur gradually, the quantity of sugars absorbed being approximately 1 g/kg body weight · hr. It appears that the absorption rate is quite constant in the small intestine regardless of the amount (within wide limits) of sugar present or the concentration in which it is introduced. In passing through the liver, fructose and galactose are either metabolized to other compounds according to homeostatic need or converted to glucose, the usual circulating blood sugar. Within $\frac{1}{2}$ to 1 hr after a meal the blood sugar level usually reaches a maximum of approximately 130 mg/dl and then decreases in 2 to $2\frac{1}{2}$ hr to approximately 70 to 90 mg/dl. The ability to handle a carbohydrate load may be determined by a glucose tolerance test. The blood glucose level is followed over time after the ingestion of 100 g of glucose. A typical glucose tolerance curve for a normal adult is shown in Fig. 7-4. Curves for the diabetic are discussed in case 2 at the end of this chapter. If for any reason the blood glucose level exceeds approximately 160 to 180 mg/dl, the sugar is not completely reabsorbed from the glomerular filtrate. Thus the kidney threshold is exceeded and glucosuria results.

Recognizing that the total blood volume in a 70-kg man is approximately 6 liters and that 100 g of glucose is absorbed in about $1\frac{1}{2}$ hr, it follows that the equilibrium level of 70 to 90 mg/dl is reached through either the storage or use of some of the carbohydrate. The process of storage involves the synthesis of glycogen, while that of use primarily involves glycolysis or an alternative oxidative process, the pentose phosphate shunt. The metabolic pathways will now be considered separately.

INTERCONVERSION OF D-GLUCOSE, D-GALACTOSE, AND D-FRUCTOSE

A typical meal containing starch, sucrose, and lactose places a load of D-galactose and D-fructose on the liver. These sugars must be converted to D-glucose. The

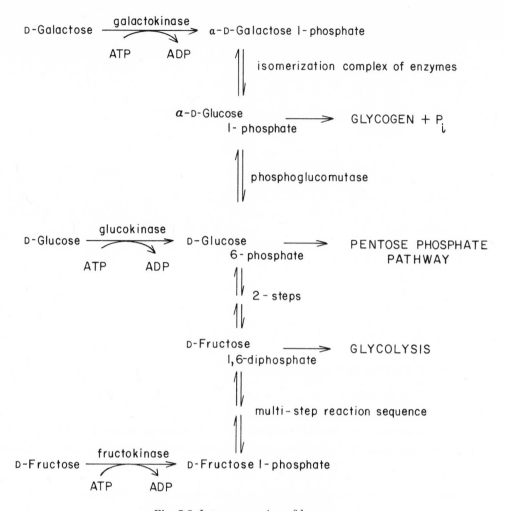

Fig. 7-5. Interconversion of hexoses.

postabsorptive blood levels of these sugars decline rapidly, approaching zero in 1 to 2 hr. The overall isomerization reactions are summarized in Fig. 7-5.

There are several enzymes called kinases, for example, glucokinase, fructokinase, and galactokinase, that catalyze the phosphorylation of hexoses. Each kinase (Fig. 7-5) is specific for its particular hexose. In addition, there are hexokinase isozymes in various tissues that are less specific and that catalyze the phosphorylation of D-glucose, D-fructose, D-mannose, and D-glucosamine to give the corresponding 6-phosphate esters. Each hexokinase is different in its Michaelis constant (K_m) for each sugar and for ATP. The data for brain hexokinase are given in Table 7-1. The hexokinases are subject to product inhibition, whereas glucokinase is not. Furthermore, the hexokinases are constitutive enzymes, being present at a constant level in the cells, whereas glucokinase is induced in the liver by insulin.

The conversion of both D-galactose and D-fructose to D-glucose 1-phosphate or D-glucose 6-phosphate involves several intermediate steps. In each case, homeostatic regulatory mechanisms determine the ultimate fate of the sugars and, as seen in Fig. 7-5, the point at which the intermediate enters some other pathway.

Table 7-1.   Brain hexokinase (K_m values)

SUGAR	K_m (molar)
D-Glucose	8×10^{-6}
D-Fructose	2×10^{-3}
D-Mannose	5×10^{-6}
D-Glucosamine	8×10^{-5}

For example, if the cells of the body are "charged" with glucose and its interme-diary metabolites, excess carbohydrate available to the cells will be converted to glycogen for storage. The D-galactose would therefore be converted to D-glucose 1-phosphate and not to D-glucose 6-phosphate in the liver cell. If, on the other hand, carbohydrate were needed immediately for energy production, the D-galactose would be converted to D-glucose 6-phosphate or even to D-fructose 1,6-diphosphate.

Utilization of D-glucose

Blood glucose penetrates the liver cell membrane without the need for insulin to be present. In contrast, glucose entry into muscle and adipose tissue cells is greatly enhanced by the presence of insulin. How insulin enhances glucose entry into these cells is not clear. In liver, insulin induces the synthesis of the enzyme glucokinase, which is absent or present only in very low levels in the starved state and in diabetes mellitus. Glucokinase is present only in hepatic tissues. Liver, muscle, and adipose hexokinases appear to be unaffected by insulin.

In the preprandial state the blood glucose concentration is approximately 5 mM. The tissues take up glucose from the extracellular fluids as it is available and as insulin is available for the muscle, adipose, and some other tissues. The rate of glucose uptake is controlled in part by the feedback inhibition of the hexo-kinase. After eating and the concomitant rise in blood glucose levels to 7 to 10 mM, the glucokinase ($K_m = 1 \times 10^{-2}$ M glucose) becomes effective. Glucokinase does not exhibit feedback inhibition, and it acts to reduce the blood glucose con-centration by converting the sugar to D-glucose 6-phosphate even in the presence of a high glucose 6-phosphate concentration. The higher K_m value allows for a higher glucose 6-phosphate concentration before the enzyme is saturated. There-fore, at times of elevated blood glucose the catalytic activity of glucokinase is less affected than hexokinase by the glucose load and can better return conditions toward the normal. As the glucose concentration in blood falls, the contribution of the glucokinase to the homeostatic mechanisms is reduced. However, as noted in Fig. 7-4, there is a short hypoglycemic period produced by the residual effects of the insulin and glucokinase. During this adjustment period, glucose release from the liver is stimulated by a second hormone, glucagon, and the blood glucose level fluctuates until it becomes stabilized in a dynamic equilibrium. With the return to normal glucose levels the hexokinases, with their greater affinity (lower K_m) for glucose, take over the phosphorylation of D-glucose as it enters the cell.

Experiments have shown that in starvation or diabetes, glucokinase is present in the liver in low concentrations, if at all. Administration of insulin to a diabetic animal results in the biosynthesis of glucokinase, but the maximum concentration is not attained for several days. Similarly, a high-carbohydrate diet would return the glucokinase concentration in a starved animal to normal in a few days. Thus under normal circumstances the liver responds to insulin by maintaining fairly constant levels of glucokinase, levels that do not fluctuate widely at mealtimes.

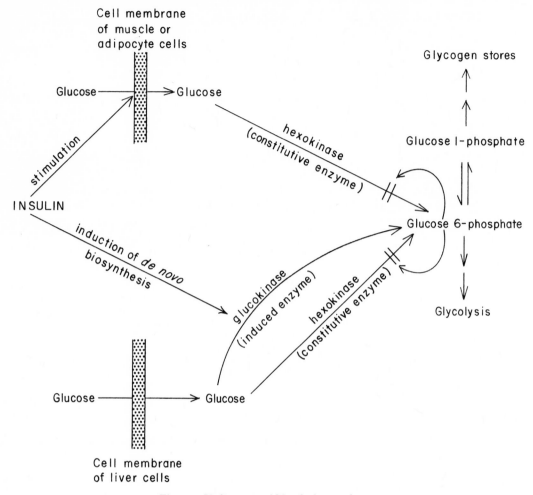

Fig. 7-6. Utilization of blood glucose by tissues.

This characteristic is also recognized when carrying out a glucose tolerance test; if necessary, the patient is primed with a carbohydrate diet for 1 or 2 days before the test. The initial steps in the utilization of blood glucose by the various tissues are summarized in Fig. 7-6.

Utilization of D-fructose

The utilization of D-fructose from the diet illustrates the many ways in which an intermediate may be used in metabolism, depending on the homeostatic needs of the tissues. In the liver the D-fructose is converted to the 1-phosphate, which is cleaved into two triose fragments by *aldolase,* as shown in Fig. 7-7. The reactions catalyzed by aldolase occur with a small change in free energy and have an equilibrium constant of nearly one. The enzyme forms a Schiff base intermediate between an ϵ-amino group of a lysyl residue and the dihydroxyacetone phosphate. This intermediate may either transfer the dihydroxyacetone phosphate to water, and so drive the reaction in the direction of cleavage, or be transferred to an aldose, for example, D-glyceraldehyde. In the reversible reaction, where there may be alternate aldose receptors, any of the aldoses may become condensed with the dihydroxyacetone phosphate residue for which the enzyme is highly specific.

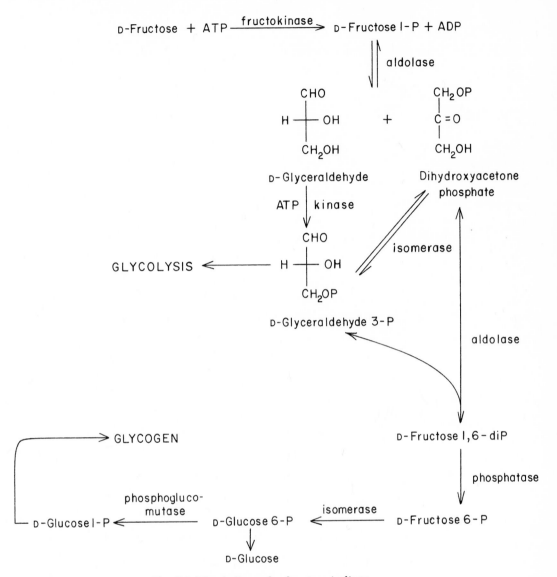

Fig. 7-7. Metabolism of D-fructose in liver.

In the liver the liberated D-glyceraldehyde, which is derived from carbon atoms 4, 5, and 6 of the D-fructose 1-phosphate, is phosphorylated with ATP and a triose kinase to D-glyceraldehyde 3-phosphate, an alternate aldose acceptor for the dihydroxyacetone phosphate. Furthermore, the triose phosphates are reversibly interconverted by *triose phosphate isomerase,* an important enzyme in the catabolism of D-glucose.

The condensation of the triose phosphates produces D-fructose 1,6-diphosphate, the 1-phosphate group of which is specifically hydrolyzed by D-fructose 1,6-diphosphate 1-phosphatase. The D-fructose 6-phosphate so formed is isomerized to D-glucose 6-phosphate, in which the phosphate group must be changed to position 1 if the sugar is to be converted to glycogen. The position of the phosphate ester group can be transferred through the catalytic action of a mutase, in this case a *phosphoglucomutase.* This enzyme requires the presence of a coenzyme

(D-glucose 1,6-diphosphate) and functions by the formation of an intermediate phosphoenzyme. This activated enzyme contains an O-phosphorylated seryl residue that may donate the phosphate to either position 1 or 6 of a D-glucose phosphate. The 1,6-diphosphate formed in this manner transfers one of its phosphates back to the seryl group. Which of the two phosphate groups is transferred to the enzyme depends on the relative concentrations of the two phosphate esters. The sequence of events is represented below, starting with D-glucose 6-phosphate.

$$\text{D-Glucose 6-P} + \text{Enz-P} \rightleftharpoons \text{D-Glucose 1,6-diP} + \text{Enz}$$

$$\Updownarrow$$

$$\text{D-Glucose 1-P} + \text{Enz-P}$$

Phosphoglucomutase was one of the first enzymes proved to function by formation of a protein seryl-phosphate intermediate.

Evidence for this apparently circuitous route of fructose metabolism in liver to give D-glucose has been gained by isotopic labeling of D-fructose and determination of the position of the label in the resulting D-glucose. A ^{14}C label at C_1 of the D-fructose subsequently appears at positions C_1 and C_6 of D-glucose.

In the short time during which D-fructose circulates in the blood before removal by the liver, adipose tissue and muscle are probably the main extrahepatic tissues that make use of it. The pathway for the conversion of D-fructose to D-glucose in extrahepatic tissues is shown in Fig. 7-8. Because hexokinase has a relatively high K_m for D-fructose, this pathway is not a major one unless a high level of fructose is maintained in the blood. The hexokinase of muscle directly produces D-fructose 6-phosphate, which is isomerized to D-glucose 6-phosphate. Muscle does not contain D-glucose 6-phosphatase; therefore once a hexose has entered the muscle cell it cannot be released into the blood as glucose to maintain the circulating sugar levels.

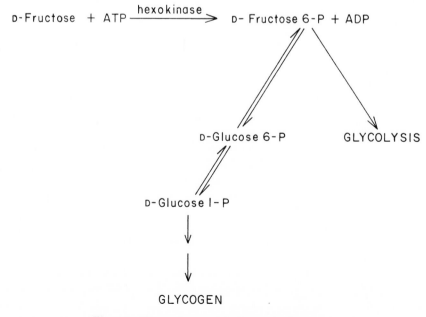

Fig. 7-8. Alternative metabolism of D-fructose.

As will be seen later, the metabolism of D-fructose is closely interrelated with glycolysis (p. 322) and the pentose phosphate pathway (p. 331). Since all of these pathways occur in the cytoplasm, the common branch points in these schemes (for example, the reactions involving D-glucose 6-phosphate) will be subject to metabolic controls.

Utilization of D-galactose

D-Galactose is readily converted to D-glucose in the liver by the metabolic process shown in Fig. 7-9. The transformation is rapid. After ingestion of about 40 g of D-galactose a healthy adult will show a maximum concentration in the blood after 1 hr. All of the D-galactose will have disappeared from the blood within 2 hr of ingestion.

The conversion of D-galactose in the liver occurs with the formation of D-galactose 1-phosphate. Next, the D-galactosyl residue is exchanged with D-glucose from uridine diphosphate glucose (UDPG), and uridine diphosphate galactose (UDPGal) is formed.

$$\text{D-Galactose 1-phosphate} + \text{UDPG} \rightleftharpoons \text{UDPGal} + \text{D-glucose 1-phosphate}$$

This exchange of sugar residues is catalyzed by D-galactose 1-phosphate uridyl transferase. The D-glucose 1-phosphate so produced is degraded by glycolysis,

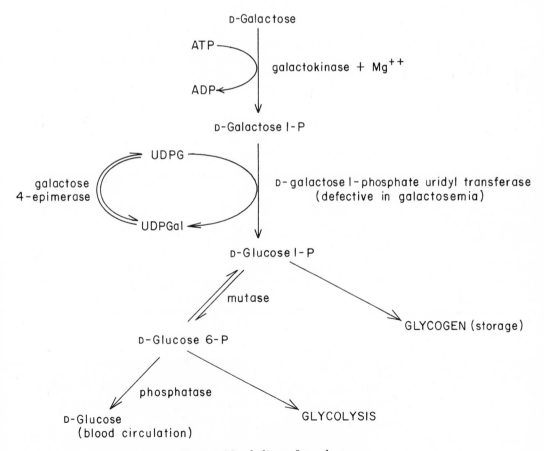

Fig. 7-9. Metabolism of D-galactose.

passed into the circulation as the free sugar, or converted to glycogen for storage. The UDPGal is epimerized back to UDPG (Fig. 7-9).

Galactosemia

There are several pathologic conditions that prevent the utilization of D-galactose derived from dietary lactose. Two such diseases involve a congenital deficiency in the metabolism of D-galactose to D-glucose. Von Reuss was the first to report such a disorder when he described an 8-month-old child with what are now recognized as characteristic symptoms of galactosuria and a cirrhotic liver. (The latter condition may also have been related to the fact that the infant was fed cognac.) In subsequent cases it was noted that dramatic relief occurred if the patient was given a galactose-free diet.

The most common type of galactosemia is caused by a defect in an autosomal recessive gene that results in the reduction or absence of D-galactose 1-phosphate uridyl transferase. Although the heterozygote will show a reduction in this enzyme, the level usually is sufficient to maintain health unless continuously challenged by high-galactose diets. The homozygote, however, because of the absence of uridyl transferase enzyme, must be recognized at birth so that milk, which contains lactose, can be excluded from the diet. Fortunately the erythrocytes reflect the enzymatic status of these patients, and cord blood can be used to assay D-galactose 1-phosphate uridyl transferase levels. Such a test is particularly important in families with a history of this disease. Within several weeks or months after birth the consequences of galactosemia (a high level of galactose 1-phosphate in all tissues) are evident and irreversible. A galactose-free diet avoids these difficulties (case 4), and the galactose that is necessary for the biosynthesis of cell membranes, cerebrosides, glycoproteins, and other cell components can be formed from glucose 1-phosphate, as shown in Fig. 7-10. The UDPGal that is formed is used directly in the biosynthetic processes.

Some capacity to metabolize ingested D-galactose may develop later in the life of a galactosemic person, caused in part by the development of the enzyme UDPGal pyrophosphorylase, a UTP:galactose 1-phosphate uridyl transferase that catalyzes a reaction analogous to that shown in Fig. 7-10.

$$UTP + Galactose\ 1\text{-}P \rightleftharpoons UDPGal + PP_i$$

A less common congenital form of galactosemia is caused by a deficiency in galactokinase.

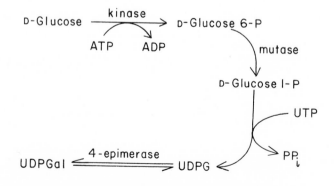

Fig. 7-10. Biosynthesis of D-galactose.

In galactosemia the increased concentration of galactose in blood and the aqueous humor leads to greater entry of the sugar into the lens. There it is converted partly to the corresponding sugar alcohol, galactitol (dulcitol), which eventually causes cataract formation.

NUCLEOSIDE DIPHOSPHATE SUGARS AND CARBOHYDRATE BIOSYNTHESES

The nucleoside diphosphate sugars are an important group of carbohydrate derivatives that are involved in some of the pathways mentioned earlier in the text, for example, bilirubin diglucuronide biosynthesis and D-galactose metabolism. Several nucleosides have been found in these sugar derivatives, but the most common is uridine (Fig. 7-11).

The nucleoside diphosphate sugars may be considered to be activated forms of the sugars, as they contain the necessary energy to form a glycosidic bond. Since the free energy of formation of a glycosidic bond is positive, the reaction must be coupled to a chemical driving force. The following is an example of the coupled reactions involved in disaccharide biosynthesis.

$$Sugar_1 + ATP \longrightarrow Sugar_1 - P + ADP$$

$$Sugar_1 - P + UTP \longrightarrow UDPSugar_1 + PP_i$$

$$H_2O + PP_i \longrightarrow 2 P_i$$

$$UDPSugar_1 + Sugar_2 \longrightarrow Sugar_1 \underline{\quad} Sugar_2 + UDP$$
$$(disaccharide)$$

$$Sugar_1 + Sugar_2 + ATP + UTP + H_2O \longrightarrow Disaccharide + UDP + ADP + 2 P_i$$

Most glycosidic bonds are synthesized in this manner. The acceptor of the glycosyl residue from the UDP sugar may be a substance to be excreted, for example, bilirubin, a growing polysaccharide chain, or another simpler sugar derivative.

One of the more interesting systems described recently for the biosynthesis

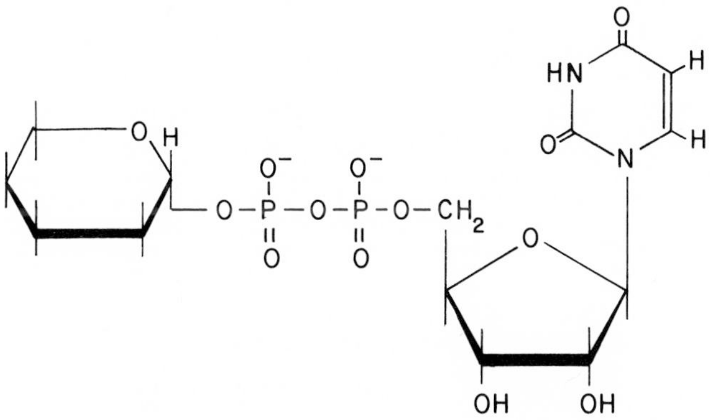

sugar–pyrophosphate–D–ribose–uracil

Fig. 7-11. Uridine diphosphate sugar.

of disaccharides is that of the lactose synthetase system, which is a β-D-galactosyl transferase. The galactosyl residue is transferred from uridine diphosphate galactose to glucose to form lactose (p. 298) but only in the presence of α-lactalbumin.

$$\text{UDP-Galactose} + \text{D-Glucose} \rightarrow \text{Lactose} + \text{UDP}$$

In the absence of α-lactalbumin the β-D-galactosyl residue will transfer to N-acetyl-D-glucosamine groups, either as the free sugar or as nonreducing terminal residues in biopolymers (p. 338) to form the same type of $1 \rightarrow 4$-β-D-galactosyl linkage as in lactose. The catalytic enzyme subunit is found in the mammary gland, liver, and small intestine. The modifying regulatory protein α-lactalbumin is synthesized only by the mammary gland under the hormonal control that is in effect during lactation. Prolactin, the hormone that stimulates milk production, is discussed further in Chapter 13.

CONTROL OF GLYCOGEN METABOLISM

The synthesis (glycogenesis) and breakdown (glycogenolysis) of glycogen occur through different metabolic paths, which are delicately controlled in an interrelated manner to ensure the following:

1. Maintenance of blood sugar levels from the liver and, to a smaller extent, kidney glycogen stores. All other tissues either have no D-glucose 6-phosphatase or possess a physiologically insignificant amount and cannot hydrolyze the intracellular D-glucose 6-phosphate to D-glucose. Without such an enzyme the cells must metabolize all the carbohydrate within the cell, and none can get out to maintain the blood glucose level.

2. Intracellular availability of D-glucose 1-phosphate for glycolysis and ATP production.

3. Relief of a hyperglycemic state by glycogen biosynthesis. There is, however, a limit to the amount of glycogen that can be stored in normal tissues. When this limit is exceeded, the excess glucose is converted to fat, which has no obvious limit as to its storage.

Metabolic interrelationships

Before proceeding, it should be emphasized that except for the isolation of reactions by subcellular compartmentalization, the pathways of biochemistry are primarily conveniences for discussion rather than isolatable metabolic gears in a cellular machine. For example, the D-glucose 6-phosphate present in the cytoplasm is involved by coupled reactions in all the metabolic pathways of D-glucose. A change in this concentration affects all the pathways. This interrelationship is also supported by *normal* human homeostatic controls, which do not call for the simultaneous stimulation of oppositely directed metabolic pathways. Thus the synthesis of liver glycogen does not occur while the liver is responding to hypoglycemia. Similarly, the synthesis of D-glucose from noncarbohydrate sources (gluconeogenesis) is not stimulated during hyperglycemia, when the body is trying to reduce the glucose levels. These concepts are illustrated by the pathways of glycogenolysis and glycogenesis.

Glycogenesis

The breakdown and synthesis of glycogen occur at the nonreducing terminal ends of the branched D-glucose polymer (Fig. 7-2). The chemical changes are shown in Figs. 7-12 and 7-13, in which a two-branched segment of the molecule is used for purposes of illustration. During synthesis the outer chains of glycogen

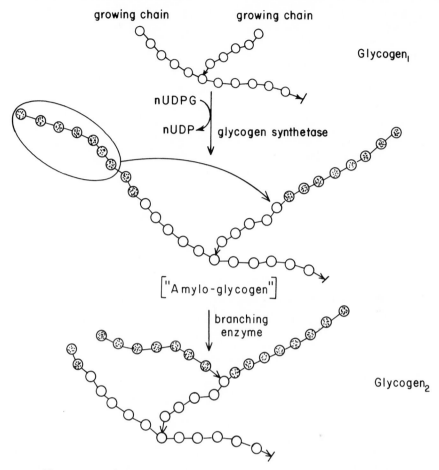

Fig. 7-12. Glycogen synthesis pictorialized. Glycogen₁ and glycogen₂ represent molecules of differing molecular weight.

are elongated by the transfer of D-glucosyl residues from UDPG. The enzyme responsible for this biosynthesis, *glycogen synthetase*, is closely bound to the particulate glycogen stores in the tissue. The reaction is essentially irreversible and results in the synthesis of $(1 \rightarrow 4)$-α-D-glucosidic linkages. Glycogen itself is the most efficient acceptor, but smaller $(1 \rightarrow 4)$-α-D-glucose oligosaccharides can also serve this purpose. As a result of the sequential addition of glucosyl residues to the nonreducing terminal residues, the chains are longer than those in the original acceptor, which is most commonly glycogen. In the absence of any other enzyme an "amylose type" of glycogen would result. However, when any chain has grown more than approximately eleven glucose units from the last branch point, a *branching enzyme* (α-D-glucosyl 4:6 transferase, also called oligo-1,4 $\rightarrow$ 1,6-glucantransferase) transfers segments of approximately seven glucose units in length from the outer chains onto position 6 of an α-D-glucosyl residue in another chain of the glycogen molecule. A $(1 \rightarrow 6)$-α-D linkage is thus formed, and a new branch point is produced. The oligosaccharide segments are transferred in a single step and not by sequential transfer of single α-D-glucosyl units. The reaction is irreversible and produces new chains, all of which can be elongated subsequently by the glycogen synthetase reaction.

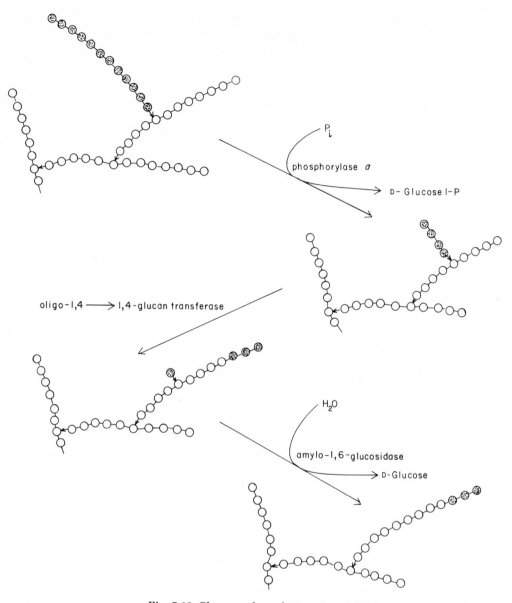

Fig. 7-13. Glycogen degradation pictorialized.

Glycogenolysis

Glycogenolysis occurs by a different pathway. In the presence of phosphorylase *a*, inorganic phosphate (P_i) cleaves the nonreducing terminal glucose residues of glycogen one at a time to produce D-glucose 1-phosphate. The reaction is analogous to hydrolysis in that the elements of phosphate ($H—OPO_3H$), instead of water ($H—OH$), are added to the cleaved glycosidic bond; the reaction is thus described as *phosphorolysis*. The action is terminated near the branch point, three to four D-glucosyl residues away from it on the average. The intermediate product is called a limit dextrin. A second enzyme (oligo-1,4 → 1,4-glucantransferase) transfers a segment of all but one glucosyl unit from the chain stubs that are linked to a glucosyl residue at position 6. The segments are transferred to the nonreducing terminal residues of the other chain stubs. The remaining chains are two to three units longer than those present in the limit dextrin, to which are attached single D-glucosyl residues. These single units, the original branch points in the glycogen, are now hydrolyzed by amylo-1,6-glucosidase to D-glucose. The branch point that was stopping the further action of the phosphorylase *a* has been removed, and phosphorolysis can proceed to the next branch point. This reaction sequence is represented in Fig. 7-13.

The combination of the amylo-1,6-glucosidase and the oligo-1,4 → 1,4-glucan-transferase thus removes the blockage to further action of phosphorylase *a*. Together the transferase and glucosidase enzymes have been referred to as the *debranching enzyme.*

In summary, the synthesis and degradation of glycogen occur by different paths. Depending on the process affecting it, the glycogen molecule becomes smaller or larger, but rarely if ever is it completely degraded; so that even in starved animals the glycogen stores are never completely depleted. A nucleus of glycogen remains to act eventually as an acceptor so that new glycogen can be built up when adequate supplies of carbohydrate are provided. Approximately 85% of the D-glucose produced by glycogen degradation is in the form of the 1-phosphate, while 15% is in the form of free sugar.

Control of glycogen metabolism in different tissues

Nearly every cell is capable of glycogen metabolism. The enzymes of different tissues may vary in terms of cofactor requirements, quaternary structure, and inhibition, but they catalyze their respective reactions similarly.

Muscle

There are two forms of glycogen synthetase in muscle. One form, synthetase D (dependent), requires the presence of D-glucose 6-phosphate. The activity of the other form, synthetase I (independent), is not increased by this sugar phosphate. Synthetase D has a low affinity (high K_m) for UDPG, which increases only slightly in the presence of D-glucose 6-phosphate, although the V_{max} increases by fortyfold to fiftyfold. The I form shows no similar increase in V_{max}.

The I form of glycogen synthetase is converted to the D form by ATP and a *protein kinase* (Fig. 7-14). The conversion reaction results in the phosphorylation of the hydroxy group of an L-seryl residue in the I protein. The reverse reaction, D form to I form, is brought about by a phosphoprotein phosphatase.

A similar interconversion of phosphorylases occurs, phosphorylase *b* being converted to phosphorylase *a* in the presence of ATP and phosphorylase *b* kinase. Phosphorylase *b* is a dimer with pyridoxal phosphate bound to each subunit. Two of the phosphorylase *b* dimers combine to form a tetramer, the *a* form of phosphorylase.

$$2 \text{ Phosphorylase } b + 4\text{ATP} \xrightarrow{\text{Kinase}} \text{phosphorylase } a + 4\text{ADP}$$
$$\text{(inactive)} \qquad\qquad\qquad\qquad \text{(active)}$$

L-Seryl-phosphate residues, introduced by ATP and the kinase, may play some role in the stabilization of the quaternary structure. The pyridoxal phosphate does not appear to have any role in the catalytic function but may be required to maintain the active native state of the protein.

Phosphorylase *a* is converted to the *b* form by a specific phosphatase that removes four phosphate ester groups per mole of *a* form. Phosphorylase *b*, normally an inactive enzyme, becomes active in the presence of AMP, an effector that is not required by phosphorylase *a* for activity.

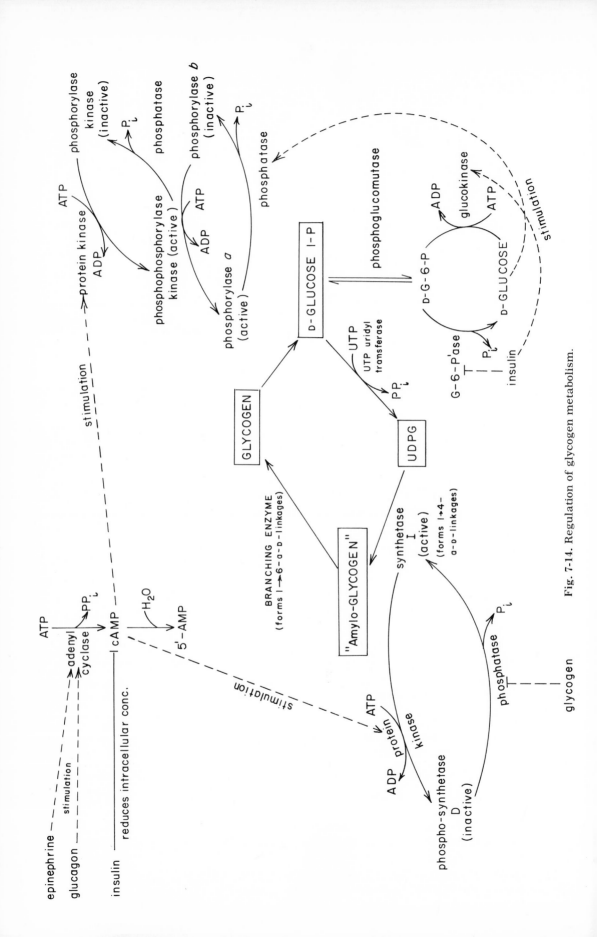

Fig. 7-14. Regulation of glycogen metabolism.

Liver and heart

Liver glycogen synthetase appears to be similar to the corresponding muscle enzyme; it exhibits D and I forms, the former being activated by D-glucose 6-phosphate. Similarly, the interconversion of liver phosphorylase *a* and *b* occurs, but the inactive phosphorylase *b* is not activated by AMP and does not form a tetrameric *a* form when activated.

The heart has three phosphorylase isozymes, one of which is similar in most respects to the form found in skeletal muscle. The two others are AMP independent.

Cyclic AMP and adenyl cyclase

There is an inactive-active pair of enzymes for both the anabolic (glycogen synthetase) and catabolic (phosphorylase) paths in glycogen metabolism. The interconversion of each pair is initiated through the activation of protein kinase by cyclic 3',5'-AMP (cAMP), which is formed from ATP by the enzyme adenyl cyclase, or adenylate cyclase, as shown in Fig. 7-15.

Although a detailed description of cAMP and a consideration of its role in many hormonal mechanisms will be presented in Chapter 13, the general concept of its action must be included here if glycogen metabolism is to be understood. As a result of the appropriate hormonal interactions, adenyl cyclase forms cAMP, which activates protein kinase. This enzyme activates another enzyme that activates yet another, so that the resulting *cascade effect* is multiplicative. Thus if one hormone molecule results in the formation of x molecules of protein kinase, each of which activates y molecules of the next enzyme, etc., the final effect of each mole of hormone is the production of $(x) \times (y) \times \cdots$ moles of enzyme to act on the substrate concerned, for example, phosphorylase *a* on glycogen.

It is necessary to have an effective terminator of such a powerful cascade; this is present in the form of a phosphodiesterase that hydrolyzes the cAMP to 5'-AMP. The synthesis and catabolism of glycogen are regulated exquisitely by intracellular cAMP concentration, which is controlled hormonally (Fig. 7-14).

Added to these hormonal controls are those of enzyme activity, which are exercised either by metabolites, for example, D-glucose 6-phosphate and UDPG, or by cofactors, for example, AMP. The physiologic consequences of these interacting controls can best be illustrated by a consideration of hypoglycemia and hyperglycemia (Fig. 7-14). The hormone and metabolite effects on blood glucose are summarized in Table 7-2.

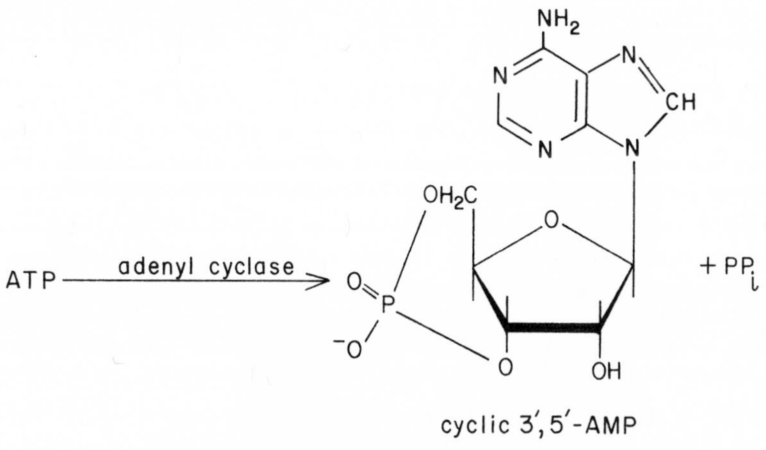

$$ATP \xrightarrow{\text{adenyl cyclase}} \text{cyclic 3',5'-AMP} + PP_i$$

Fig. 7-15. Adenyl cyclase and cyclic AMP.

Table 7-2. Factors causing hypoglycemia or hyperglycemia

HORMONE OR METABOLITE		ENZYME ACTIVITY CHANGED		BLOOD GLUCOSE
Factors causing hypo-glycemia				
Glucagon	↓	Adenyl cyclase	↓	↓
Insulin	↑	Glucose 6-phosphatase	↓	↓
		Glucokinase	↑	↓
cAMP	↓	Protein kinase	↓	↓
Glucose 6-P	↑	Glycogen synthetase D	↑	↓
Factors causing hyper-glycemia				
Glucagon	↑	Adenyl cyclase	↑	↑
Epinephrine	↑	Adenyl cyclase	↑	↑
Insulin	↓	cAMP system	↑	↑
AMP	↑	Phosphorylase *b*	↑	↑
cAMP	↑	Protein kinase	↑	↑

Enzymatic response to hypoglycemia

In response to a hypoglycemia in which the circulating level of insulin is abnormally low, glucagon is secreted from the pancreas and stimulates adenyl cyclase in the liver to produce cAMP, which in turn stimulates protein kinase. This latter enzyme shuts off any synthetase I by converting it to the D form, which is relatively inactive in the absence of D-glucose 6-phosphate. Protein kinase has a cascading effect down the phosphorylase mechanism to bring the active form, phosphorylase *a*, into action as quickly as possible, to give D-glucose 1-phosphate. This product is converted to the 6-phosphate, the intracellular concentration of which does not greatly increase since it is hydrolyzed to D-glucose by the D-glucose 6-phosphatase, present at physiologic concentrations only in liver and kidney.

The muscle and other extrahepatic tissues have a reduced glycogen content resulting from the low blood glucose levels. The levels of ATP may be low and those of AMP may be high. A high level of AMP stimulates phosphorylase *b*, and glycogenolysis can still be effected to provide D-glucose 6-phosphate for glycolysis and ATP production. If concentrations of both glucose 6-phosphate and ATP increase significantly in the cell, they competitively inhibit the AMP stimulation of phosphosphorylase *b*, and glycogenolysis will be controlled at the cofactor and substrate levels. Epinephrine can provide the additional stimulus for glycogenolysis when required.

Enzymatic response to hyperglycemia

During hyperglycemia the concentration of glucagon in the blood is low, but the circulating insulin concentration is elevated. This increases the rate of the transport of glucose across the extrahepatic cell membranes and also stimulates the de novo biosynthesis of glucokinase in the liver. The diurnal variations of insulin concentrations caused by eating probably do not have great or rapid effects on the activity of glucokinase in the liver cells but rather serve to maintain it approximately at maximum. By unknown mechanisms insulin reduces the intracellular concentration of cAMP; consequently, the amounts of phosphorylase *a* and synthetase D are reduced. The rates of formation of the *b* and I forms as a result of the action of the corresponding phosphatases are likewise increased. All tissues are directed toward glycogen synthesis, which demands the production of UDPG from UDP, ATP, and D-glucose (Fig. 7-10). In the liver, UDPG inhibits phosphorylase, whereas high levels of UDP inhibit the synthetase. The regulation of glycogen synthesis is therefore delicately balanced by the levels of UDPG, UDP, ATP, and glycogen itself, which inhibits the synthetase D phosphatase.

Role of D-glucose 6-phosphate and AMP

Muscle contraction requires the continued presence of ATP, which is produced by the oxidative catabolism of D-glucose, lipids, and α-L-amino acids, or is formed from two molecules of ADP, as shown below.

$$2ADP \xrightarrow{\text{adenylate kinase}} ATP + AMP$$

This transphosphorylation reaction not only generates ATP for further muscular contraction but also raises the concentation of AMP, which results in the activation of phosphorylase *b*. The continuous degradation of glycogen is thus possible without hormonal intervention so long as ADP is being formed and violent activity is not demanded.

The use of the α-D-glucose 1-phosphate for glycolysis involves its conversion to the 6-phosphate. If D-glucose 6-phosphate accumulates, glycogen synthetase D will be activated, and the glucose will be laid down again as glycogen. Such an occurrence in the muscle is quite possible, for example, on the occasion of an excessive release of epinephrine that subsequently is not needed. The false alarm would stimulate phosphorolysis of glycogen in the muscle, and the cell could be flooded with the D-glucose phosphates.

Abnormal glycogen metabolism

The confidence with which one can describe the intricate balance of glycogen metabolism in vivo results in large measure from the careful description of inherited diseases of glycogen metabolism by clinicians. Although diseases of glycogen metabolism are rare, examples involving each enzyme in the pathways

Table 7-3.   Glycogen storage diseases*

NAME	TYPE	PRINCIPAL TISSUE AFFECTED	ENZYME DEFICIENCY
von Gierke's disease	I	Liver and kidney	Glucose 6-phosphatase
Pompe's disease	II	Liver, heart, and muscle	1,4-α-D-Glucosidase (lysosomal)
Limit dextrinosis	III	Liver and muscle	Amylo-1,6-glucosidase or oligo-1,4 → 1,4-glucan-transferase
Amylopectinosis	IV	Liver	"Branching" enzyme
McArdle's disease	V	Muscle	Phosphorylase
Hers' disease	VI	Liver	Phosphorylase
	VII	Muscle	Phosphofructokinase
	VIII	Liver	Phosphorylase kinase
	IX†	Liver	Glycogen synthetase

*The disease entities listed here are those for which enzyme deficiencies have been identified.
†A deficiency in glycogen results.

have been studied. The structure of the glycogen has been described in each case, as have the enzyme levels in the affected tissues. The biochemical aspects of these glycogen storage diseases are summarized in Table 7-3, and an example is described in case 5.

GLYCOLYSIS

Glycolysis is the degradative pathway whereby D-glucose is oxidized to pyruvate or lactate. It is linked to glycogen metabolism through D-glucose 6-phosphate. Glycolysis is concerned with the following:

1. Generating ATP, in the course of which the glucose molecule is partially oxidized

2. Producing pyruvate

3. Forming intermediate compounds for other biochemical processes, for example, glycerol 3-phosphate for triglyceride and phospholipid biosynthesis, 2,3-diphosphoglycerate in erythrocytes, pyruvate for L-alanine biosynthesis, etc.

Glycolysis can proceed either under aerobic conditions, when the supply of oxygen is adequate to maintain the necessary levels of NAD^+, or under anaerobic (hypoxic) conditions, when the level of NAD^+ cannot be maintained by way of the mitochondrial cytochrome system and is dependent on the temporary expedient

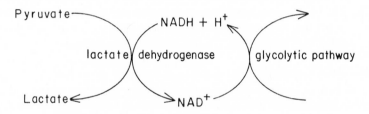

Fig. 7-16. Regeneration of NAD^+ by pyruvate.

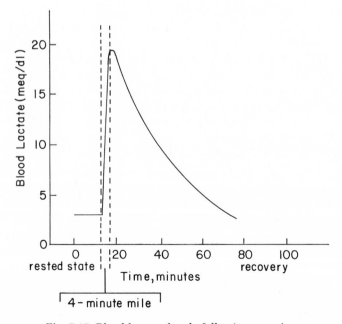

Fig. 7-17. Blood lactate levels following exercise.

of converting pyruvate to lactate (Fig. 7-16). Anaerobic glycolysis, the temporary reliance on pyruvate, is the means by which the body awaits the return of adequate oxygen. Therefore, the condition is called an *oxygen debt.*

As was noted in Chapters 1, 4, and 5, the maintenance of certain oxygen and carbon dioxide levels in cells is essential to their normal function. Abnormal situations do arise, however, when a stress is placed on the body. Such a stress might be a high-energy demand, for example, violent exercise or the hyperventilation of encephalitis (case 3, Chapter 4), when the rate at which oxygen is brought to the cell is not sufficient to keep pace with the oxidative catabolic reactions producing ATP. Since these oxidation reactions are linked to oxygen through $NAD^+/NADH$ and the cytochrome system and since they cannot proceed unless $NADH + H^+$ is converted to NAD^+, an emergency step involving pyruvate is called into play. This results in the conversion of pyruvate to lactate. When the lactate level in the blood increases, the pH falls, and the expected signs of rapid breathing and exhaustion set in. Variations in blood lactate levels accompanying changes in physical activity are illustrated in Fig. 7-17. The lactate that is produced and released into the blood is converted back to pyruvate in the liver when oxygen is made available.

Pathway of glycolysis

The pathway of glycolysis is summarized in Fig. 7-18. All the reactions leading to the formation of pyruvate and lactate are catalyzed by enzymes present in the cytoplasm. (The various enzymatic reactions in glycolysis have been considered previously, pp. 307 to 311.) The phosphorylation of the D-glucose that has been absorbed by the cell from the interstitial fluid or produced by the debranching enzyme system in glycogenolysis forms D-glucose 6-phosphate. D-Glucose 6-phosphate is also produced from D-glucose 1-phosphate. The cellular concentration of the 6-phosphate acts as a product inhibitor of hexokinase and activates glycogen synthetase D. Therefore, mechanisms are present that limit the entry of unnecessary glucose into glycolysis.

Following the isomerization of D-glucose 6-phosphate into D-fructose 6-phosphate, a second phosphorylation occurs to produce D-fructose 1,6-diphosphate. At this point the sugar is irreversibly committed to glycolysis. It is reasonable to assume that this irreversible step is carefully controlled. A large cellular energy charge (p. 215) and a high concentration of citrate in the cytoplasm, which increase respiratory chain activity and hence ATP production, represent negative allosteric effectors for phosphofructokinase (PFK). Glycolysis will not proceed until the levels of ATP and citrate are lowered, that is, until ADP and AMP are proportionately increased.

The D-fructose 1,6-diphosphate is split by aldolase to form the two triose phosphates, which are interconverted by triose phosphate isomerase. At equilibrium this reaction favors dihydroxyacetone phosphate to approximately 95%. However, in glycolysis the D-glyceraldehyde 3-phosphate is continuously phosphorylated oxidatively to 1,3-diphospho-D-glycerate, and the equilibrium is probably never established.

The oxidation of 3-phospho-D-glyceraldehyde is another example of conservation of the energy derived from chemical oxidations by conversion of ADP to ATP. The reaction is represented in simple terms by Fig. 7-19, in which an —SH group of the enzyme is shown to react with the aldehyde group to form a thiohemiacetal. The NAD^+ coenzyme that is part of the enzyme complex oxidizes the thiohemiacetal to a reactive thioester, which undergoes phosphorolysis with inorganic

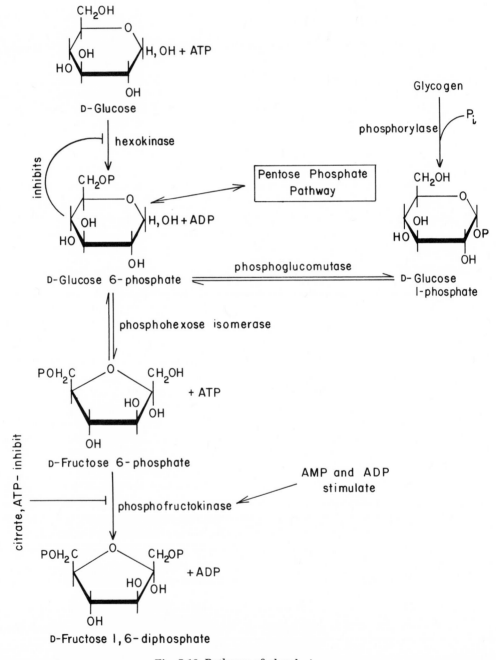

Fig. 7-18. Pathway of glycolysis.

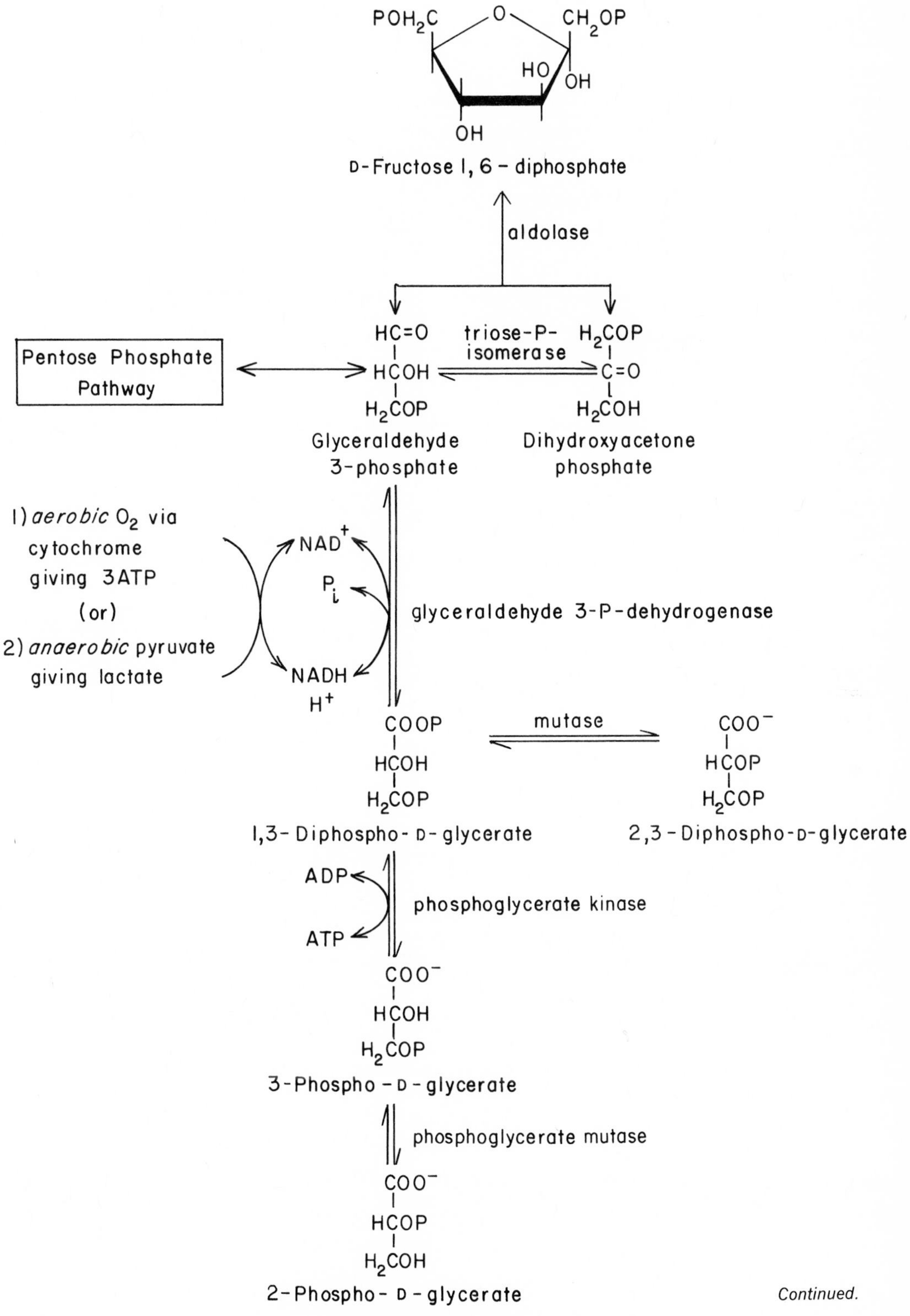

POH$_2$C   O   CH$_2$OP

HO OH

OH

D-Fructose I, 6 – diphosphate

aldolase

HC=O   triose-P-   H$_2$COP

           isomerase

| Pentose Phosphate Pathway |

HCOH   C=O

H$_2$COP   H$_2$COH

Glyceraldehyde   Dihydroxyacetone
3-phosphate   phosphate

1) *aerobic* O$_2$ via
   cytochrome
   giving 3ATP
      (or)
2) *anaerobic* pyruvate
   giving lactate

NAD$^+$

P$_i$

glyceraldehyde 3-P-dehydrogenase

NADH
H$^+$

COOP   mutase   COO$^-$

HCOH   HCOP

H$_2$COP   H$_2$COP

I,3-Diphospho-D-glycerate   2,3-Diphospho-D-glycerate

ADP

phosphoglycerate kinase

ATP

COO$^-$

HCOH

H$_2$COP

3-Phospho-D-glycerate

phosphoglycerate mutase

COO$^-$

HCOP

H$_2$COH

2-Phospho-D-glycerate

Continued.

Fig. 7-18, cont'd. Pathway of glycolysis.

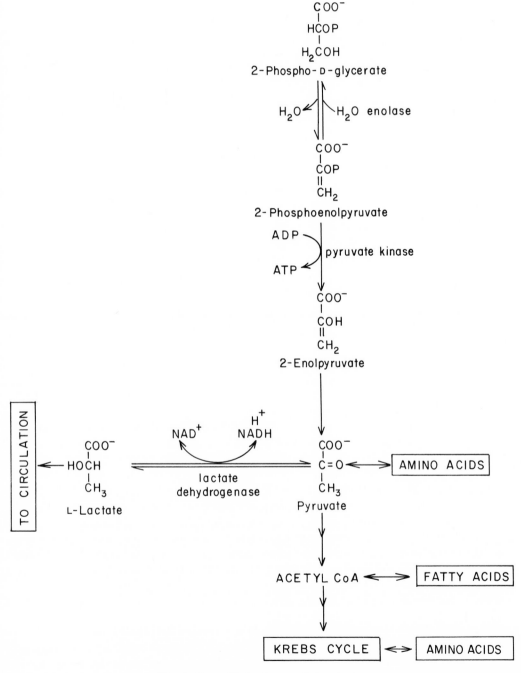

Fig. 7-18, cont'd. Pathway of glycolysis.

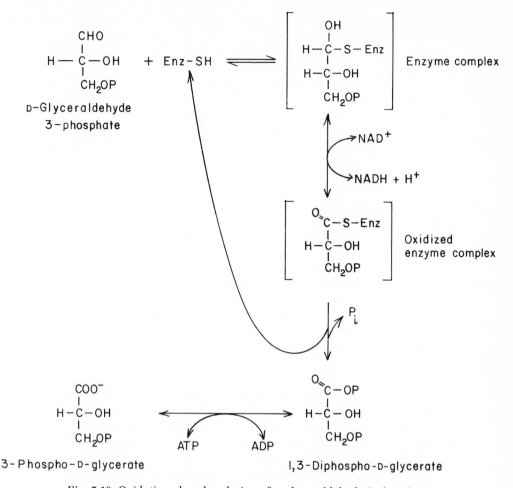

Fig. 7-19. Oxidative phosphorylation of D-glyceraldehyde 3-phosphate.

phosphate to liberate the enzyme and form a reactive 1,3-diphospho-D-glycerate ($\Delta G^{o\prime}$ for hydrolysis is approximately -12 kcal/mole). The phosphate ester can be transferred to ADP to form ATP.

Alternatively, in the erythrocyte the 1,3-diphospho-D-glycerate may be isomerized to the 2,3-diphosphate, a molecule that plays an important role in the transport of oxygen by hemoglobin (p. 81).

Glycolysis proceeds as phosphoglycerate mutase catalyzes the formation of 2-phospho-D-glycerate, which by the elimination of water gives 2-phosphoenolpyruvate. 2-Phosphoenolpyruvate is another molecule with a high negative free energy of hydrolysis ($\Delta G^{o\prime}$ is approximately -15 kcal/mole). It reacts with ADP to form ATP and pyruvate. Several metabolic routes are available to pyruvate (Chapter 6).

Regulation and energetics

Several characteristics of the glycolytic pathway should be noted.

1. Some of the reactions are reversible and are also found in the pathway of gluconeogenesis. The physiologically irreversible reactions are those for which metabolic controls have been established; for example, as was seen previously,

hexokinase is inhibited by D-glucose 6-phosphate. PFK is allosterically inhibited by ATP and citrate and stimulated by AMP, ADP, or P_i.

2. One effect of the ATP inhibition of PFK is that the cell will not activate glycolysis to produce *excess* ATP.

3. The $\Delta G^{0\prime}$ for the oxidation of one mole of D-glucose to two moles of pyruvate is approximately -120 kcal. Part of this is conserved as ATP through the mitochondrial oxidation of NADH and the substrate-level phosphorylation. The steps involving ATP may be summarized as follows:

AEROBIC GLYCOLYSIS	ATP/MOLE D-GLUCOSE
1. Phosphorylation of D-glucose	-1
2. Phosphorylation of D-fructose 6-P	-1
3. 2(1,3-diP-Glycerate → 3-P-glycerate)	$+2$
4. 2(Phosphoenolpyruvate → pyruvate)	$+2$
5. 2(NADH + H$^+$ → NAD$^+$)	$+6$
	$+8$

The $\Delta G^{0\prime}$ of hydrolysis of the eight moles of ATP is equal to approximately -56 kcal. Therefore, of the energy available (-120 kcal/mole D-glucose), slightly less than 50% is conserved as ATP for subsequent use. It should be noted, however, that much of the energy of the D-glucose remains in the two moles of pyruvate, since the total oxidation of D-glucose is associated with a $\Delta G^{0\prime}$ of -686 kcal/mole. As was demonstrated in Chapters 5 and 6, the metabolism of one mole of pyruvate to acetyl CoA yields three moles of ATP, and the oxidation of acetyl CoA in the Krebs cycle and respiratory chain adds another twelve moles of ATP. Therefore, the two moles of pyruvate produced from one mole of D-glucose corresponds to thirty moles of ATP, giving a total of thirty-eight moles of ATP/one mole of D-glucose metabolized completely to $CO_2 + H_2O$. Thus the overall energy conservation as ATP is about 40%. Your car should be so efficient!

4. During periods of anaerobic glycolysis the ATP from mitochondrial oxidative phosphorylation is not available. Consequently, the net gain in ATP per mole of D-glucose converted to two moles of L-lactate is reduced to two moles. Although this is low in comparison to aerobic glycolysis, in times of stress it is better than nothing and may suffice. If the stress is not too prolonged, the total loss of calories is not serious. However, prolonged anaerobic glycolysis results in loss of lactate in the urine, which results in a wasting of calories.

5. The interplay among the glycolytic and other pathways is extensive. Depending on the total metabolic state, other pathways may pass intermediates into a segment of the glycolytic pathway for conversion to pyruvate or glucose. The reverse may also occur; for example, while other tissues can phosphorylate glycerol, dihydroxyacetone phosphate from glycolysis is the only source of glycerol 3-phosphate in adipose tissue (Chapter 8).

GLUCONEOGENESIS

D-Glucose is essential to the proper function of most cells; it is an absolute necessity for the nervous system and the erythrocytes. In the event that adequate amounts of D-glucose are not provided in the diet, the blood glucose level falls and the body responds by synthesizing D-glucose from noncarbohydrate precursors. This process is called gluconeogenesis.

The carbon source for gluconeogenesis is a number of glucogenic precursors that are derived principally from L-amino acids (Fig. 7-20). How the precursors are converted to 2-phosphoenolpyruvate (PEP) was discussed in Chapter 6 and

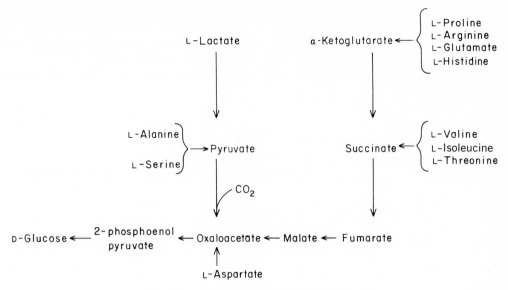

Fig. 7-20. Carbon sources of gluconeogenesis.

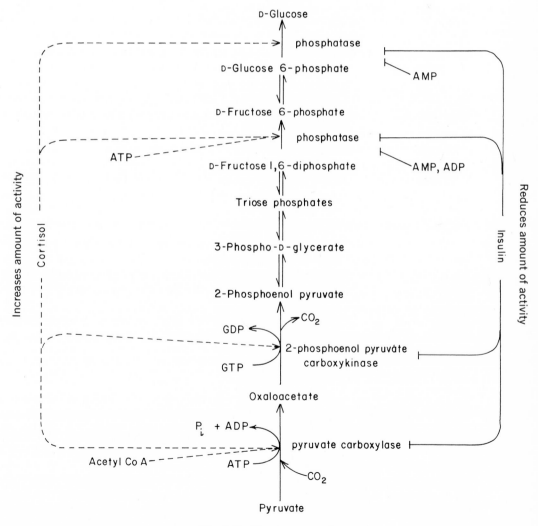

Fig. 7-21. Gluconeogenesis.

illustrated in Fig. 6-16. The conversion of amino acids to pyruvate and to other Krebs cycle intermediates will be further discussed in Chapter 9.

The formation of PEP from pyruvate in the gluconeogenic pathway is not a simple reversal of the corresponding reaction in glycolysis. Two other steps are mediated by enzymes that differ from those in glycolysis. One is the hydrolysis of D-fructose 1,6-diphosphate to D-fructose 6-phosphate, catalyzed by fructose-1,6-diphosphate 1-phosphatase. The other has been discussed previously and involves the enzyme glucose 6-phosphatase. The pathway is summarized in Fig. 7-21, which makes no attempt to show the subcellular compartmentalization of the reactions (see Fig. 6-16).

As gluconeogenesis provides glucose during periods when circulating blood glucose levels are low, it is not surprising to find that this pathway occurs principally in the liver and kidney. It does not occur to any physiologic extent in muscle, which has no glucose 6-phosphatase and cannot support blood glucose levels.

Regulation of gluconeogenesis

Pyruvate carboxylase is found in the mitochondria of liver and kidney, whereas the pyruvate kinase occurs in the cytoplasm. Without such a compartmentalization of these enzymes, an energy-demanding and futile cycling of pyruvate might occur. As shown in Figs. 7-21 and 6-16, mitochondrial pyruvate carboxylase catalyzes the carboxylation of pyruvate to oxaloacetate, which is transported to the cytoplasm via malate and then converted to PEP. These steps involve the use of one mole each of ATP and GTP and two moles of nucleotide coenzymes. The PEP may then be available for reconversion to pyruvate by pyruvate kinase, with the subsequent regeneration of ATP. However, this cycle is pointless, as there is no need to effect the hydrolysis of one mole of GTP.

Gluconeogenesis is regulated primarily by four key enzymes: pyruvate carboxylase, PEP-carboxykinase, D-fructose 1,6-diphosphate 1-phosphatase, and D-glucose 6-phosphatase. The first enzyme is stimulated by acetyl CoA (an absolute requirement) and inhibited by ADP. The 1-phosphatase is strongly inhibited by AMP and ADP, while the D-glucose 6-phosphatase is subject to product inhibition by P_i and D-glucose.

At a slower rate of control the biosynthesis of the four key enzymes is influenced by insulin and the glucocorticoid hormone cortisol. Insulin represses the synthesis of all four enzymes, while cortisol induces their de novo synthesis. These effects seem reasonable when the physiologic conditions that necessitate gluconeogenesis are considered. If for some reason there is an inadequate supply of glucose, for example, starvation, high-fat diet, or hormonal imbalance, glucose must be provided by amino acid metabolism from protein catabolism, which is also stimulated by cortisol. Most of the energy for gluconeogenesis is probably derived from fatty acid oxidation, the fatty acids having been produced by lipolysis of triglycerides, which is also potentiated by cortisol. There is some evidence that fatty acids have a stimulating effect on gluconeogenesis, but further research is needed to confirm this in vivo. As the glucogenic amino acids enter the gluconeogenesis pathway (Fig. 7-20), the Krebs cycle is involved to a significant extent. Malate, which is produced from oxaloacetate in the mitochondria using $NADH,H^+$ in the Krebs cycle, is translocated from the mitochondria. After moving into the cytoplasm, malate is oxidized back to oxaloacetate with $NADP^+$ in a reaction that is catalyzed by another malate dehydrogenase. Therefore, the ratios of $NAD^+/NADH,H^+$, $NADP^+/NADPH,H^+$, and ATP/ADP as well as the concentrations of Krebs cycle acids and acetyl CoA all play a role in the control of gluconeo-genesis. Fatty acid oxidation in the mitochondria requires NAD^+ and results in increased $NADH,H^+$ and acetyl CoA. Furthermore, the levels of Krebs cycle acids are increased by amino acid transamination. All these processes favor gluconeo-genesis.

HORMONAL INTERACTION TO CONTROL GLUCOSE METABOLISM

High concentrations of insulin in the blood act to suppress the release of glucagon. On the other hand, low concentrations of insulin in the blood are ac-companied by high glucagon concentrations, which is particularly relevant in conditions of insulin insufficiency, such as diabetes mellitus, severe burns, and hemorrhagic shock. Thus in the absence of a normal secretion of insulin in response to hyperglycemia, the balance of the insulin and glucagon hormones is lost. The blood concentration of glucagon is also reduced by somatostatin, a hor-mone that inhibits the release of growth hormone. So it is that in patients with high circulating concentrations of growth hormone, a condition found in acro-megaly, treatment with somatostatin concomitantly reduces the levels of glucagon and blood glucose. Taken together with the stimulation of gluconeogenesis by cortisol, the interplay of hormones in the control of glucose metabolism is of considerable importance in the understanding of metabolism and the treatment of disease.

PENTOSE PHOSPHATE PATHWAY (HEXOSE MONOPHOSPHATE SHUNT)

An important branch point in carbohydrate metabolism is the oxidation of D-glucose 6-phosphate by the pentose phosphate pathway (PPP). This pathway is the source of D-ribose for nucleotide synthesis and of NADPH, the biochemical reducing agent for many anabolic reactions.

The pathway is composed of two parts. The first involves the oxidation of the reducing carbon of D-glucose 6-phosphate to CO_2 by two equivalents of $NADP^+$ to give D-ribulose 5-phosphate and two equivalents of $NADPH + H^+$ (Fig. 7-22).

D-Glucose 6-phosphate dehydrogenase is widely distributed in tissues and under physiologic conditions uses $NADP^+$ as the coenzyme. The oxidation to the 6-phospho-D-gluconolactone is followed by its irreversible hydrolysis to 6-phospho-D-gluconate, which is oxidized and decarboxylated to give D-ribulose 5-phosphate. The 6-phospho-D-gluconate dehydrogenase also requires $NADP^+$, and the reaction proceeds through the intermediate formation of a 3-keto-D-gluconate 6-phosphate; a dephosphorylated isomer, 3-keto-L-gulonate, is important in another carbohy-drate pathway to be discussed on p. 337.

The second part of the pathway entails the reorganization of the pentose phosphate product back to D-glucose 6-phosphate. This step is difficult to demon-strate unless the PPP is assumed to commence with six moles of hexose phos-phate, as shown below.

For part 1:   6 Hexose P + $12NADP^+$ + $6H_2O$ → 6 pentose P + 12NADPH + $12H^+$ + $6CO_2$

For part 2:   4 Pentose P $\xrightarrow{\text{transketolase}}$ 2 heptose P + 2 triose P

2 Heptose P + 2 triose P $\xrightarrow{\text{transaldolase}}$ 2 hexose P + 2 tetrose P

2 Pentose P + 2 tetrose P $\xrightarrow{\text{transketolase}}$ 2 hexose P + 2 triose P

2 Triose P $\xrightarrow[\text{+ (phosphatase)}]{\text{aldolase}}$ hexose P + P_i

Sum of parts 1 and 2:   Hexose P + $12NADP^+$ + $7H_2O$ → $6CO_2$ + 12NADPH + $12H^+$ + P_i

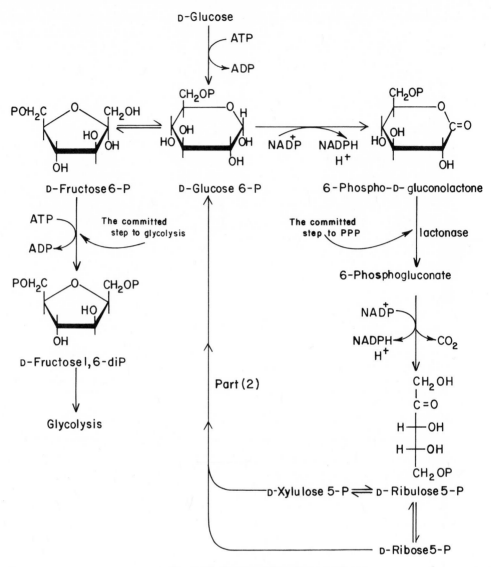

Fig. 7-22. Pentose phosphate pathway.

Part 2 is presented to show that the rearrangement of pentose to hexose is possible and that the triose phosphates are common substrates in the glycolytic and gluconeogenic pathways. The two enzymes that have yet to be discussed, transketolase and transaldolase, function by transferring a two- or three-carbon fragment from a "donor" molecule to an "acceptor." Aldolase, discussed previously, has an analogous action.

Transketolase

Transketolase transfers a —CO · CH$_2$OH group according to the reaction:

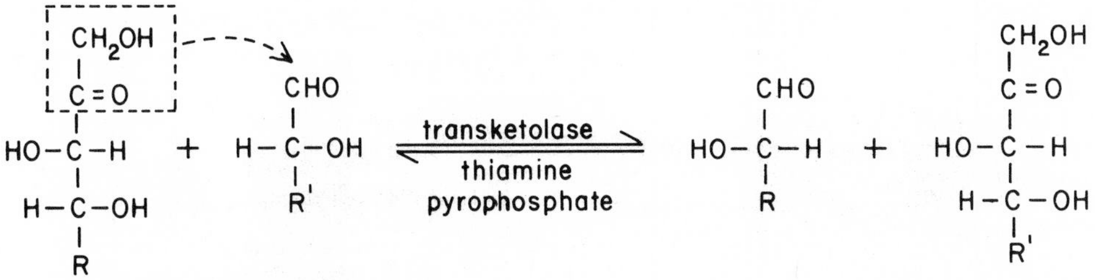

```
┌─────────┐
│ CH₂OH   │ ----→
│ |       │         CHO                                          CHO          CH₂OH
│ C=O     │         |                                            |            |
└─────────┘         |            transketolase                   |            C=O
  |          +   H—C—OH    ⇌ ─────────────→     HO—C—H    +   HO—C—H
HO—C—H              |            thiamine            |            |
  |                 R'         pyrophosphate          R            H—C—OH
H—C—OH                                                             |
  |                                                                R'
  R

   Donor            Acceptor
```

The thiamine pyrophosphate functions in the transfer in much the same manner as in pyruvate decarboxylase (Chapter 6), but in this case an intermediate α,β-dihydroxyethyl thiamine pyrophosphate is formed. The reaction is reversible. The donor molecules have an L configuration at C$_3$, and the hydroxy groups on positions 3 and 4 are in a D-*threo* configuration: that is, they are *trans*. Thus, as shown below, D-xylulose 5-phosphate donates the two-carbon fragment to D-ribose 5-phosphate to form a seven-carbon sugar (D-sedoheptulose 7-phosphate) as well as D-glyceraldehyde 3-phosphate.

```
┌─────────┐                                          CH₂OH
│ CH₂OH   │ ---→                                      |
│ |       │        CHO                                C=O
│ C=O     │        |                                  |                 CHO
└─────────┘        H—C—OH                           HO—C—H               |
  |         +      |                ⇌                 |            +    H—C—OH
HO—C—H             H—C—OH                           H—C—OH              |
  |                |                                  |                 CH₂O P
H—C—OH             H—C—OH                           H—C—OH
  |                |                                  |
CH₂O P            CH₂O P                            H—C—OH
                                                      |
                                                    CH₂O P
```

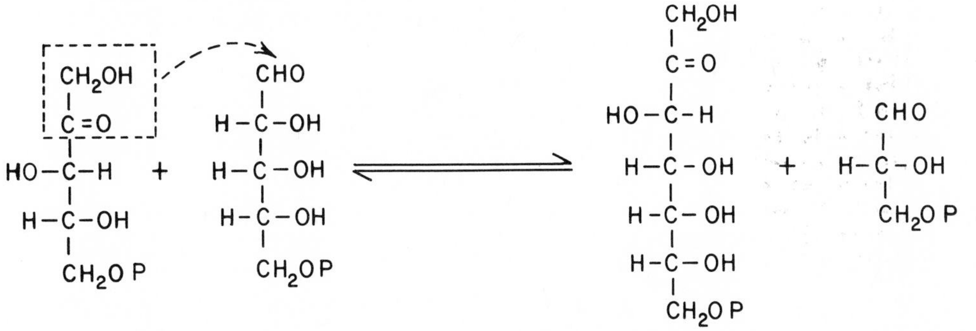

D-Xylulose 5-P   D-Ribose 5-P   D-Sedoheptulose 7-P   Glyceraldehyde 3-P

Transaldolase

Transaldolase catalyzes the reaction between D-sedoheptulose 7-phosphate and D-glyceraldehyde 3-phosphate to produce D-fructose 6-phosphate and a tetrose (D-erythrose 4-phosphate). The transfer, shown below, involves a three-carbon fragment but does not have any identified coenzyme or cofactor.

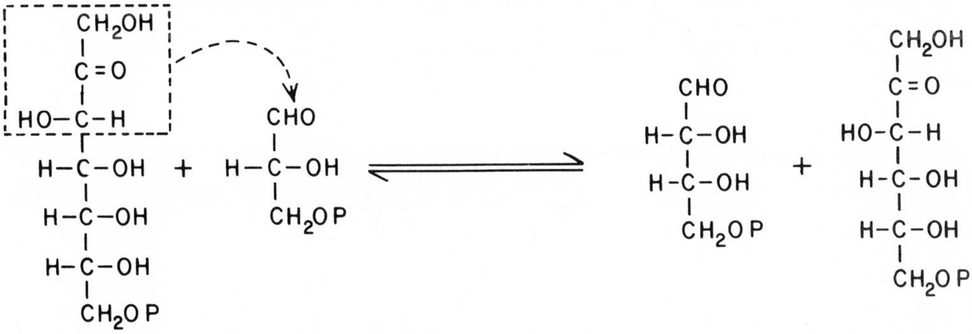

```
┌─────────┐
┊ CH₂OH   ┊ ---→
┊ |       ┊
┊ C=O     ┊         CHO                       CHO             CH₂OH
┊ |       ┊         |                         |               |
┊ HO—C—H  ┊         H—C—OH                     H—C—OH          C=O
└─────────┘   +     |             ⇌           |         +   HO—C—H
  |                 CH₂O P                     H—C—OH          |
H—C—OH                                         |               H—C—OH
  |                                            CH₂O P          |
H—C—OH                                                         H—C—OH
  |                                                            |
H—C—OH                                                         CH₂O P
  |
CH₂O P
```

D-Sedoheptulose 7-P   D-Glyceraldehyde 3-P   D-Erythrose 4-P   D-Fructose 6-P

The combination of these reactions, as shown in Fig. 7-22, results in the oxidation of D-glucose to CO_2 and H_2O, with the formation of twelve moles of NADPH + H^+/one mole of hexose. If these reduced nucleotides could be oxidized with NAD^+ to form $NADP^+$ and NADH, they would be equivalent to (12×3) or thirty-six moles of ATP by reoxidation of NADH in the repiratory chain. Thus the conservation of energy in the PPP is similar to that in glycolysis and the Krebs cycle.

Regulation of the pentose phosphate pathway

Several points pertaining to PPP regulation should be made.

1. The PPP is of minimal importance in muscle but is significant in erythrocytes, liver, adipose tissue, and kidney.

2. PPP reactions occur in the cytoplasm.

3. PPP is a major source of NADPH in the cytoplasm, the other sources being NADP-linked malate and isocitrate dehydrogenases.

4. In each pathway there is an irreversible reaction that, once passed, commits the molecule to its fate. In the PPP it is the lactonase-catalyzed step; in glycolysis it is the formation of D-fructose 1,6-diphosphate.

5. The regulation of the PPP occurs mainly in the dehydrogenation of D-glucose 6-phosphate. The enzyme concerned, D-glucose 6-phosphate dehydrogenase, is increased in amount when the diet contains excess carbohydrate. A tenfold increase is noted in liver when the subject passes from a starved state to one in which excess dietary carbohydrate is being converted to fatty acid. This coarse regulation coexists with a finer control of enzyme activity by NADPH acting as a competitive inhibitor, $K_i = 7$ μM. The inhibition is over 90% when the ratio of the concentrations of NADPH to $NADP^+$ is above 10. The inhibition is reversed by oxidized glutathione (GSSG), which acts specifically and not by oxidizing NADPH to $NADP^+$. These recent observations by Krebs go far to explain why the D-glucose 6-phosphate dehydrogenase is active in liver cells where the concentration of NADPH is often 100 times that of $NADP^+$. The cellular concentration of D-glucose 6-phosphate is approximately 1 mM, so that it places little constraint on the dehydrogenase activity, for which its K_m is in the micromolar range.

6. Depending on the needs of the cell, the conversion of pentose phosphate to hexose phosphate (part 2 of the pathway) may not be obligatory. The reactions may favor the formation of glycerol-3-phosphate for biosynthesis of phosphoglycerides or triglycerides or the conversion of triose phosphates to 2,3-diphospho-D-glycerate in the erythrocyte where the PPP is very significant.

The functioning of the pentose phosphate pathway in the red blood cell is important for the maintenance of its viability. The normal erythrocyte has a life span of around 120 to 135 days in the circulation. When mature, the erythrocyte is unable to synthesize protein and is devoid of mitochondria. It requires energy for the synthesis of simple compounds, for example, glutathione, coenzymes, and ATP, and this energy is derived from the metabolism of glucose. The maintenance of ATP levels is achieved by anaerobic glycolysis with the formation of lactic acid, since oxidative phosphorylation through the respiratory chain is not possible. The formation of NADPH by the pentose phosphate pathway provides for the reduction of methemoglobin, the Fe^{3+} form of hemoglobin that does not bind oxygen but is continuously being formed, and for the maintenance of glutathione in its reduced form.

Glutathione functions as a reducing agent to maintain other molecules in the reduced form, for example, enzymes with an essential —SH group at their active site. As a result, the glutathione is oxidized to G—S—S—G, from which the reduced form, GSH, is regenerated by NADPH. This reaction is catalyzed by *glutathione reductase.*

$$G—S—S—G + NADPH + H^+ \rightarrow 2GSH + NADP^+$$

The $NADP^+$ is converted back to NADPH in the pentose phosphate pathway.

It is therefore clear that the supply of ATP and the maintenance of the proper redox state of the erythrocyte is dependent on the catabolism of glucose. The interrelationship of these pathways may be summarized as shown below.

The normal erythrocyte hexokinase level is low compared to the other enzymes, so that the initial step, which is common to both pathways, is usually rate limiting. The relative rate of the two pathways is influenced by blood pH, [NADP]/[NADPH]

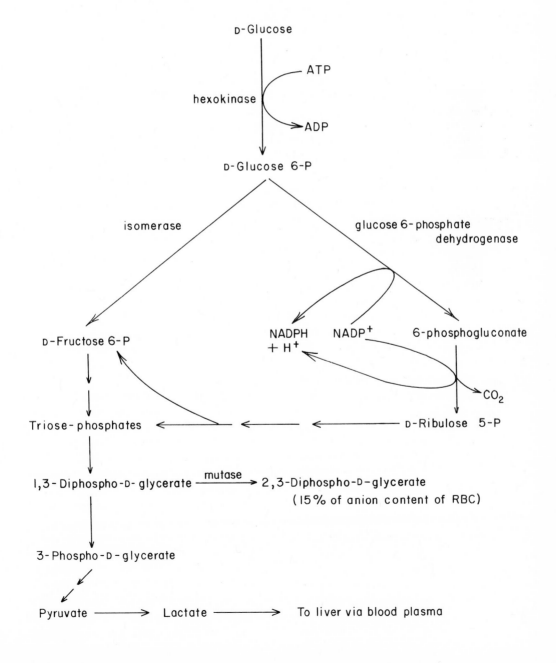

ratio, glucose 6-phosphate dehydrogenase activity, and the rate of GSH oxidation.

Crossover between these pathways is achieved through the recycling of the pentose phosphates, involving transaldolase and transketolase, to form the fructose phosphates and glyceraldehyde-3-phosphate. It is conceivable, therefore, that the pentose phosphate and glycolytic pathways run in series to generate NADPH and ATP. The substrate phosphorylations would occur in glycolysis after the formation of triose phosphate. Such a serial sequence also permits the maintenance of 2,3-diphospho-D-glycerate, which is important in oxygen transport (Chapters 2 and 4).

GSH deficiency in red blood cells results in serious impairment of essential metabolic processes and produces hemolysis. Probably less than 1% of the glutathione is normally present as G—S—S—G. It follows therefore that the pentose phosphate pathway must function efficiently, requiring an adequate level of enzymes, including glucose 6-phosphate dehydrogenase. The level of this enzyme is normally reduced in older erythrocytes. Several types of inherited deficiency of this enzyme have been recognized, all of which result in sensitivity to certain drugs that induce hemolysis. The breakthrough in the understanding of this clinical condition came with the in-depth study of the hemolytic effects of the 8-aminoquinoline antimalarial agent primaquine. Subsequently, many more drugs have been found to act like primaquine. The administration of these drugs to a glucose 6-phosphate dehydrogenase–deficient patient results in a rapid fall in the average GSH level. Since the deficiency is greatest in the older cells, it is cells in this group that hemolyze, which explains the higher percentage of circulating young cells that are found in the blood of such patients. This is also consistent with the shorter half-life of these patients' cells and a larger percentage of reticulocytes (immature erythrocytes).

D-GLUCURONATE AND POLYOL PATHWAYS

In studying intermediary metabolism it is most important to think in terms of the total organism and homeostasis. Even though a very small segment of metabolism in a particular tissue and cell may be under discussion, it should not be forgotten that there is a dynamic equilibrium in each cell that is constantly affected by many other aspects of metabolism. Thus after a segment of intermediary metabolism is understood in detail, it should be considered in relation to the other parts.

It is difficult to identify many of the metabolites of D-glucose catabolism with any one pathway; this problem is illustrated by a series of reactions that are particularly important in the understanding of pentosurias and in the metabolism of vitamin C.

D-Glucuronate, formed from D-glucose by way of UDPG, is metabolized by reduction of the aldehyde group to form L-gulonate. It is this hexonic acid that undergoes lactonization and oxidation to form ascorbic acid in many mammals. Man, however, is deficient in the L-gulonolactone oxidase and thus is absolutely dependent on diet to supply this vitamin. The metabolic fate of ascorbic acid depends greatly on the species, and it is not possible to extrapolate animal data to humans. Man does not have the lactonase enzyme, unique to the guinea pig, that permits ascorbic acid and dehydroascorbic acid to enter the catabolic process by way of 2,3-diketo-L-gulonate. In man, little if any ascorbic acid is catabolized to CO_2, but there is evidence that some (less than 10%) of the ascorbic acid is cleaved to oxalate and a four-carbon intermediate. When an excess of ascorbic acid is ingested, there are mechanisms that prevent its accumulation in the body.

First, there is a limited intestinal absorption of the vitamin, and second, an efficient urinary excretion of any excess occurs.

The alternate pathway of L-gulonate is oxidation to 3-keto-L-gulonate, which undergoes decarboxylation to give L-xylulose in much the same manner as was seen in the PPP for D-ribulose, except that the coenzyme is NAD^+. The reactions to this point are shown below, with the original C_1 of the D-glucuronate identified by an asterisk to clarify the interconversions of the sugars.

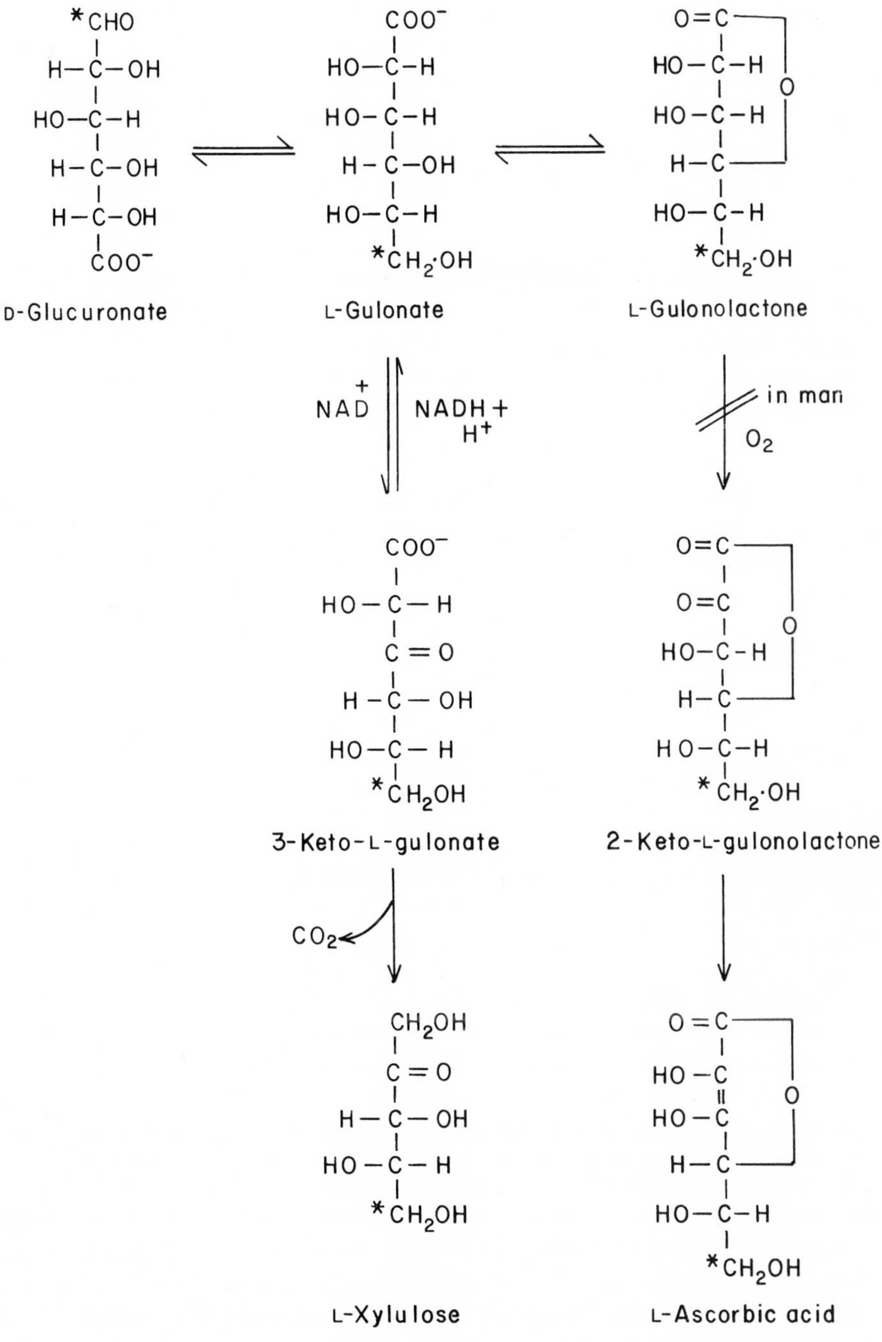

D-Glucuronate

L-Gulonate

L-Gulonolactone

3-Keto-L-gulonate

2-Keto-L-gulonolactone

L-Xylulose

L-Ascorbic acid

Through the action of two stereospecific xylitol dehydrogenases, L-xylulose is converted to D-xylulose through the intermediate pentitol xylitol, which is a symmetric molecule.

```
     CH2OH                      CH2OH                      CH2OH
      |                          |                          |
      C=O       NADPH         HO—C—H        NAD+            C=O
      |        ⇌               |           ⇌               |
   H—C—OH       NADP+       H—C—OH          NADH         HO—C—H
      |                          |                          |
   HO—C—H                     HO—C—H                     H—C—OH
      |                          |                          |
     CH2OH                      CH2OH                      CH2OH

   L-Xylulose                  Xylitol                   D-Xylulose
```

Each dehydrogenase has a different coenzyme, and that which is linked to $NADP^+/NADPH$ is absent or deficient in pentosuria. As a result, L-xylulose is excreted in the urine. This genetic defect does not appear to involve any serious physiologic consequences. However, the test for reducing sugars in the urine should be checked to differentiate the excretory material from D-glucose.

D-Xylulose can be phosphorylated with ATP and a kinase to D-xylulose 5-phosphate, which enters the PPP.

A reaction analogous to that for the xyluloses is the interconversion of D-glucose and D-fructose by way of D-glucitol (sorbitol).

```
   *CHO                       *CH2OH                     *CH2OH
    |                          |                          |
  H—C—OH                     H—C—OH                       C=O
    |                          |                          |
 HO—C—H        NADPH        HO—C—H        NAD+          HO—C—H
    |          ⇌               |          ⇌               |
  H—C—OH        NADP+        H—C—OH        NADH         H—C—OH
    |                          |                          |
  H—C—OH                     H—C—OH                     H—C—OH
    |                          |                          |
   CH2OH                      CH2OH                      CH2OH

  D-Glucose                  D-Glucitol                 D-Fructose
```

This polyol pathway may be the source of D-fructose in cerebrospinal and seminal fluid. The pathway has been described in many tissues, including those of the peripheral nerves, and it may be important in diabetics where episodes of hyperglycemia may be common (see case 2, p. 354, for further discussion).

BIOPOLYMERS CONTAINING CARBOHYDRATE

Aside from glycogen and blood glucose, carbohydrate is present in the body as a part of the glycolipids and glycoproteins. In the latter, carbohydrate may be present as oligosaccharide groups attached to a protein core, for example, fibrinogen, the immunoglobulins, the blood group substances, and the mucous secretions. Some glycoproteins are composed of polysaccharide-protein complexes, for example, the mucopolysaccharides, the chondroitin sulfates, dermatan sulfate, and keratan sulfate. The larger carbohydrate groups are present in molecules that have more of a structural function. By contrast, the function of the glycoproteins that contain oligosaccharides is related to molecule or cell recognition. For example, the immunochemistry of the blood group substances involves only a few sugar residues at the nonreducing termini of the oligosaccharide groups. Thus blood group B substance is converted to blood group H substance simply by removing its terminal residues of D-galactose. Examples of glycoproteins are given in Table 7-4.

In abbreviated structural representations of complex carbohydrates, the suffix A (for acid) is added to the symbol for the monosaccharide to represent the uronic acids, for example, D-GA (or Glc A) is D-glucuronic acid, and L-IdoA is L-iduronic acid. The N-acetyl amino sugars have NAc added as a suffix to the monosaccharide symbol; for example, D-Gal NAc is N-acetyl-D-galactosamine. Similarly, S is added for sulfate. In arrays that represent the structures of the polysaccharides, the positional numerals and anomeric prefixes are added at the appro-

Table 7-4. Distribution and function of some glycoproteins

PRESUMED FUNCTION	NAME
Structural	Collagen
	Bacterial cell wall peptidoglycans
	Proteoglycans (mucopolysaccharides)
Food reserve	Plant pollen allergens
	κ-Casein
Enzyme	Ribonuclease B
	Prothrombin
Transport	Ceruloplasmin
	Transferrin
Hormone	Thyroglobulin
	Erythropoietin
Plasma and body fluids	α-Acid glycoprotein
	Fibrinogen
Immune systems	γ-Globulins
	Blood-group substances

Table 7-5. Structure of mucopolysaccharides

MUCOPOLYSACCHARIDE	REPEATING UNIT	OCCURRENCE
Hyaluronic acid	→ 4)-β-D-GA-(1 → 3)-β-D-GlcNAc-(1 →	Vitreous tissue, joint fluids, skin, umbilical cord
Chondroitin 4-sulfate	→ 4)-β-D-GA-(1 → 3)-β-D-GalNAc-(1 → 4-sulfate	Cartilage, skin, bone
Chondroitin 6-sulfate	→ 4)-β-D-GA-(1 → 3)-β-D-GalNAc-(1 → 6-sulfate	Cartilage, nucleus pulposus, skin
Heparin	→ 4)-α-L-IdoA*-(1 → 4)-α-D-GlcNSO$_4$-(1 → 2-sulfate 3,6-disulfate	Lung, spleen, liver, muscle
Keratan sulfate	→ 3)-β-D-Gal-(1 → 4)-β-D-GlcNAc-(1 → 6-sulfate	Cornea, nucleus pulposus, cartilage
Dermatan sulfate†	→ 4)-α-L-IdoA*-(1 → 3)-β-D-GalNAc-(1 → 4-sulfate	Skin, lung

*D-Glucuronic acid (unsulfated) also present.
†Older nomenclature; dermatan sulfate is also known as chondroitin sulfate B and β-heparin.

priate places. The numbers are separated by arrows pointing from the glycosyl group to the hydroxyl group where the glycosidic linkage is made. These are illustrated in Table 7-5, Fig. 7-24, and Fig. 7-25.

Glycoproteins

The covalent linkage between the carbohydrate groups and the polypeptide chain in glycoproteins may be of two types: (1) through the amide group of L-asparaginyl residues, for example, ribonuclease B, and (2) to the hydroxyl group of L-seryl (submaxillary mucin), L-threonyl, hydroxy-L-lysyl (collagen), or hydroxy-L-prolyl (plant cell wall glycoprotein) residues. In the first type the linkage is most frequently made to an N-acetyl-D-glucosamine unit, whereas those of the second type are more varied; N-acetyl-D-galactosamine, D-galactose, and D-xylose are common linkage units. These linkages are shown in Fig. 7-23. The other carbohydrate units are glycosidically linked as oligosaccharides or polysaccharides, which are attached to the sugar in the linkage areas as shown in Fig. 7-23.

The number of carbohydrate groups attached to each protein varies; for example, only one is attached in ribonuclease B, several are attached in the immunoglobulins, and many are attached in the epithelial mucins and the blood group substances. The proportion of carbohydrate in glycoproteins varies widely. Collagen usually contains less than 1% carbohydrate; ribonuclease B, 8%; fetuin, 20%; serum α-acid glycoprotein, around 38%; submaxillary mucins, around 50%; blood group substances, more than 80%; and the mucopolysaccharides, approaching 100%.

An important and ubiquitous family of sugars found in glycoproteins and glycolipids are the sialic acids. These are derivatives of neuraminic acid where R may be an acetyl (CH_3CO-) or glycolyl ($HOCH_2CO-$) group and some of the hydroxyl groups may also be acetylated. The biosynthesis of neuraminic acid is catalyzed by an aldolase that condenses N-acetyl-D-mannosamine 6-phosphate with phosphoenolpyruvate, giving the 2-keto sugar acid.

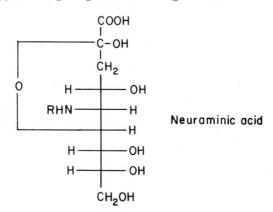

Neuraminic acid

N-Acetyl-D-mannosamine + ATP $\xrightarrow{\text{kinase}}$ N-acetyl-D-mannosamine 6-phosphate + ADP

N-Acetyl-D-mannosamine 6-phosphate + phosphoenolpyruvate $\xrightarrow{\text{aldolase}}$
N-acetyl neuraminic acid 9-phosphate + P_i

N-Acetyl neuraminic acid 9-phosphate $\xrightarrow{\text{phosphatase}}$ N-acetyl neuraminic acid + P_i

Through the formation of the "activated" cytidine monophosphate N-acetyl neuraminic acid (CMP—NANA), the sialic acid residues are transferred to

L-Asparaginyl-N-acetyl-
β-D-glucosaminyl linkage

L-Seryl-β-D-xylosyl linkage

Fig. 7-23. Carbohydrate-protein linkages in glycoproteins.

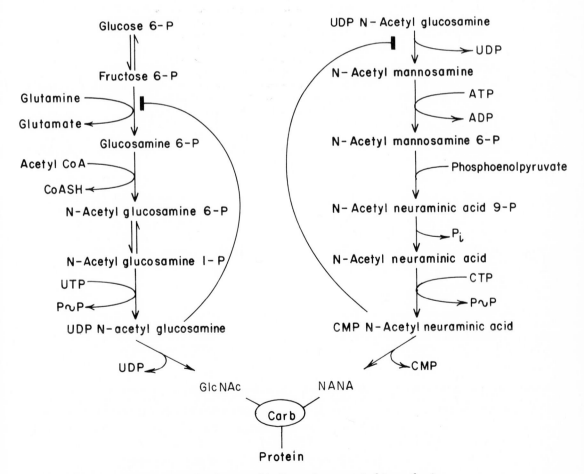

Fig. 7-24. Feedback control in liver glycoprotein biosynthesis.

terminal sugar residues, such as galactose in the biopolymers and may serve as terminators to any further glycosyl-chain growth.

$$\text{N-Acetyl neuraminic acid} + \text{CTP} \xrightarrow{\text{transferase}} \text{CMP—NANA} + \text{P} \sim \text{P}$$
(NANA)

$$\text{Protein} - (\text{Sugar})_x\text{-Gal} + \text{CMP—NANA} \xrightarrow{\text{sialyltransferase}} \text{Protein} - (\text{Sugar})_x\text{-Gal-NANA} + \text{CMP}$$

The sialic acid residues in carbohydrate groups of biopolymers are hydrolyzed in vivo by neuraminidase.

$$\text{Protein} - (\text{Sugar})_x\text{-Gal-NANA} + \text{H}_2\text{O} \xrightarrow{\text{neuraminidase}} \text{Protein} - (\text{Sugar})_x\text{-Gal} + \text{NANA}$$

The turnover of plasma glycoproteins may well be controlled by their desialylation. Thus a serum glycoprotein such as ceruloplasmin is rapidly removed from the circulation by the reticuloendothelial system after the terminal galactosyl residues in the carbohydrate groups have been exposed by hydrolysis of the blocking NANA residues with neuraminidase.

The biosynthesis of the carbohydrate group is preceded by the biosynthesis of the polypeptide chain on the ribosome, which is followed by the sequential transglycosylation of each sugar residue. Except for CMP-NANA (see above), donors of the sugar units are the nucleotide diphosphate sugars. The first transglycosylase must be specific for the amino acid residue to which the sugar will be linked. Each transglycosylase is specific for the sugar residue that is transferred. To varying degrees each enzyme is also specific for the sugar residue that acts as the acceptor. Although there is no template for the synthesis, the specificities of the transglycosylases can properly order the sugar residues. Additional control is provided in the biosyntheses of the nucleotide phosphate sugars. CMP-NANA inhibits the formation of N-acetyl-D-mannosamine from UDPGlcNAc, which inhibits the formation of D-glucosamine 6-phosphate from D-fructose 6-phosphate. This is illustrated in Fig. 7-24 where the nucleotide phosphate sugars are shown transferring the glycosyl group to the growing carbohydrate group on the protein.

An alternate mechanism of biosynthesis involves a carbohydrate-lipid intermediate, which is probably a dolichol-diphosphate oligosaccharide. The oligosaccharide is then transferred to a protein acceptor. It is possible that the sequence of reactions shown at the top of the opposite page occurs. The interplay of the two pathways for the glycosylations of proteins is complicated, and it is possible that dolichol intermediate may be important in the glycosylation of proteins in a hydrophobic membrane.

Glycoproteins in membranes

For many years the functions of several glycoproteins and particularly the carbohydrate groups were an enigma. It is now increasingly clear that the carbohydrate alters the physical properties of some glycoproteins, for example, the viscosity of the mucins. It is known that the glycoproteins are important, biologically active components of cell membranes. As will be discussed in greater detail in Chapter 10, the glycoproteins of cell membranes may be considered to be embedded in a lipid bilayer. The glycoprotein molecule is exposed on one or both surfaces of the membrane, and in all examples studied the carbohydrate is present on the exterior surface. One of the best studied examples is that of glycophorin, one of several glycoproteins in the human erythrocyte membrane. It comprises about 5% of the total protein of the membrane and itself consists of about 40%

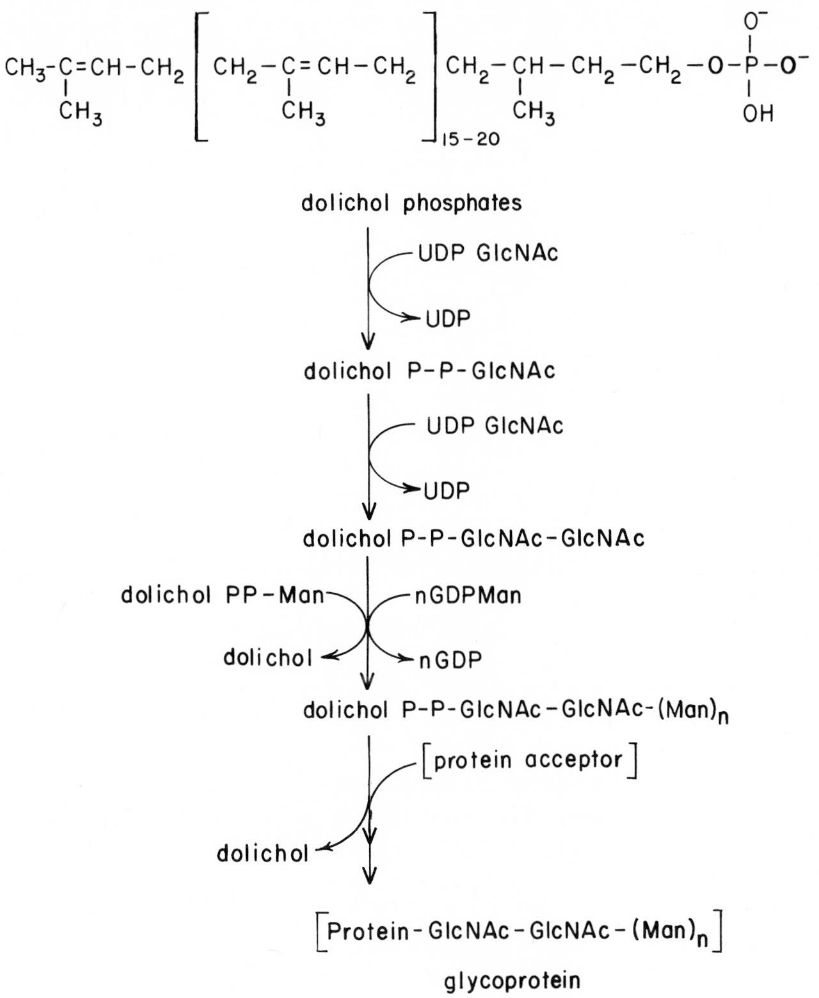

$$CH_3-\underset{\underset{CH_3}{|}}{C}=CH-CH_2\left[CH_2-\underset{\underset{CH_3}{|}}{C}=CH-CH_2\right]_{15-20}CH_2-\underset{\underset{CH_3}{|}}{CH}-CH_2-CH_2-O-\underset{\underset{OH}{|}}{\overset{\overset{O^-}{|}}{P}}-O^-$$

dolichol phosphates

UDP GlcNAc

UDP

dolichol P-P-GlcNAc

UDP GlcNAc

UDP

dolichol P-P-GlcNAc-GlcNAc

dolichol PP-Man — nGDPMan

dolichol — nGDP

dolichol P-P-GlcNAc-GlcNAc-(Man)$_n$

[protein acceptor]

dolichol

[Protein-GlcNAc-GlcNAc-(Man)$_n$]

glycoprotein

protein and 60% carbohydrate; the latter includes 25% N-acetyl neuraminic acid. The molecule extends through the membrane, the N-terminal segment in the external environment carrying all the carbohydrate in sixteen oligosaccharide groups, fifteen of which are linked through the hydroxyl group of seryl or threonyl residues and one of which is in β-asparaginyl linkage. The C-terminal segment of glycophorin appears to be exposed to the cytoplasm of the erythrocyte and contains a high proportion of hydrophilic amino acids. The segment of the protein within the membrane lipid bilayer is hydrophobic in composition.

The structures of the oligosaccharides in glycophorin are similar to those identified to carry blood group determinants (see below). Glycophorin is representative therefore of those glycoproteins that have important roles in membrane biology, the carbohydrate groups being identified with receptors for some peptide hormones and viruses, blood group antibodies, some other immunoglobulins, and phytohemagglutinins. The accessibility to the external environment of carbohydrate groups in normal cell membranes enables them to react with specific plant proteins, called lectins, such as concanavalin A. This ability to react is increased in the transformed, malignant cell, the meaning of which is the subject of much research at present. Similarly, the distribution of the surface glycoproteins on the cell plasma membrane changes in the various stages of cell division for reasons that are not presently understood.

The glycolipids (see Chapter 10) that have ABH and Le blood group activities are minor constituents of the erythrocyte membrane. However, as expected, the antigenic determinant groups are the same as those noted in the glycoprotein blood group substances.

The second human blood group system, the MN system, relates to the antigenic glycoprotein, glycophorin, found only in the erythrocyte membrane. As noted for the A, B, O(H) system, a common precursor may be proposed.

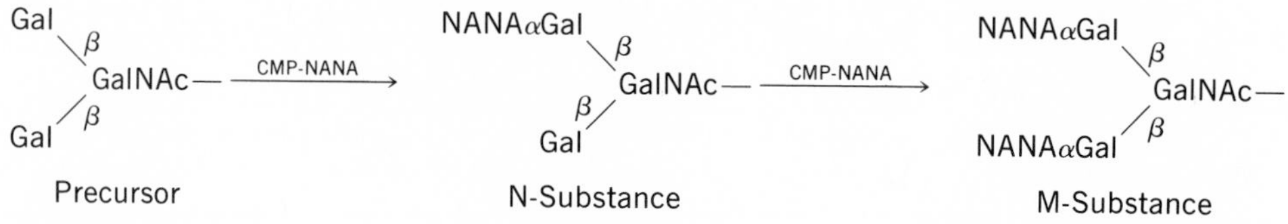

$$
\begin{array}{ccc}
\text{Gal} & \text{NANA}\alpha\text{Gal} & \text{NANA}\alpha\text{Gal} \\
\searrow\beta & \searrow\beta & \searrow\beta \\
\quad\text{GalNAc}- \xrightarrow{\text{CMP-NANA}} & \quad\text{GalNAc}- \xrightarrow{\text{CMP-NANA}} & \quad\text{GalNAc}- \\
\nearrow\beta & \nearrow\beta & \nearrow\beta \\
\text{Gal} & \text{Gal} & \text{NANA}\alpha\text{Gal} \\
\text{Precursor} & \text{N-Substance} & \text{M-Substance}
\end{array}
$$

The M-antigen can be converted to N-substance by mild hydrolysis using neuraminidase.

Blood group substances

Human erythrocyte membranes contain antigenic substances, over 100 of which can be classified into approximately fifteen blood groups. The A, B, O(H), and Lewis (Le) systems are known to involve glycoproteins. Antigenic substances present in saliva, gastric mucin, cystic fluids, and other body secretions can also be characterized by their blood group properties. Thus the sera or secretions of an individual who belongs to group A would agglutinate the erythrocytes of B and AB types. Carbohydrate accounts for 80% to 90% of these molecules, and

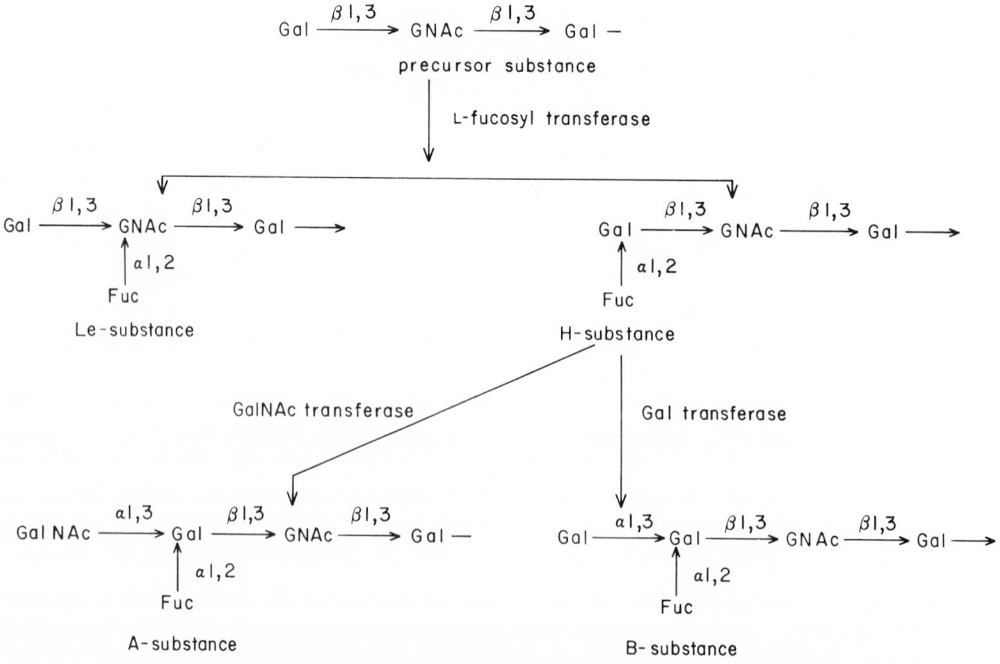

Fig. 7-25. Possible pathways for biosynthesis of blood group substances.

blood group specificity is determined by the sugar residues close to the non-reducing termini. The existence of precursor structures that have similar antigenic determinants to those in the polysaccharide from the cell wall of a type XIV pneumococcus has been proposed. The pathways of biosynthesis follow the lines indicated in Fig. 7-25. Similarly, a series of transglycosylations to a Gal $\xrightarrow{\beta 1,4}$ GNAc $\xrightarrow{\beta 1,3}$ Gal precursor has been postulated to provide a parallel family of molecules with the same blood group activities. Since every individual maintains the same inherited blood group during life, it is clear that the transglycosylations cannot be random; otherwise, the blood type of each person would not be constant.

Proteoglycans (mucopolysaccharides)

Proteoglycans (mucopolysaccharides) occur in many tissues and fluids. The proteoglycans of the ground substances and tissues, for example, cartilage, bone, cornea, and synovial fluid, are polysaccharides that are linked to protein through a xylosyl serine linkage (Fig. 7-23). The polysaccharide portion of the proteoglycans may be classified into one of six types (Table 7-5). The simplest is *hyaluronic acid*, present in synovial fluid of joints, in vitreous humor of the eye, and on cell membranes. The structure is a repeating unit of a disaccharide, composed of D-glucuronic acid and N-acetyl-D-glucosamine.

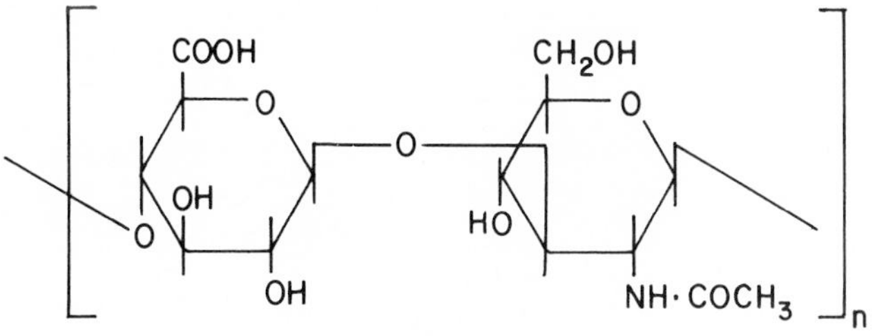

Hyaluronic acid

$$\left[\rightarrow 4\right)-\beta\text{-}D\text{-}GA\text{-}(1\rightarrow 3)-\beta\text{-}D\text{-}GlcNAc\text{-}(1\rightarrow\right]$$

Hydrolysis of hyaluronic acid with hyaluronidases of mammalian origin gives rise to a tetrasaccharide consisting of two of the repeating units. A more complex reaction occurs when bacterial hyaluronidase is used, giving rise to an unsaturated disaccharide in which a double bond is formed between positions 4 and 5 of the glucuronyl residue of the repeating unit. A similar unsaturated disaccharide is formed from the mucopolysaccharides chondroitin 4-sulfate, chondroitin 6-sulfate, and dermatan sulfate, except that N-acetyl-D-galactosamine replaces the N-acetyl-D-glucosamine.

The *chondroitin sulfates* are the most abundant mucopolysaccharides in the body. The repeating sugar unit is the same for each of them (see Table 7-5), but the sulfate is either esterified to the 4-hydroxyl or the 6-hydroxyl of the N-acetyl-D-galactosamine residue. Both polysaccharides are hydrolyzed to their corresponding tetrasaccharide units by mammalian hyaluronidase.

Keratan sulfate always accompanies the chondroitin sulfate in adult mammalian cartilage. It is composed of repeating units of $\rightarrow$ 3)-β-D-Gal-(1 $\rightarrow$ 4)-β-D-GlcNAc-(1 $\rightarrow$ with a sulfate ester at position 6 of the hexosamine residue as well as to varying degrees on the galactose.

Heparin and dermatan sulfate both contain L-iduronic acid residues. D-Glucuronic acid is a minor component of dermatan sulfate but is more significant (about 50%) in heparin. Heparin is not present in connective tissue but can be isolated from lung where it occurs with dermatan sulfate and a less well-defined mucopolysaccharide called *heparan sulfate.* Heparan sulfate contains more N-acetyl groups and thus less N-sulfate groups than heparin, and it also has less O-sulfate. Unlike heparin, heparan sulfate has little if any blood anticoagulant activity. It would appear that heparin and heparan sulfate are at opposite ends of a spectrum of a family of mucopolysaccharides.

The biosynthesis of the mucopolysaccharides may be illustrated (Fig. 7-26) using chondroitin 4-sulfate as an example. The polysaccharide is built on a D-xylose unit that is attached to the protein through the hydroxyl side chain of an L-seryl residue. Two D-galactosyl units are then added. Following this, D-glucuronosyl residues and N-acetyl-D-galactosaminyl residues are added alternately. Finally, the carbohydrate chain is sulfated in certain positions by 3'-phospho-adenosine 5'-phosphosulfate, a molecule performing a function similar to that of ATP in phosphorylation. Depending on the type of sulfate transferase, the

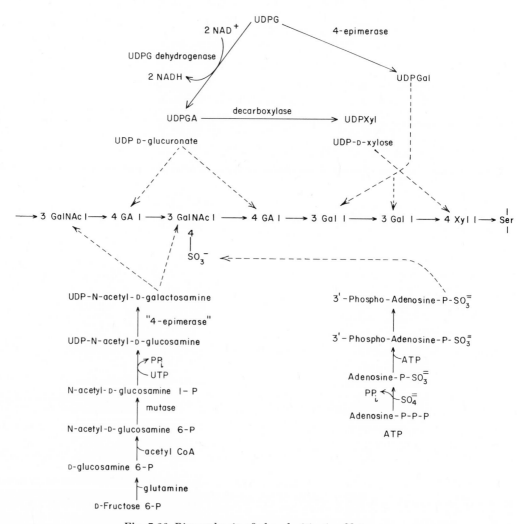

Fig. 7-26. Biosynthesis of chondroitin 4-sulfate.

sulfate ester may be located in various positions of the chondroitin sulfate. The most common species are the 4- and the 6-chondroitin sulfates, in which the sulfate group is at either position 4 or 6. Every glycoprotein, each with its particular carbohydrate groups, is synthesized in this general manner. The generation of the nucleoside diphosphate sugar precursors for the biosynthesis of chondroitin 4-sulfate is shown in Fig. 7-26. The dashed lines indicate the points of insertion of the appropriate residues as chain elongation proceeds. In every case except hyaluronic acid the linkage of the oligosaccharide to the protein is similar to that shown for chondroitin 4-sulfate (Fig. 7-26), and the biosyntheses are presumed to be similar.

Although some of the proteoglycans have a long biologic half-life, the normal turnover requires the proteolysis of the protein core, which may involve the lysosomal cathepsin enzymes. The mucopolysaccharide chains are hydrolyzed to smaller oligosaccharides by enzymes such as hyaluronidase, followed by the sequential removal of terminal sugars by glycosidases; for example, α-L-iduronidase hydrolyzes terminal α-L-iduronic acid residues.

Abnormalities of mucopolysaccharide metabolism

It should be reemphasized at this point that the carbohydrate groups in the biopolymer molecules are synthesized by the sequential addition of single mono-

Table 7-6.   Mucopolysaccharidoses

SYNDROME	TYPE	CLINICAL CHARACTERISTICS	BIOCHEMICAL FINDINGS	GENETICS
Hurler's	I	Severe mental retardation, skeletal deformities, marked corneal opacity, marked somatic changes	Dermatan sulfate* and heparan sulfate† in urine and tissues, dermatan sulfate in fibroblasts	Autosomal recessive
Hunter's	II	Moderate mental retardation, marked skeletal deformities, no corneal clouding, early deafness, marked somatic changes	Dermatan sulfate and heparan sulfate in urine and tissues, dermatan sulfate in fibroblasts	X-linked recessive
Sanfilippo's (two types exist)	III	Severe mental retardation, mild skeletal changes, corneal clouding questionable	Heparan sulfate in urine and tissues, dermatan sulfate in fibroblasts	Autosomal recessive
Morquio's	IV	Severe skeletal deformities, marked spondylepiphyseal dysplasia, no mental retardation, corneal opacities may occur	Keratan sulfate and chondroitin sulfates in urine	Autosomal recessive
Scheie's	V	Mild skeletal changes, no mental retardation, severe corneal opacity	Dermatan sulfate in urine, α-L-iduronidase deficiency	Autosomal recessive
Maroteaux-Lamy	VI	Severe skeletal deformities, gross corneal opacity, no mental retardation	Dermatan sulfate in urine	Autosomal recessive

*Dermatan sulfate is also known as chondroitin sulfate B and β-heparin.
†Heparan sulfate is also known as heparitin sulfate, heparin monosulfate, and N-acetyl heparin sulfate.

saccharide units from "activated" derivatives. In the case of glycogen, UDPG is the activated sugar donor. The transfers are catalyzed enzymically, each type of sugar unit having its particular transglycosylase. The absence of any one of these enzymes results at the very least in a faulty biopolymer. If the defect is compatible with life, the individual will have a particular genetic disease, for example, a glycogen storage disease or a mucopolysaccharidosis.

Fortunately, these genetic diseases are relatively rare. The mucopolysaccharidoses are summarized in Table 7-6, which also notes the tissues that are affected by deposits of particular mucopolysaccharides. The deposits are often present in skin fibroblasts, which can be grown in tissue culture. By this method a deficiency of α-L-iduronidase was found in Hunter's and Scheie's syndromes; the addition of this enzyme to the tissue culture media caused the catabolism of the dermatan and heparan sulfate to return to normal, and the fibroblasts grew normally. Similarly, corrective factors in the form of deficient hydrolases were identified for the other mucopolysaccharidoses: Hunter's, iduronate sulfatase; Sanfilippo's, heparan N-sulfatase or N-acetyl-α-D-glucosaminidase; Maroteaux-Lamy, N-acetyl-D-galactosamine 4-sulfatase; and Morquio's, N-acetyl-D-galactosamine 6-sulfatase. These enzymes, when added to the culture fluid, are presumably taken up into the lysosomes, which contain the stored, excessive amounts of the corresponding mucopolysaccharides. Normal catabolism can then ensue. While such findings suggest the potential value of enzyme therapy, many technical problems have precluded its use.

REFERENCES

Ashwell, G., and Morell, A. G.: The role of surface carbohydrates in the hepatic recognition and transport of circulating glycoproteins, Adv. Enzymol. 41:99, 1974.

Dickens, F., Randle, P. J., and Whelan, W. J., editors: Carbohydrate metabolism and its disorders, vols 1 and 2, New York, 1968, Academic Press, Inc. *A comprehensive statement of the material on carbohydrates with a good discussion of the disease states.*

Elbrink, J., and Bihler, I.: Membrane transport: its relation to cellular metabolic rates, Science 188:1177, 1975.

Fritz, I. B., editor: Insulin action, New York, 1972, Academic Press, Inc. *A modern treatise on the many roles proposed and proved for insulin.*

Ginsburg, V.: Enzymatic basis for blood groups in man, Adv. Enzymol. 36:131, 1972.

Holmes, R.: Carbohydrate digestion and absorption, J. Clin. Pathol. 24(suppl.):5, 1971. *Short review of the subject.*

Hsu, A. F., Baynes, J. W., and Heath, E. C.: The role of a dolichololigosaccharide as an intermediate in glycoprotein biosynthesis, Proc. Natl. Acad. Sci. U.S.A. 71:2391, 1974.

Larner, J.: Intermediary metabolism and its regulation, Englewood Cliffs, N.J., 1971, Prentice-Hall, Inc. *An authoritative statement on the regulation of carbohydrate metabolism and its integration with other pathways.*

Marchesi, V. T., et al.: Molecular features of the major glycoprotein of the human erythrocyte membrane, Fed. Proc. 32:1833, 1973.

Neufeld, E. F.: Inherited disorders of lysosomal metabolism, Annu. Rev. Biochem. 44:357, 1975.

Owen, O. E., and Reichard, G. A., Jr.: Fuels consumed by man: the interplay between carbohydrates and fatty acids, Prog. Biochem. Pharmacol. 6:177, 1971. *A review article that will best be appreciated after the lipids have been studied.*

Pastan, I.: Cyclic AMP, Sci. Am. 227:97, 1972. *A recent statement in simple terms of the ubiquity of this compound.*

Robinson, S.: Physiology of muscular exercise. In Mountcastle, V. B., editor: Medical physiology, ed. 13, vol. 2, St. Louis, 1974, The C. V. Mosby Co. *A good discussion of the relationship of carbohydrates to work demands.*

Schachter, H., and Roden, L.: The biosynthesis of animal glycoproteins: metabolic conjugation and hydrolysis (editor, W. H. Fishman), vol. 3, New York, 1973, Academic Press, Inc.

Stanbury, J. B., Wyngaarden, J. B., and Fredrickson, D. S., editors: The metabolic basis of inherited disease, ed. 3, New York, 1972, McGraw-Hill Book Co. *A classic reference to inborn errors of metabolism.*

Clinical examples

CASE 1: HYPOGLYCEMIA

A girl, X. Y., now age $5\frac{1}{2}$ years, had been admitted to her community hospital at 6 months of age because of "pneumonitis." Fever, dyspnea, severe metabolic acidosis, and hepatomegaly were noted. She responded to therapy with intravenous fluids and antibiotics, but the liver enlargement persisted.

The family history is of interest. A male sibling had died previously at age 6 months. He had seemed to be a normal child until 24 hr before death, when he developed unexplained severe metabolic acidosis and hepatomegaly. Necropsy revealed only the presence of "severe fatty changes" in the liver. Four other siblings are alive and healthy. There is no evidence of consanguinity.

On her first admission years earlier the patient had been alert and height and weight were normal. Her abdomen was protuberant, and a soft liver edge was easily palpable 8 cm below the right costal margin. Serum albumin, globulin, and cholesterol levels were normal, as were the SGOT and SGPT. Fasting hypoglycemia was easily elicited, and her blood glucose was 8 mg/100 ml after an 8 hr overnight fast. Glucagon (1 mg intramuscularly) given at that time caused no rise in blood glucose levels; however, when the same dose of glucagon was given 1 hr after a feeding, the blood glucose levels rose by 60 mg/100 ml in 30 min. A diagnosis of a limit dextrinosis variety of glycogen storage disease (type III) was entertained, and a liver biopsy was performed. The glycogen content was 1.4% of the wet weight, a level that is not excessive. Levels of hepatic glucose 6-phosphatase, phosphorylase, amylo-1,6-glucosidase, acid phosphatase, α-glucosidase, and phosphoglucomutase activities were normal. Histologic examination revealed ballooning of the hepatic cells by fat-containing vacuoles without inflammatory or fibrotic changes. This picture was very similar to that found at necropsy in the liver of the deceased sibling.

The child was discharged without a specific diagnosis on a regimen of frequent feedings. Over the next 2 years she was frequently admitted to the hospital for episodes of symptomatic hypoglycemia and severe metabolic acidosis, which were often associated with an intercurrent infection. Blood glucose concentrations below 10 mg/100 ml and arterial pH below 7.15 were noted repeatedly.

After several admissions to the hospital, she was finally discharged on a diet that excluded all fructose, sucrose, and sorbitol. On a routine hospital evaluation at $5\frac{1}{2}$ years of age she was noted to be above the fiftieth percentile for both height and weight. The physical examination was completely normal. The liver was no longer palpable. Neurologic examination was completely within normal limits, and she was doing well in kindergarten.

At $5\frac{1}{2}$ years of age the effect of fasting was investigated. Following a 12 hr overnight fast the values noted in Table 7-7 were seen. These changed precipitously over the next several hr of continued fasting. The glucose tolerance test was normal, as were the plasma insulin levels. Both fructose and glycerol tolerance tests produced hypoglycemia.

Hepatic enzyme activities were found to be normal *except* for the absence of fructose 1,6-diphosphate 1-phosphatase.

Table 7-7. Effect of fasting on patient X. Y.

TIME (hr)	12	18	21
D-Glucose (mg/dl)	56.0	35.0	10.0
pH	7.4	–	7.17
Total CO_2 (meq/liter)	25.0	–	9.3

Biochemical questions

1. What explanation can be provided for the fasting hypoglycemia in patient X. Y.?
2. Why was the diagnosis of type III glycogen storage disease dismissed?

3. Aside from glucose, what other sugars or sugar alcohols would relieve the fasting hypoglycemia? Explain.
4. What are the the causes of metabolic acidosis?

Case discussion

The patient X. Y. presents with a history and symptoms that are interesting to compare with the normal person and those with enzyme deficiencies in the metabolism of D-fructose. The metabolism of D-fructose from the blood occurs in liver, kidney tubular epithelial cells, and intestinal mucosa. The pathway is summarized in the following scheme. In the normal person D-fructose is removed from the blood faster than D-glucose. The half-life of D-fructose is about 20 min compared to about 45 min for blood glucose. This results in a more rapid fall in blood inorganic phosphate (a hypophosphatemia) and an elevation of blood lactate, pyruvate, α-ketoglutarate, and citrate, demonstrating a more rapid utilization of D-fructose. In the liver this proceeds principally by way of D-fructose 1-phosphate, as is also the case in the kidney and intestine.

D-Fructose 1-phosphate may play an important role in the control of the D-fructose metabolism. It is known that D-fructose 1-phosphate inhibits fructose kinase completely at 0.01 M, which may be a high intracellular concentration in the normal state but not so where enzyme deficiencies in the pathway oc-

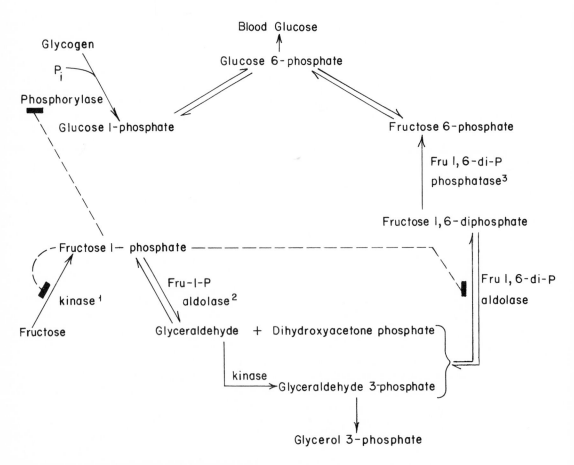

[1]Essential fructosuria deficient enzyme (type 1)
[2]Hereditary fructose intolerance deficient enzyme (type 2)
[3]Fructose intolerance for patient X. Y. deficient enzyme (type 3)
P_i Intracellular inorganic phosphate

cur, such as in the type 1 fructosuria in particular.

In type 1 fructosuria the time for the return of blood fructose concentrations to zero after fructose ingestion is usually greater than 2 hr as opposed to the normal of less than 1 hr. Similarly, D-fructose 1-phosphate inhibits (K_1 0.9×10^{-2} M) the aldol cleavage of D-fructose 1,6-diphosphate to the triose phosphates in a liver homogenate of a patient with type 2 fructosuria, although it is interesting to note that no such inhibition is found with crystalline normal aldolase. Again, D-fructose 1-phosphate completely inhibits crystalline muscle phosphorylase at 8×10^{-2} M concentration but could not be shown to inhibit glycogen breakdown in the liver of persons with type 2 fructosuria. These examples serve to illustrate the care with which in vitro experimentation should be interpreted in in vivo situations.

That the severe hypoglycemia produced on starvation is not responsive to glucagon is critical, so much so that fructose tolerance tests should be carried out with great caution and with IV glucose at hand. This applies also to patients with type 2 fructosuria. The symptoms of hypoglycemia are not seen in type 1 fructosuria. It may be noted that in all of these patients the glucose and galactose tolerance tests are normal unless there is cirrhosis of the liver or other mixed inborn errors of metabolism.

In the case of patient X. Y. it is noted that in the starved state there is no response to the hypoglycemia by glucagon but that, after feeding, the injection of glucagon produces an increase in blood glucose. This is interpreted to mean that the glycogenolysis cf the liver (and kidney) glycogen stores, which are replenished after a glucose meal, is normal. Starvation for 8 to 12 hr depletes these stores in patient X. Y. In normal persons the blood glucose is maintained by gluconeogenesis from glucogenic amino acids, but this is impaired in patient X. Y. by the fructose 1,6-diphosphatase

deficiency. For these reasons the hypoglycemia could not be relieved by feeding sorbitol or glycerol.

Patient X. Y. has type 3 fructosuria, in which a pileup of metabolites from D-fructose 1,6 diphosphate, back through the triose phosphates and D-fructose 1-phosphate. Excessive accumulation of D-fructose 1-phosphate has been noted in the proximal kidney tubules in type 2 fructosuria with some deterioration in the kidney function and loss of the ability to acidify the urine (fructose-induced renal tubular acidosis). This situation is reversible when fructose is removed from the diet and in the present case may explain in part the severe metabolic acidosis.

The hypoglycemic episodes will be associated with reduced amounts of secreted insulin but elevated concentrations of glucagon in the blood. The glucagon will act to mobilize the triglyceride pool in the adipose tissues, producing an increase in glycerol and free fatty acid concentration in the blood. These will be metabolized in the liver, with the reduced glucose availability in this hypoglycemic state resulting in the conversion of incoming free fatty acids partly to ketone bodies and triglycerides. The glyceryl residue of these triglycerides will be derived in part from the phosphorylation of blood glycerol and the glycerol 3-phosphate from the fructose 1-phosphate (see p. 309). These reactions lead to excess fatty deposit in the liver ("fatty liver") and hepatomegaly, without any associated inflammation or fibrosis (thus the normal SGOT and SGPT).

Associated with the impaired glucose metabolism, the free fatty acids will be oxidized at an increased rate producing excessive ketone bodies and an acidosis, not unlike the situation in diabetes mellitus.

REFERENCE

Baker, L., and Winegrad, A.: Fasting hypoglycemia and metabolic acidosis with deficiency of hepatic fructose 1,6-diphosphate 1-phosphatase activity, Lancet 2:13, 1970.

CASE 2: DIABETES MELLITUS
AND OBESITY

D. M., a 24-year-old, 5' 10" graduate student, was seen in a diabetes outpatient clinic. His chief complaints were fatigue, weight loss, and increases in appetite, thirst, and frequency of urination. He had considered himself well until 6 months prior to his visit. At that time he tired easily, and he tended to fall asleep in class, especially after the noon meal. He had attributed this to a heavy work load, the stress of a difficult research project, and rather high living. The symptoms continued and during the last 2 months he had lost approximately 15 pounds from his normal weight of 160 pounds. His appetite was good; in fact, it was excessive. For the last 3 to 4 weeks he had been excessively thirsty and had to urinate every few hr. He began to get up three to four times a night to urinate. With the continued weight loss, muscle weakness, fatigue, and frequent urination, it finally became apparent to the patient that his life-style was not the cause of these problems. He presented himself to the student health service and was referred to the diabetes outpatient clinic.

His family history was significant in that his maternal grandfather had had diabetes mellitus and had died at the age of 50 years of a myocardial infarction. The patient had one older sister, 40 years of age, who was obese and had recently been diagnosed as having adult-onset diabetes.

Since the age of 15 years the patient had experienced frequent episodes of weakness, during which he had felt tremulous and faint. He had found that eating a candy bar relieved the symptoms, which always occurred several hours after a meal. A routine physical examination at the age of 18 years revealed a trace of sugar in his urine. His physician obtained a postprandial blood sugar, which was normal at 90 mg/dl. Follow-up urinalysis revealed no glycosuria. The earlier analysis was attributed to a contaminated urine jug and dismissed.

The patient's physical examination was essentially within normal limits. Laboratory studies revealed a normal blood count. Urinalysis revealed a specific gravity of 1.040 and was 4+ positive for glucose using a Clinitest tablet; the glucose content in a 24 hr analysis was 35 g. The urine also showed a moderate amount of ketone bodies as determined by an Acetest tablet. A glucose tolerance test was performed, and a blood sample was analyzed for other substances as indicated at the bottom of the page.

Biochemical questions

1. What is the basis for the symptoms of the patient? Explain.
2. What role does the polyol pathway (see p. 338) play in the disturbances of metabolism?
3. Could the symptoms and laboratory values be the result of inadequate insulin levels? Explain.
4. Would you deduce from the electrolyte

	D. M.	NORMAL
Glucose tolerance test		
Fasting	160 mg/dl	60-100 mg/dl
1/2 hr	240 mg/dl	< 180 mg/dl
1 hr	325 mg/dl	< 180 mg/dl
1 1/2 hr	305 mg/dl	< 140 mg/dl
2 hr	285 mg/dl	< 120 mg/dl
Electrolytes		
Na$^+$	140 meq/liter	136-145 meq/liter
K$^+$	4.1 meq/liter	3.5-5.0 meq/liter
Cl$^-$	98 meq/liter	100-106 meq/liter
Total CO_2	25 meq/liter	24-30 meq/liter
BUN	16 mg/dl	10-20 mg/dl
Creatinine	0.9 mg/dl	0.7-1.5 mg/dl

values that the patient was acidotic, dehydrated, or suffering from another type of imbalance?

5. On entering the hospital, what would you expect D. M.'s tissue glycogen levels to be? Explain.

Case discussion
Diabetic signs

In his glucose tolerance test, D. M. demonstrated an inability to handle a normal glucose load as well as hyperglycemia that exceeded the kidney threshold values. Such findings might have been expected in light of the earlier urinalysis and the familial history of diabetes.

Evidence is now overwhelming that the potentiality to develop diabetes mellitus is inherited. Although it had remained latent for a number of years, there had been earlier indications of the disease in this patient at the ages of 15 and 18 years. At 24 years of age his physical examination was normal, but he reported an increased appetite and excessive fluid intake and fluid loss. These symptoms suggested that his energy stores were being wasted and that frequent urination was required for their excretion. In short, his tissues were starving in the presence of excessive amounts of circulating glucose. His muscle mass was producing approximately normal levels of creatinine.

The major forms of diabetes are identified as *juvenile-onset* diabetes and *adult-onset* diabetes. The juvenile form is characterized by an abrupt presentation and requires insulin to control the hyperglycemia. The adult-onset diabetes produces milder symptoms, develops gradually, and can usually be controlled by diet. These patients are often obese. Either form can be found at any age despite their names.

Glucose and insulin response in blood

The *essential* sign of diabetes mellitus is sustained hyperglycemia. Diabetes involves many other biochemical components. These abnormalities often are attributed to an inadequate secretion of insulin. As noted in the discussion of obesity, which is often associated with adult-onset diabetes, the change in the plasma levels of insulin is abnormal in response to carbohydrate loads in both obese and diabetic subjects. When plasma insulin levels are followed in relation to blood glucose concentrations, it is found that the overt diabetic has a slower, reduced secretion of insulin in response to glucose intake. The obese person has a nearly normal glucose tolerance, but it is achieved with very high insulin levels. The obese, mild diabetic individual shows a slower but still high rise in plasma insulin, which is associated with a hyperglycemia and a slow recovery. These findings are summarized in Fig. 7-27. Thus it would appear that the concentration of circulating insulin, measured by radioimmunoassay, is not always indicative of physiologically active hormone.

It has been clearly demonstrated that cells reponding to insulin have receptor molecules in their plasma membranes to which insulin binds specifically. The concentration of these receptors on the cell surface can be determined by the use of radioactively labeled insulin by measuring the amount of label that is bound under standard conditions. There is good evidence that the insulin does not enter the cell but mediates its influence indirectly (see Chapter 13).

By such quantitative methods it is found that the number of receptor sites on a cell in the obese person with glucose intolerance is reduced. As the obesity is corrected, the concentration of receptors increases. These findings help to explain why obese individuals are refractory to the action of insulin (Fig. 7-27).

The hyperglycemia noted after the ingestion of 100 g of glucose by this patient, who is not obese, was more severe than that in Fig. 7-27. Therefore, one might conclude that his circulating levels of insulin were low. In spite of his hyperglycemia and glucosuria (urine loss of 35 g glucose/24 hr), his cells were deficient in glucose and were living essentially on a low-carbohydrate, high-fat energy supply, which resulted in a ketosis and compensated metabolic acidosis. There were no physical signs of acidosis, but no plasma pH was recorded. The total

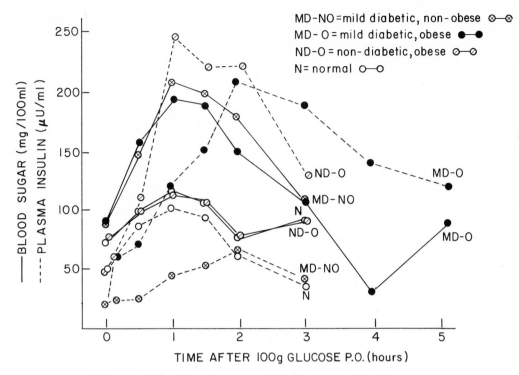

Fig. 7-27. Glucose tolerance tests—blood sugar and insulin responses.

CO_2 was within normal limits but the serum chloride was slightly changed. The Na^+ and K^+ were within normal limits, indicating minimal dehydration. The water concentration mechanisms were functioning properly, as seen by the high urine specific gravity. A more serious condition of diabetic ketosis and acidosis is discussed in Chapter 4 (case 1).

When the patient entered the hospital, the level of circulating insulin was insufficient to prevent glucosuria but not so low that it precipitated a coma. He felt faint at times, and it seems reasonable to assume that glucose was not being maintained at adequate levels in the extrahepatic tissues, where the rate of transport across the cell membranes is abnormally low in the absence of insulin.

Many tissues do not require insulin to assimilate glucose so that the glucose concentrations within these cells approach that in the blood. Some of this glucose may be reduced to sorbitol, which may then be oxidized to D-fruc-

tose, both of which may stay within the cell after the blood glucose has been returned to normal. These reactions have been noted earlier (p. 338) and the enzymes, particularly the NADPH-linked sorbitol dehydrogenase, which produces sorbitol from glucose, is found in high concentration in lens epithelium, the Schwann cell in peripheral nerve, the papilla in kidney, and the islets of Langerhans in the pancreas. These localizations identify with areas of pathology in the diabetic, such as cataract formation and diabetic neuropathy. Thus the glycogen levels in these tissues were low, inducing a state that resembled starvation. While glucose entered the liver cells, it could not be metabolized at a sufficient rate because of the low level of glucokinase. Hexokinase is competitively inhibited by a high level of D-glucose 6-phosphate. To the extent that it can function, glycogen stores would be maintained.

Hyperglycemia was controlled in this patient through regular injections of insulin.

REFERENCES

Cahill, G. F., Jr., and Owen, O. E.: Some observations on carbohydrate metabolism in man. In Dickens, F., Randle, P. J., and Whelan, W. J., editors: Carbohydrate metabolism and its disorders, New York, 1968, Academic Press, Inc., p. 497.

Hales, C. N.: Actions of hormones in the regulation of glucose metabolism. In Campbell, P. N., and Greville, G. D., editors: Essays in biochemistry, vol. 3, New York, 1967, Academic Press, Inc.

CASE 3: PEDIATRIC GASTROENTERITIS

R. D., a 6-week-old girl, was born after a normal pregnancy and had weighed 3.2 kg at birth. Her parents and two older siblings were reported to be in good health. She had been breast fed for 4 weeks and her weight gain had been normal. The composition of human milk is as follows:

Fat	3.5%-4.0%	Total	0.2%
Lactose	7.0%	minerals	
Total protein	1.35%	Water	85%-88%
Lactalbumin	0.70%	kcal/100	66
Casein	0.5%	ml	

Her stools during this period were described as "loose" but were not considered by her mother to be abnormal. At 4 weeks of age, breast feeding was discontinued and a common baby formula was substituted. As a result of poor initial formula preparation, the child developed a viral gastroenteritis and after several days exhibited fussiness, watery diarrhea, and vomiting. At the age of 6 weeks she was admitted to the hospital.

Physical examination at the time of admission revealed only a moderate degree of dehydration (dry mucous membranes, slight depression of the anterior fontanel, and sunken appearance of the eyes). Skin turgor seemed normal, and body weight was 3.4 kg. Urinalysis yielded a 1+ reaction for reducing substance and no reaction for glucose.

The infant was hydrated with intravenous fluids over a period of 24 hr and then with fluids orally for an additional 24 hr. During this time the diarrhea subsided and her weight increased to 3.7 kg. She was then fed formula X (67 kcal/100 ml). Within 24 hr her stools became watery and were passed at the time of each feeding.

On the fourth day of hospitalization, formula Y was substituted for formula X.

The number of stools decreased to two or three daily; they contained no reducing substance and became semiformed in character. The infant gained 40 g/day during the next 4 days. However, because of the inconvenience and expense of formula Y, a single feeding of formula X was once again given as a trial. The infant demonstrated some fussiness but did not pass a stool. A second feeding of formula X 4 hr later was followed by several explosive, watery stools. The stool pH was 5.0, and it gave a positive reaction for reducing substance. Data on formulae X and Y are given below. Both formulae contain 67 kcal/100 ml.

Biochemical questions

1. Was there any significant difference between the breast milk and the baby formulae?
2. Was the gastroenteritis related to the diarrhea and vomiting?
3. What might have caused the explosive, acid, watery stools containing reducing substances?
4. How did the gastroenteritis affect the digestion of carbohydrates?

Case discussion

At 6 weeks of age, R. D. suffered from a disorder that was not present during the first month of life when her development was normal. This period coincided with breast feeding, which indicated that the infant had been able to handle all of the nutrients in her mother's milk. The deficiency may have developed gradually, but this is unlikely unless for some reason the mother was producing abnormal milk. Again, the nutritional adequacy of the milk is indicated by the infant's normal weight gain.

It is reasonable to conclude that this problem was not inherited but rather

COMPONENTS	FORMULA X*	FORMULA Y*
Cow milk protein	x	
Casein hydrolysate		x
Fat		
Corn oil	x	x
Coconut oil	x	
Carbohydrates		
Lactose	x	
Sucrose		x
Arrowroot starch		x
Major constituents (gm/100 ml)		
Protein	1.7	2.2
Fat	3.4	2.6
Carbohydrate	6.6	8.5
Minerals	0.4	0.6
Caloric distribution (% of calories)		
Protein	11.0	15.0
Fat	48.0	35.0
Carbohydrate	41.0	50.0
Minerals per liter		
Calcium (mg)	670.0	1000.0
Phosphorus (mg)	517.0	700.0
Sodium (meq)	11.0	17.0
Potassium (meq)	28.0	26.0
Chloride (meq)	15.0	23.0
Magnesium (mg)	62.0	
Iron (mg)	12.0	10.0
Vitamins per liter		
A (IU)	2643.0	1586.0
Thiamine (μg)	710.0	486.0
Riboflavin (IU)	1100.0	1903.0
Niacin (mg)	4.4	4.2
Pyridoxine (μg)		529.0
Pantothenate (mg)		3.4
C (mg)	55.0	32.0
D (IU)	440.0	423.0
E (IU)		5.3

*An x indicates presence of a component.

caused by the effects of viral gastroenteritis. The evidence of dehydration was to be expected in light of the excessive water losses through vomiting and diarrhea. Thus the principal question concerned the infant's intolerance of formula X. The following aspects were noted.

1. Intravenous feeding followed by oral feeding of simple solutions reduced the diarrhea and produced a weight gain through rehydration.
2. Each feeding of formula X caused diarrhea.
3. Substitution of formula Y relieved the problem and produced an acceptable weight gain.
4. Feeding of formula X once again produced diarrhea. This points rather directly to a relation between some component of formula X and the diarrhea.

There were a number of differences between formulae X and Y, as indicated by the summary of their compositions. However, further study of the case history helped limit the possibilities. When the infant was admitted to the hospital, urinalysis indicated the presence of a reducing substance or substances that were not glucose. Thus a diabetic type of overload could be ruled out. Reducing substances were also present in the watery stools induced by formula X, but these disappeared from the stools when formula Y was substituted. The most common reducing substances are the carbohydrates, which were present in formula X as lactose and in formula Y as sucrose. Lactose is a reducing substance; sucrose is not. It would seem therefore that the gastroenteritis had caused at least a temporary inability to digest lactose.

Normally the digestion of carbohydrates is initiated in the mouth, where the salivary α-amylase enzyme acts on starch to initiate the partial hydrolysis to simpler materials called starch dextrins (glucose, maltose, and oligosaccharides). This process is shown below. Digestion of carbohydrates ceases in the stomach as a result of the extreme gastric acidity that is produced by hydrochloric acid secretion. When the gastric contents are emptied into the small intestine, bile and pancreatic juice neutralize the hydrochloric acid. Digestion of the starch dextrins then continues as a result of the secretion of pancreatic α-amylase. At this stage all the digestible carbohydrate, not including cellulose

and similar roughage, has been cleaved into smaller fragments. These fragments include monosaccharides, principally glucose and perhaps some fructose; disaccharides, principally maltose, sucrose, and lactose; and a small amount of the α-limit dextrins of starch. Final hydrolysis of the disaccharides occurs in the brush border of the intestinal mucosa. Several disaccharidases (maltase, sucrase, lactase, and isomaltase) are present in the membranes of these brush border cells. The resulting monosaccharides are then actively transported by carrier systems, probably located in the brush border membranes, into the cell for eventual passage into the portal system and thence to the liver. Some portion of the monosaccharides resulting from the disaccharidase activity are not immediately picked up by the carrier system; these leak back into the lumen, from which they are completely reabsorbed before reaching the colon.

Viral gastroenteritis damages the mucosa, and some of the brush border cells have a significant proportion of their disaccharidases destroyed. How-

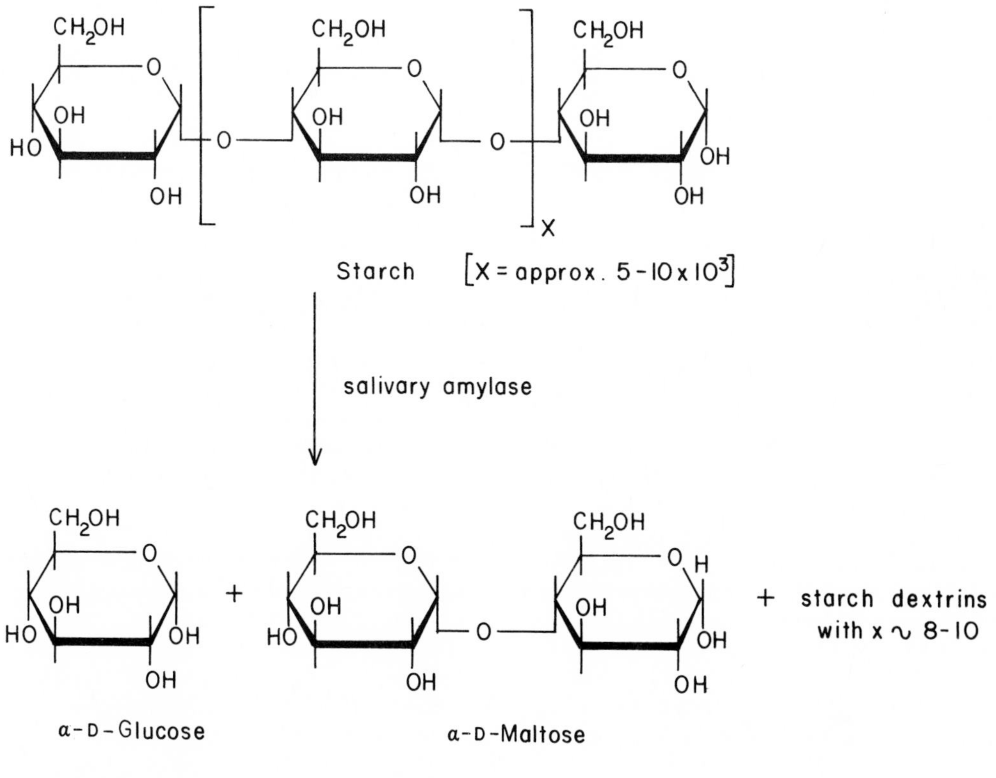

Starch $[X = \text{approx}. 5-10 \times 10^3]$

salivary amylase

α-D-Glucose $\qquad$ α-D-Maltose $\qquad$ + starch dextrins with x $\sim$ 8-10

ever, since there are at least four different maltases, serious deficiencies of maltose hydrolysis are rare. It is generally accepted that there is only one type of sucrase. However, mucosal damage does not usually affect sucrose hydrolysis, probably because a high level of sucrase is normally present. In contrast, lactase activity is sensitive to mucosal damage, and lactose hydrolysis is significantly reduced. Because of the damage to the microvilli in the intestinal mucosa, some lactose is absorbed without hydrolysis, while a large proportion passes on to the colon.

These facts explain many of the symptoms exhibited by R. D. when she was fed formula X. The lactose that was absorbed intact could not be hydrolyzed anywhere in the body other than the mucosa, and was thus excreted intact in her urine. This produced the 1+ reaction for reducing substances.

The infant was fed 120 kcal/kg body weight each day. Thus, based on a weight of 3.5 kg and 67 kcal/100 ml of formula X, her daily intake was:

$$\frac{120 \times 3.5}{67} \times 100 \text{ ml/day} = 627 \text{ ml/day}$$

Assuming five equal feedings per day, this would be equivalent to about 125 ml per feeding. Formula X contains 6.6 g

lactose/100 ml, so that $6.6 \times \dfrac{125}{100}$ g or about 8 g of lactose is fed at each meal. A large proportion of this lactose would pass into the colon, where the bacterial flora would ferment part of it to CO_2 and lactic acid. The presence of these extra low molecular weight materials in the colon would draw water in from the blood, producing an osmotic diarrhea and discomfort ("fussiness"). This explains explosive (CO_2), watery (osmotic), acidic (lactic acid) stools containing reducing substances (lactose); therefore, the advisability of maintaining the infant for some time on a formula that is lactose free is apparent.

Mucosal disaccharidase deficiencies, except maltase deficiency, are well-described clinical entities. The most common is lactase deficiency, which varies greatly among different age and racial groups. Primary lactase deficiency is genetically determined. Secondary lactase deficiency, such as that described in this case, can be precipitated by a number of disturbances of the gastrointestinal tract, including tropical sprue.

REFERENCES

Appleton, H., and Higgins, P. G.: Viruses and gastroenteritis in infants, Lancet 1:1297, 1975.

Gray, G. M.: Intestinal digestion and maldigestion of dietary carbohydrates, Am. Rev. Med. 22:391, 1971.

CASE 4: GALACTOSEMIA

The patient, a boy, was the first child of healthy parents without known consanguinity. Delivery was normal and birth weight was 3780 g. From the third day of life he developed an increasing degree of icterus and at the same time became indolent and difficult to feed. No blood group incompatibility could be demonstrated. At 6 days of age he had a serum bilirubin value of 29.5 mg/100 ml, and his weight was 15% below birth weight. He was readmitted to the hospital on the seventh day after birth. Muscular tonus was increased, and the patient later developed convulsions. Between the seventh and ninth days, exchange blood transfusion was performed three times, but the serum bilirubin concentration

still remained high. On the ninth day of life the boy began vomiting, liver enlargement was noted, and the cerebral symptoms became accentuated.

A positive test for reducing sugars in urine had been obtained already on the sixth day after birth. A repeated test was performed on the seventh day, and this was positive, while at the same time a Clinistix test, specific for D-glucose, was negative. Hereditary galactosemia was then suspected, and special tests that were performed on the eighth day of life confirmed the diagnosis.

Hemoglobin	20 g/dl
SGOT	299 Karmen-Ordell units
SGPT	202 units (normal after 7th day)

Bilirubin (max)   32.1 mg/dl (at 7th day)
Galactose 1-P uridyl   0 (normal 22-31
  transferase   units/g Hb)
  (in erythrocytes)

Milk feeding was stopped on the ninth day, and it was replaced by intravenous glucose administration. From the tenth day of life, a galactose-free diet was introduced. With this treatment the patient improved dramatically.

Biochemical questions

1. What are the biochemical effects of galactosemia?
2. Is there an alternate source of tissue galactose for patients on a galactose-free diet?

3. Would a mother who is homozygous for galactosemia be able to produce lactose in her milk?
4. What would a D-galactose tolerance curve have looked like if it had (unfortunately) been given to the patient?
5. What are the interpretations of the laboratory results?
6. Studies also indicate galactitol in the urine. By what mechanisms do you think this might occur?
7. What differences would have been noted if the deficient enzyme had been galactokinase?

REFERENCE

Dahlquist, A., Jagenburg, R., and Mark, A.: A patient with hereditary galactosemia, Acta Pediatr. Scand. 58:237, 1969.

CASE 5: VON GIERKE'S DISEASE

The patient was a 12-year-old girl who had a grossly enlarged abdomen. She gave a history of frequent episodes of weakness, sweating, and pallor that were eliminated by eating. Her development had been somewhat slow; she sat at 1 year of age, walked unassisted at 2 years, and was doing poorly in school.

Physical examination revealed blood pressure 110/58 mm Hg; temperature 38° C; weight 22.4 kg (low); and height 128 cm (low). The patient had clear lungs and a normal heart. Slight venous distention was present over the prominent abdomen. The liver was enlarged, firm, and smooth and was descended into the pelvis. The spleen was not palpable, nor were the kidneys. The remainder of the physical examination was within normal limits except for "poor musculature."

Laboratory findings for a fasting blood sample were:

	Patient	Normal values
Glucose (mg/dl)	50	70-100
Lactate (mg/dl)	59	5-18
Pyruvate (mg/dl)	3.8	0.4-0.8
Free fatty acids (mM)	1.6	0.3-0.8
Triglycerides (mg/dl)	3145	150
Total ketone bodies (mg/dl)	40	3
pH	7.25	7.35-7.44
Total CO_2	12	24-30

A liver biopsy specimen was obtained through an abdominal incision. The liver was huge, buff-colored, and firm but not cirrhotic. Histologically, the hepatic cells were bulging and dilated. The portal areas were compressed and shrunken. No inflammatory reaction was present. Stain for carbohydrate revealed large amounts of positive material in the parenchymal cells that was removed by digestion with salivary amylase. Glycogen content was 11 gm/100 gm liver (normal up to 8%), and lipid content was 20.2 gm/100 gm liver (normal less than 5%). Hepatic glycogen structure was normal.

Results of enzyme assays performed on the liver biopsy tissue are shown below:

Enzyme	Patient (units per gm liver N)	Normal range (units per gm liver N)
Glucose-6-phosphatase	22	214 ± 45
Glucose-6-phosphate dehydrogenase	0.07	0.05-0.13
Phosphoglucomutase	27	25 ± 4
Phosphorylase	24	22 ± 3
Fructose-1,6-diphosphatase	8.4	10 ± 6

Biochemical questions

1. What is a normal structure for hepatic glycogen?

2. What changes in this structure would accompany a deficiency in branching enzyme?
3. Which other tissues might be expected to accumulate excessive amounts of glycogen?
4. Explain the reasons for the fasting hypoglycemic episodes.
5. To what might be ascribed (a) the elevated free fatty acid, (b) the ketonemia, and (c) the metabolic acidosis?
6. What would you predict would be the result of continuous parenteral feeding? Explain.
7. What is the nature of the acidosis?

REFERENCES

Fischer, E. H., et al.: Phosphorylase and the control of glycogen degradation, Curr. Top. Cell Regulation 4:211-251, 1971.
Folkman, J., et al.: Portacaval shunt for glycogen storage disease: value of prolonged intravenous hyperalimentation before surgery, Surgery 72:306-314, 1972.
Howell, R. R.: The glycogen storage diseases. In Stanbury, J. B., et al., editors: The metabolic basis of inherited disease, New York, 1972, McGraw-Hill Book Co., pp. 149-173.

CASE 6: DIABETES MELLITUS

A 61-year-old man presented himself for periodontal checkup with signs of severely ulcerated and inflamed mucosal surfaces of the lips, severe fissuring of the tongue, generalized edema of the gingiva, extensive calculus, and nine new carious lesions of the teeth. The possibility of uncontrolled diabetes mellitus was recognized and supported by laboratory tests: urine had a specific gravity of 1.030 and contained albumin, glucose (Dextrostix Reagent Strips), and occasional casts. The serum blood glucose was 232 mg/dl. There was no evidence of diabetes in the family history. The patient was referred to his physician for therapy before dental treatment was pursued. He returned to the clinic after a few months, by which time his blood glucose was said to be 175 mg/dl. He was placed on an 1800 kcal liquid diet for 24 hr preoperatively.

Biochemical questions

1. What other biochemical tests would be needed to confirm the diagnosis of diabetes mellitus? Explain.
2. Differentiate between juvenile- and adult-onset diabetes (note the absence of diabetes in the family).
3. Does the requirement for insulin change when the body is under stress, such as during infections? Explain.
4. What might be an appropriate diet for this patient?
5. From the most recent theories of caries formation, what dietary components should be reduced? Salivary amylase activity is generally increased in the diabetic.
6. What is the specificity of the glucose test strips that depend upon glucose oxidase?
7. The urine specific gravity (1.030) was above normal (1.015 to 1.025), yet some tissue edema was noted. How could these observations be reconciled?
8. Is the frequency of supragingival calculus higher in the diabetic? What might be the possible explanation?

REFERENCES

Burket, L. W.: Oral medicine, ed. 6, Philadelphia, 1971, J. B. Lippincott Co.
Gabbay, K. H.: The sorbitol pathway and the complications of diabetes, New Engl. J. Med. 288: 831, 1973.
Kjellman, O., et al.: Oral conditions in 105 subjects with insulin-treated diabetes mellitus, Swed. Dent. J. 63:99, 1970.
Lavine, M. H.: Diagnosis and management of the diabetic patient, Oral Surg., Oral Med., Oral Pathol. 24:16, 1967.

CASE 7: TYPE IV GLYCOGEN STORAGE DISEASE – ENZYME THERAPY

A girl, J. V., appeared normal at birth but developed signs of liver dysfunction, muscular weakness, and persistent infections at the age of 3 months. Enzyme assays confirmed that branching enzyme was absent in leukocytes and liver. The clinical condition of the child was deteriorating, and six doses (total 0.5 g) of purified α-glucosidase from *Aspergillus niger* was given IV over a period of 3 days. The liver glycogen at the commencement of the enzyme therapy was 6% of the net weight of tissue. A biospsy sample of liver immediately after the last enzyme injection (3 days) showed little change in glycogen, but a postmortem sample (13 days) contained 1% glycogen. The glycogen accumulated in all tissues, in the lysosomes, and in the cytoplasm. Two brothers of J. V. had died earlier from the same disease. The parents of these children were second cousins.

Biochemical questions

1. What is the nature of the abnormal glycogen that is associated with this disease?
2. What carbohydrate hydrolase enzymes are usually present in lysosomes?
3. α-Glucosidase hydrolyzes normal glycogen and amylopectin equally well. What is the pattern of hydrolysis of glycogen by α-glucosidase, which has a pH optimum at 4.7 to 4.8 and is 20% of maximum in the physiologic pH range?
4. Describe the mechanism whereby α-glucosidase would enter a cell.
5. In what tissue would the α-glucosidase be concentrated after IV injection?
6. What explanation might be given for the fact that extralysosomal glycogen is hydrolyzed before that in the lysosome?
7. What are some of the associated problems that can result from the intravenous injection of an enzyme?
8. What could be deduced concerning the genetics of this case?

REFERENCES

Fernandes, J., and Huijing, F.: Branching enzyme–deficiency glycogenesis, Arch. Dis. Child. 43: 347, 1968.

Huijing, F., et al.: α-Glucosidase administration: experiences in two patients with glycogen storage disease compared with animal experiments, Enzyme Therapy in Genetic Disease, Birth Defects: Original Article Series IX:191, 1973.

Mercer, C., and Whelan, W. J.: The fine structure of glycogen from type IV glycogen storage disease, Eur. J. Biochem. 16:579, 1970.

Neufeld, E.: Inherited disorders of lysosomal metabolism, Annu. Rev. Biochem. 44:357, 1975.

CASE 8: HEREDITARY FRUCTOSE INTOLERANCE

A child had nausea, vomiting, and symptoms of hypoglycemia: sweating, dizziness, and trembling. It was reported that these attacks occurred shortly after eating fruit or cane sugar. This was resulting in a strong aversion to fruits, and the mother was therefore providing large supplementations of multivitamin preparations. The child was below normal weight and was an only child. The child had been breast-fed, during which time none of the above symptoms were evident. The clinical findings included a normal glucose tolerance test, reducing substances in the urine that did not react positively with glucose test papers in which glucose oxidase was used as the basis for test, and some cirrhosis of the liver. A fructose tolerance test was ordered, using 3 g fructose per M^2 surface, given IV in a single, rapid push. Within 30 min the child displayed the symptoms of hypoglycemia. Blood glucose analysis confirmed this and revealed that the hypoglycemia was greatest after 60 to 90 min. Fructose concentrations reached a maximum (60 mg/dl) after 15 min and gradually decreased to zero in 2½ hr. Inorganic phosphate concentrations fell by 50%, and SGOT and SGPT elevations were noted after 1½ hr. The urine was positive for fructose.

Biochemical questions

1. Explain why the fructose concentration in the blood remained elevated for an extended period.
2. What evidence exists to suggest that the aldolase for D-fructose 1-phosphate and D-fructose 1,6-diphosphate is the same protein?
3. Explain the elevations noted for SGOT and SGPT in the fructose tolerance test.
4. In the face of severe low intracellular inorganic phosphate concentrations produced by rapid D-fructose assimilation, what explanations can be proposed for the hypoglycemia that is not reponsive to glucagon?
5. Why are the symptoms of hypoglycemia not found in essential fructosuria (type 1)?
6. What would be the consequences of a deficiency of phosphofructokinase in the patient?

REFERENCES

Froesch, E. R.: Essential fructosuria and hereditary fructose intolerance. In Stanbury, J. B., et al., editors: The metabolic basis of inherited disease, New York, 1972, McGraw-Hill Book Co.

Lai, C. Y., and Horecker, B. L.: Aldolases: structure and function, Essays Biochem. 8:149-178, 1972.

ADDITIONAL QUESTIONS AND PROBLEMS

1. *Genetic defects*. What would be the consequence of a genetic defect that severely reduced or eliminated the activity of the glycogen phosphorylase kinase? (See Morishita, Y., et al., Biochem. Biophys. Res. Comm. 54:833, 1973.)
2. *Malabsorption*. How would you check for an impairment in the absorption of sugars, for example, in celiac disease (nontropical sprue)? What diet would you recommend for such a patient?
3. *Hyperkalemia and insulin therapy*. A case diagnosed as hyperkalemic paralysis with Addison's disease is reported by R. G. Van Dellen and D. C. Purnell (Mayo Clin. Proc. 44:904, 1969). The disease was also associated with diabetes mellitus. Dramatic improvement was noted after the patient had been given fluid therapy, sodium bicarbonate, and insulin. What was the biochemical basis for this treatment?

CHAPTER 8

LIPID METABOLISM

OBJECTIVES

1. To describe the structure of the lipid components of mammalian tissues, exclusive of the sterols and steroids
2. To discuss the biosynthesis and catabolism of the commonly occurring lipids in mammalian cells
3. To discuss lipid transport
4. To describe the biochemistry of diseases associated with abnormalities in lipid transport or metabolism, including hyperlipoproteinemias and the lipid storage diseases

Lipids are organic compounds that are poorly soluble in water but quite soluble in organic solvents such as ether, benzene, or chloroform. In the human body, lipids function as structural components of cell membranes, as storage forms of energy, as metabolic fuel, and as emulsifying agents. Four of the vitamins are lipids. In addition, the prostaglandins, substances that stimulate smooth muscle contraction and function in intracellular regulatory processes, are lipid derivatives. The transport of lipids through the blood plasma is an extremely important subject from the standpoint of health, for abnormalities in these processes are thought to be a major factor in the development of coronary artery disease. Obesity results from the storage of excessive amounts of lipid in the body. Such common diseases as diabetes mellitus, obstructive jaundice, pancreatitis, and hypothyroidism have associated plasma lipid transport abnormalities. In addition, there are a number of rare inherited diseases, known as the lipid storage diseases or lipidoses. All health scientists should have at least some acquaintance with the metabolic defects responsible for these diseases in terms of fundamentals of lipid chemistry and metabolism.

CLASSIFICATION OF LIPIDS

The lipids contained in the body can be classified according to their chemical structure into five groups, as shown in Table 8-1. Fatty acids are constituents of a variety of lipids, many of which are present in substantial amounts. The fatty acid portions of these more complex lipids serve as a major energy source. Prostaglandins are intracellular regulatory substances that modify the cell's response to external stimuli. They are derived from certain polyunsaturated fatty acids. Most of the lipids in a cell are in the form of glyceryl esters. These include the acylglycerols, which are either metabolic intermediates or storage forms of fatty acid, and phosphoglycerides, which are the main structural lipids of cell membranes. Sphingolipids, derivatives of the fatty alcohol sphingosine, also are membrane components. The sterol derivatives, including cholesterol, bile acids, steroid hormones, and vitamin D, are so important in terms of human health that they are discussed separately in Chapters 10 and 13. Aspects of cholesteryl ester

Table 8-1. Classification and functions of lipids*

LIPID	FUNCTION
Fatty acids	Metabolic fuel, building blocks for other lipids
Prostaglandins	Intracellular modulators
Glyceryl esters	
Acylglycerols	Fatty acid storage, metabolic intermediates
Phosphoglycerides	Membrane structure
Sphingolipids	
Sphingomyelin	Membrane structure
Glycosphingolipids	Membranes, surface antigens
Sterol derivatives	
Cholesterol	Membrane and lipoprotein structure
Cholesteryl esters	Storage and transport
Bile acids	Lipid digestion and absorption
Steroid hormones	Metabolic regulation
Vitamin D	Calcium and phosphorus metabolism
Terpenes	
Dolichols	Glycoprotein synthesis
Vitamin A	Vision, epithelial integrity
Vitamin E	Lipid antioxidant
Vitamin K	Blood coagulation

*For additional discussion of fat-soluble vitamins, see Chapter 1; for additional discussion of prostaglandins, see Chapter 13; for additional discussion of sterols, see Chapter 10.

Table 8-2.. Fatty acid classification according to chain length

TYPE	NUMBER OF CARBON ATOMS
Short chain	2-4
Medium chain	6-10
Long chain	12-26

Table 8-3. Major fatty acids in humans

DESCRIPTIVE NAME	SYSTEMATIC NAME	CARBON ATOMS	DOUBLE BONDS	POSITION OF DOUBLE BONDS*	UNSATURATED FATTY ACID CLASS†
Acetic		2	0		
Butyric		4	0		
Caprylic	Octanoic	8	0		
Palmitic	Hexadecanoic	16	0		
Stearic	Octadecanoic	18	0		
Oleic	Octadecenoic	18	1	9	ω 9
Linoleic	Octadecadienoic	18	2	9, 12	ω 6
Linolenic	Octadecatrienoic	18	3	9, 12, 15	ω 3
γ-Linolenic	Octadecatrienoic	18	3	6, 9, 12	ω 6
Arachidonic	Eicosatetraenoic	20	4	5, 8, 11, 14	ω 6

*In the Δ numbering system, only the first carbon of the pair is listed: that is, 9 means position 9, 10.
†In the ω numbering system, only the first double bond from the methylene end is listed and, as above, only the first carbon of the pair is written.

metabolism that relate to their fatty acid moieties, however, are presented in this chapter. Terpenes, the final class, include the dolichols and three fat-soluble vitamins, A, E, and K. These isoprene derivatives are present in only very small quantities but have separate, extremely important metabolic functions.

Fatty acids

Fatty acids are compounds represented by the chemical formula R—COOH, where R stands for an alkyl chain composed of carbon and hydrogen atoms. One method of classifying fatty acids is according to their chain length, that is, the number of carbon atoms that they contain. An arbitrary but widely accepted classification is listed in Table 8-2. Most of the fatty acids present in the blood and tissues of man are of the long-chain variety. The names and structural properties of the most important fatty acids are given in Table 8-3. Almost all of the fatty acids in natural products contain an *even* number of carbon atoms.

General nomenclature

The carbon atoms of an acid are numbered (or lettered) either from the *carboxyl group* (Δ numbering system) or from the carbon atom furthest removed from the carboxyl group (ω numbering system) as follows:

| ω – terminus | $CH_3- CH_2 - CH_2 - CH_2 - CH_2 - CH_2 - CH_2 - CH_2 - CH_2 - COOH$ | carboxyl terminus |

Δ numbering	10	9	8	7	6	5	4	3	2	1
ω numbering	1	2	3	4	5	6	7	8	9	10
Letter designation	ω	$\omega-1$				δ	γ	β	α	

Greek letters also are used to indicate the various carbon atoms. The α-carbon is adjacent to the carboxyl group, and the ω-carbon atom is that furthest from the carboxyl group.

Fatty acids are abbreviated in the Δ nomenclature by listing the carbon number and position of double bonds ($C_x \Delta_y$). Thus palmitoleic acid is abbreviated as either 16:1 or $C_{16}\Delta_9$. The number after the Δ in this classification system signifies the position of the double bond relative to the carboxyl end. For example, in $C_{16}\Delta_9$ the double bond is nine carbon atoms away from the carboxyl group; that is, it is between carbon atoms nine and ten, counting the carboxyl carbon atom as carbon atom number one. In the ω-numbering system, palmitoleic acid would be referred to as 16:1ω7. This indicates that the acid has sixteen carbon atoms and one unsaturated bond that is located seven carbon atoms away from the ω-carbon atom. Since palmitoleic acid contains sixteen carbon atoms, the double bond that is seven atoms away from the ω-carbon also is nine atoms away from the carboxyl carbon.

$$CH_3-CH_2-CH_2-CH_2-CH_2-CH_2-CH=CH-CH_2-CH_2-CH_2-CH_2-CH_2-CH_2-CH_2-COOH$$

Palmitoleic acid

All three numbering systems are in current usage and therefore it is necessary to become familiar with each of them. The ω-designation is used in discussing fatty acid lengthening and desaturation reactions. The Δ-designation is used for the prostaglandins and in fatty acid β-oxidation. The letter designations also are used in β-oxidation.

Ionization

The pK_a of the fatty acid carboxyl group is about 4.8. Therefore, almost all (99.9%) of the fatty acid molecules present in body fluids are ionized; that is, the fatty acid is present as an *anion*. This imparts detergent-like properties to the long-chain fatty acids when they exist in unesterified form in biologic fluids, for the ionized carboxyl group interacts with aqueous media, whereas the hydrocarbon tail seeks a nonpolar environment.

Saturation

Fatty acids are either saturated or unsaturated. In a saturated fatty acid the alkyl chain does not contain any double bonds.

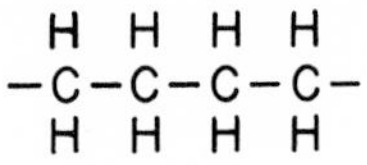

$$\begin{array}{cccc} H & H & H & H \\ -C- & C- & C- & C- \\ H & H & H & H \end{array}$$

Unsaturated fatty acids contain one or more double bonds. Those that contain one unsaturated bond are known as *monoenoic* or *monounsaturated* fatty acids; those that contain two or more unsaturated bonds are known as *polyenoic* or *polyunsaturated*.

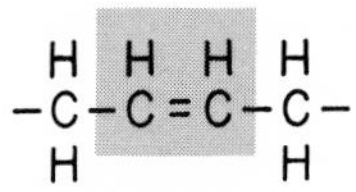

$$\begin{array}{cccc} H & H & H & H \\ -C- & C= & C- & C- \\ H & & & H \end{array}$$

A monoenoic acid

$$\begin{array}{ccccccc} H & H & H & H & H & H & H \\ -C- & C= & C- & C- & C= & C- & C- \\ H & & & H & & & H \end{array}$$

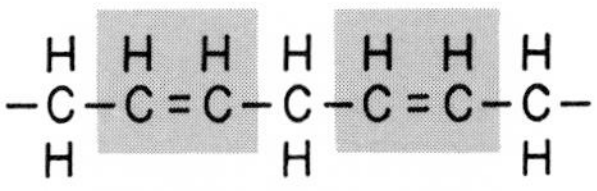

A polyenoic acid

Mammals and plants contain both monoenoic and polyenoic fatty acids, whereas all of the fatty acids containing double bonds that are present in bacteria are monoenoic. Plant and fish fats contain more polyenoic fatty acids than animal fats. The double bonds in a polyenoic fatty acid are neither adjacent nor conjugated, for this would make the structure too easily oxidizable. Rather, the double bonds are two carbons apart; this provides somewhat greater protection against oxidation, referred to as either *auto-* or *peroxidation*. Rancid fats are those that contain an appreciable amount of peroxidized fatty acid.

The unsaturated fatty acids are divided into four classes, and these are designated according to the ω numbering system.

Class	Parent Fatty Acid
ω 7	Palmitoleic acid
ω 9	Oleic acid
ω 6	Linoleic acid
ω 3	Linolenic acid

Each class is made up of a family of fatty acids, and all members of that family can be synthesized biologically from the parent fatty acid. For example, arachidonic acid ($20:4\omega6$) is synthesized from the parent of the ω 6 class, linoleic acid ($18:2\omega6$). A fatty acid of one class, however, cannot be converted biologically to another class, that is, no member of the oleic acid class (ω 9) can be converted to either linoleic acid or any other member of the ω 6 class.

The hydrocarbon chain of a saturated fatty acid usually is extended because this linear, hinged conformation is the minimum energy form. By contrast, unsaturated fatty acids have rigid bends in their hydrocarbon chains because the double bonds do not rotate, and a 30° angulation in the chain is produced by each of the *cis* double bonds that are present. In general, human cells contain at least twice as much unsaturated as saturated fatty acids, but the composition varies considerably among the different tissues and depends to some extent on the type of fat contained in the diet.

    Iodine number.   Fat from natural sources is almost always composed of a mixture of fatty acids. The extent of unsaturation of a lipid material is often estimated by the amount of I_2 that it takes up under standardized conditions. The iodine number of a fat, expressed as grams of I_2 absorbed by 100 g of fat when exposed to either IBr or ICl, is a measure of the unsaturation of the lipid. Since a saturated fatty acid has no double bonds, it takes up no I_2 and its iodine number is zero. Oleic acid, which is monoenoic, has an iodine number of 90, whereas linoleic, the corresponding dienoic acid, has an iodine number of 181. Natural fats that have a preponderance of saturated fatty acids have an iodine number of about 10 to 50; those that contain an abundance of polyunsaturated fatty acids have an iodine number of 120 to 150. Fats with low iodine numbers, for example, lard and butter, are solids at room temperature; fats with high iodine numbers, for example, corn oil and safflower oil, are liquids at room temperature. The iodine number still is in use clinically in designating the fatty acid saturation of diets that are prescribed. With the widespread availability of gas-liquid chromatography, however, much more precise information about the fatty acid composition of foods can be obtained so that the iodine number might be expected to become obsolete.

Isomerism

    The presence of double bonds restricts rotation of the alkyl chain. This allows for isomerism around the double bond, recognized by a *cis* or *trans* configuration.

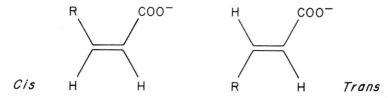

Naturally occurring unsaturated fatty acids in mammals are all of the *cis* configuration. *Trans* fatty acids, however, are present in the body of people who eat the ordinary western diet. They arise as the result of catalytic hydrogenation of fats, a process used in the manufacture of certain commercially prepared foods such as margarine and peanut butter. Hydrogenation is used to produce some solidification of these fat-containing foods in order to make them easier to use. In the hydrogenation process, some of the naturally occurring *cis* double bonds are converted to the *trans* configuration. When ingested by man, *trans* fatty acids are either oxidized or incorporated into structural lipids. Although their presence in the body does not appear to produce any ill effects, the question of whether they might be injurious in a subtle way has not yet been resolved.

    Another form of isomerism that can occur in unsaturated fatty acids concerns the *position* of the double bond in the alkyl chain. For example, the double bond can be present in position 9,10 in one monounsaturated acid; in another monounsaturated acid with the same chemical formula the double bond may be in position 11,12. The mammalian organism is quite specific in its requirement for certain positional isomers and synthesizes only the 9,10 isomer.

Another kind of isomerism that can occur in fatty acids results from the presence of a branch in the alkyl chains.

$$C-C-C-C-C-C-COOH$$

Straight chain

$$C-C-C-\overset{\overset{\displaystyle C}{|}}{C}-C-COOH$$

Branched chain

The fatty acids that are normally present in mammalian tissue are almost all of the straight-chain variety. However, branched-chain fatty acids or their precursors may be ingested in certain foods, for example, *phytanic acid* in butter and *phytol* in green vegetables. In these situations the body degrades the branched-chain acids, and they do not accumulate to any appreciable extent under ordinary conditions. In at least one disease, *Refsum's disease*, a serious neurologic defect results from the accumulation of phytanic acid in the tissue.

$$CH_3-\overset{\overset{\displaystyle CH_3}{|}}{CH}-CH_2-CH_2-CH_2-\overset{\overset{\displaystyle CH_3}{|}}{CH}-CH_2-CH_2-CH_2-\overset{\overset{\displaystyle CH_3}{|}}{CH}-CH_2-CH_2-CH_2-\overset{\overset{\displaystyle CH_3}{|}}{CH}-CH_2-COOH$$

Phytanic acid

This twenty-carbon atom acid has a sixteen-carbon atom chain with methyl groups at C_3, C_7, C_{11}, and C_{15}.

Hydroxy fatty acids. The central nervous system contains some fatty acids that have a hydroxyl group attached to the alkyl chain. As exemplified by *cerebronic acid*, the hydroxyl group in these acids is attached to the α-carbon atom.

$$CH_3-(CH_2)_{21}-\overset{\overset{\displaystyle OH}{|}}{\underset{\underset{\displaystyle H}{|}}{C}}-COOH$$

Cerebronic acid

$$CH_3-(CH_2)_5-\overset{\overset{\displaystyle OH}{|}}{\underset{\underset{\displaystyle H}{|}}{C}}-CH_2-CH=CH-(CH_2)_7-COOH$$

Ricinoleic acid

Other hydroxy fatty acids that occur in natural products can have the hydroxyl group attached at other locations. In *ricinoleic acid*, the main fatty acid in castor oil, the hydroxyl group is attached to C_{12}. Castor oil is a cathartic, and this effect is somehow produced by its ricinoleic acid content.

Essential fatty acids

In addition to obtaining fat from the diet, humans can biosynthesize certain fatty acids, including the saturated, monoenoic, and polyenoic varieties. However, mammals cannot synthesize all of the necessary types of polyenoic fatty acids. Those polyenoic fatty acids that cannot be synthesized must be obtained from the diet—they are termed essential fatty acids. Two of the four classes of unsaturated fatty acids, the linoleic (ω 6) and linolenic (ω 3) acid classes, cannot be biosynthesized by mammals but are synthesized by plants. One of these, linoleic acid, is essential for health, probably because the longer and more highly unsaturated members of its family are precursors of certain prostaglandins. As long as adequate amounts of linoleic acid are available, the mammal can synthesize from it the other members of the ω 6 class that are required. In practice, each member of the ω 6 class is known as an essential fatty acid. However, the only member of the class that actually must be provided in the diet is linoleic acid; hence in the strict sense only linoleic acid is truly essential to man.

In animals a disease state termed *essential fatty acid deficiency* occurs if essential fatty acids are excluded from the diet for long periods of time. The full syndrome has never been produced in man, but the chemical changes that precede the symptoms do occur in situations where humans are placed on fat-free diets for prolonged periods. Certain of the prostaglandins are synthesized from derivatives of linoleic acid (Chapter 13). Lack of these specific prostaglandins may in some way cause the essential fatty acid deficiency syndrome. Essential fatty acids also may be required to maintain normal membrane function and to obtain proper coupling in oxidative phosphorylation. However, it is intriguing to note that while the intact animal requires essential fatty acids, several mammalian cell lines in culture proliferate and appear to function normally even though they are deprived of essential fatty acids. Like the intact organism, these cell lines cannot synthesize any fatty acids of the ω 6 class and contain almost none of these acids when they are omitted from the culture medium.

Fatty acid derivatives

The complex lipids can be considered as being fatty acid derivatives. In most cases the fatty acid is attached covalently to an alcohol, forming an ester linkage. However, amide and ether linkages also occur in certain of the complex lipids.

Cholesteryl esters

Since the sterols are the subject of Chapter 10, only brief mention of the cholesteryl esters will be made at this point. Cholesteryl esters contain fatty acid esterified to the 3-β-hydroxyl group of the steroid ring system. The fatty acid moiety usually is of the long-chain variety and often is unsaturated. Cholesteryl esters are storage forms of cholesterol. They occur in intracellular lipid droplets and in the plasma lipoproteins. Most of the cholesterol that accumulates in the arterial intima in the disease process, atherosclerosis, is esterified. This will be discussed in great detail in Chapter 10, which presents the structural formulae of cholesterol and its esters.

Acylglycerols (glycerides)

Fatty acid esters of glycerol, the acylglycerols, are commonly known as glycerides. The fatty acid moiety in lipid esters is known as an acyl group. The class of glyceride depends on the number of glycerol alcohol groups that are esterified. Three general types of glycerides occur, *monoglycerides*, *diglycerides*, and *triglycerides*. In monoglycerides the single acyl group may be linked to either the primary or secondary alcohol groups. Therefore, two forms of monoglyceride can occur, a 1- or 2-monoglyceride. The numbering of the glycerol carbon atoms is shown for the monoglycerides. This numbering scheme applies to all of the acylglycerols, including the phosphoglycerides.

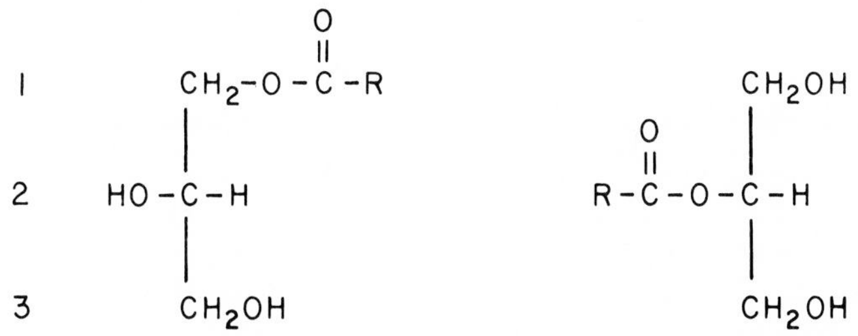

1 – Monoglyceride   2 – Monoglyceride

Monoglycerides are important in digestion and as metabolic intermediates. Two types of diglycerides also occur, depending on whether the acyl groups are attached to the 1,2- or 1,3-alcohol groups of the glycerol moiety.

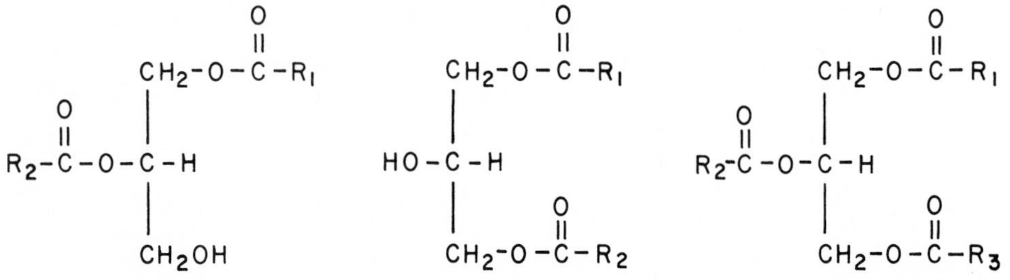

$$
\begin{array}{ccc}
\overset{O}{\overset{\|}{\text{CH}_2\text{-O-C-R}_1}} & \overset{O}{\overset{\|}{\text{CH}_2\text{-O-C-R}_1}} & \overset{O}{\overset{\|}{\text{CH}_2\text{-O-C-R}_1}} \\
\underset{}{\overset{O}{\overset{\|}{\text{R}_2\text{-C-O-C-H}}}} & \text{HO-C-H} & \overset{O}{\overset{\|}{\text{R}_2\text{-C-O-C-H}}} \\
\text{CH}_2\text{OH} & \overset{O}{\overset{\|}{\text{CH}_2\text{-O-C-R}_2}} & \overset{O}{\overset{\|}{\text{CH}_2\text{-O-C-R}_3}}
\end{array}
$$

| 1, 2 – Diglyceride | 1, 3 – Diglyceride | Triglyceride |

In man, diglycerides occur almost exclusively as metabolic intermediates. Triglycerides are the most prevalent of the acylglycerols, for they are quantitatively the major storage and transport form of fatty acids. The fatty acid residues in a given triglyceride molecule may occur in many combinations, for example, palmitic + 2 oleic; 2 palmitic + linoleic; palmitic + stearic + oleic; and many other combinations. Hence while triglyceride is considered to be a "single" species in everyday usage, it actually represents a family of molecules of varying fatty acid composition.

Phosphoglycerides

Acylglycerols that contain phosphoric acid esterified at the C_3-hydroxyl group are termed phosphoglycerides. They form bilayers when dispersed in an aqueous solution and, in this form, are the main structural components of cell membranes as described in Chapter 10. Phosphoglycerides have the general structure:

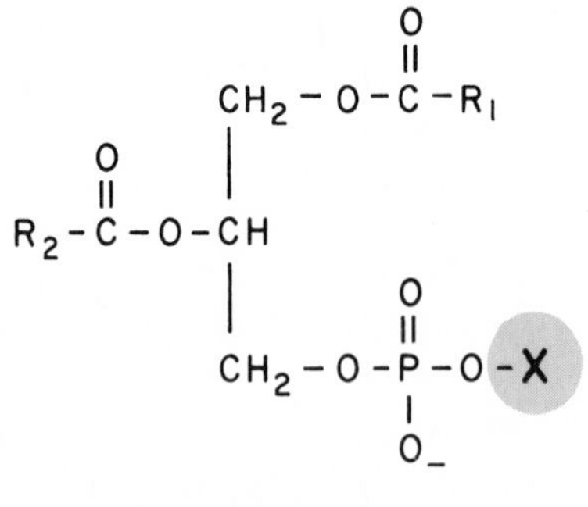

$$
\text{X} = (\text{CH}_3)_3\overset{+}{\text{N}}\text{-CH}_2\text{-CH}_2\text{-}
$$

(*Choline*)

$$
\text{X} = \text{NH}_2\text{-CH}_2\text{-CH}_2\text{-}
$$

(*Ethanolamine*)

General formula for a diacylphosphoglyceride

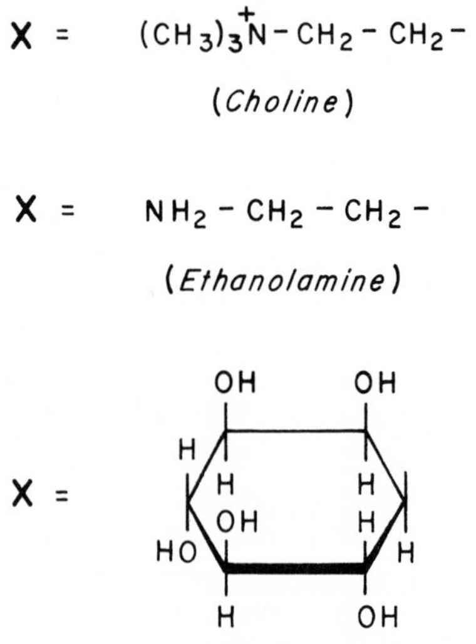

$$
\text{X} =
$$

(*Inositol*)

where X represents a group derived from an alcohol, such as *choline, ethanolamine, serine, inositol,* and *glycerol,* giving rise to *phosphatidyl choline, phosphatidyl ethanolamine,* etc. *Phosphatidic acid* is the name of a class of phosphoglycerides in which the group represented as X in the general structural formula for a diacylphosphoglyceride is a hydrogen. Hence it is the simplest form of diacylphosphoglyceride. The common name for phosphatidylcholine is lecithin. As with glycerides, the fatty acid residues of a phosphoglyceride usually vary. Hence there is actually a family of lecithins, for example, palmitoyl-oleyl-lecithin and stearoyl-linoleyl-lecithin. In general, the acids present in position 1 are more saturated than those contained in position 2, as is implied in these examples. For the sake of simplicity, each phosphoglyceride class is usually considered as a single entity without subclassifying as to fatty acid composition, for example, lecithin. The phosphoglycerides are important structural components of cellular membranes.

Stereospecific numbering. Phosphoglycerides are derivatives of glycerol phosphoric acid. Carbon atom 2 of this glycerol derivative is asymmetric, and the structure can be considered as either D-glycerol 1-phosphate or L-glycerol 3-phosphate according to the conventions discussed in Chapter 2. The latter configuration has been selected by convention. Therefore, as illustrated in the general structural formula for phosphoglyceride above, the phosphate is written as being at position 3 of the glycerol moiety, and the hydroxyl group attached to carbon atom number 2 is projected to the left. This convention is known as the stereospecific numbering (sn) system.

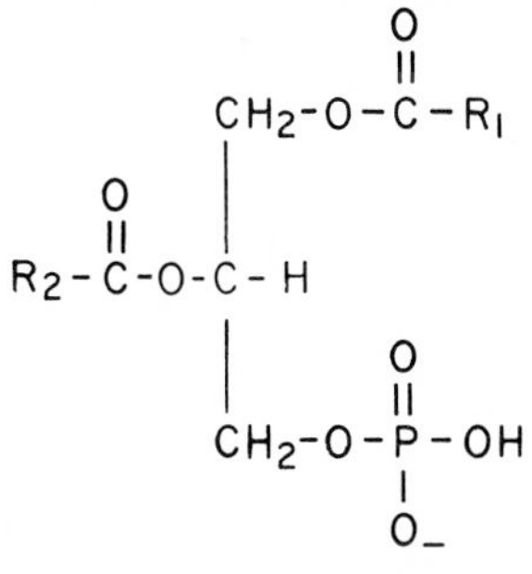

Phosphatidic acid

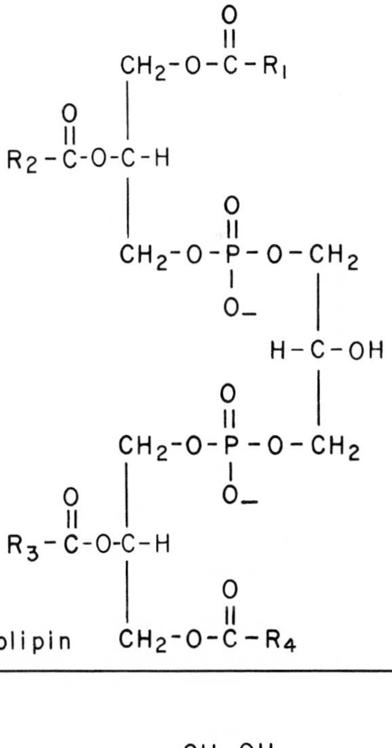

Cardiolipin

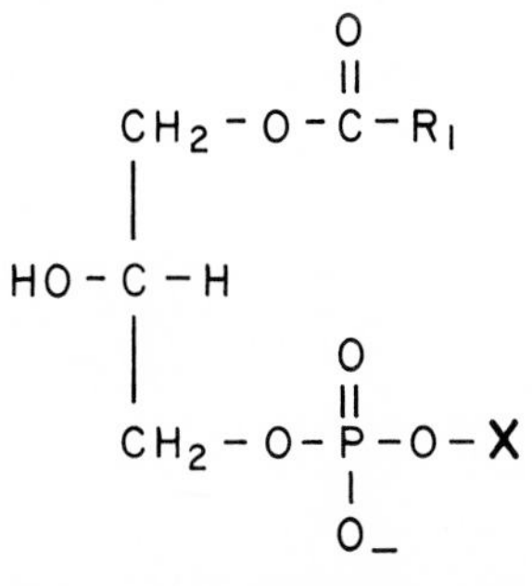

1-Acyl lysophosphoglyceride

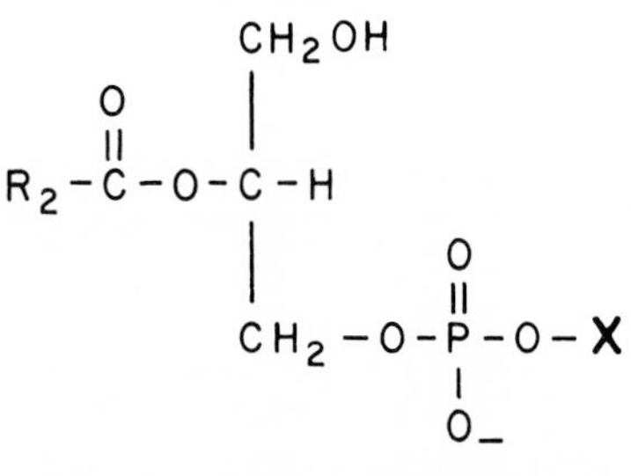

2-Acyl lysophosphoglyceride

Cardiolipin.   These are complex phosphoglycerides, present in the inner mito-chondrial membranes and in chloroplast membranes. Cardiolipins are composed of two molecules of phosphatidic acid joined together by a glycerol bridge. Like the phosphatidic acids, cardiolipins are more acidic than the diacylphospho-glycerides.

Lysophosphoglycerides.   In a lysophosphoglyceride one of the two acyl groups is removed. There is no acyl group in position 2 of a 1-acyl lysophosphoglyceride; the acyl group of position 1 is absent in a 2-acyl lysophosphoglyceride. Several types of phosphoglycerides have lyso derivatives, the most common of which are *lysolecithin* and *lysophosphatidylethanolamine*. Lysophosphoglycerides are ex-tremely damaging to cellular membranes because they act as a detergent and are lytic. Yet there is always a small amount of lysolecithin in blood plasma that arises primarily from the metabolism of the lecithin contained in the plasma lipoproteins.

Alkyl ethers and plasmalogens.   In alkyl ethers the hydrocarbon chain present at position 1 is attached to the glycerol moiety through an *ether*, not an ester, linkage. Plasmalogens are alkyl ether phosphoglycerides in which the ether-linked alkyl chain contains a double bond between C_1 and C_2; it is an *alk-1-enyl ether* or a vinyl ether, with the double bond in the *cis* configuration. Ethanolamine is the most prevalent base in the plasmalogens. Alkyl-ether linkages also occur in struc-tures that are similar to triglycerides. Here, too, the ether linkage occurs at posi-tion C_1; acyl groups are esterified at the positions C_2 and C_3. The hydrocarbon chain contained in ether linkage usually contains sixteen carbon atoms; it is derived from palmitic acid. Alkyl ethers and plasmalogens are present in particu-larly high concentration in cancer cells.

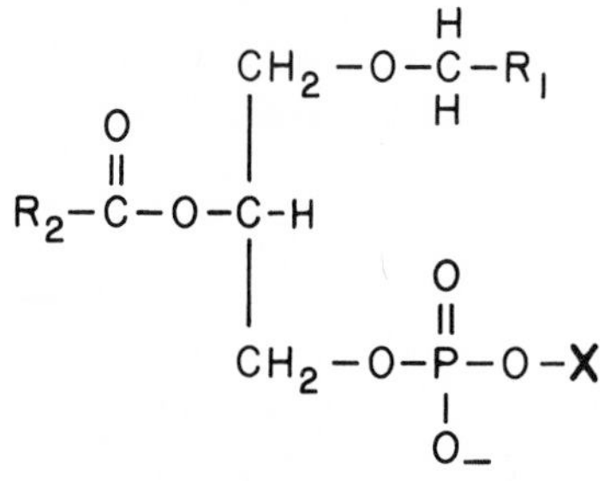

Alkyl ether phosphoglyceride 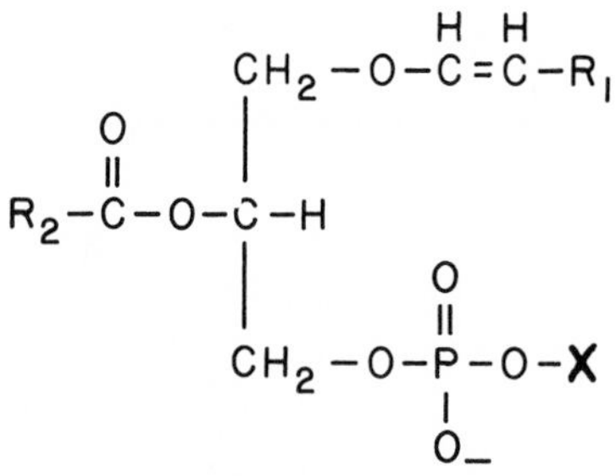   Plasmalogen

Sphingosine derivatives

Another series of lipids that occur in man are derivatives of sphingosine. This eighteen-carbon atom, dihydric alcohol contains an amino group at C_{17}. When written in the form shown below, sphingosine bears some resemblance to a glycerol moiety containing a *trans*-alk-1-enyl residue.

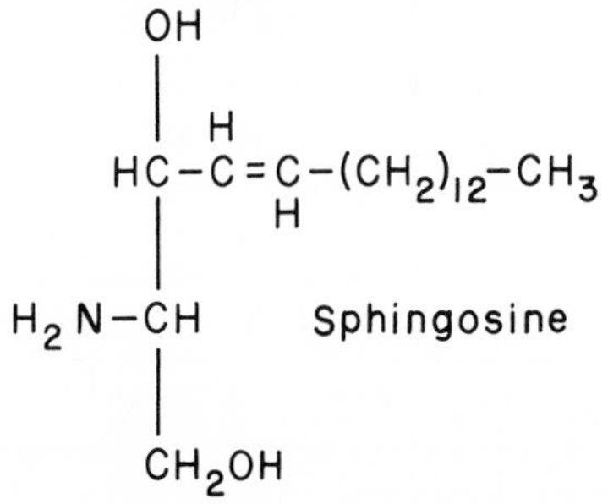

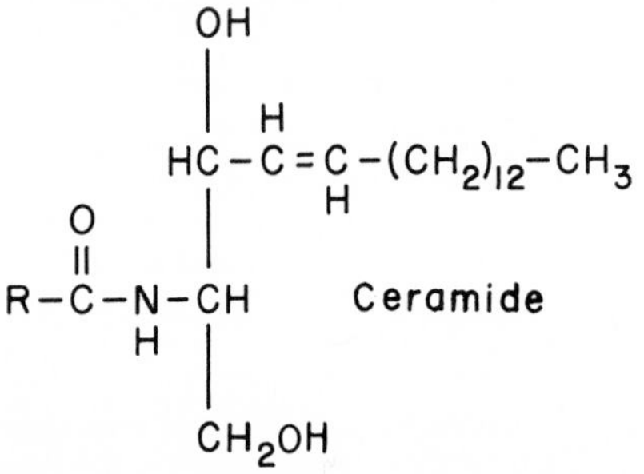

Dihydrosphingosine is a similar long-chain base that does not contain any double bonds. Fatty acids, usually containing eighteen or more carbon atoms, can attach to the amino group in amide linkage. Sphingosine containing an amide-linked fatty acid is known as *ceramide*. There are three types of ceramide derivatives in human tissue, *sphingomyelin, galactosylceramides*, and *glucosylceramides*.

Sphingomyelin. Because it contains phosphorus, sphingomyelin usually is considered as being a member of the phospholipid class. In sphingomyelin a phosphorylcholine residue is linked to the terminal hydroxyl group of ceramide. The structure is abbreviated as Chln-P-Cer. It is a major lipid component of many biologic membranes; for example, it accounts for 22% to 30% of the phospholipid in the human erythrocyte membrane.

Galactosylceramides. Ceramides that contain a D-galactose residue at the terminal hydroxyl group are known as galactosylceramides. The galactose is linked to this hydroxyl through a β-D-glycosidic bond, and these compounds are abbreviated as Gal-Cer.

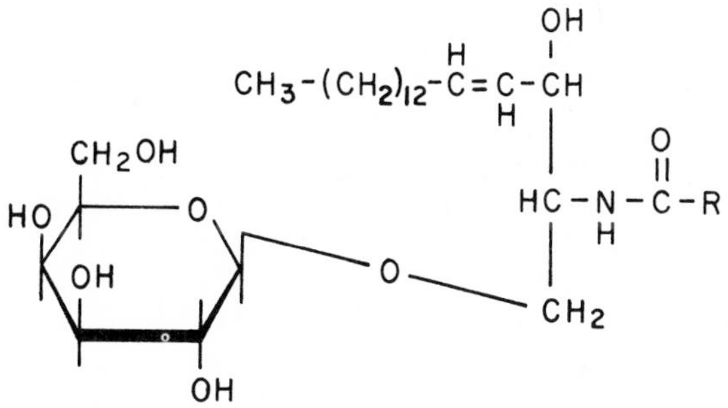

Galactosylceramide

Galactosylceramides do not contain phosphorus and are therefore not phospholipids. Ceramides that contain one monosaccharide residue are known as cerebrosides. These glycolipids are present in the brain and peripheral nervous system, particularly in the myelin sheath. A *sulfatide* is formed when a sulfate group is attached to one of the hydroxyl groups of the galactose residue, usually at position 2. This form of sulfatide is abbreviated as S-Gal-Cer.

Glucosylceramides. Ceramides that contain a D-glucose residue at the terminal hydroxyl group are known as glucosylceramides. Those that contain only the single glucose residue also are called cerebrosides. They are abbreviated as Glc-Cer. The glucose residue is attached through β-D-glycosidic linkage.

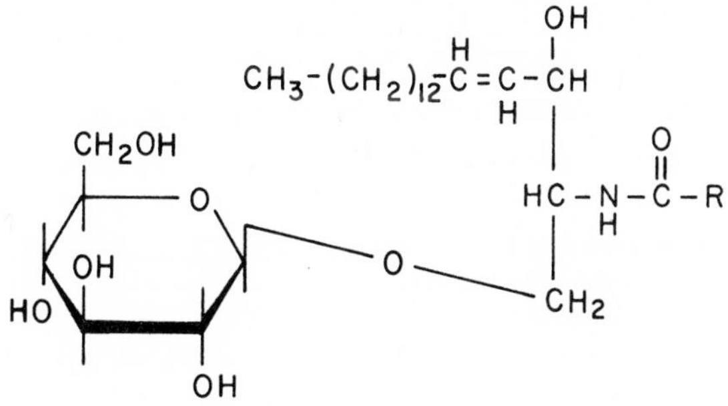

Glucosylceramide

Additional monosaccharide moieties can be attached to the terminal glucose residue, giving rise to the more complex glycosphingolipids, whose basic structures are presented below in abbreviated form.

$$\text{Gal I} \xrightarrow{\beta} \text{4 Glc} - \text{Cer} \qquad \text{Ceramide dihexoside}$$

$$\text{Gal I} \xrightarrow{\alpha} \text{4 Gal I} \xrightarrow{\beta} \text{4 Glc} - \text{Cer} \qquad \text{Ceramide trihexoside}$$

$$\text{GalNAc I} \xrightarrow{\beta} \text{3 Gal I} \xrightarrow{\alpha} \text{4 Gal I} \xrightarrow{\beta} \text{4 Glc} - \text{Cer} \qquad \text{Globoside}$$

$$\text{GalNAc I} \xrightarrow{\beta} \text{4 Gal I} \xrightarrow{\beta} \text{4 Glc} - \text{Cer} \qquad \text{Ganglioside (GM}_2\text{)}$$
$$\overset{3}{\underset{\uparrow}{}}$$
$$\text{2 NANA}$$

Each of these complex glycosphingolipids has a glucose residue attached directly to the ceramide. The globosides and gangliosides contain N-acetylgalactosamine (GalNAc), and the gangliosides contain N-acetylneuraminic acid (NANA). These glycosphingolipids are contained in cell membranes, with their carbohydrate residues projected out into the surrounding aqueous environment. Gangliosides are particularly prevalent in the gray matter of the brain and at nerve synapses. The glycosphingolipids present at the surfaces of cells are involved in cell-cell recognition and are antigenic, accounting for certain of the blood-group substances. The presence of glycosphingolipid and glycoprotein antigens on the surface of cells makes it necessary to match blood or tissue types before carrying out either a blood transfusion or a tissue transplantation.

Genetic diseases that are caused by accumulation of one of the sphingolipids occur in humans. These diseases result from the absence of, or a defect in, a hydrolase needed for the degradation of one of the sphingolipids. For example, cerebrosides accumulate in *Gaucher's* disease and ceramide trihexoside accumulates in *Fabry's* disease. The particular ganglioside illustrated above, GM_2, is the glycosphingolipid that accumulates in *Tay-Sachs* disease. This group of diseases is known as either the *lipid storage diseases* or *lipidoses*.

Plants and bacteria contain glycolipid structures in which a monosaccharide is attached in glycosidic linkage to a 1,2-diglyceride. This class of glycolipid, which is known as the glycosylacylglycerols, is not present in mammals.

Terpenes

The *terpenes* are polymers of the five-carbon atom isoprene unit. They include vitamins A, E, and K (Chapter 1) and the dolichols, polyprenols that contain many isoprene units (Chapter 7).* Dolichyl phosphate, the main dolichol in mammalian cells, is involved in glycoprotein synthesis. It transfers a carbohydrate unit, either a single monosaccharide residue or an oligosaccharide chain, to the protein core that is contained in the lipid phase of membranes. The lipid structure of dolichol somehow facilitates movement of the carbohydrate unit through the membrane lipid phase to the proper location where it can be added to the glycoprotein backbone.

---

*The lipid dolichols should not be confused with dulcitol (galactitol), the reduced alcohol form of galactose.

ANALYTIC METHODS IN LIPID CHEMISTRY

The initial problem in dealing with lipids is to remove them without degradation from the tissue or body fluid that must be analyzed. A commonly used method is to extract the material with a chloroform-methanol mixture. The extracted lipid is then isolated in the chloroform solution by separating the chloroform-methanol mixture into two phases. This is accomplished by addition of a dilute salt or acid solution to the chloroform-methanol extract. The lipids contained in the chloroform solution may be either saponified or separated by thin-layer chromatography.

Saponification

When it is necessary to isolate the fatty acids or sterols present in a given lipid, the material is treated with an alcoholic potassium hydroxide solution. The ester bonds are saponified in the reaction:

$$R-\overset{\overset{\displaystyle O}{\|}}{C}-O-R' + KOH \longrightarrow R-\overset{\overset{\displaystyle O}{\|}}{C}-O^- \; K^+ + R'OH$$

After saponification, the mixture is extracted with an organic solvent such as hexane while it remains alkaline. Any sterols that are present are extracted from the alkaline mixture into the hexane. The saponification mixture then is acidified and extracted again with the organic solvent. Acidification converts the fatty acids from the soap to the un-ionized form and permits their extraction into the organic phase. The material extracted by organic solvents after acidification is called the saponifiable fraction.

Thin-layer chromatography

The lipids contained in a tissue extract may be separated into their individual components, for example, triacylglycerols, cholesteryl esters, and phospholipids. Thin-layer chromatography (TLC) is a simple method for accomplishing this separation. An adsorbant, usually silica gel, is applied in a thin layer to a firm backing such as a glass plate. A binder such as calcium sulfate sometimes is included to promote adherence of the gel to the support. In order to obtain good separations, it is necessary to dry or *activate* the layer of adsorbant prior to use of the plate. The lipid mixture is applied near one edge of the gel, and the plate is placed in an equilibrated tank containing the appropriate solvent mixture. Solvent systems containing hexane and ether often are employed for separation of cholesteryl esters, fatty acids, and acylglycerols. More polar solvent systems containing chloroform and methanol commonly are used to separate the phosphoglycerides and sphingolipids. The solvents move up the thin-layer plate through capillary action, and the extent to which a given lipid migrates depends on its polarity relative to the given solvent mixture. After the adsorbant dries, the lipids can be made visible by exposure to iodine vapor or dyes such as rhodamine 6G that form fluorescent complexes with lipids. The areas of the adsorbant that contain the lipids can be collected, and the lipid material can be removed by elution with chloroform or another appropriate solvent.

Gas-liquid chromatography

Another powerful separation technique that is ideally suited to lipid chemistry is gas-liquid chromatography (GLC). This can be employed to separate mixtures of fatty acids or sterols, but it also has wide application in many other areas, for example, amino acid, drug, and pesticide separations. Therefore, GLC has become an important diagnostic aid in toxicology and forensic pathology. The inert support placed inside the column is coated with a material such as ethyleneglycol succinate that is a liquid at the temperature at which chromatography is performed, usually about 175° C. Derivatives of the test material that are more volatile usually are prepared before the sample is chromatographed. Fatty acids are converted to methyl esters prior to chromatography, and sterols to trimethylsilyl ethers. The sample is applied in a hydrocarbon solvent, often hexane, to a thin column that usually is about 2 m in length. A carrier gas such as nitrogen is passed over the column. The lipid derivatives partition in the liquid and gas phases. The time during which they are retained on the column depends in the case of fatty acid methyl esters on their chain length and degree of unsaturation. The column effluent is passed through a detector such as a flame ionization device that causes the organic molecules to ionize at the temperature of the hydrogen flame. An amplifier and recorder are connected to the detector, and the amount of lipid present is recorded as the area under a curve. When the sample passes through the hydrogen flame detector, it is combusted and is no longer suitable for further analysis. It is also possible to divide the vapor stream from the separating column so that samples of the eluate may be recovered for other uses.

DIGESTION AND ABSORPTION OF DIETARY FAT

The only absolute dietary lipid requirement is that about 1% of the caloric intake be in the form of essential fatty acid. However, in practice, between 20% and 40% of the dietary caloric intake is lipid. Most of the dietary lipid intake is triglyceride, but small amounts of phosphoglycerides, cholesteryl esters, and cholesterol also are ingested. These lipids must be emulsified in the intestinal lumen, digested by hydrolytic enzymes, and absorbed into the intestinal mucosal cells.

Emulsification of dietary lipids

Emulsification of the lipids present in the aqueous chyme occurs in the duodenum where lipids interact with bile. The constituents of the bile that produce emulsification are the conjugated bile acids, phosphatidylcholine, and cholesterol. A complete discussion of bile is presented in Chapter 10. Emulsification functions to get the poorly soluble dietary lipids into mixed micelles.

Micelles are aggregates formed in aqueous solution by a substance composed of both polar and nonpolar groups. Lecithin is an example of a compound that forms micelles. It contains nonpolar fatty acid chains as well as the polar phosphorylcholine group and is therefore *amphipathic*. The nonpolar components orient themselves inside the aggregate while the polar groups are on the outside, where they intereact with the surrounding water molecules. *Mixed micelles* are micelles made up of more than one compound. For example, the lipid constituents of the bile – lecithin, cholesterol, and conjugated bile acids – exist together in aggregates that are mixed micelles. Dietary lipids such as triglycerides and cholesteryl esters are not amphipathic and do not form micelles. However, the mixed micelles composed of the bile lipids are able to take up these very nonpolar materials. In other words, the mixed micelles offer triglycerides and cholesteryl esters a suitable nonpolar environment within the interstices of the micellar structure and in this way function to disperse these dietary lipids in the aqueous intestinal chyme. In micellar form these lipids can be acted on by the digestive enzymes. After hydrolysis, the products diffuse from the micelle to the intestinal mucosal cell membrane.

Hydrolytic enzymes

Three hydrolytic enzymes are produced by the exocrine pancreas and secreted into the duodenum. These are pancreatic lipase, cholesterol esterase, and phospholipase A.

Pancreatic lipase. Pancreatic lipase catalyzes the partial hydrolysis of triglycerides containing long-chain fatty acids.

$$\text{Triglyceride} + 2H_2O \rightleftharpoons 2\text{-Monoglyceride} + 2\text{ Fatty acid}$$

Almost all of the ordinary dietary triglycerides are of the long-chain fatty acid variety. Pancreatic lipase is specific for the fatty acid residues at positions 1 and 3 of the glyceryl moiety. Digestion of triglyceride largely stops at the 2-monoglyceride because the pancreatic triglyceride lipase exhibits very low activity toward this substrate. Both the 2-monoglyceride and the released fatty acids can pass through cell membranes, and they are absorbed by diffusion into the mucosal cells of the jejunum and ileum.

A protein cofactor that activates the pancreatic lipase, called *colipase*, is produced by the pancreas. Since the lipase acts at interfaces between water and the triglyceride molecules, interfacial adsorption of the enzyme is an important step in the catalytic process. Colipase binds to the mixed micelle containing the ingested triglycerides and facilitates adsorption of the lipase to the complex, thereby activating triglyceride hydrolysis.

Cholesterol esterase. Most of the dietary cholesterol is in unesterified form. Any cholesteryl ester that is present is emulsified by the bile and then hydrolyzed by the pancreatic cholesterol esterase.

$$\text{Cholesteryl ester} + H_2O \rightleftharpoons \text{Cholesterol} + \text{Fatty acid}$$

The molecular weight of the monomeric form of this enzyme is 65,000. The active form of the enzyme is polymeric and has a molecular weight of about 400,000. Polymerization is reversible and occurs in the presence of cholic acid, one of the main forms of bile acid. Thus the enzyme is activated when it comes into contact with bile. Fatty acid transfer occurs through a serine hydroxyl group, and this is fully exposed only when the enzyme is in its polymeric form. The released cholesterol and fatty acids pass from the micelles to the intestinal cell membrane through diffusion.

Phospholipase A. The phosphoglycerides present in the diet are digested by pancreatic phospholipase A_2. This enzyme catalyzes the hydrolysis of one of the fatty acid residues of the phospholipid, forming a 1-acyl lysophosphoglyceride.

$$\text{Phosphoglyceride} + H_2O \rightleftharpoons \text{1-Acyl lysophosphoglyceride} + \text{Fatty acid}$$

It is not known at present whether the lysophosphoglycerides enter the mucosal cells or are degraded further by *lysophospholipases*, enzymes that remove the remaining fatty acid residue from lysophosphatides. In terms of fulfilling caloric needs, phospholipids and cholesteryl esters serve only a minor role relative to triglycerides. Therefore, when the digestion of lipids is considered clinically, one usually refers only to pancreatic lipase and triglyceride hydrolysis.

Absorption and reesterification

Absorption from the mixed micelles into the mucosal cells is a passive process that occurs through diffusion. The main function of the intestinal mucosa in terms of lipid metabolism is to resynthesize the absorbed fatty acids and 2-monoglyceride into triglycerides, for the dietary long-chain fatty acid is absorbed into the body only after it is reconverted into triglyceride. This is an energy-dependent process requiring four moles of ATP/one mole of triglyceride that is synthesized. There appear to be two reasons for this seemingly wasteful expenditure of energy: first, the intact triglyceride cannot efficiently diffuse to or through the mucosal cell membrane, and second, the organism is given the opportunity of using endogenous fatty acids to alter the fatty acid composition of the triglyceride that is resynthesized and actually enters into the body. Triglycerides are synthesized in the intestine through the *2-monoglyceride pathway*.

$$\text{2-Monoglyceride} + 2\text{ FA} \sim \text{SCoA} \longrightarrow \text{Triglyceride} + 2\text{ CoASH}$$

The enzymes involved are a *monoglyceride acyl CoA acyltransferase* and a *diglyceride acyl CoA acyltransferase*.

Secretion and utilization of dietary triglycerides

Lipid droplets made up almost entirely of triglycerides accumulate in the mucosal cells. They are released into the lymph in the form of a lipoprotein, the *chylomicron*. This will be described in detail on p. 382. Chylomicrons are secreted into the lymph and pass from it into the venous blood. Their triglyceride content is removed through the action of a hydrolytic enzyme, *lipoprotein lipase*, and the released fatty acids are taken up by the tissues. Much of the absorbed dietary triglyceride is deposited in the liver and adipose tissue.

If the plasma triglyceride concentrations are measured in a healthy person who has fasted overnight and then eaten a meal containing 50 to 100 g of triglyceride, a relationship similar to that shown in Fig. 8-1 will be observed. Prior to eating, the plasma triglyceride concentration is less than 150 mg/dl. After eating, this gradually rises to a peak within 3 to 4 hr and then returns to normal within 6 to 8 hr. The triglyceride increment above fasting is caused mainly by absorbed chylomicrons. Thus development of hypertriglyceridemia is normal after a fatty meal; prolonged and excessive hypertriglyceridemia after eating fat is abnormal.

Medium-chain triglycerides

The information presented thus far applies only to the absorption of dietary long-chain triglycerides, those containing long-chain fatty acids. These are the

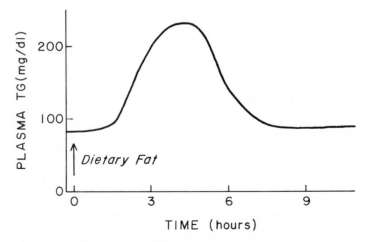

Fig. 8-1. Normal triglyceride absorption study. TG on ordinate refers to triglycerides.

triglycerides present in the ordinary foods that we eat. Recently medium-chain triglycerides have been used in certain therapeutic diets. These are triglycerides, prepared synthetically, that contain medium-chain fatty acids instead of the usual long-chain acids. The medium-chain triglyceride preparations that are currently in clinical use contain eight- and ten-carbon atom fatty acids almost exclusively. Medium- and long-chain triglycerides are digested and metabolized differently. Much of the ingested medium-chain triglyceride is absorbed intact either onto the mucosa cell villi or actually into the cells, and lipolysis is mediated by an intestinal cell microsomal lipase rather than by the pancreatic lipase secreted into the intestinal lumen. The medium-chain triglyceride is completely degraded to fatty acid and glycerol, another difference between it and the ordinary dietary triglycerides. Finally, the medium-chain fatty acids that are produced are *not* reesterified and secreted into the lymph in the form of chylomicrons. Instead, they pass directly into the portal vein as fatty acids, bind physically to plasma albumin, and are delivered directly to the liver as fatty acids by the portal circulation.

The differences in digestion and metabolism between the medium-chain and the ordinary dietary triglycerides are shown schematically in Fig. 8-2. These differences make medium-chain triglycerides useful as a caloric supplement in those disease states in which one or more of the metabolic reactions required for the utilization of ordinary dietary long-chain triglyceride is defective. In *abetalipoproteinemia*, for example, chylomicrons cannot be synthesized because of an inherited inability to make one of its protein constituents, apolipoprotein B. This

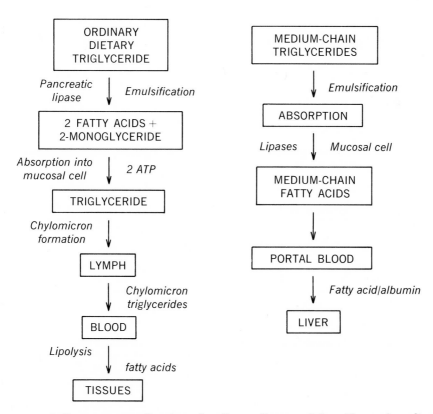

Fig. 8-2. Differences in utilization of ordinary dietary triglycerides and medium-chain triglycerides.

severely limits the quantity of long-chain fatty acid that can be absorbed. Use of medium-chain triglycerides as a dietary supplement enables the patient to obtain some dietary fat even though chylomicron synthesis is defective.

LIPID TRANSPORT

Lipids must be carried through the blood plasma from one tissue to another. An overview of this process, known as lipid transport, is presented in Fig. 8-3. The transport process serves three main purposes. Dietary triglyceride must be transported from the intestine to other tissues in the body. Triglyceride formed in the liver must be deposited for storage in adipose tissue. Finally fatty acid stored as triglyceride in the adipose tissue must be taken to other tissues in the form of free fatty acids in metabolic states, such as fasting, when these tissues require the stored fat as a source of energy.

Lipoproteins

The medium of the circulatory system is an aqueous solution, plasma, in which lipids are poorly soluble. In order to overcome this difficulty, a group of lipid transport macromolecules evolved, the plasma lipoproteins. These large complexes are made up of physically combined lipid and protein. They have a micellar structure, with triglycerides and cholesteryl esters contained in a hydrophobic core, surrounded by amphipathic lipids and proteins. The hydrophilic protein and lipid components serve to carry the nonpolar lipids through the aqueous environment, much as a boat can carry someone who cannot swim across a river. The properties of the four main classes of plasma lipoproteins are listed in Table 8-4, and the composition of these lipoproteins is given in Table 8-5. The compositions are average values, for each of the four lipoprotein classes are made up of a spectrum of molecules that vary in size and lipid content.

Plasma lipoproteins are separated either by discontinuous density gradient

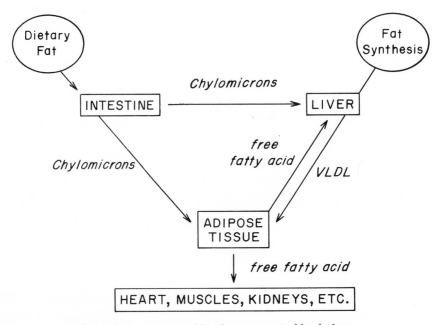

Fig. 8-3. Integration of lipid transport in blood plasma.

ultracentrifugation or electrophoresis. This has given rise to two terminologies, one based upon ultracentrifugal and the other on electrophoretic separation. For example, one of the classes of plasma lipoproteins is called either low-density lipoproteins or β-lipoproteins. In effect, both terms refer to the same class of lipoproteins.

Apolipoproteins. The protein components of the plasma lipoproteins are called apolipoproteins, abbreviated *apo-*. Letters are used to designate the various apolipoproteins, for example, apo-A, apo-B, etc. Although it was originally thought that each class represented a single apolipoprotein component, it was found subsequently that some were heterogenous. This has given rise to the use of a Roman numeral to distinguish the several members within a class, such as apo-A-I,

Table 8-4. Properties of plasma lipoproteins

LIPOPROTEIN	DENSITY RANGE	FLOTATION UNITS (S_F)*	MAJOR LIPIDS	ELECTRO-PHORETIC MOBILITY†
Chylomicron	<0.94	>400	Triglyceride	0
Very low density (VLDL)	0.94-1.006	20-400	Triglyceride	pre-β
Low density (LDL)	1.006-1.063	0-20	Cholesteryl esters	β
High density (HDL)	1.063-1.21		Phospholipids	α_1

*Flotation at density 1.063.
†Agarose gel electrophoresis, pH 8.6, in a barbital buffer.

Table 8-5. Average lipid and protein composition of plasma lipoproteins

COMPONENT	AVERAGE COMPOSITION (%)			
	CHYLOMICRONS	VLDL	LDL	HDL
Protein	2	9	21	50
Triglycerides	84	54	11	4
Cholesterol	2	7	8	2
Cholesteryl esters	5	12	37	20
Phospholipids	7	18	22	24

Table 8-6. Main apolipoproteins contained in the plasma lipoproteins

LOW-DENSITY LIPOPROTEINS (LDL)	VERY LOW–DENSITY LIPOPROTEINS (VLDL)	HIGH-DENSITY LIPOPROTEINS (HDL)
B (> 95)*	B (40)	A-I (65)
		A-II (25)
	Arginine-rich	
	C-I	C-I
	C-II	C-II
	C-III†	C-III

*Numbers in parentheses are the percentage that the particular apolipoprotein contributes to the total protein of the fraction. The values that are omitted are not known as yet with certainty.
†Apo-C-III has been subdivided further into apo-C-III-0, apo-C-III-1, and apo-C-III-2. These varients have 0, 1, and 2 sialic acid residues, respectively.

apo-A-II, etc. In addition, one class of apolipoproteins retains a descriptive name, the *arginine-rich apolipoprotein*, because it contains 11% arginine residues. The distribution of the main apolipoprotein components in three of the lipoprotein classes is listed in Table 8-6. Chylomicrons are not listed in this table because their apolipoprotein composition is uncertain at this time. It is known, however, that they do contain apo-B.

Apo-B and apo-A-I have primarily structural functions, organizing the lipids and proteins in the lipoprotein complex. Catalytic functions have been observed for several of the C-apolipoproteins. Apo-C-I is an activator of lecithin-cholesterol acyltransferase, and apo-C-II is an activator of lipoprotein lipase. These enzymes are involved in the removal of lipids from the lipoprotein complexes.

Chylomicrons. Chylomicrons are synthesized in the intestine. They contain chiefly triglycerides, and approximately 98% of their dry weight is lipid. They therefore have an extremely low density (<0.94). They are very large molecules, with a molecular weight as high as 1×10^9. Chylomicrons refract light, and they give the plasma a milky appearance when they are present in excessive concentrations. They are secreted into the lymph and enter the blood plasma via the thoracic duct. Their major function is to transport dietary fat, primarily in the form of triglyceride, into the body. The other dietary lipid that is transported in chylomicrons is cholesteryl esters.

Very low–density lipoproteins. Very low–density lipoproteins (VLDL) are also very large complexes. They contain about 90% lipid, from 50% to 65% of which is triglyceride. VLDL's are synthesized primarily in the liver and serve to transport triglycerides from the liver to other tissues, especially to adipose tissue. Because of their electrophoretic mobility, VLDL's also are known as pre-β lipoproteins.

Low-density lipoproteins. Most of the cholesterol contained in the blood plasma of a normal human after an overnight fast is present in low-density lipoproteins (LDL). These lipoproteins contain only a single apolipoprotein, apo-B. LDL's are formed in the plasma during VLDL catabolism. They are a product of VLDL and therefore can be thought of as a VLDL remnant. There is some question at present, however, as to whether all of the LDL contained in human plasma is a catabolic product derived from VLDL. Since LDL's migrate in the β-region on electrophoresis, they are also called β-lipoproteins.

Intermediate-density lipoproteins. Intermediate-density lipoproteins (IDL) are formed in the plasma during the conversion of VLDL to LDL. They contain amounts of triglyceride and cholesterol that are intermediate between those of VLDL and LDL. Like VLDL, they float in the ultracentrifuge at a density of 1.006, but like LDL, they migrate in the β-region on electrophoresis. Normally, the conversion of VLDL to LDL proceeds so efficiently that appreciable quantities of IDL do not accumulate in the blood plasma obtained after an overnight fast. Quantities of IDL do accumulate, however, in one of the lipoprotein diseases discussed below.

High-density lipoproteins. High-density lipoproteins (HDL) are synthesized in the liver. They probably act as catalysts, facilitating the catabolism of VLDL and chylomicrons. The low molecular weight C-apolipoproteins are transferred between HDL and VLDL, and lipid transfer also occurs. It is likely that HDL provides the protein activators required to allow lipoprotein lipase and lecithin-cholesterol acyltransferase to operate on the VLDL. In addition, HDL may take up lipids released from the VLDL during the action of these enzymes. Two subfractions of HDL have been identified, HDL_2 and HDL_3. In clinical practice, however, HDL is usually considered as a single fraction.

Lipoprotein lipase

The enzyme that hydrolyzes the triglycerides contained in the circulating chylomicrons and VLDL is lipoprotein lipase. It probably acts on or within the capillary endothelial cell. Lipoprotein lipase is present in many tissues, including adipose tissue, mammary gland, and heart. In the adipose tissue the activity of lipoprotein lipase is increased by administration of the hormone insulin. Lipoprotein lipase has specificity for fatty acids esterified at positions 1 and 3 of the triglycerides. The enzyme is activated by apolipoprotein CII, one of the low molecular weight proteins present in VLDL and HDL.

When an injection of heparin is given, lipoprotein lipase activity is released into the blood plasma. This appears to be a pharmacologic action of heparin, and it is unlikely that heparin is a physiologic cofactor for the enzyme. The lipolytic activity that is released is made up of two different enzymes that hydrolyze triglycerides. One of the lipases is derived from the liver. It is resistant to inactivation by protamine and is more active against triglycerides contained in VLDL than in chylomicrons. The second enzyme is of extrahepatic origin and is inhibited by protamine. It is active against both chylomicrons and VLDL and is the component primarily responsible for the hydrolysis of the triglycerides taken into the body in the form of chylomicrons. After heparin injection, from 45% to 95% of the lipolytic activity that appears in the blood plasma is protamine resistant, that is, the liver lipase.

Lipoprotein separations

Plasma lipoprotein analyses are made routinely in clinical laboratories by electrophoresis on a paper, cellulose acetate, or polyacrylamide gel support. The separation is done at a pH of 8.6 using a barbital buffer, and the lipid-containing regions are made visible with a stain such as oil red 0. Lipoproteins can be isolated from plasma in bulk quantities by preparative ultracentrifugation. Another method of lipoprotein separation involves the addition of heparin in the presence of a divalent cation. Before this is done, chylomicrons and VLDL are removed by a preliminary centrifugation. The addition of heparin and $MnCl_2$ will precipitate LDL; HDL remains in solution.

Proteolipids

A lipoprotein can be isolated from myelin, the membrane material of nerve axons, by extraction with organic solvents. This lipoprotein is known as a proteolipid. Unlike the plasma lipoproteins, proteolipid is insoluble in aqueous media and is soluble in lipid solvents. Proteolipid contains 8% protein and 92% lipid. The molecular weight of its protein component is 36,000. Like the plasma lipoproteins, proteolipid also has a micellar structure.

Free fatty acids

From 90% to 98% of the fatty acid contained in the blood plasma is in the form of fatty acid esters such as triglycerides, cholesteryl esters, and phospholipids. These fatty acid esters are contained in the plasma lipoproteins. A small quantity of fatty acid, however, is present in unesterified form as a physical complex with plasma albumin. This fraction is known as the plasma *free fatty acid* (FFA). There are two sources of the FFA that is present in the plasma. The main source is the adipose tissue, for FFA is the form in which the fatty acids stored in the triglycerides are released into the blood. In addition, some FFA is generated during the hydrolysis of the triglycerides contained in chylomicrons and VLDL when these lipoproteins deliver their fat load to recipient cells. FFA is transported to the heart, skeletal muscles, liver, and other tissues. There the FFA is either oxidized as a source of energy or incorporated into the esterified lipids of the tissues. The brain appears to be the only organ that does not utilize plasma FFA, for the FFA cannot cross the blood-brain barrier. Although plasma FFA is present in relatively small amounts, it has a half-life of only 1 to 2 min. Therefore, it has an extremely rapid turnover and is very important as a source of circulating lipid.

Integration of lipid transport

An overview of the lipid transport process is presented in Fig. 8-3. Dietary fat is digested and absorbed by the intestine. After resynthesis into triglyceride, it is moved into the lymph and then into the blood in the form of chylomicrons. Most of the dietary triglyceride is taken up by the adipose tissue and liver. In addition to taking up circulating fat, the liver synthesizes fat from carbohydrates; the excess fat accumulates in the liver as triglyceride, and this is secreted into the blood in the form of VLDL. Much of the VLDL triglyceride is delivered to the adipose tissue, where it is stored. Fat is mobilized from the adipose tissue as FFA, which is then transported through the blood as a complex with plasma albumin. LDL's are remnants of VLDL's; they are formed from VLDL during its breakdown. Although the exact role of HDL is still unknown, it probably assists in the catabolism of VLDL and chylomicrons by providing specific apolipoproteins and taking up some of their lipids.

Hyperlipoproteinemias

An abnormality very common in humans is the presence of excessive amounts of VLDL or LDL in the plasma following a 12- to 14-hr period of fasting. This gives rise to the condition termed *hyperlipidemia*, an excess of plasma lipid. This is synonymous with hyperlipoproteinemia, a term that gives a more correct picture of the actual metabolic abnormality because the plasma lipids actually are present in the form of lipoprotein complexes. Certain forms of hyperlipoproteinemia are associated with an increased incidence of atherosclerosis, the pathologic process that causes coronary heart disease and other serious illnesses. The main forms of hyperlipoproteinemia are listed in Table 8-7, and the qualitative abnormalities seen on plasma lipoprotein electrophoresis are illustrated schematically in Fig. 8-4.

Hyperlipoproteinemias are classified into the six types listed in Table 8-7. These are clinical syndromes and do not represent single genetic defects. A possible exception is type I, a hyperchylomicronemia that most likely results from a deficiency of the extrahepatic, protamine-sensitive lipoprotein lipase. Type I is

Table 8-7. Typing of hyperlipoproteinemias

TYPE	LIPOPROTEIN ABNORMALITY	MAJOR PLASMA LIPID ELEVATION	MINOR PLASMA LIPID ELEVATION
I	Chylomicrons	Triglycerides	Cholesterol
IIa	LDL (β)	Cholesterol	
IIb	LDL (β) and VLDL (pre-β)	Cholesterol and triglycerides	
III	Intermediate β	Cholesterol and triglycerides	
IV	VLDL (pre-β)	Triglycerides	Cholesterol
V	VLDL (pre-β) and chylomicrons	Triglycerides	Cholesterol

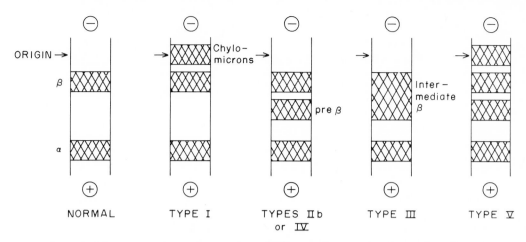

Fig. 8-4. Diagrammatic representation of plasma lipoprotein electrophoretic patterns. Plasma samples are obtained after the subject has fasted for 14 hr. The abnormality in type I is the presence of chylomicrons. In types IIb and IV, an excess of VLDL (pre-β) is present. LDL (β) also is elevated in type IIb. In type III, IDL (intermediate β) appears as a broad band covering the β and pre-β regions. This is known as a broad β band. In type V, both chylomicrons and VLDL (pre-β) are elevated. The pattern for type IIa is not shown because it appears normal qualitatively. Quantitative analysis of a type IIa sample would reveal, however, that LDL (β) is elevated.

manifested clinically as an inability to clear dietary fat from the blood at an adequate rate. Another possible exception is type IIa, which is manifested clinically by an increase in LDL in the plasma. Recent evidence suggests that type IIa may result from a defect in the high affinity cell surface receptor for LDL. Because of this, LDL removal from the plasma is slowed, leading to its abnormal accumulation. Moreover, because cholesterol from LDL cannot enter the cell in adequate amounts in spite of high circulating LDL concentrations, feedback inhibition of cholesterol synthesis does not occur (Chapter 10). This leads to continued cholesterol synthesis in spite of a sufficient supply being already present in the body and further contributes to the hypercholesterolemia. Even in this situation, however, it is not yet certain that all cases expressed clinically as type IIa are caused by a single gene defect for the LDL receptor. The metabolic abnormalities responsible for the other types of hyperlipoproteinemias are not as yet known.

ADIPOSE TISSUE

Fat in the form of triglyceride is the major storage form of energy in man. Of the three main nutrients, only fat can be stored in large quantities. This is because of the fact that the body contains specialized mesenchymal cells, *adipocytes*, that are devoted solely to the function of storing fat. Recent studies in rats indicate that the number of adipose cells in the body is determined by nutrition in infancy. Weanling rats that are overfed develop excessive numbers of adipocytes. This predisposes them to obesity in later life. Obesity, excessive accumulation of fat in the body, is associated with an increase in the number of adipocytes as well as an increase in the size of the individual adipocytes resulting from packing with triglyceride. It is therefore both a hypertrophy and a hyperplasia of the adipose tissue. Through mechanisms that as yet are not understood, the presence of excessive numbes of adipocytes signals the body to synthesize more triglycerides so that they can be filled, leading ultimately to an excess of total stored fat in the body.

As opposed to fat, there is no true storage site for protein or its constituent amino acids in the body. In times of severe need such as starvation, tissues and plasma proteins are catabolized in order to supply essential amino acids, but this is clearly a pathologic situation. There also is no specialized cell for glycogen storage; it is simply crowded into the cytoplasm of most cells, particularly in the liver and muscles. Because of this, the amount of glycogen that can be stored is quite limited, on the order of a 0.5 kg total. A primary metabolic drive in the body in times of abundant food supply is to convert excess calories from carbohydrate into fat so that they can be kept in reserve in the large and expandable adipose storage depots.

Triglyceride accumulation in adipocytes. The glyceryl moiety of triglycerides is derived from the glucose that is delivered to the adipocyte through the blood. Some fatty acid that is incorporated into the triglycerides is synthesized within the adipocyte from glucose. The remainder is delivered through the blood in the form of triglyceride contained in either chylomicrons or VLDL. In both cases the lipoprotein triglycerides must be hydrolyzed by lipoprotein lipase so that their fatty acid content can enter into the adipose cell.

Fatty acid mobilization from adipose tissue. In order to leave the adipocyte, the fat stored as triglycerides must be hydrolyzed. The fatty acid is released into the blood and is transported from the adipose tissue to other organs as FFA in a physical complex with albumin. Glycerol also is released from the adipose tissue when the lipolysis occurs. The glycerol produced by lipolysis cannot be utilized by adipocytes because they lack the enzyme glycerol kinase. Therefore, the glycerol is released and transported to the liver, an organ that contains sufficient amounts of glycerol kinase to efficiently metabolize glycerol (see Chapter 7). FFA is mobilized in large amounts from adipose tissue during periods of fasting, anxiety, or physical exertion. In this way the tissues of the body are ensured a constantly available circulating supply of fat either from the diet after eating or from the adipose tissue in the postabsorptive state or in stressful situations. Conversely, fat is accumulated in the adipose tissue when food is plentiful and the individual is calm and resting. The general aspects of adipose tissue function and lipid secretion are summarized in Fig. 8-5. Fasting, sympathetic nervous system discharge (norepinephrine), or release of many different hormones into the circulation (epinephrine, adrenocorticotropic hormone [ACTH], growth hormone, or glucagon) rapidly stimulates FFA release from the adipose tissue. Norepinephrine, epinephrine, ACTH, and glucagon activate the adipocyte triglyceride lipase. These hormones combine with receptor sites on the cell membrane and activate *adenyl cyclase*, a membrane-bound enzyme that converts ATP to 3′,5′-cyclic adenosine monophosphate (cAMP) (see Chapter 13). cAMP activates a protein kinase that in turn catalyzes an ATP-induced phosphorylation of the triglyceride lipase and transforms it from an inactive to an active form. The hydrolysis catalyzed by triglyceride lipase is the rate-limiting step in triglyceride catabolism. Hydrolysis of the resulting diglyceride and subsequently monoglyceride is mediated by a single enzyme, *monoglyceride lipase.*

The other hormones that stimulate lipolysis, for example, thyroxine and growth hormone, act more slowly. They probably act by increasing the synthesis of a regulatory protein rather than by activating existing adenyl cyclase. Glucocorticoids also stimulate lipolysis, but they facilitate the action of other fat-mobilizing hormones and do not exert a direct lipolytic effect. Since the intracellular triglyceride lipase is activated by hormonal effects, it is also known as the *hormonesensitive lipase.* It is entirely distinct from the heparin-sensitive lipase that is involved in moving triglyceride fatty acid into the adipocyte.

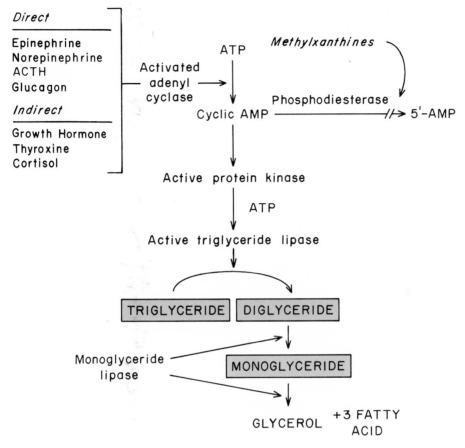

Fig. 8-5. Fatty acid mobilization from adipocytes.

High levels of glucose and insulin in the blood stimulate triglyceride accumulation in adipose tissue. When this occurs, the cAMP content in the adipocyte is reduced. Conversely, low blood glucose and insulin concentrations remove this stimulus and indirectly enhance the mobilization of fatty acid from the adipocyte. In addition, methylxanthines such as caffeine and theophylline enhance fatty acid mobilization from adipose tissue. These substances inhibit phosphodiesterase, the enzyme that inactivates cAMP. As a result, the cAMP level in the cell increases, maintaining the triglyceride lipase in active form. Prostaglandins E_2 and H_2, insulin, and nicotinic acid inhibit fatty acid mobilization by unknown mechanisms. Since the adenyl cyclase reaction and cAMP are central to the understanding of many hormonal mechanisms, these subjects will be dealt with more comprehensively in Chapter 13.

FATTY ACID OXIDATION

Fatty acids are a major fuel for man and other mammals. In the presence of O_2, fatty acids are catabolized to CO_2 and H_2O, and approximately 40% of the free energy produced in this process is conserved by the cell to form ATP. The remainder of the energy is released as heat. Fatty acid catabolism occurs in the mitochondria by a process known as β-oxidation. In this process two-carbon fragments are successively removed from the fatty acid in the form of acetyl CoA. The metabolism of the acetyl CoA fragments to CO_2 and H_2O in the Krebs cycle and the mechanism of ATP formation are the subjects of Chapters 5 and 6. Therefore, only the catabolism of fatty acid to the stage of acetyl CoA, the *β-oxidation pathway*, will be considered here. In β-oxidation the fatty acid is catabolized from the carboxyl end. Two hydrogen atoms are removed from the β-carbon atom, C_3 in the chain, and a keto group is then formed. Therefore, it is the β-carbon atom that is oxidized, giving rise to the term β-oxidation. Cleavage between the α- and β-carbon atoms occurs, and the two-carbon atom fragment, composed of the original carboxyl carbon and α-carbon atoms, is released as acetyl CoA. A single β-oxidation sequence producing one mole of acetyl CoA provides the cell with five moles of ATP. The mole of acetyl CoA that is released provides the cell with additional high-energy phosphate bonds equivalent to twelve moles of ATP if oxidized in the Krebs cycle to CO_2 and H_2O. An overview of β-oxidation is presented in Fig. 8-6, using a fatty acid containing eight carbon atoms for illustrative purposes. Although four acetyl CoA units are produced, only three β-oxidation sequences are required because the final oxidation produces two acetyl CoA units. By analogy, the oxidation of a sixteen-carbon atom fatty acid will produce eight acetyl CoA units but will require only seven β-oxidation cycles. Each of the individual reactions of the β-oxidation process will now be described.

Acyl CoA synthetase

The first step in the β-oxidation pathway is *activation* of the fatty acid, that is, formation of the acyl CoA thioester by combination with CoASH. The enzyme, acyl CoA synthetase, is also known as thiokinase or acyl CoA ligase.

$$FA + CoASH + ATP \xrightarrow{Mg^{++}} FA{\sim}S{-}CoA + AMP + PP_i$$

Notice that ATP is degraded to AMP and pyrophosphate, and the energy released is utilized to form the high energy ($\sim$) thioester bond. An acyladenylate complex, AMP $\sim$ FA, is an intermediate in the acyl CoA synthetase reaction. This also contains a high-energy bond, the acylphosphate ester, and serves to transfer the

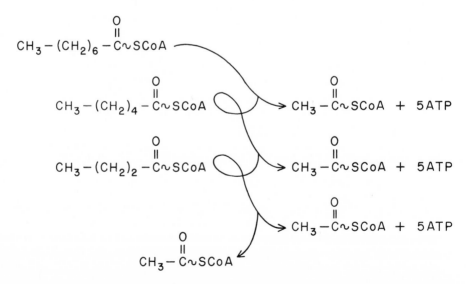

Fig. 8-6. β-Oxidation cycle. For simplicity, this is illustrated using octanoic acid (8 : 0).

activated acyl group to CoASH. Acyladenylate intermediates occur in other important biologic reactions such as protein synthesis, and they will be discussed in detail in subsequent chapters. At least three separate acyl CoA synthetases are present: a short-chain fatty acid thiokinase that activates acetate and propionate, a medium chain thiokinase that activates fatty acids containing four- to ten-carbon atoms, and a long-chain thiokinase that activates fatty acids containing twelve or more carbon atoms. The short- and medium-chain activating enzymes are exclusively mitochondrial in location; the long-chain activating enzyme is located in the endoplasmic reticulum as well as in the outer mitochondrial membrane. The pyrophosphate that is formed in this reaction is split into two moles of inorganic phosphate through the action of inorganic pyrophosphatase. Since this product is decomposed irreversibly, the activation reaction is favored in the left-to-right direction as written.

A GTP-linked acyl CoA synthetase is present in the matrix of the mitochondria. Unlike the ATP-linked synthetase, the products of the GTP reaction are GDP and inorganic phosphate, not the mononucleotide and pyrophosphate.

$$FA + CoASH + GTP \rightleftharpoons FA \sim SC_oA + GDP + P_i$$

The GTP-linked synthetase activates any free fatty acid that is generated inside the mitochondria.

Acylcarnitine formation

Acyl CoA synthesis can occur in the endoplasmic reticulum or in the outer mitochondrial membrane. However, the acyl CoA so formed cannot penetrate through the inner mitochondrial membrane to the site of the fatty acid β-oxidation enzyme system. In order to cross this barrier the acyl group is transesterified from CoASH to carnitine. This reaction is catalyzed by L-*carnitine acyltransferase*, an enzyme associated with the inner mitochondrial membrane.

L-Carnitine

Acylcarnitine

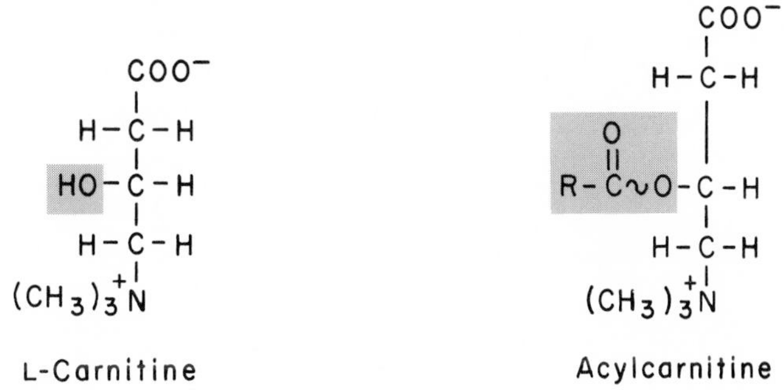

Outside:

$$FA \sim S - CoA + L\text{-carnitine} \longrightarrow FA \sim O\text{-carnitine} + CoASH$$

Inside:

$$FA \sim O\text{-carnitine} + CoASH \longrightarrow FA \sim SCoA + Carnitine$$

The enzyme is both a *transferase* and a *translocase*, for the acyl group that is bound to it as the carnitine ester is able to cross the inner mitochondrial membrane. Note that this reaction is reversible, indicating that the ester bond in acyl carnitine is a high-energy bond. Once across the membrane, a different carnitine acyltransferase located on the inner surface of the inner mitochondrial membrane catalyzes the reverse reaction, and the acyl group is transesterified to intramitochondrial CoASH. Through these reactions, the fatty acyl group is brought to the site of the β-oxidation enzymes, the mitochondrial matrix.

β-Oxidation sequence

Once the acyl CoA is inside the mitochondrion, four successive reactions occur. They, together with the enzymes catalyzing them, are listed below.

First dehydrogenation — acyl CoA dehydrogenase:

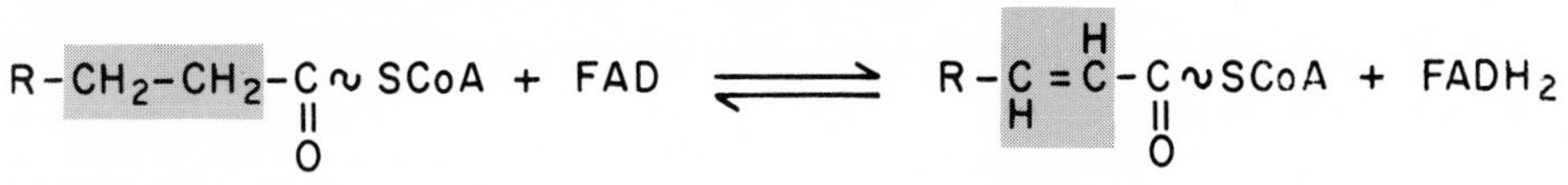

$$R-CH_2-CH_2-C \sim SCoA + FAD \rightleftharpoons R-\underset{H}{\overset{H}{C}}=C-C \sim SCoA + FADH_2$$

$$\Delta_2 \; \textit{trans} - \text{Enoylacyl CoA}$$

Hydration — enoylhydratase:

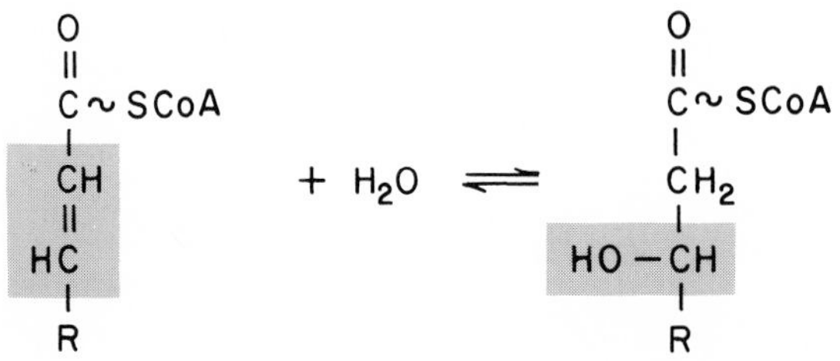

$$\begin{array}{c} O \\ \| \\ C \sim SCoA \\ | \\ CH \\ \| \\ HC \\ | \\ R \end{array} + H_2O \rightleftharpoons \begin{array}{c} O \\ \| \\ C \sim SCoA \\ | \\ CH_2 \\ | \\ HO-CH \\ | \\ R \end{array}$$

$$\text{L}-\beta-\text{Hydroxyacyl CoA}$$

Second dehydrogenation — L-β-hydroxyacyl dehydrogenase:

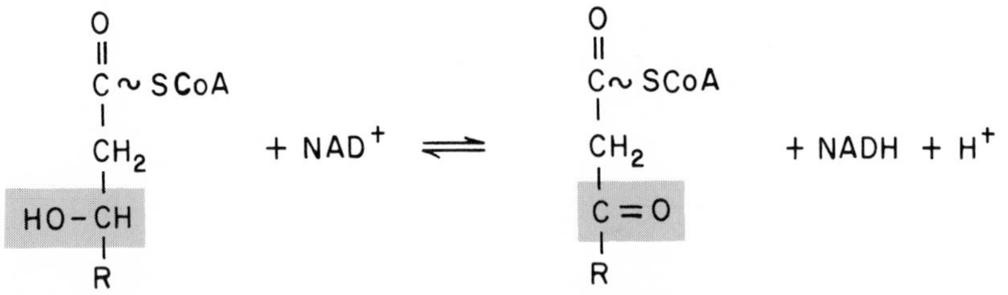

$$\begin{array}{c} O \\ \| \\ C \sim SCoA \\ | \\ CH_2 \\ | \\ HO-CH \\ | \\ R \end{array} + NAD^+ \rightleftharpoons \begin{array}{c} O \\ \| \\ C \sim SCoA \\ | \\ CH_2 \\ | \\ C=O \\ | \\ R \end{array} + NADH + H^+$$

$$\beta-\text{Ketoacyl CoA}$$

Thiolytic cleavage — thiolase:

$$R-\underset{O}{\overset{\|}{C}}-CH_2-\underset{O}{\overset{\|}{C}} \sim SCoA + CoASH \longrightarrow R-\underset{O}{\overset{\|}{C}} \sim SCoA + CH_3-\underset{O}{\overset{\|}{C}} \sim SCoA$$

$$\text{Acyl CoA} \qquad \text{Acetyl CoA}$$

In the first dehydrogenation step, a Δ_2-*trans*-enoylacyl CoA is formed. FAD is reduced in this reaction, and it in turn transfers one electron to the electron-transferring flavoprotein (FP_2), which moves the electron into the respiratory chain at the level of coenzyme Q. The second step involves hydration of the unsaturated bond, forming an L-β-hydroxyacyl CoA. Enoylhydratase, the enzyme catalyzing this reaction, is specific for the Δ_2 unsaturated bond. However, it is not specific for the *trans* configuration and, in certain instances, can operate quite well on a Δ_2-*cis* double bond. A similar β-hydroxyacyl intermediate occurs in fatty acid synthesis, but it is of the D configuration, not the L configuration as in the oxidation pathway. The dehydrogenase catalyzing the third reaction, forming the β-ketoacyl CoA, is specific for the L stereoisomer. NAD^+ is reduced in this reaction, and it subsequently transfers the reducing equivalents into the initial segment of the respiratory chain. The final reaction, thiolytic cleavage, forms acetyl CoA and an acyl CoA that is two carbon atoms shorter than the original acyl CoA that entered this β-oxidation sequence. This reaction is highly exergonic as written from left to right and serves to pull the β-oxidation sequence in the proper direction. The reaction has been written above with a one-way arrow from left to right. The acyl CoA that is generated reenters the β-oxidation cycle, and the process is repeated until the entire chain is degraded to acetyl CoA. Therefore, these reactions can be thought of in terms of a β-oxidation cycle, as shown in Fig. 8-6.

The $FADH_2$ that is formed in the first step is oxidized in the electron transport chain, yielding 2ATP. The NADH formed in the third step is also oxidized in the electron transport chain, yielding 3ATP. In other words, if an eight-carbon atom fatty acid is oxidized by β-oxidation, four acetyl CoA units would be generated and 15ATP would be formed in the β-oxidation process itself.

In three situations, auxiliary reactions are required in order to carry out the β-oxidation sequence. The situations involve the oxidation of monoenoic acids, polyenoic acids, and fatty acids with odd numbers of carbon atoms.

Enoyl CoA isomerase. When a monoenoic fatty acid undergoes β-oxidation, an intermediate acyl CoA contains a double bond in the incorrect location. Furthermore, the double bond is in the *cis*, not the *trans*, configuration. For example, consider the β-oxidation of oleic acid, $C_{18}\Delta_9$-*cis*. Recall that all naturally occurring unsaturated fatty acids have double bonds in the *cis* configuration. After three β-oxidation cycles, the acyl CoA product is a Δ_3-*cis*-enoyl CoA.

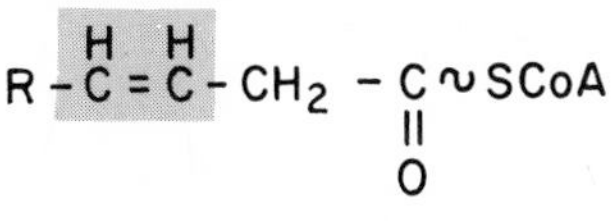

$$R - \overset{\displaystyle H}{\underset{}{C}} = \overset{\displaystyle H}{\underset{}{C}} - CH_2 - \overset{}{\underset{\displaystyle \underset{O}{\|}}{C}} \sim SCoA$$

$$\Delta_3 \; cis - \text{Enoyl CoA}$$

However, the required enoyl intermediate for the next reaction in the β-oxidation sequence must have a double bond in position Δ_2. The potential difficulty is solved by the enzyme Δ_3-*cis*-Δ_2-*trans*-enoyl CoA isomerase. This enzyme catalyzes the isomerization of the double bond to position Δ_2. In shifting positions the double bond is transformed to a *trans* configuration, producing the correct intermediate for β-oxidation, Δ_2-*trans*-enoyl CoA. β-Oxidation then proceeds normally to the ω-end of the fatty acid.

β-Hydroxyacyl CoA epimerase. Another problem arises when polyenoic fatty acids under β-oxidation; for example, consider linoleic acid, $C_{18}\Delta_{9,12}$-*cis,cis*. The problem of the 9-*cis* double bond is handled by the enoyl CoA isomerase, just as it was for the monoenoic acid $C_{18}\Delta_9$-*cis* as described above. However, the 12-*cis* double bond poses another difficulty. After the β-oxidation involving the first double bond, the remaining acyl CoA has the original 12-*cis* double bond in the proper Δ_2-position but not in the *trans* configuration, which is ordinarily the intermediate. Enoylhydratase, the enzyme that catalyzes the hydration of the

Δ_2-enoyl CoA, can utilize the *cis* isomer. However, the incorrect β-hydroxyacyl intermediate is formed, for it is the D rather than L isomer. The next enzyme in the sequence, β-hydroxyacyl CoA dehydrogenase, cannot utilize the D isomer. Fortunately, an epimerase is available to convert the D isomer to the correct L form, β-hydroxyacyl CoA epimerase. Following epimerization, β-oxidation proceeds normally to the ω-end of the acid.

Unsaturated fatty acids.   Because the enoyl CoA isomerase and β-hydroxyacyl CoA epimerase enzymes are present, unsaturated fatty acids can be oxidized just as readily as saturated fatty acids. For each unsaturated bond present, however, two fewer ATP will be generated because the FAD-linked acyl CoA dehydrogenase step (p. 390) is not required. Stearic acid, the eighteen-carbon atom saturated acid, will undergo eight β-oxidations to yield 40ATP. Oleic acid, $C_{18}\Delta_9 cis$, also undergoes eight β-oxidations but yields only 38ATP from this process. Linoleic acid, $C_{18}\Delta_{9,12}$-*cis,cis* also undergoes eight β-oxidations but yields only 36ATP.

As noted earlier some *trans* unsaturated fatty acids are taken into the body when commercially prepared hydrogenated fats are eaten, for example, margarine, peanut butter, etc. These *trans* fatty acids are oxidized just as readily as the corresponding *cis* unsaturated fatty acids. The mechanism through which a Δ_3*trans*-enoyl CoA, such as would be generated when elaidic acid ($C_{18}\Delta_9 trans$) is oxidized, has not been worked out as yet, but it appears to pose no special problem in terms of β-oxidation.

Odd-carbon fatty acids.   Acids containing an odd number of carbon atoms also undergo normal β-oxidation. However, the ω-terminal fragment that is generated by the β-oxidation process is *propionyl CoA*, not acetyl CoA. Propionyl CoA is oxidized in a special manner as is described in Chapter 6. First, it is carboxylated to a four-carbon atom intermediate, methylmalonyl CoA. In the next step *methylmalonyl CoA* is isomerized to *succinyl CoA*, a reaction that requires vitamin B_{12} in coenzyme form. Succinyl CoA enters the Krebs cycle and is oxidized.

Other types of fatty acid oxidation.   There are other pathways for fatty acid oxidation: α-*oxidation*, in which only one carbon atom is removed at a time from the carboxyl end, and ω-*oxidation*, in which the ω-terminus is oxidized to a carboxyl group to form a *dicarboxylic acid*. The substrate for both of these pathways is the fatty acid, not the acyl CoA derivative. α-Oxidation is a microsomal process that involves NAD^+ and ascorbate. An α-hydroxy fatty acid is an intermediate. ω-Oxidation also is a microsomal process, but it requires NADPH and involves the microsomal mixed-function oxidase that contains cytochrome P_{450}. Under ordinary conditions, relatively little fatty acid is oxidized by either of these pathways.

Energy yield of fatty acid oxidation.   Fatty acid oxidation is one of the main pathways through which cells derive energy for ATP synthesis. Unless otherwise specified, fatty acid oxidation is taken to mean β-oxidation to produce acetyl CoA, followed by complete combustion of the acetyl CoA in the Krebs cycle to produce CO_2 and H_2O. There are four steps in which energy is either utilized or trapped. These are illustrated for the complete oxidation of palmitic acid, the commonly occurring sixteen-carbon atom saturated fatty acid.

$$\text{Palmitate} + \text{CoA} + \text{ATP} \xrightarrow{\text{Mg}^{2+}} \text{Palmitoyl CoA} + \text{AMP} + \text{PP}_i$$

$$\text{PP}_i + \text{H}_2\text{O} \longrightarrow 2\text{ P}_i$$

$$\text{Palmitoyl CoA} + 7\text{ CoA} \longrightarrow 8\text{ Acetyl CoA}$$

$$8\text{ Acetyl CoA} + 16\text{ O}_2 \longrightarrow 16\text{ CO}_2 + 16\text{ H}_2\text{O} + 8\text{ CoA}$$

Two high-energy phosphate bonds are utilized to activate the palmitate group, one in the acyl CoA ligase reaction and the other when the pyrophosphate that is formed in this reaction is hydrolyzed. The initial expenditure by the cell, however, is well rewarded. In the β-oxidation cycle, the palmitoyl CoA is converted to eight acetyl CoA units. This requires seven β-oxidations, and each β-oxidation sequence produces 5ATP. Therefore, 35ATP are gained from the β-oxidation reactions. Furthermore, twelve high-energy phosphate bonds are formed when each acetyl CoA unit is oxidized in the Krebs cycle. Actually, 11ATP and 1GTP are formed, but this can be considered as 12ATP since GTP and ATP are interconvertible. Therefore, the combustion of the eight acetyl CoA units produced will yield 96ATP. The total ATP formed in the complete oxidation of palmitoyl CoA is 131. Since two high-energy bonds were broken in the overall process needed to form palmitoyl CoA, the net yield is one hundred twenty-nine high-energy phosphate bonds, considered as 129ATP. This represents about 40% of the energy that is derived when palmitic acid is burned to CO_2 and H_2O in a bomb calorimeter. Although 40% efficiency may appear wasteful, it is not so, for the remainder of the energy serves to maintain the body temperature at 37.5° C.

KETONE BODY OXIDATION

Three metabolic products, *acetoacetate*, D-*β-hydroxybutyrate*, and *acetone*, are referred to as ketone bodies. They are produced by the liver when excessive amounts of fat are being oxidized and glucose availability is limiting, for example, in starvation and diabetic ketoacidosis. The mechanism of ketone body synthesis is discussed in Chapter 10. Ketone bodies are secreted into the blood by the liver and are taken up by the brain, heart, and skeletal muscles as a source of energy. Acetoacetate is utilized in these tissues by either of two activation reactions.

$$\text{Acetoacetate} + \text{Succinyl CoA} \rightleftharpoons \text{Acetoacetyl CoA} + \text{Succinate}$$

$$\text{Acetoacetate} + \text{CoASH} + \text{ATP} \rightleftharpoons \text{Acetoacetyl CoA} + \text{AMP} + \text{PP}_i$$

The first is a transacylase involving succinyl CoA; the other is an acyl CoA synthetase reaction linked to ATP. Acetoacetyl CoA that is formed in these reactions undergoes β-oxidation.

β-Hydroxybutyrate is activated through an acyl CoA synthetase reaction that also is ATP linked. The resulting β-hydroxybutyryl CoA is oxidized to acetoacetyl CoA through the action of a NAD^+-linked dehydrogenase, and the acetoacetyl CoA is utilized as noted above. Acetone, the third ketone body, cannot be further metabolized by the tissues.

The amounts of ketone bodies that are utilized by the tissues is proportional to their arterial concentration until this exceeds 70 mg/dl. Above this concentration the oxidation process is saturated, the concentration in the glomerular filtrate exceeds the maximum tubular reabsorption rate, and large amounts of ketone bodies are excreted in the urine. This condition is known as *ketonuria*. Acetone also is excreted by the lungs when the arterial ketone body concentration is high, and this odor is easily detected in patients suffering from *ketosis*.

FATTY ACID SYNTHESIS

Man synthesizes fatty acids using acetyl CoA derived from carbohydrate. This pathway is known as complete or de novo synthesis, and it occurs in the cell cytoplasm. In addition, acetate fragments can be added to an existing fatty acid by a process of *chain elongation*. This reaction occurs in both the mitochondria and microsomes. Finally, *cis* double bonds can be introduced into fatty acids by a process of desaturation, a reaction that occurs in the microsomes.

Source of acetyl CoA

Acetyl CoA is generated within the mitochondrion. Pyruvate that is derived from glucose is the main source of the acetyl CoA used for fatty acid synthesis. As already described, acyl CoA derivatives cannot cross the inner mitochondrial membrane. In certain instances the acetyl group may be transferred to carnitine and exit from the mitochondrion as an *acetyl carnitine* ester by a process that is similar to fatty acid entrance into the mitochondrion. However, present evidence indicates that this is not the major mechanism for the efflux of mitochondrial acetyl CoA. Instead, most of the acetyl CoA leaves the mitochondrion in the form of *citrate*, the initial intermediate of the Krebs cycle. In the cell cytoplasm, citrate is cleaved to acetyl CoA and oxaloacetate in an ATP-requiring reaction catalyzed by *citrate lyase*.

$$Citrate + ATP + CoASH \longrightarrow Acetyl\ CoA + Oxaloacetate + ADP + P_i$$

One acetyl CoA unit enters the synthetase pathway directly, but the remainder must be first converted to *malonyl CoA*.

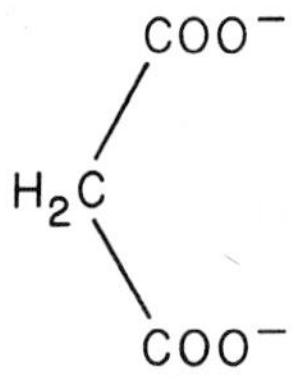

malonate 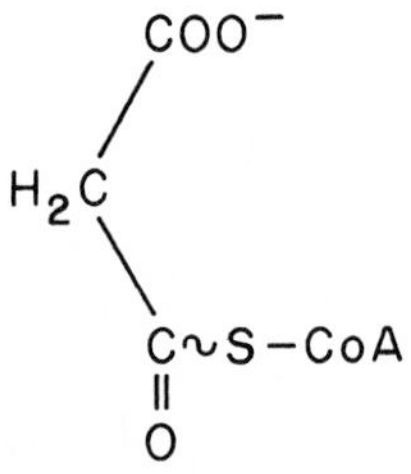   malonyl CoA

Carboxylation of acetyl CoA

The rate-limiting reaction in fatty acid synthesis is the carboxylation of acetyl CoA to malonyl CoA.

$$acetyl{\sim}SCoA + HCO_3^- + ATP \longrightarrow malonyl{\sim}SCoA + ADP + P_i$$

This reaction is catalyzed by an enzyme-complex, acetyl CoA carboxylase, that contains *biotin* and utilizes *bicarbonate*. This allosteric enzyme is activated by citrate so that the fatty acid–synthesizing system actually is primed for acetyl CoA by the compound in which the acetate moiety is delivered. An intermediate in the carboxylation reaction is CO_2 linked covalently to enzyme-bound biotin, a *carboxybiotin* complex. This is attached to a lysine residue of the enzyme. See the example at the top of p. 395.

Acetyl CoA carboxylase is activated by conversion from a monomer with a molecular weight of 4×10^5 to a polymer having a molecular weight of approximately 6 to 8×10^6. Citrate causes polymerization; palmitoyl CoA causes the polymer to disaggregate. The enzyme consists of two components. One contains two proteins, a biotin carboxyl carrier protein (BCCP) and a biotin carboxylase. The second component is a transcarboxylase that catalyzes the transfer of CO_2 from BCCP to acetyl CoA.

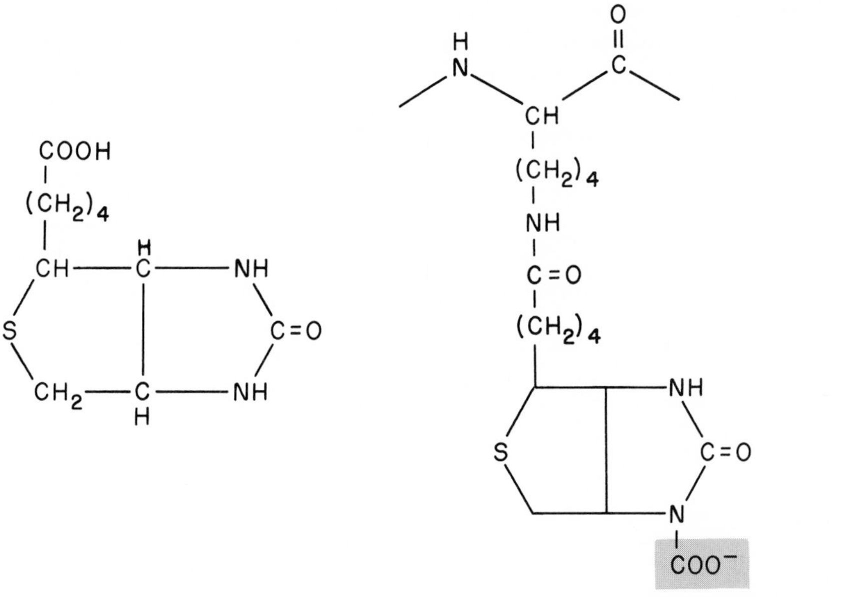

Biotin   N-carboxy biotin enzyme complex

The partial reactions are as follows:

$$BCCP + HCO_3^- + ATP \longrightarrow BCCP-CO_2 + ADP + P_i$$

$$BCCP-CO_2 + Acetyl\ CoA \longrightarrow Malonyl\ CoA + BCCP$$

The net result is the carboxylation of acetyl CoA to malonyl CoA.

Acyl carrier protein

From the bacterium *Escherichia coli* a soluble protein with a molecular weight of approximately 8500 can be isolated that serves as the carrier for the fatty acid during de novo biosynthesis. This is called the acyl carrier protein (ACP). The protein contains a 4'-phosphopantetheine group esterified to a hydroxyl group of serine that is located thirty-six residues from the N-terminal end. This provides the sulfhydryl group to which the fatty acid chain is attached as it is synthesized. A 4'-phosphopantethiene group also is the functional group of CoASH.

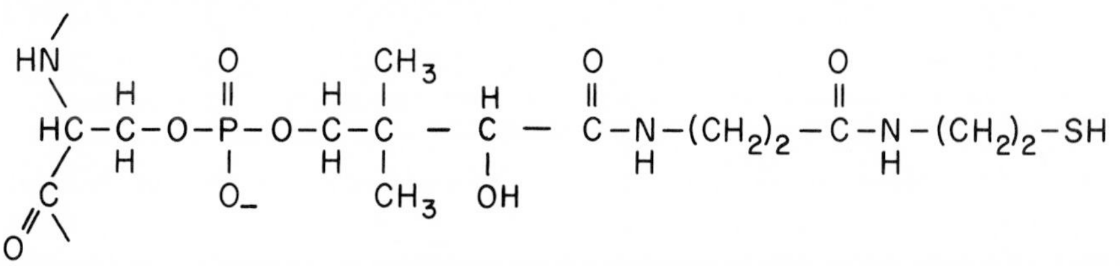

In addition to ACP, six enzymes are needed for de novo fatty acid synthesis. These, together with ACP, exist as a complex. In mammalian tissues the *fatty acid synthetase complex* is tightly associated and has a molecular weight of 4.8×10^5. Although the mammalian fatty acid synthetase has been separated into two large polypeptides, neither ACP nor any of the six enzymes can be isolated from it in active form. However, there is every reason to suspect that the mammalian complex functions in a similar manner and contains ACP or a very similar sulfhydryl-containing protein carrier. In order to facilitate communication, the de novo synthetic pathway in man will be discussed in terms of ACP even though, in a rigorous sense, it might better be described in terms of a synthetase protein-sulfhydryl group.

Fatty acid synthetase

A schematic representation of the fatty acid synthetase complex is presented in Fig. 8-7. Initially the acetyl group of acetyl CoA is transferred to the sulfhydryl group of ACP. The reaction is catalyzed by the *acetyl CoA–ACP–transacylase.*

$$\text{Acetyl} \sim \text{SCoA} + \text{ACP-SH} \rightleftharpoons \text{Acetyl} \sim \text{S-ACP} + \text{CoASH}$$

Next, the acetyl group originally attached to this sulfhydryl group is momentarily transferred to a sulfhydryl group contained on the condensing enzyme, one of the enzymatic activities contained in the fatty acid synthetase complex, abbreviated CE.

$$\text{Acetyl} \sim \text{S-ACP} + \text{CE-SH} \rightleftharpoons \text{Acetyl} \sim \text{S-CE} + \text{ACP-SH}$$

This frees the sulfhydryl group of ACP to accept the next incoming group, a

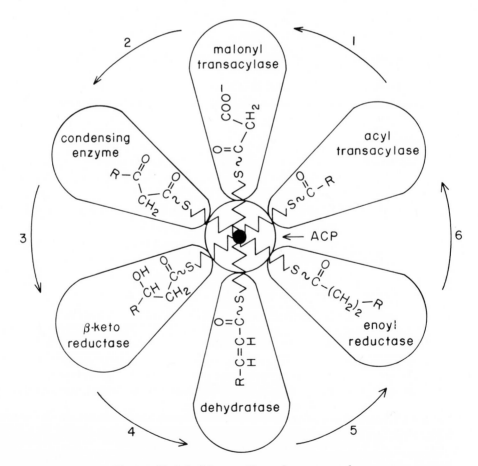

Fig. 8-7. Model of fatty acid synthetase complex.

malonyl CoA. The malonyl transfer is catalyzed by *malonyl CoA:ACP acyltransferase*.

$$\text{Malonyl} \sim \text{SCoA} + \text{ACP-SH} \rightleftharpoons \text{Malonyl} \sim \text{S-ACP} + \text{CoASH}$$

The acetyl residue then is transferred from its temporary site on the condensing enzyme to condense with the malonyl residue. In this reaction the free carboxyl group of malonate is discharged as CO_2 so that the product contains four rather than five carbon atoms.

$$\text{Acetyl} \sim \text{S-CE} + \text{Malonyl} \sim \text{S-ACP} \longrightarrow \text{Acetoacetyl} \sim \text{S-ACP} + \text{CE-SH} + CO_2$$

The condensation of the acetyl and malonyl groups is catalyzed by the *condensing enzyme*. The process that drives the condensation reaction is the decarboxylation of malonate. Therefore, the expenditure of ATP in the acetyl CoA–carboxylase reaction is not wasteful, for the energy is used subsequently for the condensation reaction.

The product of these reactions, *acetoacetyl ACP*, is reduced to *butyryl ACP* by the remaining enzymes in the fatty acid synthetase complex, *β-ketoacyl ACP reductase*, *enoyl ACP dehydratase*, and *crotonyl ACP reductase*.

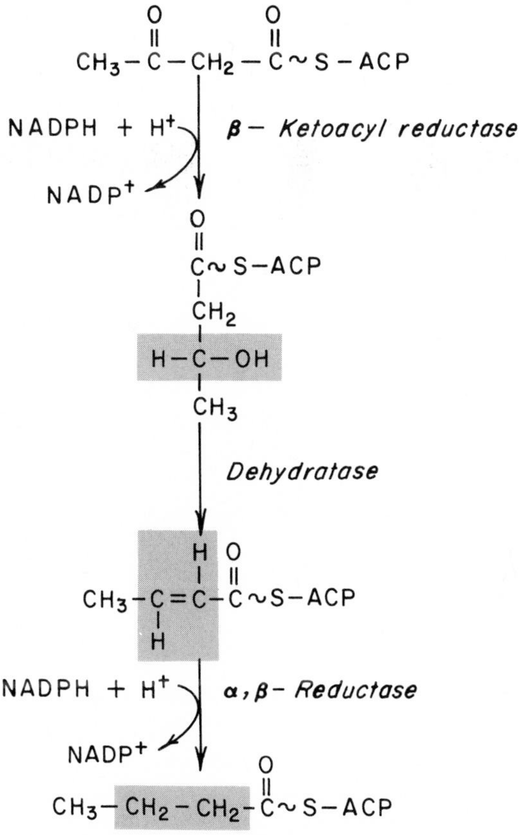

The β-hydroxyl intermediate, unlike that involved in β-oxidation, is of the D-configuration. The Δ_2 double bond of the crotonyl S-ACP intermediate is in the *trans* configuration. NADPH is the reducing agent for both the acetoacetyl-ACP and crotonyl-ACP intermediates.

Following this series of events, the entire sequence is repeated. First, the butyryl group is transferred to the sulfhydryl group of the condensing enzyme, just as was the original acetyl group. Malonyl CoA then adds once again to the free sulfhydryl group of ACP, and the butyryl group on the condensing enzyme condenses with the malonyl-ACP residue. In this condensation CO_2 is discharged, and the product now has six carbon atoms, with carbon atom 5 being a carbonyl group. The same reduction pathway described above utilizing two NADPH units takes place, forming a six-carbon–atom saturated acyl-ACP intermediate. This sequence is repeated, with a two-carbon unit from malonyl CoA being added, followed by reduction of the carbonyl group, until a sixteen-carbon atom chain is built. In each case the acyl residue leaves the ACP-sulfhydryl group and moves to the condensing enzyme momentarily so that malonate can add and then condense with the ACP-linked malonate residue, which becomes the carboxyl end of the growing fatty acid. In this way the original acetate residue remains as the ω-end of the growing fatty acid, and each malonate residue that is added successively becomes the carboxyl end.

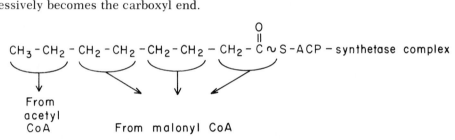

De novo fatty acid synthesis always results in a product containing an even number of carbon atoms because the acetate primer contains two carbon atoms and each malonate unit that condenses with it donates two carbon atoms to the chain. Fatty acid synthesis usually stops after the chain is sixteen carbon atoms long. This occurs partly because the enzyme that removes the acyl group from ACP, a *deacylase*, exhibits maximal activity with sixteen-carbon–atom fatty acids. It also occurs because the binding site on the condensing enzyme to which the acyl chain is transferred from ACP cannot hold chains that contain more than fourteen carbon atoms very well. Therefore, the affinity of this site for the next longer acyl group, the sixteen-carbon–atom palmitoyl chain, is so low as to allow the deacylase to cleave it from ACP before transfer to the condensing enzyme can occur. Much of the NADPH needed for the reductions is obtained from the initial reactions of the pentose-phosphate pathway (hexose monophosphate shunt), in which D-glucose 6-phosphate is oxidized (Chapter 7). The remainder of the NADPH is supplied by the oxidation of malate in the cytoplasm, catalyzed by the *malic enzyme*.

Regulation of de novo biosynthesis

There are two types of control mechanisms that regulate the complete synthesis of fatty acids by the *de novo* pathway. One is *long-term control*. This involves changes in the content of the fatty acid biosynthetic enzymes within the cell. The *production* of enzymes involved in fatty acid de novo synthesis, including acetyl CoA carboxylase, the fatty acid synthetase complex, citrate lyase, glucose 6-phosphate dehydrogenase, and the malic enzyme, is stimulated in the liver when glucose is fed. In an insulin-deficient diabetic animal, the production of these enzymes is stimulated when insulin is administered. By contrast, the content of these enzymes in the liver decreases during starvation or in diabetes associated with insulin deficiency. These enzymes have a relatively short half-life, so that their content in a cell is dependent on the rate at which they are synthesized. In this way, the cell has large quantities of these enzymes needed for conversion of carbohydrate into fatty acid only when carbohydrate and ATP are available. Because it requires enzyme production, long-term control exerts its effect relatively slowly, requiring several hr to manifest itself completely.

There is also a *short-term control* mechanism that acts on acetyl CoA carboxylase, the rate-limiting enzyme in fatty acid de novo biosynthesis. It involves modulation of the activity of acetyl CoA carboxylase. Citrate activates the enzyme by causing aggregation, while long-chain acyl CoA inactivates it by causing it to disaggregate (p. 394). Even more exquisite control may occur through variations in the composition of the acyl CoA to which acetyl CoA carboxylase is exposed, for the saturated acyl CoA's are much more inhibitory than the polyenoic acyl CoA's. Since short-term control involves modulation of the activity of already existing enzyme, its effects manifest themselves very rapidly, that is, within min. Therefore, short-term control can be thought of as the fine tuning of this biosynthetic process.

Differences between synthesis and oxidation

Although some overall similarities exist between fatty acid synthesis and oxidation, the details of these processes really are very different. In order to more clearly understand each of these very important metabolic pathways, it is helpful to summarize their main differences. These are listed in Table 8-8 as a review.

Chain elongation

A second pathway for fatty acid synthesis involves the addition of acetate units to the *carboxyl end* of an existing acyl chain. In the mitochondria the acetate group is added in the form of acetyl CoA. In the microsomes, malonyl CoA provides the two-carbon atom fragment. These pathways are called chain elongation. The newly formed β-keto acid is then reduced, and a saturated fatty acid is the

Table 8-8. Main differences between fatty acid de novo biosynthesis and fatty acid oxidation

PARAMETER	FATTY ACID OXIDATION	FATTY ACID SYNTHESIS
Intracellular location	Mitochondria	Cytoplasm
Intermediates	Acetyl CoA	Acetyl CoA, malonyl CoA
Thioester linkage	CoASH	ACP-SH*
Coenzymes for electron transfer	FAD, NAD$^+$	NADPH
Configuration of β-hydroxy intermediate	L	D
Bicarbonate dependence	No	Yes
Energy state favoring the process	High ADP	High ATP
Citrate activation	No	Yes
Acyl CoA inhibition	No	Yes
Highest activity	Fasting, starvation	Carbohydrate fed

*As noted on p. 395, the mammalian fatty acid synthetase complex does not contain a readily dissociable ACP, but it probably has a similar structure as part of a tightly associated complex.

final product. The chain elongation pathway requires the activated form of fatty acid, acyl CoA, as the substrate. It can elongate either saturated or unsaturated fatty acids. More than one elongation can take place so that fatty acids containing up to twenty-six carbon atoms can be formed. In each elongation step, the fatty acid is lengthened by two carbon atoms. This is true even when malonyl CoA is the elongating agent, for just as in the de novo synthetic pathway, CO_2 is released when the malonyl CoA condenses with the acyl CoA substrate. Therefore, the elongation mechanism also leads to the production of fatty acids having an *even* number of carbon atoms.

Desaturation

Double bonds also can be introduced into fatty acids by a desaturase enzyme system requiring O_2 and NADH. A *cis* configuration around the double bond results. This reaction occurs in the endoplasmic reticulum. Cytochrome b_5, cytochrome b_5 reductase, and a cyanide-sensitive factor that is tightly bound to the membrane are required for the desaturation reaction. The mammalian desaturase utilizes fatty acids containing sixteen or more carbon atoms and inserts the initial double bond in position 9,10. Thus *palmitic* acid can be converted to *palmitoleic* acid ($16:1\omega7$), and *stearic* acid can be converted to *oleic* acid ($18:1\omega9$). Additional double bonds can be inserted to make a polyunsaturated fatty acid. Why then is linoleic acid essential? The reason is the specificity of the mammalian desaturases. These enzymes can insert a second double bond between position 9,10 and the carboxyl group, but *not* between position 9,10 and the ω-end. Hence it is impossible to form the position 12,13 unsaturation that is required for linoleic acid, that is, eighteen carbon atoms with double bonds in positions 9,10 and 12,13. On the other hand, it is possible for mammals to form a double bond beyond position 9,10. Suppose palmitoleic acid is made from palmitic acid, and an acetate unit is added by the chain elongation pathway. This acid now has eighteen carbon atoms. Since the new two-carbon unit always is added to the carboxyl terminus, the double bond is now in position 11,12, that is, the new product remains in the ω 7 class and is $18:1\omega7$.

$C_{16}\Delta_9$
or
$16:1\omega7$

CH$_3$(CH$_2$)$_5$ —CH=CH —CH$_2$ —CH$_2$—CH$_2$—CH$_2$—CH$_2$—CH$_2$—CH$_2$—COOH

10 9   8   7   6   5   4   3   2   1

Elongation | *Acetyl CoA*

$C_{18}\Delta_{11}$
or
$18:1\omega7$

CH$_3$(CH$_2$)$_5$ —CH=CH — CH$_2$ — CH$_2$—CH$_2$—CH$_2$—CH$_2$—CH$_2$—CH$_2$ —CH$_2$—COOH

12 11   10   9   8   7   6   5   4   3   2   1

Desaturase | O_2, *NADPH*

$C_{18}\Delta_{8,11}$
or
$18:2\omega7$

CH$_3$(CH$_2$)$_5$ —CH=CH — CH$_2$ —CH=CH— CH$_2$ —CH$_2$—CH$_2$—CH$_2$—CH$_2$—CH$_2$—COOH

12 11   10 9 8   7   6   5   4   3   2   1

If the $C_{18}\Delta_{11}$ acid were further desaturated, the second double bond must be between the first double bond and the carboxyl end and, in addition, two carbon atoms distant from any other double bond (see p. 366). Therefore, the resulting acid would be $C_{18}\Delta_{8,11}$, and the double bonds would *not* be in the proper position to form linoleic acid, $C_{18}\Delta_{9,12}$. For similar reasons, linolenic acid, $C_{18}\Delta_{9,12,15}$ cannot be synthesized in mammals because the Δ_{12} and Δ_{15} double bonds cannot be inserted in the proper positions. However, when they are supplied in the diet, linoleic and linolenic acids can be further elongated and desaturated. For example, linoleic acid ($18:2\omega6$) taken in the diet can be further desaturated and elongated to form other members of the essential fatty acid family, for example, γ-linolenic ($18:3\omega6$) and arachidonic ($20:4\omega6$) acids.

Mammalian tissues require polyunsaturated fatty acids. When the usual diet containing plant foods is eaten, most of the polyunsaturated fatty acids are either linoleate or elongation and desaturation products of the linoleate class (ω 6-class). If a fat-deficient diet is eaten, the body will become depleted of these polyunsaturated fatty acids, eventually leading to a condition known as essential fatty acid deficiency (p. 368). In such a situation the body attempts to compensate for the lack of ω 6–fatty acids by synthesizing polyunsaturated fatty acids de novo. Palmitic acid ($16:0$) made by de novo biosynthesis is elongated to stearate ($18:0$), which is desaturated to oleate ($18:1\omega9$). Oleic acid then is desaturated and elongated and the major product that accumulates is an eicosatrienoic acid of the ω 9-class. When this polyenoic fatty acid, $20:3\omega9$, appears in appreciable quantities in the tissue, it is a sign of essential fatty acid (ω 6-class) deficiency. Notice that all of these fatty acids are of the ω 6-class. This again illustrates the point that the unsaturated fatty acids within a given class, for example, ω 6, are interconverted but that there is no interconversion between the different classes, that is, an ω 6–fatty acid is not converted to an ω 9–fatty acid.

Retroconversion. In addition to inserting double bonds into fatty acids, mammalian tissues can hydrogenate unsaturated bonds already present in a fatty acid. This process, a biohydrogenation known as retroconversion, has been observed in the liver and testes. It usually is associated with the removal of two carbon atoms from the carboxyl end of the fatty acid. Thus $C_{22}\Delta_{4,7,10,13,16}$ is converted to $C_{20}\Delta_{5,8,11,14}$ by a combination of hydrogenation of the original Δ_4 double bond and cleavage of carbon atoms 1 and 2 of the original twenty-two-carbon atom fatty acid. Little detail is known about these reactions. The evidence available to date indicates that the reduction occurs prior to loss of the two-carbon atom fragment. In the above example, $C_{22}\Delta_{4,7,10,13,16}$ is converted to a $C_{22}\Delta_{7,10,13,16}$ intermediate, which then has the carboxyl-terminal acetate fragment removed. With the release of the carboxyl-terminal acetate group, each unsaturated bond is located two carbon atoms closer to the carboxyl end, and the resulting acid is $C_{20}\Delta_{5,8,11,14}$, arachidonic acid.

PHOSPHOGLYCERIDE METABOLISM

Phosphoglycerides can be synthesized and degraded in human tissues. For simplicity we shall concentrate on only one phosphoglyceride, lecithin, and discuss the pathways for its synthesis and catabolism. In general, the pathways for metabolism of the other phosphoglycerides are quite similar to those of lecithin.

Lecithin synthesis

Phosphatidyl choline, commonly known as lecithin, can be synthesized in four ways. Two of these are complete or de novo synthetic pathways, one is a fatty acid addition or acylation pathway, and the last is a condensation pathway. One of the de novo pathways involves *cytidine diphosphate choline* (CDP choline) and 1,2-diglyceride, as shown in Fig. 8-8. The 1,2-diacylglycerol intermediate in this pathway can be converted to either phosphatidyl choline through interaction with

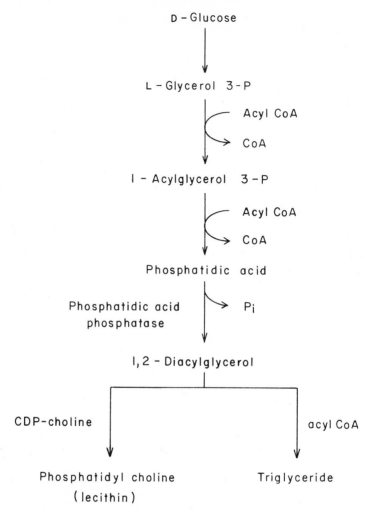

Fig. 8-8. Pathways for de novo biosynthesis of lecithin and triglyceride. This pathway for triglyceride synthesis occurs mainly in the liver and adipose tissue.

CDP choline or, alternatively, to triglyceride through interaction with acyl CoA. Therefore, this pathway also is a major route for triglyceride synthesis as is described on p. 406.

The second de novo synthetic pathway utilizes phosphatidic acid as the key intermediate instead of 1,2-diacylglycerol. As illustrated in Fig. 8-9, this pathway also employs a cytidine derivative, in this case *cytidine diphosphate diacylglycerol (CDP diglyceride)*. The structures of the two cytidine intermediates are compared in Fig. 8-10. They are quite similar, cytidine monophosphate (CMP) being linked to *phosphorylcholine* in one case and to phosphatidic acid in the other. In the CDP choline pathway, the CMP is not actually incorporated into the phosphoglyceride intermediate; it transfers its phosphorylcholine residue to the diacylglycerol. In the phosphatidic acid pathway, CMP actually is incorporated into a phosphoglyceride intermediate. *Phosphatidyl ethanolamine* and *phosphatidyl serine* are intermediates in the phosphatidic acid pathway, and often they are products of this pathway. *Phosphatidyl inositol* and *cardiolipin* also are

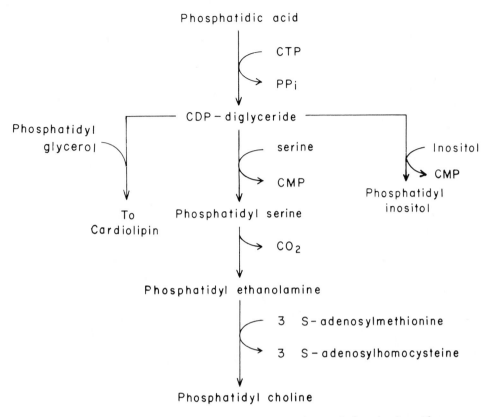

Phosphatidic acid

CTP

PPi

CDP – diglyceride

Phosphatidyl glycerol

serine

Inositol

CMP

CMP

Phosphatidyl inositol

To Cardiolipin

Phosphatidyl serine

CO_2

Phosphatidyl ethanolamine

3 S–adenosylmethionine

3 S–adenosylhomocysteine

Phosphatidyl choline

Fig. 8-9. Phosphatidic acid pathway for biosynthesis of phosphoglycerides.

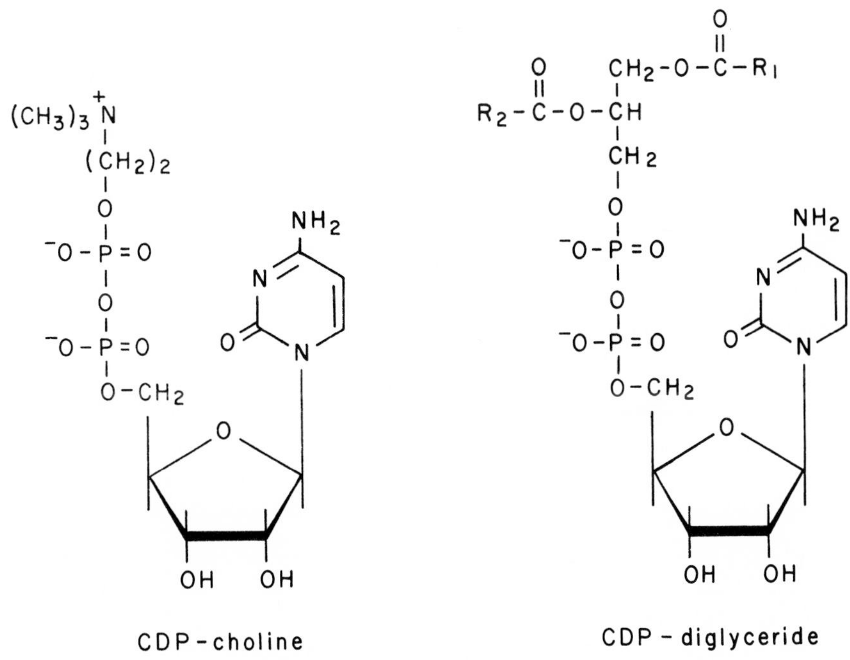

CDP–choline

CDP – diglyceride

Fig. 8-10. Structures of CDP-choline and CDP-diglyceride.

synthesized from the CDP diglyceride intermediate in the phosphatidic acid pathway. *S-Adenosylmethionine*, the methyl donor that is involved in the conversion of the ethanolamine to the choline phosphatide, will be discussed in detail in Chapter 9.

A third pathway for lecithin synthesis begins with lysolecithin. Therefore, it is not a complete synthesis pathway, for much of the phosphoglyceride structure already is formed. An acyl residue is added by reaction with acyl CoA in the presence of an acyltransferase. This acylation pathway probably serves an important function in membrane structure, for it is likely that fatty acids are constantly being hydrolyzed from lecithin and subsequently being replaced by acylation in the course of normal membrane function. In this way, the cell has the opportunity to quickly alter the fatty acyl group composition of its membrane glycerophosphatides. This may permit rapid regulation of enzymes or carriers located within the membrane, for phospholipid fatty acid chains probably interact with these proteins. Such regulation has been demonstrated for the ouabain-sensitive, Na^+,K^+-ATPase present in the plasma membrane. The types of fatty acid groups inserted into the lysolecithin are controlled by the specificity of the acyltransferase. In this pathway it is usually the fatty acid of position 2 that is removed and replaced. Acyltransferases specific for position 2 have greater specificity for polyenoic fatty acids, so that the fatty acyl group inserted into lysolecithin in this pathway usually is polyunsaturated. Phospholipase A_2 catalyzes the hydrolysis of the fatty acid ester from position 2. This enzyme is dependent on calcium. Therefore, changes in intracellular calcium content may regulate the operation of this pathway. It is likely that this is one of the mechanisms whereby calcium regulates cellular

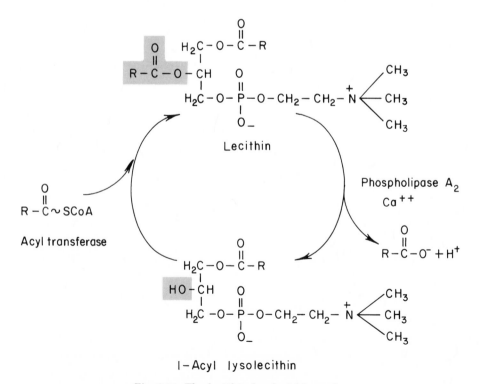

Fig. 8-11. The lecithin-lysolecithin cycle.

function. In this context, it is considered likely that the phospholipase A_2 reaction, because it releases primarily polyenoic fatty acids within membranes, provides the fatty acids utilized for prostaglandin syntheses (p. 651). The cyclic nature of the lecithin-lysolecithin interconversion is illustrated in Fig. 8-11. Other glycerophosphatides besides lecithin undergo a similar reaction cycle.

The fourth pathway for lecithin synthesis involves the condensation of two lysolecithin molecules. One of these accepts an acyl residue from the other.

$$2 \text{ Lysolecithin} \longrightarrow \text{lecithin} + \text{glycerylphosphorylcholine}$$

Phosphoglyceride degradation

Lecithin is degraded to *glycerylphosphorylcholine* through the action of phospholipases. Two phospholipases are present; A_1, which removes the acyl residue from position 1 of the glycerol backbone, and A_2, which removes the acyl group at position 2. As described above, phospholipase A_2 is most important in terms of the acylation pathway of lecithin synthesis.

A *lysophospholipase* removes the remaining fatty acid from lysolecithin to form glycerylphosphorylcholine, which is subsequently degraded. Additional phospholipases have been isolated; these remove either the phosphorylcholine or the choline residue from lecithin, but their importance in mammalian metabolism is questionable.

Alkyl ether and plasmalogen metabolism

The alkyl ether linkage (p. 372), which sometimes occurs in phosphoglycerides and acylglycerols, always is located in position 1. It is formed by exchange of the acyl group of *1-acyl-dihydroxyacetone phosphate* with a *fatty alcohol*. The fatty alcohol is formed by reduction of the corresponding fatty acid. The resulting *1-alkyl-dihydroxyacetone phosphate* is reduced to a *1-alkyl-glycerol-3-phosphate*, and this intermediate is converted to phosphoglycerides through the pathways illustrated in Figs. 8-8 and 8-9. Plasmalogens (p. 372) are synthesized from the 1-alkyl-phosphoglycerides by dehydrogenation of the alkyl chain in position 1,2. The alkyl ether bond is more stable than the corresponding ester bond. Therefore, the presence of alkyl ether derivatives may provide additional stability to cellular membrane structure.

The alkyl ether bond is cleaved during the degradation of these compounds by a microsomal enzyme system that requires oxygen and reduced pteridine. One of the products is a fatty aldehyde.

SPHINGOLIPID METABOLISM

Sphingolipids are derived from sphingosine (p. 372), which in turn is synthesized from palmitoyl CoA and serine. *Pyridoxal phosphate*, a B vitamin derivative, is needed for sphingosine synthesis. NADPH is required in this reaction sequence. Dihydrosphingosine, which does not contain any unsaturated bonds, is an intermediate in this pathway. It is oxidized by a flavoprotein enzyme to sphingosine. *Ceramide* (p. 373) is formed by N-acylation of sphingosine. A long-chain acyl CoA is the acyl group donor in ceramide synthesis.

$$\text{Sphingosine} + \text{FA} \sim \text{SCoA} \rightleftharpoons \text{Ceramide} + \text{CoASH}$$

As shown in Fig. 8-12, ceramide is the key intermediate in sphingolipid synthesis. This is the pathway employed for the synthesis of *sphingomyelin* (Chln-P-Cer) and the cerebrosides *galactosylceramide* (Gal-Cer) and *glucosylceramide* (Glc-

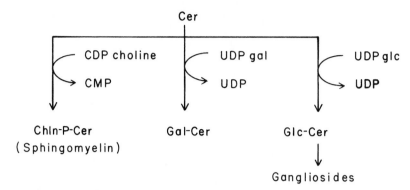

Fig. 8-12. Sphingolipid synthesis through ceramide (Cer) pathway. Choline is abbreviated as Chln.

Cer). *Gangliosides* are formed from glucosylceramides by the addition of hexose residues to the glucose.

An alternative pathway for sphingolipid synthesis has been described in which the phosphorylcholine or galactose residue is added to sphingosine instead of ceramide. The resulting sphingosine derivative is then N-acylated by acyl CoA. *Psychosine* is the galactosylsphingosine intermediate that is formed in this pathway.

Sulfatides are synthesized from galactosylceramides or glucosylceramides by sulfation of the hydroxyl group at position 3 of the hexose. The sulfate donor is 3'-phosphoadenosine-5'-phosphosulfate (PAPS).

Sphingolipid degradation

Sphingomyelin is degraded by *sphingomyelinase*, an enzyme that removes the phosphorylcholine residue. Cerebrosides and gangliosides are hydrolyzed by hexosidases, which remove one sugar residue at a time from the nonreducing terminus of the carbohydrate chain. These enzymes all are associated with the lysosomes of the cell.

Lipid storage diseases. These diseases, which are also known as the *lipidoses*, result from inherited deficiencies of a sphingolipid catabolic enzyme. For example, sphingomyelinase is deficient in Niemann-Pick disease, and the β-D-glucosidase that hydrolyzes glucosylceramide is deficient in Gaucher's disease. This results in intracellular accumulation of a sphingolipid, often leading to death in early childhood.

GLYCERIDE METABOLISM

The two pathways for triglyceride biosynthesis have been presented previously. Fig. 8-8 illustrates the synthetic pathway involving L-glycerol 3-phosphate. Phosphatidic acid and 1,2-diacylglycerol are intermediates along this route. A third fatty acid residue is transferred to the 1,2-diglyceride from acyl CoA to form the triglyceride. Adipose tissue and liver, two of the major sites of triglyceride synthesis, employ this pathway. The other major site of triglyceride synthesis, the intestinal mucosa, employs the 2-monoglyceride pathway that was discussed in the context of intestinal fat absorption (p. 377). In this pathway, two acyl residues in the form of acyl CoA are esterified at positions 1 and 3 of the 2-monoglyceride, forming the triglyceride.

Table 8-9. Common forms of lipid storage disease

DISEASE	LIPID ACCUMULATION*	ENZYME DEFICIENCY
Gaucher's	Glc$\downarrow$Cer	Glucosylceramide β-D-glucosidase
Niemann-Pick	Chln-P$\downarrow$Cer	Sphingomyelinase
Krabbe's	Gal$\downarrow$Cer	Galactosylceramide β-D-galactosidase
Metachromatic leukodystrophy	Gal-Cer $\rightarrow\downarrow$ OSO$_3$	Sulfatide sulfatase
Ceramide lactoside lipidosis	Gal$\downarrow$Glc-Cer	β-D-Galactosidase
Fabry's	Gal$\downarrow$Gal-Glc-Cer	α-D-Galactosidase
Tay-Sachs	GalNAc$\downarrow$Gal-Glc-Cer (GM$_2$) NANA	β-D-Hexosaminidase A
Farber's	Cer	Ceramidase

*The abbreviations used are as follows: Cer, ceramide; Glc, glucose; Gal, galactose; GalNAc, N-acetyl galactosamine; NANA, N-acetyl neuraminic acid; Chln, choline. The arrow indicates the bond that normally is cleaved by the enzyme.

Lipases

Triglycerides are hydrolyzed by lipases, three of which already have been discussed. *Pancreatic lipase* is secreted in the pancreatic juice and digests dietary triglycerides within the intestinal lumen to the 2-monoglyceride intermediate. *Lipoprotein lipase*, the heparin-releasable enzyme, is present in adipose tissue and capillary endothelial cells; it mediates the hydrolysis of triglycerides in chylomicrons and VLDL. *Hormone-sensitive lipase* is present in adipocytes and hydrolyzes the triglycerides stored in adipose tissue. A *lysosomal lipase* that degrades triglycerides taken into cells during phagocytosis or pinocytosis is contained in many different cells. These enzymes degrade the triglyceride to the stage of the diglyceride. Another enzyme, *monoglyceride lipase*, hydrolyzes both the diglyceride and the resulting monoglyceride, completing the degradative process.

REFERENCES

Brady, R. O., Pentchev, P. G., and Gal, A. E.: Investigations in enzyme replacement therapy in lipid storage diseases, Fed. Proc. 34:1310, 1975. *A very brief review of interesting new developments that may have clinical importance.*

Fulco, A. J.: Metabolic alterations of fatty acids, Annu. Rev. Biochem. 43:215, 1974. *A recent, thorough review.*

Goodridge, A. G.: Hormonal regulation of the activity of the fatty acid synthesizing system and of the malic enzyme concentration in liver cells, Fed. Proc. 34:117, 1975. *A brief review of long-term control of fatty acid biosynthesis.*

Scanu, A. M., Edelstein, C., and Keim, P.: Serum lipoproteins. In Putman, F. W., editor: The plasma proteins, structure, function and genetic control, ed. 2, vol. 1, New York, 1975, Academic Press, Inc. *A brief up-to-date review.*

van den Bosch, H.: Phosphoglyceride metabolism, Annu. Rev. Biochem. 43:243, 1974. *A detailed review.*

Volpe, J. J., and Vagelos, P. R.: Saturated fatty acid biosynthesis and its regulation, Annu. Rev. Biochem. 42:21, 1973. *A detailed review.*

Wood, R., editor: Tumor lipids—biochemistry and metabolism, Champaign, 1973, American Oil Chemists' Society. *The latest information about lipids in cancer, written by leading workers in this field.*

Clinical examples

CASE 1: ENDOGENOUS HYPERTRIGLYCERIDEMIA (TYPE IV HYPERLIPOPROTEINEMIA)

On routine physical examination it was noted that a 36-year-old man had a fasting plasma cholesterol concentration of 250 mg/dl (normal range, 120 to 240 mg/dl), and a plasma triglyceride concentration of 500 mg/dl (normal range, 10 to 150 mg/dl). He was otherwise asymptomatic and was judged to be healthy except for this defect. He also was about 10 kg above his ideal weight. The patient was quite upset because he had read that middle-aged men with hyperlipidemia have an increased incidence of coronary artery disease; he therefore requested immediate and vigorous treatment. Initially his physician recommended an eucaloric diet containing 15% protein, 15% fat, 70% carbodrate, and 300 mg cholesterol. Although the patient rigorously followed this diet, there was no improvement in the plasma lipid concentrations; in fact, they became worse. This puzzled the physician, and he admitted the patient to the hospital and ordered that he receive a fat-free formula diet for 2 weeks. Still no improvement in plasma lipid levels occurred. Since the treatment had been unsuccessful to this point, a radical change in dietary therapy was instituted. Instead of restricting fat and cholesterol, dietary carbohydrate was reduced and replaced with foods rich in fat (triglyceride and cholesterol). The new dietary composition was 20% protein, 55% fat, 25% carbohydrate, and 600 mg cholesterol. A marked reduction in plasma lipids was observed within 48 hr; the plasma cholesterol remained in the range of 225 mg/dl and triglycerides in the range of 200 mg/dl. These values were considered to be acceptable, and no drug treatment was recommended. The patient was discharged on a high-fat, low-carbohydrate diet.

Biochemical questions

1. What is hyperlipoproteinemia?
2. How would a plasma lipoprotein elec- trophoretic analysis have aided in the diagnosis of this problem?
3. What are the dangers of a long-term, fat-free diet?
4. Describe the type of hyperlipoproteinemia in this case and explain the improvement in plasma lipid concentrations on a high-fat diet.
5. Explain the persistence of elevated plasma triglyceride and cholesterol concentration in spite of a fat-free diet. What is the mechanism of the hypercholesterolemia in this case?

Case discussion

Hyperlipidemia is defined as any condition in which, after a fast of 12 hr, the plasma cholesterol concentration is greater than 240 mg/dl, the plasma triglyceride concentration is greater than 150 mg/dl, or both. It is presently defined in terms of plasma lipid concentrations because they can be easily measured in the clinical laboratory. However, the plasma lipids are components of plasma lipoproteins. Hyperlipidemia is a defect in lipoprotein metabolism; it actually is a hyperlipoproteinemia. The hyperlipoproteinemias are classified in two ways. The first classification is in terms of primary versus secondary etiology. A primary hyperlipoproteinemia is a disease entity of itself; it is not simply the manifestation of a well-defined illness. In contrast, a *secondary* hyperlipoproteinemia is a symptom of another disease that happens to have, as one of its signs, a plasma lipid abnormality. Some of the diseases that produce a secondary hyperlipoproteinemia include diabetes mellitus, hypothyroidism, and biliary obstruction. Since we are told that the patient was otherwise healthy, we must assume that his hyperlipoproteinemia is primary.

A suggested scheme that makes these complicated metabolic disorders a bit easier to comprehend is presented in Fig. 8-13. All of the types are considered to be

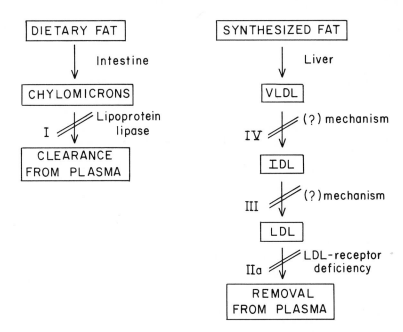

Fig. 8-13. Schematic representation of the metabolic abnormalities in hyperlipoprotein-emias. Roman numerals represent types of hyperlipoproteinemia and are placed to illustrate probable location of metabolic defect.

defects of lipoprotein catabolism. In type IV there is a defect in VLDL (pre-β-lipoprotein) catabolism, and VLDL accumulates in the plasma. The defect in type III involves the conversion of IDL to LDL, and the former accumulates in the plasma. Type IIb is a combination of type IIa and IV defects, and both LDL and VLDL accumulate. Type V is a combination of types I and IV, and both chylomicrons and VLDL are elevated.

The patient has a major elevation in plasma triglyceride and a minor elevation in plasma cholesterol. Therefore, he may have any of the types except type IIa hyperlipoproteinemia. As shown in Fig. 8-4, *lipoprotein electrophoresis* would distinguish between the possible types, for a normal electrophoretic pattern contains only β- and α-lipoproteins as major bands. Although some VLDL (pre-β-lipoprotein) normally is present, this usually appears as a trace amount on electrophoresis. A chylomicron band would be present in type I, and an excessive amount of pre-β-material would appear in types IIb and IV. Both chylomicrons and pre-β-lipoproteins would be seen in type V.

Because of the correlation between hyperlipidemia and atherosclerosis, many people believe that dietary fat is unhealthy and should be avoided. While this is basically correct for patients with types I, IIa, IIb, III, and V hyperlipoproteinemia, this probably is not true for most of the population. Fat supplies calories in a concentrated form. Moreover, polyenoic fatty acids of the *linoleic* family (ω 6) are essential components in the diet; they cannot be synthesized by man. Prostaglandins are synthesized from these essential fatty acids, and a deficiency in dietary fat eventually could produce a prostaglandin deficiency. The fat-soluble vitamins (A, D, E, and K) occur in the lipid portion of foods. Omission of fat from the diet may eventually produce deficiencies in fat-soluble vitamins unless vitamin replacement therapy is instituted.

Since this patient had a large elevation in plasma triglyceride concentration that was not helped by reduction in dietary fat, the type I and V syndromes can be excluded. Likewise, the plasma cholesterol concentration is not high enough for this to be a case of the type

IIb syndrome. As noted in Table 8-7, the high triglyceride but only minor cholesterol elevation is more consistent with type IV than type III. Therefore, it was reasoned that the patient had the type IV syndrome. In order to support this supposition, plasma lipoprotein electrophoresis was carried out. As shown in Fig. 8-4, type IV can be distinguished from type III by this method. Electrophoresis established that the lipoprotein pattern was consistent with that expected in type IV. This led to the diagnosis of type IV hyperlipoproteinemia, an excess of VLDL in the plasma. VLDL's are the triglyceride-rich lipopoteins synthesized in the liver. Much of the fatty acid used by the liver for the synthesis of triglycerides is derived from dietary carbohydrates. In this way, the liver converts excess calories into fat, the major storage form of energy in the body. The VLDL then transports the newly formed triglyceride to the adipocytes for storage. It is not surprising in view of this explanation of the hyperlipidemia that the low-fat diet did not help this patient, for the source of the circulating triglyceride was primarily dietary carbohydrate, not dietary fat. For as yet unexplained reasons, the patient was sensitive to dietary carbohydrate, the result being a plasma VLDL elevation. Recent work suggests that fructose may be more apt to cause hypertriglyceridemia than glucose. This has led to the view among some experts that sucrose, the disaccharide that contains fructose, should be reduced in the diet of patients who have carbohydrate inducible hypertriglyceridemia. It is recommended that starches replace sucrose in these diets since starch does not contain fructose.

The hypercholesterolemia, which is mild in this case, is secondary to the elevation in VLDL. Remember that VLDL functions primarily to transport triglyceride. Yet it contains cholesterol that is a structural part of the lipoprotein complex. In this situation there is no basic abnormality in cholesterol metabolism; plasma cholesterol is elevated simply because it is a component of the "package" in which the triglyceride is carried.

REFERENCES

Fredrickson, D. S., and Levy, R. I.: Familial hyperlipoproteinemia. In Stanbury, J. B., Wyngaarden, J. B., and Fredrickson, D. S., editors: The metabolic basis of inherited disease, ed. 3, New York, 1972, McGraw-Hill Book Co.

Macdonald, I.: Effects of dietary carbohydrates on serum lipids, Prog. Biochem. Pharmacol. **10:** 216, 1973.

CASE 2: HYPERCHYLOMICRONEMIA (TYPE I HYPERLIPOPROTEINEMIA)

A 15-year-old boy had a long history of abdominal complaints, including bouts of abdominal pain so severe that narcotics were required for relief. These bouts were intermittent, occurring every 6 months on the average. On one occasion abdominal surgery (an exploratory laparotomy) was performed, and the patient's appendix was removed. However, this did not correct the problem. The patient had recently felt well until he suddenly developed another episode of abdominal pain. His mother stated that the illness came on 4 hr after he had eaten a meal consisting of pork chops, spare ribs, fried potatoes, milk, and ice cream topped with a generous serving of whipped cream. No one else in the family had been made ill by this meal.

Examination of the patient indicated an acute abdominal emergency. The patient was brought to the hospital at 8:00 AM, 14 hr after his last meal. On arrival a blood specimen was drawn. Within 15 min the laboratory technician reported that valid results could not be obtained from the blood plasma because it was loaded with fat. The plasma was "milky," but on centrifugation for 30 min at 15,000 rpm, it cleared considerably and there was a thick band of "cream" located at the top of the specimen.

Biochemical questions

1. What kind of lipid abnormality would you suspect in this patient?
2. What chemical and electrophoretic studies should be obtained on the plas-

ma sample to aid in making the diagnosis?

3. What kind of a diet would be helpful in treating this disease?
4. Would you recommend that this boy's diet be supplemented with medium-chain triglycerides?

Case discussion

One would suspect the presence of type I hyperlipoproteinemia in this case. The thick band of "cream" that floated to the top of the centrifuge tube after relatively low-speed centrifugation at plasma density indicates the presence of lipoproteins of a density < 1.006. Since the illness followed a meal containing a large amount of fat, it is likely that these lipoproteins are chylomicrons of intestinal origin, which are very rich in dietary fat. VLDL's are smaller particles than chylomicrons and require longer periods of centrifugation at higher forces for flotation at density 1.006. Type I hyperlipoproteinemia is a familial disease that usually manifests itself in childhood or early adolescence. In contrast, types IV and V, which would have a similar blood plasma picture, usually become manifest later in life and in many cases are secondary to diseases such as diabetes mellitus or alcoholism. The age of the patient, the previous history of recurrent abdominal pain, and the persistence of severe hyperlipoproteinemia after a fat-rich meal strongly suggest a diagnosis of hyperchylomicronemia.

Three types of laboratory studies would support this diagnosis. First, plasma lipid determinations should be run. A large elevation in triglyceride concentrations would be expected, for chylomicrons are composed of 85% to 90% triglyceride. Lipoprotein electrophoresis should be done to demonstrate the presence of chylomicrons (Fig. 8-4). These lipoproteins are removed from the blood within 4 to 6 hr after a fatty meal in normal people (Fig. 8-1), and their presence in the plasma after 14 hr of fasting is grossly abnormal. If the defect were of type IV, an excessive pre-β-lipoprotein band would be visible and no chylomicron band would be seen. Conversely,

if the defect were of type V, both a chylomicron band and an excessive pre-β-lipoprotein band would be observed. Therefore, lipoprotein electrophoresis should provide a definitive diagnosis in this case. In order to confirm the diagnosis, the patient should be given an intravenous injection of heparin to test for lipoprotein lipase activity because the metabolic defect in type I hyperlipoproteinemia is a deficiency of lipoprotein lipase. Lipase activity is measured in the plasma using an artifically prepared radioactive triglyceride emulsion as the substrate. It must be remembered that heparin injection releases both the hepatic lipase as well as the extrahepatic or true lipoprotein lipase (p. 383). Patients with hyperchylomicronemia are deficient only in the extrahepatic enzyme, the protamine-inhibited form. Therefore, the lipase assay should be carried out with and without the addition of protamine. A normal patient will exhibit lower lipase activity in the sample containing the added protamine, indicating that some of the postheparin plasma lipolytic activity is lipoprotein lipase. Patients with type I hyperlipoproteinemia will have little or no protamine-inhibited lipolytic activity; all of their heparin-releasable activity will be caused by the liver lipase. This test demonstrated the absence of the extrahepatic lipoprotein lipase in this 15-year-old patient.

Type I hyperlipoproteinemia is an exogenous hypertriglyceridemia. The problem results from an inability to clear dietary fat from the blood plasma because of a lipoprotein lipase deficiency. Therefore, the treatment involves a drastic reduction in dietary fat intake. A diet rich in carbohydrate and protein and low in fat should be recommended for this patient.

Medium-chain triglycerides in the form of a dietary supplement would be useful in this situation. Medium-chain triglycerides are digested and absorbed differently from the long-chain fatty acid triglycerides contained in normal foods (Fig. 8-2). Medium-chain fatty acids are not resynthesized into triglycerides by the intestinal mucosa. Instead, they are absorbed as free fatty acid in the portal blood. Since they bypass the chylomicron and lipoprotein lipase steps, medium-chain fatty acid utilization would not be affected by the enzymatic deficiency in this disease. Therefore, fat calories could be supplied to this patient in the form of medium-chain triglycerides, enabling

the carbohydrate content of the diet to be reduced to a more palatable level.

REFERENCE

Krauss, R. M., Levy, R. I., and Fredrickson, D. S.: Selective measurement of two lipase activities in postheparin plasma from normal subjects and patients with hyperlipoproteinemia, J. Clin. Invest. 54:1107, 1974.

CASE 3: GLUCOSYLCERAMIDE LIPIDOSIS (GAUCHER'S DISEASE)

A 34-year-old woman was admitted to the hospital because of easy bruising and excessive bleeding. Examination of her abdomen revealed a firm, nontender mass in the left upper quadrant that was judged to be an enlarged spleen. A mass also was present in the right upper quadrant of the abdomen, probably an enlargement of the liver. Examination of the blood revealed *pancytopenia*, a decrease of all of its cellular elements. Coagulation tests indicated that a prolonged bleeding time was the only abnormality. Examination of the bone marrow revealed the presence of Gaucher's cells.

Biochemical questions

1. What are the lipidoses?
2. How is the finding of Gaucher's cells in the bone marrow related to the metabolic defect in this disease?
3. What biochemical tests can be performed to identify individuals suffering from a form of lipidosis? Will these tests detect the carrier state as well as the patient with overt disease?

Case discussion

The lipidoses are a group of lipid storage diseases that occur in man. Each is characterized by the accumulation of a particular lipid in the cells of the reticuloendothelial system and central nervous system. All of the lipid storage diseases result from an inherited deficiency of the enzyme that catalyzes the degradation of the lipid that accumulates. The lipids that are involved are sphingolipids, including cerebrosides, gangliosides, sulfatides, or sphingomyelin. Table 8-9 on p. 407 lists the most common lipid storage diseases, the lipid

that accumulates in each, and the enzymatic deficiency.

Gaucher's disease. Gaucher's disease is one of the most common forms of lipid storage disease. The most severe types occur in infancy and lead to early death. However, as is illustrated by this case, milder forms of the disease do occur in the adult. Gaucher's disease is characterized by the accumulation of a β-glucosylceramide. This is a result of a deficiency or absence of β-glucosylceramide glucosidase. The lipid accumulates in the tissues, causing splenic and hepatic enlargement as in this patient. Enlarged cells also appear in the bone marrow resulting from glucosylceramide accumulation, and these are diagnostic for this disease. The enlarged spleen acts as a trap for blood cells, filtering them out of the blood and thereby producing the *pancytopenia*. Bruising and bleeding resulted from the fact that the blood *platelets* are one of the cellular elements filtered out by the enlarged spleen. Platelets are important in the initiation of blood coagulation, and bleeding often results when there is a platelet deficiency.

One can test for the activity of this glucosidase and most of the other enzymes that are involved in the various lipid storage diseases by examining blood *leukocytes*. These cells are readily available for study, for all that is required is a small sample of venous blood. In contrast, the tissues where the lipid accumulation occurs are often difficult or impossible to biopsy for biochemical study. Leukocytes are isolated from the blood by centrifugation. After disruption, their contents are incubated with the appropriate substrate, that is, glucosylceramide if Gaucher's disease is suspected. The enzyme in question, a β-D-glucosidase, catalyzes the removal of the β-D-linked glucose residue from the ceramide. Therefore, the amount of glucose that was released would be analyzed; this would be a measure of the cellular β-D-glucosidase activity. Leukocytes obtained from a normal individual are used as a reference. This test is important in making a definitive diagnosis as well as in detecting the *carrier state*, for less than the normal amount of the enzyme is present in the cells of carriers. In this way one can determine whether apparently normal people are carriers of these diseases, information that is crucial when married couples with a familial history of a lipid storage disease contemplate parenthood. Similar diagnostic enzyme assays can be made relatively early in pregnancy with cells isolated from the amnionic fluid. This can provide early definitive evidence as to whether or not the fetus has inherited the lipid storage defect, information that is vital in determining whether the possibility of aborting the fetus should be considered.

REFERENCES

Bersohn, J., and Grossman, H. J.: Lipid storage diseases, New York, 1971, Academic Press, Inc.

Fredrickson, D. S., and Sloan, H. R.: Glucosyl ceramide lipidoses: Gaucher's disease. In Stanbury, J. B., Wyngaarden, J. B., and Fredrickson, D. S., editors: The metabolic basis of inherited disease, ed. 3, New York, 1972, McGraw-Hill Book Co.

CASE 4: CARNITINE DEFICIENCY

A 19-year-old girl was referred to a university medical center because of easy fatigability and very poor exercise tolerance. Careful neurologic examination revealed some muscle weakness in her extremities. Several muscle biopsies were performed. Microscopic examination indicated that the muscle was filled with vacuoles containing lipid. Chemical measurements indicated that these muscle specimens contained greatly elevated amounts of triglycerides, but only one-sixth as much carnitine as biopsy specimens obtained from other patients who did not have any primary muscle disease.

Biochemical questions

1. What is the main intracellular function of carnitine?
2. Would you expect that fatty acid β-oxidation might be impaired in this patient?
3. Would you expect that the oxidation of pyruvate (derived from glucose) might be impaired in this patient?
4. How might the carnitine deficiency account for the triglyceride accumulation in the muscles?

Case discussion

L-Carnitine is involved in the β-oxidation of long–chain fatty acids. Its structure is shown on p. 389. It is a part of the mitochondrial translocase system that transports long-chain fatty acyl groups across the inner mitochondrial membrane. In order that β-oxidation of fatty acids occur, the fatty acyl chain must be translocated into the mitochondrial matrix, the site of the β-oxidation system. Fatty acids are activated to acyl CoA thioesters either in the endoplasmic reticulum or the outer mitochondrial membrane. Acyl CoA thioesters, however, cannot cross the inner mitochondrial membrane. The acyl group is transferred across as the acylcarnitine ester, being first transesterified to carnitine and then transesterified back to CoA after crossing the membrane in the form of a carnitine ester. This patient's muscles were deficient in carnitine and therefore she could not efficiently transport long-chain fatty acyl groups into her mitochondria for oxidation. Fatty acid oxidation is a major source of energy for many tissues, including the skeletal muscles. The patient's muscular symptoms and intolerance toward exercise are explained by her inability to derive sufficient energy from fatty acid oxidation.

As opposed to fatty acid oxidation, one would not expect any defect in glucose oxidation in this patient. Like fatty acids, glucose is converted to acetyl CoA prior to oxidation in the Krebs cycle. The carbon atoms from glucose enter the mitochondria in the form of pyruvate, and the pyruvate is converted to acetyl CoA after it already has crossed the inner mitochondrial membrane. Therefore, the carnitine-dependent step is not involved in glucose oxidation, and a carnitine deficiency would not be expected to impair glucose oxidation. In fact, in order to replace some of the energy ordinarily derived from fatty acid β-oxidation, more than the usual amounts of glucose probably were oxidized via the Krebs cycle in this patient.

It is likely that triglycerides accumulated in this patient's muscles because of the defect in fatty acid oxidation.

There was no impairment in the activation of fatty acid in the muscle, so that long-chain acyl CoA could be formed. Not having the oxidative pathway to enter, it is likely that more than the usual amounts of fatty acyl groups were diverted into triglycerides. Muscle, as well as other tissues, store fatty acids as triglycerides. Therefore, it is expected that any excess of fatty acyl groups within the muscle might be incorporated into this storage form.

REFERENCE

Engel, A. G., and Angelini, C.: Carnitine deficiency of human skeletal muscle with associated lipid storage myopathy: a new syndrome, Science 179:899, 1973.

CASE 5: OBESITY

A 19-year-old woman sought medical help because she was 30 kg overweight. Most of her excess weight was in the form of adipose tissue triglycerides. A dietary history revealed that her diet was extremely poor. Much of her caloric intake was carbohydrate—candy, cookies, cake, soft drinks, and beer; her dietary fat intake was actually quite moderate.

Biochemical questions

1. How is it possible to form excess amounts of triglyceride in the body if a diet contains predominantly carbohydrate?
2. How does acetyl CoA generated inside the mitochondria reach the cytoplasm for use by the fatty acid de novo biosynthetic pathway?
3. Why is bicarbonate required for fatty acid synthesis?
4. What is the rate-limiting enzyme in fatty acid de novo biosynthesis?
5. If this patient were deficient in biotin, would this deficiency interfere with fatty acid biosynthesis?

6. How might the carbohydrate ingested by this patient supply the reducing equivalents needed for fatty acid biosynthesis?
7. How would VLDL and lipoprotein lipase be involved in the conversion of dietary carbohydrate to triglycerides for storage in this patient's adipose tissue?
8. Are there any differences between starch and sucrose in terms of their hyperlipidemic action?
9. Devise a test that would indicate whether this patient could mobilize the triglyceride that is stored in her adipose tissue.

REFERENCES

Albrink, M. J.: Dietary carbohydrates in lipid disorders in man, Prog. Biochem. Pharmacol. 8:242, 1973.

Nestel, P. J.: Triglyceride turnover in man: effects of dietary carbohydrates, Prog. Biochem. Pharmacol. 8:125, 1973.

Zakim, D.: Influence of fructose on hepatic synthesis of lipids, Prog. Biochem. Pharmacol. 8:161, 1973.

CASE 6: STEATORRHEA

A 61-year-old man had severe pancreatic damage as a result of pancreatitis. He began to have difficulty digesting the fat in his diet and noted the onset of fatty stools (steatorrhea). He had lost

5 kg in 2 months despite an adequate dietary intake. He was hospitalized and fed a low-fat diet that was supplemented with a medium-chain triglyceride mixture. His symptoms improved, and he began to gain weight.

Biochemical questions

1. How could pancreatic damage interfere with the utilization of dietary fat?
2. Why was the medium-chain triglyceride mixture tolerated by this patient even though he could not tolerate ordinary dietary fats?
3. Which of the following carbohydrates might be difficult for this patient to digest — sucrose, lactose, or starch?
4. If this patient were found to have a low percentage of linoleic acid ($18:2\omega6$) in his plasma, how might this finding be explained?
5. Why would the patient *not* be expected to have a low percentage of palmitic acid ($16:0$) in his plasma even if the linoleic acid content was reduced?
6. Assuming that the palmitic acid content of the body was adequate but linoleic acid was reduced, what prediction can be made concerning the amount of oleic acid and its derivatives (ω 9-class) present in this patient's blood and tissues?

REFERENCES

Borgström, B.: On the interactions between pancreatic lipase and colipase and the substrate, and the importance of bile salts, J. Lipid Res. 16:411, 1975.
Wilson, F. A., and Dietschy, J. M.: Differential approach to clinical problems of malabsorption, Prog. Gastroenterol. 61:911, 1971.

CASE 7: ANGINA PECTORIS

A 36-year-old man sought medical help because of intermittent, severe anterior chest pain that radiated to his left arm. This pain was brought on by mild exertion such as a brisk walk or climbing a flight of stairs. An exercise electrocardiagram revealed changes indicative of myocardial ischemia, and the possibility of a coronary artery bypass graft was considered. It was therefore necessary to perform a diagnostic cardiac catheterization and coronary artery angiogram. During the course of the catheterization a number of blood samples were obtained from both the left ventricular chamber and the coronary sinus in order to measure arteriovenous differences in certain metabolites. The free fatty acid concentration in the arterial plasma was 0.89 ± 0.06 μeq/ml; that in the venous samples was 0.47 ± 0.05 μeq/ml. This demonstrated that free fatty acid was being utilized by the myocardium.

Biochemical questions

1. Assume that some of the newly incorporated fatty acid was oxidized as a source of energy. What reactions are involved in producing acetyl CoA for the Krebs cycle from the free fatty acid that entered the myocardial cell?
2. How much ATP would be formed if one mole of stearic acid were to be converted completely to acetyl CoA by the myocardium?
3. What are the major uses of the ATP formed by this process in the myocardial cell?
4. Some of the newly taken up free fatty acid was not oxidized by the myocardium. Into what compounds and cellular structures might this fatty acid have been incorporated?
5. How is carnitine related to fatty acid utilization?
6. What effect would the following substances be expected to have on the plasma free fatty acid concentration if they were infused into this patient? *a*, Epinephrine, *b*, glucagon, *c*, glucose plus insulin, *d*, theophylline, and *e*, prostaglandin E.
7. How is cyclic AMP related to the epinephrine and theophylline effects on fatty acid mobilization from adipose tissue?
8. Would changes in the plasma free fatty acid concentration be expected to have any effect on the transport of drugs such as sodium warfarin, coumarin, or clofibrate that might be prescribed for this patient at some later date?

REFERENCES

Gorman, R., Hamberg, M., and Samuelsson, B.: Inhibition of basal and hormone-stimulated adenylate cyclase in adipocyte ghosts by the prostaglandin endoperoxide prostaglandin H_2. J. Biol. Chem. 250:6460, 1975.

Most, A. S., Brachfield, N., Gorlin, R., and Wahren, J.: Free fatty acid metabolism of the human heart at rest, J. Clin. Invest. 48:1177, 1969.

Spector, A. A.: Fatty acid binding to plasma albumin, J. Lipid Res. 16:165, 1975.

CASE 8: LIPOGRANULOMATOSIS (FARBER'S DISEASE)

A 9-month-old girl was admitted to the pediatric unit of a hospital because of poor weight gain and psychomotor retardation. Although the child appeared well at birth and for several months thereafter, at 5 months of age she had begun to develop these problems, which had become progressively worse. Physical examination, in addition to confirming the nutritional failure and psychomotor retardation, indicated the presence of subcutaneous nodules, hepatomegaly, and splenomegaly. In spite of vigorous supportive therapy she deteriorated rapidly and expired 3 weeks after admission. Tissue specimens were obtained for histologic and chemical analysis during the postmortem examination. Large quantities of lipid-staining material were observed in many tissues, and this was demonstrated chemically to be ceramide.

Biochemical questions

1. In what kinds of lipids is ceramide found?
2. How is sphingosine related to ceramide?
3. Are fatty acids present in ceramide?
4. Based on our current knowledge concerning the pathologic mechanisms causing lipid-storage diseases, would you expect the cause of this problem to be excessive synthesis of ceramide?
5. How is pyridoxal phosphate related to ceramide?
6. What amino acid is involved directly in ceramide synthesis?
7. Lipogranulomatosis is an inherited disease. Therefore, how can we account for the fact that serious problems were not noted until the patient was 5 months old?

REFERENCE

Sugita, M., Dulaney, J. T., and Moser, H. W.: Ceramidase deficiency in Farber's disease (lipogranulomatosis), Science 178:1100, 1972.

CASE 9: TYPE III HYPERLIPOPROTEINEMIA

A 29-year-old man was referred to a cardiologist because of a history of heart and vascular problems in his immediate family. His father had suffered from poor circulation in his legs for several years and had recently been operated on to have grafts inserted in his femoral arteries. During this hospitalization, the father was noted to have hyperlipidemia. The patient had three brothers, all older than himself. One died of a myocardial infarction at age 37 years; another had lipid accumulations (xanthoma) in the palmar creases of his hands and on his elbows and buttocks; and a third brother was asymptomatic. The patient stated that he felt well and was eating the usual American diet. After an overnight fast, blood was drawn for lipid determination. The plasma total cholesterol was 340 mg/dl and the plasma triglyceride concentration was 380 mg/dl. This test was repeated once, and similar values were obtained, indicating the presence of hyperlipidemia.

Biochemical questions

1. What are intermediate density lipoproteins (IDL)?
2. What is the precursor of IDL, and to what lipoproteins are IDL's catabolized?
3. What enzyme is involved in the hydrolysis of plasma triglycerides?
4. What functions have been attributed to high density lipoproteins (HDL)?

5. How does the apolipoprotein composition differ among the various classes of plasma lipoproteins?
6. Would a plasma lipoprotein electrophoresis be helpful in diagnosing the type of hyperlipoproteinemia present in this patient?
7. Can the triglycerides present in plasma lipoproteins such as IDL be synthesized completely from carbon atoms derived from dietary carbohydrates?

REFERENCES

Utermann, G., Jaeschke, M., and Menzel, J.: Familial hyperlipoproteinemia type III: deficiency of a specific apolipoprotein (Apo E-III) in the very low density lipoproteins, FEBS Lett. 56:352, 1975.

Zelis, R., Mason, D. T., Braunwald, E., and Levy, R. I.: Effects of hyperlipoproteinemias and their treatment on the peripheral circulation, J. Clin. Invest. 49:1007, 1970.

ADDITIONAL QUESTIONS AND PROBLEMS

1. Would removal of the gallbladder have a profound effect on lipid digestion? Explain.
2. A male infant failed to gain weight, and the presence of cystic fibrosis, a disease involving the exocrine pancreas, was suspected. Little triglyceride but lots of fatty acid was found in the feces. Does this laboratory observation support the working diagnosis? Explain.
3. A patient responded in the following way to an injection of heparin when his plasma triglyceride levels were grossly elevated. Within 10 min the triglycerides were reduced slightly and the plasma free fatty acid concentration was greatly increased. However, after 2 hr the plasma free fatty acid concentration reverted to normal. Explain the mechanism of these responses.
4. Low plasma albumin concentrations often accompany a kidney disease known as nephrosis. What effect would this have on lipid transport?
5. Many cancer cells have a high plasmalogen content. What type of lipid is this? What subcellular structures might be influenced by this lipid difference?
6. How would you determine the fatty acid composition of the plasma triglycerides of a patient? Assume that you are given a fresh plasma specimen with which to work.
7. How would you separate the lipoproteins contained in the blood plasma obtained from a patient?

CHAPTER 9

AMINO ACID METABOLISM

OBJECTIVES

1. To explain the normally occurring interconversions of the amino acids
2. To analyze deviations in these reactions produced by some diseases
3. To consider the mechanisms by which the amino acids are utilized for energy
4. To show the pathways by which the nitrogen produced in amino acid catabolism is converted to urea
5. To demonstrate how urea production and its elimination affect certain pathologic conditions
6. To describe the biosynthesis of pharmacologically active amines from amino acids

Dietary protein deficiency is the most serious nutritional problem in the world. While it has devastating effects in underdeveloped countries, protein deficiency is not uncommon even in highly developed countries, particularly among pregnant and lactating women, newborns, the poor, the aged, and those who habitually avoid foods rich in protein. In the strictest sense it is not protein itself that is required but rather the amino acids released by hydrolysis of the protein. For this reason it is important to understand how the dietary amino acids are involved in the biosynthesis of other amino acids and how they serve as precursors of a variety of nitrogen-containing metabolites.

Measurement of several of the nitrogenous constituents of blood and urine are useful in the diagnosis and treatment of certain diseases and disorders. Evidence of nitrogen imbalance is characteristic of growth and pregnancy as well as starvation, malnutrition, febrile diseases, burns, trauma, and surgery. Difficulties in protein amino nitrogen metabolism may lead to ammonia intoxication in hepatic failure. Several less common aminoacidurias, the result of genetic defects, may lead to serious metabolic derangements. Nitrogen metabolism also involves the synthesis and degradation of creatine and heme porphyrins; the latter were considered in Chapter 3. Substances derived from these metabolites are easily measured, and determination of changes in their concentrations is useful in diagnosis and treatment.

Amino acids are also important in the synthesis of several biologically active amines, for example, the neurotransmitters, norepinephrine, serotonin, and acetylcholine. The metabolism of these amines is central to our understanding of the biochemical basis for the action of a host of pharmacologic agents.

DIETARY PROTEIN REQUIREMENTS

In the broadest sense, dietary proteins serve two functions: the carbon skeletons of their amino acids can be oxidized to yield energy, and their carbon and nitrogen atoms are used for the biosynthesis of several nitrogen-containing cellular constituents. While biologic energy requirements can be met by the oxidation of lipids and carbohydrates, amino acids are indispensable for protein bio-synthesis, and it is therefore necessary to eat sufficient protein-containing foods of adequate biologic value.

As demonstrated earlier, amino acids are required for the synthesis of the por-phyrins, particularly heme, which turns over rapidly (Chapter 3), the choline and ethanolamine moities of phospholipids (Chapter 8), and the glycosamines of con-nective tissue (Chapter 7). The synthesis of biogenic amines and nucleic acids also require substantial amounts of protein nitrogen. However, the major require-ment is for the synthesis of the body's proteins.

Unlike carbohydrates and lipids, proteins and amino acids are not thought of as stored by the body, even though some of the body's proteins are mobilized during fasting or starvation. The carbon skeletons of the amino acids can be either burned for energy or converted into glycogen or triglycerides, both of which are stored. However, the nitrogen portions of the amino acids, even though they are needed for many biosynthetic reactions, are not stored. Fig. 9-1 summarizes the metabolism of proteins. The amino acid pool shown in this diagram should not be

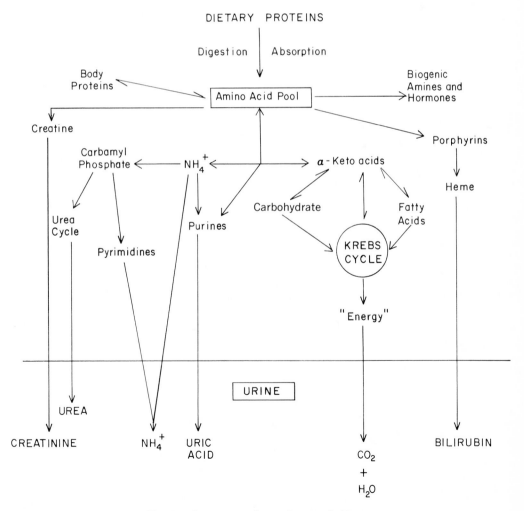

Fig. 9-1. Summary of protein metabolism.

regarded as a storage place for substantial amounts of amino acids. On the contrary, it is merely a convenient way of representing the circulating amino acids in the blood and the small amounts present within the cells as intermediates.

During starvation, amino nitrogen is derived first from plasma proteins, especially albumin. Other rapidly metabolizing tissues, for example, liver, pancreas, and intestinal mucosa, also tend to lose their proteins rapidly. Muscle is slower to provide nitrogen, but because it is approximately 60% protein it represents the largest reservoir. When fat supplies are exhausted the body may lose as much as 6% of its protein mass per day for energy.

Digestion of proteins

Dietary proteins are digested to their constituent amino acids by proteolytic enzymes and peptidases in the gastrointestinal tract. Some small proteins and several peptides are directly absorbed from the intestine, but only a very low concentration of peptides is found in either peripheral or portal blood. Most of the protein digestion products circulate as amino acids so that measurements of plasma amino nitrogen are a good index of the amount of amino acids in the blood. After a short fast, plasma amino nitrogen is about 5 mg/dl, whereas shortly after a meal rich in protein, plasma levels may increase to 8 mg/dl.

With the exception of the intestinal peptidases, all of the proteolytic enzymes are activated by conversion from inactive and larger protein precursors called zymogens. The first zymogen that comes into play is pepsinogen, which is produced in the chief cells, the secretory cells located in the mucosa of the stomach wall. The secreted pepsinogen is activated autocatalytically by the low pH of the stomach contents. Under the acid conditions in the stomach, a small amount of pepsinogen is cleaved into pepsin and inert peptides (Fig. 9-2). The pepsin produced is an active proteolytic enzyme that cleaves the remainder of the pepsinogen molecules and converts them to pepsin.

The hormone *gastrin* is secreted from the distal end of the stomach, through a stimulus initiated by hydrochloric acid. Gastrin then triggers the chief cells to release pepsinogen. In a like manner the pancreas releases its digestive juice when stimulated by *secretin,* a hormone produced in the duodenum in response to chyme (the complex contents of the stomach). The acid chyme is made alkaline in the duodenum by the bile and pancreatic juice. The latter contains the zymogens chymotrypsinogen, trypsinogen, and procarboxypeptidase, which are activated as shown in Fig. 9-3 (see also Chapter 3).

The resulting proteolytic enzymes have some remarkable specificities for cleaving protein chains at certain amino acid residues. These specificities are summarized in Table 9-1. The products of these enzymes are free amino acids, dipeptides, and small peptides. The residual peptides are hydrolyzed in the intestinal mucosal cells by *aminopeptidase* and *dipeptidases.*

A clear distinction should be made between the process of enzyme activation and that of enzyme synthesis. An increase in the concentration of a functional enzyme is accomplished in both cases but by different mechanisms. Activation depends on the existence of an inactive precursor protein that can be quickly converted to the active enzyme by a simple hydrolytic reaction, for example, removal of part of the peptide chain. Enzyme synthesis refers to the formation of new enzyme molecules from amino acids, that is, new or de novo synthesis. In this case an agent, for example, a hormone, stimulates the synthesis from the amino acids of new, fully active enzyme molecules. The body often uses enzyme activation when the response must be fast, as in digestion or blood clotting. Enzyme synthesis or induction is usually a slower process.

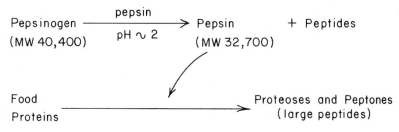

Fig. 9-2. Activation of pepsinogen.

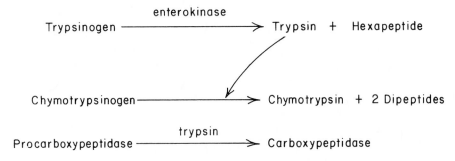

Fig. 9-3. Activation of pancreatic zymogens.

Absorption of amino acids and peptides

For a long time it was thought that the digestion of proteins must go to completion, all the way to free amino acids, before the hydrolysis products can be absorbed into the intestinal mucosa. Now it is clear that a substantial amount of small peptides, in addition to amino acids, are absorbed by stereospecific transport systems. The peptides, however, are rapidly hydrolyzed by peptidases located in the absorptive cells so that only amino acids are found in the portal blood.

The uptake mechanism for peptides is completely independent of that for amino acids. There are at least three stereospecific transport systems for the amino acids in human kidney and intestine. One system transports neutral and aromatic amino acids, another transports the basic amino acids and cystine, while the third is specific for glycine and proline. Each system transports amino acids that are structurally similar. For example, in Hartnup disease (case 3, p. 464) the transport system for the neutral and aromatic amino acids is defective. In spite of this, most Hartnup patients have only minor health problems, since essential neutral and aromatic amino acids are absorbed as peptides from the intestine by separate transport systems.

There is still much to learn about peptide transport. Important questions still unanswered are: what is the quantitative significance of peptide uptake compared to amino acid uptake; is intramembrane transport of peptides biochemically linked to their hydrolysis; and how many mucosal peptide transport systems exist?

γ-Glutamyl cycle

We know more about the transport of amino acids. According to a scheme proposed by Meister, amino acids are transported as dipeptides of glutamic acid. The important feature of this transport system is that the tripeptide glutathione serves as a donor of a γ-glutamyl group that is transferred to the amino group of the amino acid selected for transport. All of the common protein amino acids ex-

Table 9-1. Specificity of some proteolytic enzymes

ENZYME	OCCURRENCE	pH OPTIMUM	MAJOR SITE OF ACTION
Trypsin	Intestine	7.5-8.5	Arginyl, lysyl bonds
Chymotrypsin	Intestine	7.5-8.5	Aromatic amino acyl bonds (Phe, Trp, Tyr)
Pepsin	Stomach	1.5-2.5	Wide range of specificity
Carboxypeptidase	Intestine	7.5-8.5	C-Terminal amino acid
Aminopeptidase	Intestinal mucosa		N-Terminal amino acid

cept proline can serve as substrates for the membrane-bound γ-glutamyl transferase, the enzyme that catalyzes the formation of the dipeptide. Several small peptides may also be transported as γ-glutamyl peptides. This transport scheme is called the γ-glutamyl cycle or the Meister cycle since the amino acids as well as all substrates are regenerated by the action of five cytoplasmic enzymes. The cycle is illustrated in Fig. 9-4.

In the first step the amino acid is bound to a specific membrane site, probably under the influence of a membrane-associated protein that recognizes amino acids having common structural features. After the amino acid is bound, the γ-glutamyl transferase of the membrane catalyzes the transpeptidation of a γ-glutamyl residue from glutathione to the amino acid forming a γ-glutamyl amino acid and cysteinylglycine, the remaining portion of gluthathione. This is shown in the top portion of Fig. 9-4. As a consequence of this enzymatic reaction the γ-glutamyl dipeptide is transported into the cell. Once inside the cell, the cytoplasmic enzyme γ-glutamyl cyclotransferase (reaction 2, Fig. 9-4) promotes a sort of internal transpeptidation that releases the amino acid and cyclizes the γ-glutamyl group so that it forms a peptide linkage with itself. The cyclized glutamyl compound is called 5-oxoproline. This compound, also called pyroglutamic acid, is quite stable so that to convert it to glutamate requires the hydrolysis of an ATP in a reaction mediated by the enzyme 5-oxoprolinase (reaction 3, Fig. 9-4).

In order to complete the γ-glutamyl cycle, the glutathione consumed in the initial reaction must be reformed. However, glutathione is synthesized from γ-glutamylcysteine. There is no known enzyme that can utilize the cysteinylglycine dipeptide formed in the initial translocation reaction of the γ-glutamyl cycle. Consequently, the latter dipeptide is hydrolyzed by a peptidase (reaction 1, Fig. 9-4) to yield cysteine and glycine. This reaction plus that catalyzed by 5-oxoprolinase yields the component amino acids glutamate, cysteine, and glycine needed for the synthesis of glutathione. The formation of each of the two peptide bonds in glutathione requires the hydrolysis of an ATP. In the first reaction, γ-glutamylcysteine is the product of a reaction mediated by γ-glutamylcysteine synthetase (reaction 4, Fig. 9-4; see also case 4, p. 466). In the second reaction the glycyl moiety is added under the influence of glutathione synthetase (reaction 5, Fig. 9-4) regenerating gluthathione so that the cycle is completed. The transport of amino acids via the γ-glutamyl cycle is quite costly, three ATP's being required for each amino acid.

It should be emphasized that the γ-glutamyl cycle may not account for all of the amino acids transported into or out of cells. Undoubtedly, the peptide transport systems described earlier are at least as important, and other amino acid transport

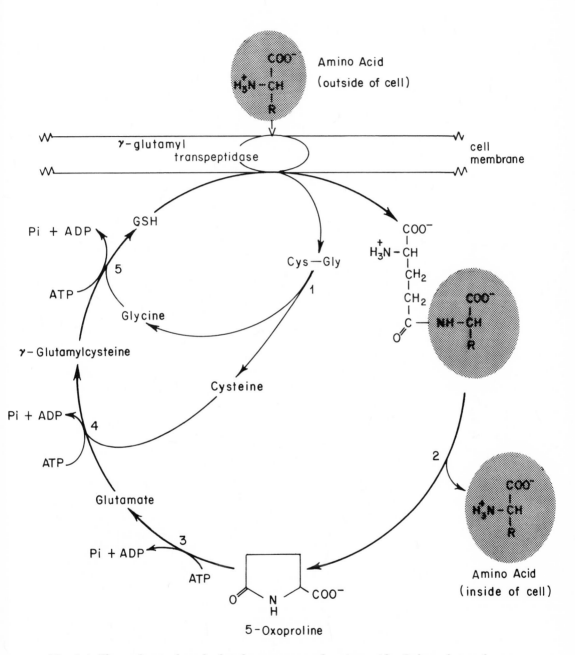

Fig. 9-4. The γ-glutamyl cycle for the transport of amino acids. Only γ-glutamyl trans-peptidase is located in the membrane; all of the other enzymes are found in the cytosol. The numbers refer to the following enzymes: *1*, peptidase, *2*, γ-glutamyl cyclotransferase, *3*, 5-oxoprolinase, *4*, γ-glutamylcysteine synthetase, and *5*, GSH (γ-glutamylcysteinyl-glycine, glutathione) synthetase.

systems also exist. Clearly, proline, which is not a substrate for γ-glutamyl transferase, must be transported by a separate mechanism. Possibly other membrane
proteins function with the γ-glutamyl transferase to give it specificity, and these
may be the proteins mutated in diseases such as Hartnup disease (case 3, p. 464).

Genetic abnormalities of the γ-glutamyl cycle

Even though the concept of the γ-glutamyl cycle has been known for only a
few years, examples of human genetic defects in three of the enzymes have been
reported. A patient with a defect in γ-glutamylcysteine synthetase (reaction 4,
Fig. 9-4) shows symptoms of hemolytic anemia (see case 4, p. 466), probably because the defective enzyme prevents the synthesis of glutathione. This is required
not only for amino acid transport but also for several other glutathione-requiring
reactions. For example, glutathione in the red cell is important in maintaining
the integrity of erythrocyte membranes.

Another genetic abnormality occurs in patients with 5-oxoprolinuria. The
enzymatic defect in these patients is for glutathione synthetase (reaction 5,
Fig. 9-4); γ-glutamyl cysteine, the substrate for this enzyme, is overproduced.
Like other γ-glutamyl amino acids, γ-glutamylcysteine is converted to 5-oxopro-
line. The third genetic abnormality in the γ-glutamyl cycle is associated with the
membrane-bound γ-glutamyl transferase. Patients with this inborn error of
metabolism excrete excessive amounts of glutathione in their urine (see case 7,
p. 468).

BIOSYNTHESIS OF NONESSENTIAL AMINO ACIDS

In most diets the mixtures of amino acids are not present in the proportions
required by the body. Consequently, it is necessary to rearrange them metabolically. The human body can accomplish this unless the diet is badly balanced.
Amino acids released to the portal blood from the gut already show some changes
in their relative compositions. Alanine levels in the portal blood exceed the relative amounts of alanine ingested, whereas glutamate and aspartate content in
portal blood is significantly lowered from the amounts ingested. Thus a large
amount of the ingested nitrogen is delivered to the liver as alanine. Alanine is also
derived from the muscle during fasting or starvation. Muscle proteins are broken
down and the amino groups of their branched chain amino acids transferred to
pyruvate to yield alanine. Recall from Chapter 1 that the amino acids are grouped
according to whether they are essential or nonessential components of the diet.
Since man is able to synthesize the nonessential amino acids, it is important to
examine the biochemical reactions involved.

Metabolism supporting amino acid biosynthesis

In previous chapters a number of metabolic pathways, including the Krebs
cycle, the glycolytic pathway, and the pentose phosphate pathway, have been described. While these metabolic schemes are important in producing the energy
needed to maintain life, several of their intermediates are also used in biosynthetic
reactions. Consequently, it is convenient to consider first those amino acids that
are synthesized by reactions leading from these familiar pathways.

Synthesis of amino acids from Krebs cycle intermediates

Several amino acids can be synthesized from Krebs cycle intermediates via
transamination reactions (Chapter 6). These reactions are shown in Fig. 9-5. Note
that the donor of amino groups in these transformations is always glutamate;
thus glutamate must be provided in order that a net synthesis of amino acids using

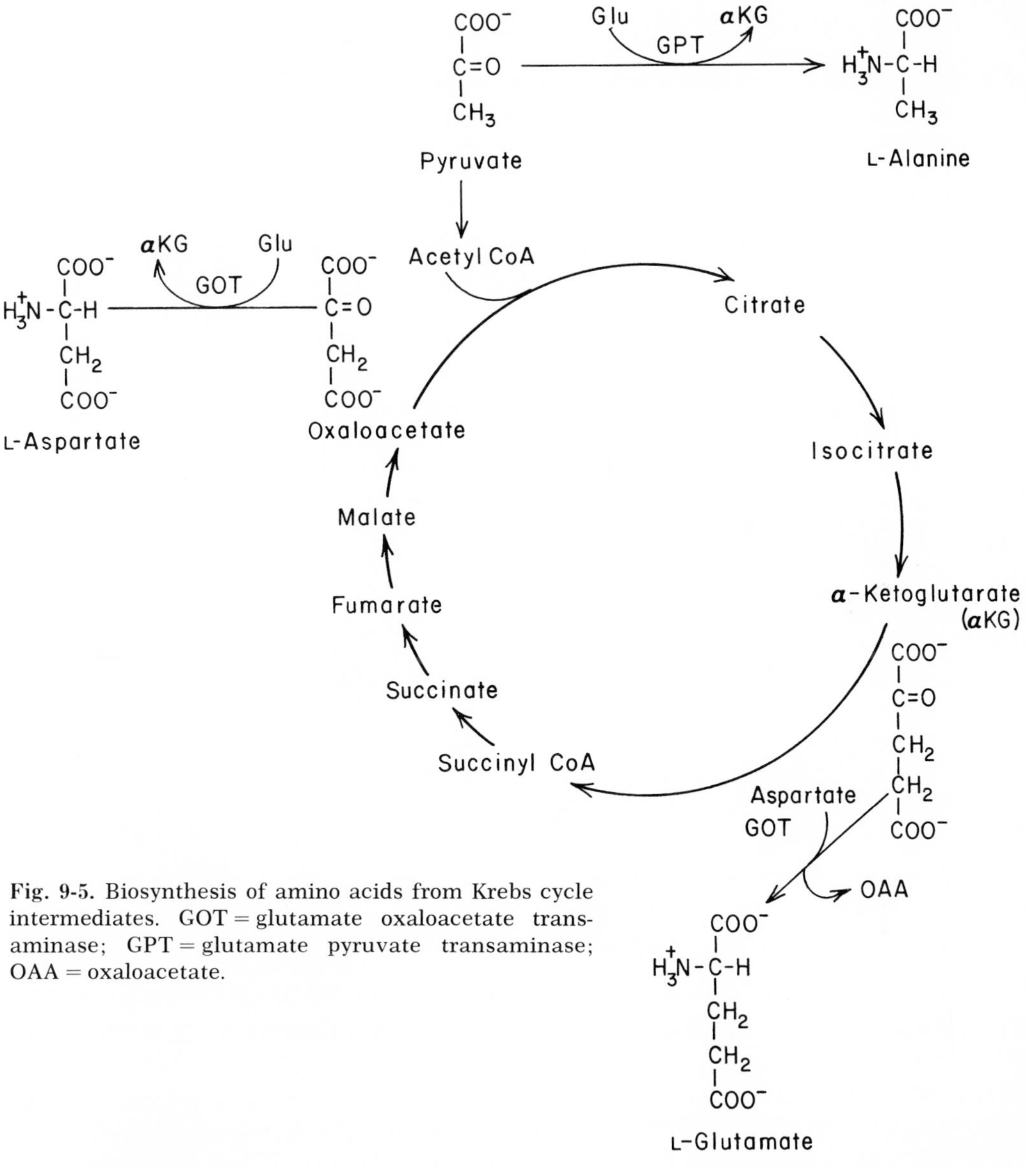

Fig. 9-5. Biosynthesis of amino acids from Krebs cycle intermediates. GOT = glutamate oxaloacetate transaminase; GPT = glutamate pyruvate transaminase; OAA = oxaloacetate.

transamination reactions may be achieved. Glutamate dehydrogenase, a widely occurring enzyme, accomplishes this by catalyzing the reaction of α-ketoglutarate with ammonium ion to form glutamate, as shown below.

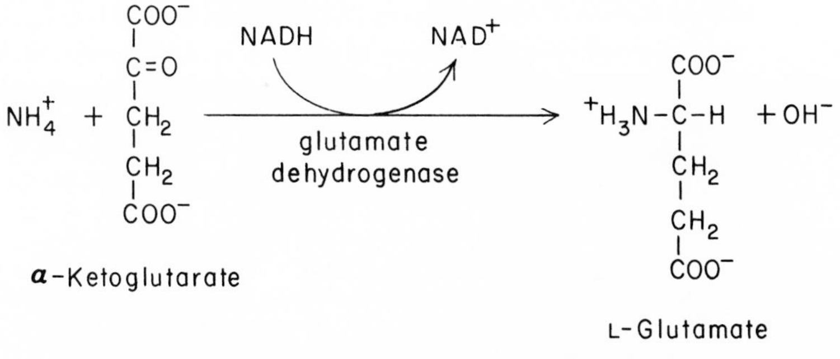

The transaminases shown in Fig. 9-5 are specific enzymes that catalyze reversible reactions. Thus amino acids enter the Krebs cycle and keto acids are drawn out of the cycle by the same reactions. Transaminases derived from different tissues may have different substrate specificities. The mechanism of transamination and the significance of these enzymes in clinical diagnosis were discussed in Chapter 6.

The amides of aspartate and glutamate (asparagine and glutamine) can both be synthesized by man. Despite the similarity of the products, the reactions involved are quite different. The enzyme glutamine synthetase catalyzes the formation of a γ-glutamyl phosphate intermediate; ADP is the other product.

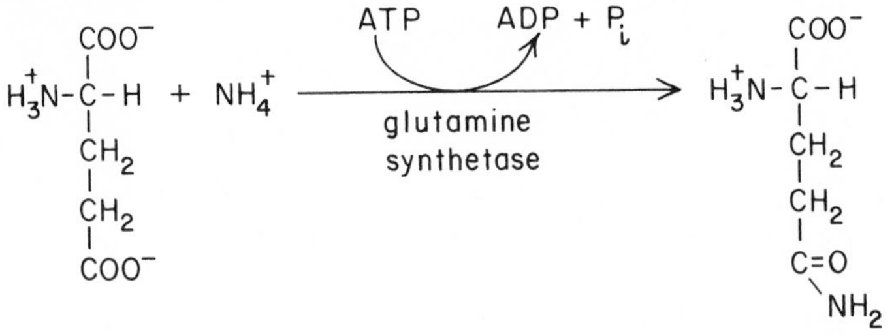

L-Glutamate   L-Glutamine

This high-energy mixed anhydride provides the driving force for the transfer of the acyl group to ammonium ion with the formation of a rather stable amide group. Asparagine synthetase works by a different mechanism. The intermediate is an asparaginyl adenylate that is formed from asparagine and ATP. Pyrophosphate is a product of the reaction. Reaction of the enzyme-bound activated acyl intermediate with ammonium ion yields asparagine and AMP.

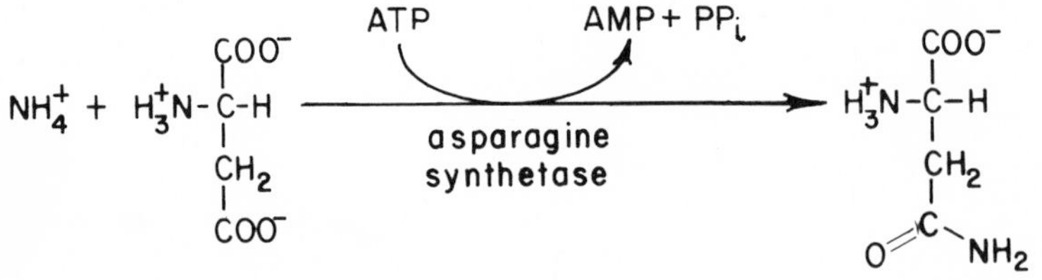

L-Aspartate   L-Asparagine

Monosaccharide metabolism — synthesis of serine and glycine

L-Serine is made from 3-phosphoglycerate, an intermediate in the glycolytic pathway that can be derived from 3-phosphoglyceraldehyde, which is also an intermediate of the pentose phosphate pathway. There are two separate routes for the synthesis of serine from 3-phosphoglycerate. One pathway is shown at the left in Fig. 9-6; the other is shown at the right. Note that the pathway on the right contains phosphorylated intermediates, whereas that on the left does not. The existence of two separate pathways and the presence of substantial amounts of serine in the diet are consistent with the function of serine as an intermediate in several metabolic transformations, some of which have already been described. Thus serine serves as a precursor for ethanolamine, choline, betaine, glycine, cysteine, sarcosine, pyruvate, and sphingosine.

Glycine.   Serine is the precursor of glycine. The hydroxymethyl group of serine is transferred to tetrahydrofolate (THF) to form 5,10-methylene THF, water, and glycine, as shown following.

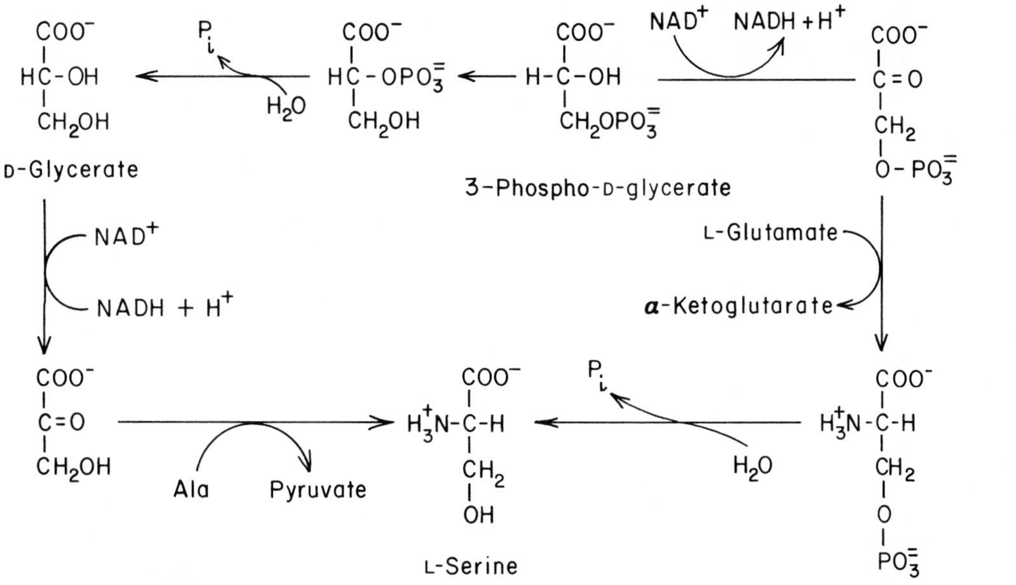

Fig. 9-6. Synthesis of serine by the glycolytic pathway.

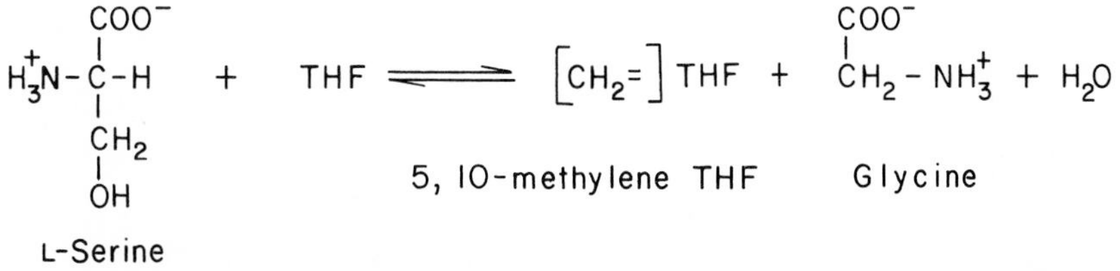

A more complete treatment of the metabolism of folic acid derivatives is given in Chapter 11.

Glycine can also be synthesized from glyoxalate. Cytoplasmic liver enzymes decarboxylate serine to give ethanolamine, which is transaminated to glycolaldehyde. Oxidation of this aldehyde yields glyoxalate. Several transaminases can then produce glycine from glyoxalate. These transformations can be summarized as follows:

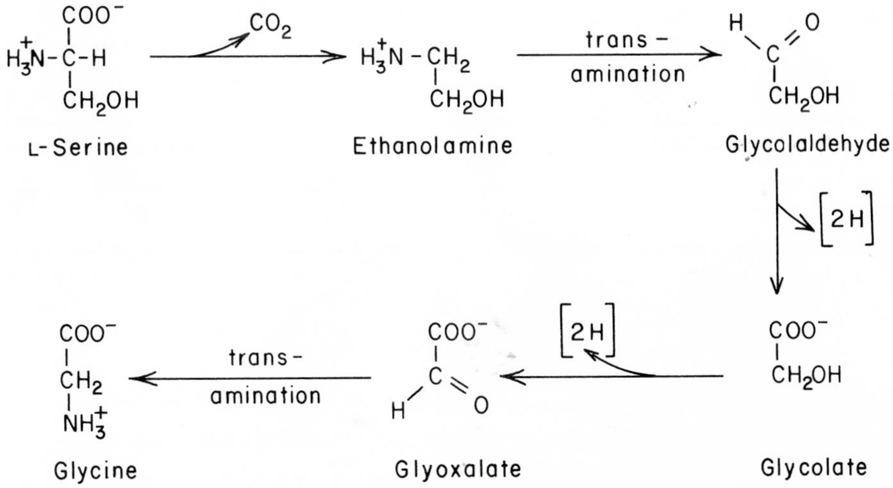

Blockage of the metabolism of glyoxalate produces a rare genetic disease, type I primary hyperoxaluria, which is characterized by excessive excretion of oxalate derived from the oxidation of the accumulated glyoxalate.

Methyl group transfer

In order to understand the biosynthesis of cysteine from methionine, it is necessary to describe in some detail the function of methionine in the transfer of methyl groups. It is by these reactions that homocysteine, a more immediate precursor of cysteine, is produced.

The transfer of one-carbon units other than carbon dioxide is generally achieved either through a derivative of methionine or by a coenzymatic form of the vitamin folic acid. Nucleotide biosynthesis requires several one-carbon transfers that are mediated by folic acid coenzymes (Chapter 11). The methionine transfer reactions are more involved with amino acid metabolism.

The methionine adenosyl transferase of liver catalyzes the formation of S-adenosylmethionine (SAM) from methionine and ATP (Fig. 9-7). The reaction is unusual in that transfer of the adenosyl group splits off the phosphates of ATP as a single triphosphate. The product, SAM, is the principal methyl donor in the body. The methionine methyl group in SAM can be transferred to amino or hydroxy groups on a variety of acceptor molecules. Some of these acceptors and their products are listed in Table 9-2.

Reactions of homocysteine transmethylation. The other product of all of the methylation reactions mentioned in Table 9-2 is S-adenosylhomocysteine (SAH). A hydrolase cleaves this molecule to yield homocysteine:

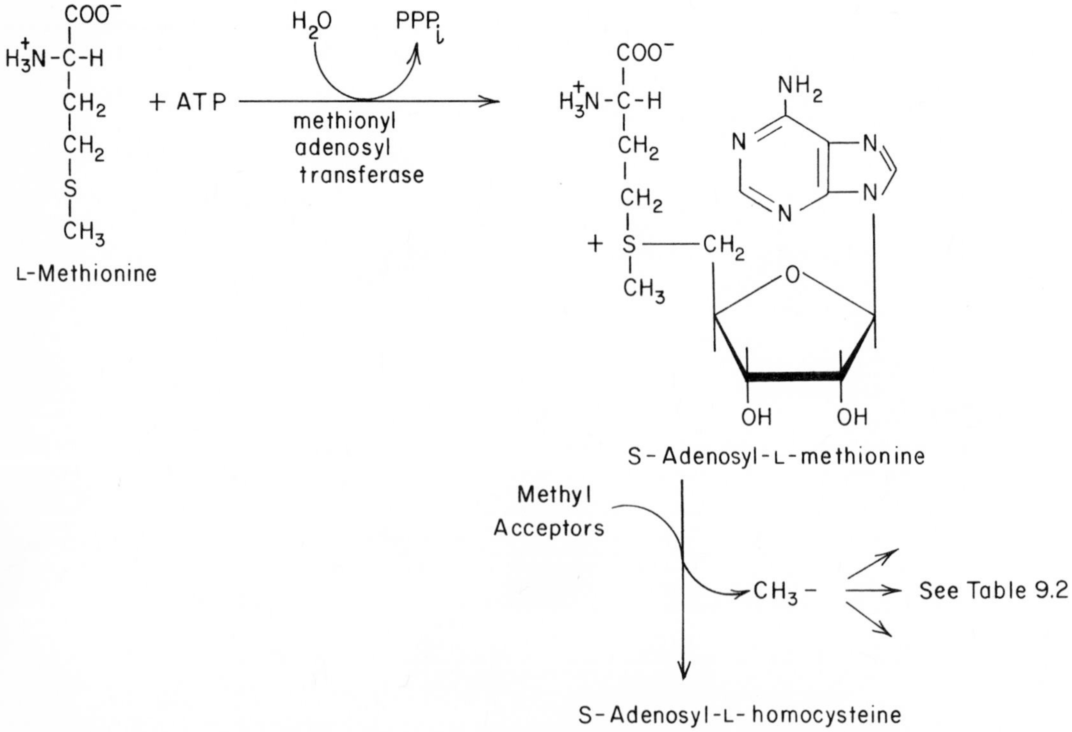

Fig. 9-7. Synthesis and utilization of S-adenosylmethionine.

$$H_2O + \text{S-Adenosyl-L-homocysteine} \rightleftharpoons \text{L-Homocysteine} + \text{Adenosine}$$

which can itself act as a methyl acceptor but in a slightly different way. The transmethylation reaction shown in Fig. 9-8 utilizes betaine as the methyl donor and homocysteine as the methyl acceptor. Betaine is derived from the oxidation of dietary choline. Based on the information in Fig. 9-8, one might predict that the essential amino acid methionine could be synthesized if sufficient choline and homocysteine were contained in the diet. This proves to be true, as man is able to synthesize all but the homocysteine portion of methionine.

Homocysteine can be converted to methionine by another pathway. This reaction is important in the metabolism of both folate and vitamin B_{12}, a relationship that will be discussed in Chapter 11. For present purposes the reaction may be illustrated by the equation below.

$$\text{L-Homocysteine} + \text{5-Methyl THF} \xrightarrow[\text{coenzyme}]{\text{corrinoid}} \text{L-Methionine} + \text{THF}$$

Table 9-2. Methyl transfer reactions

DONOR	ACCEPTOR	PRODUCT
SAM	Guanidoacetate	Creatine
SAM (3 moles)	Phosphatidylethanolamine	Phosphatidylcholine
SAM	Ribosomal and transfer RNA	Methylated RNA
SAM	DNA	Methylated DNA
SAM	Norepinephrine	Epinephrine

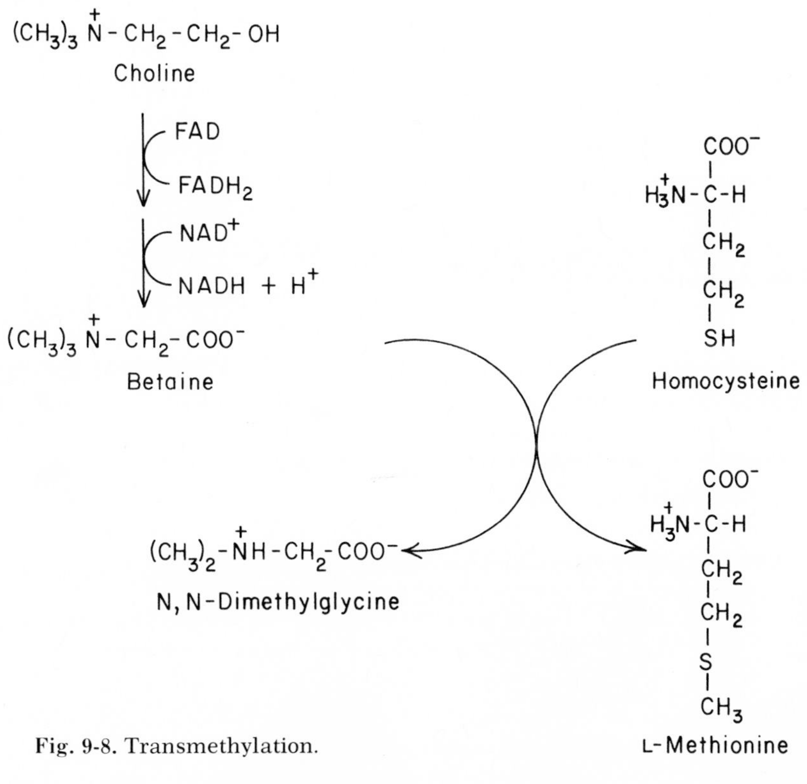

Fig. 9-8. Transmethylation.

The enzyme that catalyzes the reaction is 5-methyl tetrahydrofolate (THF) homocysteine transmethylase, sometimes called 5-methyl THF methyltransferase. The corrinoid coenzyme is a derivative of vitamin B_{12}.

Biosynthesis of amino acids from dietary essential amino acids

Cysteine synthesis. The sulfhydryl group of cysteine is derived from methionine. More specifically, it is the homocysteine portion of the molecule that furnishes the sulfhydryl group. The carbon skeleton of cysteine is derived from serine. The two amino acids condense under the influence of *cystathionine synthetase*, a pyridoxal phosphate-containing enzyme.

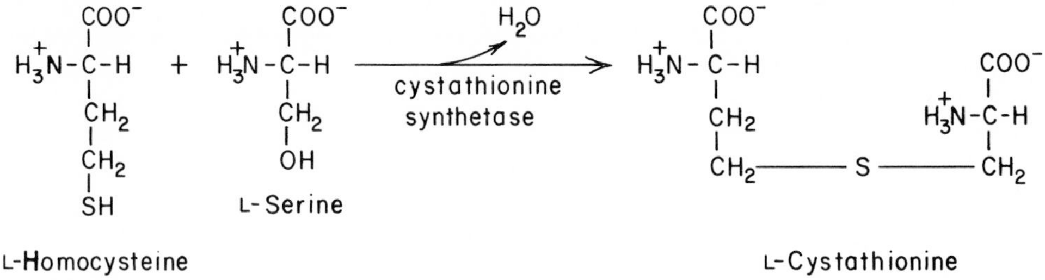

Cystathionase, the enzyme that catalyzes a hydrolytic reaction to give homoserine and cysteine, also contains pyridoxal phosphate.

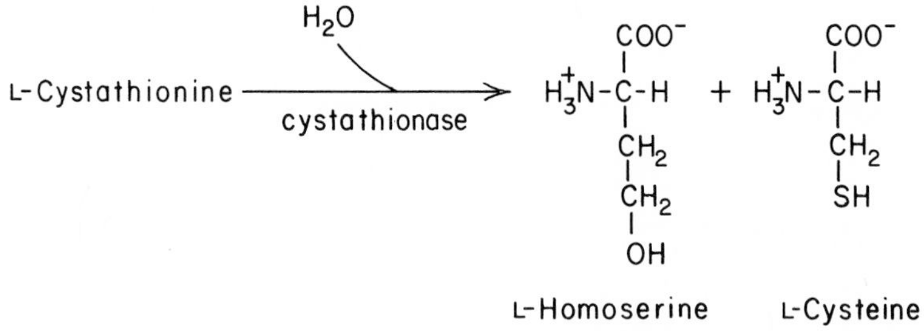

Certain genetic diseases in man are caused by deficiencies of the enzymes involved in cysteine synthesis. In homocystinuria, so called because large amounts of homocystine are found in the urine, the patient's cystathionine synthetase is defective. The homocysteine that accumulates is oxidized to form the disulfide compound homocystine, the structure of which is analogous to that of cystine. In *cystathioninuria* the defect involves the step that cleaves cystathionine to produce cysteine; consequently, large amounts of cystathionine are found in the blood and urine. The genetic defect cystathionuria is interesting in that an active enzyme protein is produced, but the protein has a much reduced affinity for its essential coenzyme, pyridoxal phosphate.

Cysteine, besides being required for the synthesis of proteins, is also used to make the tripeptide glutathione (γ-glutamylcysteinylglycine) and the amine taurine.

Tyrosine. Given enough of the essential amino acid phenylalanine, man can synthesize adequate amounts of tyrosine. The enzyme phenylalanine hydroxylase is an oxygenase that requires tetrahydrobiopterin, a pteridine coenzyme similar to folic acid. This is the enzyme involved in *phenylketonuria*, a disease caused by an autosomal recessive gene carried by 2% of the population. It occurs at an incidence of one in 10^4 births and is the most common of the aminoacidurias. Abnormal phenylalanine metabolites, for example, phenylpyruvate, O-hydroxyphenylacetate, phenyllactate, and phenylacetyl glutamine, as well as phenylalanine itself, are found in the patient's urine. Mental retardation associated with the disease is lessened by placing the phenylketonuric infant on a diet low in phenylalanine. The cause of the mental retardation is not clear, but it may be related to the fact that high levels of phenylalanine inhibit the enzyme that decarboxylates 5-hydroxytryptophan to form serotonin, a biogenic amine (p. 455). Consequently, impaired serotonin biosynthesis would be expected in the phenylketonuric infant, and it is an established fact that serotonin levels are low in these individuals.

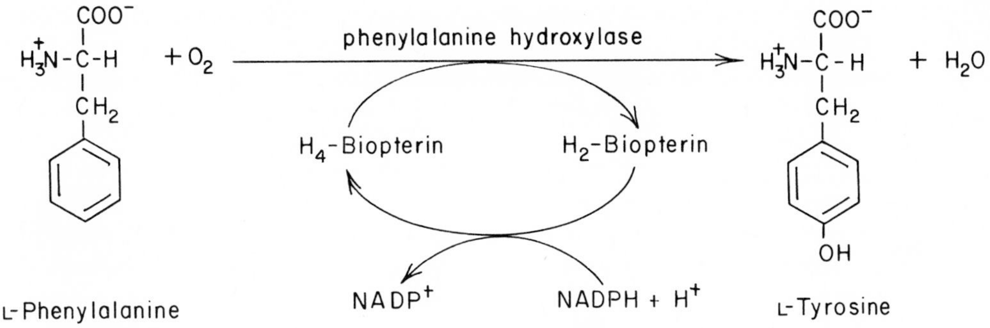

Tyrosine has a number of important functions in biosynthesis. It serves as a precursor of thyroxine (Chapter 13) and of the melanin pigments. Later in this chapter its role in the synthesis of norepinephrine and epinephrine will be discussed.

L-*Proline.* L-Proline is synthesized from arginine according to the scheme in Fig. 9-9. The first reaction in this series is catalyzed by arginase and results in the conversion of arginine to urea and ornithine (p. 437). The delta amino group of

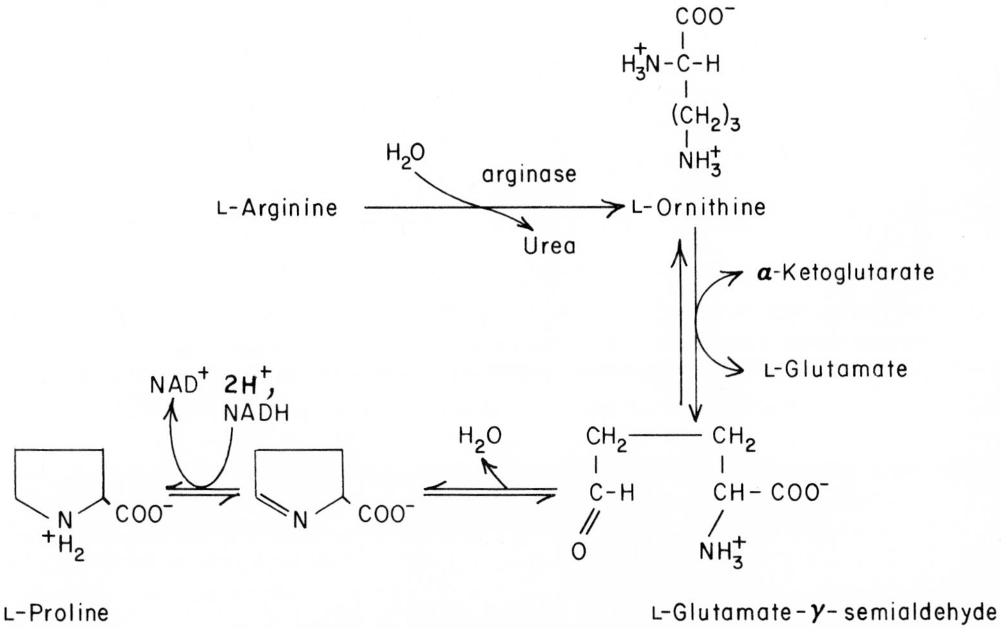

Fig. 9-9. Biosynthesis of proline.

ornithine is transaminated to form glutamate-γ-semialdehyde. The formation of a
Schiff base closes the ring, and a subsequent reduction yields L-proline.

Arginine and histidine. While both arginine and histidine are considered by
many authorities to be nonessential to the adult human, their status is not so clear
as that of the other nonessential amino acids. Both amino acids are required for
the rapid growth of infants. As will be seen later, arginine can be made from ornithine, a urea cycle intermediate. Glutamate is the precursor of the major carbon
skeleton. It is oxidized to glutamate γ-semialdehyde, which is then transaminated
to give ornithine. Apparently infants cannot obtain enough arginine through this
pathway.

It is very likely that histidine is an essential amino acid for adult humans as
it is for rats. The determination of the essential amino acids for rats was made by
measuring the drastic weight losses that occurred when the animals were denied
a single essential amino acid. Obviously humans could not be used for such an
experiment. Instead, the nitrogen equilibrium of healthy volunteer subjects was
measured. When these young men were denied histidine, their nitrogen output
equaled their nitrogen input, and histidine was scored as nonessential. More than
likely these subjects were able to mobilize enough of the histidine derivatives
carnosine and anserine, which are stored in substantial quantities in muscle, to
satisfy their needs for histidine.

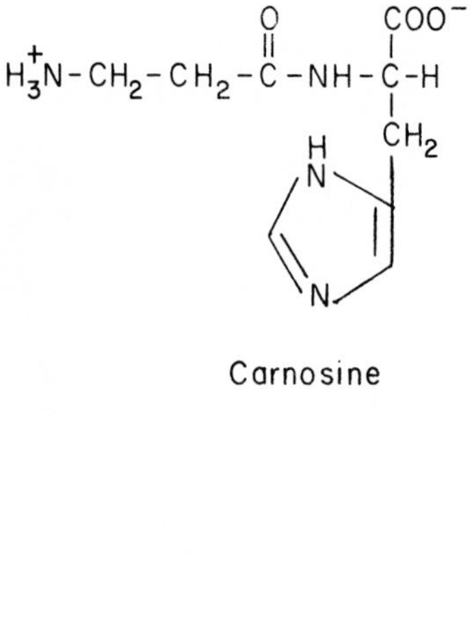

Carnosine

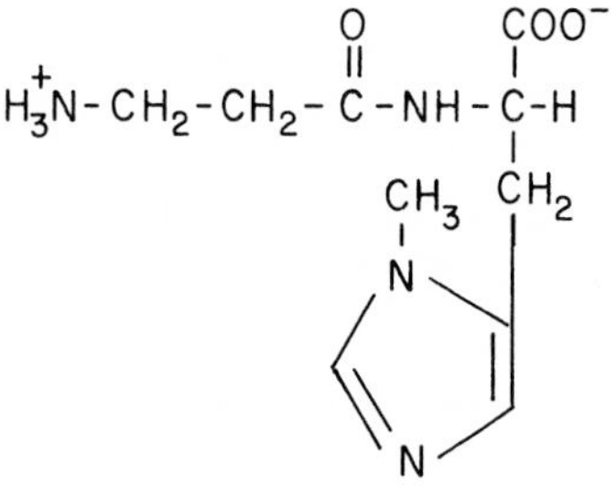

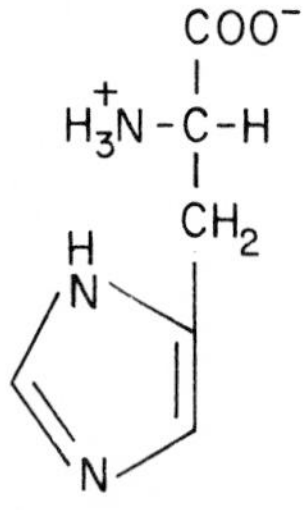

L-Histidine

Anserine

Carnosine, which is synthesized from dietary histidine, can replace histidine in
the diet of rats over a short period of time. Since the human experiments were
performed only over a short time period, it is possible that histidine is also essential
for man.

Summary of the biosynthesis of amino acids

A listing of amino acids is given in Table 9-3 together with the precursors that
provide most of the carbon atoms used in their biosynthesis. Important or unusual
features of the biosynthetic pathways as well as the names of human diseases
associated with genetic defects in these pathways are listed in the right-hand
column. Note that two essential amino acids are listed. Both can be synthesized
by the human under certain circumstances. Methionine can be synthesized if the

Table 9-3. Summary of the biosynthesis of amino acids in animal cells

AMINO ACID	PRECURSOR	DISTINGUISHING FEATURES OF PATHWAYS
Alanine	Pyruvate	By transamination
Glutamate	α-Ketoglutarate	By reductive amination
	α-Ketoglutarate	By transamination
Glutamine	Glutamate	ADP + P_i are products
Aspartate	Oxaloacetate	By transamination
Asparagine	Aspartate	AMP + PP_i are products
Serine	3-Phospho-D-glycerate	Phosphoserine intermediate
	3-Phospho-D-glycerate	Hydroxypyruvate intermediate
	Glycine	Requires 5,10 methylene THF
Glycine	Serine	Requires THF
	Serine	Glyoxalate intermediate (type 1 hyperoxaluria)
Methionine	Homocysteine	Requires 5-methyl THF
	Homocysteine	Requires betaine
Cysteine	Serine and homocysteine	Cystathionine intermediate (cystathionuria, homocystinuria)
Tyrosine	Phenylalanine	Biopterin coenzyme (phenylketonuria)
Proline	Arginine	Glutamate γ-semialdehyde intermediate
Arginine	Glutamate	Ornithine intermediate

diet contains homocysteine, and arginine can be synthesized as in the urea cycle, but it cannot be made in sufficient amounts by young animals and is essential for them.

Because the essential amino acids cannot be synthesized in the human, their biosynthesis will not be considered.

AMINO ACID CATABOLISM

The degradation of either dietary or biosynthesized amino acids usually involves an initial separation of the α-amino group from the rest of the molecule. Because metabolism branches at this early point, it is convenient to consider the fate of the amino groups prior to the transformations of the carbon skeleton.

Fate of nitrogen atoms

While the major end products of nitrogen metabolism in man are eliminated in both the feces and the urine, most of the nitrogen that represents metabolic end products is found in the urine. The relative concentrations of these end products in normal urine are given in Table 9-4. It is clear that most of the nitrogen is excreted as urea. However, the most immediate product of amino acid catabolism is not urea but ammonium ions.

Amino acid deamination

There are many ways in which amino groups may be removed from individual amino acids, but there are two general types of reactions that all amino acids undergo: oxidative deamination and transamination. Both processes produce keto acids, but oxidative deamination produces ammonium ions directly, whereas in transamination the amino group is passed to another keto acid. We have already

Table 9-4. Nitrogen-containing components of normal urine

END PRODUCT	EXCRETED NITROGEN (%)
Urea	86.0
Creatinine	4.5
Ammonium	2.8
Uric acid	1.7
Other compounds	5.0

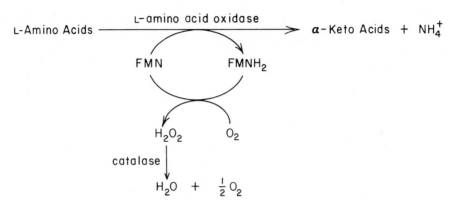

Fig. 9-10. Reaction catalyzed by L-amino acid oxidase.

seen one example of oxidative deamination, that involved in the glutamate dehydrogenase reaction (p. 425). When linked to a glutamate-requiring transaminase, the reaction catalyzed by glutamate dehydrogenase can produce ammonium ions from almost all of the amino acids. Such a coupled reaction is shown below.

$$\text{Most Amino Acids} + \alpha\text{-Ketoglutarate} \rightleftharpoons \alpha\text{-Keto acids} + \text{L-Glutamate}$$
$$\text{L-Glutamate} + NAD^+ \rightleftharpoons NADH + H^+ + \alpha\text{-Ketoglutarate} + NH_4^+$$

Sum: Most Amino Acids $+ NAD^+ \rightleftharpoons NADH + H^+ + \alpha$-Keto acids $+ NH_4^+$

The glutamate oxaloacetate transaminase catalyzes the transfer of amino groups to oxaloacetate to yield aspartate; this is important since the nitrogen atoms of urea are derived not only from ammonium ions but from aspartate as well.

The L-amino acid oxidases of kidney and liver are of lesser importance in the initial catabolism of amino acids. These enzymes have a broad specificity for many amino acids, but their relative rates of oxidation are slow. The tightly bound coenzyme of the L-amino acid oxidase from kidney is flavin mononucleotide (FMN). Fig. 9-10 illustrates the reaction catalyzed by L-amino acid oxidase.

Ammonium ion metabolism

Virtually all of the body tissues produce ammonia, which is present predominantly as ammonium ions. These ammonium ions arise primarily from the catabolism of amino acids. During digestion they are also released within the intestine as a result of bacterial action but are then removed on passage through the liver.

Thus the concentration of ammonium ions in the portal blood is quite high, while the ammonium concentration in the peripheral blood is maintained at a very low level. Normal concentrations of ammonium ions in plasma are 0.025 to 0.04 meq/liter, although higher levels are often found in severe hepatic disease. Regardless of their source, ammonium ions (as ammonia) are exceedingly toxic to the central nervous system and must be eliminated.

The brain detoxifies ammonium ions by converting them to glutamine. Indeed, brain is a rich source of the enzyme glutamine synthetase. Other tissues are also active in glutamine synthesis, and the high levels of glutamine found in the blood after ingestion of foods rich in protein are thought to represent a form of ammonia storage and transport. The circulating glutamine can be hydrolyzed to glutamate and ammonium ions by glutaminase in the kidney. In Chapter 4 we saw how this reaction was important in acid-base balance. Alternatively, the glutamine of the blood may be hydrolyzed by liver glutaminase to yield the ammonium ions used by that organ for urea synthesis. Most of the ammonia is transported to the liver by glutamine, and the plasma concentration of this amino acid is almost twice that of any other single amino acid (approximately 7.5 mg/dl). The blood ammonium level is normally many times lower (0.04 to 0.07 mg/dl).

Most ammonium ions in the portal blood are detoxified in the liver; there they are converted to urea, a form of nitrogen much less toxic to the central nervous system. If the urea-synthesizing system fails as a result of a malfunctioning liver or portal obstruction, ammonium ions pass into the systemic circulation and ammonia intoxication results. Ammonia intoxication produces blurred vision, tremors, slurred speech, and ultimately coma and death. However, this condition, which is called hepatic coma, is complex and may be precipitated by many other factors.

Urea synthesis

The first step in the formation of urea is the synthesis of *carbamyl phosphate* from carbon dioxide, ammonium ions, and ATP. Carbamyl phosphate is not only a key intermediate in the synthesis of urea but also is involved in the synthesis of pyrimidines. However, the carbamyl phosphate used for pyrimidine synthesis is produced by a slightly different mechanism (Chapter 11). The carbamyl phosphate used in urea synthesis is produced in the mitochondria in a reaction catalyzed by the enzyme carbamyl phosphate synthetase. Acetyl glutamate is a cofactor.

1. $$CO_2 + NH_4^+ + 2ATP \xrightarrow[\text{acetyl glutamate (mitochondrial)}]{\text{carbamyl phosphate synthetase}} \underset{\substack{\text{Carbamyl} \\ \text{phosphate}}}{\overset{NH_2}{\underset{O}{\overset{\mid}{\underset{\mid}{C=O}}}} {PO_3^=}} + 2ADP + P_i$$

The reaction is essentially irreversible; therefore, the uptake of ammonium ions and the synthesis of urea are assured. Carbamyl phosphate synthetase and ornithine transcarbamylase are the only enzymes of the urea cycle found in the mitochondria. As might be expected, there appears to be no difficulty in transporting amino acids and ammonium ions into and out of the mitochondria.

In the second step, carbamyl phosphate, a high-energy mixed anhydride, condenses with ornithine to form citrulline and inorganic phosphate.

2.

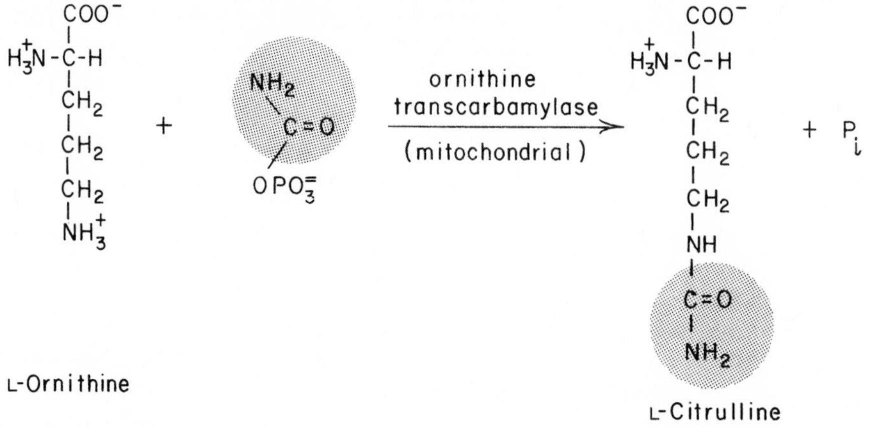

$$
\begin{array}{c}
\text{COO}^- \\
| \\
\overset{+}{\text{H}_3\text{N}}-\text{C-H} \\
| \\
\text{CH}_2 \\
| \\
\text{CH}_2 \\
| \\
\text{CH}_2 \\
| \\
\overset{+}{\text{NH}_3}
\end{array}
\quad + \quad
\begin{array}{c}
\text{NH}_2 \\
| \\
\text{C=O} \\
| \\
\text{OPO}_3^=
\end{array}
\quad
\xrightarrow[\text{(mitochondrial)}]{\text{ornithine transcarbamylase}}
\quad
\begin{array}{c}
\text{COO}^- \\
| \\
\overset{+}{\text{H}_3\text{N}}-\text{C-H} \\
| \\
\text{CH}_2 \\
| \\
\text{CH}_2 \\
| \\
\text{CH}_2 \\
| \\
\text{NH} \\
| \\
\text{C=O} \\
| \\
\text{NH}_2
\end{array}
\quad + \quad \text{P}_i
$$

L-Ornithine

L-Citrulline

In an ATP-requiring reaction, citrulline condenses with aspartic acid to form *argininosuccinate*, which in turn is cleaved to yield *arginine* and fumarate. The fumarate produced in the cytoplasm by this reaction can be converted to malate by a cytoplasmic fumarase and the malate transported to the mitochondria by the malate-isocitrate translocase system (p. 268).

$$
\begin{array}{c}
\text{COO}^- \\
| \\
\overset{+}{\text{H}_3\text{N}}-\text{C-H} \\
| \\
\text{CH}_2 \\
| \\
\text{CH}_2 \\
| \\
\text{CH}_2 \\
| \\
\text{NH} \\
| \\
\text{C=O} \\
| \\
\text{NH}_2
\end{array}
\quad + \quad
\begin{array}{c}
\text{COO}^- \\
| \\
\overset{+}{\text{H}_3\text{N}}-\text{C-H} \\
| \\
\text{CH}_2 \\
| \\
\text{COO}^-
\end{array}
\quad
\xrightarrow[\text{synthetase}]{\text{ATP} \quad \text{AMP + PP}_i \quad \quad \text{H}_2\text{O} \quad \text{argininosuccinate}}
$$

L-Citrulline

L-Aspartate

L-Argininosuccinate

$$
\begin{array}{c}
\text{COO}^- \\
| \\
\overset{+}{\text{H}_3\text{N}} - \text{C-H} \\
| \\
\text{CH}_2 \\
| \\
\text{CH}_2 \\
| \\
\text{CH}_2 \\
| \\
\text{NH} \quad \text{COO}^- \\
| \quad \quad | \\
\text{C-NH-C-H} \\
\| \quad \quad | \\
\text{NH} \quad \text{CH}_2 \\
\quad \quad | \\
\quad \quad \text{COO}^-
\end{array}
$$

3.

$$
\begin{array}{c}
\text{COO}^- \\
| \\
\overset{+}{\text{H}_3\text{N}} - \text{C-H} \\
| \\
\text{CH}_2 \\
| \\
\text{CH}_2 \\
| \\
\text{CH}_2 \\
| \\
\text{NH} \\
| \\
\text{C = NH} \\
| \\
\overset{+}{\text{NH}_3}
\end{array}
$$

arginino-
succinase

$$
\begin{array}{c}
\text{HC-COO}^- \\
\| \\
{}^-\text{OOC-CH}
\end{array}
$$

Fumarate

↓
↓

Glucose

L-Arginine

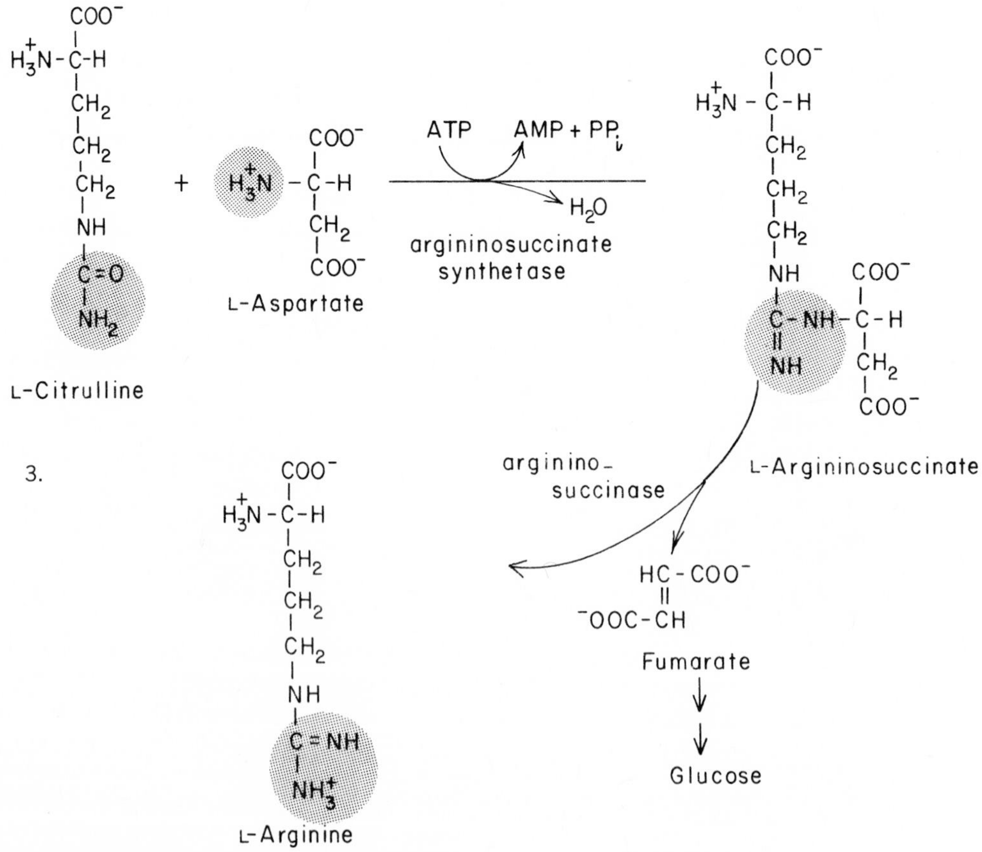

Finally, the hydrolytic cleavage of arginine by the enzyme arginase yields urea and ornithine, and the latter recycles as an acceptor of another molecule of carbamyl phosphate.

4. L-Arginine $+$ H_2O $\xrightarrow{\text{arginase}}$

$$
\begin{array}{c}
COO^- \\
| \\
H_3^+N-C-H \\
| \\
CH_2 \\
| \\
CH_2 \\
| \\
CH_2 \\
| \\
NH_3^+
\end{array}
$$

L-Ornithine

$+$

$$
\begin{array}{c}
NH_2 \\
| \\
C=O \\
| \\
NH_2
\end{array}
$$

Urea

These four reactions can be combined to form a cycle in which urea is produced from ammonium ions and carbon dioxide and in which all of the other intermediates are regenerated. Fig. 9-11 summarizes the reactions and represents them as a urea cycle.

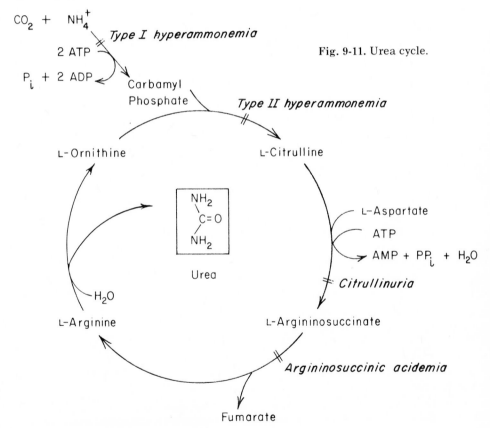

Fig. 9-11. Urea cycle.

Inherited diseases affecting the urea cycle

Hyperammonemia is sometimes associated with inherited abnormalities of the urea cycle enzymes. In congenital hyperammonemia type I the defective enzyme is carbamyl phosphate synthetase, and in type II it is ornithine transcarbamylase. The accompanying symptoms of episodic vomiting, psychomotor retardation, and stupor respond to a restricted protein diet. The glutamine concentration of the blood is often elevated in these patients, and in hyperammonemia type II, pyrimidine metabolites appear in the urine. Two other inherited diseases, *citrullinuria* and *argininosuccinic acidemia,* are caused by defects in the enzymes that ordinarily utilize the named metabolites.

Fate of carbon atoms

The carbon skeletons of the amino acids may be utilized for the production of energy. In early studies in which individual amino acids were fed to diabetic or fasting animals it was found that the carbon atoms of a particular amino acid might give rise to either ketone bodies (ketogenic amino acid) or glucose and glycogen (glycogenic amino acid). Under certain experimental conditions an amino acid appears to be "glycogenic," whereas under others it is "ketogenic." The detailed metabolism of the amino acids is probably best studied by enzymatic experimentation. Nevertheless, to emphasize the metabolic functions of the amino acids in the whole animal it is convenient to group the amino acids according to this scheme, recognizing that the conditions of the experiment or of the patient can greatly influence metabolism.

Most amino acids are glycogenic; only *leucine* is completely ketogenic. Five of the amino acids, *isoleucine, lysine, phenylalanine, tyrosine,* and *tryptophan,* are considered both ketogenic and glycogenic, or as neither significantly glycogenic nor ketogenic. Again, such findings depend on the conditions of the experiment.

Initial metabolic breakdown of amino acids

The initial steps in the metabolic breakdown of the amino acids separate the carbon-containing chains from the amino groups by quite simple reactions involving either transaminations or oxidative deaminations. The disposition of the carbon chains is accomplished by a series of rather complex enzymatic reactions that lead to the formation of either ketone bodies or glycogen, depending on the particular amino acid. A summary of the total effect achieved by metabolism of the amino acids that result from the breakdown of proteins is presented in Fig. 9-12. The Krebs cycle and the reactions leading to and from it are the dominant features of this diagram. The names of the intermediates that can be produced by catabolism of the amino acids are in boxes. Those portions of the amino acids that form Krebs cycle intermediates can be thought of as glycogenic, as oxaloacetate can be converted to phosphoenolpyruvate (PEP) and then to glucose. The names of the ketogenic amino acids are italicized. Recall that several of these are both glycogenic and ketogenic. Those portions that are ketogenic can give rise to either acetoacetate or acetyl CoA. Acetoacetate can also be biosynthesized from acetyl CoA. Whether an amino acid is glycogenic or ketogenic merely indicates that it has a potential to form either glycogen or ketone bodies. Of greater significance is the fact that either of these substances can be quickly mobilized for energy production. Thus amino acids are not only important in the synthesis of other nitrogen-containing metabolites and in the synthesis of proteins but they are also important sources of energy.

Catabolism of ketogenic amino acids

Leucine. Leucine is transaminated, converted to a CoA thioester, decarboxylated, oxidized, carboxylated, and finally hydrated to form 3-hydroxy-3-methylglutaryl CoA (HMGCoA). A summary of the reactions is shown in Fig. 9-13. The 2-ketoisocaproate dehydrogenase complex is analogous to the pyruvate and ketoglutarate dehydrogenase complexes (Chapter 6) in that thiamine pyrophosphate, lipoic acid, CoA, FAD, and NAD^+ are all involved. This complex is defective in

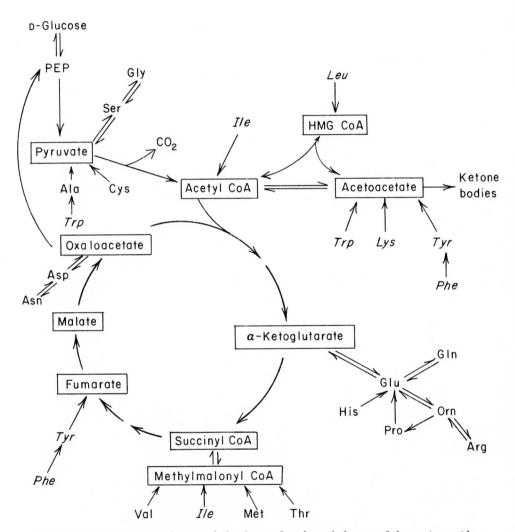

Fig. 9-12. Summary of the metabolic fates of carbon skeletons of the amino acids.

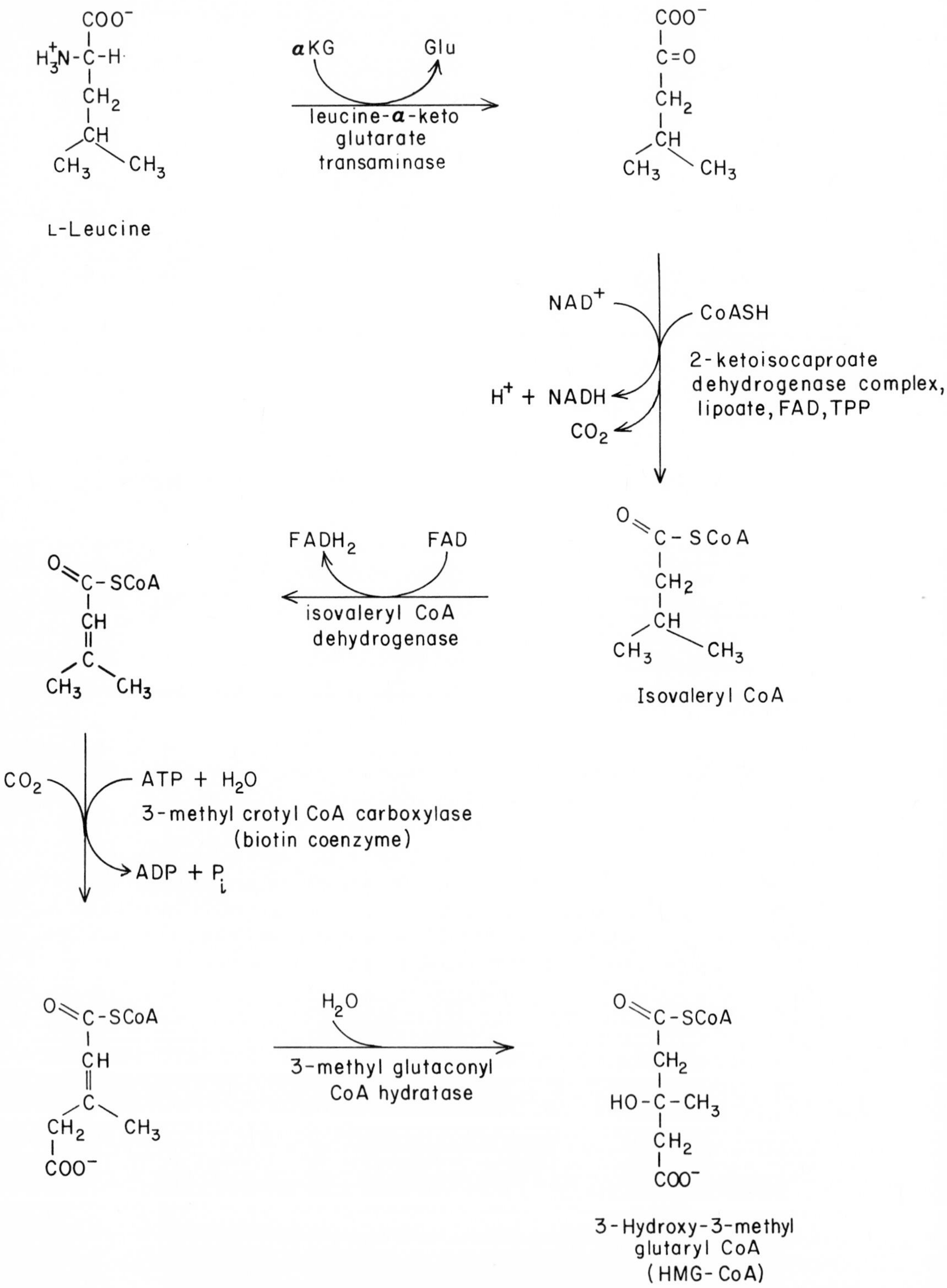

Fig. 9-13. Catabolism of leucine.

individuals with *maple syrup urine disease,* in which the urine takes on a characteristic maple syrup smell as a result of the accumulated keto acids.

HMGCoA is a precursor of cholesterol (Chapter 10), so it might be thought that leucine could be used for the synthesis of the steroids. Actually, leucine catabolism occurs in the mitochondria, whereas sterol synthesis is cytoplasmic; consequently, leucine produces acetoacetate and acetyl CoA exclusively. Acetoacetate is a ketone body, but any amino acid that can form acetyl CoA may be considered ketogenic; because the acetoacetyl CoA thiolase reaction is somewhat reversible, the following sequence is possible:

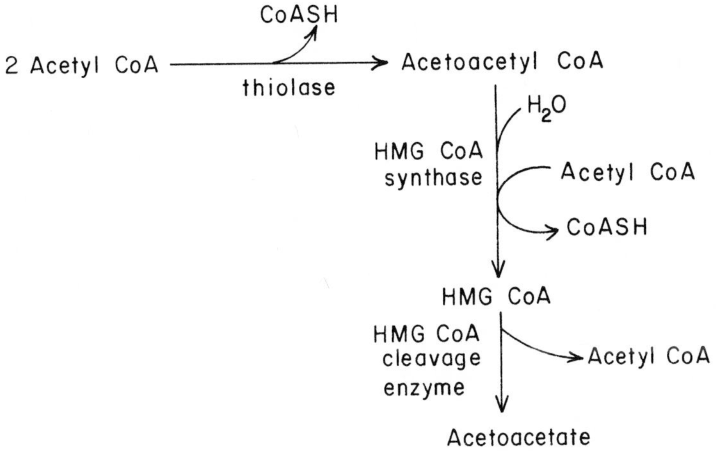

Thus it is possible to derive a net synthesis of the ketone body acetoacetate from any amino acid that gives rise to acetyl CoA.

Catabolism of amino acids that are both ketogenic and glycogenic

Isoleucine. Isoleucine is catabolized by a series of reactions similar to those for leucine. As shown in Fig. 9-14, the reactions prior to the formation of the unsaturated acid are entirely analogous to those for leucine (Fig. 9-13). The enzymes are similar for each pathway, but only the 2-ketoisocaproate dehydrogenases are identical. However, in this case the precursor for acetyl CoA formation arises from the action of the thiolase rather than a cleavage enzyme. The propionyl CoA can then be carboxylated in a reaction mediated by a biotin-containing enzyme (Chapters 6 and 8) to form methylmalonyl CoA, which can be isomerized to succinyl CoA, enter the Krebs cycle, and ultimately yield pyruvate, glucose, or glycogen.

Phenylalanine and tyrosine. Phenylalanine is converted to tyrosine by the enzyme phenylalanine hydroxylase (p. 431); thus its further catabolism is the same as that of tyrosine.

Tyrosine is deaminated under the influence of the enzyme tyrosine transaminase (tyrosine aminotransferase), which uses α-ketoglutarate as the amino acceptor (Fig. 9-15). The synthesis of this enzyme can be induced in animal liver by the glucocorticoids, steroid hormones that stimulate protein catabolism. The mechanism of this induction is poorly understood but may involve either an increase in the synthesis of the messenger RNA or a posttranscriptional activation of existing messenger RNA. Under the influence of an oxidase the keto acid product of a tyrosine transaminase action, *p*-hydroxyphenylpyruvate (Fig. 9-15), is hydroxylated, decarboxylated, and its side chain rearranged. The product of the

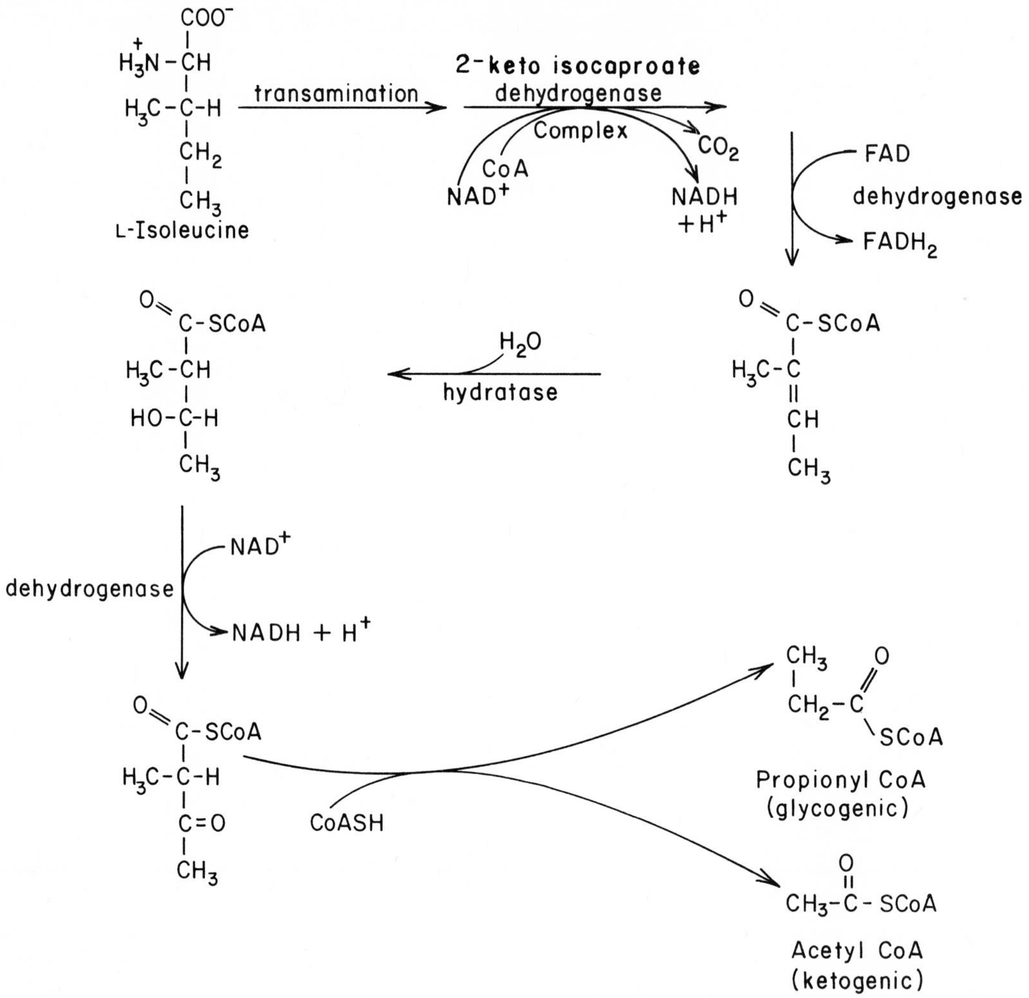

Fig. 9-14. Catabolism of isoleucine.

reaction is homogentisate, a hydroquinone. Another oxidation catalyzed by homogentisate oxidase opens the phenyl ring, giving rise to maleoylacetoacetate and ultimately acetoacetate and fumarate. Homogentisate is excreted in the disease *alkaptonuria.*

In his book *Inborn Errors of Metabolism* published in 1909 the English physician Garrod proposed that some human diseases were inherited; 30 years before the one gene–one enzyme concept of Beadle and Tatum, he correctly predicted that alkaptonuria resulted from a defect in an enzyme that splits the benzene ring. We now know that the defective enzyme is homogentisate oxidase. Homogentisate accumulates, and black oxidation products are formed when the patient's urine is exposed to air. This coloring occurs more rapidly in the presence of alkali. The disease itself is rather benign by comparison to other genetic disorders. In middle age, arthritis usually occurs with the accumulation of the black pigment in the connective tissues.

Lysine. Lysine is one of the few amino acids not deaminated in its initial catabolic reaction. There are two pathways for its metabolism in animals; both have

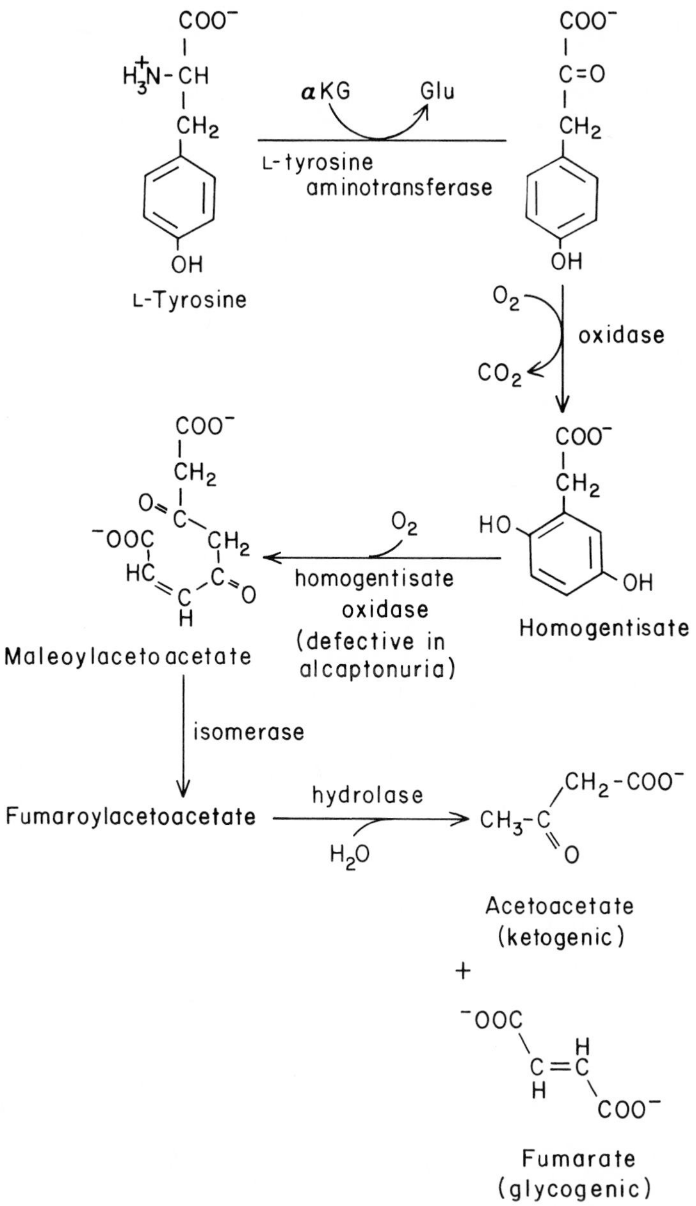

Fig. 9-15. Catabolism of tyrosine.

in common a modification of the ϵ-amino group before deamination. There is still some question as to the relative importance of the pathways, and some of the intermediates are omitted in Fig. 9-16.

Both routes ultimately yield α-aminoadipate. The oxidative decarboxylation of this substance is catalyzed by an enzyme complex similar to the pyruvate and ketoglutarate dehydrogenase complexes. Further decarboxylation and oxidation yield crotonyl CoA, which is an intermediate in fatty acid oxidation. This intermediate yields acetoacetyl CoA and eventually acetyl CoA or acetoacetate. The metabolism of lysine is still poorly understood, and although the pathways in

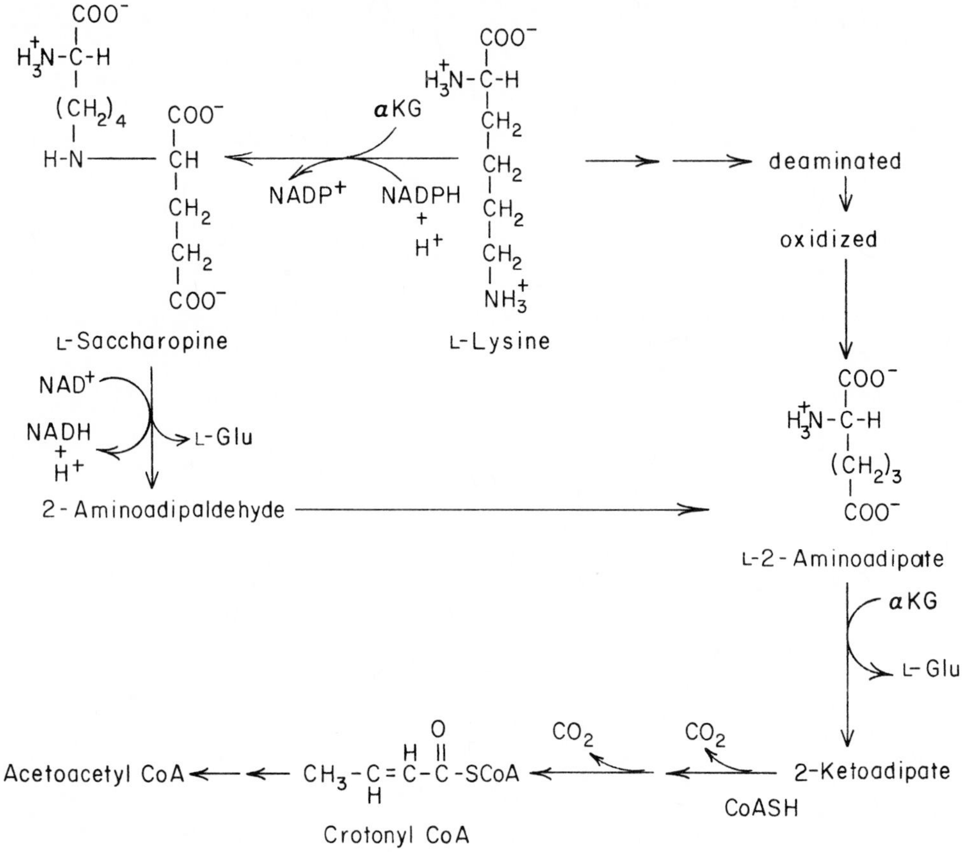

Fig. 9-16. Catabolism of lysine.

Fig. 9-16 show it to be a ketogenic amino acid, feeding experiments have shown that it is only slightly involved in producing either ketone bodies or glycogen. However, the pathway through saccharopine is probably important, as patients with hyperlysinemia also accumulate saccharopine.

Tryptophan. Tryptophan has a complex metabolism that can give rise to both alanine (glycogenic) and acetyl CoA (ketogenic). The initial reaction (Fig. 9-17) is an oxidation catalyzed by the enzyme tryptophan pyrrolase. Like tyrosine transaminase, the synthesis of this enzyme is induced by administration of glucocorticoids. The tryptophan side chain is cleaved off to give alanine and 3-hydroxyanthranilate. The latter compound is decarboxylated and reduced to α-ketoadipate. The subsequent reactions that yield acetoacetyl CoA are the same as those shown for lysine (Fig. 9-16).

Tryptophan can also be converted to NAD^+ (Fig. 9-17). More than 50 years ago it was observed that the symptoms of pellagra, the disease caused by a deficiency of niacin, could be successfully treated with tryptophan. The recommended dietary allowance for niacin depends on the degree of tryptophan conversion to NAD^+; that is, in a normal person, 60 mg of tryptophan is approximately equivalent to 1 mg of niacin. During pregnancy there appears to be an increased conversion of tryptophan to niacin. However, if a person is deficient in one vitamin, it is often the result of eating a diet so poor that it may be lacking in sufficient

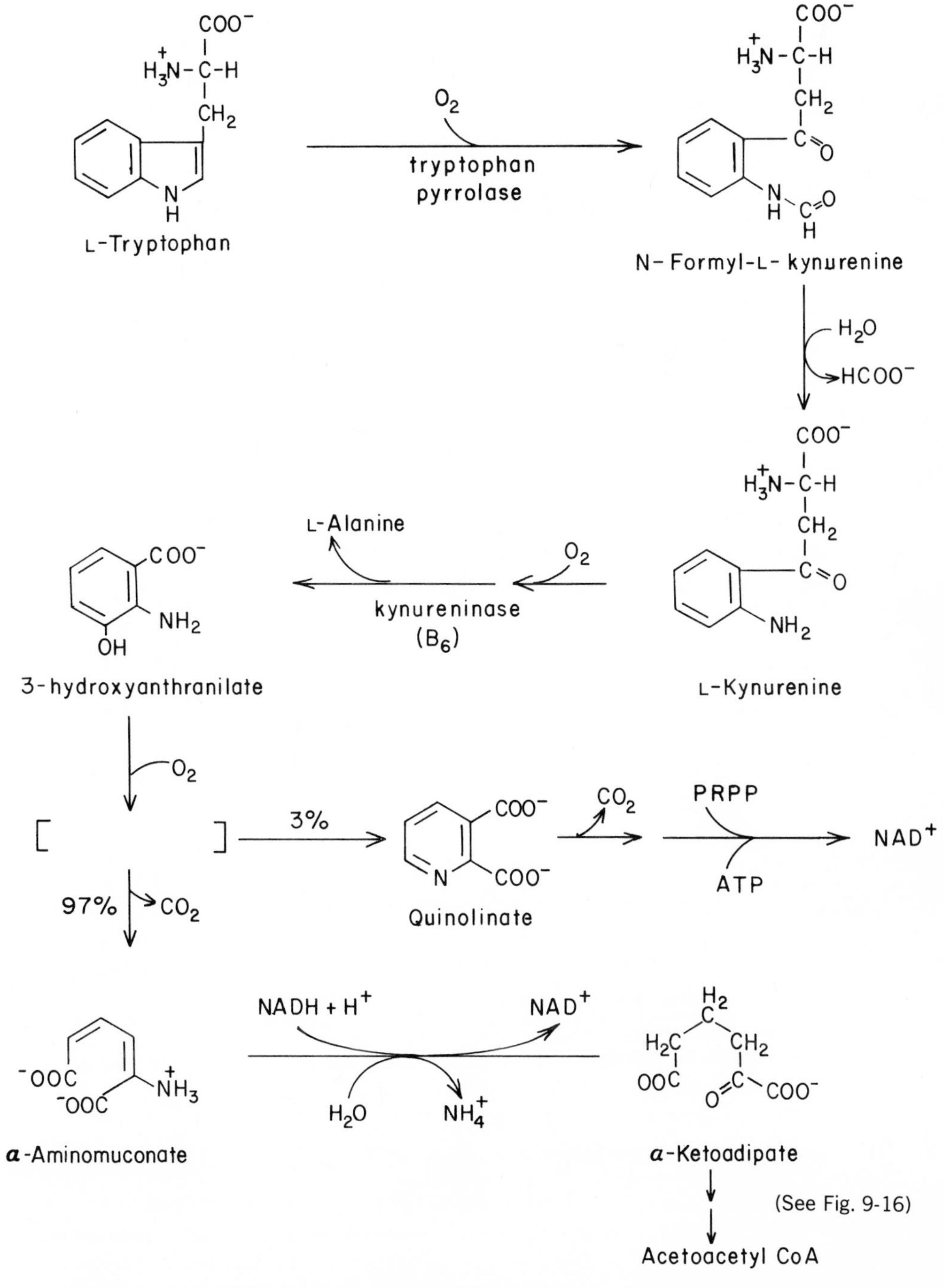

Fig. 9-17. Catabolism of tryptophan. Several intermediates after 3-hydroxyanthranilate have been omitted. The brackets indicate that one is a precursor of NAD+.

amounts of other vitamins. For example, an individual with pellagra could be eating a diet low in niacin but not in tryptophan. However, the person is unable to synthesize niacin from tryptophan if he is also deficient in vitamin B_6, since the enzyme kynureninase, which catalyzes a reaction leading to NAD$^+$ synthesis (Fig. 9-17), requires a vitamin B_6–derived coenzyme.

   The presence of excessive tryptophan metabolites in the urine occurs with several diseases; for example, tuberculosis, diabetes, Hodgkin's disease, leukemia, cancer of the bladder, and multiple myeloma may disrupt tryptophan metabolism, and the urinary metabolites are then very sensitive indicators of disease. This does not mean that defects in tryptophan metabolism cause these diseases. In Hartnup disease the inability to absorb tryptophan can lead to pellagra-like symptoms. In carcinoid syndrome, a malignancy of the enterochromaffin or argentaffin cells that produce serotonin (p. 455), excessive amounts of the tryptophan metabolites involved in the synthesis of serotonin are excreted, as well as serotonin itself.

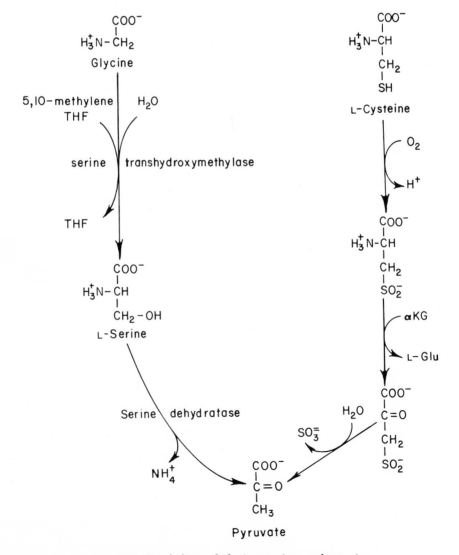

Fig. 9-18. Catabolism of glycine, serine, and cysteine.

Catabolism of glycogenic amino acids

We have already seen how *alanine, aspartate,* and *glutamate* can be metabolized via Krebs cycle enzymes by reversible transamination reactions (Fig. 9-5).

Glutamine and asparagine. Glutamine and asparagine can be converted to glutamate and aspartate by hydrolytic reactions that yield ammonia. The ammonia produced by the kidney in response to metabolic acidosis originates from glutamine and is released by the enzyme glutaminase. However, the amide nitrogen of glutamine is not always released simply in the form of ammonium ions. The nitrogen can be transferred enzymatically to a variety of acceptors in a number of biosynthetic reactions; for example, the formation of amino sugars and the synthesis of the purine and pyrimidine rings require transfers of amide nitrogen from glutamine.

Glycine, serine, and cysteine. Glycine, serine, and cysteine can all be metabolized to pyruvate, as shown in Fig. 9-18. The enzyme serine dehydratase causes the direct conversion of serine to pyruvate. Glycine can be converted to serine by the transfer of a hydroxymethyl group from 5,10-methylene THF. The sulfur atom of cysteine is first oxidized, the product transaminated, and the sulfite residue is hydrolyzed off to give pyruvate.

Threonine. Threonine dehydratase, an enzyme similar to the corresponding one for serine, produces α-ketobutyrate, as shown in Fig. 9-19. Ketobutyrate can be oxidatively decarboxylated to yield propionyl CoA by an enzyme complex similar to the pyruvate dehydrogenase complex. Propionyl CoA (Chapter 6) can be carboxylated to methylmalonyl CoA, which in turn is isomerized to succinyl CoA. Succinyl CoA can enter the Krebs cycle and give rise to pyruvate.

Methionine. Methionine is another amino acid that produces α-ketobutyrate, and thus it has the same fate as threonine. The reactions on p. 428 show the steps by which methionine can donate a methyl group and produce homocysteine, which

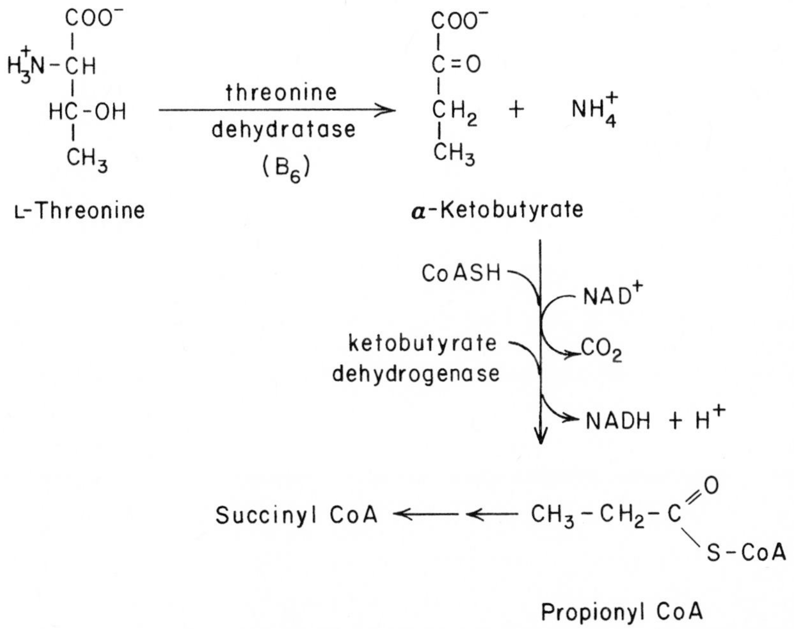

Fig. 9-19. Catabolism of threonine.

in turn can be converted to homoserine. Homoserine is then deaminated and dehydrated in much the same way as threonine (Fig. 9-19) to yield α-ketobutyrate and thus propionate.

Arginine.   As seen previously (Fig. 9-9), arginine is converted to glutamate-γ-semialdehyde, an intermediate that can be oxidized to glutamate in an NAD^+-dependent reaction.

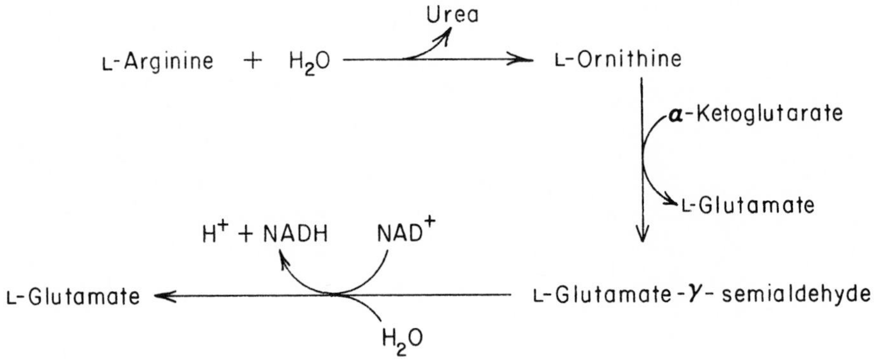

Proline.   Proline undergoes two oxidations. The first is mediated by a mitochondrial flavoprotein (FAD) to form the unsaturated compound shown below. The second oxidation uses NAD^+ to open the ring and form glutamate.

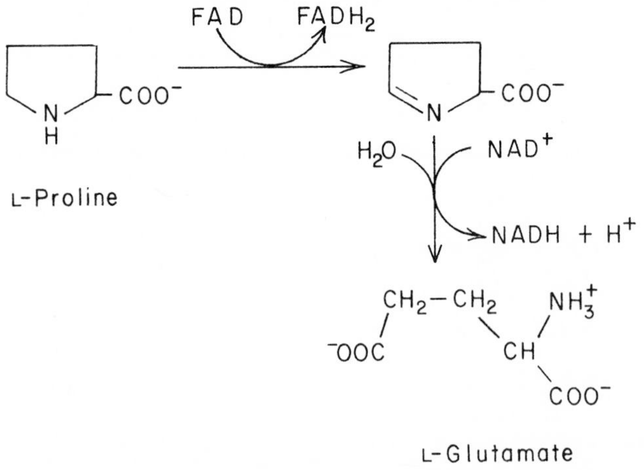

Valine.   Valine, like the other branched-chain amino acids, is first transaminated and then decarboxylated and oxidized. In this case the final product is methylmalonyl CoA, which can be isomerized to succinyl CoA (Fig. 9-20).

In Figs. 9-13, 9-14, and 9-20 one sees that the branched-chain amino acids (leucine, isoleucine, and valine) undergo a similar oxidative decarboxylation after an initial transamination to the keto acids. The enzyme complexes that catalyze these reactions appear to be similar to pyruvate dehydrogenase. The 2-ketoisocaproate dehydrogenase functions with keto acids derived from either leucine or isoleucine, and a defect in this complex causes maple syrup urine disease. Although the dehydrogenase of the valine pathway is not affected by the disease, keto acids derived from valine are found in the urine; the accumulated keto acids derived from leucine and isoleucine very likely inhibit the keto-valine dehydrogenase.

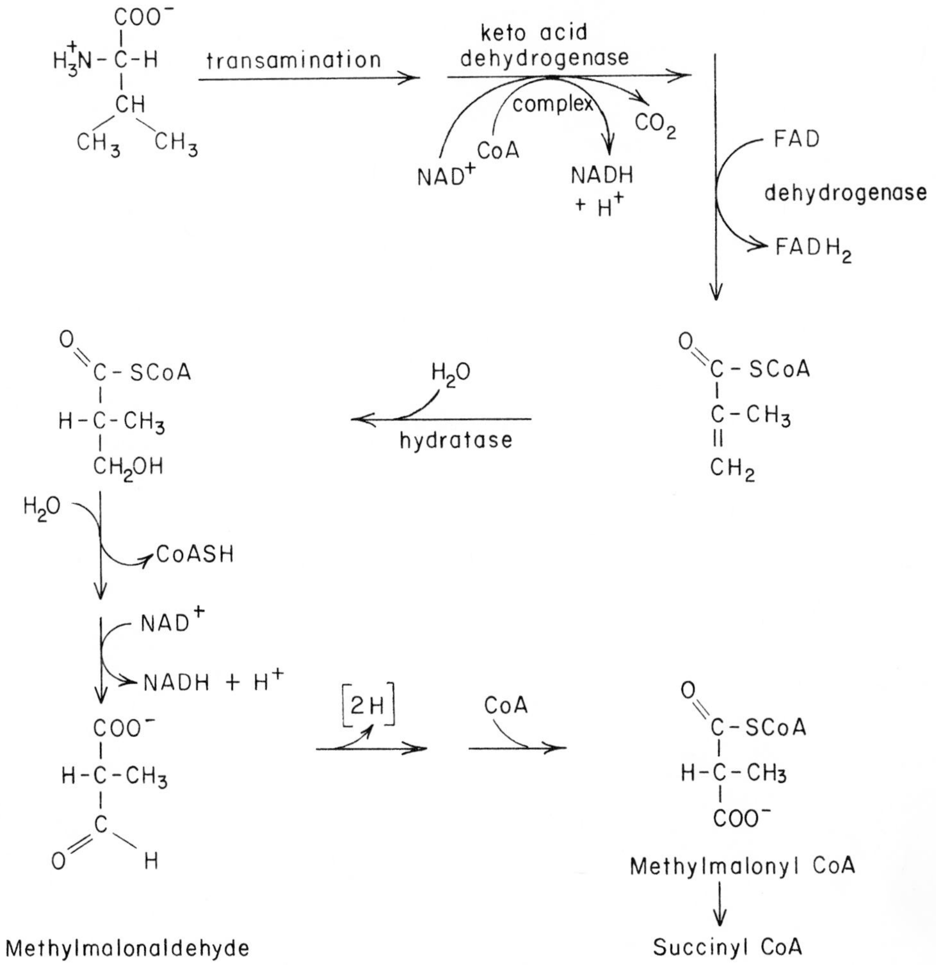

Fig. 9-20. Catabolism of valine.

Histidine. The breakdown of histidine is shown in Fig. 9-21. The enzyme his-tidase deaminates the amino acid to produce the unsaturated intermediate uro-canate. Hydration, which causes the opening of the imidazole ring, produces formiminoglutamate. Perhaps the most distinctive reaction in the pathway is the subsequent transfer of the formimino group to THF, producing glutamate.

The genetic disease histidinemia is associated with a defect in histidinase, the first enzyme of this series. Victims of the disease have difficulty with speech and the development of language skills. However, the relationship of these symptoms to histidine metabolism is unknown. Treatment by restricting dietary histidine lowers the blood levels of the amino acid but also impairs growth and development.

AMINO ACIDS AS PRECURSORS OF METABOLITES
Amine synthesis

Animals contain a number of amine compounds derived chiefly from the amino acids. Although the concentrations of these amines are low, their actions are very important.

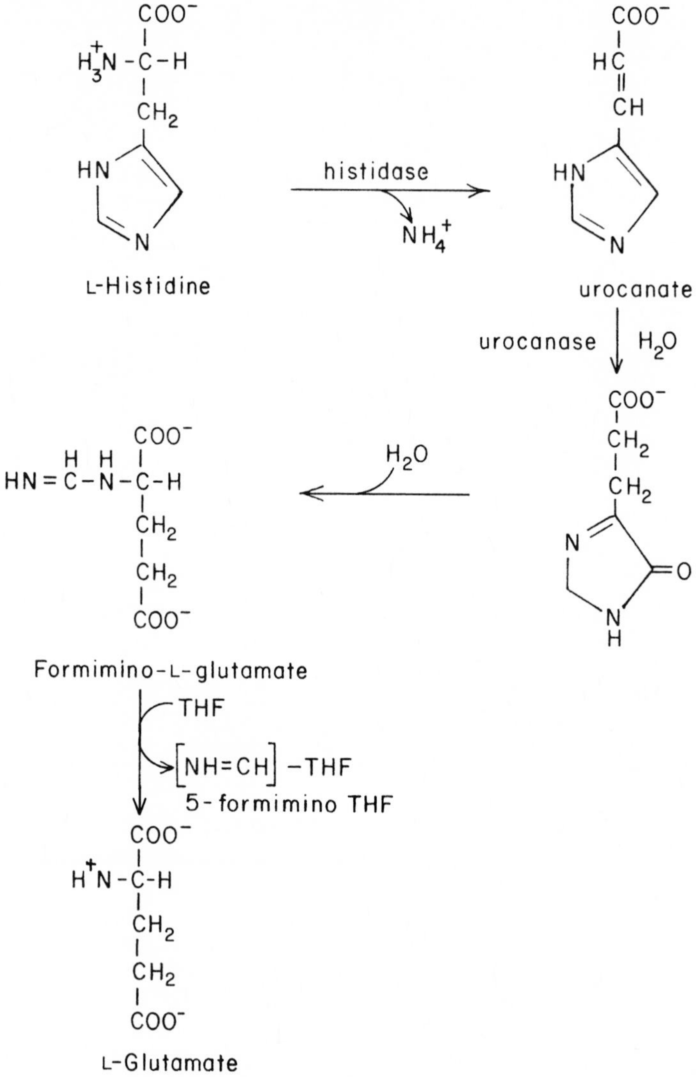

Fig. 9-21. Catabolism of histidine.

Nervous system amines

Some of the most important amines derived from amino acids are those found in the nervous system; many of these are neurochemical transmitters. A nerve can be thought of as transmitting a stimulus from one part of the body to another by means of a moving wave of ions. When the wave reaches the end of the neuron, it causes the release of a transmitter that migrates to a receptor cell, where it triggers the propagation of another wave.

Acetylcholine. *Cholinergic* receptors such as the motor end plates of muscles are sensitive to acetylcholine. Once the desired response has been provoked the

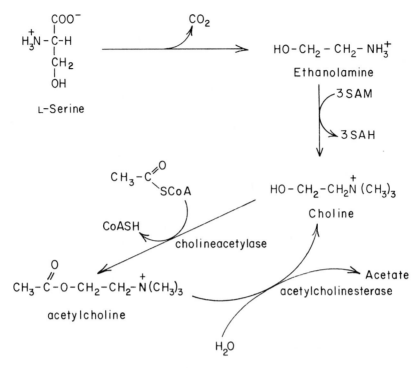

Fig. 9-22. Synthesis and hydrolysis of acetylcholine. SAM = S-adenosyl-L-methionine; SAH = S-adenosyl-L-homocysteine.

acetylcholine must be removed. The enzyme acetylcholinesterase accomplishes this. Fig. 9-22 summarizes the biochemical reactions involved in the synthesis of choline from serine and methionine and shows the reactions involved in the cycling of acetylcholine and choline, which occurs during the functioning of the amine.

Catecholamines.   Catecholamines are derivatives of:

$$HO-\!\!\left\langle\!\!\bigcirc\!\!\right\rangle\!\!-CH_2-CH_2-NH_3^+$$
$$\overset{|}{OH}$$

3,4-dihydroxy phenethylamine (dopamine)

and include norepinephrine, epinephrine, and dopamine. They are produced in the brain, sympathetic nerve endings, chromaffin cells of peripheral tissues, and the adrenal medulla. The latter is an endocrine gland that stores the hormones epinephrine and norepinephrine in chromaffin granules (named for their affinity for dichromate). These hormones permit the body to react to stress by effecting increases in cardiac output, blood flow to muscles, lungs, and brain, and cellular metabolism.

The amino acid precursor of the catecholamines is tyrosine. Fig. 9-23 shows the pathway of epinephrine synthesis from tyrosine. The first reaction is catalyzed by tyrosine hydroxylase, an enzyme similar to phenylalanine hydroxylase in its mechanism but with a different substrate specificity. The product of the reaction is 3,4-dihydroxyphenylalanine. In the older literature this compound was called *dioxy*phenyl*a*lanine and was abbreviated dopa. Although the chemical nomen-

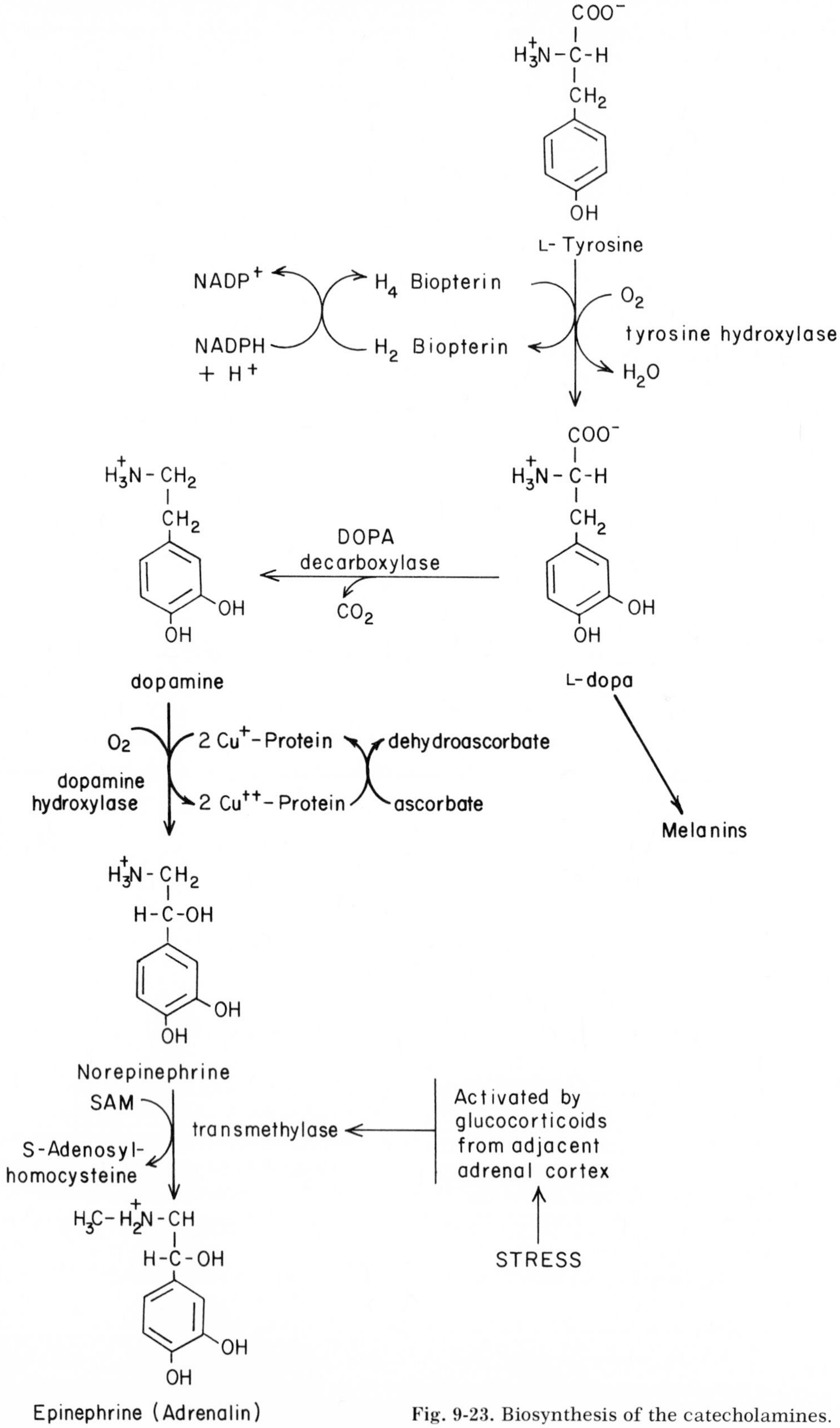

Fig. 9-23. Biosynthesis of the catecholamines.

clature has been improved, the popular abbreviation has been retained. L-Dopa has been used recently with remarkable success to treat the symptoms of Parkinson's disease. It serves as a precursor of both the melanin skin pigments and dopamine (Fig. 9-23). Some of the nerves of the central nervous system are extraordinarily rich in dopamine. The adrenal medulla contains an enzyme that hydroxylates the side chain of dopamine to give norepinephrine, a transmitter that functions at the terminals of sympathetic nerves as well as with the nerves of the central nervous system (adrenergic nerves). The hormone epinephrine is formed

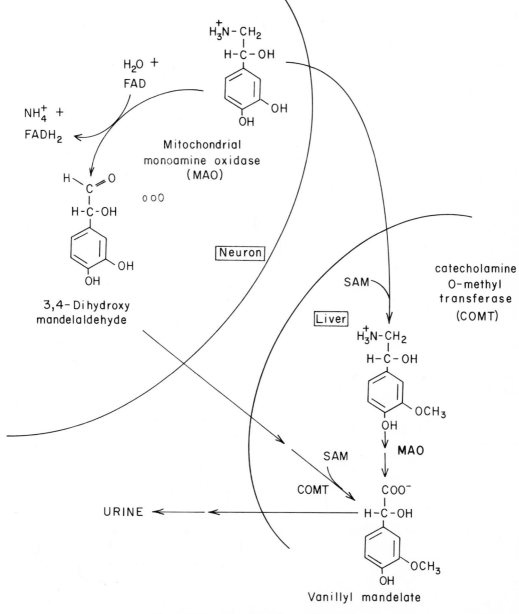

Fig. 9-24. Inactivation of norepinephrine.

through methylation of the amino group of norepinephrine; S-adenosylmethionine is the methyl donor (Fig. 9-23).

Norepinephrine is stored in chromaffin granules, the organelles located near the synapses of adrenergic nerves. The transmitter is discharged by a process of exocytosis, a kind of reverse pinocytosis. Implicit in the action of norepinephrine is its inactivation, so that after transmission has occurred the synapse is restored. The immediate termination of action is probably nonenzymatic; perhaps the transmitter is bound to a substance that renders it inactive. Shortly after it is taken up by the receptor cell, norepinephrine is oxidized by a mitochondrial enzyme, monoamine oxidase (MAO) (Fig. 9-24). This enzyme is not, however, concerned with the termination of the adrenergic impulse. MAO is a flavin enzyme that produces 3,4-dihydroxymandelaldehyde. This aldehyde is transported through the blood to the liver, where it is further oxidized to the corresponding acid and where position 3 of the ring is O-methylated to form 3-methoxy-4-hydroxymandelate. The trivial name for this substance is vanillylmandelate.

Some of the norepinephrine goes directly into the blood and is taken up by the liver, where it is O-methylated by the cytoplasmic enzyme catecholamine O-methyl transferase (COMT), which uses S-adenosylmethionine as the methyl donor. The methylated epinephrine is then oxidized by liver enzymes to vanillylmandelate, the form ultimately excreted in the urine (Fig. 9-24). Both MAO and COMT are widely distributed in the body; the highest concentrations are found in the liver and kidney. Norepinephrine that is rapidly released by adrenergic nerves is first methylated by COMT, whereas that which is more slowly released is oxidized first by MAO. In either case, as shown in Fig. 9-24, the ultimate product is vanillylmandelate.

Epinephrine acts as a hormone to stimulate oxygen consumption and increase blood glucose and lactate levels through the mediation of cAMP. Epinephrine probably cannot penetrate into the target cells, but this is not necessary for the stimulation of adenyl cyclase.

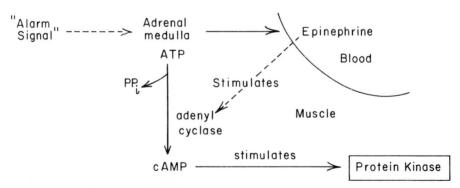

Action of epinephrine

Serotonin (5-hydroxytryptamine).   Serotonin is found in cells of the central nervous system, where it affects transmission either directly as a transmitter or indirectly in a more obscure way. It is also produced in the intestinal mucosa for unknown reasons. The drug LSD probably competes with serotonin, since LSD toxicity can be treated by serotonin administration. Biosynthesis of the amine, like that of norepinephrine and epinephrine, begins with a hydroxylation reaction followed by a decarboxylation. Although phenylalanine hydroxylase isolated from liver can synthesize the intermediate 5-hydroxytryptophan from tryptophan, a specific tryptophan hydroxylase functions in the brain. Decarboxylation of 5-hydroxytryptophan yields serotonin, as shown below.

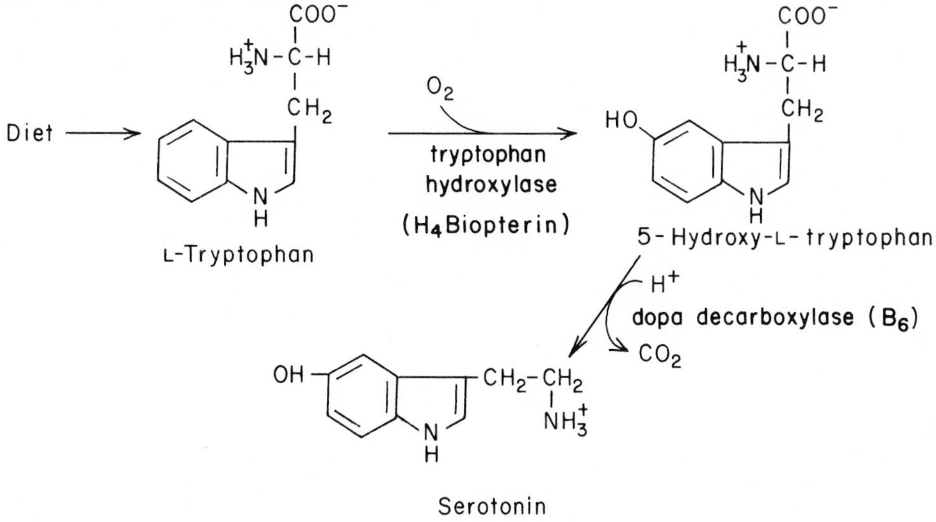

γ-*Aminobutyrate.* The decarboxylation of glutamate produces γ-aminobutyrate (GABA); it is present in very high concentrations in the brain, where it appears to inhibit synaptic transmission.

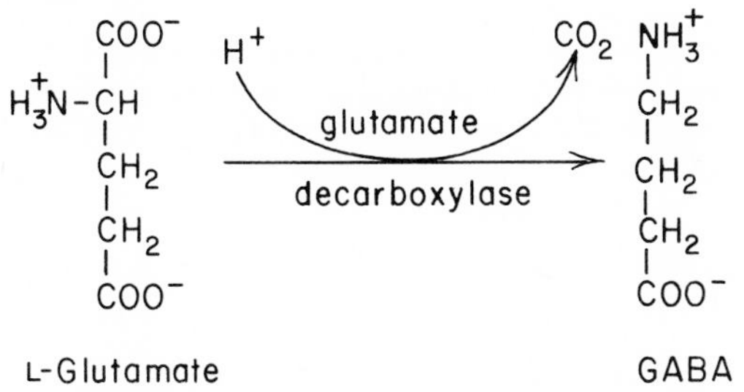

It has been suggested that several other amino acids or their derivatives serve as neurotransmitters. Probable exitory transmitters include aspartate and glutamate, whereas glycine, like GABA, may be an inhibitor. (For further discussion of neurotransmitters and the drugs that affect their action, see Chapter 13.)

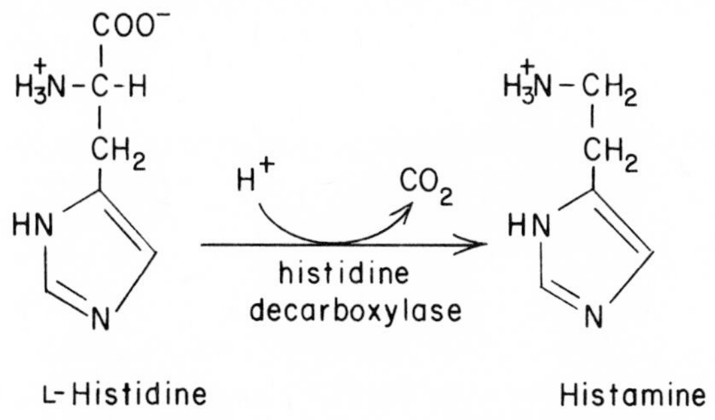

Histamine. Decarboxylation of histidine produces histamine, which is made and stored in the mast cells as well as in many other cells of the body. It is released in anaphylaxis and as a consequence of allergies, although drugs and chemicals can also trigger release. Circulating histamine is inactivated in the liver by methylation and oxidation reactions. The amine causes expansion of capillaries, probably by constricting the smaller veins that lead from them. This causes a local edema and an increase in the volume of the vascular bed. A drop in blood pressure results that, if severe enough, may induce shock. *Antihistamines* are compounds that, because of their structural similarity to histamine, can prevent the physiologic changes produced by histamine released during allergic reactions. Diphenylhydramine, a typical antihistamine, is shown below.

$$CH-O-CH_2-CH_2NH^+ \quad \overset{CH_3}{\underset{CH_3}{|}}$$

Diphenylhydramine

Polyamines

Putrescine, or decarboxylated ornithine, is a four-carbon diamine that is the precursor of *spermidine* and *spermine,* polyamines found in association with nucleic acid and present in high concentrations in semen. Their functions are still unknown. It can be seen in Fig. 9-25 that methionine and ornithine are the amino acid precursors of spermine. Notice that S-adenosylmethionine does not donate methyl groups in this reaction; instead, the main carbon chain is decarboxylated and transferred, both reactions being catalyzed by the same enzyme. The other product of the reaction is 5'-methylthioadenosine (for the sake of simplicity, its structure is not shown in Fig. 9-25).

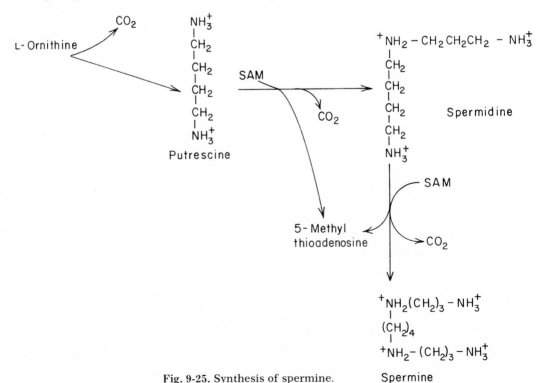

Fig. 9-25. Synthesis of spermine.

Creatine synthesis

Glycine and arginine are involved in the synthesis of creatine, an important constituent of muscle. Information on the rate of creatine turnover and its excretion as creatinine is useful in many clinical situations. When the muscles are at

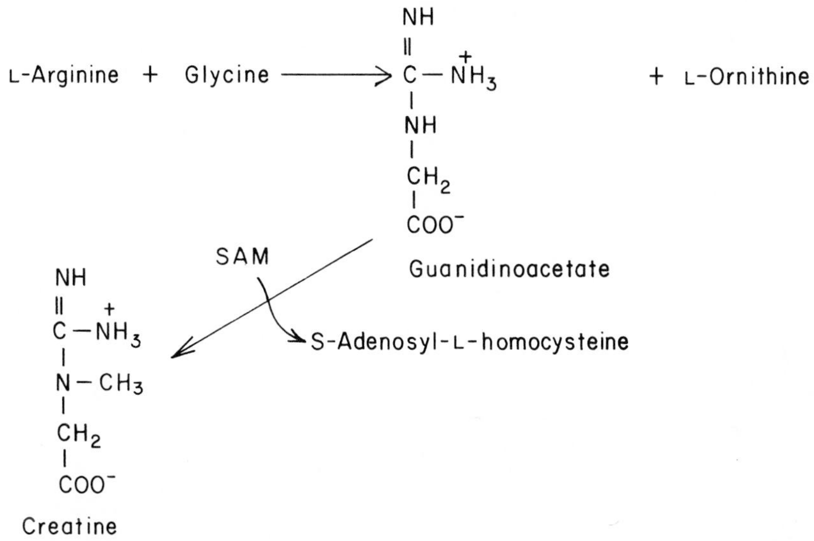

$$L\text{-Arginine} + \text{Glycine} \longrightarrow \underset{\substack{\text{Guanidinoacetate}}}{\overset{\displaystyle \underset{\displaystyle \parallel}{NH}}{C}-\overset{+}{N}H_3} + L\text{-Ornithine}$$

$$
\begin{array}{c}
NH \\
\parallel \quad + \\
C-NH_3 \\
\mid \\
N-CH_3 \\
\mid \\
CH_2 \\
\mid \\
COO^- \\
\end{array}
$$

Creatine

SAM → S-Adenosyl-L-homocysteine

rest, almost all of the creatine is phosphorylated. It is the phosphocreatine that spontaneously and nonenzymatically produces creatinine by cyclizing and splitting out inorganic phosphate.

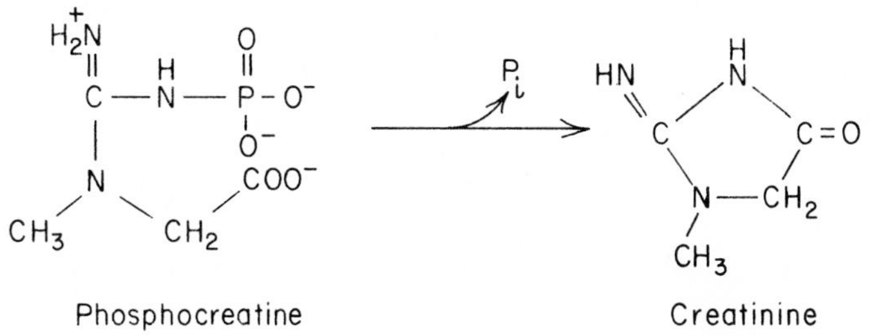

Phosphocreatine → P_i → Creatinine

REFERENCES

Beaven, M. A.: Histamine (in 2 parts), N. Engl. J. Med. 294:30, 1976. *A review of the medical aspects of histamine biochemistry, physiology, and pharmacology.*

Felig, P.: Amino acid metabolism in man, Annu. Rev. Biochem. 44:933, 1975. *Interorgan transfer of nitrogen via alanine is stressed.*

Mathews, D. M.: Intestinal absorption of peptides, Physiol. Rev. 55:537, 1975. *The importance of peptide transport is described.*

Meister, A.: Biochemistry of the amino acids, ed. 2, New York, 1965, Academic Press, Inc., vols. 1 and 2. *A somewhat dated but very comprehensive treatment of all aspects of amino acids.*

Meister, A.: On the enzymology of amino acid transport, Science 180:33, 1973. *This review describes the role of glutathione in amino acid transport.*

Munro, H. N., and Allison, J. B., editors: Mammalian protein metabolism, New York, 1964, Academic Press, Inc. *A complete presentation of a variety of topics that are of interest to students of the health sciences.*

Nyhan, W. L.: Heritable disorders of amino acid metabolism: patterns of clinical expression and genetic variation, New York, 1974, John Wiley & Sons, Inc. *This book emphasizes both the basic and applied features of diseases of amino acid metabolism, including treatment and management.*

Ratner, S.: Enzymes of arginine and urea synthesis, Adv. Enzymol. 39:1, 1973. *A complete review of the properties of these enzymes, including a discussion of their deficiencies in humans.*

Scrivner, C. R., and Rosenberg, L. E.: Amino acid metabolism and its disorders, Philadelphia, 1973, W. B. Saunders Co. *Emphasizes both biochemical and clinical aspects.*

Waterlow, J. C., and Alleyne, A. O.: Protein malnutrition in children, Adv. Protein Chem. 25:117, 1971. *A review of recent advances in this area.*

Clinical examples

CASE 1: AMINO ACID METABOLISM IN STARVATION

F. N., an obese patient, had failed to lose weight because he could not follow prescribed weight reduction diets. Consequently, he volunteered to go on a starvation diet as part of a study of amino acid metabolism during prolonged fasts (Felig and associates). Blood samples were taken and analyzed for plasma amino acids during a 5- to 6-week fast. Valine, leucine, isoleucine, methionine, and α-aminobutyrate were transiently increased during the first week but dropped below initial levels at later times. Glycine, threonine, and serine increased more slowly, but thirteen amino acids ultimately decreased. The largest decrease was seen for alanine, which rapidly dropped 70% during the first week. Total plasma amino nitrogen dropped only 12%.

Biochemical questions

1. Explain how the decreased plasma alanine concentration may be related to gluconeogenesis. What is the alanine-glucose cycle?
2. Threonine, an essential amino acid, can be converted to glycine; thus a block in glycine metabolism might allow both threonine and glycine to accumulate. As a result, folic acid was administered, and the levels of glycine and threonine were measured. Although there was no change as a result of this treatment, what was the rationale for it?
3. What is the source of the plasma α-aminobutyrate?
4. What may account for the increased plasma concentrations of the branched chain amino acids after the first 5 days of starvation?
5. It has been shown in animals that circulating branched–chain amino acids are metabolized by muscle to yield energy. What are the products of the catabolism of leucine, valine, and isoleucine? Is urea made in muscle tissue?
6. The branched–chain amino acids stimulate the production of both ala-

nine and glutamate in aminal muscle. Explain this observation, recognizing that muscle is rich in glutamate pyruvate transaminase.
7. Explain why patients with maple syrup urine disease suffer hypoglycemic episodes even though insulin levels are normal. Consider the fact that plasma alanine levels in these patients are threefold to tenfold below normal.

Case analysis*

Obese patients such as F. N. have maintained starvation diets for as long as 200 days, occasionally with serious effects. The primary need of the starving person is fuel for energy, and the initial physiologic response to the lack of food is the maintenance of the blood glucose concentration. Glucose is especially needed by the brain, which consumes about 65% of the total circulating glucose (approximately 400 to 600 calories/day). Early in starvation the liver is the most important organ in supplying blood glucose. Later the kidneys become equally important. Liver glycogen can supply only enough glucose for a few hr, and, even after a short fast, gluconeogenesis by the liver requires substrates from other tissues. These substrates are derived from glycogenic amino acids produced by protein catabolism in several tissues, but primarily in the muscle. After a few days of fasting, some of the energy requirements of the body are met by increased fat catabolism. It is the purpose of weight-reduction diets to take advantage of this phenomenon and allow excessive body fat to be converted to energy. Fat, as has been pointed out (p. 275), cannot be directly converted to glucose, but it can be used to supply the energy that would otherwise have to come from amino acids derived from protein. In this sense, fat can spare the glucose requirement by providing approximately

---

*See also case 2, Chapter 14, in which different aspects of biochemistry are considered.

16 g of the 100 to 145 g of glucose that would otherwise be needed each day. Nevertheless, a starving person would quickly use up his body protein and die within 3 to 4 weeks if the remainder of the glucose requirement continued to come from protein catabolism. Fortunately, the brain, the major consumer of glucose, adapts to the starving state by increasing its ability to use fat-derived ketone bodies for energy. The concentration of ketone bodies is normally quite low in healthy and fed individuals but increases significantly during starvation. This increased concentration of ketone bodies may trigger the brain to start utilizing them.

In spite of this, the body has needs that can be satisfied only by glucose during both the early and late stages of starvation, and these requirements can be met only by the breakdown of body protein. The liver is primarily responsible for maintaining glucose output, yet the amino acids needed for gluconeogenesis are derived mainly from the muscle. The amino acids must be transported from the muscle to the liver. Complicating this transport is the fact that much of the amino acids produced by protein breakdown are used for energy in the muscle itself. Recall that the first step in the ca-

tabolism of amino acids is a deamination or transamination reaction that removes the alpha amino group. The muscle now must accommodate a large increase in amino nitrogen. Because muscle, unlike liver, is incapable of synthesizing urea, most of the amino nitrogen is transaminated to pyruvate to form alanine. The alanine enters the blood, is taken up by the liver, the amino groups removed to form urea, and the resulting pyruvate converted to glucose. The newly synthesized glucose is secreted to the blood, taken up by the muscle, and catabolized to pyruvate as described in detail in Chapter 7. The pyruvate acts again as the acceptor for another amino group. The result of all these reactions is a cycle that transports amino groups from muscle to liver. This scheme is called the alanine-glucose cycle, or sometimes the Cahill cycle. Fig. 9-26 illustrates some features of amino acid metabolism related to this cycle.

Alanine concentration drops late in starvation as peripheral release decreases to conserve protein.

Glutamine is another important vehicle for transferring amino groups from muscle to the liver. In this case ammonium ions are coupled with glutamate to yield glutamine. The glutamine is re-

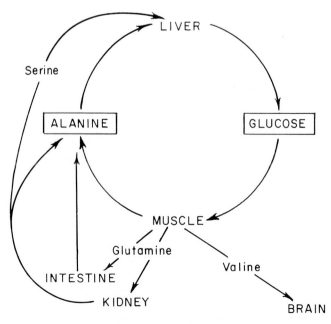

Fig. 9-26. The alanine-glucose cycle.

leased from the muscle to the blood and taken up by the kidneys and the intestine (Fig. 9-25). The intestine transports its amino groups to the liver as alanine. The kidneys also release alanine to the liver, and in addition release some amino groups as serine, another glucogenic amino acid.

Folate was administered in an effort to test whether decreased glycine utilization might lead to the accumulation of both glycine and its precursor threonine. A major function of glycine is the synthesis of serine by an enzyme that requires tetrahydrofolate (see p. 427). Consequently, it was thought that the patient might be deficient in the vitamin folic acid since he was on a starvation diet. The folate deficiency might have caused glycine to accumulate. In fact, glycine probably accumulated because of increased peripheral release that was maintained by its active reabsorption by the kidney.

Most of the branched-chain amino acids stay in the muscle where they are catabolized for energy. Leucine breakdown produces HMGCoA and ketone bodies, whereas isoleucine yields propionyl CoA and acetyl CoA. Valine, however, is released to the blood. The brain is capable of directly utilizing valine, a glucogenic amino acid that produces methylmalonyl CoA.

In this case the branched-chain amino acids exhibited a transient increase in serum concentration that peaked at 5 days. As fasting continued, the blood valine level dropped far below the initial level in the fed patient. Increased plasma concentrations of branched-chain amino acids is seen in other states as well as in starvation. For example, branched-chain amino acids accumulate in the plasma of diabetics and pancreatectomized dogs. This may be the result of lowered blood insulin. It is interesting to note that the concentration of branched-chain amino acids is usually higher in obese individuals, and we have seen that diabetic obese patients have a lower number of insulin receptors (see case 2, Chapter 7). Alternatively, the transient increase in the circulating levels of the branched-chain

amino acids may be a result of their increased release from the liver. This is known to occur in diabetic, fasted, or uremic rats.

The branched-chain amino acids stimulate the formation of glutamate as well as alanine and glutamine in experimental animals. Because muscle transaminases specific for branched-chain amino acids are much more active for α-ketoglutarate than for pyruvate, it is believed that the amino groups of these amino acids are first transferred to α-ketoglutarate to give glutamate. The glutamate is then transaminated with pyruvate to form alanine by the glutamate-pyruvate transaminase of muscle, a very active enzyme in muscle as it is in liver.

Further evidence of a coupling of the metabolism of branched-chain amino acids with alanine synthesis is obtained from patients with maple syrup urine disease. Failure to decarboxylate the branched-chain amino acids causes a threefold to tenfold decrease in the amount of plasma alanine in these patients. They also suffer from hypoglycemia even though insulin levels are normal. The lowered alanine production is a consequence of the inability of the patients to metabolize the branched-chain amino acids. If alanine is given the patients, the hypoglycemia is alleviated. The coupling of the alanine cycle to the oxidation of branched-chain amino acids is not only very important in maintaining glucose balance under a number of physiologic conditions, but it is also important for satisfying much of the energy needs of muscle. In experimental animals as much as 14% of the energy needed by the muscle may come from the oxidation of the branched-chain amino acids.

REFERENCES

Felig, P.: Amino acid metabolism in man, Annu. Rev. Biochem. 44:933, 1975.

Felig, P., Owen, O. E., Wahren, G., and Cahill, G. F., Jr.: Amino acid metabolism during prolonged starvation, J. Clin. Invest. 48:584, 1969.

Odessey, E. A., Khairallah, E. A., and Goldberg, A. L.: Origin and possible significance of alanine production by skeletal muscle, J. Biol. Chem. 249:7623, 1974.

Young, V. R., and Scrimshaw, N. W.: The physiology of starvation, Sci. Am. 225:14, October 1971.

CASE 2: GLOMERULONEPHRITIS

C. A., a 20-year-old man, was admitted to the hospital because of headache, pain in the flanks, anorexia, and passage of red-colored urine. Some of his symptoms had appeared 2 to 4 weeks earlier after a moderate bout with tonsillitis that had been untreated.

On examination, C. A. showed a mild edema around the eyes. Examination of the red-colored urine revealed hyaline casts in large number and red cell casts. The EKG showed signs of hypertrophy of the left ventricle, which was subsequently confirmed by chest x-ray films. Hemorrhages and exudates were observed around constricted retinal arteries; the optic discs were normal. Results of laboratory tests are shown below.

The patient was hospitalized and given a protein-free diet with reduced sodium and water intake (1 liter/day). His blood pressure was lowered by administration of hypotensive drugs, and after 8 days the BUN returned to normal. At this time, protein (150 g of steak/day) was added to his diet. He was then released with instructions to restrict his activity and report back to the hospital in 1 week.

Biochemical questions

1. What causes uremia? Is there a toxin involved?
2. Why was the protein in the diet lowered in this case? What levels of lipid and carbohydrate should be present in this diet? Would you recommend an all-vegetable diet? Explain.
3. If you used amino acids to replace the protein in the diet, which would you use? Why? Would this be practical?
4. If this patient's condition were left untreated, would you expect the development of alkalosis or acidosis? Explain.

Case discussion

The clinical picture of acute glomerulonephritis is characterized by a generalized edema, mild hypertension with headache and retinal hemorrhages, pain in the flanks, and oliguria. The edema is particularly noticeable around the eyes; this is because of diminished glomerular filtration. Tubular function is normal. This results in a greater retention of water and electrolytes.

C. A.'s glomerulonephritis was caused by the previous β-hemolytic streptococcal infection. It is believed that a substance derived from the microorganism becomes associated with the glomerular basement membrane, where antibodies against the substance interact to form a product that brings about the inflammation. In such patients the antistreptolysin O titer is high. Why the streptococcal infection can result in glomerulonephri-

	PATIENT	NORMAL
Blood pressure	180/111 mm Hg	120/80 mm Hg
Urine		
Specific gravity	1.010	1.003-1.030
Creatinine clearance	50 ml/min • 1.73 m²	110-150 ml/min • 1.73 m²
Serum creatinine	3 mg/dl	1-2 mg/dl
Antistreptolysin O titer	800 units	0-200 units
Serum electrolytes		
Sodium	150 meq/liter	136-145 meq/liter
Potassium	5.1 meq/liter	3.5-5.0 meq/liter
Calcium	7.2 mg/dl	9-11 mg/dl
Phosphate	5.2 mg/dl	3.0-4.5 mg/dl
Sulfate	2 meq/liter	0.8-1.2 meq/liter
Total protein	6.5 g/dl	6-8 g/dl
Albumin	4 g/dl	3.5-5.5 g/dl
Globulins	2.5 g/dl	1.5-3.0 g/dl
BUN	45 mg/dl	10-20 mg/dl
Hemoglobin	9 g/dl	14-18 g/dl
WBC	14,000/mm³	5000-10,000/mm³

tis in certain individuals but not in others is not clear.

Renal-associated hypertension. Hypertension may become very severe in an acute attack of glomerulonephritis. How the diseased kidney brings about this hypertension is fairly well understood in biochemical terms. When the blood supply to the kidney is diminished, the juxtaglomerular cells secrete renin, a highly specific proteolytic enzyme, into the bloodstream (see Fig. 4-7, p. 180). In the blood, renin catalyzes the hydrolytic cleavage of a small peptide from the α_2-globulin *angiotensinogen*. Angiotensinogen is produced in the liver and circulates in an inactive form. The decapeptide produced by the action of renin is called *angiotensin I*. It exerts no pressor activity until it is hydrolyzed in the lung to *angiotensin II*, an octapeptide. It is this octapeptide that exerts a potent pressor activity, 200 times greater than that of norepinephrine.

Angiotensin II also has a trophic effect on the adrenal cortex, causing the secretion of aldosterone. The increased levels of aldosterone result in the conservation of Na^+; this in turn leads to an increase in blood volume because water is reabsorbed in order to maintain the osmotic pressure of the plasma. Thus circulation to the kidney improves and renin secretion subsides.

In cases such as that of C. A., hypertension is often alleviated by the administration of a diuretic and α-methyl-D-dopa. Like dopa, methyldopa can be decarboxylated and oxidized. Methylnorepinephrine, the product, may function as a "false" neurotransmitter. Actually, the hypotensive properties of the drug may be related more to its inhibitory effect on dopa decarboxylase, since it clearly lowers the concentrations of serotonin, dopamine, and norepinephrine in the central nervous system. Lowering the concentrations of the catecholamines depresses vasomotor responses, and the blood pressure drops.

Glomerular filtration rate. Glomerular damage results in a decrease in the rate at which substances are excreted in the urine; renal clearance tests are designed to measure this decrease. Inulin, mannitol, and creatinine are filtered by the glomerulus but for the most part are not secreted or reabsorbed. As a result, their rate of excretion can be used to calculate the glomerular filtration rate (GFR).

$$GFR = \frac{\text{Conc. of test substance in urine} \times \text{ml of urine/min}}{\text{Conc. of test substance in blood}}$$

Because the GFR varies according to body size, values must be corrected to the body surface area of an average-size man ($1.73\ m^2$). The normal values are 110 to 150 ml/min for men and 105 to 132 ml/min for women. A constant plasma level of inulin or mannitol must be maintained during the test period; consequently, the clearance of endogenous creatinine is often measured since its serum levels are relatively constant, and it is more convenient than administering test substances.

The serum level of creatinine and the BUN are quite useful in assessing glomerular function. For the most part, the serum creatinine concentration is dependent on glomerular filtration, but the BUN may be influenced by many nonglomerular factors. A high BUN value indicates a severe problem. Nitrogen balance has a great influence on BUN, whereas under normal conditions the plasma creatinine concentration depends only on muscle mass.

Patients with uremia are unable to excrete adequately nitrogen-containing catabolites, and in the early stage of the disease they retain excessive amounts of water and electrolytes. Since the patient cannot adequately excrete nitrogenous waste materials, any excessive intake of nitrogen will increase the BUN level and exacerbate the clinical situation. Most of the nitrogen is derived from protein; consequently, it is important that the diet contain only maintenance levels of protein. Energy requirements are maintained with dietary carbohydrate and fat.

Acidosis in renal disease. The acidosis seen in patients with renal disease is usually caused by a reduced filtration rate, which leads to retention of the fixed acids of sulfate and phosphate as well as of nonprotein nitrogen. Because of the reduced filtration rate, the cations asso-

$$H_2O \; + \qquad \begin{array}{c} COO^- \\ | \\ H_3\overset{+}{N}-C-H \\ | \\ (CH_2)_3 \\ | \\ NH \qquad COO^- \\ | \qquad\quad | \\ C-NH-C-H \\ \| \qquad\quad | \\ NH \qquad CH_2 \\ \qquad\quad | \\ \qquad\quad COO^- \end{array} \qquad\longrightarrow\qquad \begin{array}{c} COO^- \\ | \\ H_3\overset{+}{N}-C-H \\ | \\ (CH_2)_3 \\ | \\ OH \qquad + \end{array} \qquad \begin{array}{c} NH_2 \qquad COO^- \\ | \qquad\quad | \\ C-NH-C-H \\ \| \qquad\quad | \\ NH \qquad CH_2 \\ \qquad\quad | \\ \qquad\quad COO^- \end{array}$$

L-Argininosuccinate   Guanidinosuccinate

ciated with these acids are slow in reaching the tubules, where they are normally exchanged. In severe acidosis, administration of sodium bicarbonate may be hazardous because the patient may be unable to handle the large sodium load. The acidosis and uremia can be corrected by either peritoneal or extracorporeal dialysis.

As the renal disease progresses, the patient loses water and electrolytes. Ordinarily there is little variation in the osmotic pressure of the plasma, since it is controlled by the kidney through the urine concentration and volume flow. However, when the solute content of the glomerular filtrate is high, osmotic diuresis occurs; that is, the extent to which the urine can be concentrated reaches a maximum. As noted previously, aldosterone is produced in response to renin-induced angiotensin to help compensate for the loss of water and sodium.

Possible involvement of a toxin. A low molecular weight substance isolated from the urine of uremic patients is toxic to cells grown in culture. It also interferes with K^+ absorption in erythrocytes, with Na^+ efflux in the toad bladder, and with other tests. This toxic substance appears to be guanidine or a guanidine derivative, perhaps the guanidinosuccinic acid that has been found in the urine of uremic patients. When guanidinosuccinic acid is injected into animals, it causes twitching, hemolysis of red blood cells, and gastroenteritis. Whether or not this compound is actually a toxin is not certain, but it is clear that it accumulates. To explain the mechanism of this accumulation, it has been suggested that high plasma levels of urea block the functioning of the urea cycle so that argininosuccinate accumulates. Because of this block, another pathway for handling arginino-guanido groups becomes operative. Argininosuccinate might be cleaved to yield guanidinosuccinate and the hydroxyl analogue of ornithine, as shown at the top of this page.

Low-protein diet. The purpose of the low-protein diet is to decrease the amount of ammonia and thus the amount of urea that the kidney must process. The source of the protein is not important so long as it is balanced in terms of all the amino acids, both essential and nonessential. A protein intake of about 0.3 g/kg of body weight · day is often sufficient, as nitrogen balance can be achieved at approximately 0.25 g/kg. Carbohydrate and lipid should be increased so that the amino acids need not be deaminated to provide carbon chains for energy production. A total of about 35 to 50 kcal/kg · day probably would be sufficient.

REFERENCES

Hoffman, W. S.: The biochemistry of clinical medicine, ed. 4, Chicago, 1970, Yearbook Medical Publishers, Inc.

Kassirer, J. P.: Clinical evaluation of kidney function—glomerular function, N. Engl. J. Med. 285:385, 1971.

Merrill, J. P.: Glomerulonephritis (three parts) N. Engl. J. Med. 290:257, 313, 374, 1974.

Merrill, J. P., and Hampers, C. L.: Uremia, N. Engl. J. Med. 282:953, 1014, 1970.

Pitts, R. F.: The role of ammonia production in regulation of acid-base balance, N. Engl. J. Med. 284:32, 1971.

Schreiner, G. E., and Maher, J. F.: Uremia, Springfield, Ill., 1961, Charles C Thomas, Publisher.

CASE 3: HARTNUP DISEASE

This disease was first classified when a 12-year-old boy, E. Hartnup, was admitted to a London hospital with a red, scaly rash and mild cerebellar ataxia. His mother thought that he was suffering from pellagra because the same symptoms in her older daughter had been diagnosed earlier as pellagra. It was clear that the boy did not have the usual *dietary*-deficiency form of pellagra, but aside from the symptoms mentioned, the only unusual aspect was an excessive excretion of free amino acids in the urine. When the older daughter had a recurrent attack of ataxia, she was brought to the hospital; her urine was also found to contain excessive amounts of amino acids. Two other siblings had the same aminoaciduria, whereas four were normal. The parents of these children were asymptomatic, but a family history revealed that they were first cousins.

Biochemical questions

1. Assuming that this disease is inherited in a simple mendelian way, what abnormality would account for the unusual amounts of aromatic and neutral amino acids in the urine?
2. Explain what is meant by two-dimensional paper chromatography. How would the analysis of urinary amino acids by this method be useful in differentiating this disease from other aminoacidurias? .
3. The parents of these children show normal amino acid excretion. On the basis of the results given, characterize the disease in terms of its dominance and whether it is X linked or autosomal.
4. What is the theoretical probability of these parents having normal, heterozygous, or homozygous children?
5. What is the relationship between the high levels of urinary excretion of the aromatic amino acids and the pellagra-like symptoms?
6. The symptoms of this disease resembling pellagra sometimes appear after patients have eaten diets rich in corn. Explain.
7. Most patients with Hartnup disease

outgrow the symptoms. It is surprising that the disease is so mild considering that the absorption of an oral load of an essential amino acid such as phenylalanine is only 25% of normal. Obviously phenylalanine and other amino acids must be absorbed in some other way or in some other form. Considering the specificity of the defect, in what other form might the amino acids be absorbed?

Case discussion

Hartnup disease is an aminoaciduria in which the genetic defect results in a reduced ability to transport several neutral and aromatic amino acids across cellular membranes. A defect in a single protein that is an essential part of a membrane transport system for this group of amino acids probably accounts for all the symptoms. While the disease is neither serious nor common, it is important because it has provided insight into certain aspects of amino acid transport, renal absorption, and protein digestion.

Amino acid transport. In patients suffering from Hartnup disease, transport in both the kidney tubule and intestine is defective. Because the disease presents a characteristic pattern of amino acid excretion, paper chromatography or amino acid analysis is used in the diagnosis. Almost all of the twenty protein amino acids can be separated by two-dimensional chromatography, a technique similar to the "fingerprinting" used to separate the tryptic peptides of hemoglobin S (case 1, Chapter 2). The chromatogram is developed with a solvent system of butanol, acetic acid, and water in one direction and with phenol-ammonia in a second direction at right angles to the first separation. This process will separate in two dimensions a mixture of amino acids that was spotted to one corner of the paper. Fig. 9-27 shows such a chromatogram. Using such a system, it was found that patients with Hartnup disease excrete five to twenty times more alanine, serine, threonine, leucine, isoleucine, valine, tyrosine, tryptophan, and phenylalanine than do controls. Several other amino acids are

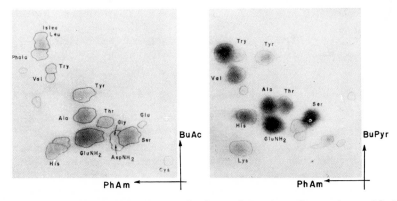

Fig. 9-27. Chromatograms of amino acids from the urine of a patient with Hartnup disease analyzed by two-dimensional paper chromatography. *Left*, BuAc is butanol:acetic acid:water; *right*, BuPyr is butanol:pyridine:water. Phenol:ammonia (PhAm), the chromatographic solvent used in the second dimension, was the same for both. (From Jepson, J. B.: Hartnup disease. In Stanbury, J. B., et al., editors: The metabolic basis of inherited disease, ed. 3, New York, 1972, McGraw-Hill Book Co.)

modestly elevated, but proline, methionine, and arginine are normal.

There seem to be three separate stereospecific transport systems for different groups of amino acids. This assumption is based on the failure of each system in three human genetic diseases. The systems are not absolutely specific, since they show some overlap for different amino acids. The transport system for the monoamino-monocarboxylate acids is defective in Hartnup disease. In *cystinuria* the transport of cystine, lysine, arginine, and ornithine is impaired, while in *glycinuria* the transport of glycine, proline, and hydroxyproline is defective.

In view of the excessive urinary losses of several essential amino acids, it is surprising that the effects of Hartnup disease are relatively benign. The explanation may be that these patients can absorb sufficient amounts of these amino acids as small peptides. It is evident that transport systems also exist for peptides and that these differ from transport systems for amino acids. Since the amino acid requirements are the highest during growth, symptoms are more apt to be evident during childhood.

Genetics. Because of lack of information, it cannot be said with certainty that the same allele is defective in every family whose members have Hartnup disease. However, every case appears to involve an autosomal recessive trait. In the case of E. Hartnup the disease was recessive, as evidenced by the fact that neither parent had symptoms. That his sister was afflicted also eliminates the possibility of the disease being X linked. If it were inherited in a strict mendelian way, half of the couple's children would be carriers, one fourth would have the disease, and one fourth would be normal. Unfortunately half of the children in this family had the disease. While these predictions are usually quite valid for large populations, they are by no means reliable for the distributions in a single family. Since the carrier state cannot be detected, it is impossible to tell if the children that do not have symptoms are normal or carriers.

Tryptophan metabolism and pellagra. The amino acid tryptophan can serve as a substrate for the synthesis of much of the body's niacin requirement. Undoubtedly the inability to transport dietary tryptophan and to reabsorb it in the kidney leads to a niacin deficiency that is responsible for the dermatologic and neurologic symptoms of pellagra. Pellagra appears sporadically as the result of infection or unusual diets (especially those consisting chiefly of corn). Corn has a very low tryptophan content, and a deficiency can occur unless adequate amounts of tryptophan are supplied by other dietary proteins. The symptoms of pellagra can be treated with niacin and a diet high in protein.

REFERENCES

Baron, D. N., Dent, C. E., Harris, E. W., and Jepson, J. B.: Hereditary pellagra-like skin rash with temporary cerebellar ataxia, constant amino aciduria, and other bizarre biochemical features, Lancet 2:421, 1956.

Jepson, J. B.: Hartnup disease. In Stanbury, J. B., Wyngaarden, J. B., and Fredrickson, D. S., editors: The metabolic basis of inherited disease, ed. 3, New York, 1972, McGraw-Hill Book Co.

CASE 4: INHERITED HEMOLYTIC ANEMIA

A 38-year-old-man had a history of anemia and intermittent episodes of jaundice since birth.

Laboratory findings included the following: hemoglobin values, 11.2 to 13.6 g/100 ml; reticulocytes, 4% to 14%; white cell count, 5500; platelet count, 160,000; and bilirubin, 1.4 mg/100 ml. The Coombs test was negative. Hemoglobin was demonstrated to be type AA by electrophoresis; hemoglobin F was 3.9%. Tests for heat-unstable hemoglobin were negative. A bone marrow examination revealed erythroid hyperplasia.

This patient's 36-year-old sister also had a history of anemia since childhood, and she had required transfusions during pregnancy. The laboratory findings in her case were similar to those for her brother.

Both brother and sister suffered from inherited hemolytic anemia, which was caused by a deficiency in the synthesis of glutathione (GSH) by the erythrocyte.

Biochemical questions

1. What is the function of GSH in the erythrocyte?
2. Explain the clinical manifestation of the rapid turnover rate of GSH in the red cells of this patient.
3. Write the series of reactions involved in GSH synthesis.
4. Hemolytic anemia resulting from GSH insufficiency might come about as a result of inherited defects in which enzymes?
5. Which enzyme is lacking in the article cited below. How was this proved? What was the assay?
6. How is this trait transmitted? Is it sex linked?

REFERENCE

Konrad, P. N., Richards, F., Valentine, W. N., and Paglia, D. E.: γ-Glutamylcysteine synthetase deficiency, N. Engl. J. Med. 286:557, 1972.

CASE 5: HEREDITARY HYPERAMMONEMIA

At 6 months of age an infant was admitted to the hospital for bronchiolitis. He weighed 6.5 kg and was eating well. Soon after discharge he began to vomit occasionally and ceased to gain weight. At the age of $8\frac{1}{2}$ months he was readmitted. There were no abnormal physical signs and the blood chemistry, urine microscopy, and x-ray examinations were normal. After 1 week he became habitually drowsy, his temperature rose to 39.4° C, his pulse was elevated at a rate of 136/min, and his liver was enlarged. All biochemical tests were normal, but the EEG was grossly abnormal. Since the infant could not retain milk given by gavage feeding, intravenous glucose was administered. He improved rapidly and came out of the coma in 24 hr. A paper chromatogram of his urine showed abnormally high amounts of glutamine and uracil, which suggested a high blood ammonium ion concentration. This suspicion was confirmed by laboratory measurement.

Biochemical questions

1. Hereditary hyperammonemia may be caused by a defect in either of two enzymes. Which are these?
2. Considering the data, which enzyme appears to be defective in this case? Explain your reasoning.

3. Why is the urine glutamine concentration elevated?
4. Although not successful in this case, the blood ammonium ion concentration can sometimes be lowered by treatment with arginine. What is the reason for this?
5. The patient was treated with a daily diet of 1.5 g of protein/kg of body weight. After 2 years on this diet, his height and weight were judged to be normal for his age. Discuss the effect of this diet on the growing child in terms of nitrogen balance.
6. Children afflicted with this disease are often alkalotic. Balance the equation for urea formation from carbon dioxide and ammonium ions. Considering the balanced equation, explain the alkalosis.
7. Offer a genetic explanation for the observation that this disease is often lethal in males but not in affected females.

REFERENCES

Alexander, G. M., et al.: Ornithine transcarbamylase deficiency, N. Engl. J. Med. 288:1, 1973.

Levin, B., Dobbs, R. H., Burgess, E. A., and Palmer, T.: Hyperammonaemia, Arch. Dis. Child. 44: 162, 1969.

CASE 6: METHYLMALONIC ACIDEMIA*

The patient was the mother of a child who had died of severe acidosis and dehydration at the age of 3 months. Diagnosis of methylmalonic acidemia was made posthumously. The patient was now pregnant and concerned about the possibility that the fetus might be affected with the same disease. At 19 weeks of gestation amniotic fluid cells were taken and cultured. The cells were assayed for propionate and succinate oxidation, vitamin B_{12} coenzyme synthesis, and methylmalonyl CoA mutase activity. Oxidation of propionate was found to be much lower than controls, whereas succinate oxidation was normal. Radioactivity from ^{57}Co-labeled vitamin B_{12} was found in methyl cobalamin but not in deoxyadenosyl-cobalamin. Methylmalonyl CoA mutase activity was measured in cell lysates, to which the necessary substrates had been added, and found to be normal. The patient's urine was found to contain large amounts of methylmalonic acid, which had not been present before pregnancy.

Biochemical questions

1. What amino acids produce propionyl CoA in the human? What would happen if a person ate a diet lacking these amino acids?
2. How can you explain the high concentration of methylmalonic acid in the urine when amniotic cells showed normal amounts of methylmalonyl CoA mutase? Assume that all metabolic defects are caused by the fetus and not by the mother.
3. Why was one B_{12} coenzyme labeled after administration of radioactive vitamin B_{12} and not the other coenzyme?
4. At 32 weeks of gestation the mother was given oral cyanocobalamin at 10 mg/day. At this time the mother was excreting approximately six times the normal amount of methylmalonate in her urine. One week after therapy was started, the methylmalonate concentration dropped to threefold above normal and by delivery at 41 weeks was almost normal. The infant was born in excellent condition; however, cultured skin fibroblasts from the infant displayed the same abnormalities as the amniotic cells. The infant was put on a low protein diet (1.5 g/kg/day) but was not further treated with vitamin B_{12}. Why was the low protein diet used, considering that the infant will be growing rapidly?
5. During the next 6 weeks, the infant cleared the vitamin B_{12} absorbed from the mother before birth. What effect would you predict this had on the methylmalonate concentration of the urine?

*Case description adapted from Ampola, M., et al.: N. Engl. J. Med. 293:313, 1975.

6. At the same time the protein in the diet was increased to 2.5 g/kg/day and vitamin B_{12} was administered intramuscularly for 11 days. Considering the enzymes involved, explain how high levels of vitamin B_{12} result in decreased excretion of methylmalonate.

7. At last report (19 months of age) the child was in the 75% percentile for weight, length, and head circumference, and development was normal. She is being maintained on a low-protein diet, and continuous vitamin B_{12} therapy has not been necessary. Do you think that this child is homozygous or heterozygous for methylmalonic acidemia? Explain.

8. Do you recommend that the low-protein diet be continued for the lifetime of this child? Explain. Should she eat several meals or three ordinary meals per day?

9. What would be the effect of a prolonged fast on this child?

10. If as an adult this girl became obese and desired to lose weight, how and for what reasons would you treat her differently from other obese patients?

REFERENCES

Ampola, M. G., Mahoney, M. J., Nakamura, E., and Tanaka, K.: Prenatal therapy of a patient with vitamin B_{12} responsive methylmalonic acidemia, N. Engl. J. Med. 293:313, 1975.

Rosenberg, L. E.: Disorders of propionate, methylmalonate, and vitamin B_{12} metabolism. In Stanbury, J. B., Wyngaarden, J. B., and Fredrickson, D. S., editors: The metabolic basis of inherited disease, ed. 3, New York, 1972, McGraw-Hill Book Co.

CASE 7: GLUTATHIONURIA*

As part of a routine screening of institutionalized individuals for defects in amino acid metabolism, a mildly retarded adult male was found to excrete abnormally large amounts of glutathione in his urine. Serum concentrations of glutathione were also abnormally high; however, both serum and renal concentration of individual amino acids were normal. γ-Glutamyl transpeptidase activity was not detected in fibroblasts cultured from a skin biopsy.

Biochemical questions

1. Describe the currently accepted role for γ-glutamyl transpeptidase in humans.

*Case description adapted from Schuhman, J. D., and associates.

2. What is the γ-glutamyl cycle? Illustrate your answer with equations.

3. Offer explanations for the lack of serious health problems in this patient. Consider enzyme stability, peptide transport, and alternative functions for this enzyme.

REFERENCES

Binkley, F., and Wiessman, M. L.: Glutathione and γ-glutamyl transferase in secretory processes (a mini-review), Life Sci. 17:1359, 1975.

Mathews, D. M.: Intestinal absorption of peptides, Physiol. Rev. 55:537, 1975.

Schuhman, J. D., et al.: Glutathionuria: inborn error of metabolism due to tissue deficiency of γ-glutamyl transpeptidase, Biochem. Biophys. Res. Commun. 65:68, 1975.

CASE 8: HOMOCYSTINURIA

Skin fibroblasts were cultured from a 9-year-old retarded boy who exhibited dislocated optic lenses and markedly elevated methionine and homocystine concentrations in both plasma and urine and undetectable plasma cystine. After 2 weeks of treatment with pyridoxine, the plasma concentrations of homocystine and methionine fell to normal. The boy was diagnosed as having a defect in cystathionine synthetase. The cultured fibroblasts were used in the purification of this enzyme, which was compared to enzyme prepared from fibroblasts cultured from normal young adult males.

Biochemical questions

1. What is the reaction catalyzed by cystathionine synthetase?

2. What is the metabolic fate of cystathionine?

3. Why is homocystine overproduced when it is not the substrate of the defective enzyme?

4. The enzyme from control extracts was purified fivefold to sevenfold with a 70% to 79% yield. Using the same technique, the mutant enzyme was purified only twofold to threefold with a 26% to 44% yield. Offer explanations for these results.

5. Pyridoxal phosphate functions as a coenzyme for a wide variety of biochemical reactions involving amino acids. List some of the different types of reactions.

6. Predict the change in the Michaelis constant of the mutant cystathionine synthetase.

7. Even at concentrations of pyridoxal phosphate that are saturating for the enzyme, the mutant synthetase has only 3% of the activity of the control enzyme. What does this say about the V_{max} of the enzyme?

REFERENCES

Kim, Y. J., and Rosenberg, L. E.: On the mechanism of pyridoxine responsive homocystinuria. II. Properties of normal and mutant cystathionine B-synthase from cultured fibroblasts, Proc. Natl. Acad. Sci. U.S.A. **71:**4821, 1974.

Rosenberg, L. E., and Scriber, C. R.: Disorders of amino acid metabolism. In Bondy, P. K., and Rosenberg, L. E., editors: Duncan's diseases of metabolism, ed. 7, Philadelphia, 1974, W. B. Saunders Co., p. 465.

ADDITIONAL QUESTIONS AND PROBLEMS

1. In calculating nitrogen balances, losses of nitrogen via flaking epidermis and sweat are sometimes ignored. How significant are these losses?

2. Hyperammonemia caused by liver damage is associated with disturbances of cerebral function. Arginine and glutamic acid have been used in the treatment of this condition with variable results. Protection from ammonia intoxication with N-carbamyl glutamate and arginine in rats suggests that these might be useful in the human disease (Proc. Natl. Acad. Sci. U.S.A. **69:**3530, 1972.) N-Carbamyl glutamate is a compound similar to N-acetyl glutamate. What is the biochemical basis for the protective effect of these amino acids on ammonia intoxication?

3. Why are amino acid analogues of limited usefulness in the treatment of cancer?

4. How are D-amino acids metabolized? Where do they occur in nature?

5. What is the origin of elevated levels of vanillylmandelate in the urine?

6. Bacitracin is a peptide antibiotic. It can be administered orally or intramuscularly. What route of administration would you predict to be more effective? Why?

7. Considering the structural similarity of diphenylhydramine (p. 456) and epinephrine, what kind of side effects might be produced by the antihistamine?

8. Thienylalanine is a growth inhibitor of some bacteria and a structural analogue of phenylalanine. Phenylalanine will competitively reverse the growth inhibition, whereas glycylphenylalanine will noncompetitively reverse inhibition by thienylalanine. Explain these data in terms of amino acid and peptide transport.

9. Extracts prepared from biopsied liver of a child with an unusual form of phenylketonuria showed only a modest decrease in phenylalanine hydroxylase activity but no detectable dihydrobiopterin reductase activity. (Kaufman, S., et al.: N. Engl. J. Med. **293:**785, 1975.) How could lack of this enzyme cause phenylketonuria? What are other reactions that might be affected? Consider reactions that use tetrahydrobiopterin as a coenzyme. What symptoms might be expected as a result of the reduced activity of these enzymes?

10. The keto forms of valine, isoleucine, and leucine were administered to a patient with hyperammonemia. Explain why the plasma ammonia level fell from 0.050 to 0.028 mM (Batshaw, M., et al.: N. Engl. J. Med. **292:**1085, 1975).

CHAPTER 10

STEROL AND STEROID METABOLISM

OBJECTIVES

1. To describe the structure of the sterols and steroids that are important in normal metabolism and in disease states
2. To discuss the biosynthesis, metabolism, and excretion of cholesterol, the bile acids, and the steroid hormones
3. To consider the structure and function of cellular membranes
4. To consider the role of cholesterol in the development of atherosclerosis and the relationship of hypercholesterolemia and dietary fat intake in this disease
5. To discuss the function of the bile and the relationship of cholesterol to gallstone formation

Cholesterol is the major sterol in the human body. It is a structural component of cell membranes and plasma lipoproteins, and it is also the starting material from which bile acids and steroid hormones are synthesized. The sterols and their derivatives are poorly soluble in aqueous solution but quite soluble in organic solvents, particularly alcohols. Therefore, these compounds are classified as lipids, as has been noted previously in Table 8-1 (p. 364).

An abnormality in either cholesterol metabolism or transport through the plasma appears to be related to the development of atherosclerosis, the form of hardening of the arteries that can lead to myocardial infarction, stroke, aneurysm, or gangrene. In addition, the gallstones that occur most commonly in inhabitants of Western nations are made up predominantly of cholesterol. Finally, many commonly used drugs, for example, cortisone derivatives and oral contraceptives, are steroids. Therefore, health science students should have a basic understanding of sterol and steroid chemistry and be familiar with the metabolic pathways of these substances in man.

STEROID CHEMISTRY

Steroids are derivatives of the perhydrocyclopentanophenanthrene ring system.

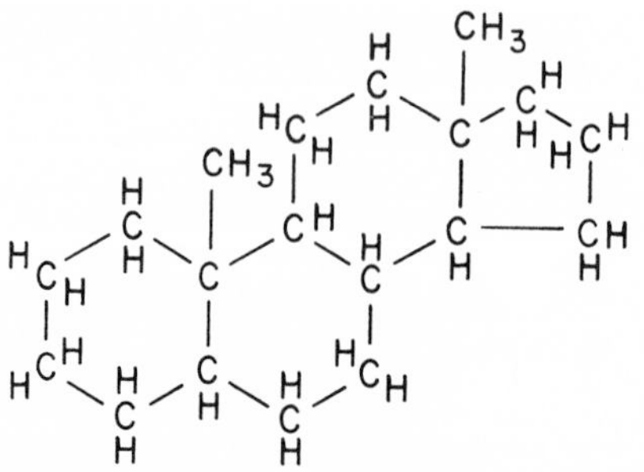

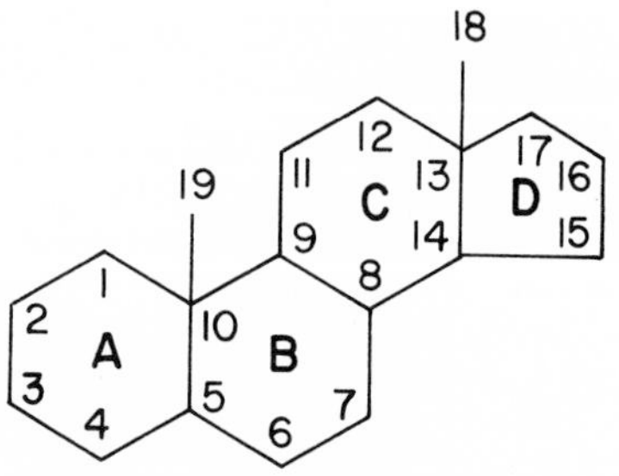

The complete structure, including carbon and hydrogen atoms, is shown at the left; the commonly used line drawing in which carbon and hydrogen atoms are omitted is shown for comparison at the right. In the line drawing the four rings are identified by the letters A to D, and the nineteen carbon atoms are identified by number. Three of the rings contain six carbon atoms, whereas the D ring contains five. An angular methyl group, C_{19}, is attached at the junction of the A and B rings, and a second angular methyl group, C_{18}, is attached at the junction of the C and D rings.

The steroid nucleus has a conformation that is approximately planar. The side groups such as the angular methyl groups at C_{10} and C_{13} that are above the plane of the rings, as shown below, are designated as being in a β-configuration. This is indicated by their connection to the ring structure with a solid line. Those side groups that project below the plane of the rings are designated as being in an α-configuration, indicated by their connection to the ring structure with a dashed line. The hydrogen atoms or hydroxyl groups that are attached to the ring carbon atoms also project either above or below the plane of the rings and are also designated as being in either an α- or a β-configuration.

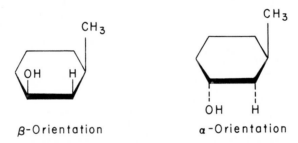

The angular methyl groups, C_{18} and C_{19}, are always in the β-configuration. Therefore, the hydrogens and hydroxyl groups that are in α-orientation in the above structures actually are in *trans* configuration relative to these angular methyl groups, which by convention are written in the β-configuration.

Sterols

Sterols are a class of steroids that contain a hydroxyl group at C_3 and an aliphatic chain of at least eight carbon atoms at C_{17}.

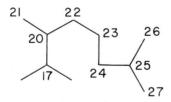

Cholesterol is the main sterol in human tissue. It has an eight-carbon atom hydrocarbon chain that is numbered 20-27 as a continuation of the steroid nucleus. Some of the cholesterol that is present in man is esterified; that is, the hydroxyl group that projects from C_3 is attached to a fatty acid residue, usually containing sixteen or more carbon atoms, in ester linkage. The hydroxyl group in cholesterol is β-oriented, as indicated by the solid line between it and C_3. There is a double bond in the B ring at position 5,6. The angular methyl groups at positions 10 and 13 as well as the eight-carbon atom hydrocarbon chain are also in a β-configuration.

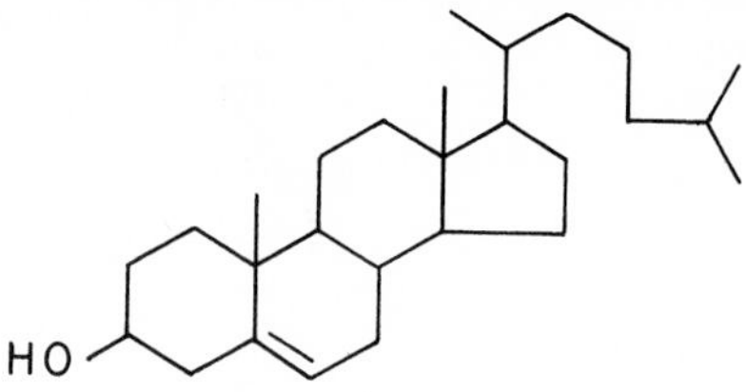

Cholesterol

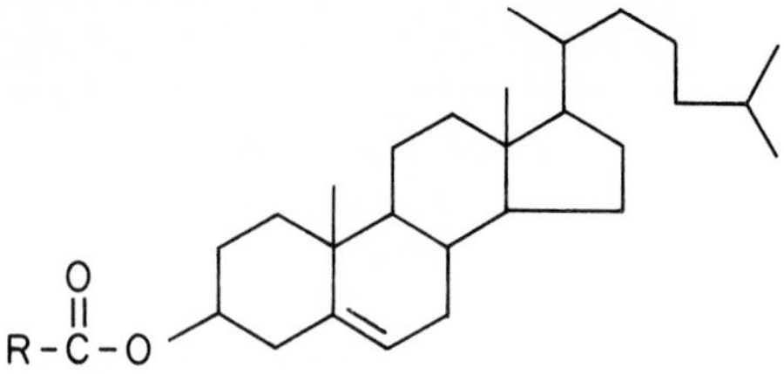

Cholesteryl ester

Foods that are derived from animal products contain cholesterol. Those foods that are particularly rich in cholesterol include eggs, dairy products such as butter, cheese, and cream, and most meats (Chapter 1). Some of the cholesterol that is contained in these animal products is in the form of cholesteryl esters. Therefore, the ordinary diet contains a mixture of cholesterol and cholesteryl esters.

The esterified fatty acid moieties of cholesteryl esters usually contain sixteen to twenty carbon atoms and in many cases are either monoenoic or polyenoic. Among the most abundant of the esters of cholesterol in human tissues are cholesteryl oleate and cholesteryl linoleate. These compounds are present in appreciable amounts in the plasma lipoproteins, particularly low-density lipoproteins (LDL) and high-density lipoproteins (HDL), and in the adrenal cortices and liver. They are the most abundant form of lipid that accumulates in the arterial wall in atherosclerotic lesions. Cholesteryl esters are the most nonpolar lipids that occur in human tissues, and they function as a storage form of sterol. Unlike cholesterol itself, cholesteryl esters do not exchange readily between cell membranes and plasma lipoproteins or among the various classes of plasma lipoproteins.

Plant sterols.   Plants do not contain cholesterol; they have other sterols that are known as *phytosterols*. The most abundant of the phytosterols is β-sitosterol. Although the ring structure of β-sitosterol is identical to that of cholesterol, the chain attached to C_{17} contains ten rather than eight carbon atoms, for there is an ethyl group attached to C_{24} of the chain.

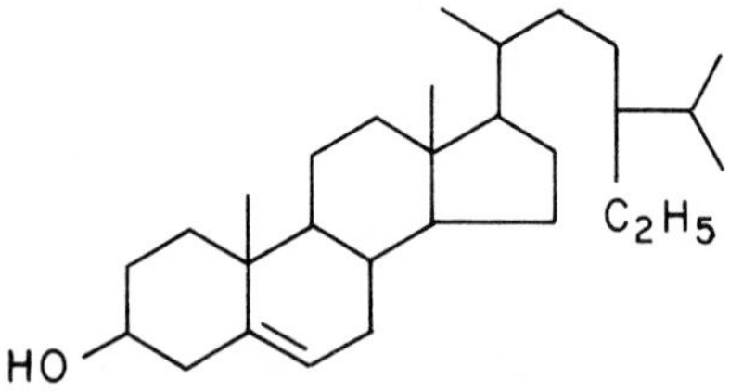

β-Sitosterol

CELL MEMBRANES

Biologic membranes separate cells from their external environment and divide the interior of the cell into compartments. They are 75 to 90 Å thick. The chemical composition of cell membranes varies widely, as is shown in Table 10-1. As a general estimate, a representative membrane can be thought of as being made up of about 50% protein, 45% lipid, and 5% carbohydrate. Approximately 10% of the membrane proteins are glycoproteins. A major exception to this general chemical composition is the myelin sheath of nerves, one of the most widely studied biologic membranes. Myelin is made up of 20% protein and 75% lipid.

Lipid composition.   The lipid composition of human erythrocyte membranes and myelin, two of the most intensively studied human membrane preparations, are

Table 10-1. Composition of cell membranes

MEMBRANE	PERCENTAGE COMPOSITION BY WEIGHT		
	Protein (%)	Lipid (%)	Carbohydrate (%)
Myelin	20	75	5
Erythrocyte	49	43	8
Hepatocyte plasma membrane	54	39	7
Outer mitochondrial membrane	50	46	4
Inner mitochondrial membrane	75	23	2

Table 10-2. Membrane lipid composition

LIPID	ERYTHROCYTE MEMBRANE		MYELIN	
	Weight (%)	Molar ratio	Weight (%)	Molar ratio
Phospholipids	69	1.0	43	1.0
Cholesterol	26	0.8	27	1.3
Glycosphingolipids	5	0.1	30	0.7

compared in Table 10-2. The main lipid components are phospholipids, cholesterol, and glycosphingolipids. There is a higher percentage of glycosphingolipids in myelin than in the erythrocyte or other membranes. The erythrocyte membrane and myelin are both *plasma membranes*, that is, they are located at the cell surface and separate the extracellular fluid from the cytoplasm. As described on pp. 102 to 105, mammalian cells also contain a number of intracellular membraneous structures. The different membranes within a cell vary in composition. For example, mitochondrial membranes contain almost no cholesterol and are the only mammalian membranes that contain cardiolipin. Membranes of different cell types also vary widely in lipid composition, as illustrated by the difference between the erythrocyte and myelin.

Phospholipid composition

As shown in Table 10-2, the main lipids in myelin and the erythrocyte membranes are phospholipids. This is true of all biologic membranes, for the basic structure of membranes in a bilayer composed of phospholipids is shown in Fig. 10-1. In this arrangement the hydrocarbon chains of the phospholipid fatty acid groups project into the center of the bilayer and are shown schematically in Fig. 10-1 as wavy lines. The hydrophilic glyceryl-phosphoryl–base components of the phospholipids are located on the outsides of the bilayer where they interact with water or other polar and charged molecules. The lipid bilayer is composed of two leaflets, the outer phospholipid leaflet that faces the extracellular fluid and the inner phospholipid leaflet that faces the cytoplasm. The membrane lipid bilayer contains a mixture of phospholipids as is shown in Table 10-3, which gives the phospholipid composition of normal human erythrocyte membranes. Both lecithin and sphingomyelin are phospholipids that contain the base choline, and it is important to note that together they comprise about 50% of the human erythrocyte membrane phospholipids.

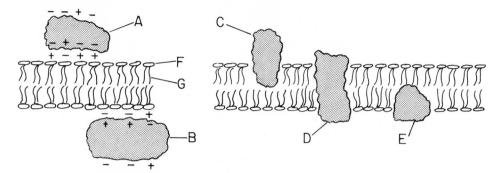

Fig. 10-1. Membrane proteins. Peripheral membrane proteins are illustrated by **A** and **B** in the left drawing. They are illustrated schematically as sitting on top of the polar head groups of the membrane phospholipid bilayer, on both the extracellular fluid and cytoplasmic surfaces. Electrostatic interactions are primarily responsible for their binding to the lipid bilayer. Integral membrane proteins are illustrated schematically by **C**, **D**, and **E** in the right drawing. They are embedded in the lipid bilayer, the primary binding forces being nonpolar interactions with the fatty acyl chains of the phospholipids in the bilayer. Integral proteins either penetrate only part of the distance through the bilayer, **C** and **E**, or completely across it, **D**. Proteins such as **D**, which span the bilayer, are known as transmembrane proteins. The phospholipid polar head groups in the lipid bilayer are illustrated as small circles, **F**, and their fatty acyl tails are illustrated by wavy lines, **G**.

Table 10-3. Phospholipid composition of the human erythrocyte membrane

PHOSPHOLIPID	PERCENTAGE OF MEMBRANE TOTAL PHOSPHOLIPIDS (%)
Lecithin	28
Phosphatidylethanolamine	26
Phosphatidylserine	13
Phosphatidylinositol	1
Phosphatidic acid	2
Sphingomyelin	25
Minor components	5

This distribution of phospholipids is quite similar in erythrocytes taken from different people and from the same person at different times during his life span.

The fatty acyl group compositions of the various classes of membrane phosphoglycerides are somewhat different. For example, about 50% of the fatty acids of the human erythrocyte lecithins are saturated, whereas only about 30% of the fatty acids in the erythrocyte phosphatidylethanolamines are saturated. Moreover, the fatty acid composition of the erythrocyte phosphoglycerides can vary in response to dietary changes. This observation is useful in determining whether a patient is adhering to certain modified diets. For example, one form of diet that is prescribed for the treatment of certain types of hyperlipoproteinemia is high in polyenoic fatty acids. If a patient who has been eating an ordinary diet is given this modified diet, the fatty acid composition of his or her erythrocytes will be altered within 2 to 4 months. The erythrocyte fatty acid composition can be measured by gas liquid chromatography (as described in Chapter 8) in order to determine whether the patient actually is following the prescribed dietary modification.

Protein composition. The subject of membrane proteins is extremely compli-

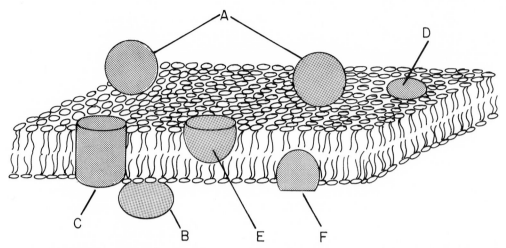

Fig. 10-2. The fluid mosaic model of membrane structure. A segment of the lipid bilayer is illustrated schematically in a three-dimensional projection. Various types of proteins are associated with the lipid bilayer. They are, **A,** peripheral proteins associated with the extracellular fluid surface of the bilayer; **B,** a peripheral protein associated with the cytoplasmic surface of the bilayer; **C,** a transmembrane integral protein viewed in cross-section; **D,** the exposed surface of an integral protein viewed from the extracellular fluid surface of the lipid bilayer; **E,** an integral protein exposed on the extracellular fluid surface that does not pass entirely through the bilayer, viewed in cross-section; and **F,** an integral protein exposed on the cytoplasmic surface that also does not pass through the entire bilayer, viewed in cross-section.

cated. As an initial approximation, they can be divided into two general types — peripheral and integral. *Peripheral* proteins are bound loosely to the membrane and can be removed by mild treatment with solutions high in ionic strength or chelating agents such as ethylenediamine tetraacetate. They are soluble in aqueous solution and do not contain tightly adherent lipid. Peripheral proteins comprise about 30% of the membrane proteins. The remainder of the membrane proteins are classified as the *integral* proteins. They are tightly bound and are removed only by such drastic treatments as with organic solvents or detergents. Lipid adheres to the integral proteins when they are removed from the membrane, and many of them appear to be true lipoprotein complexes. Integral proteins that are isolated from membranes are insoluble when they are introduced into aqueous media. Many different individual proteins make up each of these two classes of membrane proteins, and their molecular weights vary widely. The relationship of peripheral and integral proteins to the lipid bilayer of the membrane is illustrated schematically in Fig. 10-1.

FLUID MOSAIC MODEL OF MEMBRANE STRUCTURE

Singer and Nicolson have proposed the most widely accepted explanation of membrane structure, the fluid mosaic model. This model is illustrated schematically in Fig. 10-2. According to this model, the phospholipid bilayer forms the matrix of the membrane. This is composed of two rows of phosphoglycerides having their fatty acyl groups pointed toward each other and their phosphorylbase head groups oriented outward at the extracellular and cytoplasmic surfaces. Therefore, the inside of the bilayer is composed of nonpolar fatty acyl hydrocarbon chains while the outsides that interact with aqueous media contain the polar phospholipid head groups. At body temperature the inside of the lipid bilayer is in a fluid physical state analogous to an oil droplet. If the membrane is cooled, the lipids will pass from the liquid to the solid state. The temperature at which the change from the liquid to the crystalline physical state occurs is known as the *transition temperature.* The transition temperature depends primarily on the fatty acyl group composition of the phospholipid bilayer. Phospholipid molecules can move laterally within each of the leaflets of the lipid bilayer very rapidly. Thus a phospholipid molecule can exchange places with the one on either side of it within fractions of a second. In this process the phosphoglyceride remains within the same leaflet of the lipid bilayer, and it does not cross from the extracellular fluid half to the cytoplasmic half of the bilayer structure. The opposite process, movement of a phospholipid molecule from the extracellular fluid to the cytoplasmic leaflet of the bilayer, known as *flip-flop,* occurs very slowly. It is very unfavorable energetically to move the polar phospholipid head group through the central nonpolar hydrocarbon region, a process that would be required for the phospholipid to cross to the opposite side of the bilayer.

Proteins are interspersed in the lipid bilayer as shown in Fig. 10-2, producing a mosaic effect. Peripheral proteins are attached relatively loosely on both the extracellular fluid and cytoplasmic surfaces of the lipid bilayer. They are attached primarily through electrostatic interactions and hydrogen bonding with the polar head groups of the phospholipids. By contrast, the integral proteins are embedded within the lipid bilayer. Integral proteins that penetrate completely through the bilayer are known as transmembrane proteins. They probably are the transport carriers for ions as well as water-soluble substrates such as glucose. Transmembrane proteins are exposed on both the extracellular fluid and cytoplasmic surfaces of the bilayer and, in this way, can form a channel through the lipid phase. The portions of these proteins that are exposed to the aqueous environment on both surfaces of the membrane are composed of polar amino acid side chains such as glutamate and serine, whereas the central segment of the protein that exists within the lipid phase contains primarily nonpolar amino acid side chains such as valine and leucine. Some integral proteins penetrate only partially into the bilayer, being exposed on either the extracellular fluid or cytoplasmic surfaces. These proteins include certain hormone receptors and membrane-bound enzymes. The proteins are not fixed in position within the lipid bilayer. Rather, they have lateral mobility in two dimensions so that they are free to diffuse from place to place within the plane of the bilayer. The proteins also rotate about their longitudinal axes at high speeds. What the proteins cannot do, however, is tumble from one side of the lipid bilayer to the other. In other words, the surface of the protein exposed to the extracellular fluid probably does not roll over and become exposed to the cytoplasm.

Asymmetry. Recent studies using cross-linking reagents and enzymes that degrade membrane constituents have indicated that biologic membranes are asymmetric. This is true for the protein, carbohydrate, and lipid components of the membranes. Different peripheral proteins are present on the two surfaces of the lipid bilayer, and the integral proteins that only partially penetrate into the bilayer are different on the cytoplasmic and extracellular fluid surfaces. For example, the enzyme adenyl cyclase is an integral protein that appears to be located on the cytoplasmic side of the bilayer, whereas the hormone receptor that is involved in activating this enzyme is on the extracellular fluid side. Likewise, the transmembrane proteins such as the large subunit of the Na^+/K^+-adenosine triphosphatase (ATPase) are asymmetric, with the Na^+- and ATP–binding sites located on the surface exposed on the cytoplasmic side and the K^+- and ouabain- (an inhibitor) binding sites located on the region exposed at the extracellular fluid surface. The glycolipids and the carbohydrate chains of the glycoproteins are oriented so that they project out into the extracellular fluid. Furthermore, the lipid bilayer itself is asymmetric. Phosphatidylcholine and sphingomyelin are concentrated to a greater extent in the leaflet of the bilayer that faces the extracellular fluid. Conversely, phosphatidylethanolamine, phosphatidylserine, and phosphatidylinositol are concentrated in the leaflet that faces the cell cytoplasm. Studies with the human erythrocyte indicate that cholesterol is more concentrated in the extracellular fluid leaflet of the lipid bilayer than in the leaflet that faces the cell cytoplasm.

Fluidity and permeability.   At body temperature the lipid bilayer exists in a fluid state. The more fluid the bilayer, the more permeable is the membrane. Unsaturated fatty acids present in the membrane phospholipids increase the fluidity of the membrane and make it more permeable. By contrast, saturated fatty acids decrease the fluidity and permeability of the membrane. Cholesterol modulates the fluidity, decreasing it in regions of the membrane that contain a great deal of unsaturated fatty acid and increasing fluidity in the regions that are composed primarily of saturated fatty acids. Therefore, cholesterol can be considered as a modulator of membrane fluidity and therefore of permeability.

Cholesterol is inserted in the lipid bilayer between phospholipid molecules. Its hydroxyl group is oriented toward the aqueous interface and interacts with the polar head groups of the phospholipids. The nonpolar ring structure and hydrocarbon tail of cholesterol are positioned so that they interact with the hydrocarbon chains of the phospholipid fatty acyl groups. The amount of cholesterol contained in the various cell membranes differs considerably. For example, cholesterol comprises about 25% of the lipids by weight in the plasma membrane, but it is not present in the inner mitochondrial membrane. Likewise, all of the cholesterol in the plasma membrane is in the free or nonesterified form, whereas both cholesterol and cholesteryl esters are contained in the endoplasmic reticulum.

Phospholipid exchange.   Experiments with cell homogenates indicate that phospholipid molecules contained in one membrane within a cell can be transferred to another membrane. For example, radioactive lecithin is transferred between microsomes and mitochondria if these subcellular particles are incubated together. This exchange process is catalyzed by one or more proteins contained in the cell cytoplasm. The physiologic role of phospholipid exchange between membranes is presently uncertain. The de novo synthesis of phospholipids (p. 401) is thought to occur primarily in the endoplasmic reticulum. Therefore, the experimentally observed exchange reaction may be a mechanism whereby newly synthesized phospholipid molecules are transferred from their site of synthesis to other membranes within the cell. Phospholipids also exchange between plasma lipoproteins and the plasma membranes of cells. As described on p. 478, cholesterol also exchanges between lipoproteins and cell membranes.

CHOLESTEROL IN PLASMA LIPOPROTEINS

Cholesterol serves two functions in the plasma lipoproteins. One is a structural function. Because it contains both hydrophobic and hydrophilic groups, cholesterol serves to help solubilize the main nonpolar material that must be transported by the lower density lipoproteins. Together with the phospholipids and protein, it is located at or near the aqueous-lipid interfaces of the micellar structures. The 3-β-hydroxyl group interacts with the aqueous medium, whereas the sterol nucleus and side chain interact with the core of nonpolar lipids. Little unesterified cholesterol is contained in high-density lipoproteins (HDL), the lipoproteins that are rich in phospholipids and protein.

In addition to serving as a structural component, cholesterol also is one of the lipids that is actually transported by the plasma lipoproteins. Transport cholesterol is present predominantly as cholesteryl esters. These are located within the lipid core in the various lipoproteins. Much of the transport cholesterol actually arises in the plasma through the action of lecithin-cholesterol acyltransferase, an enzyme that acts in the plasma to transfer a fatty acid residue from lecithin to cholesterol (p. 486). The conversion of the free cholesterol that was required for the lipoprotein structure into the esterified form is associated with triglyceride removal and catabolism of VLDL and possibly of chylomicrons. Cholesteryl esters formed in this way must be transported to tissues where they can be taken up and hydrolyzed. This is thought to occur primarily in the liver.

Cholesterol exchange.   Unesterified cholesterol contained in human lipoproteins and the membrane of the human erythrocyte is exchangeable. In other words, if human lipoproteins containing radioactive cholesterol are mixed in an incubation flask with erythrocytes, the radioactive cholesterol will rapidly begin to distribute itself between the erythrocyte membranes and the lipoproteins. Unesterified cholesterol also exchanges between two lipoproteins, for example, between LDL and HDL. Results with subcellular fractions and intact cells in tissue culture indicate that much of the cholesterol contained in many membranes other than the erythrocyte membrane also is exchangeable. On the other hand, cholesteryl esters do not exchange rapidly between lipoprotein classes and are not transferred in large quantities from lipoproteins to cell membranes. However, cholesteryl esters probably are transferred between VLDL's and HDL's during VLDL catabolism. In this process, VLDL triglycerides are transferred to HDL. These lipid transfers as well as unesterified cholesterol transfer occur during collisions between two lipoproteins or between lipoproteins and cell membranes. Although the significance of cholesterol exchangeability is not completely understood, it serves to point out the dynamic state in which cholesterol exists in cell membranes and plasma lipoproteins.

DIETARY CHOLESTEROL
Absorption

Man can readily absorb cholesterol contained in his diet. Most people in Western societies eat between 600 and 1000 mg/day of cholesterol and absorb from 300 to 400 mg/day. When the dietary intake is relatively small, absorption is quite efficient. However, when the intake exceeds approximately 500 mg/day, cholesterol absorption becomes less efficient, and only about 30% to 35% of the intake is absorbed. If one is reasonably cautious and avoids foods that are rich in cholesterol, it is relatively easy to reduce the dietary cholesterol intake to about 400 to 500 mg/day. However, about 200 to 300 mg/day will nonetheless be absorbed under these conditions. In order to reduce absorption further, it is necessary to lower the dietary intake of cholesterol to the range of between 100 and 300 mg/day. This requires severe dietary restrictions, for the average person consumes at least 300 to 400 mg/day of cholesterol contained in meat.

Unlike cholesterol, plant sterols are absorbed very poorly by man. Indeed, feeding of large quantities of plant sterols such as β-sitosterol inhibits cholesterol absorption. This fact has been used clinically, for β-sitosterol was administered in the past to patients with *hypercholesterolemia* (excessive blood plasma cholesterol levels) in an attempt to reduce cholesterol absorption.

Digestion

Most of the cholesterol contained in the diet is present as the free sterol. Any esterified cholesterol is hydrolyzed within the intestinal lumen by an enzyme secreted in the pancreatic juice, *cholesterol esterase*. All of the dietary cholesterol is incorporated into micelles that are formed from constituents present in the bile. These micelles contain *conjugated bile acids* and phospholipids in addition to cholesterol. Emulsification is necessary because cholesterol is poorly soluble in the chyme, the aqueous medium that is present in the intestinal lumen. It must be brought into a physical state suitable for uptake by the intestinal mucosa. Hydrolysis of dietary cholesteryl esters by cholesterol esterase occurs on or within the micelle. Cholesterol is absorbed from the micelles by diffusion into the mucosal cells, where much of it is subsequently reconverted into cholesteryl esters. Cholesterol absorption occurs mostly in the ileum.

The cholesteryl esters that are synthesized in the mucosal cells, together with some unesterified cholesterol, are incorporated into large lipid-protein particles that are released into the lymph. These particles are called *chylomicrons*, and they transport cholesterol as well as other dietary lipids into the plasma from the lymphatic vessels via the thoracic duct (Chapter 8). Eventually the cholesterol is deposited in the tissues, mostly in the liver. Cholesterol digestion, absorption, and excretion are illustrated schematically in Fig. 10-3.

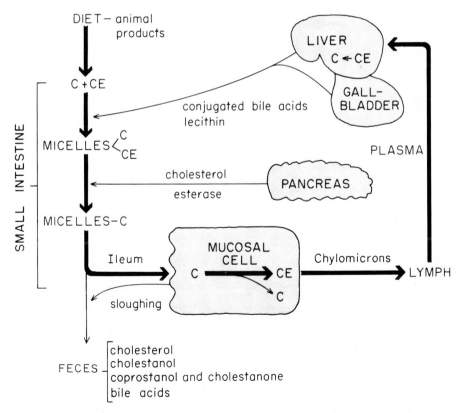

Fig. 10-3. Digestion and absorption of cholesterol. C, Cholesterol; CE, cholesteryl esters. Mucosal cells constantly degenerate and are discharged into the intestinal lumen. This is noted as "sloughing."

Excretion

Cholesterol is excreted in the feces. Cholesterol is delivered into the intestine in the bile, and additional amounts are derived from the sloughed intestinal mucosal cells. Moreover, some dietary cholesterol is excreted without being absorbed. Some of the cholesterol in the bowel is acted upon by intestinal bacterial enzymes and converted to other *neutral sterols* before excretion in the feces. In man the main neutral sterol products in the stool are coprostanol and cholestanone. Another neutral sterol reduction product that is excreted in the feces is cholestanol. It is a metabolic product of cholesterol that is formed in the liver and delivered into the intestine in the bile. Coprostanol and cholestanol are isomers, the only difference being that the hydrogen atom between the A and B rings has a β-orientation in coprostanol and an α-orientation in cholestanol.

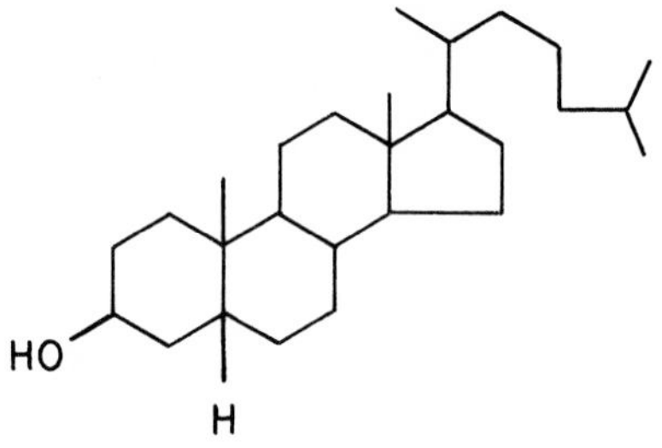

Coprostanol 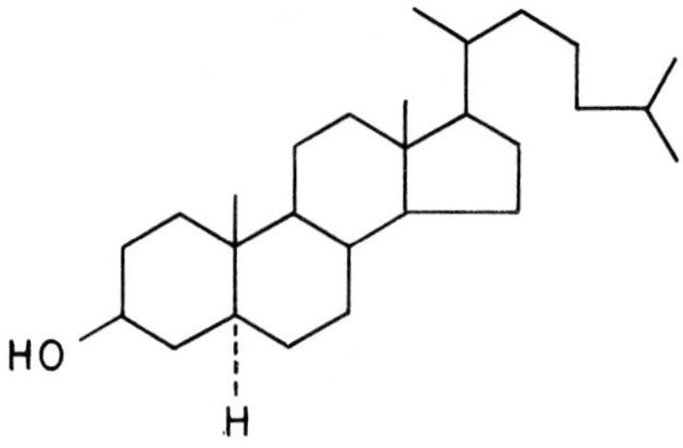   Cholestanol

BILE ACIDS

The main metabolic products of cholesterol in terms of the amounts formed are the bile acids. The bile acids in man are cholic, chenodeoxycholic, deoxycholic, and lithocholic acids.

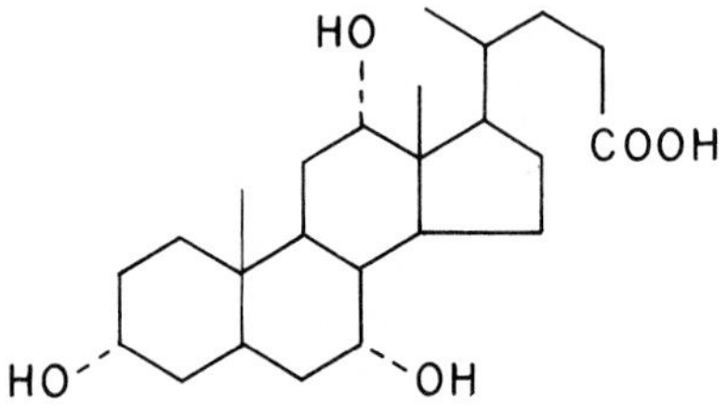

Cholic acid 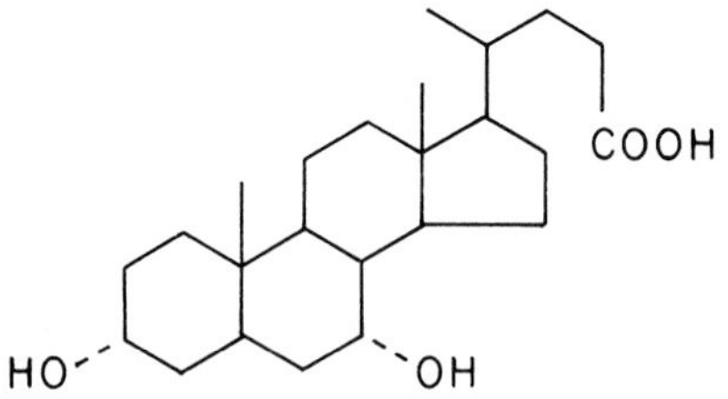   Chenodeoxycholic acid

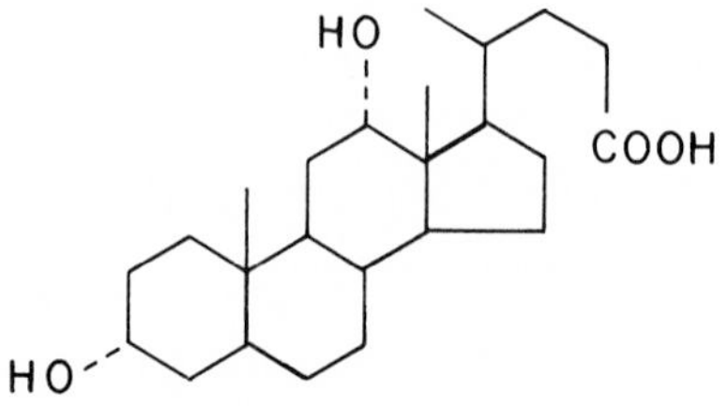

Deoxycholic acid 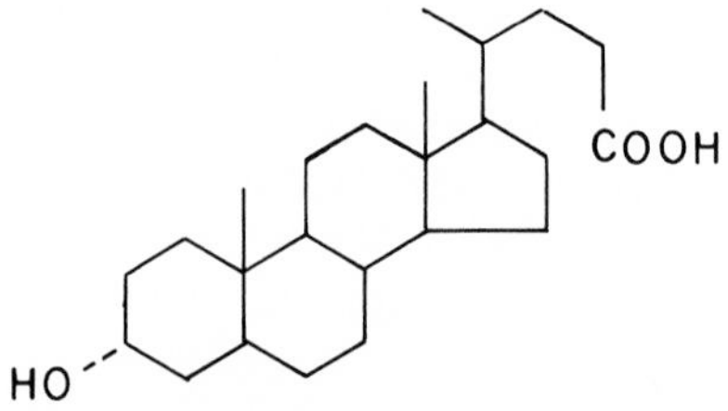   Lithocholic acid

Bile acids contain twenty-four carbon atoms; the terminal three carbon atoms of the cholesterol side chain are removed during their synthesis from cholesterol. The double bond that is present in the B ring of cholesterol is also reduced in the synthetic process. The four bile acids differ only in the number of hydroxyl groups that are attached to the steroid nucleus. Cholic acid has three hydroxyl groups, and they are attached to the carbon atoms at positions 3, 7, and 12. Chenodeoxycholic and deoxycholic acids have two hydroxyl groups, and lithocholic acid has only one. Notice that all of the hydroxyl groups are in the α-configuration.

Cholic and *chenodeoxycholic* acids are *primary* bile acids; that is, they are synthesized from cholesterol in the liver. Deoxycholic and lithocholic acids are *secondary* bile acids; they are synthesized in the intestinal lumen from the primary bile acids by the action of intestinal bacteria. Both types of bile acids are condensed with either glycine or taurine in the liver to form glyco- or tauro-conjugated bile acids. Both glycine and taurine are joined in amide linkage with the carboxyl group (C_{24}) of the five-carbon atom chain attached to C_{17} of the steroid nucleus.

Glycine conjugate   Taurine conjugate

Synthesis of the primary bile acids

The pathways for bile acid synthesis are shown in Fig. 10-4. Either one or two hydroxyl groups are added to the steroid nucleus in an α-configuration. The rate-limiting reaction in the pathway is the insertion of the first hydroxyl group, which occurs at position C_7 of the cholesterol nucleus. This reaction is catalyzed by the enzyme *cholesterol 7α-hydroxylase*. The 3-β-hydroxyl of cholesterol then is converted into a 3-α-hydroxyl group. A 3-keto group is an intermediate in this isomerization reaction. The double bond of the cholesterol B ring is reduced, and the hydrogen on C_5 introduced in this reduction is in β-orientation; that is, it is in *cis* configuration relative to the methyl group at C_{19}. Next, the terminal carbon atom of the cholesterol side chain is oxidized and then attached to CoASH in thioester linkage. The C_{24} group also undergoes oxidation, and the terminal three carbon atoms, C_{25} to C_{27}, are released as propionyl CoA. In this cleavage reaction, CoASH adds to C_{24}. One of the products, cholyl CoA, contains three hydroxyl groups, whereas the other product, chenodeoxycholyl CoA, contains only two hydroxyl groups projecting from the steroid nucleus. The CoASH at C_{24} is then replaced by either glycine or taurine, which are in amide linkage with C_{24}, the terminal carbon atom of the chain. These synthetic reactions take place in the liver, and a mixture of glycocholic, glycochenodeoxycholic, taurocholic, and taurochenodeoxycholic acids are secreted into the bile. The glycine and taurine amides of the bile acids are known as *conjugated* bile acids or bile salts. Almost all of the bile acids that are released from the liver are present in conjugated form.

Bile acid biosynthesis is regulated by the amount of bile acid that is returned from the intestine to the liver. Biosynthesis *decreases* as the bile acid return to the liver increases. Bile acid synthesis also is controlled by the amount of cholesterol that is transported from the intestine to the liver; it *increases* as cholesterol absorption increases.

Bile acid production is the most important catabolic pathway for cholesterol.

Continuous conversion of cholesterol into bile acid in the liver prevents the body from becoming overloaded with cholesterol. Excessive accumulation of cholesterol in the tissues is harmful. Unlike many other metabolites, cholesterol cannot be destroyed by oxidation to carbon dioxide and water, for mammalian tissues do not have enzymes capable of oxidizing the steroid nucleus. Therefore other regulatory mechanisms have evolved to prevent cholesterol overload. Malfunctioning of these control mechanisms leads to some of the most serious diseases known to man. These diseases result from a pathologic process that involves the wall of arteries, known as atherosclerosis (p. 511).

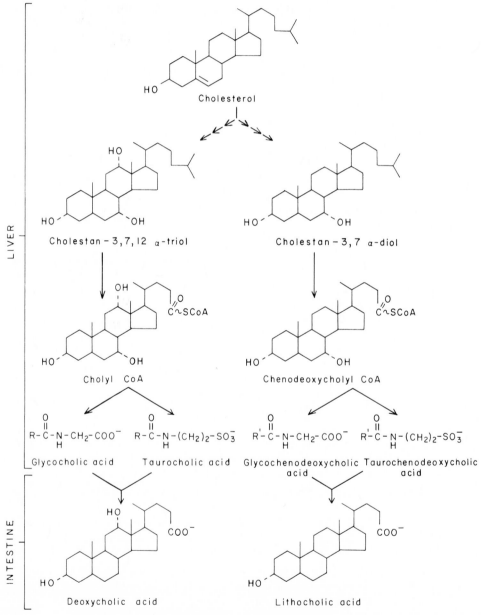

Fig. 10-4. Biosynthesis of bile acids.

Metabolism

Conjugated bile acids either pass from the liver directly into the duodenum through the common bile duct or are stored in the gallbladder when not needed immediately for digestion. They form a part of the bile that, in addition, contains phospholipids, cholesterol, salts, water, and excretory metabolites such as bilirubin. Bile that is stored and concentrated in the gallbladder is released into the intestine via the cystic and common bile ducts when it is needed to aid in the digestion of dietary lipids. The bile acids aid in the emulsification of the ingested lipids, a process that facilitates enzymatic digestion and absorption of dietary fat. As will be described in Chapter 13, the secretion of bile from the liver and the emptying of the gallbladder are processes under hormonal control. *Hepatocrinin* stimulates bile secretion by the liver, and *cholecystokinin* causes the gallbladder to empty. These hormones are synthesized by the intestine and are released when the partially digested food passes from the stomach into the duodenum.

Secondary bile acids

Deoxycholic and lithocholic acids are the secondary bile acids. They are synthesized in the intestine through the action of bacterial enzymes on the primary bile acids. Only a portion of the primary bile acids present in the intestine are converted to secondary bile acids. First, the primary acids are deconjugated; that is, the amide linkage at C_{24} is hydrolyzed and glycine or taurine is released. Next, the hydroxyl group present at C_7 is removed. In this way cholic acid (3,7,12-trihydroxy) is converted to deoxycholic acid (3,12-dihydroxy), and chenodeoxycholic acid (3,7-dihydroxy) is converted to lithocholic acid (3-hydroxy).

Enterohepatic circulation

Most of the bile acids present in the intestine, the remaining primary acids as well as the newly formed secondary acids, are reabsorbed into the portal blood. As shown in Fig. 10-5, there are three intestinal sites at which bile acid reabsorption occurs. In the jejunum and colon, absorption occurs by passive diffusion.

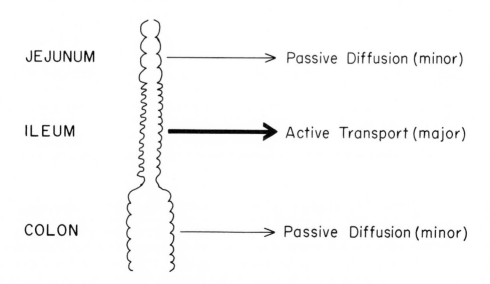

JEJUNUM   ⟶ Passive Diffusion (minor)

ILEUM   ⟶ Active Transport (major)

COLON   ⟶ Passive Diffusion (minor)

Fig. 10-5. Bile acid absorption from intestine.

In the ileum, the major site of absorption, the process is one of active transport. The bile acids are removed from the portal blood by the liver and then are resecreted into the bile. Bile that is released from the liver therefore contains all four bile acids, not only the two primary acids that actually are synthesized by the liver. The recycling of bile acids between intestine and liver is called the enterohepatic circulation of bile acids. In the portal blood the bile acids are carried as a noncovalent, physical complex by the main plasma protein, *albumin.* From 15 to 30 g/day of bile acids is passed into the intestine from the liver. Only a small fraction, approximately 300 mg/day, is lost in the feces. This quantity must be synthesized daily from cholesterol in order to maintain the size of the bile acid pool. The bile acids present in the feces are known as *acidic sterols.*

Cholelithiasis

The disease in which the gallbladder contains gallstones is called cholelithiasis. Most of the stones that form in the gallbladders of adults in Western countries are made up predominantly of cholesterol. The sterol is laid down around a central core containing protein and bilirubin. Because of an unknown defect, bile of abnormal composition is secreted by the liver, causing cholesterol to crystallize and form a stone.

Cholesterol is insoluble in the aqueous salt medium of the bile and is brought into solution by molecular association with the bile salts and lecithin. Mixed micelles made up of lecithin, bile salts, and cholesterol are formed, and the relatively large amounts of cholesterol ordinarily present in human bile are held in solution. This allows cholesterol to be transported harmlessly through the biliary tract into the intestine, including temporary storage in a solubilized form in the gallbladder. The mixed micelles that form have a limited capacity to solubilize cholesterol. The amount of cholesterol that can actually be held in micellar aggregates depends on the relative proportions of lecithin, bile salts, and cholesterol as well as on the water content of the bile.

A convenient method for describing the defective bile composition that leads to cholesterol precipitation and gallstone formation is shown in Fig. 10-6. The molar percentage composition of biliary cholesterol, lecithin, and bile salt is plotted on triangular coordinates. In any given sample of bile the molar percentage of these three components can be measured chemically. The three coordinates are then plotted in graphic form (Fig. 10-6), and their point of intersection is determined. If this point falls within the small shaded area in the lower left corner, the cholesterol present in the sample of bile will exist in soluble form, that is, in the liquid micellar state. However, if the point of intersection of the three coordinates falls outside of the shaded area, cholesterol will crystallize in the form of discrete aggregates. Cholesterol gallstones will form in bile that is incapable of completely solubilizing all of its cholesterol content. For example, a bile containing 5% cholesterol, 15% lecithin, and 80% bile salts will have all of the cholesterol in the micellar state, whereas bile containing 20% cholesterol will form gallstones independently of the relative lecithin and bile salt concentrations.

Lithogenic bile is bile with a cholesterol content that falls outside the shaded area in Fig. 10-6 and therefore is supersaturated with respect to cholesterol. Recent investigations indicate that the total amount of bile acids in the body, referred to as the bile acid pool, is reduced in size in patients with gallstones and in patients whose genetic background indicates that they have a high risk of developing this disease. It is thought that the reduction in bile acid pool size may be responsible for the formation of lithogenic bile. Normally, the liver would

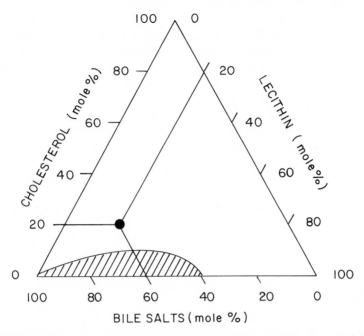

Fig. 10-6. Phase diagram of lipid component composition of bile. The shaded area represents the region in which cholesterol exists as a micellar liquid. Cholesterol forms crystals if concentrations of the three components fall outside shaded area. The point shown on the phase diagram represents a composition of 20 moles % cholesterol, 20 moles % lecithin, and 60 moles % bile salts. (Adapted from Small, D. M.: N. Engl. J. Med. 279:588, 1968.)

sense the reduced pool size and increase its synthesis of bile acids until the pool size would return to the normal value. The fact that this homeostatic regulatory mechanism does not operate properly suggests that the defect causing the formation of lithogenic bile probably occurs within the liver cell. Additional clinical investigations have shown that the ingestion of 1 to 2 g of cholesterol daily for 1 to 3 months increases the cholesterol content of the bile into the supersaturated range. These results suggest that continuous very high dietary cholesterol intake also can predispose to gallstone formation through the production of a lithogenic bile.

CHOLESTEROL METABOLISM

Cholesterol can undergo a number of metabolic reactions in man. It can be esterified, and the resulting cholesteryl esters can be hydrolyzed. It can be transformed into bile acids as described in the previous section. Conversion to bile acids is the main catabolic pathway in terms of the quantity of cholesterol that is metabolized. Cholesterol also is the substrate of steroid hormone synthesis. Although the amount of cholesterol converted into steroid hormones is small in terms of total cholesterol utilization, these reactions are of enormous importance in terms of bodily function. Cholesterol is either obtained from the diet or synthesized completely from acetyl CoA. As noted above, however, the one metabolic reaction involving cholesterol that cannot be carried out by human or other mammalian tissues is the catabolism of the sterol nucleus to carbon dioxide and water. The potential problems imposed by this limitation will become apparent when diseases associated with abnormalities in cholesterol metabolism are discussed.

Ester formation

Cholesteryl ester synthesis occurs through two separate pathways. One pathway, the cholesteryl ester synthetase, is operative in tissues. This pathway involves two reactions. First, fatty acid is activated by incorporation into thioester linkage with CoASH to form an acyl CoA. Formation of the thioester bond requires the expenditure of one ATP molecule as the source of energy. This reaction is mediated by *fatty acid–activating enzymes*, as was described in Chapter 8. The acyl CoA thioester then reacts with cholesterol through the action of an *acyl CoA cholesterol acyltransferase* (ACAT).

$$\text{Acyl CoA} + \text{Cholesterol} \xrightarrow[\text{Acyltransferase}]{\text{Acyl CoA Cholesterol}} \text{Cholesteryl ester} + \text{CoASH}$$

Major sites for cholesteryl ester synthesis by this pathway are the liver, intestine, and adrenal cortex.

The second pathway of cholesteryl ester synthesis occurs in the plasma and is mediated by the enzyme *lecithin-cholesterol acyltransferase* (LCAT). In this reaction a fatty acid contained in position 2 of lecithin is transferred directly to cholesterol without passing through either a fatty acid or acyl CoA intermediate.

$$\text{Lecithin} + \text{Cholesterol} \xrightarrow[\text{Cholesterol Acyltransferase}]{\text{Lecithin-}} \text{Cholesteryl ester} + \text{lysolecithin}$$

As noted earlier, most of the *plasma* cholesteryl esters are synthesized by this pathway. In other words, the cholesterol that is released by the liver as a constituent of plasma lipoproteins is mostly in nonesterified form. It is esterified subsequently in the plasma through the action of lecithin-cholesterol acyltransferase. This enzyme is produced in the liver and secreted into the plasma, where it functions. Apolipoprotein C-I, which is contained in very low– and high-density lipoproteins, is an activator of the enzyme. Esterification serves to trap cholesterol within the lipoprotein, preventing it from diffusing to other circulating macromolecules or into the tissues.

Ester hydrolysis

Cholesteryl esters are hydrolyzed to cholesterol and fatty acids through the action of cholesterol esterases.

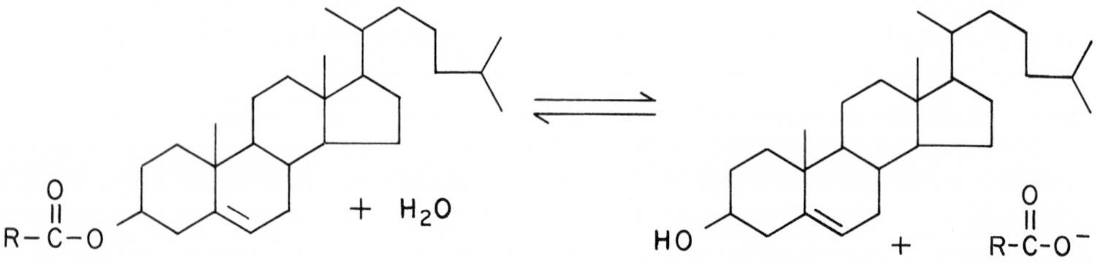

Cholesterol   Fatty acid

The reaction equilibrium is in the direction of hydrolysis. Cholesterol esterases are present in many tissues, and there is a cholesterol esterase in the pancreatic juice that functions in the digestion of dietary cholesteryl esters. The pancreatic esterase has been described in Chapter 8.

Biosynthesis

In addition to being taken in from the diet, cholesterol can be synthesized by human tissues. Cholesterol is synthesized from acetyl CoA, which can be derived from carbohydrates, amino acids, or fatty acids. The liver is the main site of cholesterol synthesis, but the intestine also is an important site of synthesis in man. In addition, cholesterol is synthesized in glands that produce steroid hormones, for example, the adrenal cortex, testes, and ovaries. All of the synthetic reactions occur in the cytoplasmic compartment of the cell, and most of the required enzymes are located in the endoplasmic reticulum.

The synthetic pathway can be thought of as occurring in three stages. In the first stage, acetyl CoA is converted into a six-carbon atom thioester intermediate, *3-hydroxy-3-methylglutaryl CoA (HMGCoA)*. The second stage involves the conversion of HMGCoA to squalene, an acyclic hydrocarbon containing thirty carbon atoms. An intermediate in this stage is the five-carbon atom isoprene unit, which occurs in two isomeric phosphorylated forms in this pathway, Δ_3-*isopentenyl pyrophosphate* and *3,3-dimethylallyl pyrophosphate*. In the third stage squalene is cyclized and converted to the twenty-seven–carbon atom sterol, cholesterol. The series of reactions from squalene to cholesterol occur while the intermediates are bound physically to a cytoplasmic protein, the *sterol carrier protein*. A schematic representation of the three stages in the cholesterol biosynthetic pathway is given in Figs. 10-7 to 10-9.

HMGCoA formation. Two molecules of acetyl CoA condense in the initial step of HMGCoA formation to produce *acetoacetyl CoA*. This reaction occurs in the cell cytoplasm.

$$2\ CH_3-\overset{\overset{\textstyle O}{\|}}{C}\!\sim\!SCoA \quad \xrightarrow[\text{Synthetase}]{\text{Acetoacetyl CoA}} \quad CH_3-\overset{\overset{\textstyle O}{\|}}{C}-CH_2-\overset{\overset{\textstyle O}{\|}}{C}\!\sim\!SCoA + CoA$$

Acetyl CoA   Acetoacetyl CoA

Remember that the acetoacetyl group is the four-carbon intermediate involved in fatty acid biosynthesis, a reaction sequence that also occurs in the cytoplasmic compartment of the cell. However, in fatty acid synthesis the intermediates are bound covalently to the acyl carrier protein component of the fatty acid synthetase

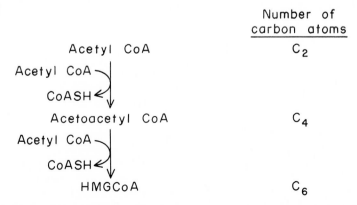

Fig. 10-7. First stage of cholesterol biosynthesis. In this sequence of reactions, acetyl CoA is converted to HMGCoA.

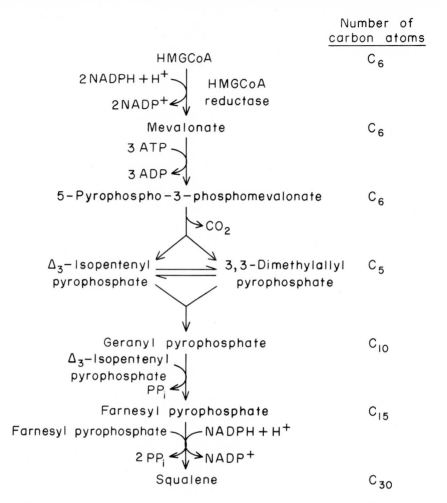

Fig. 10-8. Second stage of cholesterol biosynthesis. This sequence of reactions involves conversion of HMGCoA to squalene.

complex, not to CoASH as in the cholesterol pathway. In the next reaction, acetoacetyl CoA condenses with another acetyl CoA molecule, forming HMGCoA.

$$CH_3-\overset{O}{\overset{\|}{C}}-CH_2-C\sim SCoA + CH_3-\overset{O}{\overset{\|}{C}}\sim SCoA \xrightarrow[\substack{\text{condensing}\\\text{enzyme}}]{\text{HMGCoA}} HO-\overset{\overset{COO^-}{|}\overset{CH_2}{|}}{\underset{\underset{C\sim SCoA}{|}\underset{\|}{\underset{O}{}}}{C}}-CH_3 + CoASH$$

3-Hydroxy-3-methyl glutaryl CoA (HMGCoA)

Notice that in this reaction the acetate unit adds to the β-carbon atom of acetoacetate.

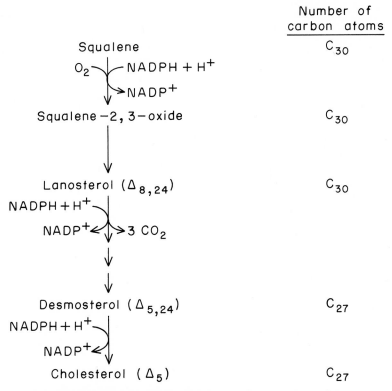

	Number of carbon atoms
Squalene	C_{30}
Squalene $-2,3-$oxide	C_{30}
Lanosterol $(\Delta_{8,24})$	C_{30}
Desmosterol $(\Delta_{5,24})$	C_{27}
Cholesterol (Δ_5)	C_{27}

Fig. 10-9. Third stage in cholesterol biosynthesis. In this sequence of reactions, squalene is converted to cyclized intermediates, which eventually are transformed into cholesterol. These reactions occur while the intermediates are bound to sterol carrier protein.

Conversion of HMGCoA to squalene. In the next step, HMGCoA is reduced to mevalonate, the thioester group being converted to an alcohol.

$$
\begin{array}{c}
COO^- \\
| \\
CH_2 \\
| \\
HO-C-CH_3 \\
| \\
CH_2 \\
|
\end{array}
+ \; 2\,NADPH + 2\,H^+ \quad \xrightarrow[\text{reductase}]{\text{HMGCoA}} \quad
\begin{array}{c}
COO^- \\
| \\
CH_2 \\
| \\
HO-C-CH_3 \\
| \\
CH_2 \\
| \\
CH_2OH
\end{array}
+ \; CoASH + 2\,NADP^+
$$

Mevalonate

This reaction utilizes two molecules of NADPH as the reducing agent and is catalyzed by HMGCoA reductase, a microsomal enzyme. It is the rate-limiting step in cholesterol synthesis. A peculiar property of this enzyme is that after it is solubilized from the microsomes it becomes labile in the presence of cold temperatures and is almost completely inactivated by storage for 30 min at $0°\,C$.

Mevalonate is converted to *5-pyrophospho-3-phosphomevalonate* in three steps, each of which requires the expenditure of one molecule of ATP. The five-carbon atom *isoprene unit* is formed from this phosphorylated intermediate.

Actually two isomeric isoprene units are formed, *isopentenyl pyrophosphate* and *dimethylallyl pyrophosphate;* they are interconvertible.

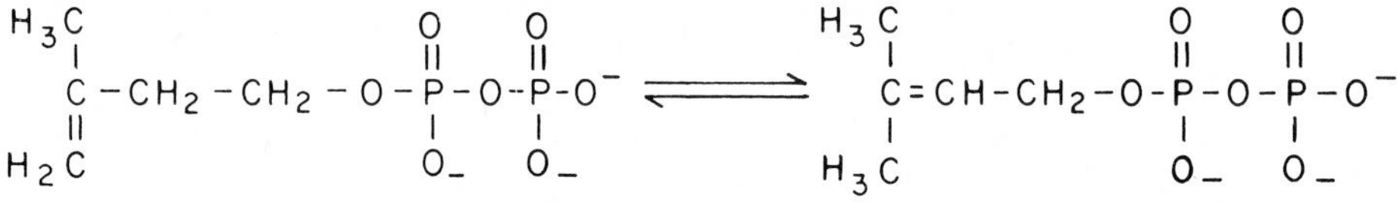

Δ_3-Isopentenyl pyrophosphate   3,3-Dimethylallyl
pyrophosphate

These isomers condense, forming a ten-carbon atom intermediate, *geranyl pyrophosphate*. Another molecule of isopentenyl pyrophosphate condenses with the newly formed geranyl pyrophosphate, forming *farnesyl pyrophosphate*, a fifteen-carbon atom intermediate. Two molecules of farnesyl pyrophosphate then combine in a series of reactions in which NADPH is utilized and both pyrophosphate groups are eliminated. The product is squalene, a thirty-carbon atom hydrocarbon that exists in a folded conformation but is not cyclized.

Conversion of squalene to cholesterol.   Squalene binds to a specific protein carrier, *sterol carrier protein,* in the cell cytoplasm. The conversion of squalene to cholesterol occurs while the intermediates are bound to this carrier. The initial reactions are shown in Fig. 10-10. First, an oxygen atom is attached to both positions C_2 and C_3 of squalene, a reaction that requires NADPH. The resulting *squalene-2,3-oxide,* an epoxide, then cyclizes to form *lanosterol,* the first ring-containing intermediate. Cyclization is accomplished by migration of the double bonds in four segments of the squalene oxide that will become the four rings of the sterol. The intermediate that is formed has methyl groups projecting from positions C_8 and C_{14} of the ring structure. In the next step two 1,2-methyl shifts occur. The methyl group originally attached to C_{14} migrates to C_{13}. This becomes the angular methyl group, C_{18}, that projects between the C- and D-rings of the steroid nucleus (see p. 491). In addition, the methyl group originally attached to C_8 migrates to C_{14} where it now projects below the plane of the ring structure. The next intermediate that is formed, lanosterol, contains thirty carbon atoms. The β-hydroxyl group is present at C_3, but there are double bonds between C_8 and C_9 and in the chain between C_{24} and C_{25}. In addition, there are two methyl groups projecting in β-orientation from C_4 as well as the one in α-orientation from C_{14}. In a series of reactions involving oxygen and NADPH, these three methyl groups are released as carbon dioxide and the double bond in the B ring is moved to position C_5-C_6, forming a twenty-seven–carbon atom intermediate, *desmosterol.*

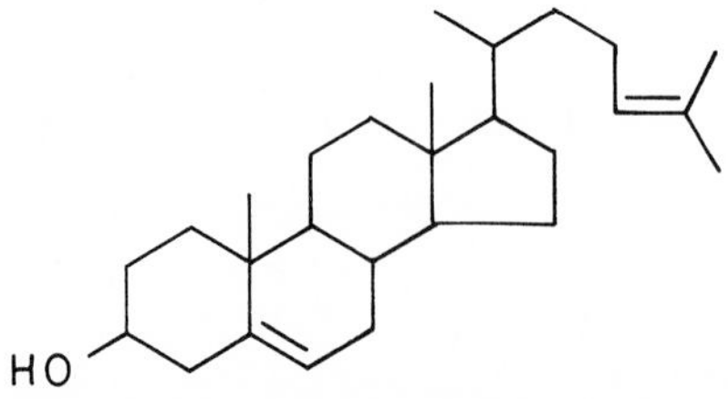

Desmosterol

Desmosterol still contains the side chain double bond at C_{24}-C_{25}, but it possesses many of the properties of the final product, cholesterol. In the last step the C_{24}-C_{25} double bond is reduced by NADPH, forming cholesterol.

In summary, it should be remembered that the synthesis of a single molecule of cholesterol represents the expenditure of considerable substrate, energy, and

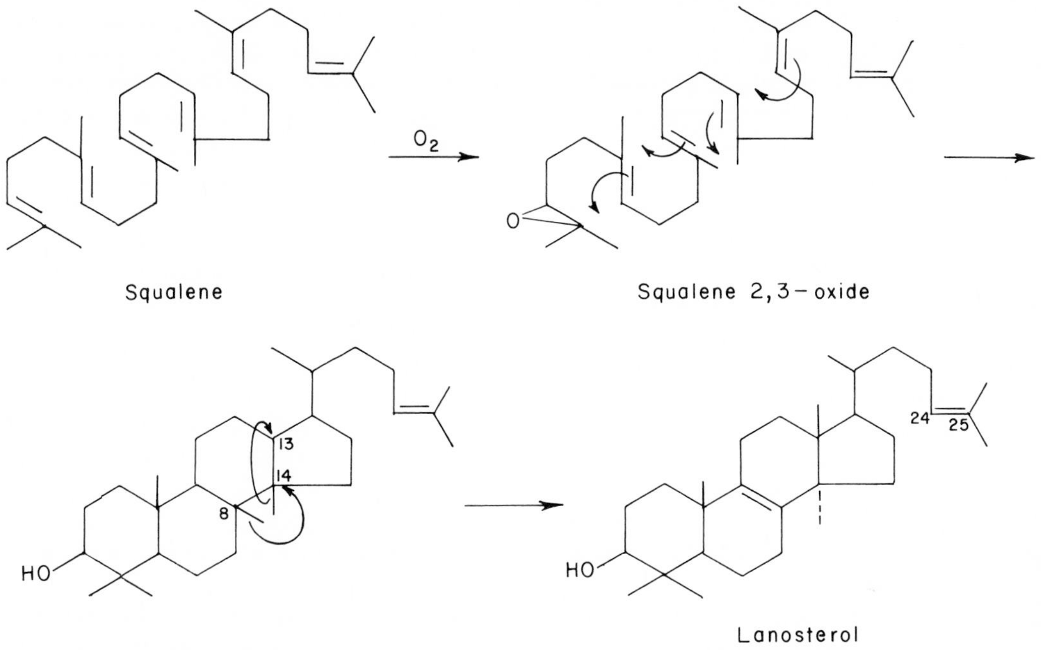

Fig. 10-10. Cyclization of squalene and its conversion to lanosterol.

reducing equivalents. The overall process requires eighteen moles of acetyl CoA, thirty-six moles of ATP, and sixteen moles of NADPH to form one mole of cholesterol. Although six molecules of HMGCoA are utilized, only twenty-seven of the thirty-six carbon atoms are retained in the final product, cholesterol. Of the thirty-six moles of ATP required, eighteen are used for acetyl CoA formation.

The mevalonate shunt

It used to be thought that once mevalonic acid was formed, the carbon atoms were committed irreversibly to cholesterol synthesis. Very recent work indicates, however, that C_2 of mevalonate can appear in fatty acids in tissues of ectodermal origin such as the brain, spinal cord, and skin. The proposed mechanism involves a dephosphorylation of dimethylallyl pyrophosphate, one of the isoprene units formed from mevalonate (p. 488). In a series of reactions involving alcohol and aldehyde dehydrogenases and a carboxylase, the carbon atoms derived from the dimethylallyl alcohol are converted to HMGCoA. This is cleaved by HMGCoA lyase to acetoacetate and acetyl CoA (p. 492), intermediates that can either be catabolized to CO_2 and H_2O or utilized for biosynthetic reactions, including fatty acid synthesis. This explains the appearance of the C_2 of mevalonate in the fatty acids of ectodermal tissues. Although the importance of this shunt mechanism in terms of the regulation of cholesterol production has not as yet been determined, it offers the potential of reducing cholesterol synthesis by drugs that might be designed to take advantage of this previously unknown pathway.

Ketone body synthesis

Ketone bodies that are released by the liver during periods of starvation or ketoacidosis supply energy to the heart, skeletal muscles, and brain. The mechanism through which these tissues oxidize ketone bodies has been described in

Chapter 8. We shall now consider ketone body synthesis, a process that occurs in the liver. Although ketone bodies are not steroids, their formation is discussed at this point because the initial steps in their synthesis are identical to those in the sterol synthetic pathway. The pathway for ketone body formation in the liver is shown in Fig. 10-11.

Ketone body formation increases markedly when fatty acid is the main oxidative substrate and carbohydrate availability is limited. Since both acetoacetate and β-hydroxybutyrate are similar in structure to intermediates in the fatty acid β-oxidation cycle, it was thought originally that the ketone bodies were derived directly from the acetoacetyl CoA intermediate in fatty acid oxidation. In other words, if excessive amounts of acetyl CoA were being presented to the Krebs cycle, the β-oxidation process was thought to stop at the four-carbon atom intermediate, acetoacetyl CoA, which was then deacylated. However, early experiments using radioactive carbon to tag fatty acids demonstrated that this was incorrect, and further studies demonstrated that ketone bodies, like cholesterol, were synthesized from acetyl CoA. Since acetyl CoA is the end product of the fatty acid β-oxidation cycle, the overall concept that ketone bodies were derived from fatty acids was basically correct, and only the route and complexity of their synthesis was somewhat surprising.

Ketone bodies are synthesized inside the mitochondria from HMGCoA using acetyl CoA generated primarily by the β-oxidation of fatty acids. Initially, two molecules of acetyl CoA condense to form acetoacetyl CoA, which then condenses with another acetyl CoA molecule to form HMGCoA. The HMGCoA is then cleaved through the action of *HMGCoA lyase* to yield acetoacetate and acetyl CoA, the latter being available to combine with another molecule of acetoacetyl CoA. The HMGCoA pathway is thought to be the major route to ketone body synthesis, but this point is not entirely conclusive. Some of the acetoacetate that is formed is released into the plasma, whereas the rest of it is reduced through the action of a NADH-dependent reductase to β-hydroxybutyrate and released from the liver in this form. When the acetoacetate concentration is elevated, some of it is spontaneously decarboxylated to acetone. The reactions to the point of HMGCoA formation are identical to those in the cholesterol synthetic pathway. In cholesterol synthesis, however, these reactions take place in the cell cytoplasm. In contrast,

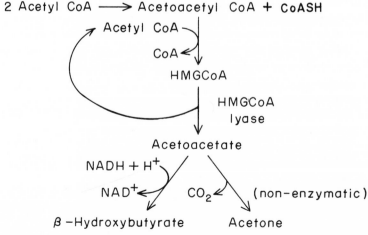

Fig. 10-11. Ketone body biosynthetic pathway.

ketone bodies are synthesized inside the mitochondria. This difference in the intracellular location of synthesis suggests that any simple concept involving a single HMGCoA pool is an oversimplification. In reality, ketone body and cholesterol synthesis are two separate pathways, each of which utilizes a different HMGCoA pool.

Regulation of cholesterol synthesis

Cholesterol synthesis is regulated by the dietary cholesterol intake, the caloric intake, certain hormones, and bile acids. Estrogen, the female sex hormone, inhibits synthesis by reducing the rate of HMGCoA formation through an effect on the HMGCoA condensing enzyme. Starvation also decreases synthesis by decreasing the availability of acetyl CoA, ATP, and NADPH. Bile acids have a direct inhibitory effect on cholesterol synthesis in the intestinal mucosa.

Feedback inhibition. Dietary cholesterol itself does not inhibit intestinal cholesterol synthesis, but it does have a strong "feedback inhibitory" effect on cholesterol synthesis in the liver. This is illustrated schematically in Fig. 10-12. Cholesterol absorbed from the diet inhibits the liver biosynthetic pathway by decreasing synthesis at the HMGCoA reductase reaction. Originally, it was thought that cholesterol directly inhibited the enzyme, perhaps as a negative allosteric effector. Since cholesterol is the end-product of the biosynthetic pathway, the regulation was termed feedback inhibition. Subsequent studies with cells in culture indicated that regulation occurred through changes in HMGCoA reductase content, not by allosteric regulation of the preexisting enzyme. Measurements made in rat liver indicate that the half-life of HMGCoA reductase is about 4 hr. Therefore, if synthesis of the enzyme is reduced or interrupted, the content within the hepatocyte decreases appreciably after a few hr. Cholesterol biosynthesis slows because of depletion of HMGCoA reductase, the rate-limiting enzyme in the pathway. Although the possibility that cholesterol exerts an additional negative allosteric effect cannot be excluded, the available evidence indicates that control is mediated primarily through changes in enzyme content.

Recent work with cultured cells indicates that certain oxygenated derivatives of cholesterol, including *7α-hydroxycholesterol*, *7β-hydroxycholesterol*, and *7-ketocholesterol*, are much more potent inhibitors of HMGCoA reductase than

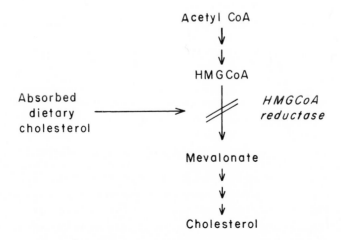

Fig. 10-12. Regulation of hepatic cholesterol synthesis.

cholesterol itself. These studies raise the important question of whether cholesterol must be converted to some oxygenated metabolic product before it exerts its negative feedback effect on HMGCoA reductase production.

Studies with skin fibroblasts taken from patients who have the homozygous from of type IIA hyperlipoproteinemia (Chapter 8), indicate that feedback inhibition of cholesterol synthesis is defective in this disease. These patients have severe hypercholesterolemia, with plasma cholesterol concentrations as high as 800 to 900 mg/dl. The fibroblast studies indicate that there is a defect in the ability of low-density lipoproteins (LDL) to bind to the surface of the cell and, as a result of this, cause feedback inhibition of cholesterol.synthesis.

Three explanations have been put forth to explain this observation. The first is the existence of a genetic defect in the high-affinity membrane receptor for LDL. The second is a defect in the mechanism whereby LDL enters the cell subsequent to binding. The third is an excessive leakage of cholesterol from the cell after it enters. Regardless of mechanism, the fact remains that the cells from the homozygous type IIA patients continue to produce cholesterol in spite of being exposed to adequate amounts of LDL, whereas normal cells exhibit feedback inhibition under these conditions. Most of the cholesterol that is removed irreversibly from the plasma of a human is in the form of LDL. It is not known at present, however, exactly how the observed cellular defect is associated with the hypercholesterolemia and LDL elevation that occurs clinically in the type IIA patient.

Circadian rhythm. Based upon studies done in animals, cholesterol synthesis also appears to vary at different times during the day. This effect, known as a circadian rhythm, occurs predominantly in the liver. Rats were given free access to food and water and kept in a room that was illuminated from 6 AM to 6 PM and darkened from 6 PM to 6 AM. Cholesterol synthesis was highest at midnight and lowest at noon. The activity of HMGCoA reductase isolated in a microsomal fraction from these livers exhibited an identical diurnal variation—highest at midnight and lowest at noon. Therefore, the circadian variation in cholesterol synthesis is secondary to changes in the activity of HMGCoA reductase, the rate-limiting enzyme in the cholesterol biosynthetic pathway. Under these conditions the animals begin to eat soon after the lights are turned off. If they are trained to eat between 9 AM and 1 PM under the same environmental conditions, the peak synthesis of cholesterol is shifted and occurs 3 hr after eating. These results suggest that the circadian rhythms that occur are related to the time that food is ingested. Animal chow usually is low in cholesterol. Therefore, in attempting to extrapolate these results to humans, one must take into account the fact that the high cholesterol content of a meal, through feedback inhibition, might blunt the expected food-induced rise in hepatic cholesterol synthesis.

Hormonal regulation. In addition to the diet, a number of hormones have regulatory actions on cholesterol synthesis. They, too, operate through effects on HMGCoA reductase activity, probably by controlling the production of the enzyme. Insulin or triiodothyronine administration increases HMGCoA reductase activity, while glucagon and cortisol decrease the enzyme. Stress induced by restraining animals for 48 hr also increases the enzyme activity. These findings, together with the dietary and circadian effects, illustrate the enormous complexity with which HMGCoA reductase and therefore cholesterol synthesis is controlled in the liver. All of this regulatory information has been obtained in rats. Although experiments on hepatic enzyme levels cannot be carried out in man, there is reason to suspect that similar diurnal, hormonal, and dietary-induced variations occur in the human.

Unlike normal hepatocytes, liver tumors, known as hepatomas, do not exhibit this feedback regulatory control, for they continue to synthesize cholesterol in spite of a dietary cholesterol load.

Regulation of cholesterol levels in man

Most of our clinical information about cholesterol regulation in man is of necessity based on measurements of the plasma cholesterol concentration. Plasma cholesterol, which is contained in lipoproteins, interacts with tissue cholesterol. Therefore, some information about tissue cholesterol synthesis and turnover can be obtained from plasma cholesterol measurements, especially in clinical research studies in which radioactive cholesterol is administered to volunteer test subjects and the plasma cholesterol specific radioactivity is followed. However, there are still many uncertainties introduced when attempts are made to extrapolate from plasma cholesterol values to tissue contents or enzymatic regulatory mechanisms.

Plasma cholesterol concentration

The normal cholesterol concentration in plasma obtained from a fasting human is considered by most clinical laboratories to be between 120 and 240 mg/dl. In young, healthy adults the mean value for plasma cholesterol is between 170 and 180 mg/dl. About 65% of the cholesterol in the plasma of normal fasting subjects is esterified. These measurements usually are made after an overnight fast of 12 to 14 hr. The plasma of a normal patient will have no chylomicrons remaining after this period of fasting and few VLDL's. Therefore, the main lipoproteins present in normal fasting plasma are LDL's and HDL's. About 60% of the total cholesterol is present in LDL's in normal fasting plasma. Males have relatively more LDL's, whereas premenopausal females have relatively more HDL's. Because LDL's contain more cholesterol than HDL's, normal males tend to have higher plasma cholesterol levels than premenopausal females.

Origins of body cholesterol. Studies with biopsy specimens of human liver clearly demonstrate that feedback inhibition occurs in man. Yet even if large quantities of cholesterol are ingested, about 60% of the plasma cholesterol is still derived from biosynthesis. Moreover, if patients are placed on a cholesterol-free diet for a prolonged period, there is a 10% to 25% decrease in the plasma cholesterol concentration. These findings lead to several important conclusions. First, although hepatic feedback inhibition occurs, it appears not to be as important a regulatory mechanism in terms of overall plasma cholesterol concentrations in the normal human as might be expected from the animal and tissue culture studies. This has led to the speculation that the intestine, an organ that does not exhibit feedback inhibition, may be a major site of cholesterol synthesis in man. Another important conclusion from these observations is that large decreases in circulating cholesterol usually cannot be achieved simply by restriction of dietary cholesterol. This can occur only through inhibition of cholesterol biosynthesis. However, a modest reduction of the plasma cholesterol level almost always can be achieved by reducing the dietary cholesterol intake. When the diet is rich in cholesterol, the plasma concentration is maintained at a level that is from 10% to 25% higher than can be achieved if biosynthesis is the only cholesterol source. Thus biosynthesis cannot completely compensate for a significant reduction in dietary cholesterol intake, and some decrease in the circulating level results.

This is a very important observation in terms of recommending dietary cholesterol restriction for patients with hypercholesterolemia as well as for the population in general. Epidemiologic studies indicate that persons with plasma cholesterol concentrations above 220 mg/dl have a greater tendency to develop atherosclerosis, the arterial disease that leads to myocardial infarction and stroke. It is currently recommended that these patients with elevated plasma cholesterol concentrations go on a low-cholesterol diet in order to lower the plasma concentration. Moreover, there is a general feeling that the entire population would benefit over the long term by ingesting less cholesterol, and diets low in cholesterol are now being recommended by many physicians for everyone, beginning even in infancy. When the information about feedback inhibition became known, serious question was raised as to whether dietary cholesterol restriction would have any benefit. It was thought that biosynthesis might compensate for the dietary reduction, leading to no reduction in the plasma content. Subsequent clinical tests indicated, as described above, that this is not the case. Indeed, many people with hypercholesterolemia have plasma cholesterol concentrations between 220 and 270 mg/dl. It is in this group that dietary cholesterol restriction probably is most helpful, for the modest reductions in plasma cholesterol concentrations that can be achieved by dietary modification are sufficient to reduce their values to 220 mg/dl or less. Although it has not been proved that such reductions actually are beneficial in preventing atherosclerosis, most experts now believe that it should be attempted as a potentially helpful preventive health measure.

Saturation of dietary fat and plasma cholesterol concentration. When humans are fed diets that are rich in saturated fatty acids, the plasma cholesterol concentration is increased. During replacement of the saturated fat with a fat that is rich in polyenoic fatty acids, such as linoleic acid, a decrease in the plasma cholesterol concentration occurs. The mechanism through which polyunsaturated fats produce this effect is not known at present. Because the plasma cholesterol concentration is lowered, diets that contain higher percentages of polyunsaturated fatty acids are considered by many to be beneficial to human health.

STEROID HORMONES
Chemistry

In addition to bile acids, cholesterol can be converted into steroid hormones. These include the twenty-one–carbon atom adrenal corticosteroids, *cortisol*, *aldosterone*, and *corticosterone*.

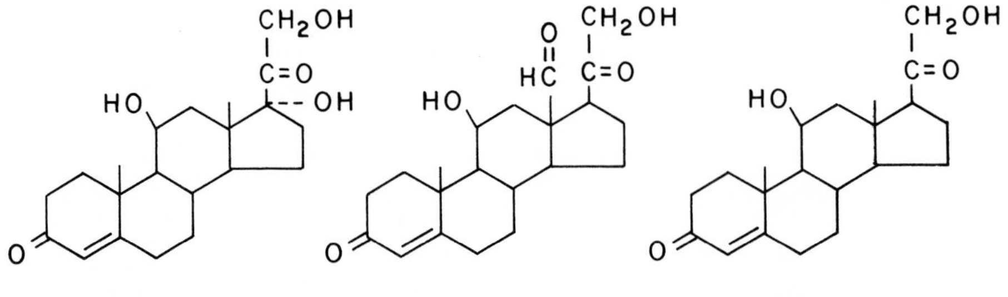

Cortisol   Aldosterone   Corticosterone

The adrenal corticosteroids retain the basic four-ring sterol structure, but the 3-β-hydroxyl group is converted to a 3-keto group and the double bond is shifted from the B ring to position 4,5 in the A ring. Each corticosteroid hormone contains an 11-β-hydroxyl group. This group is not present in cholesterol and therefore must be added during the synthesis of these hormones. Only two of the original eight-carbon atoms of the cholesterol hydrocarbon side chain remain, with C_{20} being converted to a keto group and C_{21} to an alcohol group. Cortisol contains an α-hydroxyl group at C_{17}. In aldosterone, the C_{18} angular methyl group is oxidized to an aldehyde. The aldehyde group can cyclize with the 11-β-hydroxyl group, forming a hemiacetal.

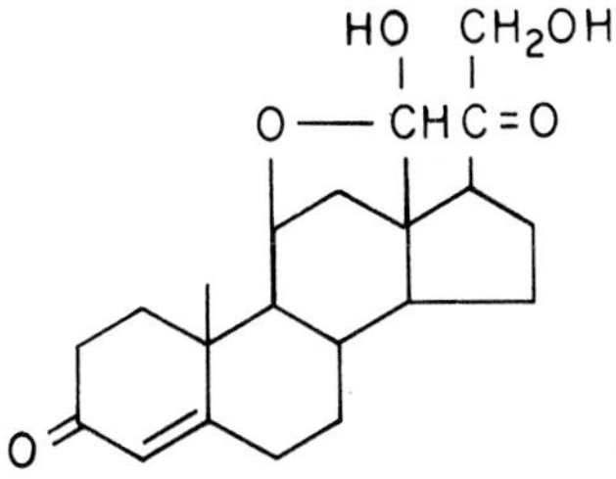

Aldosterone (hemiacetal form)

The other major steroids are the female sex hormones, *progesterone* and the *estrogens*, and the male sex hormones,, *testosterone* and *androstenedione*. The predominant estrogen is *17-β-estradiol*, but *estrone* and *estriol* also are present to some extent.

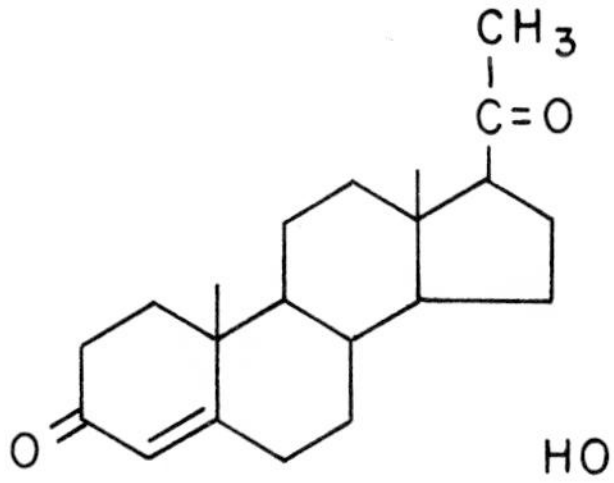

Progesterone 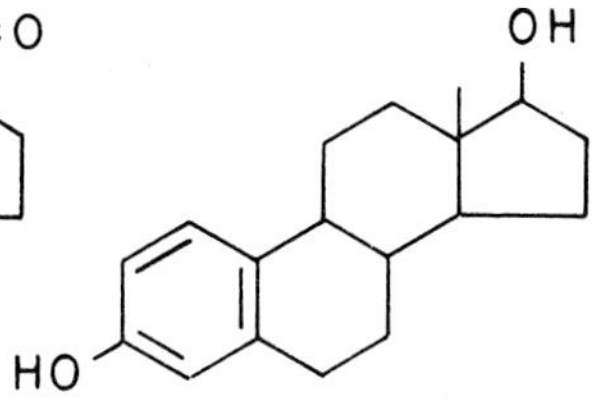   17-β-Estradiol 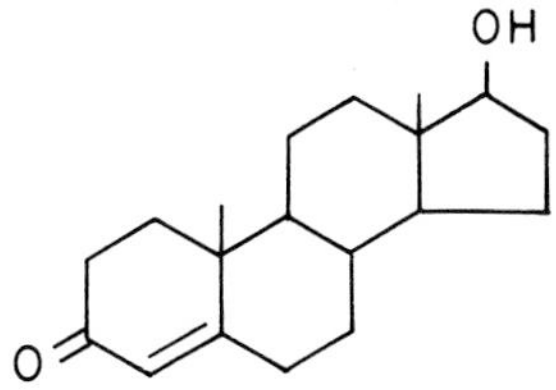   Testosterone

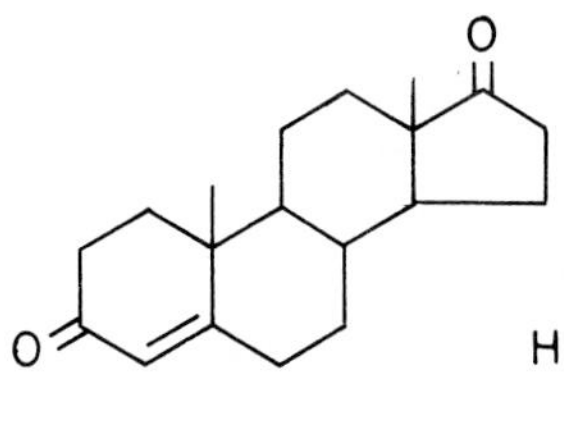

Androstenedione 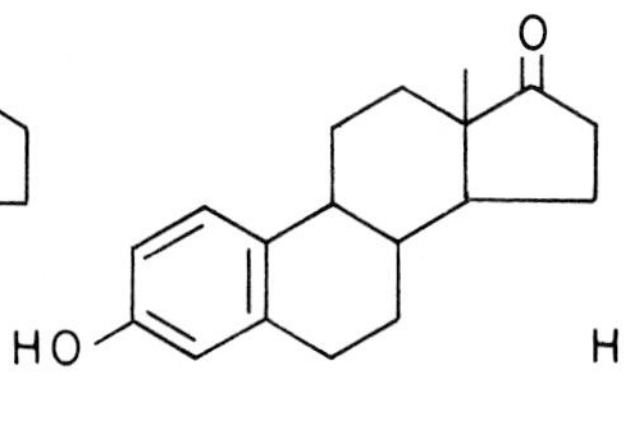   Estrone 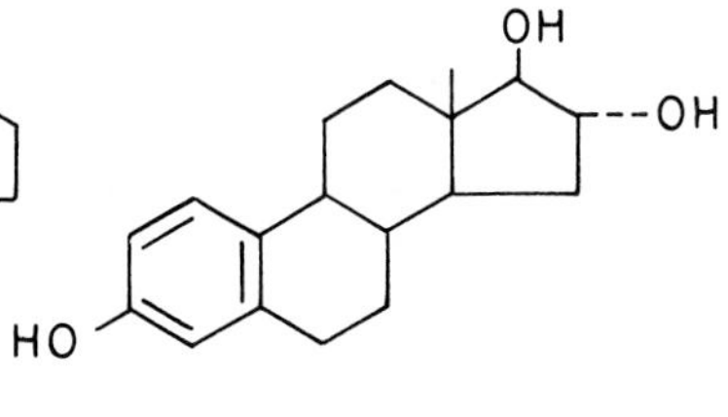   Estriol

Like the adrenal cortical hormones, progesterone contains twenty-one carbon atoms, the 3-keto group and the $C_{4,5}$ double bond. However, the terminal carbon atom of the chain, C_{21}, is a methyl group. Testosterone and androstenedione contain nineteen carbon atoms, whereas the estrogens contain only eighteen carbon atoms. The hydrocarbon side chain attached to the carbon atom at the apex of the D ring, C_{17}, is absent in the estrogens and androgens. The structural differences among these steroid hormones are summarized in Table 10-4.

Recognition of steroid hormones. It is often difficult for anyone other than a sophisticated steroid chemist to remember the variations in structure among these hormones. Yet it is important that the health scientist be able to recognize at least the major steroid hormones that are listed above. Several helpful hints may aid in recognizing these structures. First, examine the top carbon atom of

Table 10-4. Structural differences among major steroid hormones

HORMONE	CARBON ATOMS	C_3	C_{11}	C_{17}	C_{18}	C_{19}	C_{21}
Cortisol	21	C=O	OH	CH_2OH—CO + OH	CH_3	CH_3	CH_2OH
Aldosterone	21	C=O	OH	CH_2OH—CO	CHO	CH_3	CH_2OH
Corticosterone	21	C=O	OH	CH_2OH—CO	CH_3	CH_3	CH_2OH
Progesterone	21	C=O		CH_3—CO	CH_3	CH_3	CH_3
Testosterone	19	C=O		OH	CH_3	CH_3	
Androstenedione	19	C=O		C=O	CH_3	CH_3	
17-β-Estradiol	18	OH		OH	CH_3		

the D ring, C_{17}. The estrogenic and androgenic hormones do not have a carbon atom–containing side chain at this position; they have either a hydroxyl or a keto group attached to C_{17}. All of the other steroids have a two-carbon atom chain attached to C_{17}.

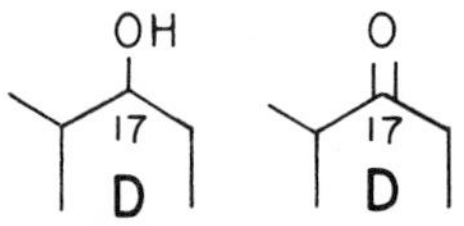

Estrogens and androgens 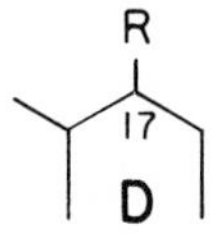   Others

Remembering this simple fact allows the separation of the steroid hormones initially into two classes, the estrogens and androgens in one group and all of the other steroid hormones in the other group.

Inspection of the A ring of the first group of hormones allows one to easily distinguish the estrogens from the androgens. The *estrogens* have a phenolic A ring. Moreover, they contain a hydroxyl group at C_3, and they do not have any angular methyl group at C_{10}, the junction of the A and B rings. In contrast, the A ring of the *androgens* has only one double bond, between C_4 and C_5, and there is a keto group at C_3. Moreover, the angular methyl group, C_{19}, is retained at the juncture of the A and B rings of the androgens.

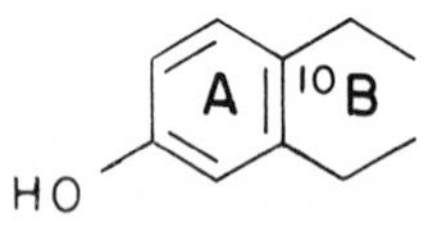

Estrogens 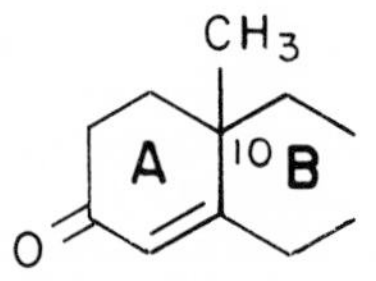   Androgens (and all others)

In all of the other steroid hormones the A ring is like that of the androgens, making the estrogens truly unique in this respect.

Having separated out the estrogens and androgens, let us now consider those steroid hormones that have a carbon atom chain attached to C_{17}, the apex of the D ring. These hormones also can be subdivided into two general classes, *progesterone* and the *corticosteroids*. The corticosteroids have a hydroxyl group attached to C_{11} of the C ring; progesterone has no substitution at this position. Moreover, the terminal carbon atom of the C_{17} chain, C_{21}, is a primary alcohol group in the corticosteroids and a methyl group in progesterone.

Corticosteroids

Progesterone

Finally, aldosterone, the main mineralocorticoid in man, can be distinguished from the other corticosteroids by simple inspection. Only aldosterone contains an aldehyde group at C_{18}, the carbon atom projecting from the juncture of the C and D rings. In the other corticosteroids, C_{18} is a methyl group. The ability to further distinguish among the various corticosteroids depends on memory.

Aldosterone

Other steroid hormones

This simplified scheme for recognition of the steroid hormones is presented in the form of a flow chart in Fig. 10-13.

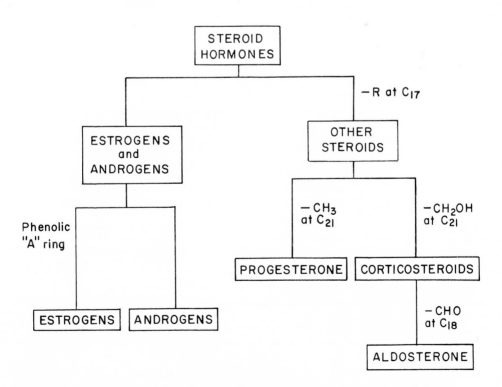

Fig. 10-13. Scheme for identification of steroid hormones into major subclasses.

Biosynthesis

Cholesterol is the precursor of all of the steroid hormones. It is converted to pregnenolone and then progesterone, and these are the initial intermediates in each of the synthetic pathways.

Pregnenolone. The initial reaction involves oxidative cleavage of cholesterol at C_{20} with the introduction of a keto group and loss of a six-carbon atom fragment. This reaction is catalyzed by desmolase, and the product that is formed is pregnenolone.

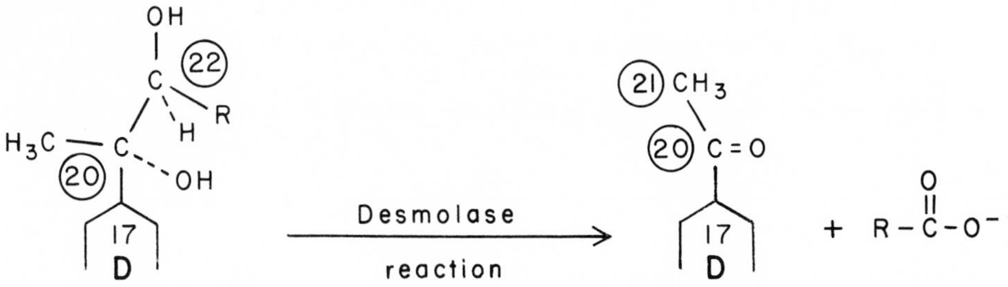

The desmolase reaction, the rate-limiting step in steroid hormone biosynthesis, requires NADPH. In the adrenal cortex this reaction is stimulated by ACTH, the pituitary hormone that causes increased glucocorticoid production and secretion. cAMP is the intracellular mediator of this hormonal signal. cAMP acts at least in part by activating a cholesterol esterase. This releases cholesterol from the cholesteryl ester storage pool and makes it available for the desmolase reaction.

Progesterone. All of the steroid hormones are derived from pregnenolone. Conversion of the 3-β-hydroxyl group to a keto group and migration of the double bond from positions 5,6 to 4,5 (B ring to A ring) produces progesterone.

In the ovarian corpus luteum, steroid synthesis stops at this point and progesterone is the hormonal product.

The pathway from cholesterol to progesterone is shown in Fig. 10-14, the shaded regions pointing out the structural modifications that occur in this series of reactions. In all other steroid-producing tissues, progesterone is merely an intermediate in the biosynthetic pathway.

Androstenedione and testosterone. In the androgen-secreting cells of the adrenal cortex, progesterone is formed from cholesterol as shown in Fig. 10-14 and then

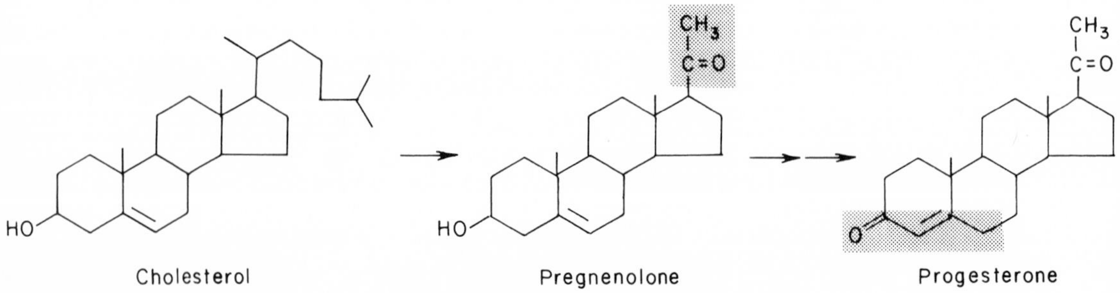

Cholesterol   Pregnenolone   Progesterone

Fig. 10-14. Progesterone synthesis.

further transformed. It is first hydroxylated at C_{17}, forming 17α-hydroxyproges-
terone. The two-carbon atom chain then is removed from C_{17}, forming androstenedione, which has only a keto group at the C_{17} position. In the testes these reactions
are carried one step further. The keto group at C_{17} is reduced to a hydroxyl group,
forming testosterone. These reactions are summarized in Fig. 10-15, with shading
again being used to point out the main chemical modifications that occur in this
sequence of reactions.

   Estrogen.   In the ovary, the series of reactions from cholesterol to testosterone
also occurs and then proceeds further. This is shown in Fig. 10-16. The A ring is
oxidized in several steps to a phenolic ring (shaded region) with loss of the C_{19}
angular methyl group and conversion of the 3-keto group to a hydroxyl group.
The product, 17-β-estradiol, is the main estrogenic hormone.

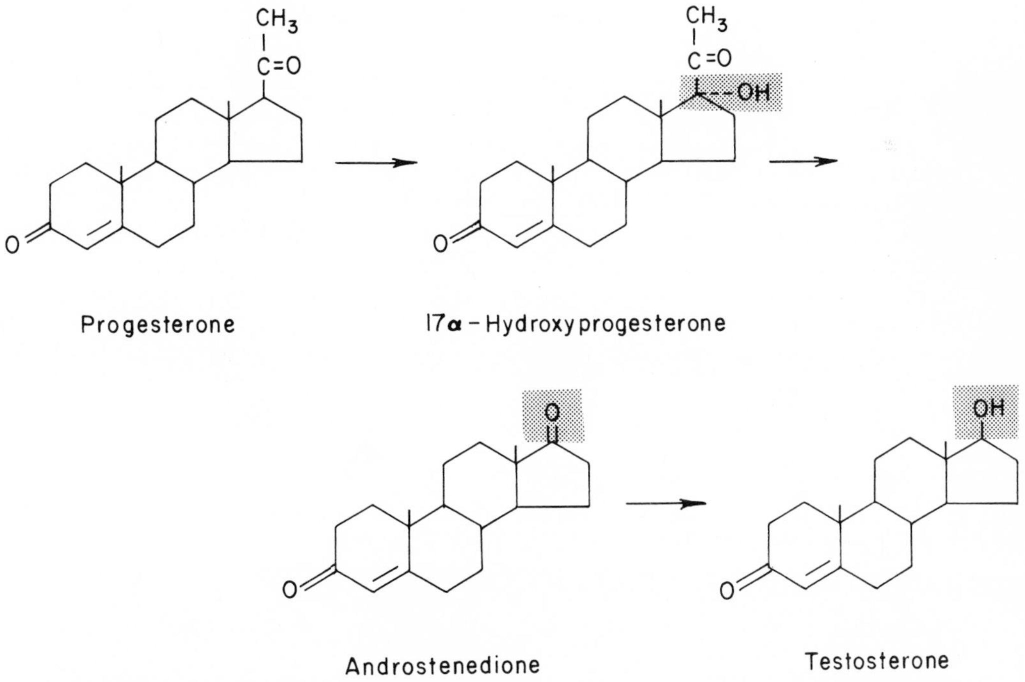

Progesterone   17α – Hydroxy progesterone

Androstenedione   Testosterone

Fig. 10-15. Conversion of progesterone to androstenedione and testosterone in androgen-
producing cells.

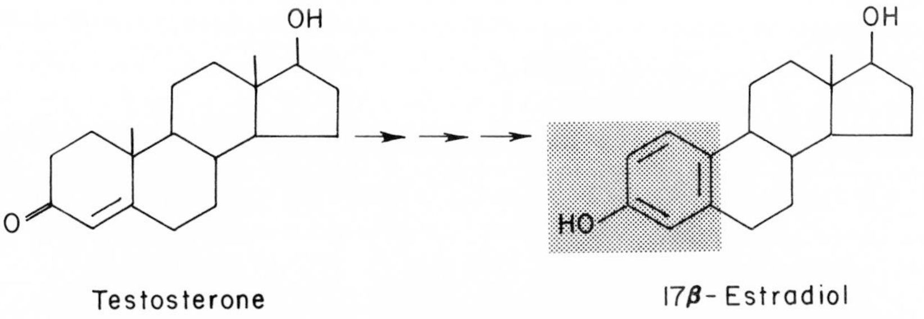

Testosterone   17β – Estradiol

Fig. 10-16. Conversion of testosterone to 17β-estradiol in the ovary.

Cortisol.   In the adrenal cortical cells that synthesize cortisol, cholesterol also is converted to 17α-hydroxyprogesterone as shown in Figs. 10-14 and 10-15. The 17α-hydroxyprogesterone that is formed is then hydroxylated in positions C_{21} and C_{11}, forming cortisol. This is illustrated in Fig. 10-17, with shading being used to point out the hydroxyl group that is inserted in each step.

Corticosterone and aldosterone.   Cells in the adrenal cortex that form the mineralocorticoids, corticosterone, and aldosterone, handle the progesterone that they

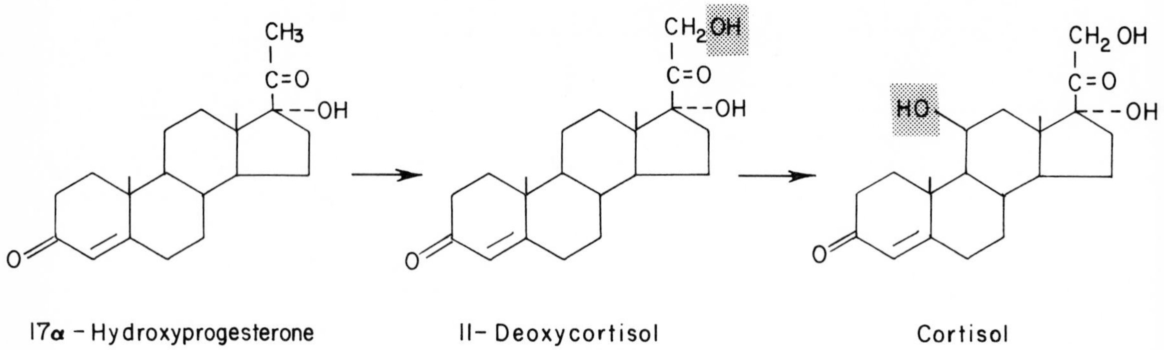

17α – Hydroxyprogesterone   ll– Deoxycortisol   Cortisol

Fig. 10-17. Cortisol formation.

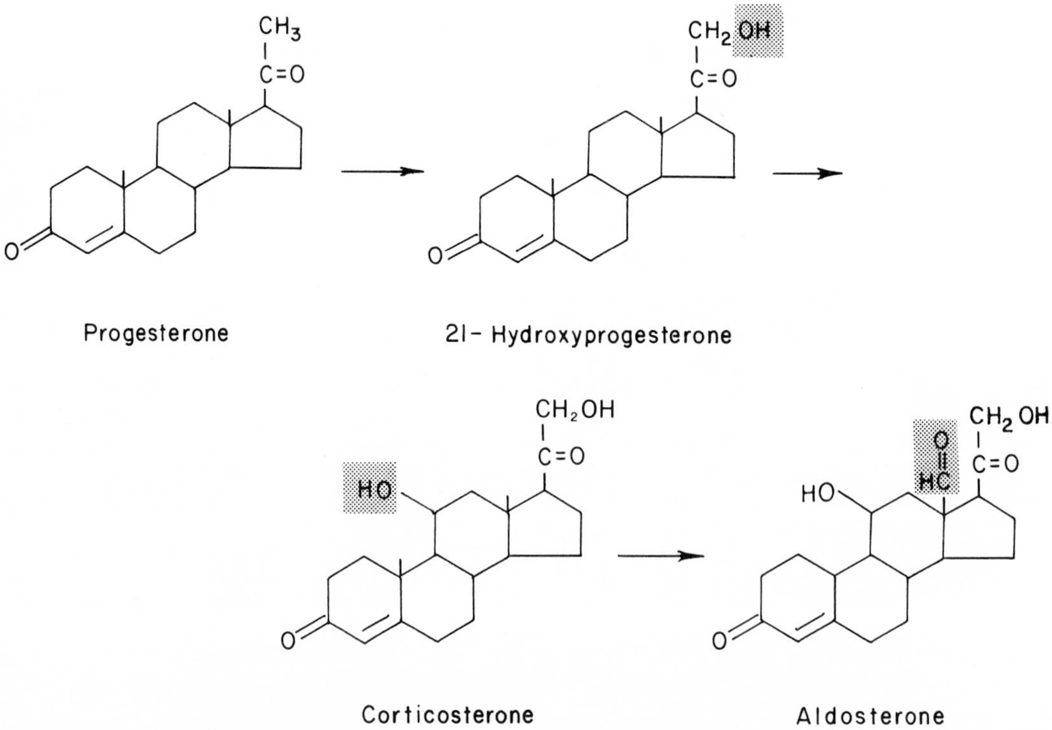

Progesterone   21– Hydroxyprogesterone

Corticosterone   Aldosterone

Fig. 10-18. Formation of corticosterone and aldosterone.

synthesize differently from the other hormone-producing cells. All of the other cells convert progesterone to 17α-hydroxyprogesterone as shown in Fig. 10-17, but the mineralocorticoid-producing cells omit this step. They hydroxylate progesterone directly in position C_{11}, forming corticosterone. Some of the corticosterone that is formed is converted to aldosterone by oxidation of the C_{18} angular methyl group to an aldehyde group. This pathway is illustrated in Fig. 10-18, with shading again being employed to point out the chemical change that occurs in each reaction.

An overview of the main steroid hormone synthetic pathways is presented schematically in Fig. 10-19. As discussed above, however, it should be remembered that no single cell carries out all of these reactions. Each hormone-producing cell performs only the segment of the scheme that leads to its product.

Alternate pathways. The biosynthetic pathway shown in Fig. 10-19 is somewhat simplified. Actually, there are several alternative pathways through which each of the steroid hormones can be synthesized. For example, consider the conversion of progesterone to corticosterone. Two hydroxyl groups must be introduced, one at C_{11} and the other at C_{21}. In Fig. 10-18 we show the hydroxylation at C_{21} as occurring first, forming 21-hydroxyprogesterone. This intermediate is then hydroxylated at C_{11}. However, the hydroxylation steps can occur in the reverse order.

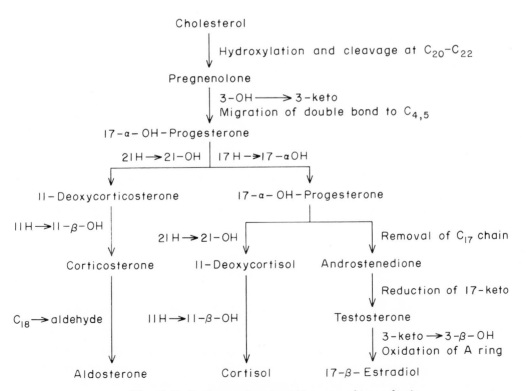

Fig. 10-19. Pathways for steroid hormone biosynthesis.

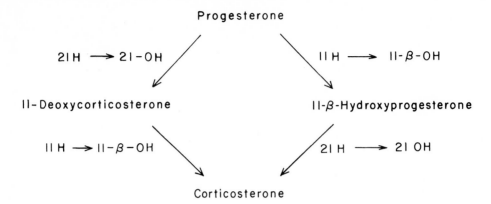

Alternative routes also occur in several other sequences, for example, the conversion of pregnenolone to androstenedione and the conversion of the latter to 17-β-estradiol.

Hydroxylation reactions

The hydroxylations of the steroid nucleus require molecular oxygen and NADPH. The enzymes, *hydroxylases*, are mixed-function oxidases. They are so called because in the reaction, one of the oxygen atoms is incorporated into the hydroxyl group, and the other is reduced to water.

$$R + O_2 + NADPH + H^+ \xrightarrow[\text{lase}]{\text{Hydroxy-}} ROH + H_2O + NADP^+$$

In the adrenal cortex the 11-β-hydroxylase is located in the mitochondria. It requires *adrenodoxin*, a nonheme iron- and sulfur-containing protein, for activity. The adrenal *21-hydroxylase* is located in the endoplasmic reticulum. There is a large amount of vitamin C, *ascorbic acid*, in the adrenal cortex. Although the exact function of ascorbic acid is unknown, it may act as a reducing agent to provide the NADPH required for the hydroxylation reactions.

Transport in plasma

Like other hormones, the steroids must be transported via the blood from their sites of synthesis to their target organs; for example, cortisol must be transported from the adrenal cortex to the liver as well as to its other sites of action. Plasma albumin can act as a nonspecific carrier for cortisol as well as other steroid hormones in the circulatory system. However, most steroid hormone transport is mediated by specific plasma steroid–carrier proteins that bind the steroid hormones much more tightly than does albumin. For example, cortisol binds tightly to a specific α-globulin of plasma, *transcortin*. This carrier protein, which has a molecular weight of 52,000, is synthesized by the liver. Corticosterone also binds to transcortin. The sex steroids, the estrogens and testosterone, are transported by a completely different plasma globulin, the *sex hormone–binding protein*.

Metabolism

Steroid hormones are for the most part converted into metabolic excretion products in the liver. The reactions that are involved are reductions of unsaturated bonds and the introduction of hydroxyl groups. In certain cases the two-carbon atom side chain of the corticosteroids may be removed and replaced at the C_{17}

position by either a keto or hydroxyl group. As an example, the pathway for cortisol inactivation in the liver is illustrated in Fig. 10-20. The shaded area in each reaction points out the chemical modification that takes place. First, the A ring and 3-keto group are reduced. Because of the asymmetry at C_5, two *tetrahydrocortisol* isomers are formed that have the five-hydrogen atom either in α- *(trans)* or β- *(cis)* orientation relative to the C_{19} methyl group. These compounds are reduced further to *cortol* through the conversion of the C_{20} keto group to a hydroxyl group. Although not shown, one should note that the C_{20} hydroxyl group also may

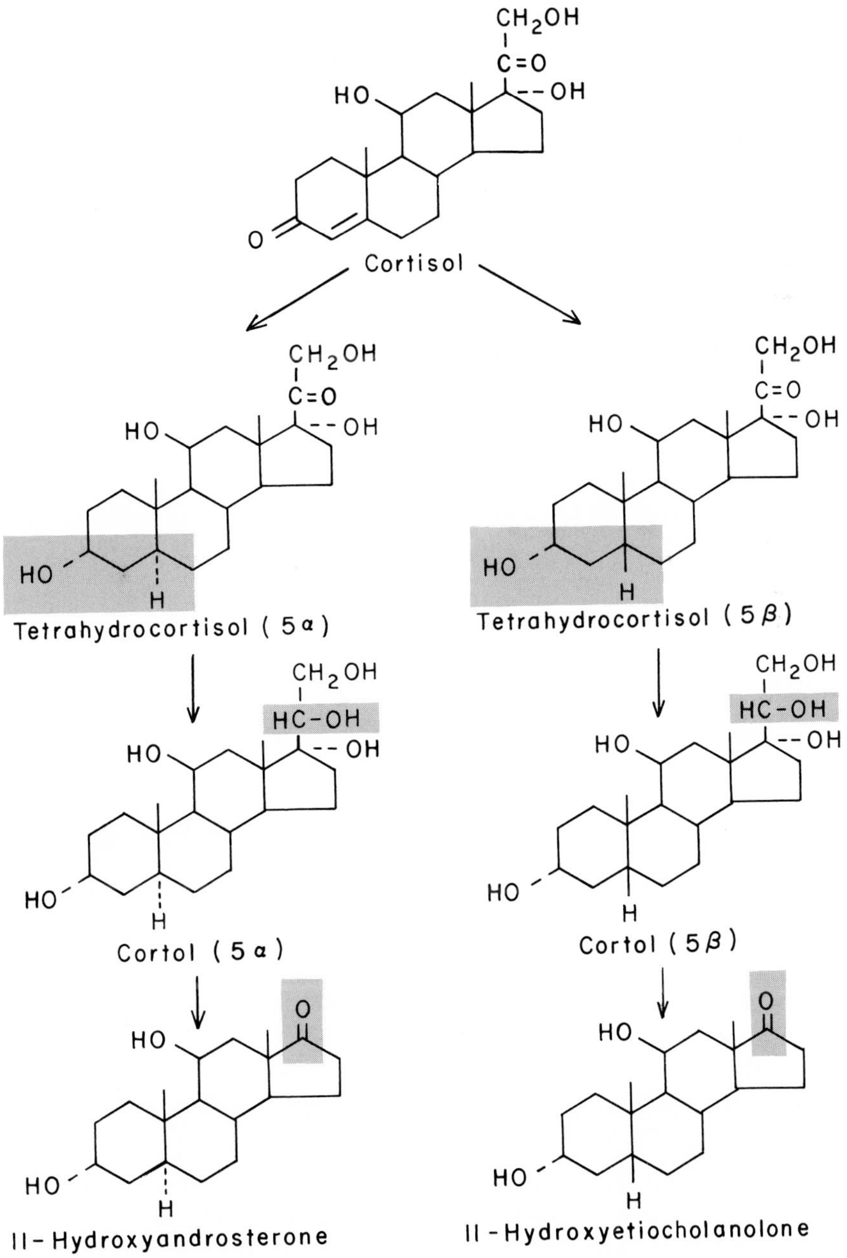

Fig. 10-20. Pathways for cortisol inactivation.

be in either α- or β-orientation, giving rise to two additional isomers. The side chain at C_{17} is then removed, giving rise to either *11-hydroxyandrosterone*, which contains an α-hydrogen atom at C_5, or *11-hydroxyetiocholanolone*, which contains a β-hydrogen atom at C_5. Note that these products actually are a special subclass of 17-ketosteroids and more precisely should be classified as *11-oxy-17-ketosteroids*. This distinguishes the cortisol metabolites from a group of substances commonly designated as 17-ketosteroids, which are excretion products of the male sex hormones.

Through a similar series of reactions, progesterone is converted to *pregnanediol* and *pregnanetriol*, its main metabolites.

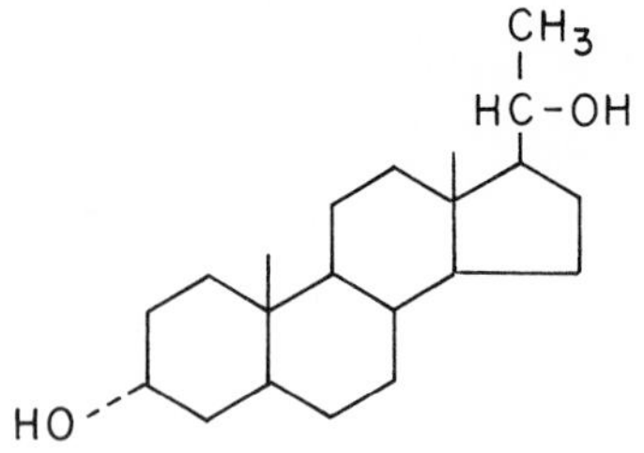

Pregnanediol 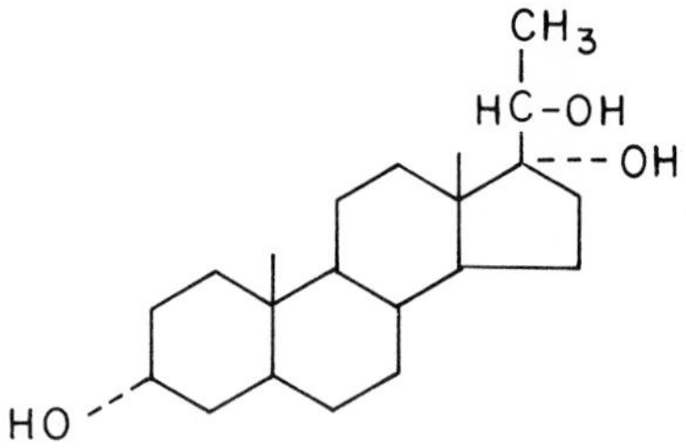   Pregnanetriol

The estrogens are inactivated in a similar manner. For example, 17-β-estradiol is hydroxylated at C_{16} to form *estriol* (p. 497). The A ring is not reduced when the estrogens are metabolized, and the hydroxyl group at C_3 remains.

The androgens are converted to either *androsterone* or *etiocholanolone*, which differ only in the orientation of the hydrogen atom at C_5; and *dehydroepiandrosterone*, which contains a β-hydroxyl group at C_3 and a double bond in the B ring. These metabolites are known as the *17-ketosteroids*.

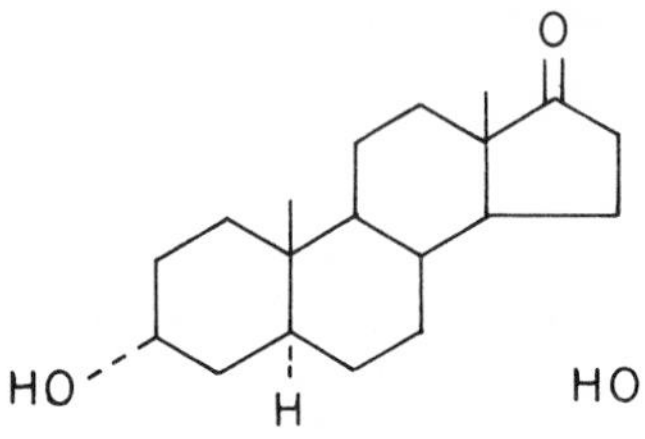

Androsterone 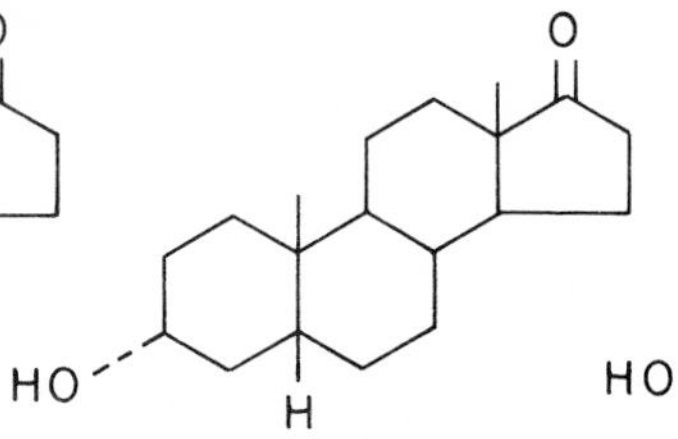   Etiocholanolone 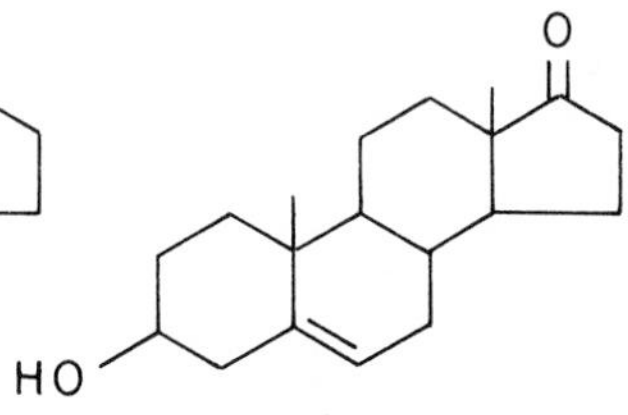   Dehydroepiandrosterone

Dehydroepiandrosterone, in addition to being a metabolite of the androgenic hormones, is produced and released in large amounts by all steroid-secreting glands, including the testes, ovaries, and adrenal cortex. The main metabolic products of the steroid hormones in man are summarized in Table 10-5.

Table 10-5. Steroid hormone metabolites in man

HORMONE	MAJOR METABOLITES
Corticosteroids	11-Oxy-17-ketosteroids, tetrahydrocortisol, tetrahydroaldosterone, cortol
Progesterone	Pregnanediol and pregnanetriol
Estrogens	Estriol
Androgens	17-Ketosteroids

Conjugation and excretion

The liver makes steroid hormone metabolites more soluble by conjugation with either glucuronic acid or sulfate. Approximately 20% of these metabolites are secreted into the bile and then excreted in the feces, and the remainder are released into the blood and subsequently excreted in the urine. These metabolites are quite soluble in plasma and are transported without binding to proteins. The steroid hormone metabolites are filtered from the plasma in the kidney and pass into the urine. From 70% to 80% of these metabolites are excreted in the urine. A typical glucuronide, *tetrahydroaldosterone-3-β-D-glucuronide*, is illustrated below. This is the main excretion product of aldosterone.

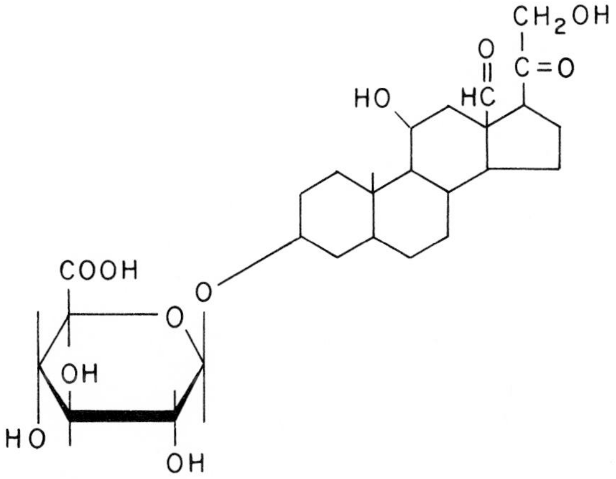

Tetrahydroaldosterone – 3 – β – D – glucuronide

UDP glucuronic acid is the glucuronyl group donor in these conjugation reactions that are catalyzed by *glucuronyl transferases*. The donor of the sulfate group in the sulfation reactions is *3′-phosphoadenosine 5′-phosphosulfate (PAPS)*.

REFERENCES

Bloch, K.: The biological synthesis of cholesterol, Science **150**:19, 1965. *A classic discussion of cholesterol biosynthesis.*

Brown, M. S., and Goldstein, J. L.: Receptor-mediated control of cholesterol metabolism, Science **191**:150, 1976. *A recent review of this new and exciting area.*

Connor, W. E., and Lin, D. S.: The intestinal absorption of dietary cholesterol by hypercholesterolemic (type II) and normocholesterolemic humans, J. Clin. Invest. **53**:1062, 1974. *An important study in humans.*

Cuatrecasas, P.: Membrane receptors, Ann. Rev. Biochem. **43**:169, 1974. *A comprehensive review of this rapidly expanding field.*

Danielsson, H., and Sjövall, J.: Bile acid metabolism, Ann. Rev. Biochem. **44**:233, 1975. *A thorough review.*

Dempsey, M. E.: Regulation of steroid biosynthesis, Ann. Rev. Biochem. **44**:967, 1975. *A brief overview describing recent advances in this field.*

Edmond, J., and Popjak, G.: Transfer of carbon atoms from mevalonate to n-fatty acids, J. Biol. Chem. **249**:66, 1974. *A description of the mevalonate shunt.*

Singer, S. J.: The molecular organization of membranes, Ann. Rev. Biochem. **43**:805, 1974. *A recent review by the investigator who described the fluid mosaic model of membrane structure.*

Singer, S. J., and Nicolson, G. I.: The fluid mosaic model of the structure of cell membranes, Science **175**:720, 1972. *Description of an alternative model for membrane structure.*

Clinical examples

CASE 1: β-SITOSTEROLEMIA

Two sisters, aged 19 and 17 years, were referred to the metabolism clinic of a university hospital because they had extensive tendon xanthomatosis. Repeated measurements of their plasma cholesterol concentrations after an overnight fast revealed values that were within the normal range, 206 and 193 mg/dl, respectively. It was thought most unlikely that plasma cholesterol could be responsible for such extensive xanthoma formation. However, no physical or biochemical abnormalities could be detected. In an attempt to explore the cause of this abnormality, the lipids from the patients' plasma were saponified and analyzed by gas-liquid chromatography. β-Sitosterol comprised about 10% of the total plasma sterols in each case, whereas it accounts for only about 0.2% in normal people. About 60% of the plasma β-sitosterol was in esterified form. From 75% to 85% of the plasma β-sitosterol was recovered in the LDL fraction. Erythrocytes obtained from these patients contained β-sitosterol, amounting to 13% of the total cell sterol. Analysis of specimens from tendon xanthomas removed from these patients demonstrated that β-sitosterol also was present in these lesions.

Biochemical questions

1. What is β-sitosterol?
2. What is the most likely origin of the β-sitosterol found in these patients' plasma and tissues?
3. Present a likely mechanism for the formation of β-sitosteryl esters in the blood plasma.
4. Would you expect the β-sitosterol in the erythrocytes to be predominantly in the free or esterified form?
5. Suggest a possible mechanism for xanthoma formation in these patients.

Case discussion

β-Sitosterol is a plant sterol. Its structure, shown on p. 472, is similar to that of cholesterol except for the presence of a two-carbon atom branch on C_{24}. This is located in the hydrocarbon tail that is attached to the sterol nucleus. β-Sitosterol is the predominant sterol in many plants, but it is not the only one. The others include campesterol and stigmasterol, which differ only slightly in structure from β-sitosterol. More detailed analyses of these patients' plasma and tissues revealed that campesterol and stigmasterol also were present, although in smaller amounts than β-sitosterol. For example, in the plasma of one of the patients, the β-sitosterol value was 27 mg/dl, campesterol was 10 mg/dl, and stigmasterol was 0.5 mg/dl.

Cholesterol is the only sterol synthesized by animal tissues. Plant products, however, are contained in the usual diet. Therefore, the most likely source of the β-sitosterol in these patients was the diet. Studies with isotopically labeled β-sitosterol revealed that these sisters absorbed about 25% of a test dose. Similar tests in normal volunteers revealed absorptions of much less than 5%. It is well known that humans absorb very little plant sterols and, in fact, high doses of plant sterols inhibit the absorption of dietary cholesterol. In the past, β-sitosterol was fed to patients with hypercholesterolemia in order to reduce dietary cholesterol absorption. Based on this information, it was concluded that these two patients with β-sitosterolemia have an inherited metabolic defect that causes them to absorb excessive quantities of plant sterols.

Most of the cholesteryl esters present in the blood plasma are formed through the action of lecithin-cholesterol acyltransferase (p. 486). This enzyme catalyzes the transfer of a fatty acid group from position 2 of lecithin to cholesterol, using lipoprotein lipids as substrates. The reaction occurs in the blood plasma. Although the enzyme is produced in the liver, it acts only after it is released into the blood. It is not known how β-sitosteryl esters are formed in the plasma. By analogy, however, there is every reason to think that they also are produced by the action of the lecithin-cholesterol acyltransferase enzyme, with the β-sitosterol contained in the plasma lipoproteins substituting for cholesterol. This actually was demonstrated by incubating radioactive β-sitosterol with plasma from these patients.

Essentially all of the cholesterol in the erythrocyte is in the free form. It is contained in the plasma membrane, where it modulates the fluidity of the fatty acyl chains of the membrane bilayer. This helps to regulate the permeability of the erythrocyte membrane as well as the function of integral proteins within the membrane. Again, we can assume by analogy that β-sitosterol replaces some cholesterol in the erythrocyte membrane simply because it is present in large quantities in these patients' bodies. It is so similar in structure that it almost certainly behaves as would cholesterol in the erythrocyte. Therefore, we would pre-dict that all of the erythrocyte β-sitosterol, like cholesterol, is in the free form.

At present, there is no definitive explanation as to why the high level of β-sitosterol in the plasma was associated with xanthomatosis. Relative to the usual plasma cholesterol concentrations, the amount of β-sitosterol in the plasma of these patients actually is low. Therefore, it is unlikely that the xanthoma formed simply through a mass-action effect. Although plant sterols were recovered in the xanthoma, they still contained primarily cholesterol. Why the presence of plant sterols caused cholesterol accumulation in the tendon sheaths when the plasma cholesterol concentration itself was not elevated also is unknown at this time.

REFERENCE

Bhattacharyya, A. K., and Connor, W. E.: β-Sitosterolemia and xanthomatosis: a newly described lipid storage disease in two sisters, J. Clin. Invest. 53:1033, 1974.

CASE 2: SPINOCEREBELLAR DEGENERATION

A 20-year-old male was admitted to a university hospital because of progressive deterioration of his gait and coordination. He had difficulty in running, fell frequently, and could not keep his balance in the dark. He also complained that his feet were numb and cold. The patient kept his legs far apart when he walked, and his feet slapped the floor with a flapping motion. Electrical measurements revealed that his sensory nerve conduction velocities were low. The plasma cholesterol concentration after an overnight fast was 103 mg/dl, and the plasma LDL content was 50% below the normal range.

Biochemical questions

1. What is the relationship of LDL to the other plasma lipoproteins?
2. How is an impulse conducted along a nerve? How might the nervous system abnormalities be related to the hypocholesterolemia?
3. How is acetylcholine thought to be involved in nerve conduction?
4. Suppose only one plasma apolipopro-tein obtained from this patient could be analyzed chemically and physically in great detail. Which one would you select in terms of possibly providing the greatest insight into the biochemical mechanism responsible for this disease?

Case discussion

Unlike most plasma lipid abnormalities, this case involves hypolipoproteinemia. The specific abnormality is a less than normal amount of LDL. The main lipid contained in the plasma LDL is cholesterol, about 60% to 70% of which is in the form of cholesteryl esters. LDL is formed in the plasma from VLDL. It is the metabolic product that remains after the triglycerides are removed from the VLDL. In this process much of the VLDL cholesterol is converted to cholesteryl esters through the action of lecithin-cholesterol acyl transferase. All of the apolipoproteins contained in VLDL except apolipoprotein B are released during the catabolic process. The molecular weight is reduced from about 1×10^7 to about

2×10^6, and what remains is a particle with a density range of 1.019 to 1.063 g/ml and S_f of 0 to 12 Svedberg units. It is rich in cholesteryl esters and apolipoprotein B (see Chapter 8).

Nerve cells are specialized so that they can conduct an impulse along their axons. Excitation along the axon is caused by a change in permeability of its membrane. The axoplasm is negatively charged with respect to the external surface of the axon membrane, the resting potential difference being −60 mV. This potential difference is maintained by an unequal distribution of ions across the axon membrane, the extracellular fluid being rich in Na^+ and the axoplasm being rich in K^+ and Cl^- ions. In the resting state, the axon membrane is very impermeable to these ions. When the nerve is stimulated, there is an increase in membrane permeability that proceeds down the axon. Each region of the axon remains permeable for less than a millisecond. The increased permeability is believed to be caused by a change in the conformation of an "active" membrane protein, leading to the creation of holes in the lipid bilayer that allows Na^+ to cross the membrane. Na^+ flows into the axon along its concentration gradient, reversing the polarization so that the axoplasm reaches a potential difference of +40 mV with respect to the outer surface of the membrane. As the impulse passes away from a region of the axon, Na^+ flows out and the original potential difference of −60 mV is reestablished. Any slight Na^+ retention in the axoplasm is removed through the action of the Na^+/K^+-ATPase, a membrane-bound enzyme (Chapter 5).

It has been postulated that acetylcholine is involved in the permeability change of the axon membrane. According to this theory, acetylcholine is stored within the membrane in a bound form during the resting state. When the impulse reaches a particular region of the axon, the acetylcholine is released and physically combines with the active protein in the membrane, causing a change in the conformation of this protein. The conformational change in this protein produces holes in the lipid bilayer through which Na^+ ions can freely pass. Acetylcholinesterase then hydrolyzes the acetylcholine. As soon as the acetylcholine concentration is reduced, the active protein returns to its original conformation and the membrane holes or Na^+-ion gates close. The acetylcholine content is then replenished by synthesis from choline and acetyl CoA. Until this is accomplished, the neuron is refractory to another stimulus.

The nervous system abnormalities can be explained by defects in nerve conduction. These, in turn, probably are caused by abnormalities in the membranes of the nerve cells. Like other membranes, those of the neuron are composed of a lipid bilayer in which proteins are embedded. Phospholipids form the matrix of the lipid bilayer. They are arranged so that their polar head groups occur on the surfaces of the bilayer, whereas their fatty acyl hydrocarbon chains project inward to form a nonpolar core. Cholesterol is intercalated between the phospholipid groups and modulates the movement of their fatty acyl tails. Therefore, cholesterol modulates the fluidity and hence the permeability of the membrane. The enzyme that is involved with ion movements across membranes, Na^+/K^+-ATPase, is an integral protein embedded in the membrane lipid bilayer. Carriers that move polar substrates into the cell by facilitated diffusion also are embedded in the lipid bilayer. Other vital substances cross the cell membrane by simple diffusion, a process that depends on the permeability characteristics of the lipid bilayer. It is not known how the observed hypocholesterolemia and reduced LDL concentration in this case are associated with or produce the nerve conduction abnormalities. We do know, however, that there is a rapid exchange of cholesterol between cell membranes and plasma lipoproteins. Therefore, it is reasonable to suggest that the less than normal plasma cholesterol concentration may have resulted in depletion of cholesterol from nerve cell membranes, leading to permeability defects or abnormalities in the function of membrane proteins. Alternatively, there may be a generalized defect in cholesterol synthesis, leading to less than normal amounts of cholesterol in both tissues and plasma lipoproteins. Another alternative is that LDL's deliver or remove substances other than cholesterol that are vital for membrane function. The lack of adequate amounts of LDL would result in membrane functional defects unrelated to cholesterol. Many other possible mechanisms can be suggested, providing hypotheses for future studies. At present, however, we can only speculate as to the molecular mechanism of this as well as most other neurologic problems.

Since LDL contains only apolipoprotein B, it is logical to study this protein if only one can be selected for analysis. Perhaps this might provide a clue to the genetic defect responsible for this disease, for if apolipoprotein B were abnormal, it might lead to the production of an abnormal, cholesterol-poor LDL. While the selection of apolipoprotein B would seem most reasonable, it certainly is not the only possible approach. The problem may involve the production of the LDL precursor, VLDL, or its normal catabolism. HDL is involved in the catabolism of VLDL, probably transferring apolipoprotein cofactors and lipids to and from the VLDL. HDL and VLDL contain a mixture of different apolipoproteins. Therefore, it is conceivable that any of these might hold the clue to this disease. Yet, when there is no obvious indication as to how to proceed, the best approach usually is the most direct one. Since the only known abnormality is in LDL, most experienced investigators probably would elect to study apolipoprotein B if they could exercise only one choice.

REFERENCE

Aggerbeck, L. P., McMahon, J. P., and Scanu, A. M.: Hypobetalipoproteinemia: clinical and biochemical description of a new kindred with "Friedreich's ataxia," Neurology 24:1051, 1974.

CASE 3: HYPERCHOLESTEROLEMIA, ATHEROSCLEROSIS, MYOCARDIAL INFARCTION, AND XANTHOMATOSIS

A 33-year-old man was referred to a university hospital because of severe intermittent chest pain. He had begun to have these pains in his late twenties, and they gradually became worse. They occurred typically after mild exertion such as brisk walking, carrying a bag of groceries up a flight of stairs, or mowing the lawn. A coronary artery angiogram was performed and revealed severe narrowing and irregularity of all three coronary arteries. A diagnosis of coronary artery atherosclerosis was made. Blood was obtained after an overnight fast on two occasions, and these samples contained 390 and 410 mg/dl of cholesterol, respectively. A careful physical examination revealed the presence of firm masses on the extensor tendons of the hands. These were thought to be xanthomas of the tendon sheaths. While in the hospital, the patient developed a severe chest pain that persisted for several hr. An electrocardiogram taken during the episode of pain showed changes that were consistent with an acute myocardial infarction. The patient was transferred immediately to the coronary care unit where his recovery from this acute illness was uneventful.

Biochemical question

1. How is hypercholesterolemia thought to be related to the development of atherosclerosis, myocardial infarction, and tendon xanthomas?

Case discussion

High plasma cholesterol concentrations, particularly in young and middle-aged men, are associated with an increased incidence of *coronary artery disease* and its end result, *myocardial infarction*. The underlying disease process is *atherosclerosis*. Fig. 10-21 illustrates the prevalent theory concerning the role of hypercholesterolemia in these diseases.

Arteriosclerosis. The general term for hardening of the arteries is arteriosclerosis. The most prevalent form of arteriosclerosis is characterized by the

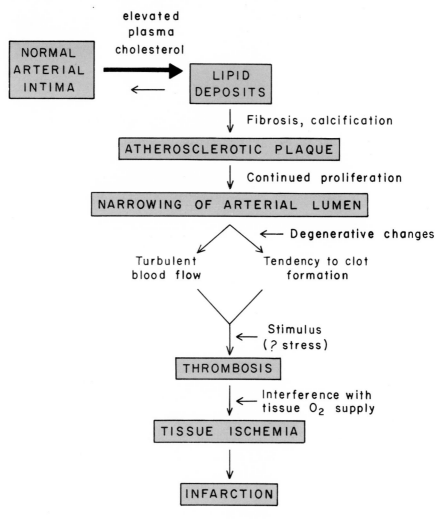

Fig. 10-21. Mechanism producing atherosclerosis and tissue infarction.

accumulation of lipid, particularly cholesteryl esters, in the arterial intima. This form of arteriosclerosis is called *atherosclerosis*. These lipid deposits occur inside cells that are thought to be derived from arterial smooth muscle. The cholesteryl esters present in these cells contain a high proportion of cholesteryl oleate. At this stage the lipid deposits are reversible, and no permanent damage to the vessel wall results. This lesion is called a *fatty streak*. Subsequently, lipids begin to accumulate extracellularly in the arterial intima. The pathologic mechanisms that are involved in converting the lesion containing predominantly intracellular lipid accumulation to one containing extracellular lipid

deposits are unknown at present. Although the predominant lipid is still cholesteryl ester, the composition of the extracellular cholesteryl esters resembles that of the plasma lipoproteins, particularly LDL. More cholesteryl linoleate than oleate is present in these extracellular lipid deposits. When the lesion progresses to this stage, it is no longer completely reversible. Yet recent studies on diet-induced atherosclerosis in monkeys indicate that some lipid can be removed from even advanced lesions so that the atherosclerotic areas actually shrink. This shrinking process is known as *regression*. Although there is no evidence as yet that human atherosclerotic lesions also can undergo regression, it is reasonable to assume that a biologic phenomenon that occurs in primates also occurs in man. Much of the currently employed medical therapy in patients with atherosclerosis is aimed at producing some degree of regression of the atherosclerotic lesions.

Hypercholesterolemia. Elevated plasma cholesterol concentrations, particularly the form resulting from elevations in LDL's (type II hyperlipoproteinemia), somehow increase the tendency toward atherosclerosis. It is thought that plasma LDL's actually deposit their lipid content in the vessel wall, causing the extracellular accumulation of cholesteryl esters. The higher the plasma LDL concentration, the greater is the tendency toward lipid deposition. What cannot be explained at present is the mechanism leading to intracellular cholesteryl ester accumulation (the initial process in fatty streak formation) and the relationship between the intracellular and extracellular cholesteryl esters in the arterial intima. Somehow, the relation between LDL elevation and arterial wall lipid accumulation will have to be tied together with the LDL binding or internalization defect noted in skin fibroblasts of patients with severe hypercholesterolemia. These points are currently under vigorous investigation. Additional facts concerning hypercholesterolemia and its role in the development of atherosclerosis are presented in the next case.

Advanced atherosclerotic lesion

Ultimately the lipid deposit triggers a fibrotic reaction in the arterial wall, and collagen is deposited to form a fibrous plaque. At this point the process is no longer completely reversible. Calcium is deposited within the plaque and the lesion grows; the result is a decrease in the diameter of the channel through which the blood flows. Subsequently, the plaque can degenerate, producing a roughened intimal surface. This, together with the increased turbulence of blood flow resulting from the narrowed arterial lumen, predisposes to clot formation within the blood channel. If the occlusion of the channel persists, the cells that are normally supplied with oxygen by the vessel will become *anoxic* and ultimately die; the death of tissue is termed an *infarction.* As described in Chapter 3, this results in the leakage of certain enzymes from the damaged cells; measurement of these enzyme levels provides an indication of whether an infarction has occurred. Atherosclerosis can occur in many different arteries, and the clinical disease that results depends on which vessels are involved.

Xanthomatosis. Xanthomas are collections of lipid, mostly cholesteryl ester, in the skin or tendon sheaths. Xanthomas are often associated with and are a sign of severe hypercholesterolemia and premature atherosclerosis.

REFERENCES

Brown, M. S., and Goldstein, J. L.: Regulation of the activity of the low density lipoprotein receptor in human fibroblasts, Cell 6:307, 1975.

Jagannathan, S. N., Connor, W. E., Baker, W. H., and Bhattacharyya, A. K.: The turnover of cholesterol in human atherosclerotic arteries, J. Clin. Invest. 54:366, 1974.

Stein, O., Weinstein, D. G., Stein, Y., and Steinberg, D.: Binding, internalization and degradation of low density lipoprotein by normal human fibroblasts and by fibroblasts from a case of homozygous familial hypercholesterolemia, Proc. Natl. Acad. Sci. U.S.A. 73:14, 1976.

CASE 4: HYPERCHOLESTEROLEMIA

A 36-year-old man was found to have hypercholesterolemia. After an overnight fast, his blood plasma cholesterol concentration on two separate occasions was approximately 330 mg/dl. He was treated with a synthetic cholesterol-free formula diet for 3 months, but the fasting plasma cholesterol level decreased only to 300 mg/dl. Subsequently, he was treated with cholestyramine, a bile acid–binding resin. This resin is not absorbed and therefore remains in the intestine. There, it binds bile acids, causing increased amounts of them to be excreted in the feces. The drug treatment was successful in lowering the fasting plasma cholesterol to the range of 230 to 250 mg/dl, a value that was considered to be acceptable for this patient.

Biochemical questions

1. Why were the cholesterol determinations made on plasma obtained after an overnight fast?
2. How is it possible for a patient to continue to have circulating cholesterol after being on a cholesterol-free diet for 3 months?
3. What connection is there between bile acids and cholesterol? Why would increased excretion of bile acids lead, in some cases, to a decrease in the plasma cholesterol concentration?
4. Why would lowering of the plasma cholesterol concentration be potentially beneficial for this patient?

Case discussion

Cholesterol is absorbed from the diet. Most of us have a dietary cholesterol intake of 500 to 1000 mg/day and absorb 200 to 500 mg/day. Any dietary cholesterol that is present in esterified form is hydrolyzed to the free alcohol through the action of cholesterol esterase, a pancreatic digestive enzyme. Hydrolysis occurs within or on the surface of micelles that are formed through mixture of the conjugated bile acids, lecithin, and cholesterol contained in the bile. Cholesterol is transferred from the micelle to the intestinal mucosal cell by passive diffusion. Most of the absorption occurs in the ileum. Once inside the mucosal cell, most of the cholesterol is reesterified and released into the lymph in chylomicrons. These lipoproteins enter the bloodstream, and the cholesteryl esters are taken up by the liver. The absorptive process may continue for 8 to 10 hr after ingestion of a fatty meal. Therefore, a person normally has transient hypercholesterolemia after eating a meal containing fat, and this is an entirely physiologic response. Hypercholesterolemia is abnormal only if it persists in the fasting state. It is for this reason that all evaluations of plasma cholesterol concentration in man should be made after an overnight fast of 12 to 14 hr in duration.

It is perfectly reasonable for a patient to have cholesterol in his blood plasma and, for that matter, in all of his tissues in spite of the fact that he has been on a cholesterol-free diet for a prolonged period. The diet is not the only source of cholesterol for man. In fact, most of the cholesterol present in the body is biosynthesized, not derived from the diet. Acetyl CoA is the substrate for cholesterol synthesis. This metabolic intermediate is produced from the catabolism of carbohydrates, fatty acids, and many of the amino acids. Therefore almost any food that man ingests will provide the necessary building blocks for cholesterol synthesis. It is true that less cholesterol is synthesized when the diet contains adequate amounts of this sterol. This occurs because the biosynthetic pathway in the liver is inhibited when the cholesterol supply to the liver is adequate. Inhibition occurs because the production of HMGCoA reductase, the rate-limiting enzyme in the cholesterol synthetic pathway, is reduced. However, considerable cholesterol synthesis occurs even when the diet is rich in cholesterol, and it is estimated that more than 50% of the plasma cholesterol is supplied by biosynthesis even when the diet is rich in cholesterol. When the diet is free of cholesterol, biosynthesis proceeds maximally. Yet the quantity that can be synthesized cannot compensate completely for what ordinarily would be obtained from the diet. Therefore, as noted in this case, one would expect some decrease in plasma cholesterol concentration when the patient is on a cholesterol-free diet. The decrease was considered to be too small to be of sufficient therapeutic value for this patient.

The main route of cholesterol metabolism from the quantitative standpoint is conversion to bile acids. About 300 mg/day of bile acids are lost in the feces. This is replaced daily, for the body is geared to maintain the size of the bile acid pool at a constant level of approximately 15 to 30 g. Replacement of bile acids is accomplished through synthesis from cholesterol in the liver. Administration of a bile acid-binding resin decreases bile acid reabsorption from the intestine and hence increases the loss of bile acids in the feces. In order to compensate for the greater losses, more cholesterol must be converted into bile acids in the liver. If the dietary intake of cholesterol is restricted to a modest level, about 300 mg/day or less, much of the cholesterol required for bile acid synthesis will have to be supplied from within the body. The capacity of the liver to synthesize cholesterol is limited, and some of the extra cholesterol that is needed for bile acid synthesis will have to be supplied from what normally would go into the plasma and tissues. Eventually a new equilibrium develops so that cholesterol is supplied at a rate that will satisfy the need for bile acid synthesis. However, the steady-state plasma cholesterol concentration that is reached in the newly equilibrated condition is often lower than that which was present prior to administration of the bile acid–binding resin.

Lowering the plasma cholesterol concentration in patients who have hypercholesterolemia is considered by many physicians to be helpful in preventing atherosclerosis and in treating existing atherosclerotic lesions. Elevations in plasma cholesterol concentrations, particularly in middle-aged males, have been correlated with an increased incidence of myocardial infarction, one of the most severe consequences of the atherosclerotic process. The exact means by which high circulating cholesterol concentrations predipose to atherosclerosis is presently unclear. One possibility is that a simple equilibrium between plasma lipoprotein cholesterol and tissue cholesterol exists. As the plasma lipoprotein cholesterol increases, more deposits in tissues such as the arterial wall. Another hypothesis is that intact lipoproteins enter into and become trapped inside the arterial wall. This is the *filtration theory*, which postulates that the protein and lipid components of the trapped lipoproteins are degraded. However, because the sterol nucleus cannot be catabolized, most of the lipoprotein cholesterol content remains with the arterial wall as cholesteryl esters, producing atherosclerosis. Independently of the mechanism, it is an empirical fact that populations whose members have high circulating cholesterol concentrations are more prone to develop atherosclerosis. Modern medical treatment is therefore aimed at lowering plasma cholesterol concentrations, particularly in individuals such as this 36-year-old man in whom a serious elevation is known to exist. Experiments in animals strongly suggest that lowering plasma cholesterol concentrations is effective in reducing the incidence of atherosclerosis and even in causing some decrease in the size of existing atherosclerotic lesions, a process that has been termed regression. We do not have proof at this time that this type of therapy is similarly helpful in human atherosclerosis. This question is being studied extensively at present, but the prevailing approach is vigorous treatment of hypercholesterolemia, as was done in this patient.

REFERENCES

Armstrong, M. L., Warner, E. D., and Connor, W. E.: Regression of coronary atheromatosis in rhesus monkey, Circ. Res. 27:59, 1970.

Glueck, C. J., Ford, S., Jr., Scheel, D., and Steiner, P.: Colestipol and cholestyramine resin: comparative effects in familial type II hyperlipoproteinemia, J.A.M.A. 222:676, 1972.

Goldstein, J. L., Hazzard, W. R., Schrott, H. G., Bierman, E. L., and Motulsky, A. G.: Hyperlipidemia in coronary heart disease I. Lipid levels in 500 survivors of myocardial infarction, J. Clin. Invest. 52:1533, 1973.

CASE 5: LECITHIN-CHOLESTEROL ACYLTRANSFERASE DEFICIENCY

A patient with familial lecithin-cholesterol acyltransferase (LCAT) deficiency was admitted to a university hospital for metabolic studies. Electron microscopy revealed structural abnormalities in all of the classes of plasma lipoproteins. These structural abnormalities were similar to those noted in patients with very severe liver damage. Analysis of the plasma lipids revealed elevated amounts of cholesterol and lecithin and a very low cholesteryl ester content. The patient had proteinuria, indicating rather severe kidney damage. Cholesterol was markedly elevated in erythrocytes. Although the total amount of phospholipids in the erythrocytes was normal, their lecithin content was considerably increased.

Biochemical questions

1. What are the two ways in which cholesteryl esters are synthesized in the body?
2. Where does LCAT operate, and what is its role in lipoprotein metabolism?
3. How might an LCAT deficiency account for the observed plasma lipid and lipoprotein structural abnormalities?
4. Why might severe liver damage produce lipid abnormalities similar to those seen in this case?
5. How might the erythrocyte lipid abnormalities be explained?
6. Assuming that the kidney damage was caused by abnormalities in the membranes of renal cells, how might this be related to the LCAT deficiency?
7. How might lipid defects in the tubular cell membranes produce alterations in tubular reabsorption?

8. How might these membrane defects also influence the operation of the Na^+/K^+-ATPase?

REFERENCES

Farras, R. N., Bloj, B., Moreno, R. D., Sineriz, F., and Trucco, R. E.: Regulation of allosteric membrane-bound enzymes through changes in membrane lipid composition, Biochem. Biophys. Res. Commun. 415:231, 1975.

Forte, T., Nichols, A., Glomset, J., and Norum, K.: The ultrastructure of plasma lipoproteins in lecithin: cholesterol acyltransferase deficiency, Scand. J. Clin. Lab. Invest. 33(Suppl. 137): 121, 1974.

Gjone, E., Torsvik, H., and Norum, K. R.: Familial plasma cholesterol ester deficiency: a study of the erythrocytes, Scand. J. Clin. Lab. Invest. 21:327, 1968.

Glomset, J. A., Nichols, A. V., Norum, K. R., King, W., and Forte, T.: Plasma lipoproteins in familial lecithin: cholesterol acyltransferase deficiency, J. Clin. Invest. 52:1078, 1973.

Huang, L., and Pagano, R. E.: Interaction of phospholipid vesicles with cultured mammalian cells I. Characteristics of uptake, J. Cell Biol. 67:38, 1975.

CASE 6: CHRONIC ADRENAL CORTICAL INSUFFICIENCY

A 26-year-old woman was referred to the endocrinology service of a university hospital because she complained of nausea, dizziness, and skin discoloration. Although she had not been feeling completely well for about 5 years, her complaints were fairly nonspecific and had been passed off as resulting from overwork and tension. She had begun to develop persistent nausea 6 months prior to admission, and she vomited intermittently. She had also become much more fatigued and complained of generalized weakness. A 5 kg weight loss had occurred during the previous 6 months. At times she felt faint when she stood up from a lying or sitting position, and her blood pressure in the sitting position was only 90/50 mm Hg (normal range varies considerably, but usually is taken as 120/80 mm Hg). She had not sunbathed since the summer and she did not use an ultraviolet tanning lamp; yet parts of her body were deeply bronzed, and there was increased pigmentation over the creases of her joints and in the palmar creases. Measurement of steroid excretion products in the urine indicated a decrease in metabolites of cortisol. The plasma cortisol determination in a blood sample taken at 9 AM was 2 μg/dl (normal range is 15 to 25 μg/dl). A diagnosis of chronic adrenal cortical insufficiency was made; after a number of additional tests to further define the exact nature of the problem, steroid replacement therapy with a cortisol analogue was begun.

Biochemical questions

1. How are steroids, which are poorly soluble in aqueous media, made available for excretion in the urine?
2. How might urine studies be used to distinguish between defects involving cortisol and the sex steroids?
3. Certain forms of adrenal cortical insufficiency result from decreased adrenocorticotropin (ACTH) production rather than from a defect within the adrenal gland. How is ACTH related to adrenal steroid production?
4. There are three steroid hydroxylase enzymes in the adrenal cortex. What carbon atoms of the steroid hormones are hydroxylated by each of these enzymes, and which of these three enzymes are involved in cortisol production?

5. Why would you *not* expect administration of a steroid hormone with a phenolic A ring to help this patient's condition?
6. If the metabolic defect in this case was located in the adrenal cortex, could it be the result of a deficiency in the 17α-hydroxylase enzyme?

REFERENCES

Bongiovanni, A. M.: Disorders of adrenocortical steroid biogenesis. In Stanbury, J. G., Wyngaarden, J. B., and Fredrickson, D. S., editors: The metabolic basis of inherited disease, ed. 3, New York, 1972, McGraw-Hill Book Co.

Frieden, E., and Lipner, H.: Biochemical endocrinology of the vertebrates, Englewood Cliffs, N. J., 1971, Prentice-Hall, Inc.

CASE 7: GALLSTONES

A 39-year-old woman consulted her physician because of intermittent abdominal distress. The discomfort usually followed the ingestion of a large meal, often one that contained greasy or fried foods. The pain was located in the upper abdomen and sometimes radiated to her chest. The patient felt bloated during these episodes and thought that she obtained some relief from belching. Occasionally she became severely nauseated and vomited during one of these acute episodes. She had not experienced any previous episodes of jaundice or gastrointestinal bleeding. Initially a diagnosis of irritable bowel syndrome was entertained, and her physician prescribed antacids and a bland diet. This treatment produced no relief. A cholecystogram demonstrated the presence of numerous gallstones in the gallbladder. A cholecystectomy was performed, and the gallstones were found to be composed predominantly of cholesterol.

Biochemical questions

1. What is the function of bile in digestion?
2. What is the metabolic relationship between cholesterol and bile acids?
3. How is cholesterol kept in the soluble state in normal human bile?
4. What physical-chemical factors cause the formation of cholesterol gallstones?
5. Can the composition of the bile be altered by changes in the dietary fat intake?

REFERENCES

Den Besten, L., Connor, W. E., and Bell, S.: The effect of dietary cholesterol on the composition of human bile, Surgery 73:266, 1973.

Holt, P. R.: The roles of bile acids during the process of normal fat and cholesterol absorption, Arch. Intern. Med. **130**:574, 1972.

Redinger, R. N., and Small, D. M.: Bile composition, bile salt metabolism and gallstones, Arch. Intern. Med. **130**:618, 1972.

CASE 8: TYPE IIa HYPERLIPOPROTEINEMIA

A 61-year-old woman was referred to a university hospital because of severe headaches, fainting spells, and severe leg cramps on walking. She was noted to have hypertension (blood pressure 200/100 mm Hg), and bruits were heard over the carotid and femoral arteries. An angiogram of the right carotid artery demonstrated a narrowed lumen. An aortogram revealed severe atherosclerosis of the lower abdominal aorta and both common iliac arteries; the left common iliac artery was almost completely obstructed. The fasting plasma cholesterol level was 385 mg/dl, triglycerides were 180 mg/dl, and the lipoprotein electrophoresis pattern showed that no abnormal lipoprotein bands were present. A diagnosis of type IIa hyperlipoproteinemia was made. Eight days after admission to the hospital a right carotid endarterectomy was performed. Ten days later a left aortofemoral bypass operation was carried out. These procedures produced significant improvement in the patient's condition. She was discharged from the hospital with instructions to take cholestyramine, a bile acid–binding drug that often is effective in reducing hypercholesterolemia.

Biochemical questions

1. What relationship is thought to exist between hypercholesterolemia and atherosclerosis?
2. What is the relationship between cholesterol and plasma lipoproteins?
3. What benefit could be gained by the use of the drug cholestyramine? Would the use of a low-cholesterol diet also be helpful for this patient?
4. Type II hyperlipoproteinemia is associated with elevated concentrations of plasma LDL's. What is the relationship of LDL to the other classes of plasma lipoproteins?
5. How is plasma LDL taken up by human cells?
6. What abnormalities have been described concerning the ability of cells to metabolize cholesterol in patients with familial hypercholesterolemia?
7. Are lysosomes involved in LDL metabolism?

REFERENCES

Brown, M. S., and Goldstein, J. L.: Familial hypercholesterolemia: defective binding of lipoproteins to cultured fibroblasts associated with impaired regulation of 3-hydroxy-3-methyl-glutaryl coenzyme A reductase activity, Proc. Natl. Acad. Sci. U.S.A. 71:788, 1974.

Goldstein, J. L., Dana, S. E., Faust, J. R., Beaudet, A. L., and Brown, M. S.: Role of lysosomal acid lipase in the metabolism of plasma low density lipoprotein, J. Biol. Chem. 250:8487, 1975.

Scanu, A. M., and Wisdom, C.: Serum lipoproteins: structure and function, Ann. Rev. Biochem. 41:703, 1972.

ADDITIONAL QUESTIONS AND PROBLEMS

1. A patient with severe hypercholesterolemia who did not respond to drug therapy was treated surgically by excision of a part of the intestine. What particular part of the intestinal tract might be selected for excision to produce the maximal decrease in cholesterol levels?
2. During a surgical procedure a sample of bile was obtained for research purposes from the gallbladder of a patient whose biliary and intestinal tracts were normal. Which bile acids would be recovered in this specimen?
3. A child who is suspected of having an adrenal tumor is noted to have an extremely high urinary excretion of androsterone and etiocholanolone. Urinary excretion of the 11-hydroxy-17-ketosteroids, cortol, and all tetrahydro derivatives are normal. In biochemical terms, what type of adrenal tumor would probably be present?
4. A female child is found to have a deficiency in steroid 21-hydroxylase activity. Could she exhibit normal sexual development? If a male child inherited this defect, could he undergo normal sexual development?
5. If the urinary pregnanediol excretion is elevated, which steroid hormone is most likely elevated in the body?
6. Would a steroid hormone with a hydroxyl group attached to position C_{11} of the steroid nucleus be expected to produce masculinization if administered to a female?
7. Would a drug that interfered with biotin-mediated carboxylation reactions be expected to directly block cholesterol biosynthesis?
8. How is the metabolism of glucose by the pentose phosphate pathway related to cholesterol synthesis?
9. If a drug blocked the synthesis of HMGCoA in the cell cytoplasm, would it be expected to interfere directly with ketone body synthesis?

CHAPTER 11

NUCLEIC ACIDS AND NUCLEOTIDES

OBJECTIVES

1. To explain how the structural features of the nucleic acids make them suitable templates for storing or expressing biologic information
2. To describe the roles of folic acid coenzymes in the biosynthesis of the purine and pyrimidine nucleotides
3. To explain how deficiencies in folate can lead to megaloblastic anemia associated with some pathologic conditions
4. To determine how certain drugs, for example, those used in cancer chemotherapy or in disease states such as gout, interfere with nucleotide biosynthesis

Biochemistry can be divided along functional lines into those aspects that relate to energy metabolism and those that relate to biosynthesis; that is, one can compare the biochemistry of homeostasis with that of growth. Central to a discussion of growth or reproduction at the molecular level is a consideration of the functions of the nucleic acids. In order to understand these functions it will be necessary to study certain of their chemical and physical properties. Because virtually all of the constituent parts of the nucleic acids are synthesized rather than provided by the diet, it will be necessary to consider both the biosynthesis of their nucleotide precursors and how these precursors are assembled into information-laden polymers. This chapter will consider only the biosynthesis of the nucleotides and the following chapter will describe how the nucleotides are used for the synthesis of the nucleic acids and how the information stored in these polymers is used for the synthesis of proteins.

One-carbon metabolism using folic acid derivatives is important for the synthesis of several cellular constituents. Some of the best examples of one-carbon metabolism are found in the synthesis of purine and pyrimidine nucleotides. The metabolic functions of folic acid and its derivatives form the basis for its use in the treatment of tropical sprue and megaloblastic anemia. Moreover, an understanding of the synthesis of folic acid compounds is important in understanding the effectiveness of certain folate derivatives and the sulfonamides in treating disease.

The usefulness of folic acid analogues in the treatment of leukemia is closely related to the vitamin's functions in the biosynthesis of the purine and pyrimidine nucleotides. Furthermore, the rather common genetic disease gout and the very rare Lesch-Nyhan syndrome appear to be similarly related to purine nucleotide metabolism, a knowledge of which is needed to understand and treat these conditions.

NUCLEIC ACIDS

In 1868 the Swiss physician Miescher described the first isolation of nucleoprotein from the nuclei of pus cells taken from bandages discarded by the surgical clinic of a nearby hospital. Later he isolated a similar substance from salmon sperm and showed that it consisted of a basic protein, *protamine*, and an acidic substance now known to be nucleic acid. Common sources of nucleic acid used by these early workers were thymus glands and yeast. When thymus nucleic acid was hydrolyzed, it yielded purine and pyrimidine bases, deoxy-D-ribose, and phosphate. Yeast nucleic acid also yielded purine and pyrimidine bases and phosphate, but its sugar was primarily D-ribose rather than deoxyribose. It was subsequently found that the thymus nucleic acid was mostly *deoxyribonucleic acid* (DNA), while the yeast nucleic acid was mostly *ribonucleic acid* (RNA). All cells contain both DNA and RNA, but some, for example, yeast and thymus, contain more of one type than the other. A particular cell may contain a whole variety of distinct nucleic acids, so that the terms DNA and RNA are generic in the same sense as the word "protein," which can refer to either a mixture of several individual substances or to a single kind of protein.

Nomenclature and hydrolysis products

Some of the first experiments performed on nucleic acids by nineteenth century chemists consisted of hydrolyzing these macromolecules in order to understand their chemical composition. Complete hydrolysis of a nucleic acid (Fig. 11-1) yields purine and pyrimidine bases, deoxyribose or ribose, and phosphoric acid.

DNA and RNA both contain the purine bases adenine (A) and guanine (G). DNA contains the pyrimidines thymine (T) and cytosine (C), but RNA contains cytosine and uracil (U). The structure of these purine and pyrimidine bases are shown in Fig. 11-2. The hydrolysis of nucleic acids by enzymes such as deoxyribonuclease (DNase) and ribonuclease (RNase) can yield *mononucleotides*. The mononucleotides contain equimolar amounts of a nitrogenous base, a sugar, and phos-

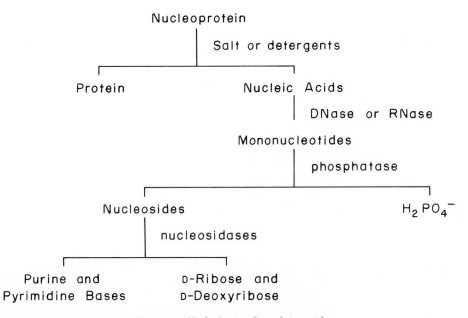

Fig. 11-1. Hydrolysis of nucleic acids.

phoric acid. Further hydrolysis by phosphatases yields inorganic phosphate and *nucleosides*. In the nucleosides the bases are linked to the pentose sugars in a β-D configuration. For example, the structures of deoxyadenosine and thymidine may be represented as follows:

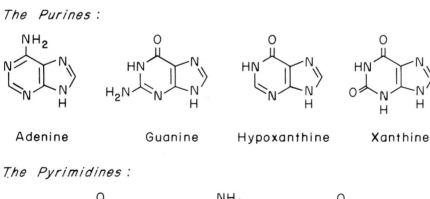

Deoxyadenosine   Thymidine

Note that all the positions on the purine and pyrimidine rings are numbered and that the positions on the sugars carry a prime symbol to distinguish them from the ring positions. The other bases are linked to either D-ribose or deoxy-D-ribose in an analogous way. The names of the nucleosides are given in Table 11-1.

The nucleosides may be either deoxyribonucleosides or ribonucleosides. Similarly, the nucleotides may be either deoxyribonucleotides or ribonucleotides. The

The Purines :

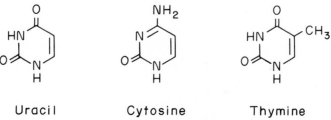

Adenine   Guanine   Hypoxanthine   Xanthine

The Pyrimidines :

Uracil   Cytosine   Thymine

Fig. 11-2. Structures of purine and pyrimidine bases.

Table 11-1. Names of nucleosides

BASE	RIBONUCLEOSIDE	DEOXYRIBONUCLEOSIDE
Adenine	Adenosine	Deoxyadenosine
Guanine	Guanosine	Deoxyguanosine
Cytosine	Cytidine	Deoxycytidine
Uracil	Uridine	Deoxyuridine
Thymine	Thymine riboside	Thymidine

Table 11-2. Alternate names of nucleotides derived from nucleic acids

	ACIDS	PHOSPHATES	ABBREVIATION*
Ribonucleotides	2'-Adenylic acid	Adenosine-2'-mono-phosphate	2'-AMP
	3'-Adenylic acid	Adenosine-3'-mono-phosphate	3'-AMP
	5'-Adenylic acid	Adenosine-5'-mono-phosphate	AMP
	3'-Guanylic acid	Guanosine-3'-mono-phosphate	3'-GMP
	3'-Cytidylic acid	Cytidine-3'-mono-phosphate	3'-CMP
	3'-Uridylic acid	Uridine-3'-mono-phosphate	3'-UMP
Deoxyribonucleotides	Deoxyadenylic acid	Deoxyadenosine-5'-monophosphate	dAMP
	Deoxyguanylic acid	Deoxyguanosine-5'-monophosphate	dGMP
	Deoxycytidylic acid	Deoxycytidine-5'-monophosphate	dCMP
	Thymidylic acid	Thymidine-5'-mono-phosphate	dTMP

*When an abbreviation does not have a numerical prefix, it is assumed to be the 5' derivative.

names of the common mononucleotides are given in Table 11-2. Polynucleotides consist of mononucleotides linked one to another by phosphodiester bonds between the 3'- and 5'-hydroxyl groups of the adjacent D-ribose or deoxy-D-ribose residues. For example, a dinucleotide would be represented as shown at the top of p. 523. By convention nucleic acids and oligonucleotides are written with the 5'-terminal at the left and the 3'-terminal at the right. A common form of abbreviation is to use a capital letter such as A, G, U, or C to represent the nucleoside (dA, dG, etc. for deoxyribonucleosides) and a lowercase p to represent the phosphate. Thus the compound shown on p. 523 would be abbreviated pUpC; 5'-AMP would be pA, 5'-ATP would be pppA, the three p's indicating the three phosphoryl groups of ATP.

Cellular location of DNA

Most of the DNA of animal cells is found in the nucleus, whereas most of the RNA is located in the cytoplasm. Nuclear DNA is the major constituent of the chromosomes, where it exists as a thin double helix with a width of only 2 nm. The double helix is folded and complexed with protein to form chromosomal strands of approximately 100 to 200 nm in diameter. Alternatively, the chromosomal strands may consist of many duplicate DNA molecules stacked together in association with proteins. Exactly how the chromosomes are assembled is not known, but the role of histone proteins in this process is undoubtedly important.

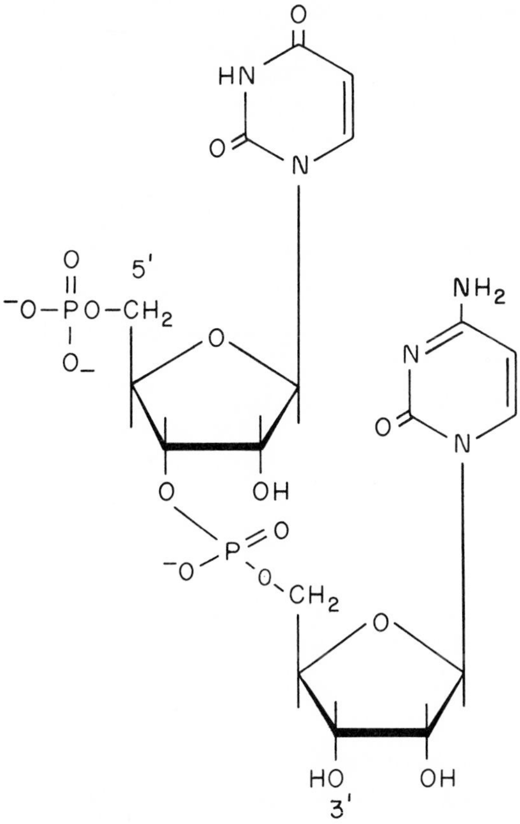

Histones and chromosome structure

The histones are the major proteins associated with the chromosomes. These basic substances can be separated by polyacrylamide gel electrophoresis into five groups. All five histone groups were found to be present in every eukaryotic cell examined, but some cells contain more of one group than of another. One of these groups, the so-called lysine-rich histones, is heterogeneous. The proteins of this group migrate in close proximity to one another on electrophoresis, but they actually consist of slightly different polypeptide chains. The other four groups of histones are apparently homogeneous proteins and have been sequenced. The sequences of histone f3 from calf, fish, and peas show a remarkable constancy. This indicates that during evolution the amino acid residues of f3 histone cannot be substantially changed without affecting its activity to the extent that it may become deleterious or lethal.

Although it is not known how the characteristic banded structure of a chromosome is related to its function, Crick has speculated that the DNA of a single chromosome consists of one or more identical strands running from one end of the chromosome to the other. These bands, which can be seen microscopically after staining with Feulgen's basic fuchsin dye, are believed to represent regions of globular DNA complexed with histones and nonhistone proteins. The DNA in these regions is unavailable for use in protein synthesis. The fibrous DNA that occurs between bands represents the only part of the chromosome utilized for the synthesis of proteins.

Mitochondrial nucleic acid

Not all of the DNA is in the nucleus; some is also contained in the mitochondria. In addition, the mitochondria contain RNA as well as the enzymes necessary for protein synthesis. Interestingly, this genetic material bears a closer resemblance to that found in bacterial cells than to that of animal cells. For example, the rather small DNA molecule of the mitochondrion is circular. Its information is contained in approximately 15,000 nucleotides that function in the synthesis of ribosomal and transfer RNA as well as in that of the messenger RNA for some of the mitochondrial membrane proteins. Because mitochondria are inherited cytoplasmically, an individual does not necessarily receive his mitochondrial nucleic acid equally from each parent. As a matter of fact, the evidence seems to favor the maternal inheritance of mitochondria.

Cellular location of RNA

RNA is found in every subcellular organelle that is capable of synthesizing protein. Most protein synthesis, and thus most RNA, occurs in the cytoplasm, although a considerable amount is present in the nucleus. The nucleolus is the site of synthesis for a large part of the cell's RNA and is particularily rich in ribosomal RNA.

Structure

Hydrolysis of nucleic acids

Both DNA and RNA can be hydrolyzed by strong acids, but only RNA can be hydrolyzed by alkali. A wide variety of hydrolysis products are possible, but heating with strong acid hydrolyzes nucleic acids to their constituent sugars, phosphate, and purine and pyrimidine bases. This process is employed to determine the base composition of a particular nucleic acid, but it sometimes causes the formation of decomposition products.

Enzymatic hydrolysis produces mononucleotides and oligonucleotides relatively free of decomposition products. The nucleases can be classified as *deoxyribonucleases* or *ribonucleases*. Some of these are exonucleases that cleave mononucleotides one at a time from the ends of a polynucleotide. Others, the endonucleases, cleave polynucleotides at internal positions. Both exonucleases and endonucleases are phosphodiesterases in that they cleave phosphodiester bonds. Other nucleases cleave only phosphomonoester bonds and are called phosphomonoesterases. Many of these hydrolytic enzymes function in nucleic acid turnover and digestion.

Nucleases are specific for cleaving phosphoester bonds at either the 3' or 5' hydroxyl of the sugar portion of the polynucleotide chain. For example, some nucleases produce 3' mononucleoside phosphates, whereas others produce 5' nucleoside phosphates. The specificity for the 3' or 5' phosphoester bond applies to both phosphodiesterases and phosphomonoesterases. Some nucleases hydrolyze only RNA and others hydrolyze only DNA, but a few can hydrolyze either nucleic acid. Deoxyribonucleases may have additional specificity for double- or single-stranded DNA. Table 11-3 lists a few common nucleases with their specificities.

After hydrolysis the purine and pyrimidine bases can be separated from each other by thin-layer, paper, or column chromatography or by electrophoresis. Estimation of the molar concentration of the separated bases is most conveniently

Table 11-3. Specificity of some nucleases

ENZYME	SOURCE	SUBSTRATE	SPECIFICITY
Ribonuclease I	Pancreas	RNA	An endonuclease that splits 5′ C—P phosphodiester bonds connected to pyrimidines
Ribonuclease T₁	*Aspergillus* (mold)	RNA	An endonuclease that splits 5′ C—P phosphodiester bonds connected to guanine residues
Deoxyribonuclease I	Pancreas	DNA	An endonuclease that splits some 3′ C—P bonds
Phosphodiesterase	Spleen	RNA or DNA	An exonuclease, starts at 5′ end, splits 5′ C—P bonds
Phosphodiesterase	Snake venom	RNA or single-stranded DNA	An exonuclease, starts at 3′ end, splits 3′ C—P bonds

done by spectrophotometry, since the bases all have spectra that show a major peak of absorption near 260 nm.

Certain endonucleases from bacteria are very useful in the study of DNA structure because of their unusual specificities. They are called restriction endonucleases. Because of their stringent specificity they are able to hydrolyze foreign DNA but not the DNA of their host; thus they can restrict the development of certain DNA viruses. The restriction endonucleases not only cleave specific nucleotide residues in DNA, they also require that the DNA have a symmetry in the bases around a given point. For example, the restriction endonuclease from *Escherichia coli,* EcoRl, requires a DNA substrate with the following sequence:

$$\downarrow$$
$$5'...\text{GAA TTC}...3'$$
$$\cdot$$
$$3'...\text{CTT AAG}...5'$$
$$\uparrow$$

Notice that the nucleotide sequence shown above is identical, if each strand were written with the 5′ end at the left; also note that they are symmetric around the middle dot. The specific sites cleaved are marked by arrows. A large number of restriction endonucleases exist that have slightly different specificities, but all require a symmetric DNA sequence of from four to six bases. Because these sequences may occur only a few times in small DNA's, judicious use of the enzymes will yield fragments of sizes that can be easily sequenced. By sequencing the fragments produced by a number of different restriction endonucleases, overlapping fragments can be arranged so as to determine the sequence of large DNA oligonucleotides.

Base composition of DNA

The base composition of DNA offers some insight into the structure and function of the hereditary material. When techniques for the quantitative separation of the purine and pyrimidine bases became available, a remarkable observation was made by Chargaff. No matter what cells were used to prepare the DNA, the molar concentration of the pyrimidines was exactly the same as that of the purines. On closer analysis it was found that the concentration of cytosine always equaled that of guanine and that the concentration of thymine always equaled that of adenine.

Table 11-4. Guanine plus cytosine content of DNA from various species*

ORGANISM OR TISSUE	GUANINE + CYTOSINE CONTENT (%)
Dictyostelium discoideum	22.0
Staphylococcus aureus	33.0
Bacillus cereus	36.0
Calf thymus	39.0
Human cell line D98	39.5
Mouse liver	40.0
Bull sperm	41.0
Mouse spleen	44.0
Escherichia coli	50.0
Brucella abortus	56.0
Pseudomonas aeruginosa	68.0
Mycobacterium phlei	73.0

*Taken in part from Handbook of biochemistry, Cleveland, Ohio, 1968, The Chemical Rubber Co.

The only exception to Chargaff's rule is the base composition of the DNA of bacteriophage ΦX174. Unlike all others, this DNA is single stranded rather than double stranded.

Although the bases of DNA occur in ordered pairs, the base composition of DNA may vary widely from one species to another. When base compositions are compared, the data are usually given as the percentage of guanine plus cytosine or sometimes as the ratio of A + T/G + C. As seen in Table 11-4, the base composition of animal DNA covers a narrow range from about 39% to 44% G + C. In contrast, the composition of DNA from microorganisms covers a very broad range. The only exception seems to be a peculiar "satellite" DNA from mice that is very rich in adenine and thymine (34% G + C). The satellite DNA is easily separated from the bulk of the DNA, but it comprises only about one tenth of the total cellular DNA.

Structure of DNA

Combining the chemical observations of Chargaff with the x-ray diffraction data of Wilkins and other workers, Watson and Crick proposed a structure for DNA in which hydrogen bonds paired adenine to thymine and guanine to cytosine (Fig. 11-3). Bases in one strand of the DNA were paired with bases in the complementary strand in an antiparallel arrangement, and both were twisted into a right-handed double helix. Antiparallel strands are arranged so that the 5' end of one strand lies next to the 3' end of the other.

$$5'\ldots p\ A\ p\ C\ldots 3'$$
$$ \| \| \|$$
$$3'\ldots T\ p\ G\ p\ldots 5'$$

The result of such an arrangement is a molecule in which the planar hydrogen-bonded base pairs are tightly stacked to form a rigid linear molecule. This structure is highly susceptible to breakage by shearing forces and makes for very viscous solutions. Much of the stability of the DNA molecule results more from the very strong interactions between the stacked bases than from the hydrogen bonds between paired bases. The bases along one strand of the helix lie on top of one another. The hydrophobic interactions between these stacked flat aromatic bases stabilize the helical structure. However, the base pairs are obviously very important for the remarkable specificity required in the replication and expression of

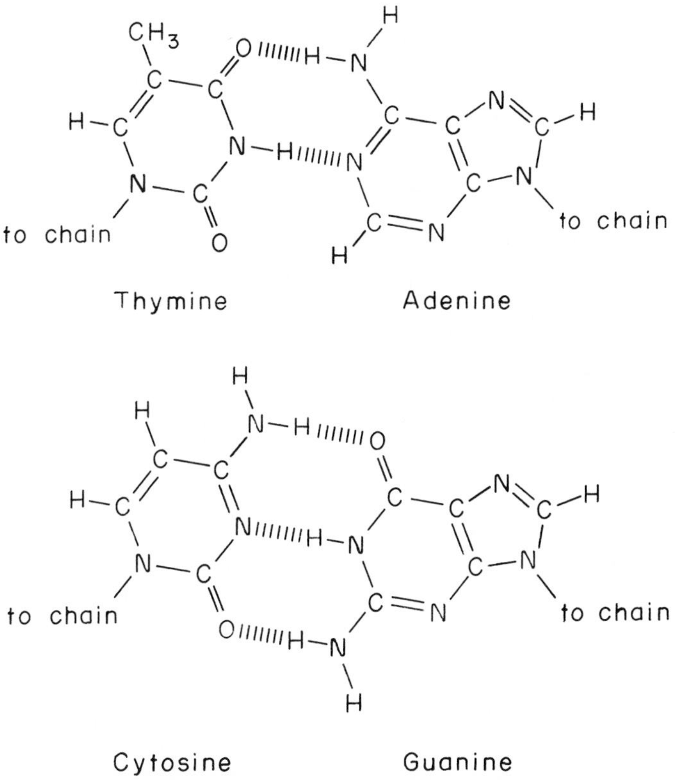

Fig. 11-3. Base pairs present in DNA.

the hereditary material. All of these forces can be disrupted by heating; consequently, DNA is like protein in that it can be denatured. When denaturation occurs, both the stacking forces and the specific hydrogen bonds between the bases are scrambled to the extent that all biologic or template activity is lost. The ordered helix is said to take on the configuration of a random coil. Coincident with denaturation is an increase in the absorption of light by the bases at 260 nm, which the powerful stacking forces in the native helix tend to suppress. When a sample of pure or homogeneous DNA is slowly heated and the absorption at 260 nm recorded, one finds that the DNA abruptly denatures when a characteristic temperature, the *melting temperature*, is reached. This is illustrated by the data in Fig. 11-4. The melting temperature (T_m) is dependent on variables such as the kind and concentration of solvents used, but it can be useful as a rough measure of the base composition of the particular DNA. The T_m is defined as the temperature at which a given DNA is midway between the helix and coil forms as judged by the absorption at 260 nm. Because the T_m is proportional to the percentages of G + C, as shown in Fig. 11-4, this measurement can be used to determine the G-C content of a DNA molecule. Denatured DNA is capable of being renatured; the hot denatured DNA will reform an intact helix if cooled slowly. However, if cooling is rapid, the DNA molecule remains denatured for the most part. This property of regaining the duplex structure is very useful because it is highly specific for complementary strands and because hybrid helices can form with complementary strands of RNA. This process, called *hybridization*, occurs when a single strand of a denatured DNA molecule is allowed to renature in the presence of single-stranded RNA that has a complementary sequence. It can be used to determine

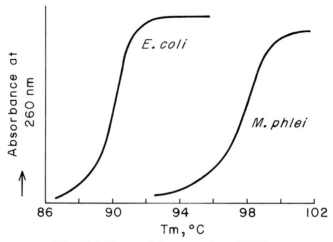

Fig. 11-4. Thermal denaturation of DNA.

whether a certain DNA molecule is capable of serving as a template for the synthesis of a partcular RNA. This will be considered further in Chapter 12 when RNA synthesis is described.

Structure of RNA

There are three types of cellular RNA. Transfer RNA (tRNA) molecules are the smallest and have a molecular weight of about 28,000. In rapidly growing cells, approximately 18% of the RNA is tRNA. Originally tRNA was called soluble RNA (sRNA), and this nomenclature is still occasionally used. However, the newer name better describes the function of these molecules, which is to transfer amino acids to the ribosomes where proteins are synthesized. Several tRNA's have been sequenced. This structural information helps elucidate the functions of tRNA and will be discussed when protein synthesis is considered in Chapter 12. Messenger RNA (mRNA) represents only 2% of the RNA content of a cell. Yet these single-stranded polymers are the templates for the synthesis of all of the cell's proteins. Their sizes vary, but generally they are quite large, that is, with molecular weights of approximately 10^6. About 80% of the cellular RNA is found in the ribosomes. This material is known as ribosomal RNA (rRNA). The ribosomes are the subcellular particles on which proteins are synthesized. Three kinds of RNA are found in the ribosomes, and they are designated according to their sedimentation coefficient, which reflects their size. Thus the ribosomes contain 28S, 18S, and 5S RNA's, with molecular weights of 1.8×10^6, 0.7×10^6, and 36,000, respectively.

Because of their large variation in size and their sensitivity to hydrolysis, RNA molecules are isolated by rather special techniques. Zone centrifugation through a steep density gradient of sucrose will separate the three ribosomal RNA's from one another. This separation can also be performed by electrophoresis on polyacrylamide gels or by absorption column chromatography. The smaller tRNA's can be isolated from the bulk of the cellular RNA by differential centrifugation and fractionation with 1M NaCl. Individual tRNA molecules specific for a single amino acid can be isolated in pure form by countercurrent distribution. Messenger RNA's, because of their large number, low concentration, and lability, are exceedingly difficult to isolate in pure form. The mRNA from purified preparations of RNA viruses, however, is easier to obtain and is often the source of the mRNA used in the laboratory.

Table 11-5. Comparisons of DNA and RNA

	DNA	RNA
Bases		
Purines	Adenine	Adenine
	Guanine	Guanine
Pyrimidines	Cytosine	Cytosine
	Thymine	Uracil
Sugar	Deoxy-D-ribose	D-Ribose
Intranucleotide linkage	3'-5'-Phosphodiester	3'-5'-Phosphodiester
Size (molecular weight)	$>2 \times 10^9$	tRNA: $\sim 28,000$
		mRNA: $\sim 10^6$
		rRNA: 0.5-1×10^6
Shape	Base-paired double-stranded helix	Single-stranded random coils (tRNA is about 75% base paired)
Associated substances	Histones, protamines, spermine	Ribosomal proteins with rRNA
Sensitivity to alkaline hydrolysis	Insensitive	Sensitive
Presence in viruses	Animal or bacterial viruses	Animal, bacterial, or plant viruses
Base distribution	Purines = Pyrimidines; A = T; G = C; contains few unusual bases	No predictable distribution except in double-stranded viral RNA; may contain several unusual bases
Cellular distribution	Mostly in nucleus; some in mitochondria	Highest in cytoplasm but much in nucleus and nucleolus

A summary of the similarities and differences in the physical and chemical properties of DNA and RNA is presented in Table 11-5.

Digestion of dietary nucleic acids

The purine or pyrimidine bases are not essential components of the diet; most mononucleotides are synthesized de novo within the human body. Dietary nucleic acids are digested by pancreatic ribonuclease and deoxyribonuclease to mononucleotides. In the intestine these are converted to nucleosides by mononucleotidases and to free bases by nucleosidases. Fig. 11-5 outlines these conversions. The tissue nucleosidases are of two types: one functions with pyrimidine nucleosides to simply hydrolyze the glycosidic bond, while the other is specific for purine mononucleosides. These enzymes catalyze a phosphorylytic cleavage of the glycosidic bond to yield D-ribose 1-phosphate. Guanosine and inosine are better substrates than adenosine for the purine nucleosidases.

The phosphate and sugars produced by the digestion of nucleic acids (Fig. 11-5) can be reutilized, as can the base adenine, but most of the other purine and pyrimidine bases are catabolized and excreted. Purines and their derivatives are efficiently converted to uric acid by an active xanthine oxidase of the intestinal mucosa. Nevertheless, small amounts of dietary nucleosides and nucleotides may be absorbed and used directly for nucleic acid synthesis, but these amounts are so low that most of the purine and pyrimidine nucleotides must be synthesized from smaller molecules.

Mononucleotides are important for all living things. They serve as precursors for the synthesis of nucleic acids and other metabolites. They also function as sources of energy for physiologic and metabolic reactions.

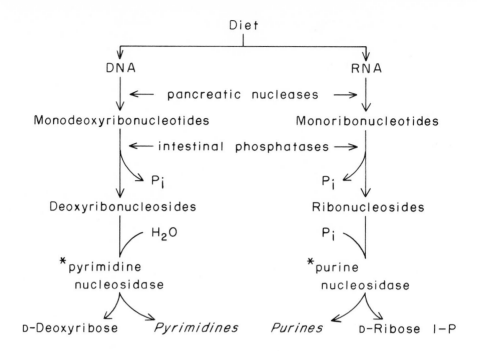

Fig. 11-5. Digestion of nucleic acids.

BIOSYNTHESIS OF PURINE AND PYRIMIDINE NUCLEOTIDES
Folic acid functions

For many years the structures of the intermediates in purine synthesis were unknown. The first intermediate was identified in a culture medium of *E. coli* that had been treated with *sulfanilamide*. This antibiotic, still useful in the treatment of some kidney infections, is highly toxic to bacteria that must synthesize their own folic acid. *p*-Aminobenzoic acid, a precursor of folic acid, will competitively reverse sulfonamide inhibitions.

The intermediate in the synthesis of purine nucleotides that accumulates in these cells is 5-aminoimidazole-4-carboxamide ribotide, as shown below.

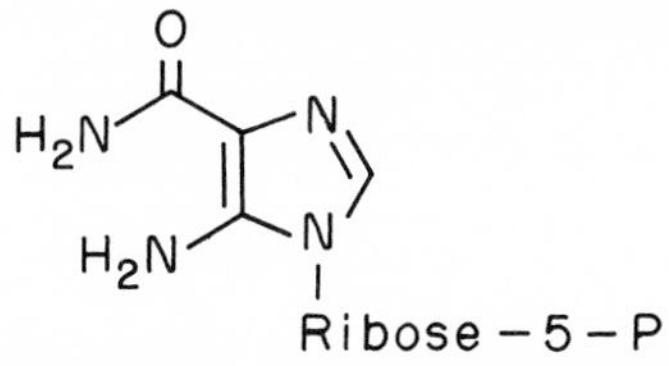

This compound is not an intermediate in folate synthesis but rather accumulates because of a decrease in the amount of a folate coenzyme required for purine synthesis. Sulfanilamide is toxic to microorganisms that synthesize folate, but because man lacks these enzymes, it is relatively innocuous to him. Recall from Chapter 1 that folate is a vitamin for man and must be supplied preformed in the diet.

Folate in nutrition. Lack of folic acid is the most common vitamin deficiency in the world today. People in underdeveloped countries or in low economic classes are most often affected, although deficiencies may be caused by infections, for example, malaria, or by hemorrhage, pregnancy, or certain drugs. In short, a deficiency may occur if the diet is poor or if an individual must utilize extraordinarily large amounts of folic acid for the synthesis of nucleic acids.

Because blood cells turn over very rapidly, folate deficiencies are usually seen first as megaloblastic anemias in which hemoglobin levels are low and the bone marrow shows abnormally high numbers of megaloblastic cells (large abnormal immature erythrocytes).

Utilization of folic acid. Dietary folic acid is converted to tetrahydrofolate (THF) by dihydrofolate (DHF) reductase, which uses NADPH as the reducing agent. DHF reductase participates in both of the reductions shown in Fig. 11-6, and it is the enzyme most sensitive to the folic acid analogues (antimetabolites) used in cancer chemotherapy. The structures of two common folate analogues are shown below.

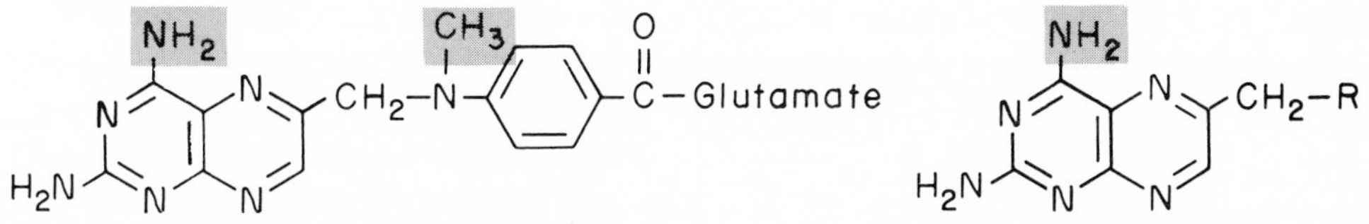

Methotrexate   *Aminopterin*

The portions of the molecules that differ from folate are shaded (compare with Fig. 11-6). Because of folate's many functions in synthesizing the precursors of nucleic acids (Table 11-6), its inhibitory analogues will slow down nucleic acid synthesis in all rapidly growing cells, including cancer cells. All of the coenzymatic forms of folic acid are derivatives of THF (Fig. 11-7). Consequently, inhibition of DHF reductase by folate analogues would be expected to affect every reaction requiring folate coenzymes.

Structural relationships between folate and its analogues. The THF derivative of the vitamin is the coenzymatic acceptor of one-carbon units. The one-carbon units come from a variety of sources and are linked to THF to form four important coenzymes, one at the oxidation level of methanol (5-methyl THF), one at the level of formaldehyde (5,10-methylene THF), and two at the level of formic acid (5,10-

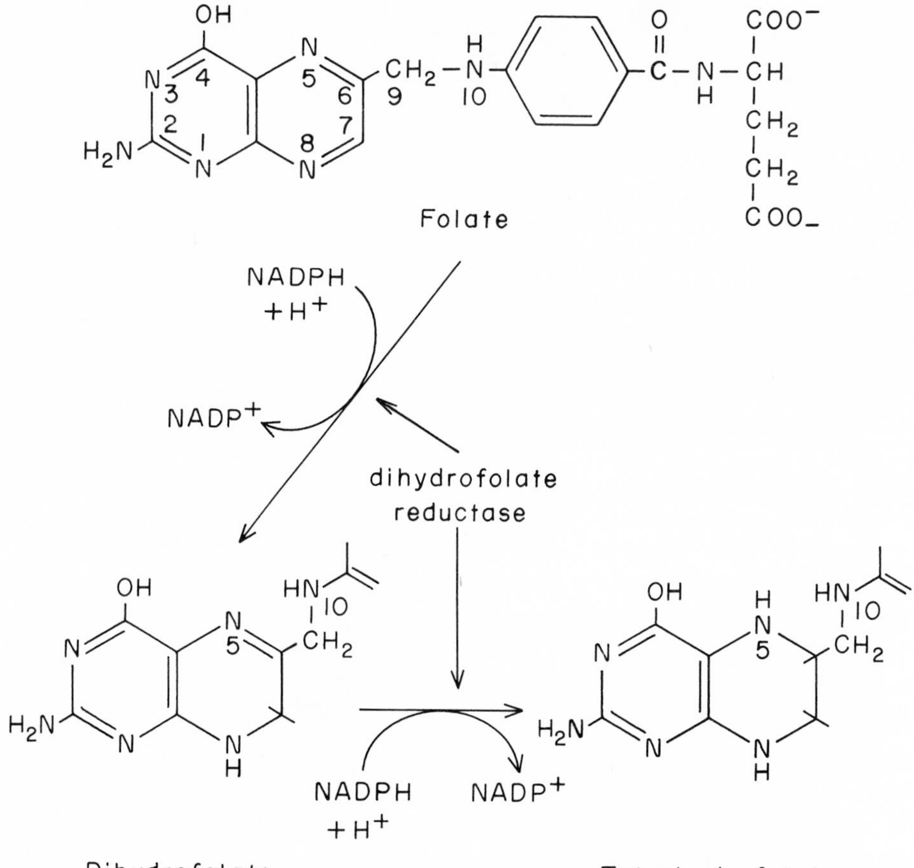

Fig. 11-6. Reactions catalyzed by dihydrofolate reductase.

Table 11-6. Some reactions where folate coenzymes function

COENZYME	REACTION		
1. 5-Methyl THF	Homocysteine	$\longrightarrow$	Methionine
2. 5,10-Methylene THF	Glycine	$\longrightarrow$	Serine
3. 5,10-Methylene THF	dUMP	$\longrightarrow$	dTMP
4. 5,10-Methenyl THF	Glycine amide ribotide	$\longrightarrow$	Formylglycine amide ribotide
5. 10-Formyl THF	Aminoimidazole carboxamide ribotide	$\longrightarrow$	Formamidoimidazole carboxamide ribotide
6. 10-Formyl THF	Met-tRNA	$\longrightarrow$	Formyl Met-tRNA

methenyl THF and 10-formyl THF). The structures of these compounds are shown in part in Fig. 11-7. A derivative of less importance, 5-formimino THF, results from histidine catabolism (Fig. 9-21). Some of the more important reactions in which folic acid coenzymes function are summarized in Table 11-6. Several of these reactions occur during the biosynthesis of the purine (reactions 4 and 5) or pyrimidine nucleotides (reaction 3).

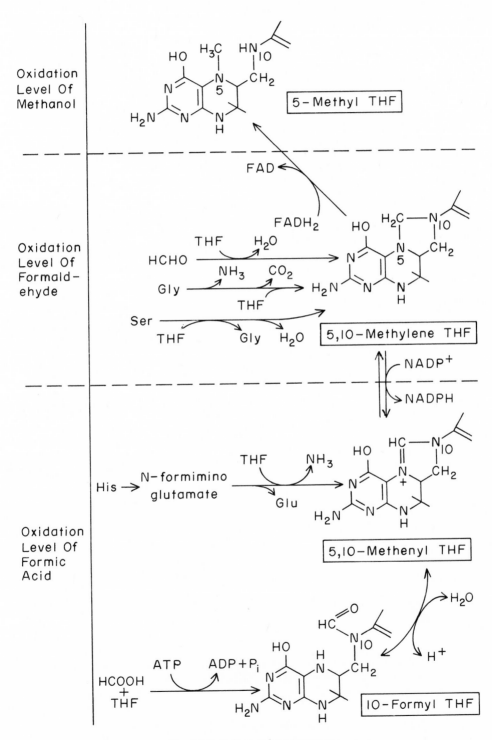

Fig. 11-7. Folic acid coenzymes.

Where B_{12} and folate meet.   5-Methyl THF is involved in the synthesis of methionine by the liver. It is formed and utilized as follows:

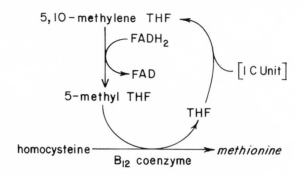

It should be recalled that although methionine is an essential amino acid, only its homocysteine portion cannot be synthesized by man.

The enzyme 5-methyl THF methyltransferase has a requirement for a methyl B_{12} coenzyme. This enzyme catalyzes the methylation of homocysteine shown in the above reaction, and it is one of the few sites where folate and B_{12} functions converge. Consequently, some investigators believe that a vitamin B_{12} deficiency results in the transformation of all of an individual's THF to 5-methyl THF. The 5-methyl THF methyl transferase reaction is the major, and perhaps the only, metabolic reaction that can utilize 5-methyl THF. If this reaction is blocked or slowed down as a result of a deficiency of vitamin B_{12}, 5-methyl THF accumulates and the THF needed for other folate-requiring reactions is not formed. Eventually, most of the folate in the body ends up as 5-methyl THF. Even in normal individuals, the major blood and tissue form of folate is 5-methyl THF; thus any impairment of its utilization could lower greatly the concentration of other folate coenzymes. This hypothesis, the *methyl-trap hypothesis,* has been offered to explain why certain anemic conditions respond to either vitamin B_{12} or folic acid and particularly to a combination of the two. For example, the administration of folate will alleviate the hematologic symptoms of the pernicious anemia caused by vitamin B_{12} deficiency but does not prevent the irreversible neurologic damage caused by a continued lack of vitamin B_{12} (see also case 3). Although the methyl-trap hypothesis offers a convenient explanation for the interrelationships of these two vitamins, the situation may be more complicated. Some evidence indicates that 5-methyl THF may participate in reactions other than methionine synthesis. Consequently, 5-methyl THF would not be expected to accumulate. Other evidence indicates that a B_{12} deficiency somehow hinders the absorption of folate derivatives and affects the rapidly growing hematopoietic cells so that megaloblastic anemia results. Therefore, it is still not known with certainty how B_{12} and folic acid metabolism are related.

Purine nucleotide synthesis

One of the first applications of isotopes to the study of biologic systems showed that the purine ring was labeled when simple, small radioactive molecules were fed or injected into laboratory animals. Considerably fewer nucleic acids were labeled when the animal was fed radioactive purines. See formula at top of opposite page. Three atoms of the ring are derived from glycine, while two carbon atoms are labeled when ^{14}C formic acid is administered. Two of the nitrogen atoms come from the amide nitrogen of glutamine, and the remaining nitrogen originates in aspartate. Finally, the carbon atom at position 6 is provided by carbon dioxide.

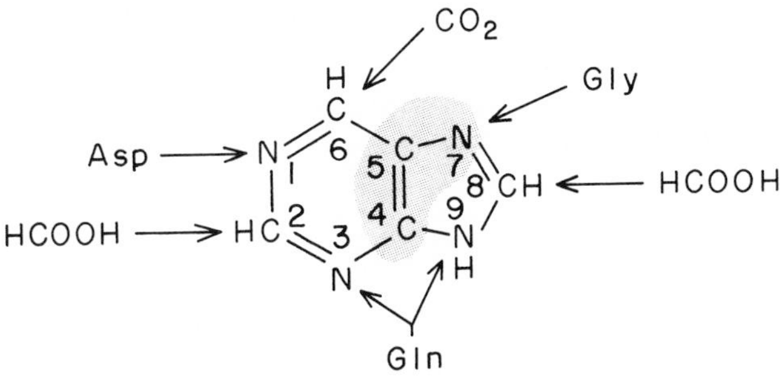

In man, all of the enzymes of purine metabolism are found in the cytoplasm. Synthesis starts with the formation of 5-phospho-α-D-ribosyl-1-pyrophosphate (PRPP).

$$\alpha\text{-D-Ribose } 5-P + ATP \longrightarrow PRPP + AMP$$

The PRPP is then aminated in the following reaction.

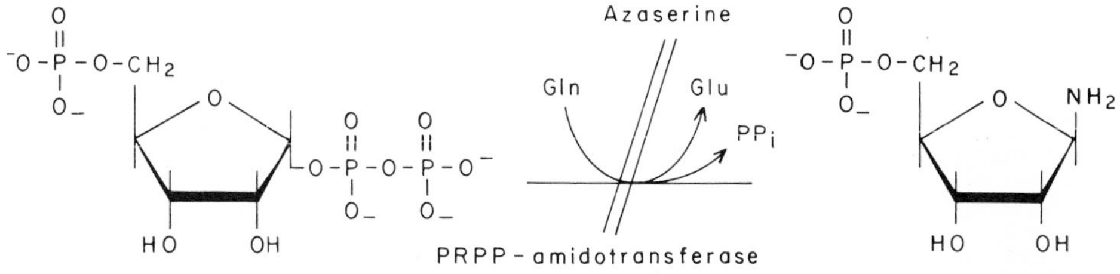

In this reaction, catalyzed by the enzyme PRPP amidotransferase, the pyrophosphate group of PRPP is displaced by the amide group of glutamine with an inversion to yield the β-D configuration found in all nucleotides. PRPP functions in several other ribosylating reactions, including pyrimidine nucleotide synthesis, NAD^+ synthesis, histidine biosynthesis, and the conversion of guanine to GMP, as shown in Fig. 11-17.

This initial step in purine nucleotide synthesis is at a branch point in PRPP utilization and thus represents the first committed, essentially irreversible step in purine nucleotide biosynthesis. As is true for many other branch points in metabolic pathways, the reaction is susceptible to feedback inhibition. The feedback inhibitors in this case are the purine nucleotides, ATP, ADP, AMP, GTP, GDP, GMP, or IMP. When the cells overproduce purine nucleotides, these same nucleotides function to limit their own biosynthesis. It has been suggested that excessive synthesis of *uric acid* (Fig. 11-15), the final catabolic metabolite of purines excreted in the urine, results from an overproduction of purine nucleotides caused by a breakdown in the feedback control of the PRPP amidotransferase. *Azaserine,* a glutamine analogue once used as an anticancer agent, blocks this reaction as well as two other glutamine-requiring reactions in purine nucleotide synthesis.

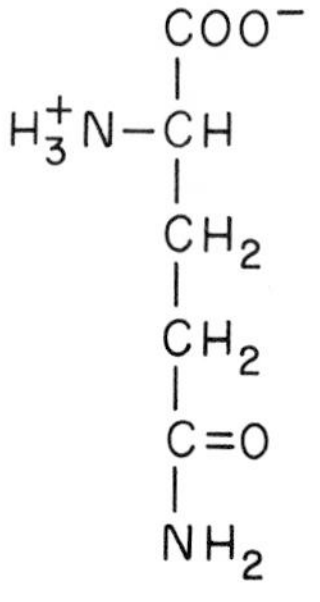

$$
\begin{array}{cc}
& \mathrm{COO^-} \\
& | \\
\mathrm{COO^-} & \mathrm{H_3^+N-CH} \\
| & | \\
\mathrm{H_3^+N-CH} & \mathrm{CH_2} \\
| & | \\
\mathrm{CH_2} & \mathrm{O} \\
| & | \\
\mathrm{CH_2} & \mathrm{C=O} \\
| & | \\
\mathrm{C=O} & \mathrm{CH_2} \\
| & | \\
\mathrm{NH_2} & \mathrm{N\equiv N} \\
& + \\
\text{Glutamine} & \text{Azaserine}
\end{array}
$$

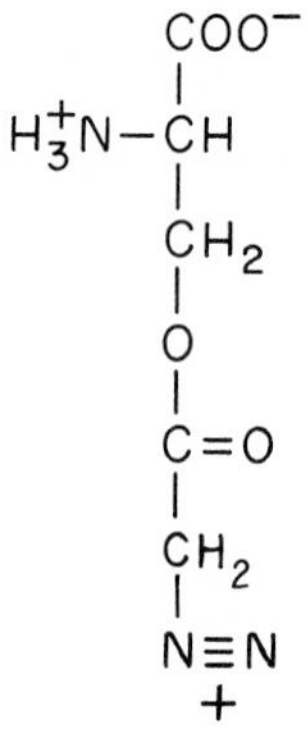

The other steps leading to purine nucleotide biosynthesis are illustrated in Figs. 11-8 and 11-9. A quick glance at the intermediates reveals that all are ribose-phosphate derivatives. This is consistent with the poor utilization of dietary purines, which in part is related to a poor conversion of the free purine bases to the ribose-phosphate derivatives.

The first atoms incorporated into the purine ring are derived from glycine. The enzyme glycinamide ribotide synthetase catalyzes the addition of glycine to 5-phosphoribosylamine. ATP provides the driving force, probably by forming a high-energy glycyl phosphate intermediate. The reaction is quite reversible, however, since the product, GAR, is a high-energy compound. GAR is formylated at the expense of 5,10-methenyl THF, which can in turn be formed from formic acid by way of 10-formyl THF. This accounts for the fact that radioactive formate will label position 8 of the purine ring.

Another amination reaction with glutamine as the nitrogen donor allows the imidazole portion of the purine ring to be closed by an enzymatic reaction driven by ATP. The subsequent carboxylation reaction occurs on an enzyme that does not appear to contain biotin, even though the precursor, 5-aminoimidazole ribotide, accumulates in biotin-deficient animals. The amination of this carboxyl group occurs in two steps in which aspartate is the nitrogen donor. The aspartate carbon atoms are released as fumarate (Fig. 11-8), and AICAR is formed. Recall that AICAR is the substance that accumulates in sulfanilamide-inhibited bacterial cells, since the subsequent reaction requires the folate coenzyme 10-formyl THF. The formylated intermediate, FAICAR, undergoes enzymatic ring closure with the formation of IMP, the purine nucleotide that is the common precursor of both AMP and GMP.

AMP synthesis requires the amination of position 6, which is accomplished in a GTP-requiring reaction in which the amino group is derived from aspartate. This reaction is similar to the amination that formed AICAR (Fig. 11-8). GMP synthesis requires an oxidation at the expense of NAD^+ followed by an amination reaction in which glutamine is the nitrogen donor. AMP and GMP can be further phosphorylated by ATP or by direct mitochondrial oxidative phosphorylation to form the purine ribonucleoside triphosphates.

Thus AMP formation occurs in a series of reactions requiring GTP, while GMP is formed at the expense of ATP. It is likely that the control of the synthesis of AMP and GMP from their common intermediate, IMP, depends on the levels of ATP and GTP; that is, the adenine nucleotides control the synthesis of guanine nucleotides,

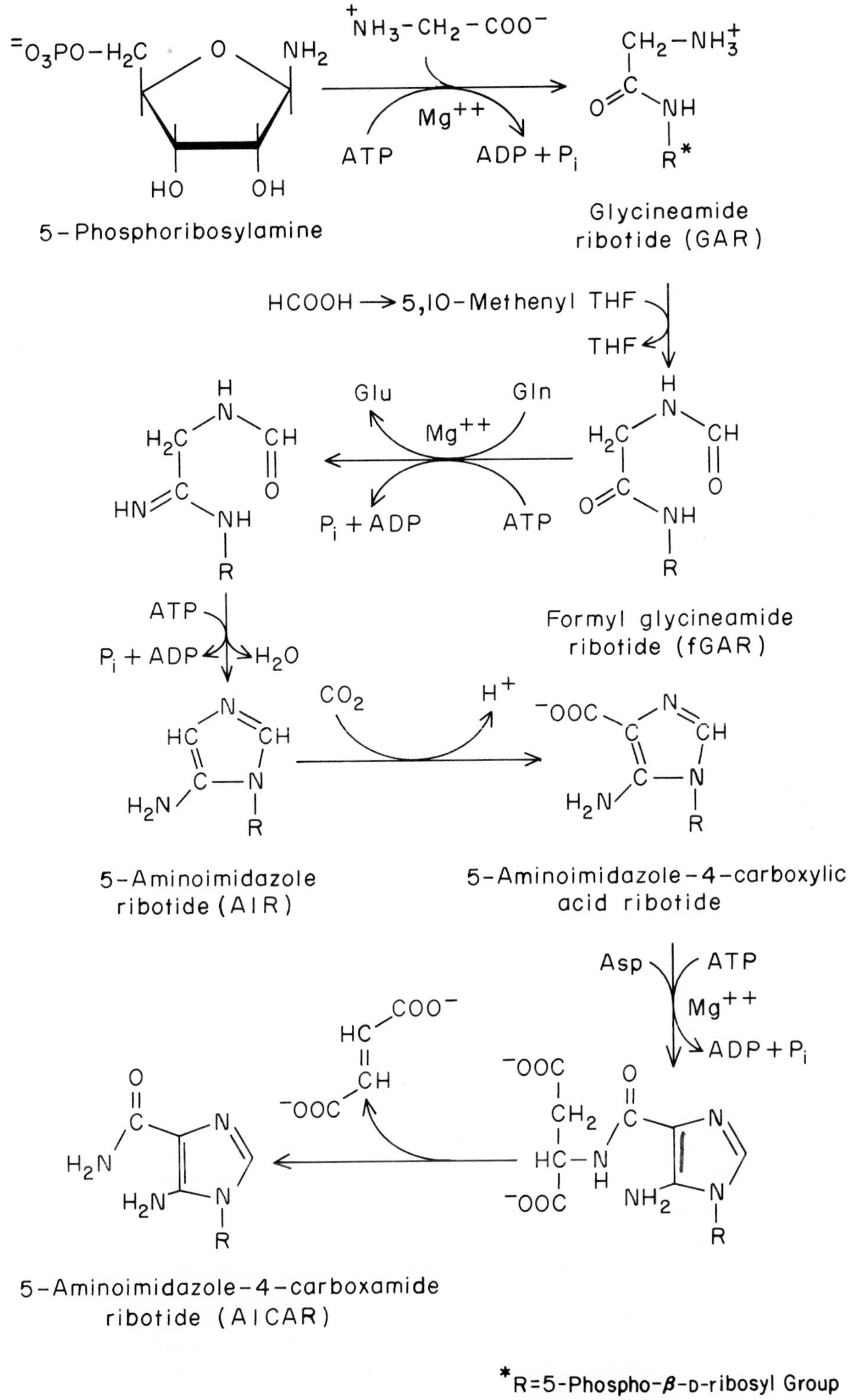
$^{=}O_3PO-H_2C$
NH_2
HO
OH
5-Phosphoribosylamine
$\overset{+}{N}H_3-CH_2-COO^-$
Mg^{++}
ATP
$ADP + P_i$
$CH_2-NH_3^+$
NH
R*
Glycineamide ribotide (GAR)
HCOOH → 5,10-Methenyl THF
THF
Glu
Gln
Mg^{++}
$P_i + ADP$
ATP
Formyl glycineamide ribotide (fGAR)
ATP
$P_i + ADP$
H_2O
CO_2
H^+
5-Aminoimidazole ribotide (AIR)
5-Aminoimidazole-4-carboxylic acid ribotide
Asp
ATP
Mg^{++}
$ADP + P_i$
5-Aminoimidazole-4-carboxamide ribotide (AICAR)
*R=5-Phospho-β-D-ribosyl Group

Fig. 11-8. Biosynthesis of purine nucleotides.

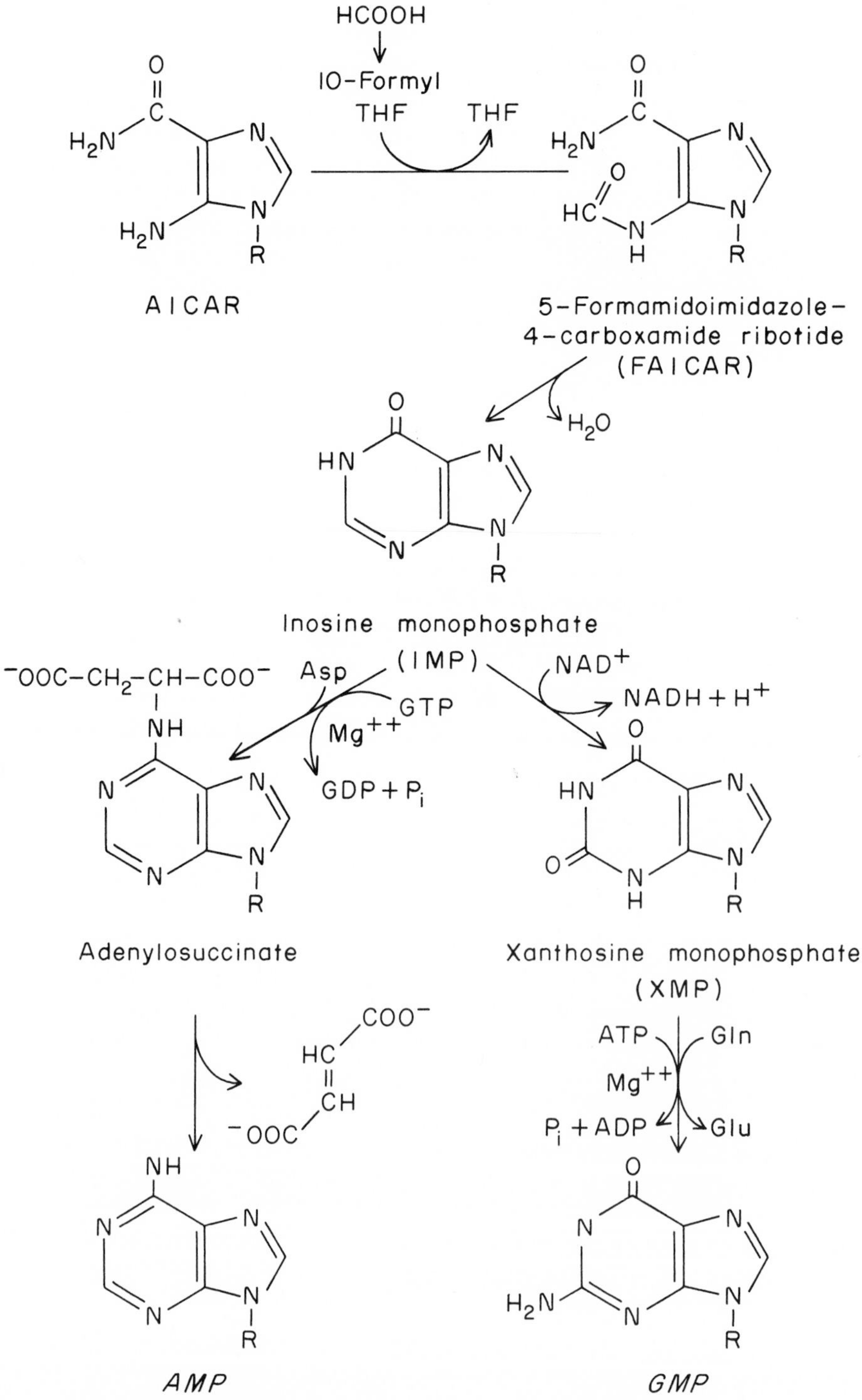

Fig. 11-9. Biosynthesis of purine nucleotides.

and the guanine nucleotides control the extent to which the adenine nucleotides are synthesized.

It should be emphasized that certain of the steps in purine nucleotide biosynthesis are indirectly sensitive to aminopterin and methotrexate, both of which are antileukemic agents (p. 531).

Pyrimidine nucleotide synthesis

Unlike synthesis of the purine nucleotides, the pyrimidine ring is formed before the β-D-ribose 5-phosphate moiety is attached. Yet both pathways are similar in that the initial reaction controls the whole sequence of the following reactions. In pyrimidine synthesis this first reaction is catalyzed by carbamyl phosphate synthetase, a cytoplasmic enzyme that uses the amide group of glutamine as the nitrogen donor. Recall that the carbamyl phosphate synthetase that functions in the urea cycle is mitochondrial and uses ammonium ion as the nitrogen donor. When UTP is increased above the steady-state concentration, carbamyl phosphate synthetase is inhibited. In *E. coli*, pyrimidine biosynthesis is controlled by aspartate transcarbamylase, an allosteric enzyme regulated in a novel way. It contains a regulatory protein subunit as well as a catalytic subunit; the catalytic subunit can function in vitro in the absence of the regulatory subunit. When both subunits exist together, as they do in vivo, the regulatory subunit, by binding CTP, can decrease the affinity with which asparate binds to the catalytic subunit, thus slowing the rate of pyrimidine synthesis.

The carbamyl phosphate used in pyrimidine formation is synthesized by enzymes quite different from those involved in the synthesis of the carbamyl phosphate of the urea cycle: the nitrogen is derived from the amide group of glutamine; acetyl glutamate is not required.

$$CO_2 + 2\,ATP + H_2O + Glutamine \xrightarrow{\substack{\text{carbamyl} \\ \text{phosphate} \\ \text{synthetase}}}$$

$$\overset{\overset{\displaystyle O}{\displaystyle \|}}{H_2N-C-OPO_3^=} + 2\,ADP + P_i + Glutamate$$

As seen in Fig. 11-10, orotic acid is an intermediate in pyrimidine nucleotide synthesis. This intermediate accumulates in the urine of individuals with a rare hereditary metabolic disease, *orotic aciduria*. Individuals who inherit this disease lack orotidylate pyrophosphorylase and orotidylate decarboxylase. An acquired form of orotic aciduria may occur in patients receiving the antineoplastic agent 6-azauridine. When converted to 6-azauridylic acid, this drug is a competitive inhibitor of the decarboxylase. The inherited disease has been treated by oral administration of uridine. In spite of the poor utilization of dietary nucleosides, it appears that enough uridine is absorbed to satisfy the patient's requirements for pyrimidine nucleotides.

Salvage synthesis

The importance of utilizing pyrimidine derivatives from the diet has not been established. Some dietary uracil can be utilized for UMP synthesis, but high concentrations are required. The enzyme uracil 5-phosphoribosyl pyrophosphorylase uses uracil and PRPP to produce UMP. While dietary uridine is better utilized in

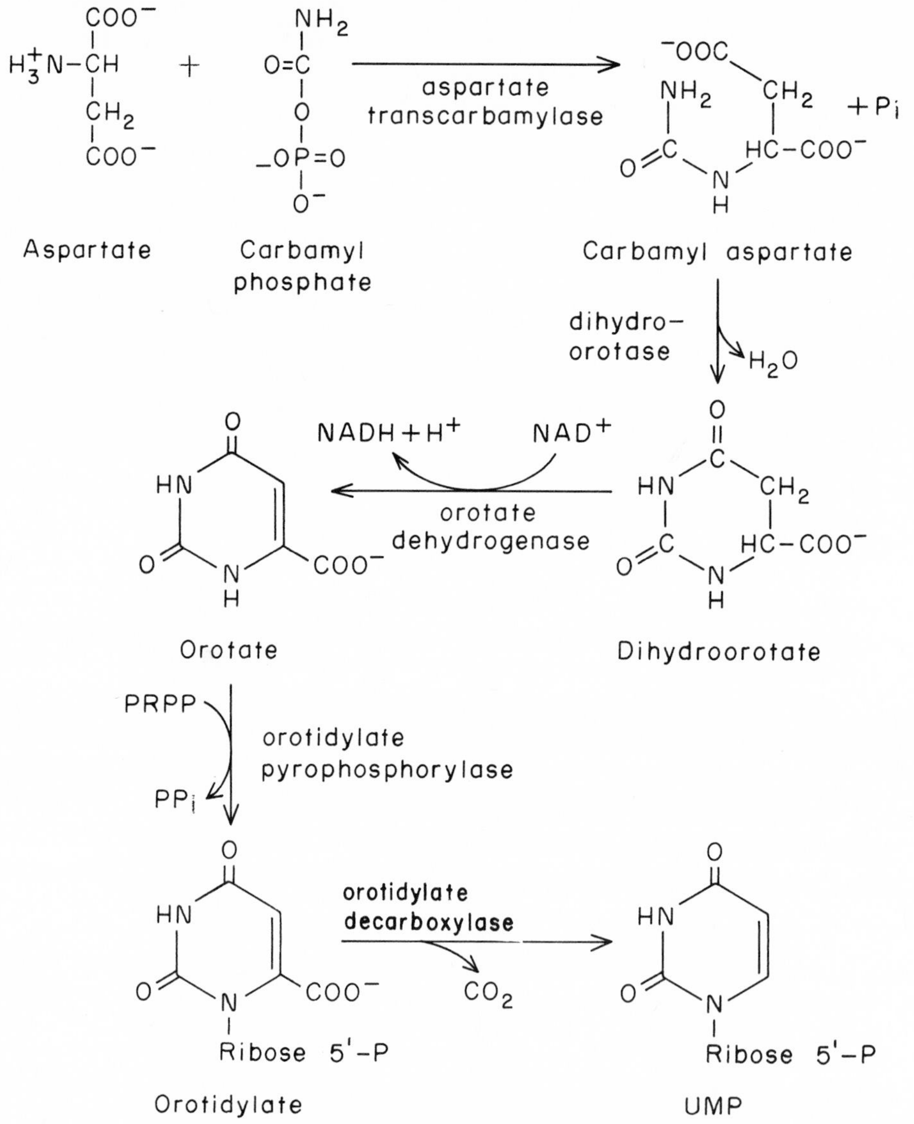

Fig. 11-10. Biosynthesis of uridine 5'-monophosphate.

its conversion to UMP by a uridine kinase, high concentrations are again required.

The first pyrimidine nucleotide synthesized is UMP, and all the others are derived from this common precursor. These transformations can be illustrated as follows:

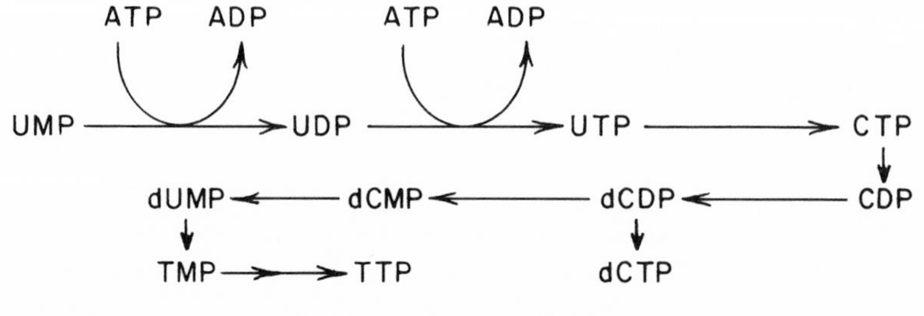

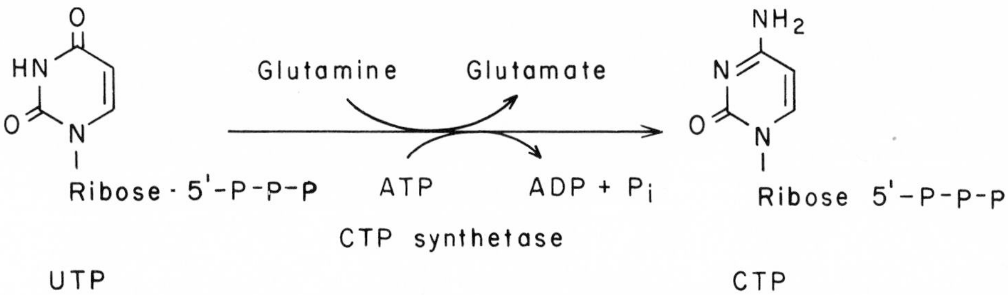

Fig. 11-11. Synthesis of cytidine triphosphate.

Kinases are required to convert UMP to UTP. The amination of UTP is catalyzed by the enzyme CTP synthetase. In animal cells the amino donor for the CTP synthetase reaction is the amide group of glutamine (Fig. 11-11), whereas the bacterial enzyme utilizes ammonium ion.

Purine and pyrimidine analogues

Several analogues of purines and pyrimidines or of their ribonucleosides or deoxyribonucleosides have been used in cancer chemotherapy. Because the fluorine atom is approximately the size of a hydrogen atom, 5-fluorouracil is usually thought of as a uracil analogue, bromodeoxyuridine as a thymidine analogue, and trifluorothymidine as a thymidine analogue. Other pyrimidine analogues useful in chemotherapy are β-D-arabinofuranosylcytosine and azauridine. Examples of purine analogues are 6-thioguanine and 6-mercaptopurine.

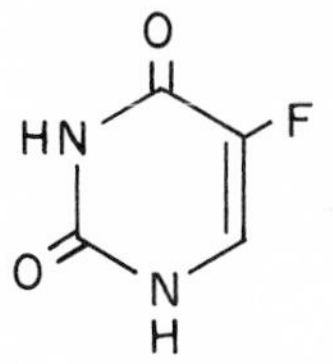

5-Fluorouracil

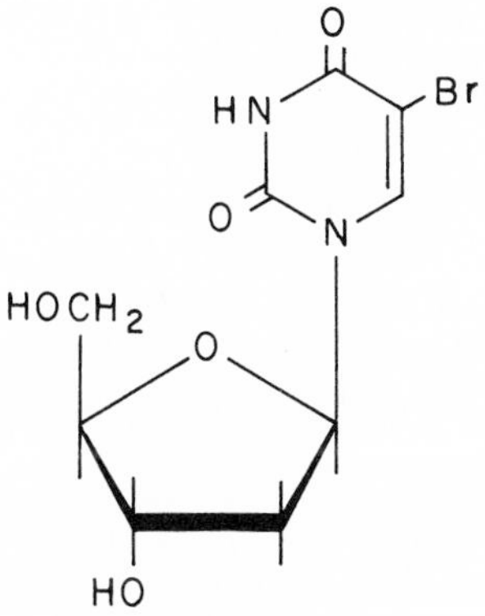

Bromodeoxyuridine

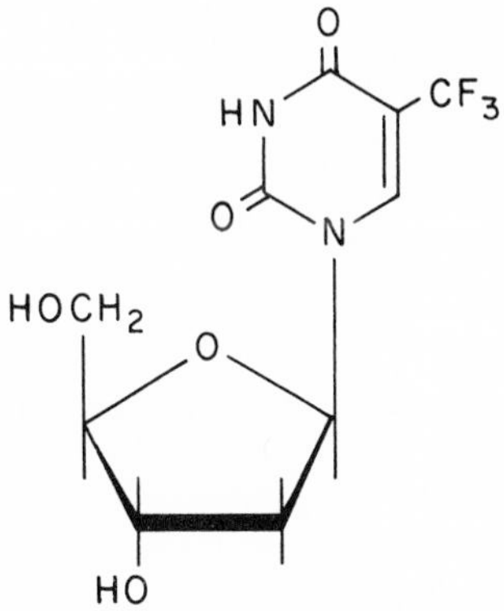

Trifluorothymidine

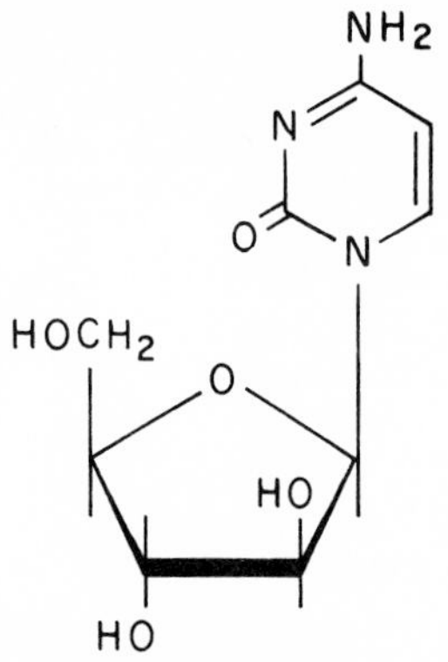

β-D-Arabinofuranosylcytosine
(Ara - C)

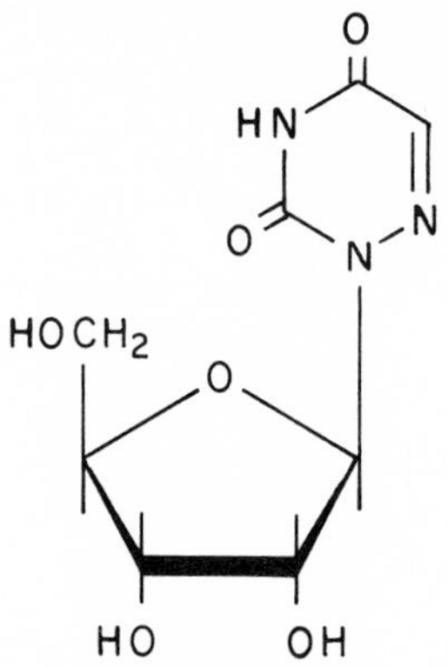

Azauridine

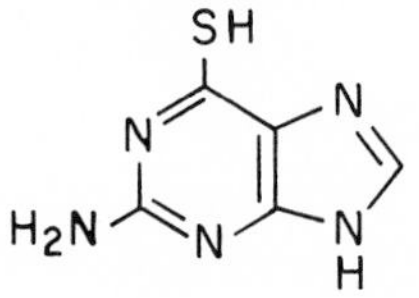

6 - Thioguanine

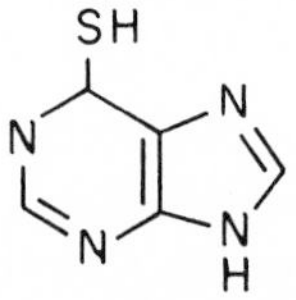

6 - Mercaptopurine

These substances are best utilized as the free bases or as nucleosides, since either the phosphate derivatives are transported poorly or the phosphate group is rapidly hydrolyzed. However, these antimetabolites must be converted to their nucleotide derivatives if they are to be toxic to the rapidly growing cancer cells. In the case of the purine analogues, conversion to the nucleotide derivative is catalyzed by the enzymes hypoxanthine-guanine phosphoribosyl transferase or adenine phosphoribosyl transferase (see case 1). In the case of the pyrimidine analogues, UMP pyrophosphorylase probably serves to attach the ribose phosphate moiety. In all cases the ribose phosphate donor is PRPP.

Deoxyribonucleotide synthesis

All of the deoxyribonucleotides are synthesized from ribonucleotides. The structures of the four deoxyribonucleotides are the same as those of the ribonucleotides except for the different sugar. In addition, thymidylic acid, which is present only in DNA, is formed by the methylation of deoxyuridine monophosphate. The formation of the deoxyribonucleotides from ribonucleotides is shown in Fig. 11-12. The enzyme ribonucleotide reductase requires ribonucleoside diphosphates as substrates. However, it has loose specificity with respect to the particular base, for ADP, GDP, CDP, and UDP are all reduced by the enzyme. The reducing agent is *thioredoxin,* a small protein that contains two free sulfhydryl groups positioned in such a way that a disulfide bond can be easily formed between them. A separate enzyme, thioredoxin reductase, uses NADPH to regenerate reduced thioredoxin.

A rather unusual ribonucleotide reductase has been found in certain lactobacilli and a few other species. These enzymes contain a cobamide coenzyme and utilize ribonucleoside triphosphates instead of the diphosphates. However, it is quite clear that ribonucleotide reductases from animal cells and *E. coli* do not contain B_{12} coenzymes. Interestingly, thioredoxin from *E. coli* can be used by the

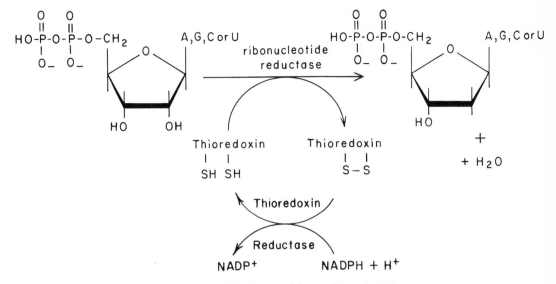

Fig. 11-12. Synthesis of deoxyribonucleotides.

ribonculeotide reductases isolated from either lactobacilli or animal cells. Ribonucleotide reductase is inhibited by hydroxyurea, a substance sometimes used in cancer chemotherapy.

Thymidylate synthesis

The precursor of thymidylic acid is dUMP, which in *E. coli* is synthesized from dUTP. However, in regenerating rat liver the enzyme dCMP deaminase is induced slightly before dTMP formation, a finding that argues for a role of this enzyme in the synthesis of thymidylic acid in animal cells. That it need be induced implies that this enzyme may be one of the rate-limiting reactions in DNA synthesis. Fig. 11-13 illustrates the sequence of reactions from dCMP to dTMP. The methylation of dUMP is catalyzed by thymidylate synthetase and requires 5,10-methylene THF as the one-carbon donor. However, this folate derivative is at the oxidation level of formaldehyde (Fig. 11-7) and must be further reduced to form the methyl group. The reducing agent in this case is THF; the oxidized product, DHF, is converted to THF by DHF reductase (Fig. 11-6).

The immediate precursors for DNA synthesis are the deoxyribonucleoside 5'-triphosphates, which are formed by phosphorylating the products of the ribonucleotide reductase–catalyzed reactions.

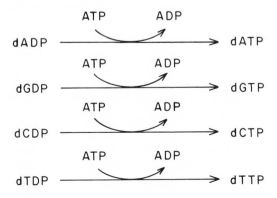

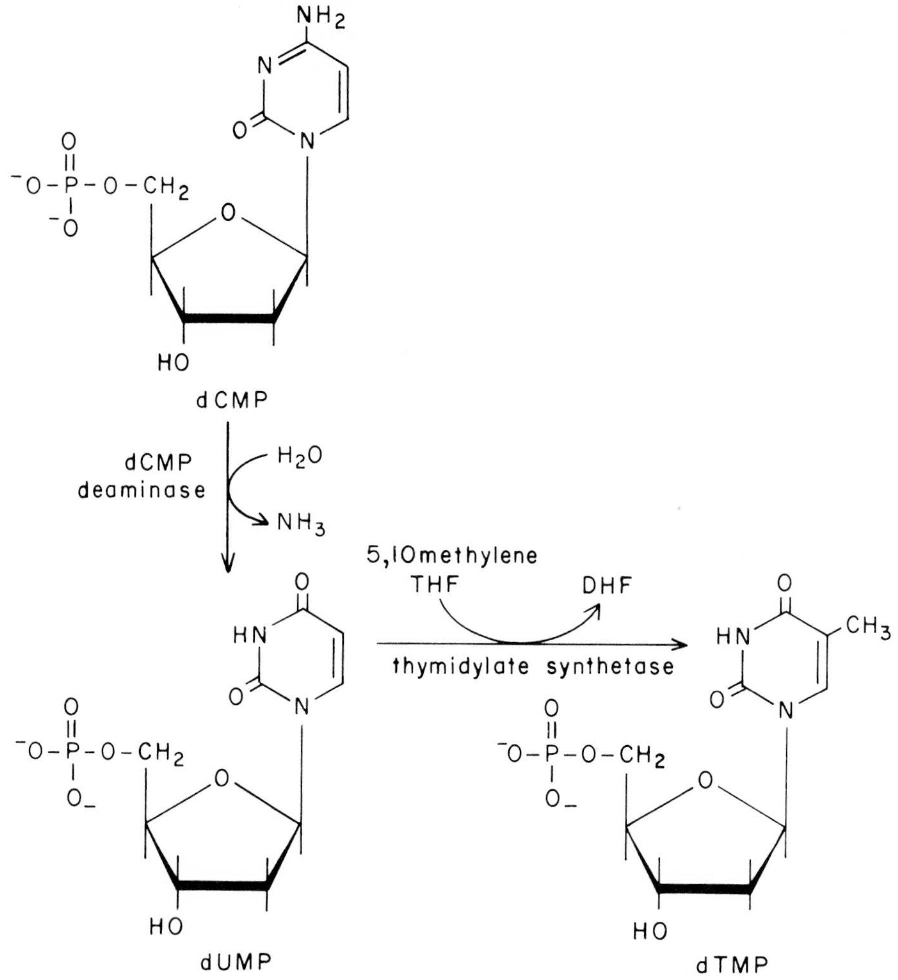

Fig. 11-13. Synthesis of thymidine 5'-monophosphate.

The enzymes catalyzing these reactions are all constitutive with the exception of thymidylate kinase, which is induced when accelerated DNA synthesis is necessary.

BIOSYNTHESIS OF NUCLEOTIDE-CONTAINING COENZYMES

Many coenzymes are derived from vitamins provided by the diet. All the B vitamins must be converted to coenzymatic forms before they can function in enzymatic reactions. Many of these coenzymes are produced in reactions involving ATP; for example, the vitamin pantothenic acid undergoes several reactions with ATP in order to form CoA. Fig. 11-14 outlines these reactions.

Similarly, ATP is required to phosphorylate riboflavin in order to form flavin mononucleotide (FMN) or riboflavin phosphate.

ATP + Riboflavin $\longrightarrow$ Riboflavin phosphate (FMN) + ADP

ATP is also required for the synthesis of flavin adenine dinucleotide (FAD), but in this transformation the AMP moiety is donated to the riboflavin phosphate to form the dinucleotide.

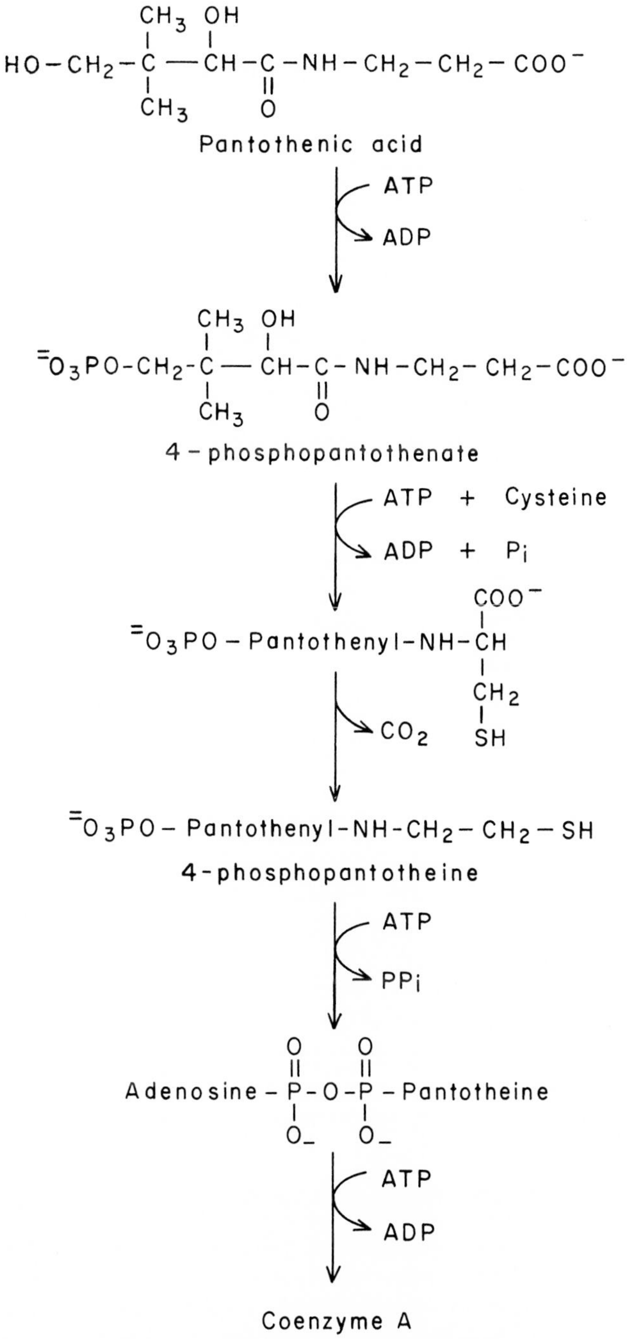

$$\text{HO}-\text{CH}_2-\underset{\underset{\text{CH}_3}{|}}{\overset{\overset{\text{CH}_3}{|}}{\text{C}}}-\underset{}{\overset{\overset{\text{OH}}{|}}{\text{CH}}}-\underset{\underset{\text{O}}{\|}}{\text{C}}-\text{NH}-\text{CH}_2-\text{CH}_2-\text{COO}^-$$

Pantothenic acid

ATP

ADP

$$^=\text{O}_3\text{PO}-\text{CH}_2-\underset{\underset{\text{CH}_3}{|}}{\overset{\overset{\text{CH}_3}{|}}{\text{C}}}-\underset{}{\overset{\overset{\text{OH}}{|}}{\text{CH}}}-\underset{\underset{\text{O}}{\|}}{\text{C}}-\text{NH}-\text{CH}_2-\text{CH}_2-\text{COO}^-$$

4-phosphopantothenate

ATP + Cysteine

ADP + P_i

$$^=\text{O}_3\text{PO}-\text{Pantothenyl}-\text{NH}-\underset{\underset{\underset{\text{SH}}{|}}{\overset{}{\text{CH}_2}}}{\overset{\overset{\text{COO}^-}{|}}{\text{CH}}}$$

CO_2

$$^=\text{O}_3\text{PO}-\text{Pantothenyl}-\text{NH}-\text{CH}_2-\text{CH}_2-\text{SH}$$

4-phosphopantotheine

ATP

PPi

$$\text{Adenosine}-\underset{\underset{\text{O}_-}{|}}{\overset{\overset{\text{O}}{\|}}{\text{P}}}-\text{O}-\underset{\underset{\text{O}_-}{|}}{\overset{\overset{\text{O}}{\|}}{\text{P}}}-\text{Pantotheine}$$

ATP

ADP

Coenzyme A

Fig. 11-14. Biosynthesis of coenzyme A.

$$ATP + FMN \longrightarrow FAD + PP_i$$

Strange as it may seem, the vitamin niacin must first be converted to nicotinic acid before NAD^+ synthesis can occur.

$$H_2O + Niacin \longrightarrow Nicotinic\ acid + NH_3$$

$$Nicotinic\ acid + PRPP \longrightarrow PP_i + Nicotinic\ acid\ nucleotide$$

$$Nicotinic\ acid\ nucleotide + ATP \longrightarrow Deamido-NAD$$

$$Deamido-NAD + Glutamine + ATP \longrightarrow ADP + P_i + Glutamate + NAD^+$$

As might be expected, $NADP^+$ is synthesized from NAD^+, and the phosphoryl group is derived from ATP; ADP is the other product. In similar reactions, thiamine pyrophosphate and pyridoxal phosphate are formed from their vitamin precursors by phosphorylations using ATP. Like fatty acids, biotin and lipoic acid must first be activated to acyladenylates. However, instead of being transferred to CoA, the vitamins are attached in covalent linkage to the ϵ-amino groups of the lysyl residues contained in their respective apoenzymes.

$$Biotin + ATP \longrightarrow Biotin-adenylate + PP_i$$

$$Biotin-adenylate + Apoenzyme \longrightarrow AMP + Holoenzyme$$

$$Lipoate + ATP \longrightarrow Lipoate-adenylate + PP_i$$

$$Lipoate-adenylate + Apoenzyme \longrightarrow AMP + Holoenzyme$$

CATABOLISM OF NUCLEOTIDES

Purines. The ribonucleotides and deoxyribonucleotides derived from the hydrolysis of nucleic acids are catabolized to form the corresponding sugar, phosphate, and purine and pyrimidine bases. In man and other primates the purine bases are catabolized to uric acid (Fig. 11-15). Exactly how the nucleic acids are broken down intracellularly has not yet been fully elucidated. However, the available evidence suggests that the pathways from the mononucleotides to uric acid follow the scheme presented in Fig. 11-15.

Adenylate deaminase catalyzes the synthesis of inosine 5'-monophosphate, and phosphomonoesterases convert IMP and GMP to their respective ribonucleosides or deoxyribonucleosides. These in turn are converted to the free bases by the enzyme purine nucleoside phosphorylase. Mechanistically, this enzyme acts like glycogen phosphorylase, as it removes a sugar 1-phosphate derivative through a phosphorolytic reaction utilizing inorganic phosphate. The bases guanine and hypoxanthine have two fates; they may be reconverted to their 5'-ribonucleotides (case 1) or they may be converted to xanthine (Fig. 11-15). The oxidation of xanthine to uric acid and the oxidation of hypoxanthine to xanthine are catalyzed by xanthine oxidase, an enzyme found in milk, liver, and the intestinal mucosa. This enzyme is a metalloflavoprotein that contains FAD, molybdenum, and nonheme iron. Electrons from the substrates are passed to molybdenum, FAD, iron, and finally to molecular oxygen, which is reduced to hydrogen peroxide. Increased levels of uric acid in the blood and urine are characteristic of the disease known

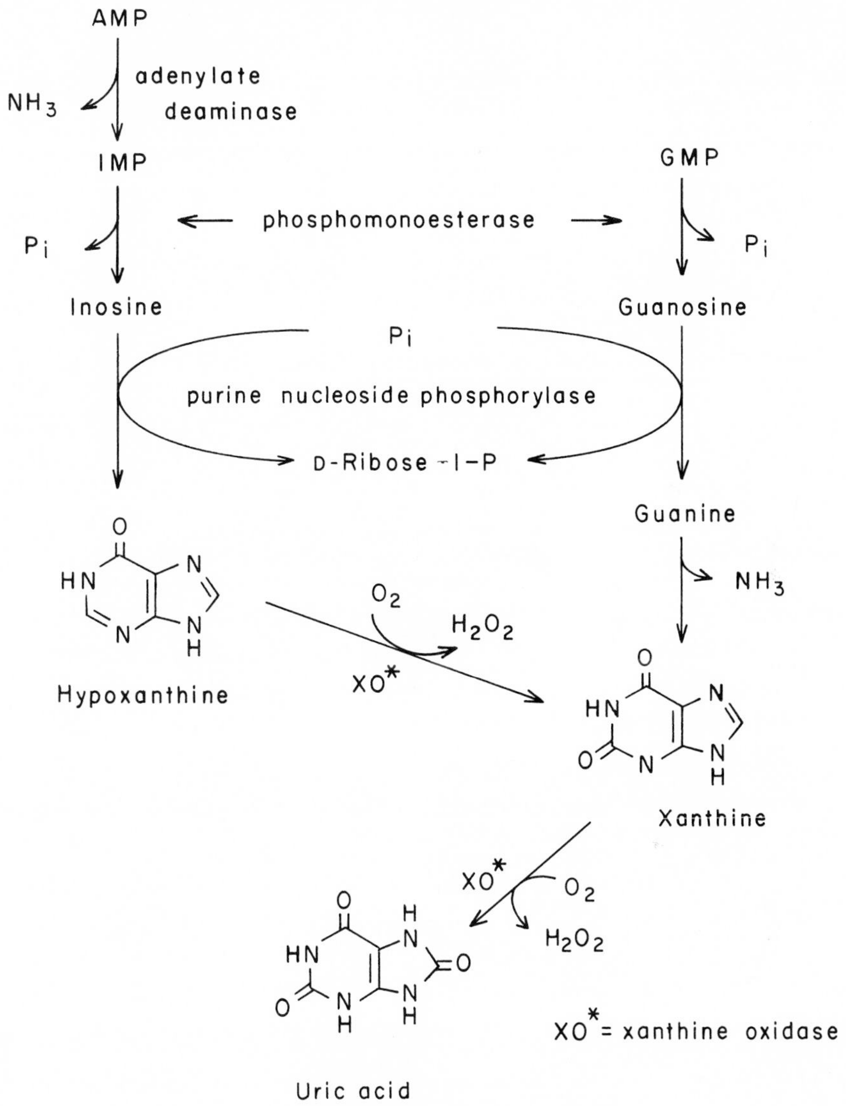

Fig. 11-15. Catabolism of purines.

as gout. However, high levels of uric acid excretion also may occur in various groups of normal individuals, for example, those with a large body surface area, those of certain ancestries, and some executives or supervisors.

Pyrimidines. The pyrimidine bases are catabolized in the liver through reduction by NADPH. As shown in Fig. 11-16, the catabolic enzymes operate on pyrimidine bases that do not contain ring amino groups. Although the diagram shows only cytosine as being deaminated, it is possible that deamination reactions also occur with cytidine and cytidylic acid. The reduced rings are opened in a hydrolytic reaction to form carbamyl derivatives. This reaction is much like the reverse of the reactions involved in pyrimidine synthesis except that the intermediates are not derivatives of orotic acid. The carbamyl groups are eliminated as ammonia and carbon dioxide, yielding β-alanine in the case of uracil or cytosine and β-amino-

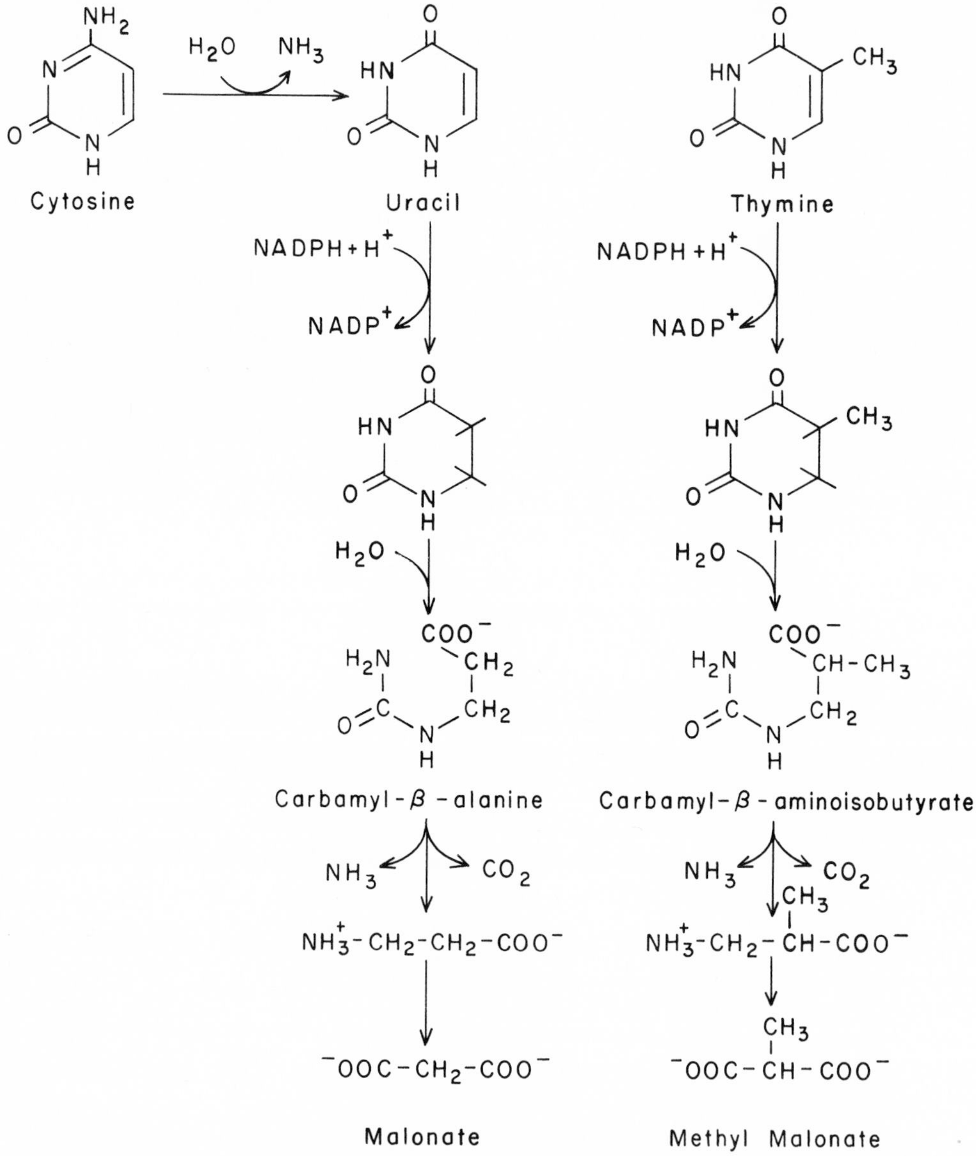

Fig. 11-16. Catabolism of pyrimidines.

isobutyrate in the case of thymine. These amino acids are transaminated to yield the corresponding aldehydes, which in turn are oxidized to malonate and methylmalonate, respectively.

When the diet is high in DNA-containing foods, the intermediate of thymine catabolism, β-aminoisobutyrate, is excreted in the urine. Some individuals are genetically constituted so that even on normal diets they excrete large amounts of β-aminoisobutyrate. They probably have a defect in one of the enzymes that convert this substrate to methylmalonate. As might be expected, excretion of β-aminoisobutyrate is also high in leukemia and in patients undergoing radiation therapy; in both conditions there is an abnormally high rate of cell destruction and therefore pyrimidine catabolism.

The excretory products derived from the pyrimidines are more soluble than those produced by purine degradation. This is probably the reason that there is no disease associated with excessive pyrimidine breakdown, which is analogous to gout.

REFERENCES

Blakley, R. L.: The biochemistry of folic acid and related pteridines, Amsterdam, 1969, North-Holland Publishing Co. *A complete exposition of almost every aspect of folic acid.*

Crick, F.: A general model for the chromosomes of higher organisms, Nature 234:25, 1971.

Davidson, J. N.: The biochemistry of nucleic acids ed. 7, New York, 1972, Academic Press, Inc. *A concise and well-referenced description of nucleic acid properties and metabolism.*

Klinenberg, J. R.: Proceedings of the Second Conference on Gout and Purine Metabolism, Arthritis Rheum. 18:(suppl.) 659-883, 1975. *Several excellent reports of recent advances in our understanding of purine metabolism in man.*

Murray, A. W.: The biological significance of purine salvage, Ann. Rev. Biochem. 40:811, 1971. *A carefully documented analysis of purine metabolism.*

Stanbury, J. B., Wyngaarden, J. B., and Fredrickson, D. S.: The metabolic basis of inherited disease, ed. 3, New York, 1972, McGraw-Hill Book Co. *Detailed accounts of all important inherited diseases with special emphasis on molecular events.*

Watson, J. D.: The molecular biology of the gene, ed. 3, New York, 1975, W. A. Benjamin, Inc. *A very readable account of information transfer in both eukaryotes and prokaryotes.*

Clinical examples

CASE 1: GOUT

R. B. P., a 61-year-old distinguished member of the medical faculty, was awarded an honorary degree in Denver 4 days before admission to the hospital. Following the award ceremony he spent the evening with friends where, as he put it, the "conviviality flowed extensively." The following morning he noticed a dull pain in the upper left flank, which worsened while he was flying to Omaha for a speaking engagement. By the time he reached the Omaha airport the pain was of such severity that he canceled his engagement and returned home. The pain continued, and he was brought to the hospital several hr after his arrival.

Physical examination revealed no overt signs of disease. Prior to this episode the patient had enjoyed good health, the only previous complaint being metatarsal arthralgia, which was presumed to be the result of golf and tennis. On admission the patient had a temperature of 39° C and a pulse rate of 90. X-ray films of the abdomen and pelvis showed no distinct abnormalities. A urine sample obtained on admission had a pH of 4.5 and was positive for protein. Microscopic examination of the centrifugal sediment from the sample revealed the presence of some fine crystalline material and numerous casts. A 24-hr urine sample was then collected; it contained 115 mg protein and 1.52 g uric acid. The serum uric acid content was then measured and found to be 11.8 mg/dl (normal = 2.5 to 7.0 mg/dl).

Biochemical questions

1. Could any useful information be obtained concerning the biochemical defect in gout by questioning this patient about a family history of the disease? Explain.
2. What foods are high in purines and pyrimidines? Is it important for a physician to know this information?
3. Are dietary purines and pyrimidines or their derivatives utilized? Are they required in the diet? Explain.

4. Would you expect a diet high in protein to be harmful to this patient? Explain.
5. What is the biochemical basis for the action of drugs used in the treatment of gout?
6. Did the overindulgence the night before precipitate the attack? If so, how is alcohol metabolism related to uric acid secretion?
7. What is the ionic form of uric acid in this patient's urine? What is its solubility?

Case discussion

Chemical manifestations. The incidence of gout is relatively high, occurring in about 0.3% of the population, while the familial incidence may be as high as 75% to 80%.

Gout is often classified into two broad types: *primary* and *secondary.* Primary gout, of which there are several subtypes, is inherited. Secondary gout is brought on by a variety of disorders such as leukemia (increase in leukocytes) and polycythemia (increase in RBC mass) or by antimetabolites used in the treatment of cancer. Gout is associated with either increased formation of uric acid or its decreased renal excretion. Primary gout is most often found in men over 30 years of age. When women are affected, the age of onset is postmenopausal. Secondary gout occurs in both sexes and at younger ages.

Most cases of primary gout are idiopathic. There are probably several genetic forms of the disease, some of which may be polygenic. However, many cases of primary gout are associated with known enzymatic defects. For example, individuals with glucose 6-phosphatase deficiency develop a glycogen storage disease; since they cannot make glucose from phosphorylated sugars, hypoglycemia develops. As a consequence, ketone bodies, such as β-hydroxybutyrate, and lactic acid build up increasing acid concentrations, tubular urate secretion is inhibited, and hyperuricemia and gout

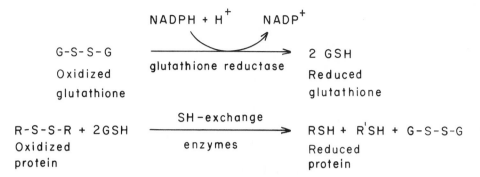

result. Whatever the cause, gout is associated with hyperuricemia (although hyperuricemia is not always associated with gout). The disease produces a painful arthritis, particularly in the joints of the extremities.

Glutathione reductase. Some persons who have an autosomal dominant variant of glutathione reductase develop gout. This variant has *increased* activity as compared to the normal enzyme. Glutathione is a widely distributed tripeptide, γ-glutamylcysteinylglycine. The peptide functions in the disulfide-sulfhydryl exchange reactions that maintain cellular proteins in a properly reduced form. It can also function indirectly to protect protein sulfhydryl groups by serving as a substrate for glutathione peroxidase, an enzyme that removes the hydrogen peroxide formed in certain oxidase reactions (Fig. 11-5). In the absence of glutathione, hydrogen peroxide accumulates. Glutathione is kept in the reduced state by the reaction shown at the top of this page, which is catalyzed by glutathione reductase. People with the superactive variant of glutathione reductase rely more heavily on the pentose phosphate pathway to provide the NADPH used by the reductase. As a result, this leads to increased production of ribose 5-phosphate, and 5-phosphoribosyl pyrophosphate is produced in excess. PRPP then drives the excessive synthesis of purine nucleotides, which must be catabolized to uric acid.

Defects in purine metabolism. It is believed that most cases of primary gout are caused by excessive purine synthesis rather than increased purine nucleotide breakdown. Consistent with this view is the observation that some gouty patients have PRPP amidotransferases (p. 535) that are resistant to feedback inhibition by purine nucleotides. The regulatory sites on this enzyme are like those of other allosteric enzymes in that they are separate from the catalytic sites. Consequently, a defect in a regulatory site as a result of a mutation could lead to the overproduction of purines as seen in gout.

Gout and Lesch-Nyhan syndrome. Another disease characterized by a tremendous overproduction of purines is the Lesch-Nyhan syndrome. This severe X-linked disease has a very early age of onset and is characterized by extremely aggressive behavior that generally leads to self-mutilation. The defect in the Lesch-Nyhan disease is associated with the enzyme hypoxanthine-guanine phosphoribosyl transferase (HGPRT). Another name for this enzyme is GMP (IMP) pyrophosphorylase. The reaction catalyzed is shown at the top of p. 552. Initially, this reaction was considered to function as a "salvage" pathway that permitted reutilization of purine bases that otherwise would be oxidized. However, the severity of the Lesch-Nyhan syndrome suggests a more important role for the phosphoribosyl transferase. Very likely the enzyme has an essential role in nonhepatic tissues where de novo synthesis of purines occurs at a very low rate. The nonhepatic tissues contain the phosphoribosyl transferases but depend on circulating purine bases or nucleosides derived from the liver. They take up the circulating purines and, through the action of HGPRT, form nucleotides. A similar enzyme, adenine phosphoribosyl transferase, produces adenine nucleotides.

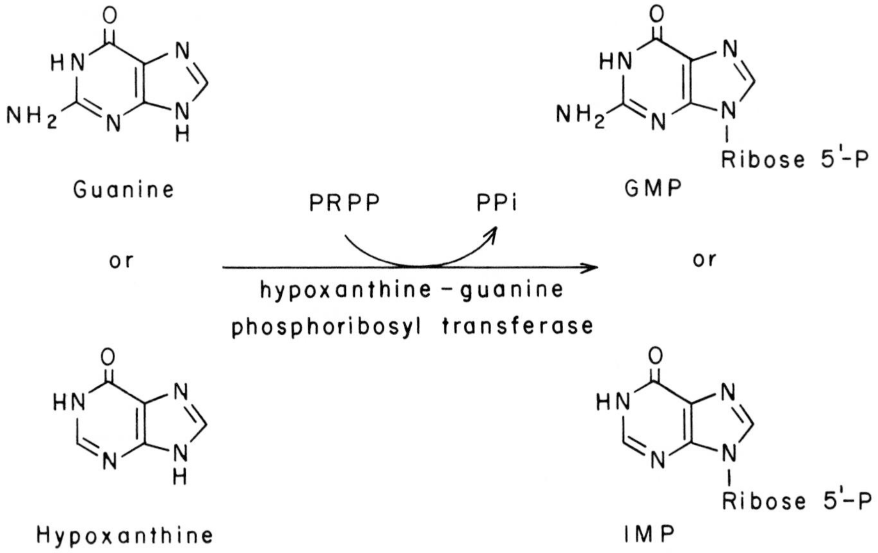

Guanine   or   Hypoxanthine

PRPP   PPi   GMP   or   IMP

hypoxanthine – guanine
phosphoribosyl transferase

Ribose 5'-P

Fig. 11-17 summarizes the catabolic interconversions of the purines as well as their ultimate conversion to uric acid. Man can deaminate adenosine to inosine but not adenine to hypoxanthine. Additionally, the purine nucleosidase favors guanosine and inosine over adenosine. Consequently, all of the catabolized purines are funneled through hypoxanthine and guanine en route to uric acid. Recall that the biosynthesis of the purine nucleotides is very sensitive to the levels of the GMP and IMP pools. This sensitivity occurs at the level of the PRPP amidotransferase reaction, the first enzyme in the purine pathway. In the Lesch-Nyhan syndrome, excessive purine synthesis might result from two effects. The levels of GMP and IMP might drop as a result of the inability of the defective HGPRT to "salvage" guanine and hypoxanthine; consequently, the PRPP amidotransferase might respond by producing excessive amounts of 5-phospho-D-ribosylamine. Alternatively, the defective HGPRT might cause PRPP to accumulate and become available for the stimulation of the PRPP amidotransferase. In either case, abnormally high levels of 5-phospho-D-ribosylamine would result, and purine production would be excessive. The available evidence seems to favor the second interpretation — that involving PRPP accumulation.

The lack of HGPRT activity in patients with the Lesch-Nyhan syndrome is virtually complete. Since some of these patients also develop gouty arthritis, and because this disease is X-linked recessive and thus limited to males, a search was made to find patients with gout but not the severe neurologic symptoms of the Lesch-Nyhan syndrome. It was thought that their levels of HGPRT might be between the normal values and those found in patients with the Lesch-Nyhan syndrome. Such patients have been found, but it is unlikely that they represent a significant percentage of the total number of patients with primary gout. This emphasizes the fact that primary gout is probably inherited in a number of ways.

Diet. Gout has been thought of as an almost fashionable disease associated with wealthy and talented people. For example, noteworthy sufferers included the Medici, Charles IV, Samuel Johnson, Benjamin Franklin, Martin Luther, John Calvin, Isaac Newton, and Charles Darwin. In the traditional view, gout is associated with rich foods and high living. A rich diet and an excessive life-style probably do contribute to the expression of the disease in genetically susceptible individuals. However, diets high in protein are consumed in such a broad spectrum of today's population that this appears to be a less important causative factor in the twentieth century than it might have been in the past.

Animal experiments have shown that

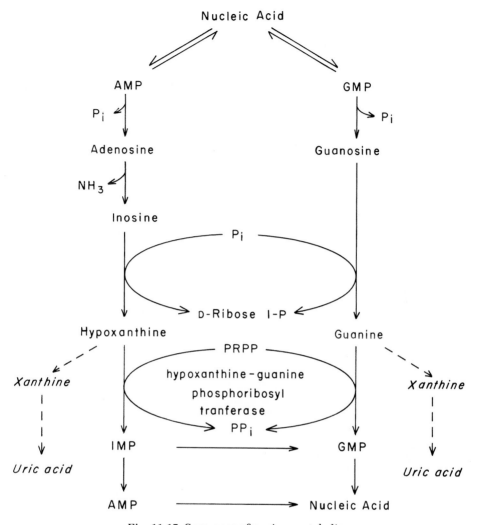

Fig. 11-17. Summary of purine metabolism.

few of the purines derived from dietary nucleic acids are incorporated into the animal's nucleic acids. Xanthine oxidase, the enzyme that converts hypoxanthine to xanthine and xanthine to uric acid, is found in the liver and intestinal mucosa. Most of the dietary purine bases are converted to uric acid by intestinal xanthine oxidase. Thus a diet that includes 4 g of yeast RNA/day will raise blood urate levels to those found in gout. A sustained intake of large amounts of food rich in nucleic acids, for example, sweet breads, liver, yeast, anchovies, kidneys, or sardines, could raise the serum urate levels to over 7 mg/dl. On the other hand, putting a patient on a purine-free diet would lower the urate level by only about 1 mg/dl, as most of the purines in the body are synthesized de novo.

In treating the disease a diet adequate but not high in protein is advised. High-protein foods might also contribute to an accelerated synthesis of purines, since several amino acids are required. Obesity and dehydration should also be avoided. Obesity, because of the excessive amount of food required to maintain body weight, contributes to the urate level. Dehydration should be avoided since urate crystallization is very dependent on concentration, and alcohol causes diuresis leading to dehydration. A high rate of alcohol metabolism might contribute to

$$CH_3CH_2OH \xrightarrow{\text{alcohol dehydrogenase}} CH_3-C\overset{\displaystyle O}{\underset{\displaystyle H}{\diagdown}}$$

Ethanol   Acetaldehyde

$$NAD^+ \qquad NADH + H^+$$

$$NAD^+$$

$$\text{Pyruvate} \xrightarrow{\hspace{3cm}} \text{Lactate}$$

$$\text{lactate dehydrogenase}$$

a lactic acidosis, thus causing the pre-cipitation of sodium urate crystals in the joints. Alcohol dehydrogenase promotes the production of NADH, which is uti-lized to form lactate from pyruvate. The resultant lacticacidemia suppresses the tubular secretion of uric acid, which is highly insoluble in its undissociated form.

Treatment of gouty arthritis. Crystals of sodium urate initiate a series of reac-tions that produce the inflammation associated with acute gouty arthritis. The first step is probably the activation of the Hageman factor (Chapter 14), a protease zymogen. In a series of poorly understood reactions the activated pro-tease causes leukocyte infiltration and phagocytosis of the urate crystals. The phagocytosis causes lysosome destruc-tion and the subsequent release of hydro-lytic enzymes, which causes destruction of tissue and inflammation. The urate crystals found in the fluid of the joints at a nearly neutral pH are composed mostly of monosodium urate, since the pK_a's of uric acid are 5.7 and 10.3. On the other hand, because the gouty patient's urine may have a pH of 4.5 to 5.0, renal stones are more likely to be uric acid crystals.

Treatment of gout may differ with the particular case, depending on whether it is acute or chronic. Since details of treat-ment are beyond the scope of this discus-sion, we will consider only a few of the drugs used. Colchicine has been used in the treatment of gout for many years. It seems to be highly specific for this condi-tion, but its mechanism of action is not well understood. The drug is known to bind to the subunits of proteins that make up the microtubules of polymorphonu-clear leukocytes. The disaggregation of the microtubules inhibits leukocyte loco-

motion, adhesiveness, and metabolism during phagocytosis. Although colchi-cine produces clinical improvement, it does not affect serum urate levels.

Another method of treatment involves lowering the serum urate levels. The dietary measures cited previously cou-pled with hypouricemic agents should lower serum urate levels to approxi-mately 6 mg/dl. Probenecid and sulfin-pyrazone are drugs that promote the excretion of uric acid by preventing tubu-lar reabsorption. Because uric acid is soluble to only 6.45 mg/dl and sodium urate to 110 mg/dl, the use of these drugs may trigger stone formation in the uri-nary tract when uric acid levels are very high.

The drug that most effectively inhibits the formation of urate is allopurinol, a competitive inhibitor of xanthine oxi-dase. Hypoxanthine and xanthine, which are more soluble than uric acid, are ex-creted during allopurinol therapy. How-ever, they are not excreted in amounts stoichiometric with the decrease in uric acid seen in gouty patients who have normal phosphoribosyl transferase lev-els. In individuals deficient in this en-zyme, hypoxanthine and xanthine are excreted instead of urate. Allopurinol, like guanine and hypoxanthine, can be converted to its ribotide by phospho-ribosyl transferase. This reaction would be expected to consume additional PRPP, and the analogue ribotide might even in-hibit PRPP amidotransferase. Further-more, the high levels of hypoxanthine resulting from inhibition of xanthine oxi-dase may cause phosphoribosyl transfer-ase to reutilize the base and thus further inhibit purine synthesis. Purine nucleo-tide synthesis is lowered regardless of the mechanism. Reducing the formation of uric acid through administration of allopurinol relieves the symptoms and decreases the possibility that uric acid kidney stones will form. However, allopurinol is a purine antagonist and must be used with caution in treating secondary gout in leukemic patients undergoing therapy with purine analogues. This is especially true when purine analogues are given orally in combination with allopurinol. The intestinal xanthine oxidase normally degrades many of the administered analogues, but when it is inhibited by allopurinol, more of the analogues can enter the bloodstream.

Another theory that explains the high urate levels in gout is based on the absence of an α_1-α_2 globulin (on electrophoresis it migrates between α_1 and α_2 globulins), which specifically binds urate and prevents it from crystallizing out of solution in the joints, where it causes gouty arthritis. The factor absent in patients with hereditary gout is presumed to be an arginase that cleaves a unique arginyl residue on the α_1-α_2 globulin to produce an ornithinyl residue essential for binding urate.

REFERENCES

Kelley, W. N., and Wyngaarden, J. B.: Enzymology of gout, Adv. Enzymol. 41:1, 1974.

Sletten, K., Aakesson, I., and Alvaaker, J. O., Presence of ornithine in the urate-binding α_1-α_2 globulin, Nature [New Biol.] 231:118, 1971.

Stanbury, J. B., Wyngaarden, J. B., and Fredrickson, D. S.: The metabolic basis of inherited disease, ed. 3, New York, 1972, McGraw-Hill Book Co.

Hypoxanthine   Allopurinol

CASE 2: TROPICAL SPRUE*

A 35-year-old man, a resident of New York City, was hospitalized 3 months after returning from a 1-month visit to Puerto Rico. He had a recent history of watery diarrhea, anorexia, 30-pound weight loss, and progressively severe weakness. He was found to have a megaloblastic anemia with pancytopenia. His serum folate concentration was 3.3 ng/ml (normal = 7 to 20 ng/ml) and serum vitamin B_{12} was 37 pg/ml (normal = 150 to 900 pg/ml). Serum concentrations of iron, albumin, and calcium were normal. A jejunal biopsy showed moderate (2+) villous abnormalities, and radiography of the small bowel showed dilatation.

Parenteral administration of 25 μg folic acid/day over a 24-day period induced a hematologic response, return of appetite, and onset of weight gain, but the diarrhea persisted. A second jejunal biopsy after 1 week of treatment showed no histologic improvement or increase in xylose absorption. Pharmacologic doses of folic acid were administered for 3 days, and the patient was found to have malabsorption of both folic acid and B_{12}. He was then treated with tetracycline for 1 week, which caused cessation of the diarrhea.

Three years passed before the patient was again hospitalized because of a 3-month history of anorexia, progressive weight loss, and weakness. He had not taken medication in the intervening period. Recently his alcoholic intake had been excessive and his dietary intake reduced. He had not returned to Puerto Rico, and he denied having diarrhea. Physical examination showed no stigmata of liver disease. Once again he was found to have megaloblastic changes in the bone marrow and a pancytopenia. His serum folate concentration was 2.8 ng/ml and serum vitamin B_{12} was 150 pg/ml. The peak serum concentration of folic acid was 60 ng/ml (tested after preliminary saturation) (normal = >40 ng/ml), the urinary excretion of ^{60}CoB$_{12}$ was 9.2% (normal = >8%), and fecal fat

---

* Case description adapted from Klipstein, F. A., and Falaige, J. M.: Medicine 48:475, 1969.

excretion was 7.3 g/24 hr. Radiographic examination of the small intestine revealed no abnormality. A biopsy showed the jejunum to be grossly convoluted, but histologic examination showed only slight abnormalities. The pharmacologic doses of folic acid and vitamin B_{12} administered during the course of absorption studies resulted in a hematologic and clinical remission.

Biochemical questions

1. What would have happened if foods rich in folic acid had been given to the patient?
2. What are folic acid polyglutamates?
3. How are the glutamate residues linked together in folic acid polyglutamates?
4. Can folic acid polyglutamates be absorbed from the intestine?
5. What reaction is catalyzed by the enzyme γ-glutamyl-carboxypeptidase?
6. Conjugase can be inhibited by an oral progestogen or estrogen. What symptoms might you expect to find in some women using oral contraceptives?
7. Cholic acid is not inhibitory, but deoxycholate is an inhibitor of conjugase. What does this tell you about the etiology of this patient's disease? What treatment does it suggest?
8. How might a toxin function in this disease? What would be its source? What might the toxin be?

Case discussion

Types of sprue. Sprue is a syndrome characterized initially by diarrhea, anorexia, and weight loss. The intestinal morphology as determined by x-ray analysis or biopsy is abnormal, and the absorption of a variety of nutrients such as glucose, fat, and vitamins is impaired. Consequently, megaloblastic anemia usually develops because of the inability to absorb adequate amounts of folate or vitamin B_{12}. Two types of sprue can be distinguished on the basis of their causative agents. One type is celiac disease of children and nontropical sprue of adults. This type is induced by the presence of a wheat gluten in the diet. Gliadin, the protein in wheat gluten that causes the disease, is composed of over 40% glu-

tamine residues. When gluten is removed from the diet, celiac disease and nontropical sprue are ameliorated. The second type, tropical sprue, does not respond to a gluten-free diet but can be treated with folate and B_{12}, usually taken in combination with an antibiotic such as tetracycline.

Megaloblastic anemia. The megaloblastic anemia seen in sprue is often associated with deficiencies of both folate and B_{12}. The anemia is characterized by the appearance of large immature erythrocytes in which the size of the nuclei and particularly the cytoplasm is increased. Maturation is arrested by a slowing of DNA synthesis in the S phase. The bone marrow accumulates abnormal precursors of the polymorphonuclear leukocytes with hypersegmented nuclei and the giant platelets that appear in the peripheral blood. Almost any defect in DNA synthesis will cause megaloblastic anemia, but the usual cause is deficiency in either folate, B_{12}, or both. The lack of these vitamins is also reflected in other fast-growing cells, especially those of the gastrointestinal tract and the epithelial cells of the mouth. Glossitis, cheilosis, and stomatitis are typical signs, and buccal smears reveal epithelial cells with large nuclei.

Absorption of B_{12} and folate. A B_{12} coenzyme is involved in the utilization of 5-methyl THF for methionine synthesis from homocysteine (p. 534). In B_{12} deficiency, 5-methyl THF accumulates in the serum. According to the methyl-trap hypothesis, all of the body's folate becomes tied up as 5-methyl THF and is thus unavailable for other THF-requiring reactions. The administration of large doses of folate will correct the hematologic symptoms of megaloblastic anemia. However, if the folate deficiency is secondary to B_{12} deficiency, the irreversible neurologic damage caused by B_{12} deficiency will proceed undetected. Megaloblastic anemia is caused by a folate deficiency. This deficiency may be related to inadequate folate intake and absorption, or it may be secondary to a B_{12} deficiency.

Vitamin B_{12} is readily absorbed when administered subcutaneously or intramuscularly. Absorption from the gastrointestinal tract is more complex. A glycoprotein secreted by the gastric mucosa, called *intrinsic factor*, binds B_{12} and enhances its absorption by the ileum. Insufficient amounts of B_{12} are transported by other mechanisms. In the absence of intrinsic factor, B_{12} cannot be absorbed unless it is given in amounts greater than those present in the ordinary diet.

Folic acid absorption is also complex. The derivatives of folate in most foods are the folate polyglutamate forms of 5-methyl THF and 10-formyl THF. Very little unmodified folate occurs in food. As many as six glutamyl residues are joined in peptide linkage through the γ-carboxyl group.

The folate polyglutamates cannot be absorbed from the intestine until all but one of the glutamate residues are hydrolyzed off. An intestinal γ-L-glutamyl carboxypeptidase is the enzyme that ordinarily converts folate polyglutamate to folate. In much of the medical literature this enzyme is referred to by the trivial name *conjugase*. This enzyme is located in the brush border cells of the intestinal mucosa. Diseases that cause degeneration of the mucosa (for example, intestinal cancer, tropical or nontropical sprue) cause loss of conjugase and with it a decreased ability to absorb folate or its derivatives. Dietary folate is utilized so poorly in tropical sprue that orally administered folate polyglutamate in amounts as high as 1.5 mg/day produces no effect. However, folate lacking all but one glutamyl residue may produce hematologic responses in patients when only 25 μg/day are given orally.

Inhibition of conjugase. Some drugs and metabolites inhibit conjugase activity so severely that serum folate levels will be drastically reduced. These drugs inhibit the enzyme directly so that their effects are seen even without mucosal degeneration. Sodium sulfobromophthalein, a substance used in a liver function test, drastically lowers serum folate levels. Oral contraceptives containing a progestogen and estrogen have the same effect. Certain bile salts also inhibit conjugase, and this may be important in patients with tropical sprue. In Chapter 10 it was noted that two bile acids, deoxycholate and lithocholate, are produced from cholate and chenodeoxycholate, respectively, by enzymes produced by intestinal bacteria. Fig. 11-18 shows the activity of conjugase in the presence of the taurine derivatives of these bile acids. Note that the two produced by bacterial enzymes inhibit conjugase more than the others.

Speculations on pathogenesis of tropical sprue. Tropical sprue is a disease that can occur in epidemic proportions, and it responds well to vigorous antibiotic

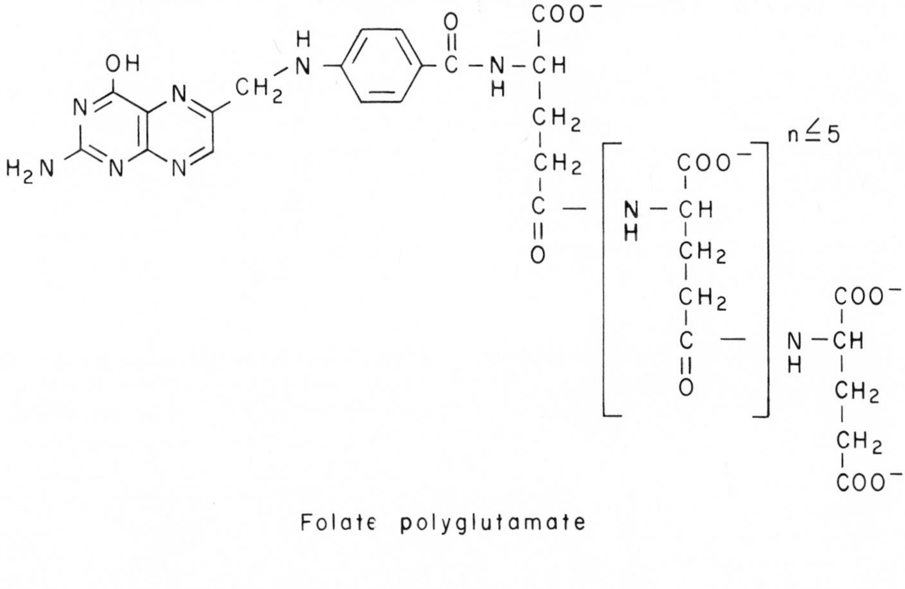

Folate polyglutamate

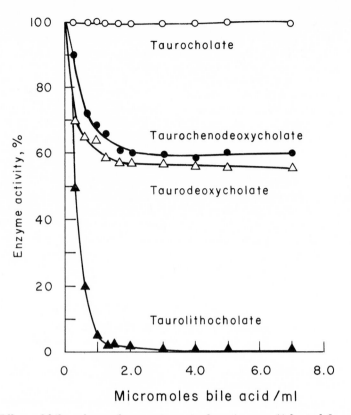

Fig. 11-18. Effect of bile salts on human intestinal conjugase. (Adapted from Bernstein, L. H., et al.: Am. J. Med. 48:574, 1970.)

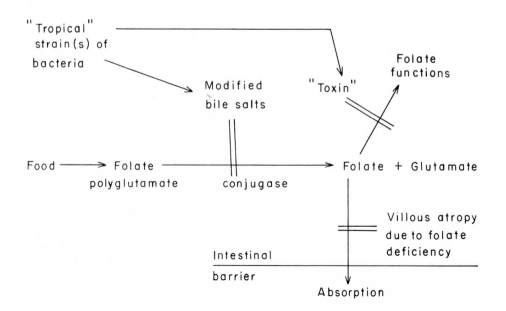

therapy. This disease, but not the gluten enteropathies, also responds to folate therapy. Some investigators believe that other factors or toxins, probably of microbial origin, are causative agents of tropical sprue. Speculation as to the chemical nature of these toxins suggests that they are antifolates or unsaturated fats. Although a microbial origin for the disease has not been proved conclusively, circumstantial evidence strongly favors it. The diagram at the bottom of the page shows the changes induced by the puta-tive microorganisms that bring about the vitamin deficiencies seen in tropical sprue.

REFERENCES

Bernstein, L. H., Gutstein, S., Weiner, S., and Efron, G.: The absorption and malabsorption of folic acid and its polyglutamates, Am. J. Med. 48: 570, 1970.

McIntyre, L. J., et al.: Cellular acquisition of folate: a compartmental model, J. Mol. Med. 1:3, 1975.

Nixon, P. F., and Bertino, J. R.: Interrelationships of vitamin B_{12} and folate in man, Am. J. Med. 48:570, 1970.

Rosenberg, I. H.: Folate absorption and malabsorption, N. Engl. J. Med. 293:1303, 1975.

CASE 3: DEFECT IN SYNTHESIS OF B_{12} COENZYMES*

A male infant, a patient at Massachusetts General Hospital from age 4 weeks until his death at age $7\frac{1}{2}$ weeks, was found to have very low serum, brain, and liver concentrations of methionine as well as methylmalonic aciduria. Homocystine and cystathionine were elevated, but other amino acid concentrations were normal. The total B_{12} content of the liver and kidneys was normal, but the concentrations of 5'-deoxyadenosylcobalamin were reduced to less than 10% of controls. Folic acid was administered on the fifteenth day of hospitalization; 5 days later the serum 5-methyl THF was found to be well above normal values. Postmortem examination of extracts prepared from the liver and kidneys showed abnormally low amounts of 5-methyl THF methyltransferase.

Biochemical questions

1. What metabolic abnormality is suggested by the low methionine and high homocystine levels? Give the reaction involved. What is the coenzyme?
2. What are two well-established enzymatic reactions in animals that require B_{12}-containing coenzymes?
3. Considering the laboratory values presented, what reaction would you expect to be blocked in this infant? Explain.

4. Are these findings consistent with the methyl-trap hypothesis? Explain.

Case discussion

The low methionine and high homocystine levels in this case suggest a defect related in some way to the synthesis of methionine from homocysteine. Homocystine, the disulfide form of homocysteine, often is present in situations where homocysteine accumulates, since the sulfhydryl group of the latter compound is very sensitive to oxidation. Because the vitamin B_{12} findings are also abnormal, it is possible that the B_{12} coenzyme–directed synthesis of methionine is disturbed. This prediction was borne out by a decreased activity of 5-methyl THF methyltransferase. The genetic defect, however, could be in either the enzyme itself or in the availability of a coenzyme essential to the reaction. Consequently, in order to fully understand the biochemical features of this case, it is necessary to describe the structure and functions of vitamin B_{12} coenzymes.

Most of the modern literature now designates the group of cobalt-containing compounds to which vitamin B_{12} belongs as *cobalamin*. The structure of a cobalamin is given on p. 22, but it is convenient to abbreviate it as:

---

* Case description adapted from Mudd, S. H., Levy, H. L., and Abeles, R. H.: Biochem. Biophys. Res. Commun. 35:121, 1969.

The four straight lines on the right and left sides, which are directed at the cobalt atom, represent bonds to four nitrogen atoms in the corrin ring structure. The corrin ring system has several structural features in common with the porphyrin ring system but also differs in many respects. The arrow at the bottom of the cobalt atom represents a coordinate covalent bond from a nitrogen atom of the dimethylbenzimidazole moiety. A corrinoid compound that contains dimethylbenzimidazole is called a cobalamin.

Vitamin B_{12}, as it occurs in vitamin capsules, is cyanocobalamin and is designated as:

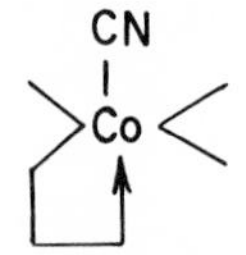

The cyanide substituent arises during the isolation of the vitamin from its natural commercial sources, usually bacteria. It is not found on cobalamin that occurs in the body, and is not toxic in the usual concentrations of cobalamin taken in vitamin capsules.

The three cobalamins found in the body are designated as:

Only methylcobalamin and adenosylcobalamin are coenzymes. These coenzymes are normally synthesized from dietary cyanocobalamin or aquocobalamin. Two enzymatic reactions have been conclusively shown to require cobalamin coenzymes in humans. One requires adenosylcobalamin, and the other uses methylcobalamin.

Adenosylcobalamin is required for the isomerization of methylmalonyl CoA to succinyl CoA. This reaction is important in utilizing the methylmalonyl CoA synthesized by the carboxylation of propionyl CoA so that odd chain fatty acids and certain amino acids, whose catabolism yields propionate, can be metabolized. The methylmalonic aciduria seen in this case is suggestive of a defect in this 5'-deoxyadenosylcobalamin-requiring reaction.

The reaction catalyzed by 5-methyl THF-homocysteine methyl transferase contains methylcobalamin as a tightly bound prosthetic group. When isolated, the holoenzyme appears to have the cobalamin present as aquocobalamin. To activate the prosthetic group it must be reduced with a reduced flavin and methylated with S-adenosylmethionine (SAM).

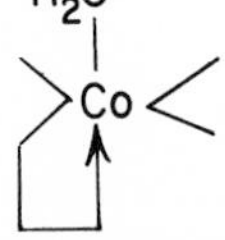

Aquocobalamin

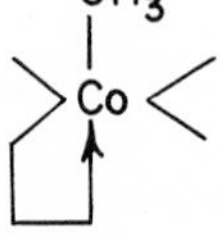

Methylcobalamin

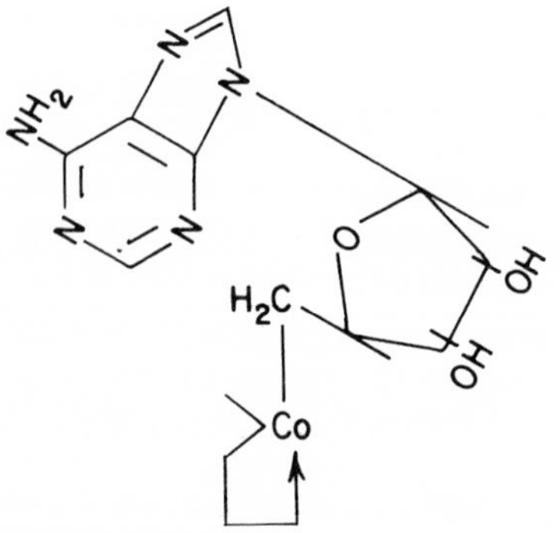

5'-Deoxy-5'-adenosylcobalamin
(Adenosylcobalamin)

$$\text{Aquocobalamin} + \text{SAM} \xrightarrow[\text{FADH}_2 \quad \text{FAD}]{} \text{methylcobalamin} + \text{S-adenosylhomocysteine}$$

The methylcobalamin prosthetic group can then transfer its methyl group to homocysteine and in turn be remethylated by 5-methyl THF

1. Methylcobalamin + homocysteine →
   methionine + reduced cobalamin
   (cob[I]alamin)

2. Cob[I]alamin + 5-methyl THF →
   methylcobalamin + THF

Although the transferase can catalyse many cycles of methyl transfer in this way, the prosthetic group occasionally reverts to an inactive form again, necessitating further reactivation by FADH$_2$, and SAM. The chemical details of this inactivation-reactivation process are obscure, but it is probably related to the very marked nucleophilic properties of the cob[I]alamin intermediate that occasionally cause the latter to undergo an accidental side reaction.

Considering these facts, the data in this case indicate a defect in the conversion of cobalamin to both coenzymatic forms of cobalamin. Total cobalamin content of liver and kidneys was normal, but 5' deoxyadenosylcobalamin was low; thus the methylmalonic aciduria resulted. Although methylcobalamin was not determined in this case, it was determined in a more recent similar case and found to be low. Thus the low activity of the 5-methyl THF methyltransferase was likely caused by the lack of methylcobalamin. The reactions involved in the conversion of aquocobalamin to cobalamin coenzymes are not well understood in the human. Likely, the defect in this case is in a reaction that produces an intermediate common to the synthesis of

both coenzymes. This intermediate, as indicated in the above equation, is probably a reduced cobalamin, and the reducing agent is likely FADH$_2$.

Even though the lack of activity of 5-methyl THF-homocysteine methyl transferase is secondary to a primary defect in the synthesis of cobalamin coenzymes, these results are in accord with the methyl-trap hypothesis. Thus liver methionine was low and both homocystine and 5-methyl THF concentrations were high because the methylcobamide–dependent transferase reaction was blocked. It is expected that 5-methyl THF accumulated, and the concentration of other folate derivatives decreased since all of the folate was "trapped" as methyl THF. If the patient had lived as long as Dillon's (1974), it is expected that the deficiency of folate derivatives other than 5-methyl THF would also have produced megaloblastic anemia.

REFERENCES

Dillon, M. J., et al.: Mental retardation, megaloblastic anemia, methylmalonic aciduria and abnormal homocysteine metabolism due to an error in vitamin B$_{12}$ metabolism, Clin. Sci. Mol. Med. 47:43, 1974.

Erbe, R. W.: Inborn errors of folate metabolism, N. Engl. J. Med. 293:753, 807, 1975.

Mudd, H. S., et al.: A derangement in B$_{12}$ metabolism leading to homocystinemia, cystathioninemia and methylmalonic aciduria, Biochem. Biophys. Res. Commun. 35:121, 1969.

Nixon, P. F., and Bertino, J. R.: Interrelationships of vitamin B$_{12}$ and folate in man, Am. J. Med. 48:555, 1970.

Rosenberg, L. E.: Disorders of propionate, methylmalonate, and vitamin B$_{12}$ metabolism. In Stanbury, J. B., Wyngaarden, J. B., and Fredrickson, D. S., editors: Metabolic basis of inherited disease, ed. 3, New York, 1972, McGraw-Hill Book Co.

CASE 4: OROTIC ACIDURIA*

J. M. R., a male infant, was delivered normally at full term. His neonatal development seemed normal. At 4 months of age a discharge from the nose and a persistent cough were noted. At 5 months, diarrhea with large, loose, pale, and foul-smelling stools prompted his being taken to the hospital. He was found to have severe anemia. The red blood cell count was 2.55 million/mm³, and hemoglobin was 6 g/dl. He was given antibiotics and transfusions. Because the diarrhea and cough persisted, he was given 250 mg ferrous gluconate, 50 mg ascorbic acid, 4 mg folic acid, 4 μg B$_{12}$, and 260 mg liver concentrate daily. The infant improved for 2 months but then contracted an upper respiratory infection and was again hospitalized.

His hemoglobin was 9 g/dl, and despite antibiotics the anemia worsened. The appearance of the erythrocytes was very abnormal. The marrow was hypercellular and contained striking abnormalities indicative of pernicious anemia, yet there was no response following treatment with B$_{12}$ and folic acid, nor did the anemia respond to pyridoxine. However, there was a partial remission following cortisone and prednisone treatment.

A prominent feature of this child's urine was a crystalline sediment. A greater number of crystals were present when the urine volume was small, and urethral obstruction developed several times. Several tests were used to finally identify these crystals as orotic acid. Chromatography of the urine using columns of Dowex-1 showed that orotic acid in amounts as high as 1.5 g were being excreted daily (normal = 1.4 mg/day).

Treatment with uracil produced no hematologic signs of remission, but the daily excretion of orotic acid dropped from 600 to 400 mg.

While partial remission was induced by prednisolone, the infant was given a yeast extract containing substantial amounts of uridylic and cytidylic acids. There was a striking reduction in the excretion of orotic acid, and the bone marrow findings were almost normal. The circulating erythrocytes also became normal. The hemoglobin level rose to 14 g/dl, and the hematocrit was 44%. The child gained weight and became active. The prednisolone dosage was gradually reduced and the nucleotides discontinued. Actually, by this time, the absorbed dose was quite variable because of diarrhea induced by the high-salt and high-phosphate content of the nucleotide preparation. The hemoglobin level again decreased and the orotic acid excretion rate increased, at which time treatment was resumed.

Biochemical questions

1. What might cause the high urinary levels of orotic acid? Consider the elevated excretion in terms of abnormalities in absorption, clearance, biosynthesis, and catabolism. What kind of defect is most likely involved in this case? Why?

2. How would you prove that the crystals in the urine sediment were orotic acid? Use several criteria. What is Dowex-1? Explain how a column of Dowex-1 would be useful in isolating orotic acid.

3. Several cases of orotic aciduria have been described since this report. Many of these patients were successfully treated with uridine rather than with the uridylate-cytidylate mixture used in this case. Why would you expect uridine to be more effective than uracil, the nucleotide mixture, or any other ribonucleoside? What adverse side effect of the nucleotide mixture is avoided by using uridine? Explain.

4. Why are the symptoms in this patient like those in patients with deficiencies in either folate or B$_{12}$? Explain in biochemical terms.

REFERENCE

Smith, L. H., Huguley, C. M., and Bain, J. A.: Hereditary orotic aciduria. In Stanbury, J. B., Wyngaarden, J. B., and Fredrickson, D. S., editors: The metabolic basis of inherited disease, ed. 3, New York, 1972, McGraw-Hill Book Co.

---

* Case description adapted from Huguley, C. M., Bain, J. A., Rivers, S. L., and Scoggins, R. B.: Blood 14:615, 1959.

CASE 5: FOLATE DEFICIENCY
IN ALCOHOLISM*

J. B., a 57-year-old man, was admitted to the hospital because of weakness and shortness of breath. The patient had lived alone for 30 years and had consumed 2 to 4 quarts of beer, 1 to 2 pints of wine, and 3 to 4 whiskey highballs daily during this time. He often went without food for several days. He was previously hospitalized for megaloblastic anemia and scurvy. On admission his hematocrit was 28% and his leukocyte count was 4450/mm³; many oval macrocytes and hypersegmented neutrophils were noted in a peripheral blood smear. The serum folate was less than 1×10^{-9} g/ml, serum B_{12} was 116×10^{-12} g/ml, serum iron was 144×10^{-6} g/dl, and the gastric juice showed normal amounts of intrinsic factor.

J. B. was placed on a hospital diet supplemented with ascorbic acid and commercial multivitamins for 7 days. At this time a bone marrow aspirate indicated megaloblastic anemia. He was transferred to a metabolic ward and put on a low-folate diet (less than 5 µg/day). By the tenth day, when the reticulocyte count had fallen to 11.8%, muscatel wine (32 oz/day) was added to the diet. On the sixteenth day the reticulocyte count was 0.7%. On the twenty-third day, intramuscular injections of folate (75 µg/day) were begun with no change in reticulocytes, leukocytes, or platelets. The folate dose was doubled on the thirty-fourth day. Several days later the reticulocytes went to 18.4%, the platelets rose from 26,000 to 140,000/mm³, and the leukocytes increased from 3000 to 6000/mm³. The wine was deleted from the diet, but the vitamin supplement was maintained. This led to another increase in the reticulocyte count.

---

*Case description adapted from Sullivan, L. W., and Herbert, V.: J. Clin. Invest. 43: 2048, 1964.

Biochemical questions

1. Is the improvement seen in alcoholics with megaloblastic anemia the result of the folate in the hospital diet or because the alcohol, which might have been acting as a hemosuppressant, has been removed from the diet? What is the minimal daily requirement for folic acid? How does this compare with the folate level when the patient was on both folate and muscatel? What can be concluded from this observation?

2. What conclusion can be made about the relationships of alcohol and folic acid to hemopoiesis?

3. Bertino et al. studied the effects of alcohol on folate metabolism in liver and bone marrow cells. They found that 1.5% (w/v) ethanol inhibited the in vitro incorporation of ¹⁴C-formate, but not of ³H-thymidine, into the nucleic acids of bone marrow cells. From these data, what can you say about the metabolic step in nucleic acid biosynthesis that is inhibited by alcohol? Where in this scheme might alcohol and folate interact with the same enzyme?

4. Bertino et al. also found that in liver extracts the enzyme THF formylase was inhibited by 1.5% ethanol. The inhibition by alcohol was competitive in respect to formate. What steps in nucleotide synthesis would be blocked as a result of this inhibition? Theoretically, how might the ethanol inhibition be overcome? Consider as many ways as you can. Would any of these be practical?

REFERENCE

Bertino, J. R., Ward, J., Sartorelli, A. C., and Silber, R.: Effect of ethanol on folate metabolism, J. Clin. Invest. 44:1028, 1965.

CASE 6: EXCESSIVE PURINE SYNTHESIS IN GOUT
Case description*

The patients were brothers, aged 30 and 35 years, who had suffered from gouty arthritis since age 20 years. High concentrations of urine and serum uric acid were successfully reduced with allopurinol. Their erythrocyte hypoxanthine guanine phosphoribosyl transferase and adenine phosphoribosyl transferase levels were normal. The content of PRPP in their erythrocytes, however, was much above normal. This was apparently the result of a mutation in the gene for PRPP synthetase that produces a novel enzyme, since the synthetase from the patients had a *greater* than normal activity. The brother's mother and father were without symptoms and responded normally to all of the tests.

Biochemical questions

1. What biochemical reactions use PRPP as a substrate? What would be the consequence of abnormally large amounts of PRPP on these reactions?

---

*Case description adapted from Sperling, O., et al.: Biochem. Med. 6:310, 1972.

2. Excessive amounts of PRPP are also produced in the Lesch-Nyhan syndrome. How do the accumulations in the Lesch-Nyhan disease differ from those in this case?
3. The increased activity of PRPP synthetase in these patients could be caused by either the presence of quantitatively more enzyme or by an enzyme that was present in the same concentration but qualitatively more efficient. What would one need to do to test these alternatives?

REFERENCES

Becker, M. A., et al.: Human phosphoribosylpyrophosphate synthetase: increased enzyme specific activity in a family with gout and excessive purine synthesis, Proc. Natl. Acad. Sci. U.S.A. 70:2749, 1973.

Sperling, O., et al.: Accelerated erythrocyte 5-phosphoribosyl 1-pyrophosphate synthesis: a familial abnormality associated with excessive uric acid production and gout, Biochem. Med. 6:310, 1972.

Sperling, O., et al.: Altered kinetic property of erythrocyte phosphoribosylpyrophosphate synthetase in excessive purine production, Rev. Eur. Etud. Clin. Biol. 17:703, 1972.

CASE 7: LESCH-NYHAN SYNDROME

At the age of 6 months, patient L. N. showed signs of slow motor development. His mother had noticed orange crystals on his diapers, but she did not report this to the child's pediatrician until questioned about it several months later when she became concerned about the boy's failure to develop and his compulsive urge to bite at his fingers and lips. Again, after questioning, L. N.'s mother revealed that she had a younger brother with similar symptoms.

The Lesch-Nyhan syndrome was suspected so the output of urinary uric acid was measured, normalized to creatinine output, and found to be at least twofold increased over normal. Serum urate concentration was 10 mg/dl, abnormally high for a boy of this age. Further laboratory tests and examination indicated that

the child also had megaloblastic anemia. Because of this unusual disease, the child was referred to the local university hospital where as part of additional testing it was discovered that L. N.'s urine contained excessive amounts of 5-aminoimidazole 4-carboxamide.

Biochemical questions

1. One of the very earliest signs of the Lesch-Nyhan syndrome, and one often overlooked, is the appearance of orange crystals in the diapers. What are these crystals?
2. Is the family history of this patient consistent with the known X-linked recessive inheritance of this disease? Explain.
3. The enzyme defective in patients with the Lesch-Nyhan syndrome is

hypoxanthine guanine phosphoribosyl transferase (HGPRT). Explain how a defect in this enzyme leads to excessive urate excretion.

4. Immunologic assay using antisera prepared against human HGPRT fails to show cross-reacting material in most patients with Lesch-Nyhan syndrome. Propose explanations for this observation considering: (a) the specificity of the antigen-antibody reaction, (b) regulation of gene expression, and (c) lability and turnover of the mutant enzyme.

5. Offer an explanation for the megaloblastic anemia and the increased excretion of 5-aminoimidazole 4-carboxamide.

6. Adenine has been used experimentally in the treatment of patients with Lesch-Nyhan syndrome. This treatment produces a major problem, the formation of very insoluble crystals of 2,8-dioxyadenine. How is adenine converted to 2,8-dioxyadenine? How might this be prevented by administering another drug?

7. Consider the possibility of using adenosine instead of adenine in treatment. Explain the differences and difficulties that might be involved.

8. Allopurinol is often used to lower the excessive amounts of urate that must be handled by the kidneys.

What is the molecular basis for this effect of allopurinol?

9. One patient with Lesch-Nyhan syndrome has been found to have an unusual kind of HGPRT. At first the erythrocyte enzyme seemed to have normal activity. It was then found that its K_m for guanine was 4.8×10^{-5} M, whereas the normal enzyme had a K_m of 5×10^{-6} M. What does this mean and what would you expect to be the consequences for the patient?

10. Erythrocytes from female carriers of the Lesch-Nyhan syndrome have normal levels of HGPRT, yet their skin fibroblasts have half the normal activity. Offer an explanation for this based on a consideration of erythrocyte precursors and the Lyon hypothesis.

REFERENCES

Gartler, S., et al.: Lesch-Nyhan syndrome: rapid detection of heterozygotes by the use of hair follicles, Science 172:572, 1971.

Ghangas, G. S., and Milman, G.: Radioimmune determination of hypoxanthine phosphoribosyl transferase crossreacting material in erythrocytes of Lesch-Nyhan patients, Proc. Natl. Acad. Sci. U.S.A. 72:4147, 1975.

Kelley, W. N., and Wyngaarden, J. B.: The Lesch-Nyhan syndrome. In Stanbury, J. B., et al., editors: The metabolic basis of inherited disease, ed. 3, New York, 1972, McGraw-Hill Book Co., p. 969.

McDonald, J. A., and Kelley, W. N. In Sperling, O., et al., editors: Purine metabolism in man, New York, 1974, Plenum Publishing Corp., p. 167.

CASE 8: ADENOSINE DEAMINASE DEFICIENCY*

A 22-month-old girl of a consanguineous mating was found to have a severe deficiency of T lymphocytes, which direct cell-mediated immune responses since antibody molecules remain tightly associated with these cells. This resulted in recurring respiratory infections. Injections of diphtheria-pertussis-tetanus (DPT) and typhoid vaccine produced only a minimal response. As part of a family study, lysates of the girl's erythrocytes were shown to lack detectable adenosine deaminase activity. Her mother and father both showed approximately 50% of the normal red cell adenosine deaminase activity.

Biochemical questions

1. How would you characterize the pattern of inheritance for adenosine deaminase?

2. What is the reaction catalyzed by adenosine deaminase?

3. One possible explanation of these data is based on the observation that adenosine is inhibitory to pyrimidine biosynthesis in cultured lymphoid cells.

---

*Case adapted from Giblett, E. R., et al.: Lancet 1972 II: 1067, 1972.

Apparently the cells most seriously affected are fast-growing cells such as lymphoid cells. Assuming this interpretation is valid, how might the patient in this case be treated? Does this interpretation suggest a strategy for cancer chemotherapy? Explain.

4. It has been reported that the histocompatibility locus (genes important in controlling the immune response) of mice is closely linked to the adenosine deaminase locus. Assuming this is true in humans, propose another mechanism to explain the findings in this case.

REFERENCES

Chan, T., et al.: Purine excretion by mammalian cells deficient in adenosine kinase, J. Cell. Physiol. 81:315, 1973.

Giblett, E. R., et al.: Adenosine-deaminase deficiency in two patients with severely impaired cellular immunity, Lancet 1972 II: 1067, 1972.

Green, H., and Chan, T.: Pyrimidine starvation induced by adenosine in fibroblasts and lymphoid cells: role of adenosine deaminase, Science 182:836, 1973.

CASE 9: CHEMOTHERAPY IN TREATMENT OF BREAST CANCER

Patient M. I. is a 45-year-old female who had had a radical mastectomy for a potentially curable breast carcinoma performed in a cancer clinic in Milan, Italy. Axillary nodes were detected histologically. This is usually a sign that the patient already has disseminated micrometastases, which at this stage are not detectable.

In the past, after surgery for breast cancer most patients were not treated by prolonged cyclic combination chemotherapy. Yet experiments in animals and other human trials indicated that this treatment might be effective. Patient M. I. volunteered to be part of a study to test this strategy, and 3 weeks after surgery she was given a combination of cyclophosphamide (orally for 14 days), methotrexate, and fluorouracil (intravenously for 8 days). After a 2-week rest without drugs, another cycle of the drug combination (CMF) was given. Drug toxicity and dosage adjustments were analyzed using a number of hematologic tests. Twelve cycles of treatment were given, and after 27 months M. I., as well as 94.7% of a large group of similarly treated women, was apparently free of disease. A control group of patients who had similar surgery but no chemotherapy showed a treatment failure of 24% after the same time period.

Biochemical questions

1. Explain how cyclophosphamide is effective in treating cancer.
2. Methotrexate is a folate analogue. How is it effective in cancer chemotherapy?
3. Trace the route that fluorouracil takes in the blood and the interconversions it undergoes to manifest its toxicity.
4. Discuss the specificity of cancer versus normal cells for the chemotherapeutic agents. What is meant by "acceptable toxicity"?
5. Why is this particular combination of drugs more effective than the use of a single drug?
6. Cancer chemotherapy is more effective in treating micrometastases rather than large tumors. Why?

REFERENCES

Bonadonna, G., et al.: Combination chemotherapy as an adjuvant treatment in operable breast cancer, N. Engl. J. Med. 294:405, 1976.

Fisher, B., et al.: 1-Phenylalanine mustard (L-PAM) in the management of primary breast cancer: a report of early findings, N. Engl. J. Med. 292:117, 1975.

Schabel, F. M., Jr.: Concepts for systemic treatment of micrometastases, Cancer 35:15, 1975.

ADDITIONAL QUESTIONS AND PROBLEMS

1. Explain why folate analogues are useful in cancer chemotherapy.

2. In cancer chemotherapy, leucovorin (5-formyl THF), a derivative that can be converted to the other coenzymatic forms of folate, is given to counteract the effects of large doses of an antifolate. Why is folic acid itself not used? Explain in terms of enzyme functions.

3. Some years ago *p*-aminobenzoic acid was thought by some to be a vitamin. If this substance were included in vitamin preparations, under what conditions would its ingestion be hazardous?

4. Why is azaserine a rather poor anticancer drug?

5. What tests would you perform to determine if an unknown viral nucleic acid were DNA or RNA and whether it were single or double stranded?

6. Cultured cells resistant to the growth inhibitory analogue 8-azaguanine were found to lack activity for hypoxanthine-guanine phosphoribosyl transferase. Explain why.

7. Explain how patients with glycogen storage disease I (glucose 6-phosphatase deficiency) develop gout in early adult life.

8. Busulfan and dacarbazine are alkylating agents used in cancer chemotherapy. How would a bifunctional alkylating agent such as busulfan compare with a monofunctional alkylating agent such as dacarbazine in reacting with DNA?

CHAPTER 12

NUCLEIC ACID AND PROTEIN BIOSYNTHESIS

OBJECTIVES

1. To explain how the structure and metabolism of the nucleic acids relate to the functioning of the genetic apparatus
2. To show how the nucleic acids ultimately control all metabolic functions
3. To describe how changes in DNA lead to the inheritance of certain diseases
4. To describe the molecular action of drugs and antibiotics that affect nucleic acid and protein synthesis
5. To explain how viruses express their genetic information and influence cellular metabolism
6. To consider the genetic basis of antibody variability

Phenomenal progress in understanding the molecular mechanisms of heredity has been made in the last two decades. The discovery that DNA was structured as a base-paired double helix provided the theoretical framework for determining how the information coded into the DNA sequence is duplicated and how it is used to direct the synthesis of RNA and protein. Already clinical medicine has taken advantage of many of these discoveries, and the future promises much more. It is safe to say that the health sciences will benefit enormously from this knowledge in the years ahead. Even now we know that the biochemistry of the nucleic acids is central to an understanding of the virus-induced diseases, the immune response, the mechanism of action of many drugs and antibiotics, and the whole spectrum of inherited diseases.

In approaching the study of the molecular mechanisms of heredity, we will deal first with the functional roles of the genetic material DNA, considering its replication, susceptibility to mutation, and its expression as the template for RNA synthesis. The mechanism of action and control of the RNA polymerase involved in RNA synthesis will form the background for a discussion of the mechanism of protein synthesis and a description of the genetic code. Next, the unique features of the synthesis of antibody and virus proteins will be considered. Finally, an attempt will be made to describe some common features of human genetic diseases in terms of this material taken as a whole.

FUNCTIONAL ROLES OF DNA
The genetic material

Proof for the functional role of DNA in the transmission of inherited characteristics was not forthcoming for more than 70 years after the discovery of the nucleic acids. That DNA was in fact the genetic material was implied by its high content not only in the chromosomes but also in many of the viruses that infect animals and bacteria. The premise was complicated, however, by the high concentration of protein in these structures and by the presence of RNA but not DNA in some viruses. Another piece of indirect evidence pointed to a role for nucleic acids as the transmitters of biologic information; the wavelengths of light in the ultraviolet region that are the most mutagenic are the same wavelengths at which nucleic acids absorb the most light energy.

One property expected of the genetic material is a constancy of amount in every cell of the body under every environmental situation. DNA, not RNA or protein, fulfills this expectation. Its content per nucleus is the same in every cell except the germ cells, which have exactly half that found in the somatic cells. Again, this is expected if progeny obtain half of their characteristics from each parent. This constancy is so dependable that the measurement of the DNA concentration in a tissue can be used to calculate the number of nuclei and thus the number of cells. This works well for diploid cells such as those of the kidney, but corrections must be made for polyploid mammalian liver or cancer cells.

Yet it was the experiments of Avery and co-workers that provided the best direct evidence that exogenous DNA is responsible for the permanent changes that can be produced in bacterial cells. DNA from one strain of cells can be used to transform a different strain of cells so that they then resemble the strain from which the DNA was derived. This process is called bacterial transformation. In the original experiment, DNA was isolated from cells of a strain of *Diplococcus pneumoniae* that contained a characteristic complex polysaccharide on their surfaces. This polysaccharide made the cells pathogenic for mice and gave a glistening, smooth appearance to colonies formed by these cells on nutrient agar. When the polysaccharide was missing, as it was in some other strains of the microorganism, the colonies were rough in appearance and the cells were harmless when injected into mice. When DNA from the smooth cells was added to rough cells, the DNA entered some of the cells and became a permanent part of those cells' genetic apparatus; subsequent generations were permanently transformed to smooth, pathogenic cells. Considering all of this evidence, there is no doubt that DNA is a carrier of genetic information.

"The central dogma"

The genetic functions of DNA are of two types: (1) DNA must be able to replicate itself so that the information coded into its primary structure is transmitted faithfully to progeny cells and (2) the information must be expressed in some useful way. The method of this expression is through RNA intermediaries, which in turn act as templates for the synthesis of every protein in the body. The relationships of DNA to RNA and to protein are often expressed in a graphic syllogism called "the central dogma." The concept was proposed by Crick in 1958 and was revised in 1970 to accommodate the discovery of the RNA-dependent DNA polymerase. Crick's original theory suggested that the flow of information was always from RNA to protein and could not be reversed, yet it allowed for the possibility of DNA synthesis from RNA. Transcription involves the transfer of information, using the four-letter language of the nucleic acids; that is, one strand of DNA serves as a template for the synthesis of a complementary RNA strand. Transcription has been shown to be reversible in a few cases. The dashed line in Fig. 12-1 represents the synthesis of DNA from information contained in the RNA of certain tumor viruses. Information flow in the direction of RNA to protein is termed translation, since the four-letter language of the nucleic acids must be

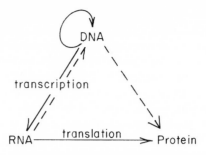

Fig. 12-1. "The central dogma."

converted to the twenty-letter language of the amino acids that make up the proteins. The process of translation is always unidirectional. Single-stranded DNA templates can be translated in the laboratory, but there is no evidence for such a function in vivo; consequently, the line between DNA and protein in Fig. 12-1 also is dashed.

DNA synthesis

The mere knowledge that DNA is the hereditary material gave no clue as to how it might reproduce itself until Watson and Crick proposed their model for the structure of DNA. In this model the two DNA strands are arranged in an antiparallel fashion and are base paired along their entire length in the form of a double helix. Recall from Chapter 11 that the molar concentration of adenine equals that of thymine, and the cytosine concentration is the same as that of guanine. Base pairing of adenine with thymine and cytosine with guanine yields a structure in which the sequence of one strand can be automatically determined if the sequence of the other strand is known. The importance of this concept in the replication of DNA and in the synthesis of RNA strands of complementary sequences was recognized immediately, but several years were required for the enzymatic studies that gave unequivocal proof.

Implicit in the functioning of the Watson-Crick DNA model is the idea that the strands of a DNA molecule must separate and new daughter strands must be synthesized in response to the sequence of bases in the mother strand. This is called semiconservative replication. Still, conservative replication, in which both strands of a daughter molecule are newly synthesized, could not be ruled out from consideration of the structure of DNA alone.

semi–conservative replication ← Double-stranded DNA → conservative replication

Maternal strands ——————
Daughter strands — — — — —

The clever experiments of Meselson and Stahl proved replication to be semi-conservative. These experiments consisted of growing *Escherichia coli* cells in a medium containing $^{15}NH_4Cl$, so that the nitrogen atoms of the purine and pyrimidine bases of the DNA were heavily labeled. Cells were then transferred to a medium containing the usual light $^{14}NH_4Cl$ and grown for one or more generations. The DNA was then prepared and separated by equilibrium centrifugation. After one generation, the progeny DNA separated in such a way that all of it appeared at a position midway between the very heavy parental DNA and the light DNA from a control culture. Because all of the DNA existed as a hybrid that contained one heavy strand and one light strand, it was clear that the DNA replicated by means of a semiconservative mechanism.

This knowledge presented an even more difficult problem: How do the double helical strands separate during DNA synthesis? In a rapidly growing cell such as *E. coli* it has been calculated that if the strands separate by untwisting, the molecule would have to rotate at 10,000 rpm, a rate that is highly improbable. Still the answer to this problem more than likely lies in an understanding of the mechanism of DNA replication at the enzyme level. We will return to this subject after first considering the enzymes involved in DNA synthesis.

DNA polymerases

DNA synthesis is a process more complex than originally thought. One reason for this is that DNA replication requires the organized functioning of a large number of different enzymes, not just a DNA polymerase. For example, replication requires enzymes that coordinate the growth of cell membranes with DNA synthesis. Other enzymes and proteins initiate the synthesis of small RNA primers that bind to single-stranded DNA. Additional enzymes are needed to remove the RNA primers from the growing deoxyribonucleotide chain, fill in the small regions vacated by the RNA primers, seal the strands together, and aid in the untwisting of the DNA helix. More than one enzyme may be required for each of these functions, and the list of functions is not meant to be complete.

The mechanisms involved in DNA synthesis are most easily developed by first considering the DNA polymerases. The most highly purified and extensively studied enzymes have been isolated from bacterial cells. Three DNA polymerases, I, II, and III, have been isolated from *E. coli*. There is still some ambiguity about the essentiality of the specific roles played by each of the polymerases. One complicating feature is that it is difficult to distinguish the DNA polymerases from enzymes that function exclusively to repair damaged DNA, primarily because some of the processes that occur during DNA replication are identical to events necessary for DNA repair. However, it is known that all of the DNA polymerases require a DNA template and all four of the deoxyribonucleoside triphosphates. Synthesis proceeds from the 5' to the 3' end of the growing polynucleotide, and inorganic pyrophosphate is a product of the reaction. These features are illustrated in Fig. 12-2 for DNA polymerase I, but they apply to other polymerases as well. Polynucleotides formed using radioactive deoxyribonucleoside triphosphates can be shown to have sequences identical to those on one strand of the DNA template and complementary to sequences on the other strand. Both strands are labeled in vitro.

One might conclude that DNA polymerase I is a nonessential enzyme since there are viable *E. coli* mutants that lack it (pol A$^-$). The enzyme, however, is unusual in that it catalyzes three separate chemical reactions. It polymerizes deoxyribonucleoside triphosphates, and it has two exonucleolytic activities, a 3' to 5' activity and a 5' to 3' activity. The pol A$^-$ mutants lack only the polymerization activity. Other mutants lacking both the polymerase and the 5' to 3' exonuclease activity are lethal. Thus the exonuclease function may be the more

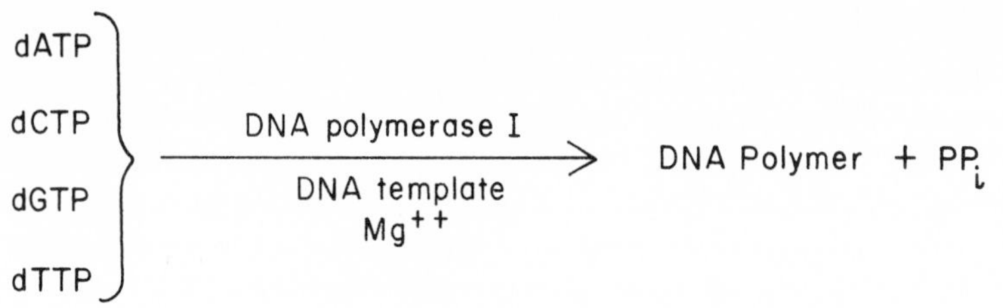

$$\left.\begin{array}{c} dATP \\ dCTP \\ dGTP \\ dTTP \end{array}\right\} \xrightarrow[\substack{\text{DNA template} \\ Mg^{++}}]{\text{DNA polymerase I}} \text{DNA Polymer} + PP_i$$

Fig. 12-2. Reaction catalyzed by DNA polymerase I.

important. This fits with the role of this enzyme in removing damaged DNA segments (DNA repair) and in removing covalently attached RNA from DNA chains. We will later see that small RNA's serve as primers of DNA synthesis.

DNA polymerase II is more likely needed for the repair synthesis of DNA. Repair synthesis requires excision of the damaged DNA, the synthesis of a fresh replacement segment complementary to the remaining single strand, and the sealing of the replacement segment to the larger polynucleotide chain. DNA polymerase II does not have 5' to 3' exonuclease activity. Mutants deficient in DNA polymerase II activity, as determined by in vitro assay, grow well; therefore, the enzyme does not seem to have an indispensible function in the cell.

DNA polymerase III has all of the enzymatic activities of DNA polymerase I. A subunit of the enzyme is the product of the dnaE gene. There are temperature sensitive mutations of this gene testifying to the importance of DNA polymerase III. A temperature sensitive mutant is one that grows at 30°C but fails to grow at 42°C, a temperature not lethal for wild-type *E. coli*. The failure to grow at the higher temperature is caused by a mutation in the gene for DNA polymerase III so that a very heat labile enzyme is produced. Since this appears to be the only mutation in this strain, DNA polymerase III is the only enzyme inactivated at 42°C; thus the enzyme is essential to the organism.

At this point it is necessary to make a distinction between the meanings of template and primer. The word template refers to the structural sequence of the polymerized monomeric units of a macromolecule that provides the pattern for the synthesis of another macromolecule with a complementary or characteristic sequence. The word primer, on the other hand, refers to a polymeric molecule that contains the growing point for the further addition of monomeric units. Glycogen is an example of a primer to which glucose units are added; however, glycogen has no template activity. Under certain circumstances DNA has both primer and template activities. For example, the addition of mononucleotides is to the 3' end of the growing DNA primer. This presents a problem with regard to how the other strand is synthesized. Biochemists have looked hard but unsuccessfully for an enzyme that can add deoxyribonucleotides onto the 5' end of DNA primers. Such a primer should contain a triphosphate on the hydroxyl group of the 5' end. Although a very active 5'-exonuclease, actually part of DNA polymerase I, has made the search for such an activated 5' end extremely difficult, a polymerase that could utilize such a primer probably does not exist. On the contrary, there is good evidence that the synthesis of both strands can be accomplished by the known DNA polymerases.

Initiation of DNA synthesis

E. coli DNA synthesis is primed by a small RNA fragment. The RNA oligonucleotide is complementary to a sequence on one of the strands of the DNA template and base pairs with this portion of the DNA molecule. Subsequently, deoxyribonucleotides are covalently attached to the RNA primer. The synthesis of the primer itself is catalyzed by RNA polymerase, the enzyme that synthesizes RNA from DNA templates. A more detailed description of this enzyme will be presented later when RNA synthesis is considered. Accessory proteins are required for the RNA polymerase to initiate synthesis at the proper spots on the DNA template. Probably a special accessory protein assures that RNA polymerase initiates the synthesis of RNA primers at the correct DNA synthesis initiation sites. RNA synthesis, unlike DNA synthesis, does not require a primer to initiate synthesis.

The precise details of the mechanism of DNA synthesis is not known with a great deal of certainty; consequently, the following description should be considered plausible but tentative in several respects. The series of steps in the sequence are illustrated in Fig. 12-3. In the first step, the DNA duplex opens slightly so that the RNA polymerase can catalyze the synthesis of a short strand of RNA (about fifty to 100 nucleotides) complementary to one of the DNA strands (Fig. 12-3, *A*). This leaves a free single-stranded region in the DNA template opposite the RNA primer to which several copies of a DNA-binding protein attach. The DNA-binding protein has been highly purified from *E. coli*. It has a molecular weight of 74,000 and a high affinity for single-stranded as opposed to double-stranded DNA. The DNA binding protein stimulates only the *E. coli* DNA polymerases, not DNA polymerases from other sources. This implies that the binding protein and the polymerase must interact. The binding protein functions by attaching cooperatively, one molecule next to another, along the entire length of the single DNA strand exposed by the RNA polymerase (Fig. 12-3, *B*). Once bound, the protein tends to hold the DNA strands apart and may even act like a wedge to actively separate the DNA strands at the replication fork.

The rigid species specificity of the DNA-binding protein suggests that it is important in directing the placement of DNA polymerase III (together with at least one accessory protein) on the DNA template. It is possible that DNA polymerase III displaces the RNA polymerase and then directs the addition of deoxyribonucleotides to the 3' end of the RNA primer (Fig. 12-3, *C*).

The opposite strand of the template is replicated in a similar manner; RNA polymerase makes an RNA primer, which is extended by DNA polymerase III (Fig. 12-3, *D*). The RNA primer contains a 5' end that still contains the triphosphate portion of the initiating purine ribonucleoside triphosphate. This is indicated by *ppp* in Fig. 12-3.

Once a DNA strand has been primed, synthesis toward the replicating fork can be visualized as continuous, although additional priming reactions would result in discontinuous synthesis, even in respect to synthesis toward the fork. Growth of the opposite strand away from the replicating fork must occur in discontinuous bursts, each burst primed by a short RNA segment. DNA synthesis visualized by electron microscopy gives the appearance of an "eye" or several eyes along a DNA template, that looks something like ——⊂⊃—— . Eyes are thought to be regions of DNA where recent synthesis has been initiated. Synthesis from an eye is bidirectional as illustrated in Fig. 12-3. Thus as the eye enlarges, DNA synthesis along either new strand may be considered continuous where the DNA polymerase is close to and moving toward the replicating fork, and it may be considered discontinuous where the DNA polymerase is close to a replicating fork but moving away from the fork. Consequently, as an eye enlarges, DNA synthesis is more or less continuous at one end of a growing strand but discontinuous at the other end of the same strand.

The end result is newly synthesized DNA that is interspersed with segments of RNA and that is discontinuous but base paired with a continuous parental strand. Subsequently, the 5′ and 3′ exonucleolytic activity of DNA polymerase I removes the RNA segment, and either DNA polymerase I or II fills the gap vacated by the RNA. DNA ligase (sometimes called polynucleotide ligase) is required to join these short pieces into phosphodiester linkage. The ligating reac-

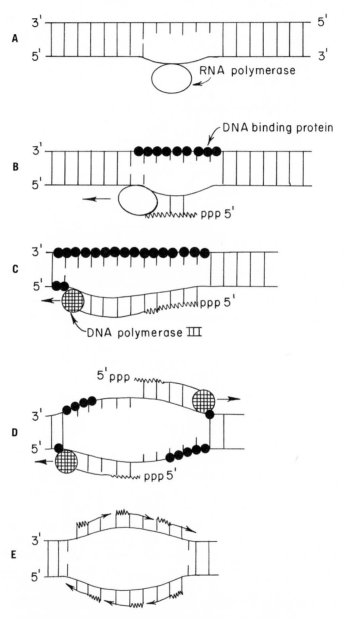

Fig. 12-3. A, Special RNA polymerase binds to DNA synthesis initiation site. B, RNA polymerase opens up DNA duplex and synthesizes RNA primer; DNA binding proteins attach. C, RNA polymerase leaves the template; DNA polymerase III attaches and adds deoxyribonucleotides to the RNA primer. D, RNA primer starts replication of the other strand. E, Both strands are replicated by making discontinuous segments.

tion shown in the diagram below requires that energy be supplied from ATP. This enzyme also occurs in animal cells.

$$ATP \;+\; ligase \;\rightleftharpoons\; ligase \sim AMP \;+\; PP_i$$

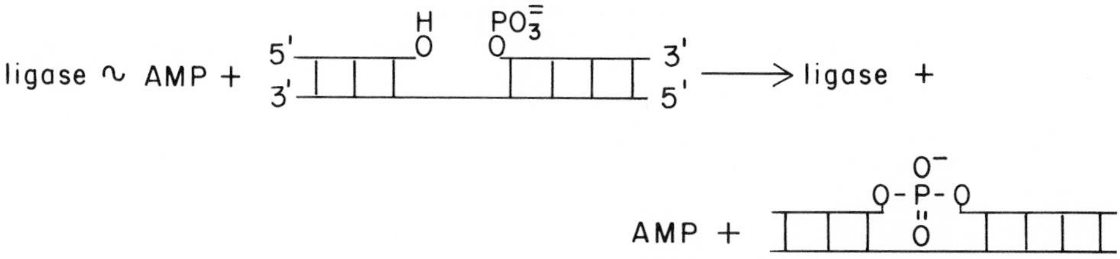

ligase $\sim$ AMP $+$ $\longrightarrow$ ligase $+$

AMP $+$

DNA ligase is not only important in DNA replication; it is used to seal deoxyribonucleotide segments in the cross-over events during gene recombination. The enzyme also functions to close breaks in segments of DNA undergoing repair and is required to join the ends of mitochondrial DNA to form their characteristic circular structure.

Armed with this information, the unwinding problem mentioned earlier can be reconsidered. One possibility is that by the alternating action of an endonuclease and DNA ligase the unwinding could be reduced to an untwisting of only a small part of the double helix at any given time. There is also evidence for a "swivelase," or untwisting protein, that releases the torque developed during the unwinding required for replication. The swivelase can be visualized as first making a single strand break upstream from the growing fork. This is not a hydrolytic cleavage but rather the breaking of the phosphodiester bond by transferring the phosphoryl group to a substituent on the swivelase so that the energy of the phosphodiester bond is conserved. The single-strand break relieves the torque as the broken strand with the enzyme still attached rotates about the unbroken strand. When the strain on the double helix is relaxed, the enzyme transfers the phosphoryl group back to the polynucleotide chain and dissociates from the DNA duplex. A swivelase would have to specifically recognize only superhelical DNA. An enzyme capable of removing superhelical twists in polyoma DNA has been isolated from the nuclei of mammalian cells.

DNA synthesis in animal cells

The replication process in mammalian cells is necessarily more complex because of the problems involved in making several chromosomes. However, DNA polymerases from a number of different animal cells have been isolated and studied. Two of these enzymes are the best candidates for synthesizing DNA in vivo. One is a soluble enzyme found in the cytosol; the other is found in the nucleus. The mechanism of action of the two enzymes is similar, but they differ in several respects. Table 12-1 compares the two with regard to these differences. Probably the cytoplasmic enzyme is the more important. The large cytoplasmic enzyme shares many functional properties with DNA polymerases II and III of *E. coli*, whereas the nuclear enzyme is much smaller, has an acidic rather than basic isoelectric point, and is relatively insensitive to agents that inhibit the large enzyme. Actually both enzymes are not unique to the nucleus or cytoplasm; they are found to some extent in both cell compartments. Nevertheless, it is difficult to rationalize why the larger, more complex DNA polymerase is found for the most part in the cytoplasm. Perhaps the large enzyme is stored in the cytoplasm until

Table 12-1. Comparison of properties of DNA polymerases from animal tissues

PROPERTY	CYTOPLASMIC ENZYME	NUCLEAR ENZYME
Molecular weight	200,000	70,000
Template specificity	Less active with native, nicked DNA template	More active with native, nicked DNA template
Deoxyribonucleoside triphosphate dependence	All four required	All four work, but single nucleotide will incorporate
Inhibition by sulfhydryl reagents	Sensitive	Less sensitive

needed for DNA synthesis during periods of rapid growth. For example, in regenerating liver the large nuclear enzyme increases in activity over twentyfold at 30 hr after partial hepatectomy. The activity of the large enzyme isolated from the cytosol also is increased, but only about threefold. There is no increase in activity of the small DNA polymerase, whether isolated from the cytosol or the nucleus. Mitochondria contain their own DNA polymerase, which shows a high specificity for the native double-stranded and circular mitochondrial DNA.

Molecular basis of mutation

On rare occasions a base may be changed or modified in the DNA sequence. As will be seen later when protein synthesis is considered, such a change in the structural gene for a protein may lead to the insertion of the wrong amino acid. If changed at a crucial position, the resulting protein will be unable to function. If the amino acid replacement occurs at a less important position, activity may be diminished or not affected at all. Mutations are responsible for dozens of known genetic diseases and undoubtedly for many more yet to be discovered. Usually these changes are subtle so they cannot be detected cytologically at the level of the chromosome. Gross chromosomal abnormalities do occur and are very important in the health sciences, but generally they are not inherited in the classic mendelian way. Rather, most are caused by nondisjunction, that is, a failure of either the egg or the sperm to receive an exact set of haploid chromosomes or of a mitotic cell to receive an exact diploid set early in development. Many others are caused by translocations that are also difficult to predict.

Mutations are caused by both chemical and physical agents, although the action of even the physical agents, for example, ionizing radiation, can usually be explained by a chemical mechanism. Regardless of the agent used to produce a mutation, none are selective in the sense that they can specifically mutate one gene and not another. Because all genes are composed of only four different types of purine or pyrimidine bases, an agent that may react specifically with only one of the four could potentially cause mutations in every gene. Mutations are essentially random events. During our evolution the selective pressures of nature eliminated an astronomic number of deleterious mutations. The smaller number of beneficial mutations gave primitive life a survival advantage over competitors and allowed for the eventual emergence of intelligent beings. Consequently, in a highly evolved species such as man, most mutations produce deleterious effects.

Chemical mutagens

Purine and pyrimidine analogues. Mutations may be produced in many ways. Bases may be deleted or new ones may be inserted; more frequently an existing base may be chemically modified so that on replication improper base pairing will cause a different base to appear at the modified position. The latter kind of mutation is called a *replacement*. When a purine is replaced by another purine or a pyrimidine by a different pyrimidine, the change is called a *transition*. A *transversion* is a change from pyrimidine to purine or purine to pyrimidine.

Many of the mutations caused by artifically produced base analogues are transitions. Mutations are produced by base analogues in one of two different ways. On entering the cell, a base analogue is converted to a nucleoside triphosphate that base pairs, perhaps incorrectly, with a DNA template and is inserted into the nucleotide chain. This is one way in which the mutation can be produced. The other requires an additional round of replication so that an improper base pair forms as a result of the previously incorporated analogue. The result is a permanently modified DNA in both cases.

As might be expected, base analogues can also inhibit DNA synthesis and cell multiplication. It is this feature that has stimulated organic chemists to create hundreds of different base analogues in the hope that some may be useful for inhibiting rapidly proliferating cancer cells. Examples of base analogues that have some usefulness in cancer chemotherapy and that are also mutagenic are 6-mercaptopurine and 2-aminopurine.

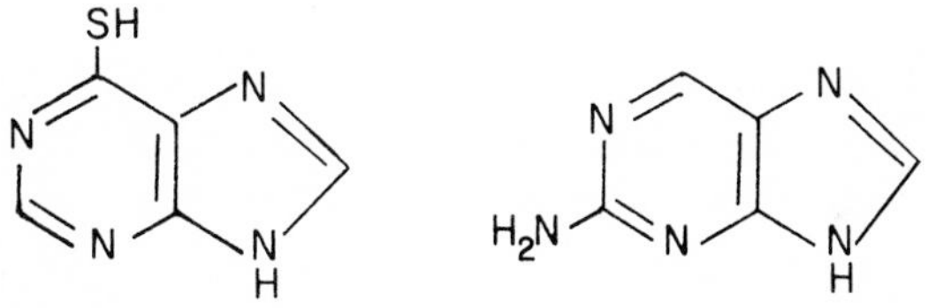

6-Mercaptopurine   2-Aminopurine

Not all analogues become active against cancer cells through incorporation into nucleic acid. As seen in Chapter 11, some analogues block the synthesis of normal purine and pyrimidine nucleotides; for example, 8-azaguanine blocks GMP synthesis and 6-mercaptopurine inhibits AMP synthesis.

Alkylating agents. Alkylating agents are also mutagenic substances that have been used in cancer chemotherapy. Alkylating agents such as nitrogen or sulfur mustards chiefly cause transversions. Bifunctional compounds such as those shown below produce cross-links between DNA strands or between a DNA strand and any other reactive group in the vicinity.

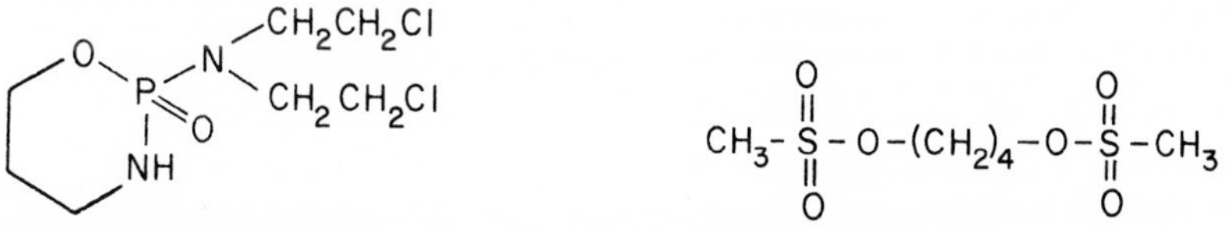

Cyclophosphamide   Busulfan

The mechanism of action of alkylating agents is complex. Adenine and guanine are easily alkylated in position 7. The reaction produces an exceedingly labile glycosidic bond. Splitting of this bond leads to depurination.

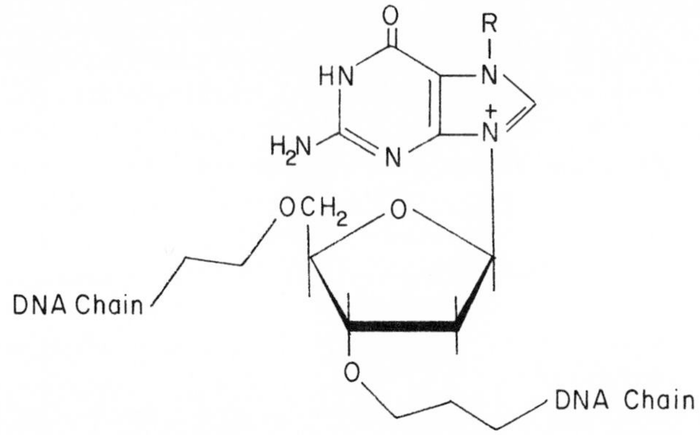

In those cases where alkylation does not lead to depurination it is more likely that the mutation will be of the transition type. On replication the position opposite the gap might be filled by any one of the four bases, a possibility that accounts for the transversions caused by these agents.

Dyes. Acridine dyes such as the antimalarial agent quinacrine (Atabrine) shown below are large planar aromatic compounds that intercalate or sandwich themselves between the stacked bases of the helix.

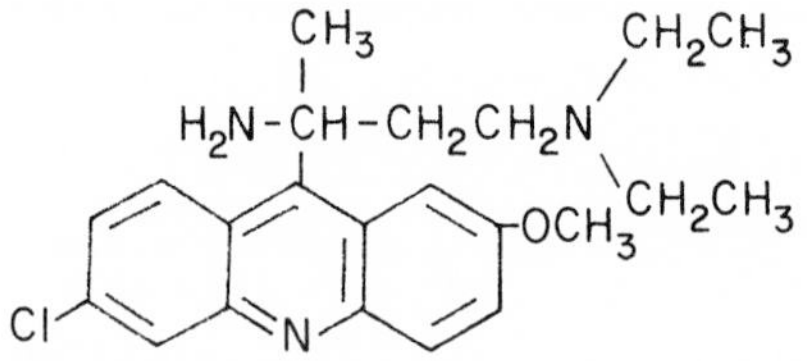

On replication, *insertion* or *deletion* of bases may occur. Chain scission and chromosome breaks are also possible. Quinacrine is useful in human cytogenetics since it intercalates significantly into the Y chromosome, making it fluoresce and rendering it easily identifiable cytologically. Detection of the Y chromosome is important in prenatal sex determination.

Other dyes present in our environment are potentially mutagenic. For example, some hair dyes have been shown to be mutagenic for *E. coli*.

Physical agents

Growing tissues are most sensitive to ionizing radiation. DNA synthesis is inhibited, yet the action of x-rays is indirect. They produce free radicals, which in turn react with DNA and thus produce point mutations or chromosomal breaks.

Large doses of ultraviolet light can damage DNA. In man this damage is confined to the skin since, unlike x-rays, ultraviolet light is easily absorbed. The chemical lesion in this case is the formation of dimers between adjacent thymine residues on the same DNA strand. Unless corrected, these dimers will stop DNA synthesis. There are enzyme systems that can repair regions of the DNA that contain thymine dimers. In the first step of the repair process, an endonuclease introduces a single-strand break on the 5′ side of the dimer. DNA polymerase I then operates to fill in the missing gap with a new polynucleotide according to the instructions in the remaining complementary strand, and an exonuclease specific for the 5′-hydroxyl group removes a part of the DNA chain, including the thymine dimer. DNA ligase seals up the break. Fig. 12-4 summarizes these reactions. While all of these steps have been worked out in *E. coli*, the same or similar reac-

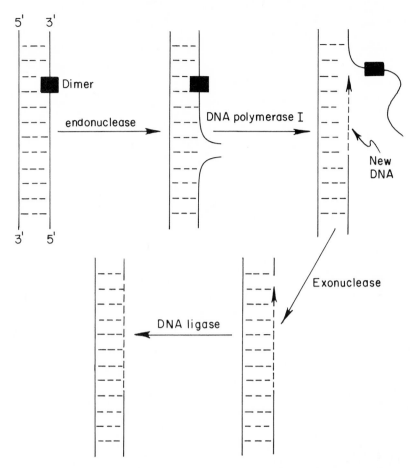

Fig. 12-4. Repair of DNA inactivated by ultraviolet light. Light causes dimerization of adjacent thymine residues that block DNA replication. The four enzymes shown are involved in removal and replacement of a portion of the DNA that contains the dimer.

tions probably occur in man. Patients with the genetic disease xeroderma pigmentosum are especially sensitive to ultraviolet light and develop skin cancer. Skin fibroblasts cultured from these patients have been shown to be defective in DNA repair.

RNA synthesis: expression of the genetic material
Heterochromatin of eukaryotic cells

Several years ago cytologists observed that some regions of the chromosomes appeared tightly packed (heterochromatin) and visible during the interphase portion of the cell cycle, whereas other regions (euchromatin) lost their visibility when the cells entered interphase. These investigators believed that the condensed heterochromatin was not expressed but that the euchromatin was expressed. This suspicion has been confirmed by the observation that the tightly packed DNA is not transcribed into RNA.

It is not known how euchromatin and heterochromatin differ chemically; their differences are not caused simply by the attachment of histone proteins since these proteins are bound to both types of chromatin.

The heterochromatin and the DNA sequences within it should not be thought of as useless, however. In embryonic life much of what later becomes hetero-chromatin does appear to be expressed. One of the most striking examples of a totally heterochromatic chromosome has already been described (Chapter 2, p. 85), that of the human X chromosome. In females one of the two X chromosomes exists as a highly condensed Barr body, the DNA of which is not expressed. There is no evidence for the total inactivation of other chromosomes; however, some observations indicate that small segments of an autosomal chromosome may be inactive in one of a pair of chromosomes. Perhaps the heterochromatic regions account for these observations.

Messenger RNA

RNA is an intermediary between DNA and the synthesis of proteins. This template RNA, or messenger RNA, is a direct reflection of the instructions "written" in DNA. The separate nucleus in animal cells and the fact that cells rich in RNA are very active in protein synthesis argue circumstantially for this inter-mediary role. Direct evidence for a role of RNA templates in protein synthesis can be obtained by in vitro experiments that show that DNA does not serve as a template for protein synthesis. There are three types of RNA: transfer RNA (tRNA or sRNA), messenger RNA (mRNA), and ribosomal RNA (rRNA). All are involved in protein synthesis. The tRNA carries the amino acids as high-energy esters to a site on the ribosome, a particle composed of about 60% RNA and 40% protein. Here the tRNA base pairs in a specific way with the messenger RNA.

RNA polymerase

In 1960 an enzyme was found in rat liver that incorporated ^{14}C from radioactive ribonucleoside triphosphates into a large polymer that had all the characteristics of RNA. This enzyme is called a DNA-dependent RNA polymerase, and it catalyzes the reaction shown in Fig. 12-5. Mechanistically, the reaction is like the DNA polymerase; addition is to the 3' end of the growing chain. A DNA template as well as all four of the ribonucleoside triphosphates are required. Some poly-nucleotide synthesizing enzymes such as polynucleotide phosphorylase can function with a single type of mononucleotide, the ribonucleoside diphosphates; these enzymes are DNA independent. The RNA product is of high molecular weight, attacked by RNase and alkali, and has a base composition like one of the strands of the DNA template. If the RNA product is heated and slowly cooled in the presence of the DNA template, a double-stranded *hybrid* is produced; one strand is DNA and the other is RNA. The RNA product will hybridize only with the kind of DNA's used as a template to form it.

Inhibitors. Several inhibitors have proved useful in studying the mechanism of transcription. The cancer chemotherapeutic agent actinomycin D inhibits the elongation of the RNA chain by binding to deoxyguanosine residues on the DNA

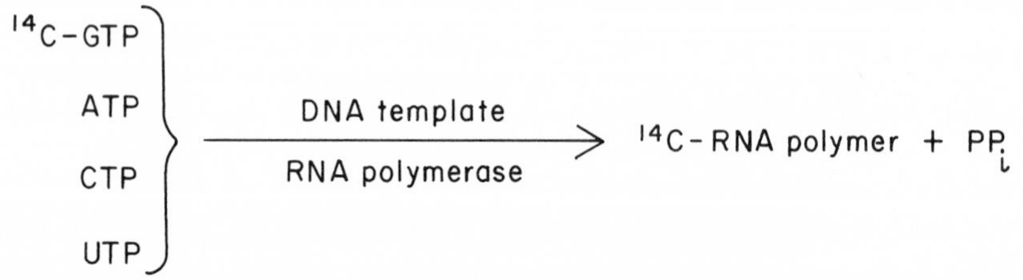

$$\left.\begin{array}{l} ^{14}C-GTP \\ ATP \\ CTP \\ UTP \end{array}\right\} \xrightarrow[\text{RNA polymerase}]{\text{DNA template}} {}^{14}C-RNA\ polymer\ +\ PP_i$$

Fig. 12-5. DNA-dependent RNA polymerase.

Table 12-2.   Properties of RNA polymerase subunits

SUBUNIT	MOLECULAR WEIGHT (daltons)	PROPERTY
α	40,000	?
β	150,000	Binds rifamycin
β'	160,000	Binds DNA
σ	90,000	Proper initiation

template in a manner that prevents movement of the RNA polymerase along the DNA molecule. Actinomycin D is not only valuable clinically but has also been widely used in laboratory experiments designed to separate the process of transcription from translation. It is generally assumed that the inhibition specifically inhibits transcription but not translation, although the drug may have significant additional effects. Proflavin, a member of the family of acridine dyes mentioned in connection with mutagenesis, also can block the elongation of RNA chains by binding to the DNA template. Proflavin is less specific than actinomycin D and binds to groups other than deoxyguanine residues.

Rifamycin and its derivatives (rifampicin) inhibit the RNA polymerases of microorganisms and mitochondria by preventing proper initiation of the synthesis of RNA chains. Rifampicin is useful in the treatment of tuberculosis since it probably inhibits the RNA synthesis of the tubercle bacilli. It is also one of the few antiviral agents. Poxviruses replicate their RNA in the cytoplasm of animal cells. The enzyme involved is induced by the virus and is sensitive to rifampicin.

Subunits.   The RNA polymerase from *E. coli* has been well characterized; it is composed of several different kinds of subunits. All of the subunits can be physically separated one from the other. Table 12-2 lists the subunits with their molecular weights and their possible functions. The *holoenzyme* is designated $\alpha_2\beta\beta'\sigma$; it has a molecular weight of about 490,000 and promotes *asymmetric* synthesis of RNA; that is, only one strand of DNA is copied. The *core enzyme* is designated $\alpha_2\beta\beta'$. It has activity in vitro with nicked DNA templates, but since it lacks the sigma subunit, initiation is faulty and often both strands of DNA are copied, yielding a double-stranded RNA, an artifact of the incomplete in vitro conditions.

Mechanism of RNA synthesis

Synthesis starts by the attachment of the holoenzyme to characteristic sites on the DNA template under the influence of the σ-subunit. These sites are called *promoter* sites. While only one of the DNA strands is read out, it is not always the same one.

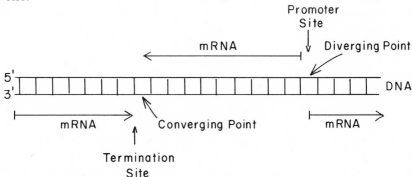

Thus synthesis on complementary strands may converge to a point or diverge from another point. It is not clear yet whether a separate kind of σ-factor is required for each strand.

The σ-subunit itself has affinity for DNA only when it is part of the polymerase. Since the β'-subunit has a high affinity for DNA, the major forces holding the RNA polymerase to the template must be between the β'-subunit and DNA. The σ-factor must ensure that binding occurs at promoter sites and nowhere else. The β-subunit interacts with rifamycin or its derivatives. Because these compounds specifically block initiation of RNA synthesis, the β-subunit must be directly concerned with forming the first intranucleotide bond. Growth of the RNA strand is at the 3' end of the new chain. The terminal 5' nucleotide is always a guanylic or adenylic acid residue. Once this bond forms, the σ-subunit comes off and awaits the termination of synthesis of the polynucleotide chain and the release of the core enzyme. The core enzyme and the σ-factor then interact, and the holoenzyme that is formed initiates RNA synthesis all over again. The σ-factor can be thought of as cycling on and off the RNA polymerase. Elongation of the polyribonucleotide results in an RNA strand that is complementary and antiparallel to the DNA strand copied.

The termination of RNA synthesis in *E. coli*, where it is best understood, may involve another protein called rho (ρ), which is, however, not a part of the RNA polymerase. Rho has a molecular weight of about 200,000. In its absence or in the absence of the DNA polynucleotides that bind it, the RNA chain is not terminated, and abnormally long RNA molecules are produced. These may contain the information for the synthesis of several proteins. In bacterial cells, long mRNA's of this type exist naturally. They are called *polycistronic* because they contain the transcripts of several genes or cistrons. There is no good evidence as yet of the presence of polycistronic mRNA's in animal cells. On the contrary, the few mammalian RNA's that have been isolated are *monocistronic*.

Posttranscriptional modification of messenger RNA

Eukaryotic messenger RNA is unusual because of its distinctive end groups, both of which are modified after the messenger RNA has been transcribed from the DNA template. An increasingly large number of virus-induced mRNA's as well as some cellular messengers have been found to have a 7-methyl guanylic acid residue at their 5' ends. The methylated guanylic acid residue is linked through a phosphodiester bond connecting its 5'-hydroxyl group to the 5'-triphosphate end of the mRNA. These "capped" messengers are resistant to some nucleases and may be essential for the translation of the messenger. The 3' end of most, but not all, animal messenger RNA's contain approximately 200 adenylic acid residues. This poly A segment, like the 5' "cap", is added posttranscriptionally by enzymes that recognize messenger RNA but not ribosomal or transfer RNA. The poly A tail probably is important in transporting mRNA from the nucleus to the cytoplasm; it is not essential for messenger translation.

Operons and control of RNA synthesis

In bacteria the synthesis of polycistronic mRNA's is often regulated in a special way. The polycistronic message is coded by a continuous stretch of DNA that contains the information for the synthesis of several proteins, all of which are required for a particular metabolic function. Such a set of genes is called an operon.

Perhaps there is more known about the lactose or *lac* operon of *E. coli* than any other; it can therefore serve as a representative example. Usually *E. coli* cells

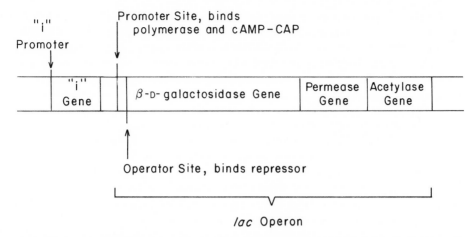

Fig. 12-6. Genes of the lactose operon. The arrangement of the genes concerned with regulation of the lactose operon are shown as they occur on the *E. coli* chromosome. The "i" gene codes for the repressor and has its own promoter site.

are grown in the laboratory in a simple medium containing glucose and a few salts. The organism can make all of its essential cellular constituents from this medium. Most strains of *E. coli* can also grow well on lactose or other galactosides; however, this requires that they induce three enzymes that are concerned with lactose metabolism. These are a β-D-galactosidase needed to hydrolyze the lactose to D-glucose and D-galactose, a D-galactoside permease to facilitate the movement of lactose into the cell, and a transacetylase whose function is not well understood. The structural genes that code for these proteins are contiguous to an *operator* site of about thirty-five base pairs and a *promoter* site for attaching the RNA polymerase. The arrangements of these genes are shown in Fig. 12-6. Next to the *lac* operon is the "i" gene, which codes for the synthesis of a *repressor* protein, a tetramer with molecular weight of about 150,000. The *lac* repressor has a high affinity for about twenty-five base pairs of the *lac* operator site and for no other operators. When it is bound to the operator, movement of the RNA polymerase along the template and into the operon is blocked. When the repressed cells are given a diet of lactose instead of glucose, lactose induces the synthesis of the proteins coded by the *lac* operon. In this role, lactose is called an *inducer*. Inducers need not always be sugars, but they are always small molecules. The inducer, either lactose or its analogues, specifically interacts with the *lac* repressor to effect its removal from the operator. The RNA polymerase proceeds then to transcribe the derepressed cistron. The control region of the *lac* operon has been sequenced. The diagram below shows the relative arrangement of the protein binding sites in the control region.

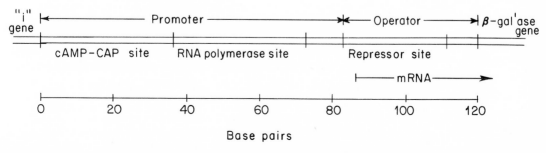

So far the sort of control described is of a negative type; that is, the expression of the *lac* operon is normally prevented by a repressor protein. Actually, the situation is more complicated. Control of the *lac* operon also has an element of positive control, and the positive controlling element is cAMP. Cells grown on glucose have low concentrations of cAMP. On induction, cAMP must bind to a protein called a catabolite gene–activator protein (CAP), a dimer with molecular weight of 45,000. The cAMP-CAP complex binds to DNA, probably at the promoter site, in such a way that the bound RNA polymerase can initiate RNA synthesis. Thus expression of the *lac* operon requires both the inducer lactose to remove the repressor protein and cAMP to activate CAP so that RNA synthesis is initiated.

Analogous to enzyme induction is the process of end-product repression. For this kind of control the end product of a metabolic sequence, for example, an amino acid, must be synthesized in excess. The excessive amount of the amino acid interacts with a repressor protein; in so doing, it *activates* the repressor to block the functioning of an operon concerned with making the enzymes used in the synthesis of the amino acid. The small molecule end product is called a *corepressor*.

Whether operons are regulated in animal cells in this way is not at all clear. Some circumstantial evidence supports such a possibility. Enzyme defects in patients with the genetic disease orotic aciduria (case 4, Chapter 11) have been interpreted according to this concept.

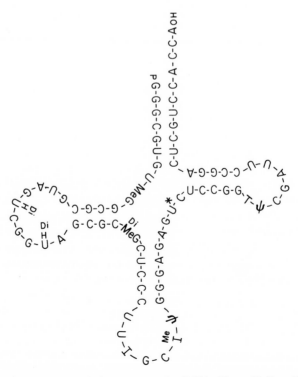

Fig. 12-7. Nucleotide sequence of alanine transfer RNA. (From Holley, R., et al.: Science 147:1462, 1965. Copyright 1965 by the American Association for the Advancement of Science.)

Transfer and ribosomal RNA synthesis

Thus far the synthesis of RNA in the simplest sense, that is, single-stranded, unmodified RNA's such as mRNA, have been considered. The rRNA's and tRNA's are more complicated, and consequently much less is known about their synthesis even though they are the most abundant and the most stable of the cellular RNA's.

Transfer RNA structure

A good deal is known about the structure of the relatively small (molecular weight 25,000 to 30,000) tRNA's. Since some amino acids can be attached to multiple species of tRNA, there are probably more than forty different types of tRNA. The reason for this will be apparent later when the genetic code is discussed. The complete nucleotide sequence of several tRNA's has been determined, and knowledge of the sequence has proved valuable.

The sequence of the yeast tRNA for alanine, as shown in Fig. 12-7, was the first reported. When the sequence is arranged to give the maximum number of base pairs, the "cloverleaf" pattern illustrated in Fig. 12-7 results. All the tRNA's that have since been sequenced can be arranged in this pattern. Other similarities have also been noted; every tRNA contains a CCA grouping at the 3′ end. It is this end of the molecule at which the amino acid is attached through its acyl group to the 3′-hydroxyl of the terminal adenosine, forming a high-energy ester.

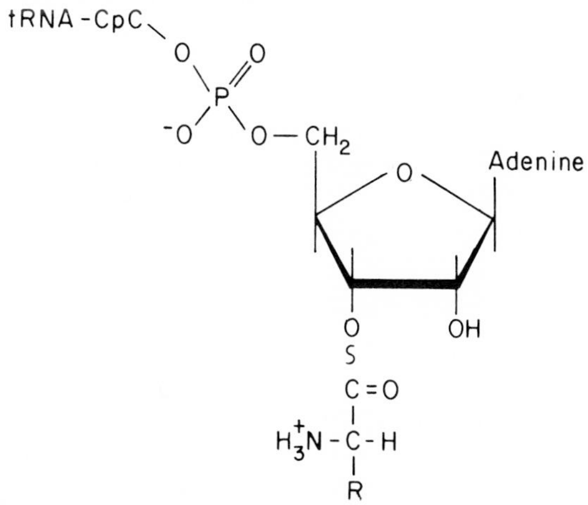

The 5′ end of almost every tRNA contains a guanine residue. There are several unusual bases in tRNA, and these proved useful in elucidating the structure. Every tRNA contains a common T-ψ-C-G tetranucleotide in the right-hand loop. T represents ribosyl thymidine, and ψ represents pseudouridine (5-ribosyl uracil). This sequence is probably part of a binding site that recognizes the ribosome or one of the enzymes involved in forming the peptide bond on the ribosome.

The *anticodon* is located midway in the molecule. It is in a single-stranded region so that it can base pair with the *codon* (a code word consisting of three nucleotides) of the mRNA in an antiparallel fashion. Note that the anticodon of yeast alanine-tRNA is I-G-C. The inosinic acid residue base pairs much like guanylic acid so that the codon G-C-C in an mRNA would specify alanine.

The synthesis of tRNA and rRNA are alike in that both are synthesized as larger precursor molecules and both contain modified or unusual bases. The large precursor molecules are first cleaved by nucleases to the proper size, and then other enzymes modify the appropriate bases. Cells contain several different methylating enzymes that use S-adenosylmethionine to methylate tRNA. Foreign nucleic acid such as that found in a virus can be overmethylated when introduced into a cell. Occasionally this acts as a defensive measure on the part of the host to eliminate foreign nucleic acid. However, one theory proposes that the transformation of cells by cancer viruses may occur because of these specific methylations. Whether the cancer causes these methylations or exists because of them is not known.

Biosynthesis and maturation of ribosomal RNA

Very little of the genetic potential of cells is used to make tRNA, perhaps one gene for each tRNA. Much more of the cell's DNA is used to make rRNA. There are as many as 100 to 1000 identical genes for rRNA in vertebrate oocytes. The site of rRNA synthesis is the nucleolus, where each multiple gene is transcribed sequentially but separately from a strand of DNA that has been amplified many times. Each nucleolus contains a single circular DNA that contains dozens of tandemly arranged genes for ribosomal RNA. Fig. 12-8 illustrates the growth of

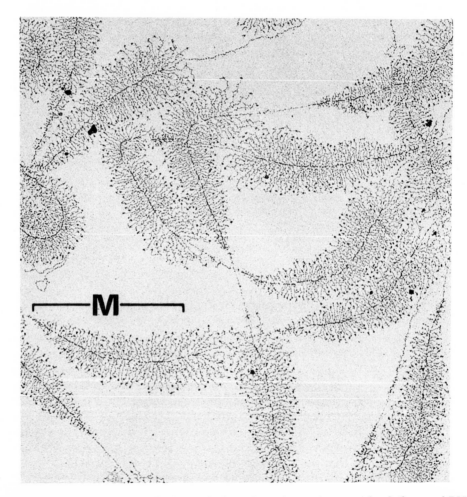

Fig. 12-8. Electron micrograph of a nucleolar core with growing strands of ribosomal RNA attached to their sequential DNA. (×25,500.) (From Miller, O. L., and Beatty, B.: J. Cell. Physiol. **74:**225, 1969.)

several strands of rRNA of varying lengths. The electron micrograph shows that the redundant genes for rRNA are separated by a short DNA segment. The length of a single gene is shown by the line designated *M*. The regular head-to-tail sequence of the fernlike pattern of RNA's emerging from the central DNA strand indicates that these genes are all read out in the same direction. The longer "leaves" on the fern indicate those RNA molecules whose synthesis is almost completed. The short strands at the other end of the matrix are RNA strands just starting to grow. Enlargements of these electron micrographs show that each growing RNA has what appears to be a molecule of RNA polymerase attached at the junction made by the RNA strand and the DNA. The 28S and 18S RNA's found in the ribosome are made as one piece of 45S RNA. Several degradative steps yield 28S and 18S RNA's. Ribosomal proteins made in the cytoplasm are transported into the nucleoli, where the 60S ribosomal subunit forms; the 40S subunit is assembled elsewhere. Like tRNA, rRNA is also methylated, but most mRNA is not.

RNA polymerases of animal cells

Although the RNA polymerase that was first described was isolated from rat liver, the polymerases from mammalian tissues are much more difficult to work with since they are tightly bound to the deoxyribonucleoprotein of the nucleus. Soluble preparations can now be obtained by treating the extracts with sound waves (sonication) under appropriate conditions. Three DNA-dependent RNA polymerases derived from rat liver have been isolated and separated from one another. RNA polymerase I is found in the nucleolus; it is involved in the synthesis of the 45S rRNA precursor. RNA polymerase II and III are present in the nucleoplasm and are more concerned with the synthesis of mRNA and tRNA. They are specifically inhibited by the antibiotic α-amanitin. A fourth RNA polymerase has been found in mitochondria. All of the mammalian polymerases have several properties in common; they are inhibited by actinomycin D, DNA is used as the template, and all four ribonucleoside triphosphates are required. However, polymerases I, II, and III have several properties that distinguish them from each other. These are summarized in Table 12-3.

PROTEIN BIOSYNTHESIS

All the different RNA's, tRNA, mRNA, and rRNA (as part of the ribosome), are involved in the synthesis of proteins. The process of protein biosynthesis is called *translation* because information must be transferred from the four-letter language of the nucleic acids to the twenty-letter language of the amino acid constituents of the proteins.

Table 12-3. Comparisons between RNA polymerases I, II, and III from rat liver

PROPERTY	POLYMERASE I	POLYMERASE II AND III
Cellular location	Nucleolus	Nucleoplasm
Optimum ionic strength	Low	High
Optimum Mn^{++}/Mg^{++} activity ratio	2	5
Sensitivity to α-amanitin	Insensitive	Sensitive; polymerase III less sensitive

Aminoacyl–transfer RNA synthetases

Amino acids themselves have no special affinity for nucleic acids; the proper amino acid must therefore be combined with the proper tRNA under the influence of an aminoacyl-tRNA synthetase. Once an amino acid is attached to a tRNA, all of the specificity in recognizing the mRNA, the ribosome, and peptide bond–forming enzymes is a property of the tRNA, not the amino acid. Thus the aminoacyl-tRNA synthetases must be highly specific both for the amino acid and for the tRNA. Only L-amino acids are recognized, not the D isomers. Neither will any of the synthetases function with peptides or amino acids without free α-amino groups.

There are twenty genetically important amino acids. These are listed with the genetic code in Table 12-5. All other amino acids in proteins are derived from the set of twenty; for example, the hydroxyproline found in collagen is formed from prolyl residues after they become part of the protein chain.

With few exceptions, there is one aminoacyl-tRNA synthetase for each amino acid; however, a synthetase can recognize all of the tRNA acceptors for a particular L-amino acid. For example, the methionyl-tRNA synthetase recognizes only methionine, but it can aminoacylate both of the tRNA's for methionine: tRNA^F, which functions to initiate the synthesis of proteins chains, and tRNA^M, which

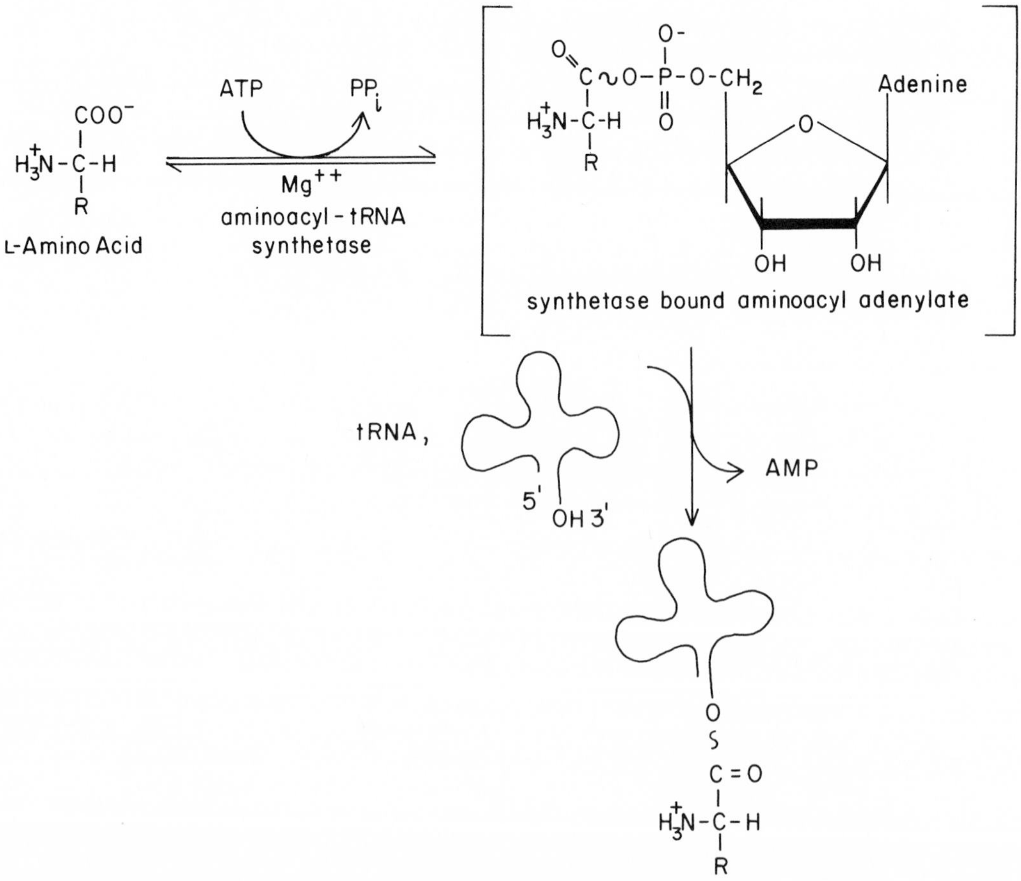

Fig. 12-9. Attachment of amino acid to tRNA.

inserts methionyl residues at internal positions in the growing polypeptide chain. These functions will be considered in some detail later. The synthetases also have a high specificity for the ribonucleoside triphosphate; only ATP will function. The reaction catalyzed is shown in Fig. 12-9.

Aminoacyl-tRNA molecules have a large, negative free energy of hydrolysis that is used to drive the synthesis of the peptide bond between two α-L-aminoacyl residues. The formation of the aminoacyl-tRNA's is coupled to the hydrolysis of ATP with the intermediate formation of aminoacyl adenylate and inorganic pyrophosphate. Transacylation to tRNA liberates AMP (Fig. 12-9). One would predict that the reaction is reversible, and this is true in vitro. However, in the cell a high concentration of aminoacyl-tRNA is maintained as a result of extremely active pyrophosphatases that convert pyrophosphate to inorganic phosphate, removing a product of the forward reaction and thus preventing reversal.

The ribosome: site of protein synthesis

Aminoacyl-tRNA's are condensed into protein on ribosomal particles, not in solution. This aspect of protein synthesis adds a new complexity since enzymatic reactions that occur in solution are much easier to study. The functional ribosome consists of two rather large particles. In animal cells these subparticles have sedimentation coefficients of 40S and 60S. In bacteria and in mitochondria the subparticles sediment at 30S and 50S. The subunits are derived from the functional particle, which is 80S in all parts of the mammalian cell except the mitochondria, where it is 70S. Even though these particles are quite large (about 200 Å in diameter), high-speed ultracentrifuges are required to sediment them. Fig. 12-10 shows the composition of a mammalian ribosome. The particle as a whole is about half protein and half rRNA. The 40S subparticle contains a single RNA that sediments at 18S and about thirty different proteins. The larger 60S

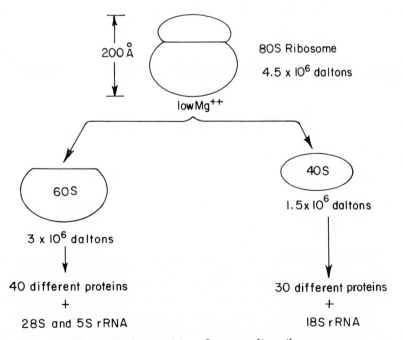

Fig. 12-10. Composition of mammalian ribosome.

subparticle contains two RNA's that sediment at 28S and 5S. The entire sequence of the 5S RNA is known, but there are few clues as to how it functions. In addition, the larger subunit contains about forty different proteins.

By lowering the magnesium ion concentration, the 80S ribosome dissociates into its constituent subparticles. This reaction is easily reversed and, as will be seen, occurs during the process of protein synthesis. Complete removal of magnesium ions breaks down the ribosome further, but strong dissociating solvents are needed to separate the RNA's from the proteins and the proteins from one another. Under special conditions the protein and RNA parts of the small and large subunits from bacteria can reassemble in vitro. Consequently, the assembly of this complex subcellular organelle resembles virus assembly in that both are processes that occur spontaneously when all the parts are present.

A single peptide chain grows from an 80S ribosome. On completion of the chain the ribosome dissociates into subparticles and becomes available for the initiation of a new protein chain.

The ribosome is a rather nonspecific protein-synthesizing particle. A special additional protein factor may give it the ability to differentiate between certain kinds of mRNA's, but the basic particle itself is capable of making any protein of the species in which it is found. Animal ribosomes will not work with peptide polymerizing factors from bacteria, but liver ribosomes appear to be the same as ribosomes from other animal tissues or organs. A smaller bacteria-like ribosome, however, is found in mitochondria. Protein synthesis in mitochondria is more like protein synthesis in bacteria since these ribosomes and their polypeptide-synthesizing enzymes are exchangeable and the protein synthesis on the mitochondrial ribosomes is sensitive to inhibition by antibiotics that specifically block bacterial but not animal protein synthesis.

The polyribosome

Only one polypeptide chain grows from each ribosome, but a strand of mRNA can accommodate many ribosomes. Such an mRNA binds ribosomes in proportion to its length and each ribosome holds a growing protein chain. The growing chain may be at one of several stages of completion. Diagrammatically such a polyribosome would appear as shown below.

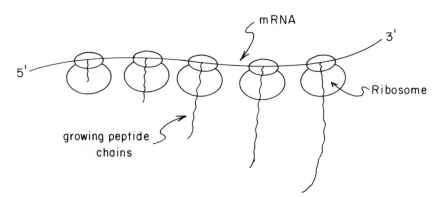

The monocistronic mRNA's of animal cells sometimes produce a very long polypeptide chain that must be cleaved after synthesis in order to become active (that is, proinsulin, trypsinogen, and blood clotting factors); some chains can be cleaved to yield more than one functional protein. Poliovirus makes a giant protein that is subsequently cleaved to form several functional proteins. In any case, it is the primary sequence of a protein that determines its tertiary structure and thus its biologically active conformation. It is very likely that the amino acid sequence of proenzymes plays an important role in forming a highly specific conformation that, on activation, yields the functional enzyme. The active enzyme might not have been able to achieve this conformation had it not existed as a proenzyme. For example, the A and B chains of insulin are difficult to cross-link in the laboratory; yet proinsulin can cross-link intramolecularly with great facility. Later it is cleaved to the A and B chains.

Protein synthesis starts by attaching a ribosome at or near the 5′ end of an mRNA. Recall that this is the end of the mRNA that is first synthesized from the DNA template. Thus in bacterial cells an mRNA molecule need not be completely synthesized before it starts to be read out. In animal cells, in which the mRNA is made in the nucleus, this linking of transcription and translation may be important only for the limited protein synthesis that occurs in the nucleus. Still exactly how mRNA is transported to the cytoplasm is not understood, and the possibility exists that translation may assist in the transport of the mRNA from the nucleus.

As a ribosome directs the addition of aminoacyl groups to a growing peptidyl-tRNA, it moves along the message, decoding it from the 5′ to the 3′ end until it has moved far enough for another ribosome to be bound onto the initiating site just vacated. Soon the messenger is completely loaded with ribosomes. These polyribosomes can be seen with the electron microscope, and the number of ribosomes on them can be counted. The number of ribosomes on a polyribosome is proportional to the size of the protein being synthesized. Thus a globin chain for hemoglobin consists of about 150 amino acid residues, and its polyribosome holds five ribosome monomers. A major chain of myosin contains about 1800 amino acid residues and its polyribosome holds sixty to 100 monomeric ribosomes.

Initiation of protein synthesis

Protein synthesis starts at the amino end of the peptide and progresses by the addition of amino acids at the carboxyl end.

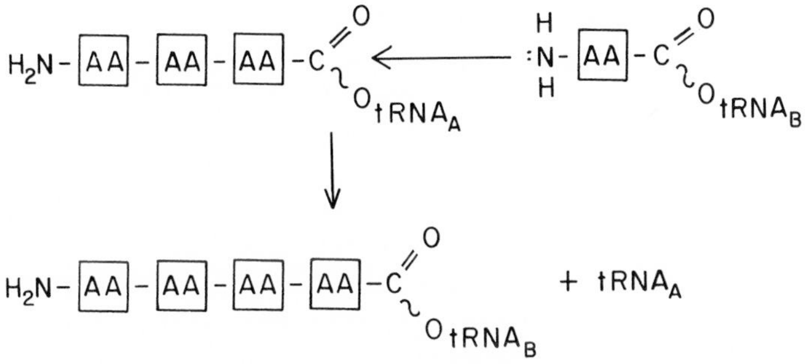

$$H_2N-\boxed{AA}-\boxed{AA}-\boxed{AA}-C\underset{O\,tRNA_A}{\overset{O}{\diagup}} \longleftarrow \quad \underset{H}{\overset{H}{:N}}-\boxed{AA}-C\underset{O\,tRNA_B}{\overset{O}{\diagup}}$$

$$\downarrow$$

$$H_2N-\boxed{AA}-\boxed{AA}-\boxed{AA}-\boxed{AA}-C\underset{O\,tRNA_B}{\overset{O}{\diagup}} \quad + \ tRNA_A$$

All protein chains start with the same amino acid, methionine, at the N-terminal position. Very often the methionine residue is cleaved off after the growing polypeptide chain has been somewhat extended; consequently, proteins isolated from cells contain amino acids other than methionine at their N-terminals. Methionine has two acceptor tRNA's, tRNA^M and tRNA^F. Only Met-tRNA^F is involved in initiation.

In bacterial cells and in mitochondria, formylmethionyl-tRNA^F initiates the synthesis of every protein chain. In animal cells it is not necessary to formylate Met-tRNA^F in order to initiate protein synthesis. All chains are started with

Met-tRNA^F, however. That the initiating tRNA in animal cells is really Met-tRNA^F and not Met-tRNA^M is indicated by the observation that the transformylase from *E. coli* can formylate the Met-tRNA^F of animal cells. The reaction catalyzed by this enzyme is illustrated below.

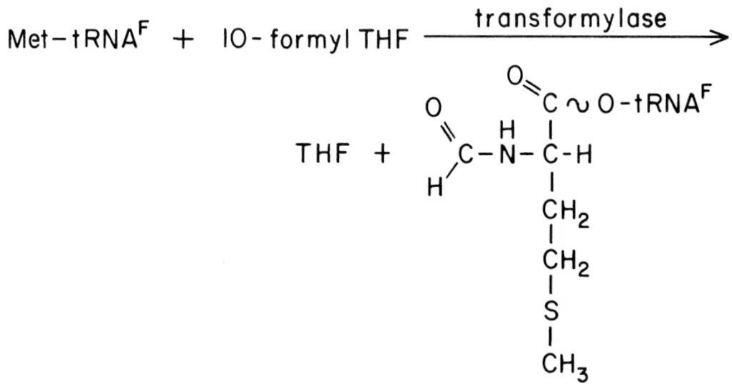

In addition to fMet-tRNA^F, the initiation of protein synthesis in bacterial cells also requires three proteins that are bound to ribosomes but are not generally considered to be structural parts of the ribosome itself. One factor, initiation factor 3, is required for the recognition of messenger RNA. Initiation factors 1 and 2 are required for positioning mRNA and fMet-tRNA on the ribosomes. A diagram of the essentials of the initiation reactions is shown in Fig. 12-11. To the extent that the analogous experiments have been done, the initiation factors from mammalian cells resemble in a functional sense those from bacteria. Because the knowledge of the initiation reactions in the bacterial system is more complete, Fig. 12-11 describes the process in bacteria.

Despite much experimentation, little is known of the early steps in protein synthesis initiation. One view proposes that messenger RNA first binds to the smaller ribosomal subunit. In this scheme the 30S subunit carries initiation factor 3 (IF-3). On subsequent addition of IF-2 and GTP, fMet-tRNA binds to an AUG sequence on the mRNA forming a stable complex. More recent experiments suggest that the 30S ribosome first interacts with fMet-tRNA, not mRNA. In a later step the messenger RNA binds. The message is bound in a functional way in a reaction that also requires the other two initiation factors. The larger subunit is added in a reaction in which GTP is split to give GDP and inorganic phosphate; initiation factor 2 acts as the guanosine triphosphatase in this reaction. The hydrolysis of GTP in this process produces a conformational change in the 50S particle so that the final initiation complex is accommodated to accept the attachment of the succeeding aminoacyl-tRNA programmed by the messenger.

Messenger RNA binding to ribosomes

A messenger RNA is attached to the ribosome so that the initiating sequence is presented in the proper reading frame. The first amino acid inserted into peptide linkage is a methionyl residue coded by an AUG or GUG codon. Not every AUG sequence in an mRNA programs the start of a new protein chain, since AUG sequences also code for methionyl residues that occur at internal positions of the polypeptide chain. If one considers AUG and GUG sequences it is obvious that several might occur within a cistron. For example, AUG occurs six times and GUG five times in the coat cistron of R17 bacteriophage RNA; yet protein synthe-

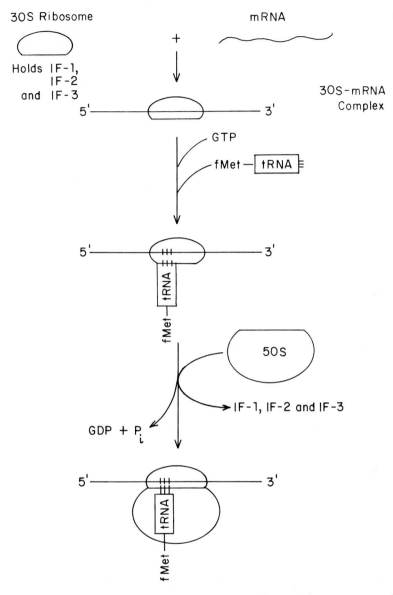

Fig. 12-11. Initiation of protein synthesis. IF-1, IF-2, and IF-3 refer to initiation factors 1, 2, and 3 from *E. coli*.

sis initiates faithfully, even in vitro, at the proper AUG codon. As one might guess, some RNA's cannot function at all as mRNA. For example, ribosomal RNA contains several AUG and GUG sequences, but because of other structural features, this RNA is not translated. The secondary structure of the RNA plays a major role in presenting single-stranded AUG codons to the ribosome. Thus formaldehyde-treated R17 RNA initiates at a few additional sites because formaldehyde reacts with the RNA to reduce its secondary structure sufficiently to generate single-stranded regions containing AUG and GUG sequences. In addition to secondary structure, other features of the RNA are important for translation. For example, ribosomes from *Bacillus stearothermophilus* translate only the minor coat cistron of formaldehyde-treated R17 RNA, whereas several other AUG segments are translated by *E. coli* ribosomes. This suggests that the ribosome recognizes primary sequences on the mRNA in addition to the initiating codons. These sequences must be located on the 5' side of the initiating codon. The sequences on the 3' side would be expected to vary considerably between different mRNA's, since it is this region that contains the code words for the protein to be synthesized. However, the complete absence of a polynucleotide chain on the 5' side of the initiating codon does not seem to prevent ribosome binding since synthetic mRNA's with AUG triplets as their 5' ends initiate translation very efficiently.

Protein chain elongation

There are two sites for tRNA attachment on the ribosome. One usually contains the peptidyl-tRNA; it is closest to the 5' end of the mRNA and is called the "P" site. After the initiation reactions shown in Fig. 12-12, fMet-tRNA is in the P site. The other site, the aminoacyl-tRNA or A site, is on the 3' side of the P site. The next aminoacyl-tRNA added to the growing chain is attached to this site under the influence of GTP and elongation factor 1 (EF-1), one of two protein elongation factors that work on the ribosome during the steps of polypeptide chain elongation. During this reaction, GTP is split to GDP and P_i.

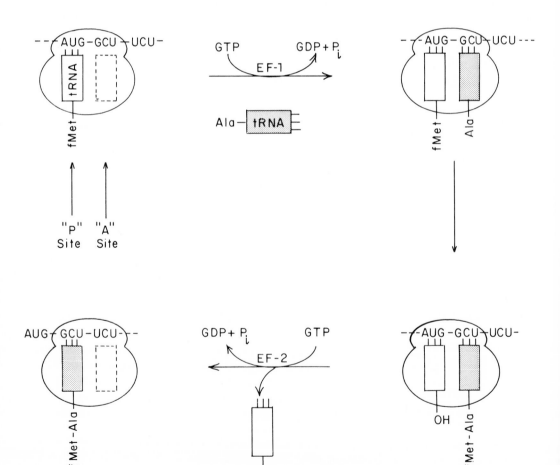

Fig. 12-12. Peptide chain elongation.

The formyl-methionyl group, or in subsequent steps the peptidyl moiety, is transferred to the nucleophilic amino group of the aminoacyl-tRNA in the A site. This reaction is catalyzed by the ribosome itself, specifically by a protein of the larger subunit. The result is a peptidyl-tRNA, now one amino acid longer, resting in the A site while the P site contains the deacylated tRNA that formerly held the peptidyl chain. This deacylated tRNA is removed in a reaction that requires elongation factor 2 (EF-2) and GTP and that is concomitant with the translocation to the P site of the new peptidyl-tRNA still hydrogen bonded to its codon on the mRNA. The cycle is now completed and a new codon is brought into apposition with the A site. These reactions are shown diagrammatically in Fig. 12-12.

Termination of protein synthesis

Termination of protein synthesis in both animal and bacterial systems requires GTP and one or more protein release factors. The release factors recognize the termination codons UAA, UAG, and UGA. Release factors 1 and 2 (RF-1 and RF-2) are found in *E. coli*; either one will function to release the peptidyl group from the tRNA bound to the ribosome, probably by assisting in the catalysis of a hydrolytic reaction. Release factor 1 recognizes the termination codons UAA or UAG, and release factor 2 recognizes UAA and UGA. It is believed that the synthesis of most proteins is terminated by the UAA signal, so that either factor might serve to stop the synthesis of most proteins. Only a single release factor has been isolated from mammalian cells. No special tRNA's are required for termination. The peptidyl transferase of the ribosome may be involved, since in the presence of the antibiotic puromycin, it can carry out chain termination in the absence of the release factor.

Antibiotics and protein synthesis

Several antibiotics recognize differences between animal and bacterial protein synthesis by specifically inhibiting one or the other. The health scientist is most interested in those antibiotics that act specifically against bacteria. Table 12-4 lists a few commonly used antibiotics and the step in protein synthesis they inhibit.

Diphtheria toxin, although it is not an antibiotic, is another substance that inhibits protein synthesis. The toxin is an enzyme that is exceedingly deadly even in small amounts. It catalyzes the rather unusual reaction shown in Fig. 12-13, which leads to the inactivation of the animal elongation factor 2. The glycosidic bond in NAD^+ that holds the ribose moiety to the pyridine ring is transferred to

Table 12-4. Antibiotics that inhibit protein synthesis

ANTIBIOTIC	STEP INHIBITED
Chloramphenicol	Ribosomal peptidyl transferase
Streptomycin	Initiation; causes misreading of code
Tetracycline	Prevents aminoacyl-tRNA attachment to ribosome
Puromycin (also inhibits animal cells)	Accepts growing peptidyl chain in place of aminoacyl-tRNA; chain terminates prematurely
Cycloheximide (inhibits only animal cells)	Ribosomal peptidyl transferase

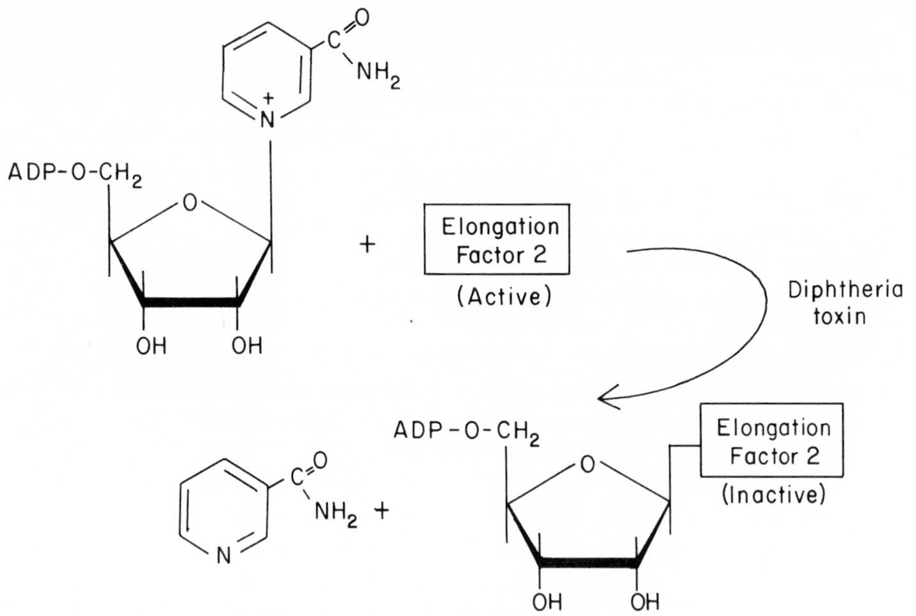

Fig. 12-13. Action of diphtheria toxin.

elongation factor 2 and in the process inactivates it. That protein synthesis is a very essential process is emphasized by the extraordinary lethal effects of this toxin.

Genetic code

Three nucleotide bases are required to specify the insertion of an amino acid into a polypeptide chain. Each series of three bases is read in a linear sequential manner without using a particular base more than once. Because there are four different bases in RNA, the maximum number of three-letter code words are 4^3, or sixty-four. Sixty-one of these words are used to specify the twenty amino acids, and essentially the same code word dictionary is used in all species tested. Consequently, the genetic code is said to be triplet, nonoverlapping, degenerate, and universal. Table 12-5 lists the codon assignments for the twenty amino acids. Notice that the first two letters of a code word are very specific but that often the same amino acid is coded regardless of the third nucleotide.

The three codons without amino acid assignments function as chain terminator signals. In a few mRNA fragments that have been sequenced, two chain terminator signals occur in sequence. Some mutations will cause a base to change so that a terminator codon is generated. This often represents a serious mutation since it results in premature chain termination; these mutations have been called "nonsense" mutations because a code word has been made for which there is no amino acid. A mutation that changes a base so that a new amino acid is now specified by the code is called a "missense" mutation. Missense mutations often result in altered or reduced enzymatic activity, and not necessarily in a complete absence of activity. Considering the degeneracy of the code, one can predict that almost one third of all base replacements will probably cause no change at all in the protein made since they will occur in the third nucleotide of the codon.

The genetic code assumes that each codon base pairs in antiparallel fashion

Table 12-5. Genetic code

FIRST POSITION (5' end)	SECOND POSITION				THIRD POSITION (3' end)
	U	C	A	G	
U	Phe	Ser	Tyr	Cys	U
	Phe	Ser	Tyr	Cys	C
	Leu	Ser	Term*	Term*	A
	Leu	Ser	Term*	Trp	G
C	Leu	Pro	His	Arg	U
	Leu	Pro	His	Arg	C
	Leu	Pro	Gln	Arg	A
	Leu	Pro	Gln	Arg	G
A	Ile	Thr	Asn	Ser	U
	Ile	Thr	Asn	Ser	C
	Ile	Thr	Lys	Arg	A
	Met	Thr	Lys	Arg	G
G	Val	Ala	Asp	Gly	U
	Val	Ala	Asp	Gly	C
	Val	Ala	Glu	Gly	A
	Val	Ala	Glu	Gly	G

*Chain-terminating codons.

with the anticodon of the tRNA's that are specific for the amino acid corresponding to the code word. It was found, however, that when purified tRNA's became available, a single tRNA could recognize several code words. For example, the tRNA for alanine, whose structure is represented in Fig. 12-7, recognizes GCU, GCC, and GCA. The anticodon for this tRNA is IGC, and the base-paired structures would be:

Notice that the nonstandard base pairing is in the third position of the codon, the position that has the least effect on specifying a particular amino acid (Table 12-5). Crick has proposed a hypothesis to account for these data. It is called the wobble hypothesis because it predicts that a "wobble" in the base pairing in the third position of the code word might account for the lessened specificity. Based on the analysis of several other tRNA's, Crick proposed the rules shown below.

THIRD POSITION OF ANTICODON IN tRNA	THIRD POSITION OF CODON IN mRNA
U or ψ	A or G
C	G
A	U
G	C or U
I	C, U, or A

Assuming that the wobble hypothesis is valid, one can see that while all sixty-one possible code words might be read when they occur in mRNA, it is not necessary to have as many as sixty-one different types of tRNA to read them.

The genetic code for the most part seems to be universal. The codon assignments established for *E. coli* are consistent with the known amino acid replacements in a large number of abnormal human hemoglobins and mutant coat proteins of the tobacco mosaic virus.

SYNTHESIS OF SPECIFIC PROTEINS

Up to this point the description of protein synthesis has been of a general nature; only those features common to the synthesis of all proteins have been considered. Many of the proteins of interest to health scientists, however, have special features in their synthesis that are peculiar only to them. Antibody proteins, for example, are synthesized in response to antigens, and virus proteins are made from instructions written in viral nucleic acids.

Immunoglobulins

Antibodies, or immunoglobulins, make up the γ-globulin fraction of the plasma. In Chapter 2 the structure of these proteins was considered, but little was said about their functions. These are special defensive proteins that are synthesized in response to exposure to a foreign material, usually a protein or a complex carbohydrate. The foreign material is called an *antigen*. The formation of antibodies affords immunity against the antigen, and this response is usually protective. For example, a human who is exposed to the rubeola virus for the first time will usually develop measles. His body then manufactures antibodies against this virus that help him to recover from the acute illness. Moreover, a clone of plasma cells, which are derived from lymphocytes, retains the memory for making antibodies rapidly against rubeola virus if the individual should ever again be challenged with this antigen. In this way, he is made immune to measles and will not develop the acute illness again, even if he should come into contact with the rubeola virus at some later date.

Humans can be protected against certain harmful diseases in a similar fashion by the administration of suitable antigens, a process known as *immunization*. Cowpox, a harmless virus, is injected in order to trigger the formation of antibodies against it. These same antibodies also attack the smallpox virus; therefore, this immunization affords protection against smallpox. In other situations, immunization is carried out by administration of inactivated bacterial toxin (tetanus toxoid), live virus in attenuated form (Sabin polio vaccine), or fully active material in an amount too low to produce serious illness (desensitization for hay fever). In these cases the individual manufactures his own antibodies against these antigens. In other situations it is possible to administer preformed antibodies (*passive immunity*) that are isolated either from a human who already has had the disease (mumps hyperimmune serum) or from an animal who has been injected previously with the antigen (rabies vaccine, tetanus antitoxin). This procedure is used when there is not enough time to allow the body to make its own antibodies, for example, for a man not previously immunized against tetanus who steps on a rusty nail. This procedure is often risky, for the recipient may develop antibodies to the injected antibodies (which themselves are foreign proteins) and become very ill (serum sickness). Finally, the immune response may actually be harmful in certain situations, for antibodies formed against a transplanted heart or kidney (foreign tissue) may lead to rejection of the transplanted organ by the new host.

We do not understand completely how the information for antibody synthesis is coded in DNA. Man makes immunoglobulins against foreign material to which he has not previously been exposed. Do we already have encoded within our DNA the information for making all possible antibodies? Alternatively, do we have partially structured DNA that can be modified slightly by specific antigen-induced recombinations? Such "plastic" genes might then recombine in such a way that they will code for a protein with enough difference in N-terminal primary structure to be specific for an individual antigen. This tremendously important question awaits a definitive answer. For the present it is sufficient to state that an antigen causes the synthesis of a soluble material, probably RNA, which directs small lymphocytes (*plasma cells*) to synthesize a specific immunoglobulin. Each clone of plasma cells makes only one species of L and H chain, that is, one antibody. The L chain is made first, is released from the ribosome, and then combines with an H chain that is still attached to its ribosome. The resulting LH complex is then released, and a carbohydrate moiety is added subsequently to each H chain.

Immunoglobulin-associated diseases. Certain diseases are characterized by an overproduction of immunoglobulins. One such disease is a plasma cell cancer, *multiple myeloma*. In this condition, L chains are overproduced by the proliferating plasma cells. These are released into the blood without combining with H chains, and they are excreted in the urine because of their relatively small size. Urinary L chain protein is called Bence Jones protein, and it has a characteristic property. Bence Jones protein is precipitated at 60° C; but on further warming, the precipitate dissolves. This simple test has been in use for many years in the diagnosis of multiple myeloma. A *monoclonal gammopathy* is a condition in which only one immunoglobulin is present in the plasma in excess. This condition results from a cancer involving one clone of plasma cells. Individuals with common allergies such as hay fever have elevated levels of IgE in their plasma. On exposure to the antigen, these people make excessive amounts of IgE.

Virus replication

Viruses are the most frequent cause of human illness. It is estimated that 60% of all illness caused by viruses is not even detected clinically. There are fifty or more different disease syndromes caused by viruses. Although some of the common virus diseases are not very serious, others such as smallpox and influenza have killed millions of people. The natural immunity processes are the best protection against viruses.

Viruses are small particles (one to 300 genes); many are the size of ribosomes. They all contain nucleic acid covered by a protein or lipoprotein coat. The nucleic acid may be either double- or single-stranded DNA or RNA, but not both. Some viruses actually carry enzymes with them as part of the virus particle. Perhaps the most interesting example of this is the reverse transcriptase that can cause the host cell to make DNA from the RNA template contained in several tumor viruses. The protein coat is probably the most important antigenic component of the virus.

A mutational event in viral nucleic acid has a high probability of affecting the coat proteins because of the small number of proteins for which the viral nucleic acid codes. Thus the amino acid sequence of the coat proteins might change considerably over a period of time, and a person's antibodies against the ancestors of the virus may no longer be able to react with the coat proteins of the present virus and inactivate it. It is for this reason that immunity against a particular virus may not be very long lasting. In general, in order that immunization against a virus be effective, the antigen should be prepared using precisely the same strain against which protection is sought. This is less of a problem in immunizing against bacteria, since their antigenic determinants are more complex and less sensitive to an abrupt change as a result of a few mutations.

Since the genes of a virus may be either RNA or DNA, viruses and the host often synthesize their nucleic acids differently. Even the smallest viruses contain the information in their nucleic acid for the synthesis of one or more coat proteins, and virtually all viruses code for the synthesis of a RNA or DNA polymerase used to replicate their nucleic acid. The synthesis of double-stranded DNA virus particles is straightforward and follows a mechanism similar to that used by host DNA. A virus particle that contains single-stranded RNA, on the other hand, uses some or all of this RNA as an mRNA when it infects the cell. A special RNA polymerase may synthesize an RNA chain complementary to the RNA chain found in the virus. The viral RNA is called the plus strand and the newly synthesized strand is called the minus strand. The minus strand serves as the template for the synthesis of dozens of plus strands. Coat proteins are made from the single-stranded plus strands. These proteins have high affinity for these same strands and interact with them to form the virus coat or capsid. Many new virus particles result, the cell lyses, the new virus particles are released, and neighboring cells become infected.

Many viruses cause the cells that they infect to produce a type of immunity substance, a glycoprotein called *interferon*. The double-stranded nucleic acid of a virus (or a synthetic polynucleotide such as poly I:C) probably causes the derepression of a host gene for interferon. The interferon mRNA is made and carried to the ribosome, and interferon proteins are produced. When these proteins are made in cells of the reticuloendothelial system, they can circulate in the blood, find their way to other cells, and there prevent the replication of virus. Interferon acts by triggering the synthesis of "antiviral" protein. It is still not clear whether "antiviral" protein inhibits virus replication at the level of transcription or translation. In any case, interferon is more active in preventing viral infections than stopping ones that are well advanced. Virtually every group of viruses is capable of eliciting interferon production. Interferon produced in response to a specific virus is active against a large number of different viruses, indicating that it must function by blocking a reaction of fundamental importance in the replication of any virus. Unfortunately, no interferon has been highly purified. Nevertheless, the induction of interferon synthesis is of considerable therapeutic interest for preventing the establishment of viral infections.

GENETIC ANALYSIS OF HUMAN DISEASE

We often think of genetically transmitted diseases as rare and of little importance. It is true that most inherited diseases occur infrequently; yet when all of the genetically transmitted diseases are considered together, they no longer seem so rare.

Data on the frequency with which genetic diseases occur in man are inaccurate for several reasons, including poor diagnosis, unusual distribution of population types, and unreported cases. Frequency data are usually reported as either the incidence or prevalence of a given disease. These terms are often confused, but they have precise meaning. The *incidence* of a disease relates to the number of cases per number of live births, whereas the *prevalence* is the number of cases in a given population at a given time.

The incidence of a disease may be difficult to determine if the disease is not physically recognizable at birth and no test is available to determine its prospective development. An example of a disease that is not physically recognizable at birth is Huntington's chorea, which rarely develops before the patient is 35 years of age. A disease for which there is no test at birth is sickle cell anemia because the β-chains of hemoglobin do not replace the γ-chains until the infant is 3 to 6 months of age. Other errors in determining the incidence of a disease are related to the chemical tests used. Some phenylketonurics escape detection because the $FeCl_3$ urine test is used instead of the Guthrie test; the latter depends on a more reliable microbiologic assay for phenylalanine metabolites. In other cases the appropriate tests are available, but the infant suffers irreversible damage before symptoms indicate that the test should have been run. An example of this problem is galactosemia.

The prevalence of a disease is even more difficult to determine because large populations are difficult to screen, especially for a rather rare disease. Furthermore, patients may refuse to be studied, or if they are cooperative they may not be aware that they carry a mutant gene; they may have a subclinical form of a disease such as myotonic dystrophy. As we have already seen, the incidence and prevalence of genetically transmitted diseases vary with race, geographic area, or sex. Table 12-6 lists frequency data for some of these diseases.

The simplest means of classifying genetic diseases is according to whether the defective gene is on the X chromosome or an autosomal chromosome and whether the disease is dominant or recessive. Furthermore, such information is often of use in counseling the family or in detecting other family members who have the disease. This classification is used in grouping the diseases listed in Table 12-6. Although many of the conditions such as color blindness are not life threatening,

Table 12-6. Incidence and prevalence of some genetically transmitted diseases*

DISEASE	INCIDENCE (1 in n live births)	PREVALENCE (1 in n people)
Autosomal recessive		
Cystic fibrosis	2000	3000
Phenylketonuria	10,000	40,000
Galactosemia (transferase deficient)		70,000
Wilson's disease		4×10^6 (U.S.A.)
		1×10^5 (Romania)
Von Gierke's disease		4×10^5
Tay-Sachs disease		6000 (Ashkenazi Jews)
		5×10^5 (Gentiles)
Sickle cell anemia		70-300 (American Negro)
Albinism		20,000
Galactokinase deficiency		1×10^5
Mucopolysaccharidosis I (Hurler's syndrome)		40,000
Autosomal dominant		
Neurofibromatosis		3300-4000
Huntington's chorea		20,000-25,000 (U.S.A.)
		3×10^5 (Japan)
Charcot-Marie-Tooth disease		5000
Achondroplasia	10,000	
X-linked recessive		
Duchenne's muscular dystrophy		6000-10,000
Color blindness		16 (American men)
		200 (American women)
Hemophilia A		20,000
Mucopolysaccharidosis II (Hunter's syndrome)		50,000

*For further information see Bergsma, E.: Birth defects: atlas and compendium, New York, 1973, The National Foundation.

they may represent situations that would lead a person to seek medical advice. It is worth noting that the most benign diseases often occur at a very high frequency.

Perhaps the best means of classifying genetic diseases is on the basis of the defective protein responsible for the condition. This has been done for almost ninety human genetic diseases, not including the hemoglobinopathies or other red cell defects. In addition, more than eighty variants of glucose 6-phosphate dehydrogenase are known, forty-nine with lowered activity. These intracistronic variants are caused by mutation at different sites in the gene responsible for programming the synthesis of the enzyme. In diseases such as Lesch-Nyhan syndrome, branched-chain ketoaciduria, Gaucher's disease, and phenylketonuria, the amount of enzymatic activity lost is in direct proportion to the severity of the disease.

Cytogenetic diseases

The cytogenetic diseases are not strictly biochemical diseases, but since they are detected as gross cytologic abnormalities of the chromosomes, they can be thought of as affecting protein and nucleic acid metabolism. Some are fairly common, and the frequency data listed in Table 12-7 emphasize the importance of these diseases and provide a basis for comparing these data with similar data on the frequency of some inherited genetic diseases (Table 12-6).

Table 12-7. Incidence and prevalence of some cytogenetic diseases

CYTOGENETIC DISEASE (chromosome abnormality)	INCIDENCE (1 in n live births)	PREVALENCE (1 in n people)
Trisomy 21 (Down's syndrome or mongolism)	770	2000-3000
XXY	800	
XYY	800	
XXX	1000	
Trisomy 18	3000	
Trisomy D	5000	
Monosomy X	5000	

REFERENCES

Becker, F. F., editor: Cancer, New York, 1975, Plenum Publishing Corp.

Chamberlin, M. J.: The selectivity of transcription, Annu. Rev. Biochem. 43:721, 1974.

Crick, F.: The genetic code III, Sci. Am. 215:55, 1966.

Dickson, R. C., et al.: Genetic regulation: the *lac* control region, Science 187:27, 1975. *Description of the nucleotide sequence of the lac promoter.*

Drake, J. W., and Koch, R. E., editors: Mutagenesis, vol. 4, Biochemical papers in genetics, New York, 1976, Hallsted Press.

Farber, E., editor: The biochemistry of disease, vol. 2, The pathology of transcription and translation, New York, 1972, Marcel Dekker, Inc.

Gefter, M.: DNA replication, Annu. Rev. Biochem. 44:45, 1975.

Haselkorn, R., and Rothman-Denes, L. B.: Protein synthesis, Annu. Rev. Biochem. 42:397, 1973.

Hood, L. J., et al.: Antibody genes, Annu. Rev. Genetics 9:305, 1975.

Kornberg, A.: DNA synthesis, San Francisco, 1974, W. H. Freeman and Co. Publishers.

Nomura, M., et al., editors: Ribosomes, Cold Spring Harbor Laboratory, N.Y., 1974.

Todaro, G. J.: Evolution and mode of transmission of RNA tumor viruses, Am. J. Pathol. 81:589, 1975.

Watson, J. D.: Molecular biology of the gene, ed. 3, New York, 1976, The Benjamin Co., Inc. *A superb account of information transfer in bacteria with applications to eukaryotic systems.*

Clinical examples

CASE 1: ACUTE LEUKEMIA

Y. B., a 5-year-old girl, was admitted to the hospital suffering from loss of appetite, weakness, pain in the joints, and fever. The child was very pale, and her spleen, liver, and lymph nodes were found to be enlarged. The laboratory findings showed a platelet count of 80,000/mm³ (normal = 250,000/mm³) and a white blood cell count of 50,000/mm³ (normal = 5000 to 10,000/mm³). Abnormal white cells were noted in a bone marrow specimen and also in the peripheral blood. Auer bodies (red-staining rods in the cytoplasm of myeloblasts or monoblasts) were not found, and there was no evidence of myelocytic or monocytic differentiation.

The diagnosis of acute leukemia was established, and therapy was started. The patient was immediately given prednisolone, 40 mg/m² · day in four divided doses, and a weekly intravenous injection of vincristine, 1.5 mg/m². After 4 weeks of treatment, hematologic and clinical remission occurred. Intensive high-dose intravenous chemotherapy was then started (Freireich and Frei, 1964). The following agents were administered on a daily schedule for 1 week: 6-mercaptopurine (1000 mg/m² · day) for 3 days followed by methotrexate (10 mg/m² · day) for 3 days and cyclophosphamide (600 mg/m²) for 1 day. After 5 weeks of therapy the child's serum uric acid level reached 15 mg/dl (normal = 1.5 to 6.0 mg/dl). Consequently, allopurinol treatment (200 mg three times a day orally) was started; the 6-mercaptopurine level was decreased to 250 mg/m² · day. Fluid intake was significantly increased. After a week the uric acid level returned to normal, the administration of allopurinol was halted, and the patient was placed on maintenance therapy of 6-mercaptopurine (50 mg/m² · day), methotrexate (20 mg/m² · week IV), and cyclophosphamide (200 mg/m² · week IV). The patient was treated with this maintenance therapy for the next 12 months.

Biochemical questions

1. What is the biochemical effect of administering high doses of corticosteroids as was done in this case?
2. What is the mechanism of action of the vinca alkaloids (vincristine), 6-mercaptopurine, methotrexate, and cyclophosphamide?
3. The above compounds have some serious side effects when high dosages are administered. Why?
4. What are the side effects of cancer therapy in a 5-year-old child?
5. Why did the serum uric acid level go up in this patient? Is this considered a good sign?
6. Why was the 6-mercaptopurine level reduced when the girl was put on allopurinol?
7. Why did the physician advise increasing the patient's fluid intake?

Case discussion

The enlarged lymph nodes and spleen seen in this case are typical of infectious mononucleosis, Hodgkin's disease, or lymphosarcoma, but abnormal white cells in the bone marrow and peripheral blood are not found in these diseases. Lymphocytosis is seen in whooping cough and infectious lymphocytosis, but the white blood cells are mature and the red blood cell and platelet counts are normal. Acute leukemia, on the other hand, has all the symptoms observed in this case. This is a disorder of the blood-forming tissue and is characterized by an abnormal proliferation of white cells. It is a neoplastic disease that strikes without regard to sex or race; it may develop at any age, although the peak incidence is from birth to 5 years of age.

Rapid and complete regression of the symptoms may be achieved through chemotherapy. At least temporary remissions are obtained in 90% of patients under 20 years of age. The objectives of the therapy are two: (1) to "induce" hematologic remission and (2) to maintain the patient in remission. The agents used vary according to whether initial remission or maintenance of remission is the goal.

Mechanism of action of drugs used. Although it is a corticosteroid, prednisone has very little salt-retaining activity. It is active, however, as an anti-inflammatory agent. The adrenocorticosteroids show dramatic effects on lymphoid tissue; they cause the nuclei to become pyknotic and disintegrate.

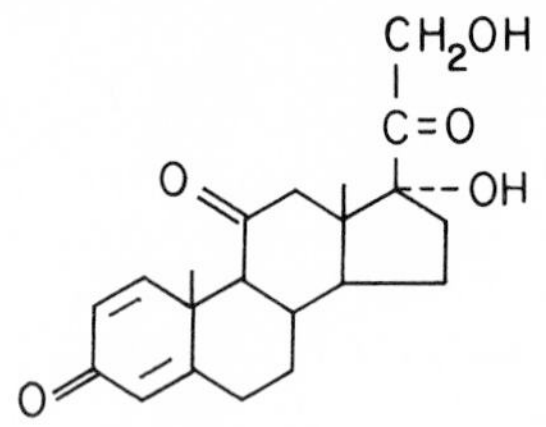

CH_2OH
|
C=O
|
---OH

Prednisone

This function may be related to their prompt (30-min) inhibition of the DNA and RNA polymerases in lymphocytes. Thus lymphocyte action in producing inflammation would be reduced, as would the number of circulating lymphocytes. The vinca alkaloids (vinblastine and vincristine) have a mode of action similar to colchicine. They prevent the assembly of the protein precursors of microtubules, including the mitotic spindles. Consequently, they cause mitotic arrest. 6-Mercaptopurine can be visualized as

an antimetabolite of either adenine or hypoxanthine. It is thought to be phosphoribosylated in a reaction with 5-phospho-α-D-ribosyl pyrophosphate to yield 6-mercaptopurine ribotide, which then functions to block the conversion of IMP to AMP. See first example below.

Methotrexate (p. 531) is an analogue of folic acid. Most antifolates are believed to act by inhibiting the enzyme dihydrofolate reductase, which catalyzes both of the transformations shown in the second example below.

Cyclophosphamide (p. 577) functions by alkylating DNA. As we have seen, reaction often occurs at position 7 of guanine, but with a bifunctional nitrogen mustard such as cyclophosphamide, alkylation can cross-link apposing strands of DNA or cross-link two moieties on the same strand. Alkylation of the phosphate group produces an unstable phosphate triester that may lead to chain scission.

The five drugs used in this case have different mechanisms of action. Therefore, it was suspected that they might produce additive effects when used in combination. Table 12-8 gives data on the percentage of patients who show signs of complete remission when each of the listed drugs is administered singly.

Because cancer chemotherapy is

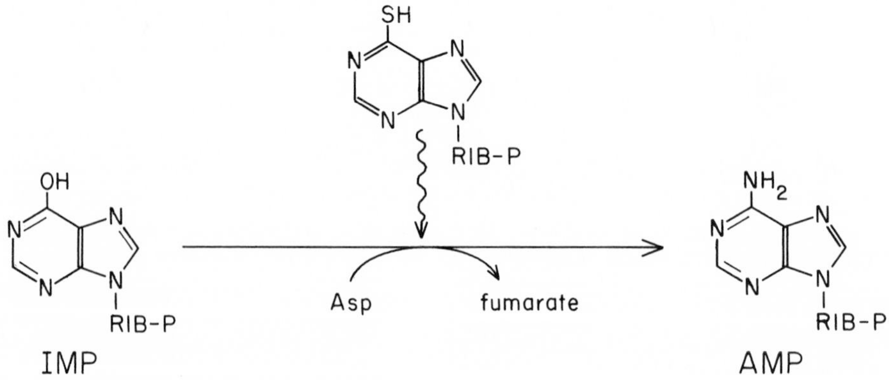

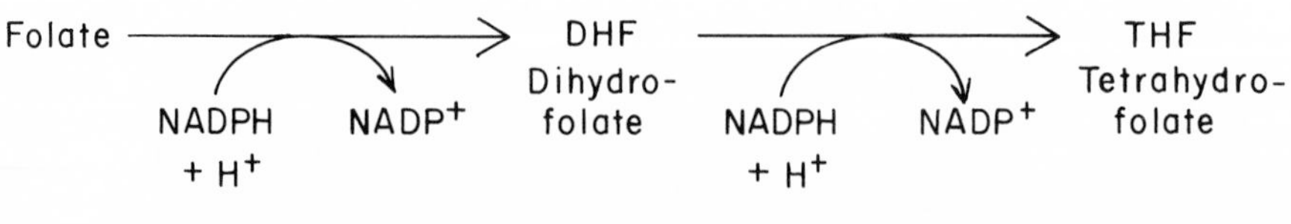

IMP → (Asp, fumarate) → AMP

Folate → (NADPH + H^+, NADP^+) → DHF Dihydrofolate → (NADPH + H^+, NADP^+) → THF Tetrahydrofolate

Table 12-8. Effectiveness of chemotherapy in inducing complete hematologic remission in patients with acute leukemia*

CHEMICAL AGENT	PATIENTS SHOWING COMPLETE REMISSION (%)	NUMBER OF PATIENTS
Prednisone	57	75
Vincristine	55	119
6-Mercaptopurine	27	43
Methotrexate	21	48
Cyclophosphamide	18	44

*From Freireich, E., and Frei, E.: Progr. Hematol. 4:187, 1964.

Table 12-9. Toxicity of chemotherapeutic agents*

CHEMICAL AGENT	TOXICITY	EFFECTIVENESS	MAJOR SIDE EFFECTS
		Most effective	
Prednisone	±		Hypertension, electrolyte imbalance
Vincristine	+		Peripheral neuropathy
6-Mercaptopurine	++		Marrow depression
Methotrexate	+++		Stomatitis, marrow depression
Cyclophosphamide	++++		Marrow depression
	Most toxic		

*From Freireich, E., and Frei, E.: Progr. Hematol. 4:187, 1964.

Table 12-10. Combination chemotherapy in acute leukemia*

DRUGS	REMISSION PREDICTED (%)	REMISSION REALIZED (%)
Vincristine + prednisone	82	85
Prednisone + 6-mercaptopurine	69	81
Methotrexate + 6-mercaptopurine	42	45
Vincristine + methotrexate + 6-mercaptopurine + prednisone	90	94

*From Freireich, E., and Frei, E.: Progr. Hematol. 4:187, 1964.

directed at such vital processes as RNA and DNA synthesis, no agent is without some toxic side effects. Interestingly, the toxicities of the compounds shown in Table 12-9 are inversely proportional to their effectiveness for inducing remission.

Combination therapy. It is now known that these drugs do act in an additive way because each inhibits nucleic acid synthesis at a different site. For example, if prednisone were administered to 100 patients, remission would occur in fifty-seven (Table 12-8). If methotrexate were combined with prednisone, 21% of the forty-three patients ($0.21 \times 43 = 9+$) refractive to prednisone alone would show a remission of the disease. Thus combining these drugs would produce remission in 66% of the patients treated. Using the data from Table 12-8, one can predict the percent remission achieved by various combinations of drugs. Table 12-10 shows the predicted remission based on the sum of each agent taken alone compared with the percent remission actually realized when patients were treated with the given combinations. The combination therapy for leukemia bears a close resemblance to antibiotic combination therapy in which combinations of antibiotics with different modes of action give additive effects. It is even possible that individual leukocytes in a large population may be resistant to one drug but not to another.

Remission maintenance therapy. For maintenance, 6-mercaptopurine and methotrexate are better than the other drugs. It seems that the order of effectiveness for maintaining remission is related more to the order of toxicity! It is possible that yet untested drugs or drugs relatively ineffective in inducing remission may be useful for maintenance, but knowledge of their effectiveness in this area is masked by their ineffectiveness in inducing remission.

The efficacy of most anticancer agents is related to their inhibition of fast-growing cells that require more nucleic acid synthesis. They do not distinguish cancer cells from other fast-growing cells, for example, normal blood-forming cells and gastrointestinal mucosa; consequently, using these drugs to treat a child for acute leukemia may significantly slow the child's growth.

The rise in serum urate is a typical finding in leukemic patients even before treatment. The additional increase on treatment can be interpreted as a positive sign or indication that the leukemic cells are dying and the patient is entering a stage of complete remission. The danger with these high levels is not the threat of gout but rather the renal problems that might occur as a result of urate crystals forming in the distal tubules. The urate levels may become as high as 80 mg/dl and are considered serious at levels greater than 15 mg/dl. Emergency therapy for severe hyperuricemia includes (1) hydration, 3 to 4 liters/day;

(2) alkalization of urine, 6 to 8 g $NaHCO_3$/day; and (3) allopurinol. Care must be taken to lower the oral dosage of 6-mercaptopurine when allopurinol is added to the regimen; allopurinol is not only a purine analogue, but when taken orally, it inhibits the intestinal xanthine oxidase that ordinarily oxidizes much of the 6-mercaptopurine to nonutilizable, excretable forms. In the presence of allopurinol, much more of the administered 6-mercaptopurine finds its way into the blood. Because this compound is generally used at doses near its toxicity range, the dosage must be reduced when allopurinol is given concomitantly.

The prognosis for a patient such as Y. B. is rather poor. The life of a few may be prolonged several years by treatment. For most, however, life is extended 6 to 12 months by treatment. In patients who remain untreated the life expectancy is 2 to 6 months. Cancer chemotherapy can actually be curative in some diseases. For example, methotrexate and other antifolates are quite effective against choriocarcinoma in women and in the treatment of Burkitt's lymphoma. Occasionally cures may also be seen in certain testicular tumors, rare cases of acute leukemia, some insulinomas, and breast cancers (case 9, Chapter 11).

REFERENCES

Aur, R., et al.: Central nervous system therapy and combination chemotherapy of childhood lymphocytic leukemia, Blood 37:272, 1971.

Freireich, E. J., and Frei, E.: Recent advances in acute leukemia, Prog. Hematol. 4:187, 1964.

Gershwin, M. E., et al.: Cyclophosphamide: use in practice, Ann. Intern. Med. 80:531, 1975.

Maugh, T. H.: Cancer chemotherapy: now a promising weapon, Science 184:970, 1974.

Pinkel, D.: Five year follow-up of total therapy of childhood lymphocytic leukemia, J.A.M.A. 216:648, 1971.

Rosenthal, S., and Kaufman, S.: Vincristine hemotoxicity, Ann. Intern. Med. 80:733, 1975.

CASE 2: BETA THALASSEMIA

M. D., a 5-year-old child of Italian descent, developed an acute upper respiratory infection. When examined, he was extremely ill with a temperature of 104° F. Respiration was rapid and his pulse was rapid and weak. He was markedly anemic. His head was somewhat enlarged, and a loud, blowing, systolic murmur was audible over the entire precordium. The abdomen was distended, ascites was present, and the spleen reached the crest of the ilium. The patient's legs were edematous. He grew rapidly worse and expired on the fourth day of his illness.

His medical history revealed that birth was normal and spontaneous; his birth weight was 4 pounds. He was one of twins, the other child weighing 5 pounds at birth. He had gained weight slowly and took food poorly.

The liver and spleen were found to be enlarged while he was still a small infant. He had been hospitalized on three different occasions for the treatment of anemia, poor appetite, and mild respiratory infections.

A twin brother and a 2-year-old brother suffered from the same type of anemia. No other type of familial illness was present.

Biochemical questions

1. What is thalassemia? What cells are primarily affected? What is its incidence?
2. How does α-thalassemia differ from β-thalassemia?
3. What is the difference between thalassemia minor and thalassemia major? Which form would you expect in the thalassemia heterozygotes? In the homozygotes? Does the child in this case have the minor or major form of thalassemia?
4. β-Thalassemia is very likely a disease(s) characterized by an abnormality in the control of protein synthesis. Describe the ways in which this control might be disturbed. Consider as many alternatives as you can and place these in the framework of what you know about the mechanisms of gene expression and protein synthesis.
5. The β-thalassemia of Ferrara is a disease in which the synthesis of the protein in question is not simply decreased; it is altogether absent. It has been shown that normal amounts of the missing β subunit are made when polyribosome fractions taken from cells of these patients are combined with the postribosomal supernatant enzyme fractions obtained from normal individuals. Can you suggest how this might happen and predict what molecular defects could account for the primary lesion in this form of thalassemia?

Case discussion

Thalassemia is a group of related inherited disorders characterized by severe anemia in early life and enlarged spleen and liver. This group of diseases represents hemoglobinopathies where the amount of hemoglobin synthesized is low. They are unlike other hemoglobinopathies such as sickle cell anemia (case 1, Chapter 2) where a specific amino acid in one of the globin chains has been changed as a result of mutation. Thalassemic hemoglobin seems normal in respect to the primary structure of the globin subunits; the abnormality is in the relative amounts of α- and β- (more properly non-α-) chains. Recall that the normal hemoglobins all contain the same α-chain, but the other non-α-chain varies. For example, in hemoglobin A, the major hemoglobin of adults, the composition is $\alpha_2\beta_2$; the composition of hemoglobin A_2 is $\alpha_2\delta_2$ and that of fetal hemoglobin F, $\alpha_2\gamma_2$. In some forms of thalassemia the rate of synthesis of each of the α, β, γ, and δ subunits is affected. In α-thalassemia the rate of synthesis of α-chains is lower than normal; in β-thalassemia the β-chains are low.

Globin genes

The genes for the human globin subunits occur in at least two sets of unlinked cistrons. Alpha subunit genes are located in a genetic region quite separate from the non-α-genes. Some of the non-α-genes on the other hand are linked one to the other, either on chromosome 2 or

4. The gene order is thought to be $G^\gamma \rightarrow A^\gamma \rightarrow Y \rightarrow \delta \rightarrow \beta$. There are at least two different forms of the normal fetal γ-subunit, G^γ having a glycyl residue at position 136 and A^γ an alanyl residue at position 136. The region designated as Y represents a hypothetical regulatory gene that controls the expression of the fetal and adult genes. During development it is assumed that the regulatory Y gene prevents expression of the γ-subunits but allows expression of the δ- and β-chains. The δ- and β-chains are known to be linked since a subunit in hemoglobin Lepore contains the N-terminal sequence of the δ-chain and the C-terminal sequence of the β-chain. Hemoglobin Lepore could arise by a crossing over between the homologous chromosomes that carry these two linked genes. The cross-over would occur between the δ-gene on one chromosome and the β-gene on the other.

Inheritance of thalassemia

All of the thalassemias are inherited in simple mendelian fashion as autosomal recessive defects. *Thalassemia major* is sometimes used to describe the disease in the homozygote and *thalassemia minor* to describe the much milder symptoms of the heterozygote. Thalassemia minor is usually not serious; however, individuals with thalassemia major, such as the boy in this case, do not survive to reproductive age. In spite of this, thalassemia is almost as prevalent in certain Mediterranean and southeast Asian kindreds as sickle cell anemia in people of African ancestry. Like sickle cell anemia, heterozygotes may be more resistant to malaria (see case discussion in case 1, Chapter 2) so that prevalence of the disease is high in regions of the world where malaria is endemic.

There is no effective treatment of thalassemia, so prevention of the disease by genetic counseling is important.

Gene expression in thalassemia

Thalassemia is fairly easy to recognize in both the homozygote and heterozygote. The low levels of a particular subunit are detected by electrophoresis of extracts of erythrocytes. Red cell precursors also show abnormal ratios of globin subunits, and because globin synthesis occurs in the precursor cells but not in the mature erythrocytes, investigators have focused attention on precursor cells such as reticulocytes in efforts to determine the nature of the genetic defects in globin mRNA or protein synthesis.

The exact molecular lesion is not known for any form of thalassemia; yet most forms of the disease appear to lack sufficient mRNA for one of the globin subunits. In the case of α-thalassemia the lack of α-mRNA is easy to explain since the α-gene is partially or completely deleted. β-Thalassemia is more complex, however, since the β-globin gene is present. Furthermore, in this disease a decrease in β-chain synthesis can be compensated by an increased synthesis of δ- and γ-globin, as in β°-thalassemia where no β-chains are made at all. Low, but detectable, amounts of β chains are made in β^+-thalassemia. In most cases of β^+- and β°-thalassemia it appears as if the β-mRNA is present in low amounts (β^+) or missing (β°). Proof for this can be obtained by hybridizing radioactively labeled DNA complementary to β-mRNA. The complementary DNA is made in vitro with the enzyme reverse transcriptase, a RNA-dependent DNA polymerase found associated with RNA tumor virus particles. The template for the reverse transcriptase is highly purified β-mRNA, and the product is a single-stranded radioactive DNA, which is complementary to β-mRNA and can be used as a probe to pick up β-mRNA.

The reason why β-mRNA is missing is unknown, but there are several possibilities. A precursor form of the mRNA may never be made, or if made it may fail to be properly modified, for example, failure to be "capped" at the 5' end or polyadenylated at the 3' end. In addition, the precursor mRNA may be improperly processed for transport to the cytoplasm. Thus the mRNA could be synthesized but because it could not be metabolized further, it would be broken down. A failure of β°-mRNA to properly initiate protein synthesis could also lead to the rapid breakdown of the mRNA and give the impression that the primary defect is in mRNA synthesis.

An unusual β°-thalassemia found in the Ferrara region of Italy may result from a defect in the translation of β-globin mRNA. Polyribosomes from the reticulocytes of patients with this disease fail to make β-globin unless supplemented with a crude soluble enzyme fraction from normal reticulocytes. Some eukaryotic protein synthesis initiation factors are found in the supernatant fraction in contrast to the mostly ribosomal location of bacterial factors. Thus it is possible that a special initiation factor, either new or modified, may be responsible for the lack of β-globin synthesis. Nevertheless, other interpretations are possible. A change in a rare tRNA essential for β-globin synthesis or a change in the affinity of an aminoacyl-tRNA synthetase for a rare tRNA would also explain the data.

REFERENCES

Conci, F., et al.: Reduced rate of β globin mRNA translation in β-thalassemia (Ferrara), Eur. J. Biochem. 32:533, 1973.

Kabat, D., and Koler, R. D.: The thalassemias: models for analysis of quantitative gene control, Adv. Hum. Genet. 5:158, 1975.

Lodish, H. F., and Nathan, D. G.: Alpha and beta-globin messenger ribonucleic acid, J. Biol. Chem. 247:7822, 1971.

Nathan, D. G.: Thalassemia, N. Engl. J. Med. 286:586, 1972.

Tolstoshev, P., et al.: Presence of gene for β-globin in homozygous β°-thalassemia, Nature 259:95, 1976.

CASE 3: XERODERMA PIGMENTOSUM

D. R. is a 45-year-old woman who has lived her entire life on a farm. She has the skin manifestations of xeroderma pigmentosum (XP) without the fairly common neurologic abnormalities. She is heavily freckled and has a lengthy history of skin cancer. In the past 5 years she has had thirty-five clinically diagnosed neoplasms excised from sun-exposed areas of her skin, and she recently developed ocular involvement requiring a corneal transplant.

Biochemical questions

1. How is DNA damaged by irradiation with UV light? What is a thymine dimer?
2. Describe the enzymatic systems involved in repairing UV-damaged DNA.
3. What is the molecular defect in XP and how is it related to a predisposition to cancer?
4. Propose a biochemical explanation for the genetic heterogeneity of XP.

Case discussion

Xeroderma pigmentosum is an autosomal recessive human skin disease characterized by sensitivity to sunlight, which causes multiple cutaneous neoplasms. It occurs in about one person in 250,000 of the general population, often in consanguineous matings, and is found worldwide in all races. Although the disease is detected by the effects it has on the exposed areas of the skin and eyes, it also can have systemic effects. For example, it can cause a number of neurologic abnormalities, including progressive mental deficiency, deafness, ataxia, and retarded growth. The neurologic manifestations are characteristic of a variant form called the De Sanctis-Cacchione syndrome. Treatment involves minimizing exposure to sunlight and the removal of the tumors when they appear, but there is no cure.

The molecular defect in XP is associated with the repair of DNA damaged by exposure to UV light. The association of this defect with a strong predisposition to cancer has attracted much attention to this otherwise rather obscure disease. The tumors of XP patients are no different from those of other individuals without the disease; only their incidence in XP patients is greatly increased. Exposure to sunlight is well correlated to the incidence of the two most common skin cancers, basal cell carcinoma and squamous cell carcinoma. Patients with XP also develop malignant melanomas. Exactly how UV light induces skin cancer is not known, but it is suspected that the mutagenicity of UV light and its ability to induce cancer may both be the consequence of UV-modified DNA. In this connection, the excessive freckling seen in XP patients may be a manifestation of the mutagenic effects of ultraviolet light. Freckles, even in normal persons, are collections of large melanocytes with melanosomes that are unusually large and dark. It is thought that each freckle represents a clone of melanocytes that arose from a single cell that had undergone a UV-induced mutation. The XP patient also has patches of hypopigmentation that could arise from mutations in pigment formation.

Repair of UV-damaged DNA

A great deal is known about how UV light modifies bacterial DNA. The 5-6 double bond of cytosine can be hydrated, and DNA-protein crosslinks may form, but the most significant lesion is the cross-linking of adjacent pyrimidine residues in DNA. Adjacent thymine residues on the same DNA strand dimerize to yield the product shown below.

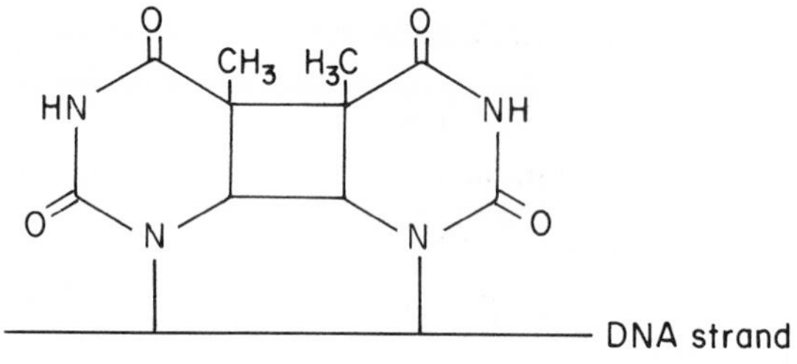

Both bacterial and animal cells have evolved enzyme systems to cope with UV-induced thymine dimers. There are at least three different enzyme systems that neutralize the effects of thymine dimers. One, called *excision repair,* was discussed on p. 579. In excision repair an endonuclease recognizes the damaged DNA and breaks the nucleotide chain close to, and on the 5′ side of, the lesion. The damaged single strand peels off the DNA duplex, and a DNA polymerase catalyzes the synthesis of a polynucleotide to replace the damaged segment. A 5′-exonuclease (perhaps part of the DNA polymerase) hydrolyzes away the damaged single-stranded segment, and a

DNA ligase seals the break, yielding a complete DNA duplex.

Another way to repair DNA strands is called *postreplication repair.* If the damaged DNA is in the process of being replicated, the DNA polymerase skips over the thymine dimer leaving a slight gap in the newly synthesized strand. This gap can be later filled by enzymes that differ in some respects from those involved in excision repair. A third method for dealing with thymine dimers is called *photoreactivation.* First discovered in bacteria, the enzyme responsible for this kind of repair has just recently been found in human cells. The photoreacting enzyme simply converts thymine dimers to monomers before they can do any harm. The enzyme binds to thymine dimers in DNA and catalyzes the formation of the thymine monomers when exposed to light of wavelengths 300 to 600 nm.

Genetic variants

There is evidence for defects in all three of the DNA repair systems from cells of patients with XP. Because there are at least five different genetic forms of the disease as detected by cell hybridization studies, it is tempting to suppose that each might be concerned with a different enzyme in one of the three repair systems. This interpretation is probably correct, but it is still not known whether components of one repair system might be shared with another. This would explain why some patients appear to be defective in more than one of the repair systems. Alternatively, a defect in a gene that regulates an operon containing two or more of the repair systems would also explain the data.

REFERENCES

Cleaver, J. E., and Bootsma, D.: Xeroderma pigmentosum: biochemical and genetic characteristics, Annu. Rev. Genet. 9:19, 1975.

Robbins, J. H., et al.: Xeroderma pigmentosum, an inherited disease with sun sensitivity, multiple cutaneous neoplasms and abnormal DNA repair, Ann. Intern. Med. 80:221, 1974.

Sutherland, B. M., et al.: Xeroderma pigmentosum cells contain low levels of photoreactivating enzyme, Proc. Natl. Acad. Sci. U.S.A. 72:103, 1975.

CASE 4: TETRACYCLINE-INDUCED HYPOPLASIA OF THE TEETH

A. B. was a 5-year-old boy who was brought to the dental clinic because of severely discolored teeth. Examination showed yellow-stained hypoplasia of the enamel. The lack of enamel on the tips of the cuspid teeth resembled the enamel notch seen in many patients who have survived erythroblastosis fetalis, a disease caused by Rh incompatibility. On questioning the child's mother, it was discovered that while his birth had been normal, the boy had been treated at 6 weeks of age for a rather serious throat infection with 500 mg tetracycline daily for 3 days.

The parents and child were tested for ABO, Rh, and Kell blood groups, and the results ruled out erythroblastosis fetalis. Unlike patients with other dental diseases that produce discoloration and hypoplasia, the teeth of this boy gave off a bright yellow fluorescence when irradiated with ultraviolet light. When sections of this patient's teeth were stained for bilirubin, the test proved negative.

Biochemical questions

1. What physical properties of tetracyline suggest that it is related to this pathologic condition?
2. How does tetracycline inhibit protein synthesis?
3. Does tetracycline inhibit protein synthesis in animals as well as bacteria?
4. Is the hypoplasia of the teeth seen in this case related to an effect of tetracycline on protein synthesis? Explain.
5. What causes erythroblastosis fetalis?

REFERENCE

Witkop, C. J., Jr., and Wolf, R. P.: Hypoplasia and intrinsic staining of enamel following tetracycline therapy, J.A.M.A. 185:1008, 1963.

CASE 5: SYSTEMIC LUPUS ERYTHEMATOSUS

A 63-year-old woman came to her physician with complaints of episodic swelling and pain of the hands, wrists, and knees. A year later an antinuclear antibody test was positive, but lupus erythematosus cell tests were negative. Still, a diagnosis of systemic lupus erythematosus (SLE) was made based on a constellation of clinical and laboratory findings. Prednisone therapy was started. Six months later she was admitted to the hospital with pleuritic pain and purpura. Serum total protein was 5.6 g/dl, and electrophoresis showed 66% albumin and 4% γ-globulin. After an additional 6 months she was admitted again with shortness of breath, chest pain, and fever. A sputum culture grew *Diplococcus pneumoniae* and coagulase-positive *Staphylococcus aureus*. Despite intensive antibiotic and supportive therapy, the patient died 1 week later. At autopsy, microscopic examination revealed a lymphoma of a mixed cell type and a pathologic lesion typical of SLE.

Biochemical questions

1. Patients with SLE produce antibodies to single- and double-stranded DNA, especially viral nucleic acids. Some patients will also show antibodies against RNA:DNA hybrids. Considering the relationship of lymphoma to SLE in this case, describe a possible role for the RNA-dependent DNA polymerase in producing these results.
2. What is the lupus erythematosus cell test?
3. Sometimes SLE develops in patients who have been receiving hydralazine hydrochloride for long periods of time for hypertension. Develop a theory considering a viral pathogenesis for the disease that encompasses these observations.
4. Symptoms of lupus are often noticed after exposure to sunlight. Can you offer an explanation based on ultraviolet light activation of a latent virus?

REFERENCES

Smith, C. K., Cassidy, J. T., and Bole, G. G.: Type I dysgammaglobulinemia, systemic lupus erythematosus and lymphoma, Am. J. Med. 48: 113, 1970.

Talal, N., and Gallo, R. C.: Antibodies to a DNA: RNA hybrid in systemic lupus erythematosus measured by a cellulose ester filter radioimmunoassay, Nature [N. Biol.] 240:240, 1972.

CASE 6: DIPHTHERIA

C. D. was a 9-year-old daughter of a migrant worker. She was hospitalized for diphtheria during an epidemic in a southwestern city. Nausea, chills, vomiting, headache, and a sore throat brought the girl to the hospital. The examining physician noted a tenacious gray membrane near the tonsils. Her leukocyte count was elevated. The patient had never been immunized against diphtheria. Antibiotic treatment with penicillin and erythromycin was begun as well as intravenous administration of diphtheria antitoxin.

Diphtheria is caused by infection with *Corynebacterium diphtheriae* strains that are lysogenic for or infected with a bacteriophage carrying the *tox* gene. The viral gene codes for the synthesis of a protein toxin that is secreted from the bacterial cell.

Biochemical questions

1. What is the mechanism of action of erythromycin?
2. Why is it necessary to administer diphtheria antitoxin, since the antibiotics will eventually kill all of the infectious bacteria?
3. Diphtheria toxin is an enzyme. What reaction does it catalyze? How does this affect protein synthesis?
4. Diphtheria toxoid is used to immunize healthy young children against diphtheria. It is made by treating diphtheria toxin with formaldehyde. What

effect would you predict formaldehyde would have on the enzymatic activity of the toxin? Why? What effect does formaldehyde have on the serologic specificity or immunogenicity of the toxin?
5. If a person is immunized, is it still possible to get a throat infection with *C. diphtheriae*? Explain.
6. Diphtheria toxin has a molecular weight of 62,000. Brief treatment with trypsin yields two fragments. The N-terminal fragment, A, has a molecular weight of 24,000 and retains enzymatic activity. The C-terminal B fragment has a molecular weight of 38,000 but is without enzymatic activity. The fragments, when assayed one at a time, do not inhibit cultured human cells, whereas the intact toxin molecule is inhibitory. Rats and mice are relatively resistant to the disease. Cultured mouse L cells are insensitive to both fragments as well as to the intact toxin; yet protein synthesis in cell-free extracts of L cells is sensitive to either the A fragment or the intact toxin. Describe a function for the B portion of the toxin that explains these data.

REFERENCES

McCloskey, R. V., et al.: The 1970 epidemic of diphtheria in San Antonio, Ann. Intern. Med. 75:495, 1971.

Pappenheimer, A. M., and Gill, D. N.: Diphtheria, Science 182:353, 1973.

CASE 7: DUCHENNE TYPE MUSCULAR DYSTROPHY

B. B. is a 5-year-old boy who was admitted to the hospital for evaluation of possible muscular dystrophy. The patient showed the first symptoms of muscle weakness in his legs at 3 years of age, when it was noted that he frequently fell while running or climbing stairs. There was a continuous progression of muscle weakness. The family history was non-contributory. Seven other siblings, including four males, were healthy.

Neurologic examination showed weakness and atrophy of the proximal parts of the lower extremities (pelvic

girdle muscles), pseudohypertrophy of the calves, and contractures of the Achilles tendons. The gait was waddling and hyperlordotic. There was no obvious weakness of the upper extremities. Deep tendon reflexes could not be elicited throughout, and plantar reflexes gave a bilateral flexor response. There was no sensory loss and no ataxia. Cranial nerves were intact. Psychologic evaluation showed an IQ of 79 (low normal value).

Electromyographic data were abnormal, indicating myopathic changes. The blood chemistry was normal; however, the serum creatine phosphokinase (CPK) was 40 μmoles of creatine formed per ml of serum per hr (normal, 3.5 ± 1.4). Serum aldolase was 95 units/ml (normal, less than 12 units/ml). The patient's mother and two sisters had normal serum concentrations of these two enzymes.

A biopsy of the vastus lateralis muscle was performed with the parent's permission. In order to complete the diagnosis a portion of the muscle sample was submitted for histopathologic examination; it showed signs compatible with muscular dystrophy. The remainder of the biopsied muscle was homogenized for a study of in vitro protein synthesis.

Biochemical questions

1. Why would you expect the serum CPK and aldolase concentrations to be high in muscular dystrophy?
2. What are the functions of these enzymes in muscle?
3. Duchenne type muscular dystrophy is an X-linked recessive disease. What is the probability that this boy's mother, father, brothers, or sisters would have the disease or be carriers of the disease?
4. The replacement of muscle tissue by connective tissue is a typical finding in this disease. Collagen synthesis in vitro is greatly increased when the heaviest polyribosomes are combined with the high-speed supernatant fraction that remains after the ribosomes have been sedimented by ultracentrifugation. What components of the supernatant fraction are involved in the synthesis of protein? What components remain bound to the ribosomes?
5. When heavy polyribosomes from a patient are mixed with the supernatant fraction from the same patient and the necessary substrates are added, almost all of the protein labeled is collagen. In the same experiment done with fractions derived from a control subject, only 15% of the protein synthesized is collagen. If the polyribosomes from the patient are mixed with the supernatant fraction from the control, only 15% of the proteins labeled are collagen. How can these data be interpreted to indicate that the genetic defect in Duchenne type muscular dystrophy is related to collagen synthesis? What other interpretations are possible?
6. A new method for detecting carriers of muscular dystrophy involves the in vitro phosphorylation of an erythrocyte protein. Phosphorylation of proteins from both carriers and patients with Duchenne type muscular dystrophy are increased. Furthermore, it is known that the gene for one of the three subunits of phosphorylase b kinase is located on the X chromosome. Another subunit is actin. Describe experiments that would test whether a defect in the X-linked phosphorylase B kinase subunit represents the molecular lesion in Duchenne type muscular dystrophy.

REFERENCES

Ionasescu, V., et al.: Identification of carriers of Duchenne muscular dystrophy by muscle protein synthesis, Neurology 23:497, 1973.

Roses, A. D., et al.: Carrier detection in Duchenne muscular dystrophy, N. Engl. J. Med. 294:193, 1976.

Tyler, F. H.: Muscular dystrophies. In Stanbury, J. B., Wyngaarden, J. B., and Fredrickson, D. S., editors: The molecular basis of inherited disease, ed. 3, New York, 1972, McGraw-Hill Book Co.

ADDITIONAL QUESTIONS AND PROBLEMS

1. Why are drugs relatively ineffective against viral diseases?
2. What is necessary for a substance to be an antibiotic?
3. In a genetically determined disease such as phenylketonuria, is the critical liver enzyme missing or just inactive? How is this related to mental retardation? How can this situation be corrected?
4. How is a dominant or recessive trait expressed at the molecular level in

terms of the protein products? In X-linked genetic diseases, why is the male usually affected and the female not?

5. What is the Philadelphia chromosome? How is it related to chronic myelogenous leukemia?*

6. Sometimes the following drugs are used in cancer chemotherapy. How do they function?

L-Asparaginase	Busulfan
Actinomycin D	Hydroxyurea
Arabinosyl cytosine	Estrogens and
5-Fluorouracil	androgens
Amethopterin	6-Thioguanine

7. Hemoglobin Constant Spring contains an extraordinarily large α-subunit. This subunit contains an additional thirty-one amino acid residues at the C-terminal end of the usual α-chain. Propose explanations for this abnormality. Consider mechanisms involving protein synthesis, chain termination, and DNA recombination.

---

*See Rowley, J. D.: Nature 243:290, 1973.

8. Several carcinogenic compounds were found to be mutagenic to a bacterial test system (Ames, B. N., et al.: Proc. Natl. Acad. Sci. U.S.A. 70:2281, 1973) but only after activation by an extract of rat liver microsomes containing NADPH. What kind of biochemical reactions are necessary to convert these compounds to mutagens?

9. Components of cigarette smoke are more mutagenic after treatment with preparations from lung similar to those used in the preceding question (Kier, L. D. et al.: Proc. Natl. Acad. Sci. U.S.A. 71:4159, 1974). The activation by the lung extract is greater if the animals have been previously treated for several days with polychlorinated biphenyls. Explain this increase in activity.

10. Some of the mutagenic substances described in questions 8 and 9 cause frameshift mutations, but only after treatment in vitro with the microsome extract. How does this observation fit with ideas about the mechanism of frameshift mutagenesis?

CHAPTER **13**

HORMONAL REGULATION
OF METABOLISM

OBJECTIVES

1. To describe the structure, biosynthesis, secretion, and mechanisms of action of the hormones present in the human body
2. To discuss the hormonal control of metabolism
3. To discuss the biochemical basis of commonly occurring human endocrine diseases

Endocrinology is the branch of biologic science that deals with hormones, hormonal regulation of metabolism, and diseases associated with hormonal abnormalities. Hormones represent one of the main communication systems of the body. Although present only in minute concentrations, they are able to turn metabolic processes on or off. In order to intelligently prescribe hormones and correct malfunctions in hormonal regulation, practitioners of the health sciences must understand how these systems operate. For example, those who prescribe or dispense oral contraceptives should know how ovulation and menstruation are controlled. The purpose of this chapter is to present the basic principles of endocrine function and control at the molecular level. This will provide a firm foundation to which the student can add the clinical and pathologic aspects of endocrinology.

NATURE OF HORMONES
Definition of a hormone

Hormones are chemical substances, present in very low concentrations in the blood, that have a regulatory effect on the metabolism of specific organs or tissues. They are secreted directly into the blood in trace amounts by specialized cells, often grouped together in a distinct anatomic structure called an endocrine gland. Hormones are transported through the blood to specific tissues called target tissues, where they exert their regulatory effects. Hormones regulate the activity or concentrations of enzymes in the target cells and thus produce alterations in their metabolism. Some hormones are *trophic;* that is, their function is to regulate the synthesis and release of another hormone by the target cell. An example of a trophic hormone is thyrotropin, the thyroid-stimulating hormone (TSH), which acts on the thyroid gland to stimulate the synthesis and release of the thyroid hormones. Other hormones produce a metabolic response in their target tissues, for example, the mobilization of fatty acid from adipose tissue by epinephrine, glucagon, and several other hormones.

The chemical structures of the hormones vary widely. Some are synthesized from amino acids, for example, the tyrosine derivatives *epinephrine* and *thyroxine.* Many of the others are polypeptides or proteins, including *vasopressin, adrenocorticotropin* (ACTH), *insulin*, and *prolactin*. Still others are steroids that are synthesized from cholesterol, for example, *testosterone, progesterone,* and *cortisol*. The structure and synthesis of the steroid hormones were discussed in Chapter 10.

Endocrine glands

Hormones are synthesized in endocrine glands, some of which, for example, the *thyroid,* are distinct gross anatomic structures. Others such as the pancreatic islets are microscopic structures that comprise only part of an organ. An endocrine gland secretes its hormone directly into the blood. This is in contrast to an *exocrine gland,* which delivers its secretory products through a duct into a body cavity. The pancreas, exclusive of the islets, is an example of an exocrine gland. This gland delivers its products, the pancreatic digestive zymogens or enzymes such as trypsinogen and ribonuclease, through the pancreatic duct into the intestine.

Neurotransmitters

Hormonal responses usually occur relatively slowly as a result of the fact that after discharge into the circulation the hormone must be concentrated at one or more distant sites of action. For example, ACTH released from the adenohypophysis acts primarily on the adrenal cortex, and parathyroid hormone released from the parathyroids acts in part on the kidneys.

When rapid signals must be sent to a tissue, the nervous system functions as the message carrier. In such cases an impulse quickly reaches the terminus of the neuronal axon that is adjacent to the specific target cell. The chemical transmitter is already present in the axon terminals at the site of action. The electrical depolarization results in the neurotransmitter making direct contact with the target cells within a fraction of a second after the initiation of the nervous impulse. The mechanism by which the neurotransmitter activates the target cell is often similar to that employed by the slower acting hormone. An example of this similarity is seen in the adipose tissue. Adipocytes receive rapid messages to mobilize fatty acids from the *sympathetic nervous system* and slower messages to the same effect from certain circulating hormones. One of the circulating fat-mobilizing hormones is epinephrine, a catecholamine released from the adrenal medulla. The neurotransmitter released at the sympathetic nerve endings is norepinephrine, which also is a catecholamine.

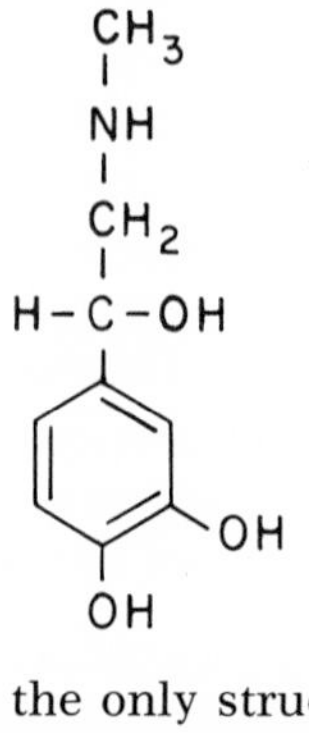

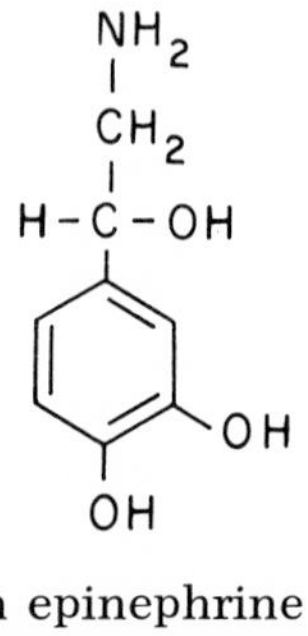

Notice that the only structural difference between epinephrine and norepinephrine is that the amino group of the former is *N*-methylated. As might be expected, these two substances activate lipolysis and initiate other metabolic responses through similar molecular mechanisms.

In addition to the adipose tissue, neurotransmitters function at other sympathetic and cholinergic nerve terminals, at the neuromuscular junction, and at synapses within the central and autonomic nervous systems. Two criteria have been established for the identification of a substance as a neurotransmitter. It must be released in appropriate amounts when the presynaptic nerve is stimulated, and it must mimic the effect of electrical stimulation of the presynaptic nerve when it is added to the postsynaptic receptor. The neurotransmitter is stored in vesicles contained in the presynaptic nerve endings. When the electrical depolarization wave reaches this region, the neurotransmitter is released by a calcium-dependent process that involves either exocytosis or passage directly through the nerve membrane. The released neurotransmitter combines with a specific receptor site, leading to transduction of the impulse and excitation of the postsynaptic cell without any electrical continuity between the two cells. It is likely that the response to the neurotransmitter in the postsynaptic cell is mediated in certain cases by cyclic nucleotides. For example, *dopamine* causes an increase in cAMP in certain postsynaptic cells, whereas *acetylcholine* produces an increase in cGMP. Although the neurotransmitter is bound tightly to the receptor, it rapidly dissociates so that the signal can be terminated quickly. Inactivation of the neurotransmitter occurs through either reabsorption and storage once again in the presynaptic nerve endings or enzymatic degradation. The enzymes needed to synthesize the neurotransmitter are made in the cell body of the presynaptic neuron. A circulation of the axoplasm appears to exist between the cell body and the nerve terminals of the axon. This enables information that additional amounts of neurotransmitter are needed to be passed to the cell body. The required enzyme is then synthesized in the cell body and sent to the axon terminals.

Neurotransmitters have been classified into two groups, established and putative. At this time, only two substances definitely have been established as neurotransmitters, *norepinephrine* and *acetylcholine*. Fourteen others are likely to be neurotransmitters but have not as yet been firmly established to have this role. They are called putative neurotransmitters and include *3,4-dihydroxyphenethylamine, dopamine, epinephrine, glutamate, γ-aminobutyrate* (GABA), *histamine, aspartate, glycine, proline, taurine, 5-hydroxytryptamine* (serotonin), *octopamine, carnosine, ATP,* and *substance P*. Most of these are amino acid derivatives. For example norepinephrine, epinephrine, and dopamine are tyrosine derivatives. Serotonin is made from tryptophan, histamine from histidine, and GABA from glutamic acid. Not all neurotransmitters, however, are amino acid derivatives. ATP is a nucleotide, and substance P is a peptide that contains eleven amino acids, including a carboxyl terminal methionine amide residue.

Norepinephrine acts primarily as the chemical transmitter at the sympathetic nerve endings, but it also is thought to function as a chemical transmitter in the central nervous system. Acetylcholine is the chemical transmitter at the endings of both the cholinergic and motor nerves as well as the chemical transmitter in the ganglia of the sympathetic nervous system. The enzyme acetylcholinesterase hydrolyzes acetylcholine after it has been released by the presynaptic cell at the postsynaptic terminal or motor end-plate. The putative neurotransmitters act primarily within the central nervous system.

Regulation of hormone secretion

The secretion of a hormone from its endocrine gland of origin is controlled through a servomechanism. In other words, the plasma concentration of the hormone itself or of a substance produced by the target tissue in response to the hormone regulates the further release of the hormone from the gland. Moreover, a hormone released from one endocrine gland often regulates the release of another hormone from a second gland, which in turn controls hormonal production in and release from the first gland. These points are illustrated by the regulation of circulating thyroid hormone levels.

The thyroid gland produces two hormones, thyroxine (*3,5,3′,5′-tetraiodothyronine*, commonly abbreviated T_4) and *3,5,3′-triiodothyronine* (T_3). These hormones are released by the gland in response to TSH, a glycoprotein that is produced in the adenohypophysis. Increased plasma concentrations of T_4 and T_3 act on the adenohypophysis to inhibit TSH release, an action analogous to a feedback-inhibition mechanism. Another level of control exists in the TSH-T_4,T_3 system; TSH release is enhanced by *thyrotropin-releasing hormone* (TRH), a tripeptide that is synthesized in the hypothalamus. TRH reaches the adenohypophysis through a portal venous system connecting the gland and the hypothalamus. TRH release is stimulated by low circulating levels of T_4 and T_3. These observations are illustrated schematically in Fig. 13-1. It should be reiterated that both effects depend on circulating T_4 and T_3 levels. When the levels are high, TSH release is depressed, and the stimulus to continued thyroid hormone release is removed. This occurs through a T_4-induced inhibition of the action of TRH on the adenohypophysis. Conversely, when the concentrations of T_4 and T_3 are low, TRH release from the hypothalamus is enhanced, which stimulates TSH release and hence thyroid hormone release.

In other cases the release of a hormone from the endocrine gland is regulated by a product of the hormonal effect and not by a second hormone, for example, parathyroid hormone release from the parathyroid glands. Parathyroid hormone acts to increase the plasma calcium concentration. When the plasma calcium level is low, parathyroid hormone release is stimulated. Alternatively, parathyroid hormone release is inhibited when the plasma calcium concentration rises.

Target tissue

More than one tissue may respond to a particular hormone. For example, while TSH primarily stimulates the thyroid to synthesize and release thyroxine, it also has a lipolytic effect in adipose tissue. For the most part, however, there is

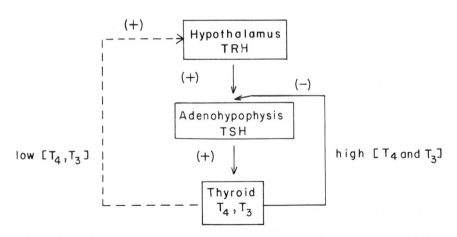

Fig. 13-1. Regulation of thyroid hormone secretion.

considerable specificity for a given target organ. Such specificity is dependent on hormone-binding receptors. The action of many hormones is initiated by a binding of the hormone to the membrane of the target cell. In these cases the *membrane binding sites* are specific for the hormone that acts on that tissue. The cell membrane may be able to bind loosely many different hormones, but only the binding sites for the specific hormones for that cell have a high affinity and are properly located for activation of the intracellular system that triggers the cell response. Epinephrine is an example of a hormone that activates its target cells through specific membrane receptor sites. The membrane receptors are situated in such a manner that they trigger the activation of an enzyme located in the membrane, adenyl cyclase, when epinephrine comes into contact with them.

Another type of target cell specificity involves *cytoplasmic binding proteins* that combine with only one hormone. These proteins allow only the hormone that acts on the particular tissue to be concentrated within that tissue. For example, uterine endometrium, vaginal epithelium, and pituitary and breast tissue contain a soluble, *estrogen-binding protein* of high molecular weight. This protein has an extremely high affinity for estradiol (affinity constant $K_a = 3 \times 10^{10} M^{-1}$), and it serves to concentrate estradiol within these tissues and to transport it into the nucleus. A similar mechanism operates with progesterone and cortisol, and it may apply to other steroid hormones.

MECHANISMS OF HORMONE ACTION

Hormones affect target tissue by regulating the enzymatic activity within that tissue. According to our present knowledge, enzymatic regulation can occur in one of two ways. Many hormones act by activating or deactivating enzymes that already exist within the target tissue. This response is mediated through a second messenger system located in the target tissue, the *adenyl cyclase–cyclic adenosine monophosphate system*. As might be expected, hormones that act in this manner produce their effects fairly rapidly. The second general method for regulation of target tissue enzymatic activity involves induction of enzyme synthesis. According to evidence that has been obtained with experimental model systems such as the hen oviduct and rat liver, the hormone must be transported into the nucleus of the target cell in order for transcription to occur. Transcription is followed by de novo synthesis of the enzyme, which is a much slower process than the activation of preexisting enzyme protein by the adenyl cyclase system. One situation is known in animal tissues where the adenyl cyclase system operates through gene regulation rather than the more common enzyme activation mechanism. This occurs in the pineal gland and is described on p. 648. At present, this action of the adenyl cyclase system should be considered as a special case and not typical of its usual regulatory action in eukaryotic cells.

Adenyl cyclase mechanism

Many hormones, including epinephrine, glucagon, TSH, ACTH, and vasopressin, act on their target organs by stimulating adenyl cyclase, an enzyme contained in the target cells. This enzyme converts ATP to *3',5'-cyclic adenosine monophosphate*, abbreviated as cAMP (see p. 623 for cAMP structure). Adenyl cyclase is located in the cell membrane, probably adjacent to the hormone-binding sites. In some way, combination of the hormone and its binding site activates the enzyme and ATP is converted to cAMP, as illustrated in Fig. 13-2. Recent experimental evidence suggests that certain prostaglandins may modulate the activity of the adenyl cyclase and in this way regulate the intracellular response to stimulation

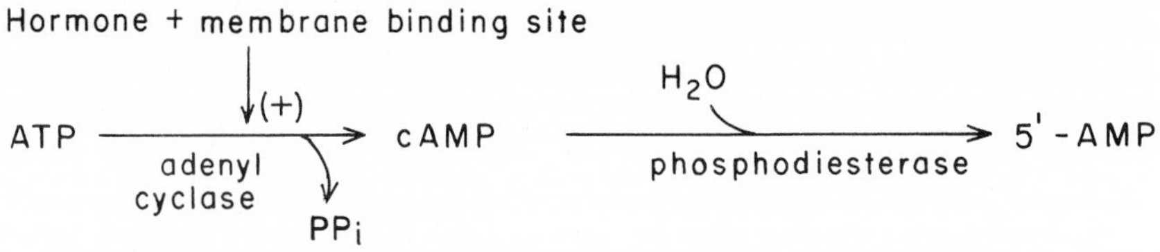

Hormone + membrane binding site

$$ATP \xrightarrow[\substack{\text{adenyl} \\ \text{cyclase}}]{(+)} cAMP \xrightarrow[\text{phosphodiesterase}]{H_2O} 5'-AMP$$

PPi

Fig. 13-2. Reactions catalyzed by adenyl cyclase and phosphodiesterase.

C_K = protein kinase catalytic subunit

RC_K = inactive protein kinase

$$RC_K + cAMP \rightleftharpoons C_K + R_K - cAMP$$

$$\text{Enzyme (inactive)} + ATP \underset{}{\overset{C_K}{\rightleftharpoons}} \text{Enzyme} - PO_4 \text{ (active)} + ADP$$

Fig. 13-3. cAMP and protein kinase reactions. R represents regulatory subunit of the protein kinase.

by a given hormone. cAMP is inactivated by conversion to 5'-*adenosine monophosphate* (5'-AMP) through the action of a *phosphodiesterase*. Methyl xanthines such as caffeine and theophylline inhibit the phosphodiesterase reaction and hence decrease the rate of cAMP breakdown. This leads to an increased cAMP concentration within the cell and therefore to an enhancement or potentiation of the cAMP effect. Recent evidence indicates that at least two different phosphodiesterases exist in mammalian cells. These enzymes have widely different K_m values for cAMP and hence probably have separate functions in metabolic regulation involving intracellular cyclic nucleotide content.

cAMP action. The effect of cAMP within the target cells is to activate *protein kinases,* enzymes that utilize ATP to phosphorylate a protein, often another enzyme, within the cell. Protein kinases contain two subunits. One subunit is *catalytic* and the other is *regulatory.* When the two are combined, the kinase is inactivated. The regulatory subunit combines with the catalytic subunit in a way that prevents the latter from functioning. cAMP binds to the regulatory subunit and causes it to dissociate from the catalytic subunit. Once the regulatory subunit no longer is attached, the catalytic subunit is free to act. It is not known at this time whether the complex of cAMP and the regulatory subunit, of itself, has any function. In prokaryotic cells, a complex of cAMP and a protein known as CAP is involved in regulating the transcription of the lac operon (see Chapter 12). Therefore, a biologic precedent exists for a role of cAMP-protein complexes, but it remains to be determined whether such complexes have any regulatory function in eukaryotic cells. The general mechanism of the cAMP-mediated phosphorylation reactions is illustrated in Fig. 13-3.

Activation of lipolysis and glycogenolysis. Lipolysis and glycogenolysis are two of the most important metabolic processes activated by cAMP-mediated phosphorylation. In the case of lipolysis, *triglyceride lipase* is the enzyme converted from the inactive to the active form through phosphorylation. This enzyme is also known as the hormone-sensitive lipase (Chapter 8). In the case of glycogenolysis,

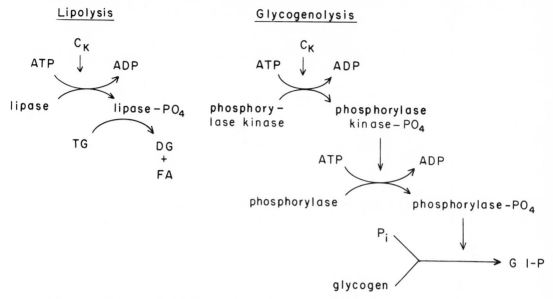

Fig. 13-4. Mechanism of lipolysis and glycogenolysis activation. TG = triglyceride; DG = diglyceride; FA = fatty acid; P_i = inorganic phosphate; G l-P = D-glucose l-phosphate.

the enzyme that is phosphorylated in the protein kinase reaction is *phosphorylase kinase*, which in turn utilizes ATP to convert phosphorylase from the inactive to the active form through phosphorylation. Activated phosphorylase then degrades glycogen (Chapter 7). The major difference between lipolysis and glycogenolysis, as shown schematically in Fig. 13-4, is that in glycogenolysis the protein kinase activates another kinase that is interposed between it and phosphorylase.

Protein kinase catalyzes the phosphorylation of a hydroxyl group of the enzyme protein, most commonly a serine hydroxyl group. In several instances, however, a threonine hydroxyl group is the site of phosphorylation. Protein kinases have broad substrate specificity; for example, the kinase obtained from reticulocytes will catalyze the phosphorylation of the appropriate serine hydroxyl group in liver phosphorylase.

Enzymatic amplification. The mechanism whereby cAMP usually acts in eukaryotic cells is an example of biochemical amplification; it permits relatively few hormone molecules to trigger a very substantial enzymatic reaction. For example, suppose that the combination of one molecule of a hormone with a cellular binding site results in the activation of one molecule of adenyl cyclase. Being an enzyme, adenyl cyclase can convert many ATP molecules to cAMP. For illustrative purposes let us assume that all of the enzymatic steps introduce a multiplication factor of 100. Therefore, 100 molecules of cAMP are formed, and in turn 100 molecules of protein kinase are activated. Each molecule of protein kinase then activates 100 molecules of either phosphorylase kinase or lipase. Each molecule of active phosphorylase kinase activates 100 molecules of phosphorylase, each of which then degrades 100 molecules of glycogen. Likewise, each activated lipase molecule degrades 100 triglyceride molecules. As shown in Fig. 13-5, the degradation of 10^6 triglyceride molecules and 10^8 glycogen molecules results from a single molecule of hormone combining with a cell receptor site.

Specificity of hormone action and cAMP. Many hormones exert their effect by activating adenyl cyclase in their respective target tissues. In each case the same

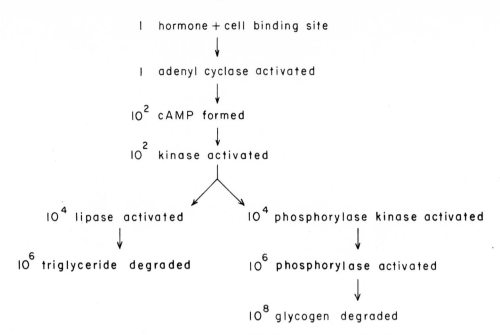

Fig. 13-5. Enzymatic amplification in cAMP mechanism. The amplification factor used in this illustration ($\times 100$) is hypothetical.

second messenger, cAMP, is formed. Adenyl cyclase is present in the plasma membranes of all human cells that have been examined, with the exception of the mature erythrocyte. In spite of the widespread occurrence of the enzyme and the presence of a common second messenger substance, the response in a given target cell is usually activated by only one hormone. That such specificity exists in the face of a common response mechanism is a result of the specificity of the hormone-binding receptors of the cell membrane. The receptors that activate adenyl cyclase in a given target cell bind only the hormone or hormones to which that cell responds. While the hormone receptors are intimately related to the adenyl cyclase and contain protein, they do not appear to be a part of the adenyl cyclase itself.

Hormone receptors. Since the specificity of the adenyl cyclase system resides in the hormone-binding receptors of the cell membrane, it is of considerable interest to isolate and characterize these receptors. The first receptor that was obtained in soluble form was a catecholamine-binding protein from canine ventricular myocardial membranes. A detergent was necessary to solubilize this receptor, and it was purified by affinity chromatography. Two proteins were obtained, one of molecular weight 4×10^4 and the other 1.6×10^5. The higher molecular weight material probably is an aggregate form of the receptor protein. The solubilized protein retains many of the binding properties of the catecholamine receptor in the intact membrane, but the association constant for epinephrine is lower. Catecholamine receptors are of two types, α and β, as will be described later in this chapter. The catecholamine receptors of the ventricular myocardium are the β-type. Therefore, the protein that has been solubilized is thought to be a *β-adrenergic receptor*.

An *insulin receptor* has been obtained in soluble form without the use of detergents. When cultured lymphocytes are incubated in a serum-free medium, material that binds insulin is released into the incubation medium. Binding of the solubilized material is abolished by digestion with trypsin, suggesting that the insulin receptor, like the cardiac β-adrenergic receptor, is a protein.

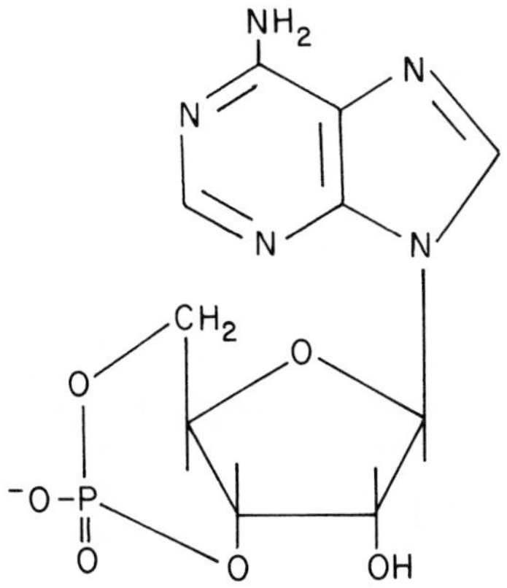

adenosine 3',5'-
monophosphate (cAMP)

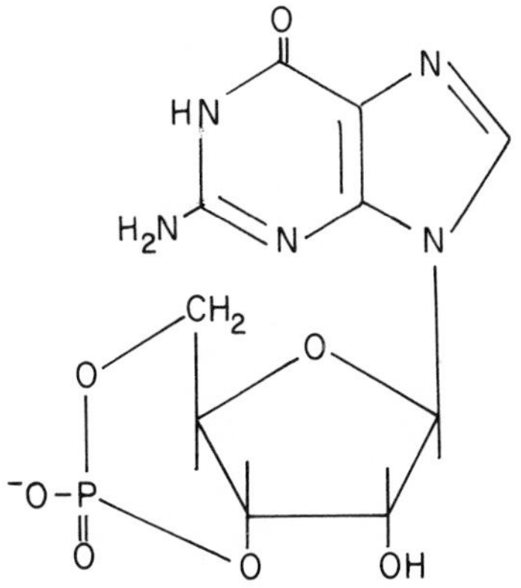

guanosine 3',5'-
monophosphate (cGMP)

cGMP. Another cyclic nucleotide, the 3',5'-phosphodiester of guanosine monophosphate, also is present in mammalian tissues. This compound is commonly abbreviated cGMP and is produced from GTP by guanyl cyclase, an enzyme that is similar to adenyl cyclase. When cell homogenates are prepared, however, most of the guanyl cyclase activity is recovered in the cytosol fraction. These guanyl cyclase preparations are stimulated by calcium ions but do not respond to hormones. One explanation is that the enzyme is loosely bound to the cell membrane in the physiologic state and that only in the membrane-bound state is it hormonally responsive. According to this interpretation, the shearing forces of the cell homogenization process cause the dissociation of guanyl cyclase from the cell membrane, and the loss of hormone responsiveness is an artifact of the preparation. By contrast, the adenyl cyclase is more tightly bound and remains associated with the cell membrane even after homogenization. Like cAMP, cGMP is degraded by phosphodiesterase. At present, cAMP and cGMP are the only cyclic nucleotides found to have a regulatory function in the eukaryotic cell.

Regulatory roles of cAMP and cGMP. The current view is that cAMP and cGMP are intracellular modulators for signals delivered from outside the cell by hormones or neurotransmitters. Both cyclic nucleotides are always present inside cells, but the cAMP concentration is ten to fifty times higher than that of cGMP. Many opposing effects within the cell are thought to be controlled by the relative concentrations of cAMP and cGMP. For example, in some rapidly dividing cells the cAMP concentration is low and the cGMP concentration is high. When proliferation ceases, the cAMP concentration rises while the cGMP concentration falls. Therefore, cGMP may be an intracellular regulator for division, whereas cAMP is an inhibitor of cell division. Similar opposing effects of the two cyclic nucleotides have been observed in other physiologic processes. Isoproterenol, a drug with properties like norepinephrine, increases myocardial contractility. This is associated with a rise in myocardial cAMP content and a lowering of its cGMP content. Conversely, acetylcholine depresses myocardial contractility, causing increased cGMP and a lowering of myocardial cAMP. A similar relationship also occurs in the pyramidal tract neurons of the cerebral cortex. Norepinephrine, which is a depressant, raises the cAMP content of the neurons. By contrast, acetylcholine excites these neurons and increases their cGMP content.

Finally, there is some evidence that insulin action is associated with an increase in the cGMP content of its target cells. Glucagon, which exerts opposite metabolic effects from insulin, is known to elevate the cAMP content of these cells. The relationship between intracellular cAMP and cGMP levels is currently a very active area of research, and much remains to be learned about their apparently opposing intracellular actions.

Induction of protein synthesis

Another general mechanism of hormonal action involves the biosynthesis of proteins, which in most cases are enzymes. This mechanism is entirely different from the usual cAMP mechanism, which almost always involves the activation of preexisting enzymes. In at least one case the hormone-induced stimulation of enzyme synthesis appears to occur through a transcriptional mechanism. Perhaps the best evidence for this comes from studies with the hen oviduct carried out independently in the laboratories of O'Malley and Schimke. *Diethylstilbestrol*, a synthetic estrogen analogue, stimulates the oviduct to synthesize a specific mRNA, which directs the production of a protein that is immunologically identical to *ovalbumin*. The mRNA is active in directing ovalbumin synthesis in a ribosomal system prepared either from the oviduct or from a rabbit reticulocyte lysate. Ovalbumin is one of the main proteins present in the egg white.

In the transcriptional mechanism, hormonal specificity is conferred by a soluble binding protein present in the cytoplasm of the target cell. The binding protein concentrates the specific hormone within the target cell, and the hormone-protein complex is transported into its nucleus. Once inside the nucleus the hormone-protein complex combines with specific sites on the chromatin. The mechanism actually amounts to a derepression of chromosomal DNA, which allows the DNA segment to be transcribed, triggering the process that leads to synthesis of the protein encoded in that particular gene.

Glucocorticoids may act through a similar mechanism. It has been known for several years that *dexamethasone*, a synthetic glucocorticoid with a cortisol-like action, induces the biosynthesis of the enzyme tyrosine aminotransferase. Much of this work has been done in the HTC line of cultured rat hepatoma cells. The results of these studies have been interpreted to show that glucocorticoids bind to cytoplasmic receptors and that the hormone-receptor complex then binds to specific chromatin sites. This causes gene transcription, and the mRNA produced directs tyrosine aminotransferase synthesis. However, earlier work with this system was interpreted to show that dexamethasone induced tyrosine aminotransferase synthesis through a posttranscriptional mechanism, that is, by affecting translation rather than mRNA synthesis. This controversy has yet to be conclusively resolved. Other posttranscriptional hormonal effects have been reported, but these observations require further confirmation.

HYPOTHALAMIC REGULATORY HORMONES

The *adenohypophysis* may be thought of as the master endocrine gland. Its hormonal secretions regulate the function of many of the other endocrine glands, including the thyroid, adrenal cortex, testes, and ovaries. As noted in Fig. 13-1, the hormones released by these glands inhibit release of the specific adenohypophyseal trophic hormone that regulates the gland. In addition, the *hypothalamus* controls the release of the trophic hormones from the adenohypophysis. This regulation is mediated through a group of nine small polypeptides known as *hypo-thalamic releasing hormones* and *hypothalamic release–inhibiting hormones.* These substances are considered by some to be factors rather than hormones, and both nomenclatures are in common usage. The substances are biosynthesized in certain groups of nerve cells, which may be thought of as neural transducer tissues, contained in the hypothalamus. The hypothalamic regulatory hormones reach the adenohypophysis through a portal venous system of blood vessels; the hormones are released from hypothalamic nerve endings, pass into the portal vessels, and are carried to the adenohypophysis. A cAMP mechanism appears to be involved in transmission of the releasing-factor message to the adenohypophyseal cells.

This neurohumoral mechanism is different from the neurosecretory mechanism that links the hypothalamus with the *neurohypophysis,* or posterior pituitary. In the latter case the actual neurohypophyseal hormones, *oxytocin* and *vasopressin,* are synthesized in the paraventricular and supraoptic nuclei, respec-

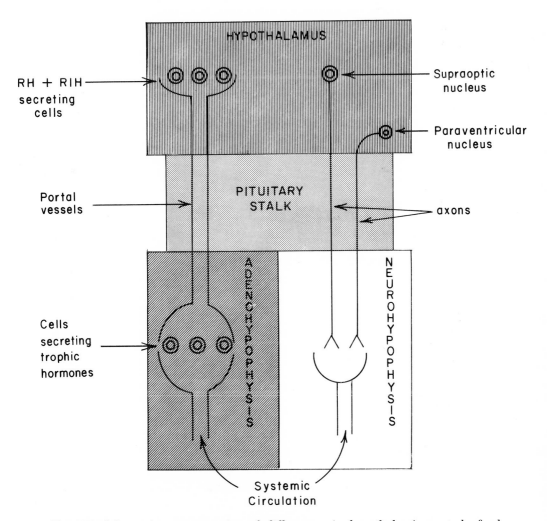

Fig. 13-6. Schematic representation of differences in hypothalamic control of adeno-hypophysis and neurohypophysis. RH = releasing hormones; RIH = release-inhibiting hormones.

tively, of the hypothalamus. These hormones are then transported down the axons of the hypothalamic-neurohypophyseal nerve tracts, the endings of which are in the neurohypophysis. Oxytocin and vasopressin are stored in the neurohypophysis and from thence released into the circulation. These differences between the hypothalamus' relationships to the two parts of the pituitary gland are illustrated schematically in Fig. 13-6.

Nine hypothalamic regulatory hormones have been identified to date and are listed in Table 13-1. The releasing hormone for luteinizing hormone (LH) is the same as that for follicle-stimulating hormone (FSH), and it has been named for both, that is, LHRH/FSHRH. The adenohypophysis secretes seven hormones. As can be seen from Table 13-1, a releasing hormone has been identified for each of them. In contrast, release-inhibiting hormones have been identified for only three of the seven adenohypophyseal hormones. The three adenohypophyseal hormones that are regulated by release-inhibiting hormones, growth hormone, prolactin, and melanocyte-stimulating hormone, have metabolic effects on their target tissues and do not cause the release of any hormone. Conversely, the function of the other four adenohypophyseal hormones, adrenocorticotropin, thyrotropic hormone, luteinizing hormone, and follicle-stimulating hormone, is to increase the synthesis and release of a hormone from their target cells. These target cell hormones, such as cortisol and thyroxine, suppress the further production and release of their respective adenohypophyseal trophic hormones. Therefore, specific release-inhibiting hormones are not required for the regulation of these four adenohypophyseal trophic hormones, for the secretions of their respective target cells serve this function.

Chemistry

All of the hypothalamic regulatory hormones characterized thus far are small peptides. The most widely studied is TRH, a tripeptide that contains glutamic acid, histidine, and proline. However, as shown in Fig. 13-7, the N-terminal gluta-

Table 13-1. Hypothalamic regulatory hormones

HYPOTHALAMIC HORMONE	ABBREVIATION	ADENOHYPOPHYSEAL HORMONE REGULATED*
Releasing hormones		
Corticotropin-releasing hormone	CRH	ACTH
Thyrotropin-releasing hormone	TRH	TSH
Luteinizing hormone– and follicle-stimulating hormone– releasing hormone	LHRH/FSHRH	LH and FSH
Growth hormone–releasing hormone	GHRH	GH
Prolactin-releasing hormone	PRH	PRL
Melanocyte-stimulating hormone– releasing hormone	MRH	β-MSH
Inhibitory hormones		
Growth hormone release– inhibiting hormone	GHRIH	GH
Prolactin release–inhibiting hormone	PRIH	PRL
Melanocyte-stimulating hormone release– inhibiting hormone	MRIH	β-MSH

*Abbreviations for the adenohypophyseal hormones are described in Table 13-2.

mate exists in a ring form, *pyrrolidone carboxylic acid,* and the carboxyl group of the C-terminal proline has an amide substitution. The structure is abbreviated as (pyro) Glu-His-Pro-NH$_2$. Another releasing hormone that has been characterized completely, LHRH/FSHRH, is a decapeptide that, in abbreviated form, is represented as:

$$\text{(pyro) Glu-His-Trp-Ser-Tyr-Gly-Leu-Arg-Pro-Gly-NH}_2$$

Like TRH, it has an N-terminal glutamate residue in pyrrolidone form, and the C-terminal residue, glycine, is present as the amide. GHRIH also has been characterized completely.

Ala-Gly-Cys-Lys-Asn-Phe-Phe-Trp-Lys-Thr-Phe-Thr-Ser-Cys
                └──────S──────────────────S──────────┘

It contains fourteen amino acid residues but has neither the N-terminal pyrrolidone glutamate residue nor the C-terminal amide. GHRIH is also called *somatostatin.* The structures of the other hypothalamic regulatory hormones are in the process of being elucidated.

In several cases the hypothalamic regulatory hormones have additional functions besides the regulation of a specific adenohypophyseal hormone. Only 31% of the TRH content actually is present in the hypothalamus, and it has been found in all parts of the brain except the cerebellum. TRH can also stimulate the release of growth hormone and prolactin. In addition, it has been observed to act as a mood elevator when administered to depressed female patients. Somatostatin also has other effects besides inhibiting growth hormone secretion. It is synthesized in the pancreas and stomach and inhibits the secretion of insulin, glucagon, gastrin, and secretin. Therefore, it is involved in the local regulation of the gastric and pancreatic secretions. Somatostatin-induced suppression of glucagon release can prevent the fasting hyperglycemia that occurs in uncontrolled diabetes mellitus. Not all of the hypothalamic regulatory hormones, however, have additional functions. For example, LHRH/FSHRH is found almost entirely in the hypothalamus, and its only known function is to cause follicle-stimulating and luteinizing hormone release.

ADENOHYPOPHYSEAL HORMONES

Seven hormones are synthesized in the adenohypophysis, that portion of the pituitary derived from the ectoderm of the primitive oral cavity (buccal compo-

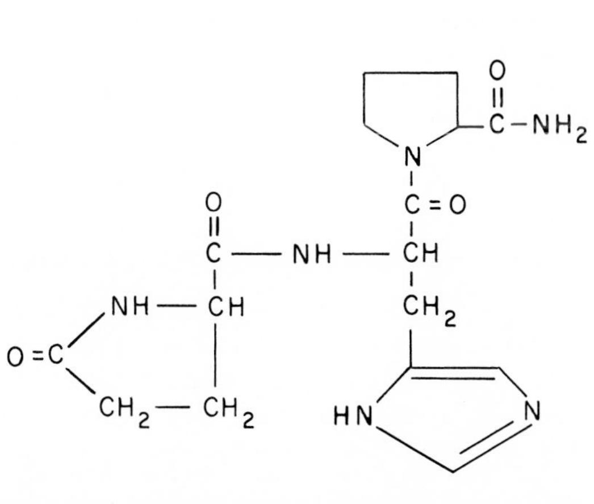

Fig. 13-7. Thyrotropin-releasing hormone (TRH).

Table 13-2. Adenohypophyseal polypeptide and glycoprotein hormones

HORMONE	ABBRE-VIATION	ADDITIONAL NAME	TARGET TISSUE	MAJOR FUNCTIONS
Growth hormone	GH	Somatotropin (SH)	Many	Protein synthesis
Lactogenic hormone	PRL	Prolactin	Mammary gland	Milk production
Follicle-stimulating hormone	FSH		In female, follicle	Maturation of follicle
			In male, testes	Spermatogenesis
Luteinizing hormone	LH	Interstitial cell–stimulating hormone (ICSH)	In female, follicle	Ovulation, formation of corpus luteum, progesterone secretion
			In male, testes	Testosterone production
Thyrotropic (thyroid-stimulating) hormone	TSH		Thyroid	Thyroxine and tri-iodothyronine production and release
Adrenocorticotropin	ACTH		Adrenal cortex	Corticosteroid synthesis and release
Melanocyte-stimulating hormone	β-MSH		Melanocytes	Pigmentation

nent). These hormones are either polypeptides or glycoproteins. Their names as well as their functions and target tissues are given in Table 13-2. The hormones vary in size from β-MSH, which contains twenty-two amino acids, to prolactin, which has a molecular weight of 3×10^4. Each hormone acts on a specific target tissue and in many cases regulates hormone production in that tissue. Secretion of these hormones is regulated by the hypothalamic regulatory hormones and in some cases by the circulating concentration of the hormone synthesized in the particular target tissue.

Growth hormone

Growth hormone (GH), which is secreted by the acidophil cells, has an anabolic effect on many tissues. The molecular weight of human GH is approximately 2.2×10^4. It is composed of a single polypeptide chain containing 188 amino acid residues. GH stimulates amino acid uptake by muscle and increases protein synthesis in several tissues. In addition, it raises the plasma glucose and free fatty acid concentrations. Therefore, GH has an anti-insulin effect. GH release into the plasma is stimulated by high plasma amino acid concentrations and is suppressed by high plasma glucose concentrations. Only human GH is active in man, making therapy with this hormone extremely costly and at times impossible. This is in contrast to other polypeptide and protein hormones such as ACTH and insulin, which are effective in man when prepared from animal products.

The most widely used test for GH secretory capacity is the *arginine tolerance test*. Arginine is the most potent amino acid stimulator of GH release. Under normal conditions the GH concentration in plasma will rise to a value greater than 5 ng/ml within 2 hr after intravenous infusion of 500 mg arginine hydrochloride/kg body weight.

Thyroid-stimulating hormone

Thyroid-stimulating hormone (TSH), also called thyrotropin, stimulates the production and release of thyroxine and triiodothyronine, the thyroid hormones. The molecular weight of TSH is 2.6×10^4. It is a glycoprotein that contains an α- and a β-chain. In addition to its main effect on the thyroid gland, TSH also acts on adipocytes to stimulate lipolysis. The actions of TSH are mediated at least in part through the adenyl cyclase–cAMP mechanism.

Adrenocorticotropin

Adrenocorticotropin (ACTH) stimulates glucocorticoid production and secretion in the adrenal cortex. It is a small polypeptide, containing only thirty-nine amino acid residues. Only the first twenty-four amino acid residues are necessary for biologic activity. The sequence of these twenty-four amino acids is identical in all of the mammalian species that have been studied to date, whereas the sequence of the last fifteen amino acid residues varies from species to species. ACTH acts through a cAMP mechanism to stimulate the hydroxylation and cleavage of the cholesterol side chain in the $C_{20\text{-}22}$ position, to form pregnenolone (Chapter 10). It also stimulates the hydrolysis of adrenal cortical cholesteryl esters, enabling the released cholesterol to be used as the substrate for steroidogenesis. Although the main action of ACTH is on the adrenal cortex, it also stimulates lipolysis in adipocytes, acting through a cAMP mechanism.

Gonadotropins

Because they act on the sex glands and regulate the production of the sex hormones, *follicle-stimulating hormone* (FSH) and *luteinizing hormone* (LH) are known as the gonadotropins. They are glycoproteins with a molecular weight of 2.8×10^4. Like TSH, they are composed of two subunits, α and β. FSH causes maturation of the ovarian follicle in the female, and it stimulates spermatogenesis in the male. LH has several actions in the female, including stimulation of ovulation, corpus luteum formation, and progesterone production. The same hormone stimulates testosterone production in the male and in this case is called interstitial cell–stimulating hormone (ICSH).

The α-chains of LH, FSH, and TSH are virtually identical. However, their β-chains differ in structure, and this imparts specificity to each hormone. The structural similarities may explain certain overlapping functions of these three adenohypophyseal trophic hormones.

Prolactin

Prolactin (PRL), a simple protein having a molecular weight of 3×10^4, stimulates milk production in the mammary gland. Human prolactin exhibits a marked diurnal variation in plasma concentration, with the highest values occurring at night during sleeping. Prolactin induces the synthesis of the two proteins that make up *lactose synthetase*, the enzyme responsible for the synthesis of lactose, or milk sugar (Chapter 7). One of these is a membrane-bound *uridine diphosphogalactosyl transferase*. The other is α-lactalbumin, a protein that modifies the specificity of the transferase to include glucose as an acceptor for the galactosyl moiety, thereby permitting lactose synthesis. In addition, prolactin causes the Golgi apparatus to proliferate and hypertrophy. The Golgi apparatus is the organelle through which lactose, protein, and the fat globules are released from the mammary alveolar cells into the collecting ducts. Before prolactin can exert its effects, the mammary cells must be primed by exposure to insulin and cortisol.

Prolactin is present in males as well as females, but its function in the male is unknown. Common amino acid sequences are present in prolactin and GH, and this may explain certain overlapping effects of these two hormones.

Recent experimental evidence suggests that prolactin may be involved in breast cancer. The relationship is not straightforward, however, for there is no simple correlation between plasma prolactin levels and the incidence of breast cancer in those females who so far have been studied.

Melanocyte-stimulating hormone

The melanocyte-stimulating hormone that occurs in man is β-MSH. In many animals it is contained in a middle lobe of the pituitary. Man does not have this structure, and human β-MSH is secreted by the same basophil cells of the adeno-hypophysis that secrete ACTH. In fact, a portion of the β-MSH primary structure is identical to the primary structure of amino acids 4 to 10 of ACTH. β-MSH is a small polypeptide containing only twenty-two amino acid residues. Its function in man is unknown, but it may act on the brain and somehow be involved in learning and memory. Administration of large quantities of β-MSH causes a deposition of pigment in the skin. As might be expected from the partial similarity in primary structure, ACTH also exerts a hyperpigmentation effect.

NEUROHYPOPHYSEAL HORMONES

The neurohypophysis, or posterior pituitary, is the storage site for two poly-peptide hormones, each containing nine amino acids. These hormones, oxytocin and vasopressin, are synthesized in the hypothalamus and transported to the neurohypophysis, where they are stored by the hypothalamic-neurohypophyseal nerve tracts. Seven of the nine amino acids in these hormones are identical, as shown in Fig. 13-8. Their N-terminal regions exist as rings made up of six amino acids linked together by a disulfide bridge. Both hormones contain a C-terminal glycinamide residue.

Oxytocin

Oxytocin is synthesized in the paraventricular nucleus of the hypothalamus. It causes uterine contraction and milk ejection from the mammary glands. In turn, stimulation of the breast nipples by the suckling infant causes oxytocin release from the neurohypophysis. Oxytocin is utilized in obstetrics, as it can induce the onset of labor when the pregnant uterus is at term. Whether the natural onset of labor is also initiated by endogenous oxytocin release is not known.

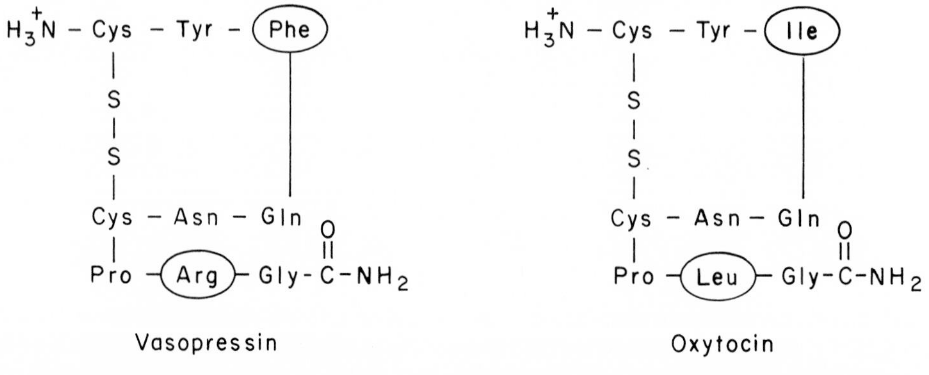

Fig. 13-8. Primary structures of vasopressin and oxytocin. Amino acid residues that differ in the two hormones are circled.

Vasopressin

Vasopressin, also known as the *antidiuretic hormone* (ADH), acts on the distal convoluted tubules and collecting ducts of the kidney, permitting passage of water molecules across the tubular wall. Hence it causes water to be reabsorbed from the urinary filtrate. Its action prevents excessive water loss in the urine and is thus antidiuretic. Vasopressin acts through the cAMP mechanism. Stimuli that cause vasopressin release include increased plasma osmolality, decrease of plasma volume, and a fall in blood pressure. The difference in action between vasopressin and the adrenocortical mineralocorticoids, which also expand the blood volume, is that vasopressin causes *water* reabsorption, whereas the mineralocorticoids cause *sodium* reabsorption. The latter, by increasing the plasma osmolality, leads to water retention in the blood. However, the water-retention effect of the mineralocorticoids is indirect, the result of their primary action on salt metabolism. The mineralocorticoids do not have a direct effect on renal water metabolism.

Neurophysins

Oxytocin and vasopressin are transported down the hypothalmic–neurohypophyseal nerve tract as physical complexes with proteins called neurophysins. The molecular weight of the neurophysins is approximately 1×10^4. Estrogen causes the release of the oxytocin-neurophysin, and nicotine or cigarette smoking causes the release of the vasopressin-neurophysin. The hormone and its neurophysin are synthesized as a single continuous polypeptide. The hormone then is split out by proteolytic cleavage and binds to the remaining protein component. Each molecule of neurophysin binds a single molecule of hormone and carries it down from the hypothalamus to the posterior pituitary, where the hormone is stored as a neurophysin complex. Both the hormone and its neurophysin are discharged during release from the neurohypophysis, but the hormone dissociates from the neurophysin and circulates in the blood plasma without being bound to any protein.

THYROID HORMONES

The thyroid hormones, thyroxine (T_4) and 3,5,3'-triiodothyronine (T_3), are synthesized in the thyroid gland from tyrosine residues present in *thyroglobulin*, a glycoprotein with a molecular weight of 6.6×10^5. The tyrosine residues of thyroglobulin are iodinated, forming either *3,5-diiodotyrosine* (DIT) or *3-monoiodotyrosine* (MIT).

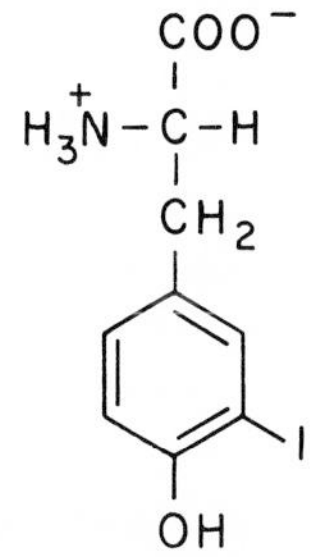

3-Monoiodotyrosine (MIT 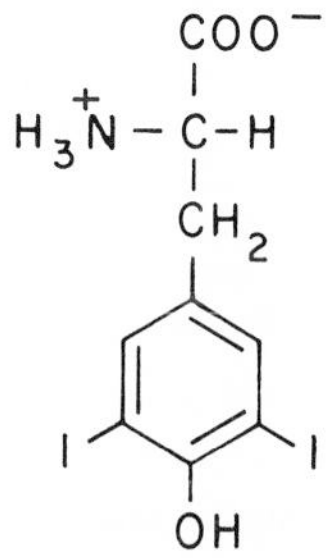)   3,5 – Diiodotyrosine (DIT)

This iodination step is blocked by the propylthiouracil group of *antithyroid drugs*. If two DIT residues combine, T_4 is formed, while a combination of DIT and MIT forms T_3. These coupling reactions occur within the thyroglobulin molecule.

Biosynthesis of thyroxine and triiodothyronine

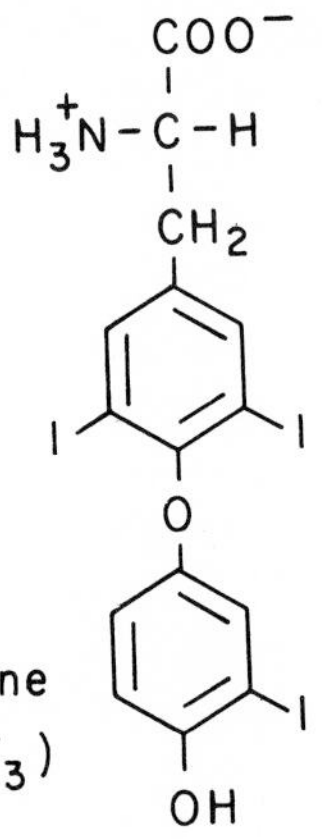

3,5,3'–Triiodothyronine
(T_3)

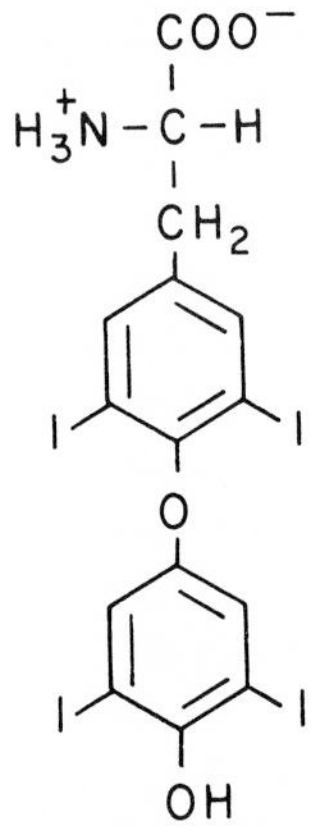

3,5,3',5'–Tetraiodothyronine
(Thyroxine, T_4)

TSH stimulates the biosynthesis and secretion of the thyroid hormones. In the first step, iodide ions are actively transported from the extracellular fluid into the glandular epithelial cells. The ability of the thyroid to concentrate iodide ions is the basis for the use of radioactive iodine (^{131}I) as a diagnostic test of thyroid function and in the treatment of hyperthyroidism and thyroid carcinoma. In the next biosynthetic step, iodide is oxidized to iodine through the action of a *peroxidase*. The tyrosine residues of thyroglobulin are then iodinated, forming MIT and DIT residues as a part of the protein. A coupling enzyme then catalyzes either the covalent linkage of an MIT residue with a DIT residue, forming an incipient T_3, or the linkage of two DIT residues, forming an incipient T_4. At this point the hormones remain as covalently linked portions of thyroglobulin, and the thyroglobulin molecule moves inside the thyroid follicle for storage. When stimulated by TSH, thyroglobulin moves from the follicle back into the cytoplasm of the glandular cell. Thyroglobulin is degraded by lysosomal proteases contained in the thyroid gland. T_4 and T_3 are released into the blood, whereas DIT and MIT are retained within the gland and deiodinated. The iodide ions that are released in this process remain inside the thyroid and can be reutilized for additional iodinations.

Peripheral triiodothyronine formation. Recent evidence suggests that under normal circumstances little of the T_3 present in the human body is actually synthesized in the thyroid. According to this view, most of the T_3 is made from T_4 at the peripheral sites of action of the hormone. T_3 is much more biologically active than T_4. In effect, T_4 actually may be a prohormone that requires conversion to T_3 at the target cell in order to exert an appreciable metabolic effect.

Plasma transport

More T_4 than T_3 is contained in thyroglobulin, and considerably more T_4 is released into the blood from the thyroid. The half-life of circulating T_4 is 6.7 days and that of T_3 is 1.4 days. T_4 and T_3 are transported in the plasma primarily by *thyroxine-binding globulin* (TBG). However, two other plasma proteins, prealbumin and albumin, can also act as carriers for these hormones. As with any other protein-bound compound, the concentrations that determine the chemical potential or biologic effectiveness of T_4 and T_3 are the unbound or free concentrations. The carrier proteins have a much lower affinity for T_3 than for T_4. Therefore, although there is ordinarily 5 to 10 μg/dl of T_4 and only 150 ng/dl of T_3 in the plasma, the normal free concentration of T_3 in the plasma ($6 \times 10^{-12} M$) is only five times less than that of T_4 ($3 \times 10^{-11} M$).

Metabolic effect

The thyroid hormones increase the metabolic rate and oxygen consumption of many tissues, but how this occurs is uncertain. Effects on oxidative phosphorylation and mitochondrial structure have been noted in certain biochemical studies. However, the hypermetabolic effect probably is not caused by uncoupling of oxidative phosphorylation or swelling of mitochondria, for thyroxine produces these changes only at concentrations that are above the physiologic range. At a given concentration, T_3 has a much greater hypermetabolic effect than T_4; the onset of its action is faster, and its duration of action is shorter. In addition to their effect as stimulators of mitochondrial respiration, the thyroid hormones activate the expression of certain genes and thereby induce protein synthesis. One of the enzymes induced by T_4 is the mitochondrial, FAD-linked glycerophosphate dehydrogenase. Since mitochondria contain their own DNA and protein-synthesizing machinery, it is thought that the thyroid hormones may promote transcription in mitochondria. Recently, binding sites for the thyroid hormones have been demonstrated in the cell nucleus, suggesting that these hormones probably also regulate nuclear transcription.

The thyroid hormones are removed from the circulation primarily by the liver and excreted in the bile, T_4 as the glucuronide and T_3 as the sulfate. There is little enterohepatic circulation of the thyroid hormones in man.

PARATHYROID HORMONE, CALCITONIN, AND VITAMIN D

Parathyroid hormone

The hormone synthesized and secreted by the parathyroid glands, parathyroid hormone, regulates calcium and phosphate metabolism as well as the release of calcium from bone. These minerals are vital for the maintenance of normal bone structure. Parathyroid hormone is a single polypeptide chain of eighty-four amino acid residues. It is synthesized as a prohormone, and the longer polypeptide chain then undergoes proteolytic cleavage to the hormonally active material. The hormone has two main sites of action—the renal tubule and bone. In the renal tubule it prevents phosphate reabsorption, thereby causing enhanced phosphate excretion in the urine. As the plasma phosphate concentration decreases, phosphate is drawn from bone in an attempt to maintain the plasma concentration. Because the phosphate in bone is in the form of hydroxyapatite, $3Ca_3(PO_4)_2 \cdot Ca(OH)_2$, calcium is released into the blood when hydroxyapatite is degraded. Parathyroid hormone also directly affects bone by increasing its release of calcium into the plasma. The net effect of parathyroid hormone action is a rise in the blood calcium concentration. Both in bone and in the proximal renal tubule, parathyroid hormone acts through a cAMP mechanism.

One of the main actions of parathyroid hormone is to stimulate the conversion of vitamin D to its hormonally active metabolite, 1,25-dihydroxycholecalciferol, in the kidney. It is likely that some actions of parathyroid hormone, such as the increase in calcium absorption from the intestine, are secondary to the production of 1,25-dihydroxycholecalciferol and actually are mediated by this vitamin D metabolite.

Calcitonin

Calcitonin is synthesized in specialized epithelial cells known as C cells contained in both the parathyroid and thyroid glands and has an effect opposite from that of parathyroid hormone. It contains thirty-two amino acid residues and, surprisingly, has several features in common with oxytocin and vasopressin.

These include the presence of an N-terminal ring structure made up of six amino acids joined together by a disulfide bridge and an amide group, prolinamide, at the C-terminal. Calcitonin lowers plasma calcium concentration by causing the deposition of calcium in bone and promoting calcium loss in the urine. Like parathyroid hormone, it also operates through a cAMP mechanism.

Vitamin D

Vitamin D also is involved in the regulation of calcium and phosphorus metabolism and the calcification of bone. It prevents *rickets*, a crippling bone deformity seen in children, and *osteomalacia* in the adult. Vitamin D has a steroid-like structure and can be synthesized in man from derivatives of cholesterol. Therefore, in a strict sense, vitamin D actually is not a true vitamin for man. Several compounds with vitamin D–like activity occur in nature, but only two are important to man. Vitamin D_3 *(cholecalciferol)* is synthesized in the skin from 7-*dehydrocholesterol* (Chapter 1), which is obtained in the diet from animal products. This reaction is catalyzed by ultraviolet light and is mediated by exposure to sunshine. Cholecalciferol is ingested as well if the diet includes certain animal products, particularly fish liver oils. The other form of the vitamin, D_2 *(calciferol)*, is ingested in foods derived from plants. It is similar in structure to cholecalciferol except that it has a double bond in the side chain at position C_{22}. Vitamin D_3 is the most prevalent form in man.

Vitamin D, together with the parathyroid hormone, increases the blood calcium concentration. The vitamin is partially activated by conversion to 25-*hydroxycholecalciferol* through a reaction requiring O_2 and NADPH that occurs in the endoplasmic reticulum of the liver. This metabolite is transported to the kidney, where it is converted in the mitochondria primarily to *1,25-dihydroxycholecalciferol*, the major metabolite. The synthesis of 1,25-dihydroxycholecalciferol in the kidney is stimulated by parathyroid hormone. The 1,25-dihydroxy derivative is transported to the intestinal mucosa, where it induces the synthesis of a specific mRNA, which in turn causes the synthesis of a *calcium-carrier protein*. This carrier facilitates the uptake of dietary calcium by the intestinal mucosa. When the plasma calcium concentration rises, the synthesis of 1,25-dihydroxycholecalciferol is inhibited. The metabolism and action of cholecalciferol are illustrated in Fig. 13-9. In general, the predominant physiologic effect of vitamin D is to facilitate bone calcification by increasing the absorption of dietary calcium, thus raising the blood calcium concentration. The main action of vitamin D, however, is to raise the plasma calcium concentration. If there is insufficient calcium available from the diet, vitamin D will have a direct effect on bone and cause calcium mobilization. This action, which also is mediated by 1,25-dihydroxycholecalciferol, requires the simultaneous presence of parathyroid hormone. The latter is released only when the plasma calcium concentration is low, ensuring that the vitamin D–induced mobilization from bone occurs only when calcium is needed in the plasma to prevent tetany.

Calcium balance and calcification

The secretion of parathyroid hormone and calcitonin is regulated by the plasma calcium concentration. The actual regulatory substance is the free (ionized) calcium that is in equilibrium with albumin-bound calcium. Low plasma ionized calcium concentrations stimulate parathyroid hormone release and inhibit calcitonin release; high ionized calcium concentrations stimulate calcitonin release and inhibit parathyroid hormone release. The influences of vitamin D, parathormone, and calcitonin on calcium balance are illustrated in Fig. 13-10.

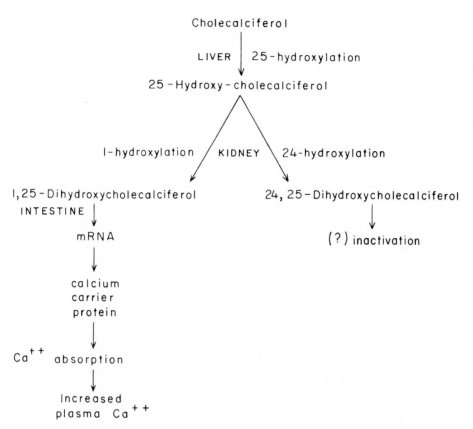

Fig. 13-9. Vitamin D metabolism and actions.

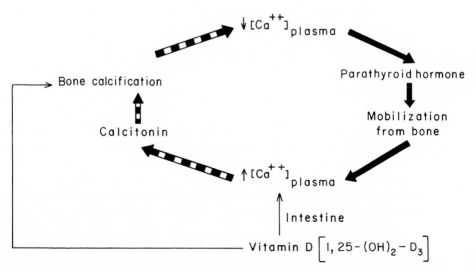

Fig. 13-10. Calcium balance. The active metabolite of vitamin D that carries out this function, 1,25-dihydroxycholecalciferol, is abbreviated as $1,25-(OH)_2-D_3$.

Recent work suggests that hydroxylation in the kidney of the vitamin D metabolites probably is the level at which calcium balance is regulated. When calcium is needed in the body, parathyroid hormone is released. This induces the 1-hydroxylase, which acts upon 25-cholecalciferol, causing 1,25-dihydroxycholecalciferol to be formed and plasma calcium and phosphate concentrations to increase. As this metabolite accumulates, it induces a 24-hydroxylase, which then diverts some of the 25-hydroxycholecalciferol into a second metabolite of vitamin D, *24,25-dihydroxycholecalciferol*. The 24-hydroxylated metabolites are funneled into the inactivation pathway. Therefore, the calcium-elevating action begins to terminate when vitamin D metabolites are diverted into the 24-hydroxylation pathway.

Through calcification the bones and teeth gain rigidity and strength. Calcium is also important for a number of enzymatic reactions, including many of those involved in blood clotting. Many of the anticoagulants that are commonly added to blood sampling tubes, for example, EDTA, citrate, and oxalate, prevent clotting by chelating calcium. Calcium also is required for muscular contraction (Chapter 5), and it must be present in the correct concentrations to permit proper transmission of nerve impulses at the neuromuscular junction. Hypocalcemia will produce tetany; hypercalcemia will lead to calcium phosphate precipitation in tissues, for example, calcium stones in the urinary tract.

PANCREATIC HORMONES

Two polypeptide hormones, insulin and glucagon, are the main products synthesized in the pancreatic islets. They have opposite effects on carbohydrate, lipid, and protein metabolism.

Insulin

Insulin is biosynthesized in the β-cells of the pancreas from the prohormone *proinsulin*. A small amount of proinsulin is released into the blood when insulin is secreted, but most of the proinsulin is converted to insulin prior to release from the pancreas through the action of a proteolytic enzyme. Proinsulin contains eighty-four amino acids in a single polypeptide chain. A segment of the chain containing thirty-three amino acids is removed, leaving two chains, A and B, having twenty-one and thirty amino acids, respectively. These are joined together by two disulfide bonds. The B chain is derived from the N-terminal end of the proinsulin, the A chain from the C-terminal end.

C-peptide.   The segment containing thirty-three amino acids is split out of proinsulin in the following manner. Amino acids 31 and 32, both arginine residues, are hydrolyzed and released. Likewise, residue 62 (lysine) and 63 (arginine) are released. The fragment that remains, containing twenty-nine amino acid residues (amino acids 33 to 61 of proinsulin), is called the connecting or C-peptide. It is released into the blood when insulin is secreted but has no insulin effect, either biologically or immunologically. The fact that the C-peptide possesses no immunologic insulin reactivity is important, for the amount of circulating insulin is measured by immunologic assay *(immunoreactive insulin)*. Proinsulin, which has very little biologic activity, does cross-react immunologically with insulin. Certain cases, for example, a pancreatic β-cell tumor that releases large quantities of proinsulin, can lead to serious error in insulin concentration measurement made by radioimmunoassay.

Insulin structure.   The synthesis of insulin from proinsulin is depicted diagrammatically in Fig. 13-11. The N-terminal end of proinsulin is connected to

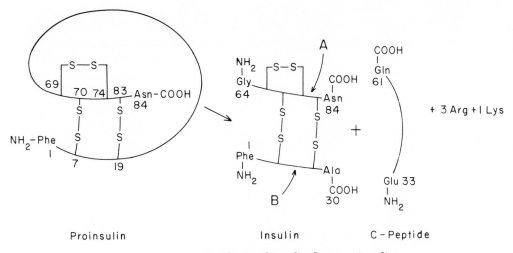

Fig. 13-11. Synthesis of insulin from proinsulin.

the C-terminal end by two disulfide bonds. One of these bonds links residue 7 to residue 70; the other links residues 19 and 83. An additional disulfide bond connects two parts of the C-terminal end, residues 69 and 74. The enzyme that converts proinsulin to insulin has a trypsin-like specificity, as it cleaves proinsulin at the arginyl and lysyl bonds. After the three arginine residues, one lysine residue, and C-peptide are split out, the N-terminal fragment of proinsulin, now a separate chain, remains joined to the original C-terminal end of proinsulin by the two disulfide bonds. The original N-terminal fragment, containing thirty amino acids, becomes the B chain of insulin. Its N-terminal residue, phenylalanine, is the original N-terminal residue of proinsulin. The remaining C-terminal fragment of proinsulin becomes the A chain of insulin. It contains twenty-one amino acid residues, and its C-terminal residue, asparagine, is the original C-terminal residue of proinsulin. The A chain retains the disulfide bond linking residues 69 and 74 of proinsulin, now residues 6 and 11 of the A chain. The disulfide bond linking these residues in the A chain is known as the intrachain disulfide bond of insulin. The B chain retains the numbering system of proinsulin, its residues being numbered from 1 to 30. However, the numbering of the A chain is changed. Residue 64 of proinsulin, glycine, becomes the N-terminal end of the A chain of insulin and is numbered 1. The numbering then continues through residue 21, which was the original C-terminal asparagine residue 84 of proinsulin.

Proinsulin is synthesized in the ribosomes of the rough endoplasmic reticulum. As the peptide chain is formed, it passes into the intracisternal space, which then narrows to form the Golgi apparatus. At this point most of the proinsulin is converted to insulin. The Golgi vesicles then combine to form the β-granules of the pancreatic islets, which are the storage sites for insulin. When it receives the appropriate stimulus, the granule merges with the plasma membrane of the β-cell, thus enabling the insulin contained inside the granule to be released into the extracellular space.

Insulin action.   The main effects of insulin are to decrease the plasma glucose and free fatty acid concentrations. Insulin also has a stimulatory effect on protein synthesis. High circulating levels of glucose or amino acids stimulate the release of insulin from the pancreas. The circulating insulin facilitates the diffusion of glucose into muscle cells and adipocytes, enhancing glucose utilization in these tissues. There appears to be a membrane barrier to glucose uptake in muscle and adipose tissue that insulin helps to overcome. Hepatocytes, on the other hand, are freely permeable to glucose, and there is no need for enhancement of glucose entry into the liver. However, insulin does have a profound influence on hepatic glucose metabolism, affecting it after diffusion of glucose across the cell membrane has taken place by inducing the synthesis of a specific *glucokinase*. This enzyme has a much higher K_m for glucose than the nonspecific hepatic hexokinase and therefore functions primarily when the circulating glucose concentration is elevated. Insulin also enhances glycogen synthesis and reduces the activity of the gluconeogenic enzymes, particularly glucose 6-phosphatase. The net effect of insulin is to stop glucose output by the liver and to stimulate glycogen and protein synthesis.

The molecular mechanism of insulin action is not completely understood at this time. However, it appears to modulate adenyl cyclase activity in some way, ultimately lowering the cellular cAMP content or effectiveness of action. Some recent evidence also suggests that insulin may raise the intracellular cGMP content and that this may explain, in part, the anti-cAMP effects that insulin exerts in many tissues.

Insulin as a pharmacologic agent. Diabetes mellitus is an extremely common disease that results in part from insufficient insulin in the blood. Many diabetics are treated with daily injections of bovine insulin. Insulin acts quickly when injected into the human, but its effect lasts for only 4 to 6 hr. Thus a diabetic treated in this manner might require several injections each day. In order to avoid this, longer acting forms of insulin are prepared by complexing it with substances such as protamine. Through the use of such preparations, the diabetic condition of most insulin-requiring patients can be controlled adequately with only one or two daily injections. Although the insulins used are animal proteins and thus foreign materials, they are so closely related in structure to human insulin that an immunologic response seldom occurs.

Glucagon

When insulin was first crystallized, the preparations were found to contain a hyperglycemic factor. Subsequently, this was identified as the polypeptide hormone glucagon, which contains twenty-nine amino acid residues and is secreted by the α_2-cells of the pancreatic islets. Unlike insulin, it consists of only a single polypeptide chain and contains no disulfide bonds. Glucagon has effects opposite to those of insulin, as it raises the blood glucose and free fatty acid concentrations. It acts directly on these processes, stimulating glycogenolysis and lipolysis through a cAMP mechanism. It also stimulates gluconeogenesis in the liver.

As will be described in the section on enteric hormones, glucagon also is secreted by α-cells contained in the stomach and duodenum. It is called "gut glucagon," and one of its probable functions is to trigger insulin release from the pancreas.

Recent studies suggest that glucagon may have a much greater role in diabetes mellitus than was previously suspected. It is likely that the fasting hyperglycemia that occurs in the more serious cases of diabetes is a result of excessive glucagon secretion or an excess of glucagon relative to the amount of circulating insulin, not insulin deficiency of itself. Lack of sufficient insulin appears to be responsible primarily for the hyperglycemia that occurs after meals. If glucagon release is inhibited by the administration of somatostatin, fasting hyperglycemia does not occur even if insulin levels are very low. Therefore, the persistent hyperglycemia that occurs in diabetes probably is a glucagon effect. Excessive ketone body production in uncontrolled diabetes also appears to be mediated by glucagon, both through mobilization of free fatty acids from the adipose tissue and direct stimulation of ketone body production in the liver.

ADRENAL HORMONES

The adrenal gland is composed of two parts, the medulla and the cortex. Epinephrine and norepinephrine are synthesized in the medulla, and three types of steroid hormones, the glucocorticoids, mineralocorticoids, and androgens, are produced by the cortex.

Epinephrine and norepinephrine

Epinephrine and norepinephrine, catecholamines, are synthesized from tyrosine in the adrenal medulla. Norepinephrine also is released at the sympathetic nerve endings. Epinephrine stimulates glycogenolysis and lipolysis, increasing the blood glucose and plasma free fatty acid concentrations. Epinephrine also increases cardiac output, systolic arterial pressure, splanchnic and skeletal muscle blood flow, and oxygen consumption. The actions of epinephrine are mediated through the cAMP mechanism. Norepinephrine, which also acts through a cAMP mechanism, increases the stroke volume of the heart, blood pressure, and the peripheral resistance of the vascular system. Catecholamines are released from the adrenal medulla, the ratio of epinephrine to norepinephrine being about $4:1$. By contrast, the sympathetic nerve endings release norepinephrine almost exclusively. The half-life of circulating catecholamines is only 10 to 30 sec, and the termination of their immediate action is caused primarily by uptake into tissues, including adrenergic nerve terminals, not enzymatic degradation.

The metabolic pathway for catecholamine synthesis, which involves L-*dihydroxyphenylalanine* (L-dopa) and *dihydroxyphenylethylamine* (dopamine), was illustrated in Chapter 9. Catecholamines are stored in intracellular vesicles called *chromaffin granules*. ATP and binding proteins, the best known of which is *chromogranin A*, are also present in these granules. All of these materials are released when catecholamines are discharged. Small amounts of the enzyme *dopamine-β-hydroxylase* are also released from the chromaffin granules when catecholamines are secreted. This enzyme is involved in catecholamine synthesis, catalyzing the conversion of dopamine to norepinephrine. Circulating levels of dopamine-β-hydroxylase are measured as a sensitive indicator of adrenal medullary and sympathetic nervous system activity.

Adrenergic receptors. There are two types of binding sites for the catecholamines on various target cell membranes, the α- and β-adrenergic receptors. As noted earlier, the β-adrenergic receptor that has been isolated in soluble form is either a protein or a protein aggregate.

The two types of receptors elicit different actions from the catecholamines that bind to them. Moreover, specific drugs are available that block both types of receptors and thus prevent the particular catecholamine responses mediated by them. The most commonly used of these drugs is *propranolol*, a β-adrenergic receptor-blocking drug. The adipose tissue contains β-receptors, and propranolol blocks the epinephrine-induced stimulation of lipolysis. Propranolol also inhibits the epinephrine-stimulated increase in blood glucose and decreases the heart rate and cardiac output. *Phenoxybenzamine* is employed clinically to block the α-adrenergic receptors. It produces vasodilation and decreases total peripheral vascular resistance, causing postural hypotension. As is evident from their structures, neither of the adrenergic blocking drugs bears much resemblance to the catecholamine hormones that normally interact with the receptors.

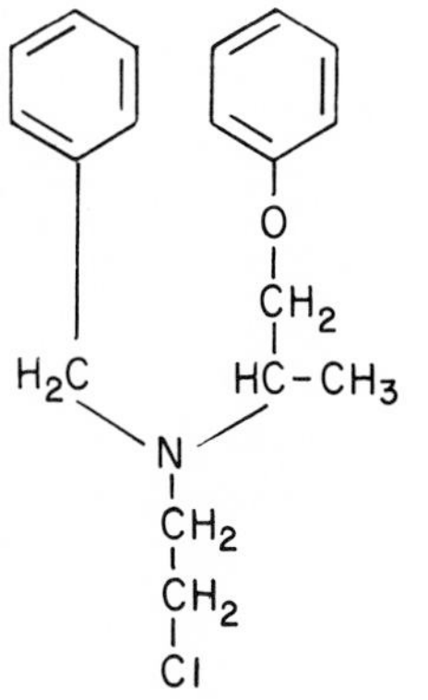

Phenoxybenzamine 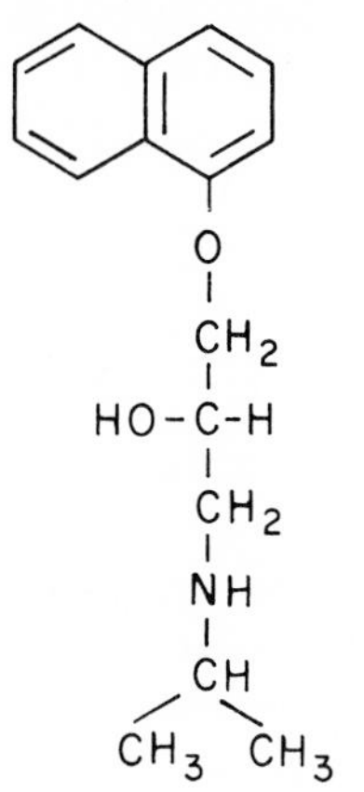   Propranolol 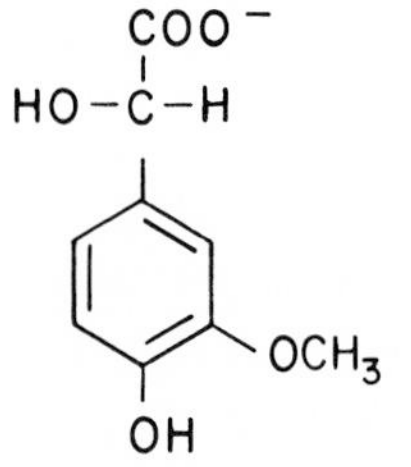   4-Hydroxy-3-methoxy-
                                                                    L-mandelic acid

Catecholamine catabolism and excretion. Two enzymes are involved in the degradation and excretion of epinephrine and norepinephrine. These are *mono-amine oxidase,* a mitochondrial enzyme that oxidizes a primary or substituted amino group to an aldehyde group, and *catechol-O-methyltransferase,* a cyto-plasmic enzyme that methylates a ring hydroxyl group. S-Adenosylmethionine (SAM) is utilized by this transferase as the methylating agent. These enzymes convert the catecholamines to several metabolites including *4-hydroxy-3-methoxymandelic acid* (VMA), the major urinary excretory product. Either urinary VMA or total urinary catecholamines are measured clinically to assess catecholamine metabolism in man. A single 24-hr urine specimen is usually sufficient for most diagnostic procedures. A false elevation of catecholamines will be observed if the subject has eaten foods rich in catecholamine metabolites, for example, bananas, or has taken drugs such as the antihypertensive agent α-methyldopa, which is a catecholamine analogue that is catabolized to similar degradation products. A normal human excretes from 2.2 to 8.5 mg of catechol-amine metabolites in the urine during a 24-hr period. VMA accounts for at least 90% of these metabolites. Therefore, measurement of VMA excretion alone is adequate to assess catecholamine metabolism in most clinical situations.

Corticosteroid hormones

As mentioned previously, three types of steroid hormones are produced by the adrenal cortex: the mineralocorticoids, which regulate electrolyte (Na^+, K^+) metabolism; the glucocorticoids, which act on protein and carbohydrate metabo-lism; and the androgenic steroids other than testosterone, which have masculiniz-ing effects. Estrogenic sex steroids are produced by the adrenal cortex only under pathologic conditions. The steroid hormones are synthesized from acetyl CoA by way of cholesterol and pregnenolone. The chemical formulae and metabolic pathways of these hormones can be found in Chapter 10.

Mineralocorticoids

Aldosterone, the principal mineralocorticoid in man, is synthesized in the zona glomerulosa of the adrenal cortex. It acts on the kidney tubule, primarily causing sodium to be reabsorbed and, to a lesser extent, enhancing the excretion of potassium and hydrogen ions. No more than 1% to 2% of the total Na^+ reabsorption occurs in exchange for K^+ and H^+, the remaining 98% being reabsorbed with Cl^- as the counterion. Reabsorption of water accompanies Na^+ reabsorption and prevents an undue increase in plasma osmotic pressure. However, the retention of water is not a direct action of aldosterone; it is merely a result of the electrolyte effect. The net result is an increase in plasma volume.

Aldosterone functions through a transcriptional mechanism: it binds to a cytoplasmic protein in the renal tubular cell and enters the nucleus as a protein complex, thus inducing DNA transcription. The resulting mRNA possesses the code for the synthesis of a protein that facilitates Na^+ reabsorption. *Deoxycorticosterone*, the major mineralocorticoid in many animals, has a lesser role in electrolyte metabolism in man.

Angiotensin II, a pressor substance that acts on the arterioles to directly increase blood pressure, also stimulates aldosterone secretion from the adrenal cortex. The increased plasma volume that results from the action of aldosterone further raises the blood pressure, which is synergistic with the direct pressor effect of angiotensin II.

Glucocorticoids

Glucocorticoids, characterized by the presence of a hydroxyl group attached to position C_{11} of the steroid nucleus, are synthesized in the zona fasciculata of the adrenal cortex. Cortisol (hydrocortisone) is the major glucocorticoid in man, but *corticosterone* also has glucocorticoid action. ACTH stimulates the synthesis of the glucocorticoids from cholesterol as well as the release of glucocorticoids from the adrenal cortex. In turn the glucocorticoids inhibit ACTH production. ACTH secretion is stimulated by *corticotropin-releasing hormone* (CRH) from the hypothalamus, whereas cortisol inhibits ACTH release. However, release of

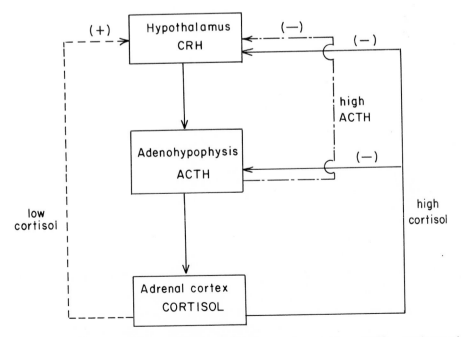

Fig. 13-12. Factors regulating cortisol production and secretion. CRH = corticotropin-releasing hormone; ACTH = adrenocorticotropin.

CRH from the hypothalamus also is inhibited by ACTH. This inhibition of the hypothalamic releasing hormone by the adenohypophyseal trophic hormone is known as short-loop feedback control. These regulatory mechanisms are illustrated schematically in Fig. 13-12.

The biosynthesis of the glucocorticoids as well as their catabolism and excretion were discussed in Chapter 10. These hormones are transported in the plasma in association with a specific α-globulin, *transcortin*. The glucocorticoids have effects on protein, carbohydrate, and lipid metabolism. These effects include (1) increased protein catabolism, (2) loss of calcium from bone resulting from the loss of osteoid matrix (osteoporosis), (3) an anti-immune effect with an associated decrease in circulating lymphocytes and eosinophilic leukocytes, (4) increased gluconeogenesis, and (5) potentiation of lipolysis.

Although the molecular action of cortisol is not known precisely, it appears that the hormone regulates enzyme synthesis by controlling the transcription of genetic information. Much evidence to this effect comes from observations concerning inducible liver enzymes such as *tyrosine-α-ketoglutarate transaminase* and *tryptophan pyrrolase*. For example, administration of *cortisone*, the 11-keto derivative of cortisol, to rats causes a large increase in tryptophan pyrrolase activity in the liver. This induction is blocked by actinomycin D, which inhibits transcription, and by puromycin, which inhibits translation. Thus the cortisone-induced increase in enzymatic activity is caused by enzyme synthesis. In the case of tryptophan pyrrolase induction, cortisone appears to function as a specific inducer. It combines with the repressor, thereby releasing the structural gene for transcription. The mRNA that is synthesized as a result of the induction then directs the synthesis of the enzyme. The mechanism of induction actually may be somewhat more complicated, for cortisone also appears to have a major post-transcriptional action in this process.

Glucocorticoids are used extensively in therapeutics. Synthetic analogues are usually administered rather than cortisol, as they can be given in smaller dosage and often are more specific in their actions.

Adrenal androgens

In addition to the main adrenal androgen *dehydroepiandrosterone, androstenedione* is also secreted from the adrenal cortex. These androgens are considerably less potent than testosterone, the androgen secreted by the testes in the male.

SEX HORMONES

Testosterone, the major male sex hormone, is synthesized in the interstitial cells of the testes, known as the Leydig cells. The female sex hormones are the estrogens and progesterone. These steroids are synthesized in the ovary, estrogen by the follicle and progesterone by the corpus luteum.

Testosterone

The synthesis of testosterone is regulated by LH, which acts on the Leydig cells through a cAMP mechanism. An elevated plasma testosterone concentration, however, inhibits the release of both LH and FSH from the adenohypophysis. Testosterone is transported in the blood by a globulin, the *sex hormone–binding protein*, that also binds estrogens. Testosterone is reduced enzymatically to 5α-*dihydrotestosterone* in its target cells, and this reduced form appears to be the hormonally active material.

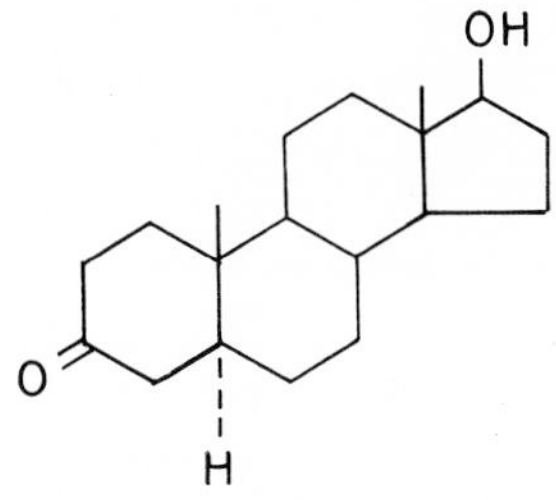

5α - Dihydrotestosterone

The active hormone binds to specific receptor proteins within the target cell. This leads to stimulation of protein synthesis by a transcriptional mechanism, producing a generalized anabolic action. Testosterone produces the male sexual characteristics. It is inactivated in the liver, being converted to *androsterone* and *etiocholanolone,* which are excreted in the urine as D-glucuronic acid derivatives. These are measured in the clinical laboratory as 17-*ketosteroids.* In the male, about 30% of the total 17-ketosteroids are derived from androgens produced by the testes, the remainder coming from androgens of adrenal cortical origin.

Estrogens

The ovary produces estrogens that induce the secondary sex characteristics of the female. In the human, *estradiol-17-β* is the major estrogen, with *estrone* and *estriol* having secondary roles. The estrogens are synthesized from cholesterol in the ovarian follicle (Chapter 10). Growth of the follicle is stimulated by FSH, but this adenohypophyseal hormone does not actually cause estrogen synthesis or release. As the follicle increases in size, it becomes capable of producing increased quantities of estrogen without any additional hormonal stimulation. Because estrogens are synthesized by the follicle, they are present in large quantities during the first half of the menstrual cycle, prior to ovulation and corpus luteum formation. Considerable amounts of estrogen also are present in the second (luteal) phase of the menstrual cycle. Estradiol is transported through the plasma by the sex hormone–binding protein, the same globulin that binds testosterone. Estrogens bind to cytoplasmic receptor proteins in the target cells and are transported as a protein complex into the cell nucleus. The complex binds to acidic proteins of the chromatin, thus freeing certain segments of DNA for transcription. The physiologic changes induced by the estrogens include maturation of the vagina, proliferation of the uterine endometrium, deposition of subcutaneous fat in the breasts and buttocks, and a peripheral vasodilatation that causes a slight lowering of body temperature. In addition, estrogens somehow promote thrombus formation and, in high dosage, lower plasma cholesterol concentration. Estriol is the main metabolite of the estrogens and is excreted in the urine.

Progesterone

Progesterone is synthesized from cholesterol by the corpus luteum in the ovary. LH from the adenohypophysis stimulates corpus luteum formation and progesterone production. The latter is mediated by a cAMP mechanism that activates the 20α-hydroxylation of cholesterol, the initial reaction in progesterone synthesis. Progesterone is the predominant female sex hormone during the second phase of the menstrual cycle (days 15 to 27), from ovulation until just prior to the onset of menstruation. The hormone is transported in the blood by both albumin and *transcortin*, the cortisol-binding globulin. Progesterone induces the synthesis of specific proteins in its target tissues, acting through a transcriptional mechanism similar to that described for the estrogens. In the chick oviduct progesterone induces the synthesis of a specific protein contained in the white of the egg, *avidin*. Binding proteins that are specific receptors for progesterone are present in the cytoplasm of its target cells. In the human, most of the actions of progesterone require the presence of estrogen, which acts either before or simultaneously with progesterone. Progesterone induces secretory changes in the uterine endometrium as well as the formation of terminal portions of the ductal system in the mammary glands. In addition, it decreases peripheral blood flow, causing body temperature to rise. A simple test to determine when and if ovulation occurs entails measurement of the increase in basal body temperature resulting from progesterone secretion. This test is commonly used in the diagnosis of different types of infertility.

The effects of the female sex hormones on the hypothalamus and adenohypophysis are complex. Both progesterone and the estrogens have some feedback inhibitory action on LH and FSH release by the adenohypophysis. However, they also have a stimulatory effect on the hypothalamus, causing LH to be released prior to ovulation.

Progesterone is inactivated and conjugated with D-glucuronic acid in the liver and excreted in the urine, primarily as *pregnanediol glucuronide*.

Because it is taken up by the liver and rapidly inactivated, progesterone cannot be administered orally. Yet the main ingredient for oral contraceptives is a progestational agent that prevents ovulation. The major breakthrough in this area was the development of synthetic progestational agents that are effective when given orally. These include derivatives of *17α-alkyl-19-nortestosterone* and *17α-hydroxyprogesterone*. To increase their effectiveness, small amounts of estrogen are usually combined with the progestational agent in some oral contraceptive preparations.

Hormonal control of the menstrual cycle

An average normal menstrual cycle in the human female requires 28 days, with menstruation occurring on days 1 to 5 and ovulation occurring on day 14. Estrogen levels remain at a baseline level of about 150 pg/ml blood plasma until day 8. They then rise sharply to a peak of about 500 pg/ml on day 13 and then rapidly decrease to baseline levels by day 15. During this initial phase of the cycle, estrogen is produced by the ovarian follicle. LH and FSH concentrations in the blood plasma also peak sharply on day 14. The progesterone concentration begins to increase after ovulation and remains elevated at a level of about 20 ng/ml until day 20, thereafter decreasing slowly to baseline values of about 2 ng/ml by day 27. Progesterone is produced by the corpus luteum. Estrogen also is produced by the corpus luteum, and an elevation in plasma estrogen levels to about 300 pg/ml occurs during the postovulatory or luteal phase of the cycle. This corresponds temporally to the progesterone elevation during the luteal phase.

Placental hormones

The placenta is the interface between the mother and fetus. One of its major functions is to produce hormones that protect the fetus and mediate many of the metabolic and structural changes that are required in pregnancy. The hormones that it produces include *human placental lactogen*, which is similar to GH; *chori-*

onic gonadotropin, which is similar to FSH and LH; estrogens; progesterone; and a TSH-like hormone.

Human chorionic gonadotropin is a glycoprotein that has a structure similar to those of TSH, LH, and FSH. It is composed of an α- and a β-subunit, the former being essentially identical in structure to its counterparts in TSH, LH, and FSH. As with these other protein hormones, it is the β-subunit that confers specificity to chorionic gonadotropin. The chorionic gonadotropin is present in the maternal plasma within 1 week after the fertilized ovum implants in the uterine wall, and it is the hormone that is measured in the ordinary test for pregnancy. Its function is to stimulate the maternal ovary to produce progesterone. The placental lactogen raises the blood glucose concentration and is responsible for causing exacerbations of diabetes that can occur during pregnancy. This protein hormone also has a prolactin-like action on the mammary gland. The TSH-like hormone has not been isolated, and the activity that is observed actually may be an effect of the chorionic gonadotropin. Progesterone is produced in large quantities by the placenta after the first 2 months of the pregnancy. The cholesterol needed for progesterone synthesis is derived from the maternal plasma. Some of the progesterone produced is taken up by the fetal adrenal and serves as a precursor for cortisol and dehydroepiandrosterone synthesis. The latter intermediate travels to the liver of the fetus where a 16α-hydroxyl group is introduced. It then passes back to the placenta where it is converted to the estrogen estriol. Therefore, the placenta requires intervention of the fetal adrenal and liver in order to produce estrogen.

ENTERIC HORMONES

At least five, and possibly as many as seven, hormones are biosynthesized and secreted by the gastrointestinal tract. These hormones function to coordinate and regulate the digestive processes. They are listed in Table 13-3. All of the enteric hormones are polypeptides.

Gastrin

Gastrin is synthesized by the mucosal cells in the gastric antrum, the distal nonacid-secreting portion of the stomach and in the upper part of the small intestine. It stimulates the parietal cells in the fundus and body of the stomach to produce acid. Gastrin is released by neural reflexes that are triggered by distention of the stomach. Its secretion is inhibited when the acid that is released

Table 13-3. Gastrointestinal hormones

HORMONE	SITE OF PRODUCTION	MAIN SITE OF ACTION	MAIN FUNCTION
Known hormones			
Gastrin	Stomach	Stomach	Acid secretion
Secretin	Duodenum	Pancreas	HCO_3^-, fluid release
Pancreozymin	Duodenum	Pancreas	Secretion of enzymes
Cholecystokinin	Duodenum	Gallbladder	Emptying
Glucagon	Duodenum	Pancreas	Insulin release
"Candidate" hormones			
Gastric inhibitory peptide	Duodenum	Stomach	Inhibits acid secretion
Vasoactive intestinal peptide	Small intestine	Small intestine	Secretion of intestinal juices

comes into contact with the gastric antral mucosa. Gastrin has effects other than the stimulation of gastric acid secretion. It also stimulates contraction of the gastroesophageal sphincter, pepsin production, gastric antral motility, pancreatic enzyme secretion, and bile flow from the liver. In addition, it inhibits absorption of water and electrolytes from the intestine and relaxes the ileocecal sphincter. Several forms of gastrin are found in the blood, including a sulfated form and a "big" form, which may be the prohormone. The main form of the hormone, *gastrin I*, contains seventeen amino acids. Certain tumors of the duodenum and pancreas secrete large quantities of gastrin, producing the *Zollinger-Ellison syndrome*, which is associated with severe peptic ulcers.

Secretin

Secretin is synthesized in the duodenal mucosa and released when hydrogen ions enter the small intestine. It acts primarily on the exocrine pancreas, causing the release of a bicarbonate-rich fluid. Secretin contains twenty-seven amino acids, fourteen of which are identical in primary structure to a sequence present in glucagon.

Cholecystokinin

Cholecystokinin is synthesized in the mucosa of the upper small intestine and released in the presence of amino acids or fatty acids. It contains thirty-three amino acids, the C-terminal pentapeptide being identical to that of gastrin. The hormone is primarily responsible for the contraction and emptying of the gallbladder. Many of the actions of cholecystokinin and gastrin are similar, but the relative sensitivity of various organs to each of the hormones varies. For example, gastrin strongly stimulates gastric acid secretion, whereas cholecystokinin has a very weak effect on it.

Pancreozymin

Pancreozymin, another duodenal hormone, has effects that are very similar to those of cholecystokinin, except that it acts primarily on the exocrine pancreas to stimulate secretion of the pancreatic digestive enzymes. There is some evidence to suggest that the structures of pancreozymin and cholecystokinin are almost identical, and in fact they may actually be the same substance.

Gut glucagon

The duodenum also produces gut glucagon, a hormone similar or identical to pancreatic glucagon. Two forms of gut glucagon occur, one with a molecular weight similar to that of pancreatic glucagon and another that is about twice as large. Both forms of gut glucagon differ immunologically from pancreatic glucagon. Gut glucagon is released in the presence of high glucose concentrations in the intestine and triggers insulin release from the pancreas. Excessive gut glucagon secretion occurs in patients who develop the dumping syndrome subsequent to subtotal gastrectomy. This may account for the hypoglycemic episodes that many of these patients suffer after eating.

Candidate hormones

Two additional hormones may be secreted by the small intestinal mucosa. They are the *gastric inhibitory peptide*, which inhibits acid secretion by the stomach, and the *vasoactive intestinal peptide*, which causes secretion of intestinal juices by the small intestine. Because the existence and exact role of these hormones is uncertain at this time, they are called candidate hormones.

HORMONE-RELATED SECRETIONS

Two substances, melatonin and the prostaglandins, are hormone-like substances. Melatonin is synthesized by the pineal gland, and it is secreted under physiologic conditions only into the cerebrospinal fluid. The prostaglandins, which actually are a group of structurally related fatty acid derivatives, function as intracellular regulatory substances.

Melatonin

Melatonin is the principal biologically active compound released by the pineal. It is an *indole* derivative that is synthesized from tryptophan by the pathway illustrated in Fig. 13-13. Other potentially important pineal products include

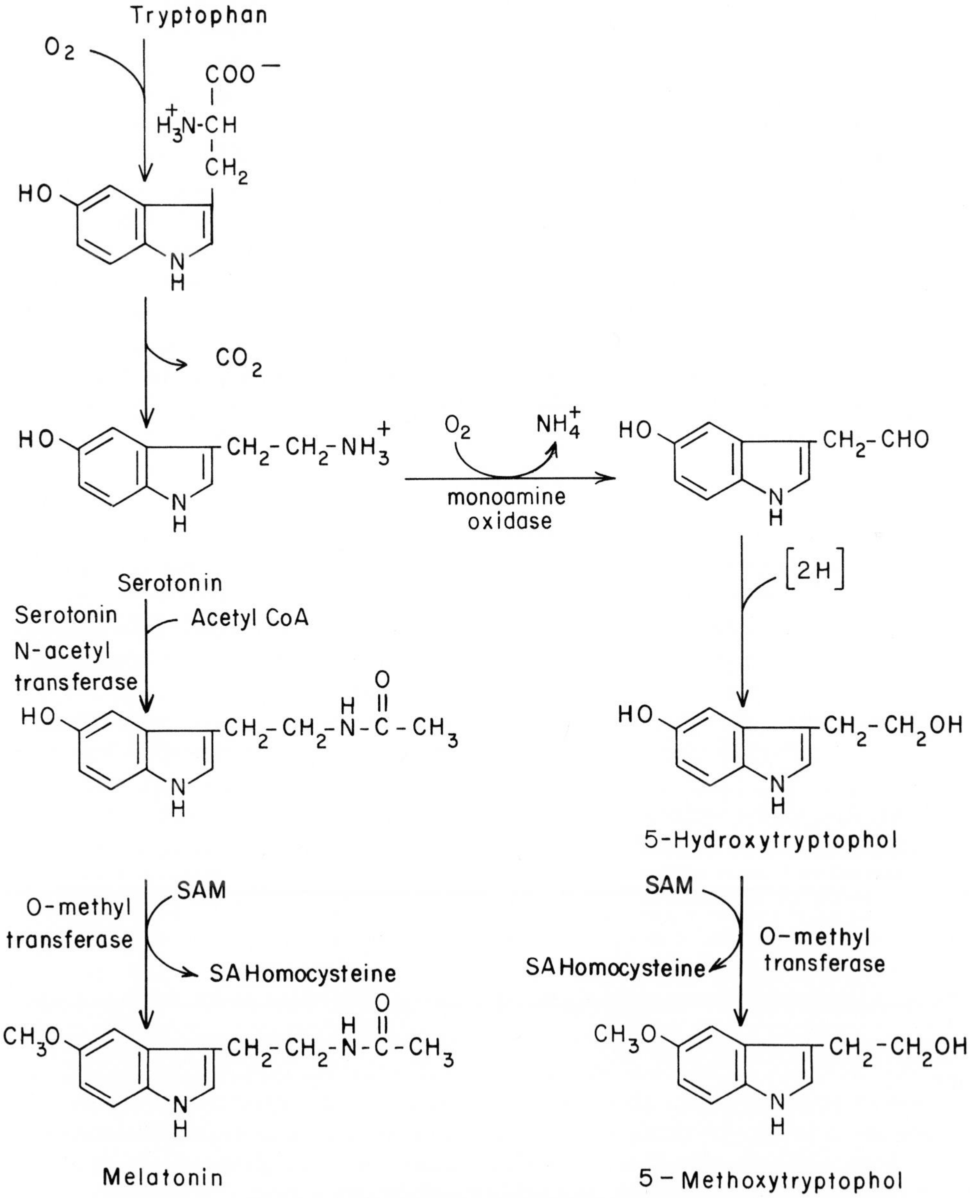

Fig. 13-13. Synthesis of pineal active metabolites.

5-*hydroxytryptophol* and 5-*methoxytryptophol*. Serotonin (5-hydroxytryptamine) and the enzymes *monoamine oxidase* and *hydroxyindole-O-methyltransferase* are involved in the pineal synthetic pathways.

The metabolic activity of the pineal is sensitive to light and darkness and undergoes cyclic variations each day. In the dark, norepinephrine is released from sympathetic nerve endings in the pineal. This stimulates adenyl cyclase activity, causing an increase in the cAMP content of the gland. Unlike the usual mechanism of cAMP action in mammalian cells, however, the cyclic nucleotide stimulates the synthesis of a new protein in the pineal, the enzyme *serotonin N-acetyl transferase*. It should be recalled that the cAMP mechanism functions in other cases by regulating the activity of an existing enzyme, although in pro-karyotic cells it does activate protein synthesis (pp. 584 and 620). The increase in serotonin N-acetyl transferase allows the conversion of serotonin to N-acetyl-serotonin and ultimately to melatonin. When light strikes the retina, the impulses are transmitted to the pineal, inhibiting the norepinephrine release. This leads to a rapid decline in serotonin N-acetyl transferase activity, and serotonin becomes the principal product of the pineal. Therefore, melatonin is secreted by the gland during periods of darkness and serotonin during exposure to light. The enzyme serotonin N-acetyl transferase also undergoes a light-dark cycle, exhibiting high activity in the pineal during darkness and essentially no activity during periods of light. Although the action of melatonin in the human is not known, animal studies suggest that it may exert an inhibitory effect on the gonads.

In addition to the light-dark cycle, there is an inherent daily variation in serotonin production by the pineal known as a *circadian rhythm*. This serotonin rhythm is controlled by impulses originating in the superchiasmatic nucleus of the hypothalamus and is a biologic clock mechanism that is unrelated to light or darkness.

Prostaglandins

Prostaglandins (PG) are twenty-carbon atom acids that have hormone-like properties. Their exact physiologic role has not been established, but they are potent smooth muscle relaxants and are involved in many important bodily functions. Prostaglandins are synthesized from polyenoic fatty acids of the linoleic and linolenic acid families. The need for prostaglandins may be the prime reason that certain fatty acids are essential to the human diet.

Prostaglandins were discovered in the 1930's when it was noted that extracts prepared from the vesicular glands of sheep or from human semen contained substances that caused contraction of intestinal or uterine strips. The extracts from the sheep seminal glands also produced a fall in systemic arterial blood pressure. The active substances were characterized as acidic lipids and named prostaglandins. It was not until the early 1960's that these compounds were identified as cyclic unsaturated fatty acids. Initially they were divided into two classes, E and F, but a more detailed classification system became necessary when the complexity of their structure was revealed.

Nomenclature. The prostaglandins are now known to be a large family of compounds that are derived from prostanoic acid, a twenty-carbon atom fatty acid that contains an internal five-carbon atom saturated ring. See prostanoic acid at top of opposite page. The seven-carbon atom chain attached to the ring at C_8 projects below the plane of the ring, as denoted by the dashed line connecting it to C_8. The eight-carbon atom chain attached at C_{12} projects above the plane of the ring, as denoted by the solid line connecting it to C_{12}.

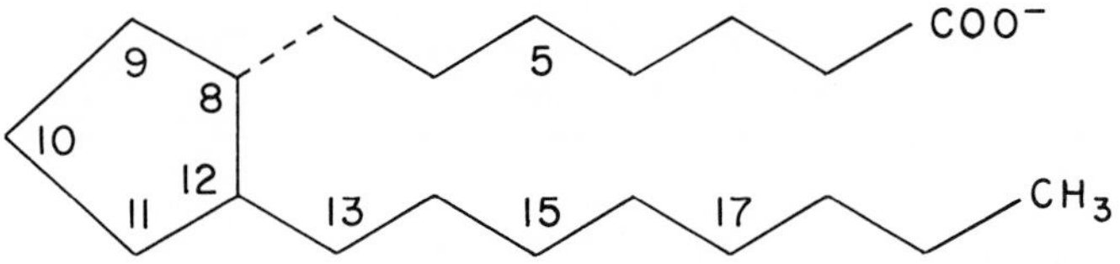

Prostanoic acid

Prostaglandins are designated by a capital letter, a number, and in some cases a Greek letter, for example, PGE_1 and $PGF_{2\alpha}$. The capital letter refers to the type of ring substitutions that are present in an individual prostaglandin molecule. Five types of rings are found in the naturally occurring prostaglandins, giving rise to prostaglandins of the A, B, E, F, and G or H series.

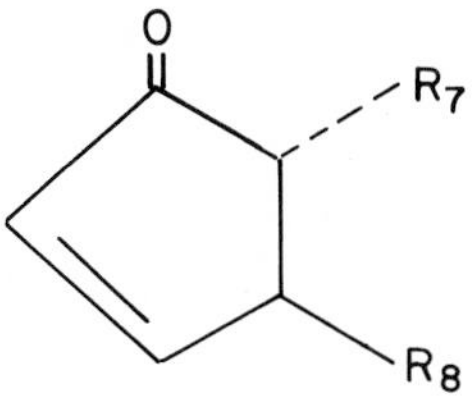

PGA 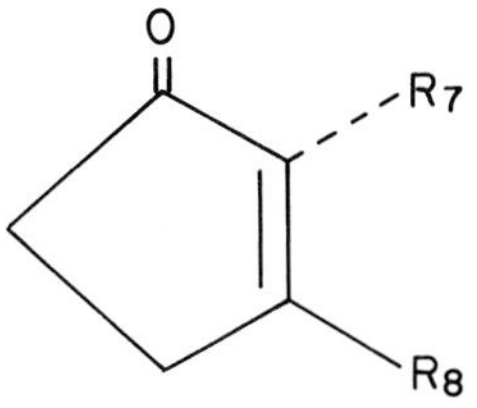   PGB 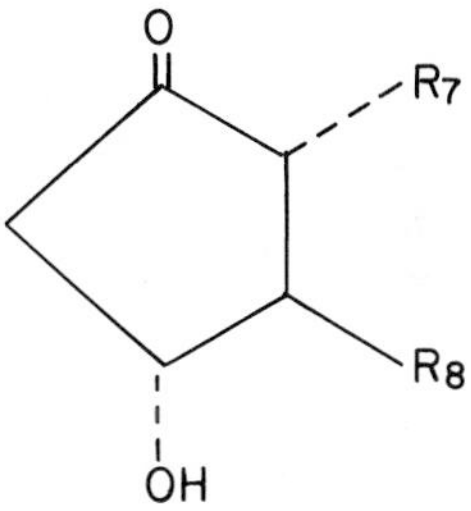   PGE

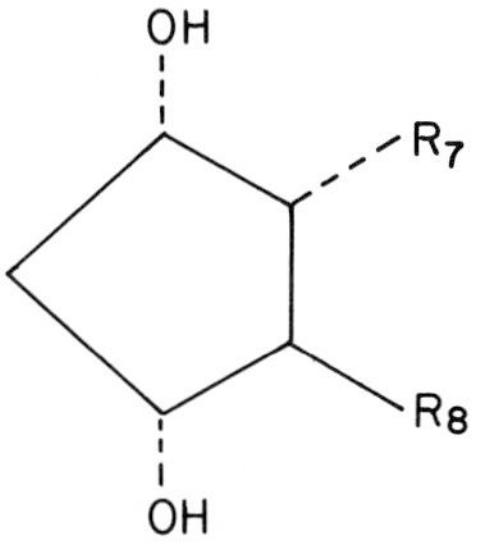

PGF 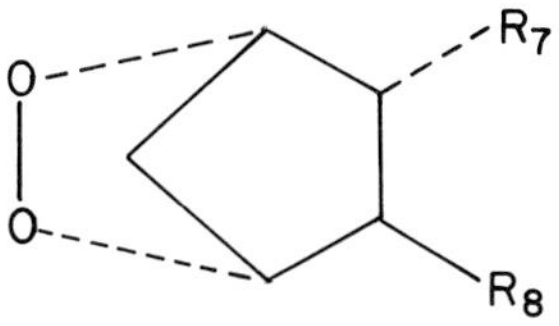   PGG or PGH

PGA and PGB have keto groups at C_9 and a double bond in the ring, which occurs at position 10,11 in the A series and at position 8,12 in the B series. PGE has a keto group at C_9 and a hydroxyl group at C_{11}. PGF has two hydroxyl groups, one at C_9 and the other at C_{11}. The nomenclature becomes somewhat inconsistent at this point because *both* PGG and PGH have the *same* ring structure, the cyclopentane endoperoxide. They differ only in the group substituted at position 15 of the side chain. PGG has a peroxide group at C_{15}, whereas PGH has a hydroxyl group. Their complete structures are shown in Fig. 13-14 in the context of prostaglandin biosynthesis.

The number denotes the unsaturated bonds that a prostaglandin contains, for example, PGE_1, PGE_2, or PGE_3. In the 1 series the double bond is at position 13,14. Double bonds are present at positions 13,14 and 5,6 of the 2 series and at positions 13,14; 5,6; and 17,18 of the 3 series. The 13,14-unsaturation is *trans*, whereas the 5,6- and 17,18-unsaturations are *cis*. In addition to the double bonds present in the hydrocarbon chains, all of the naturally occurring active prostaglandins contain a hydroxyl group at C_{15} in α-orientation.

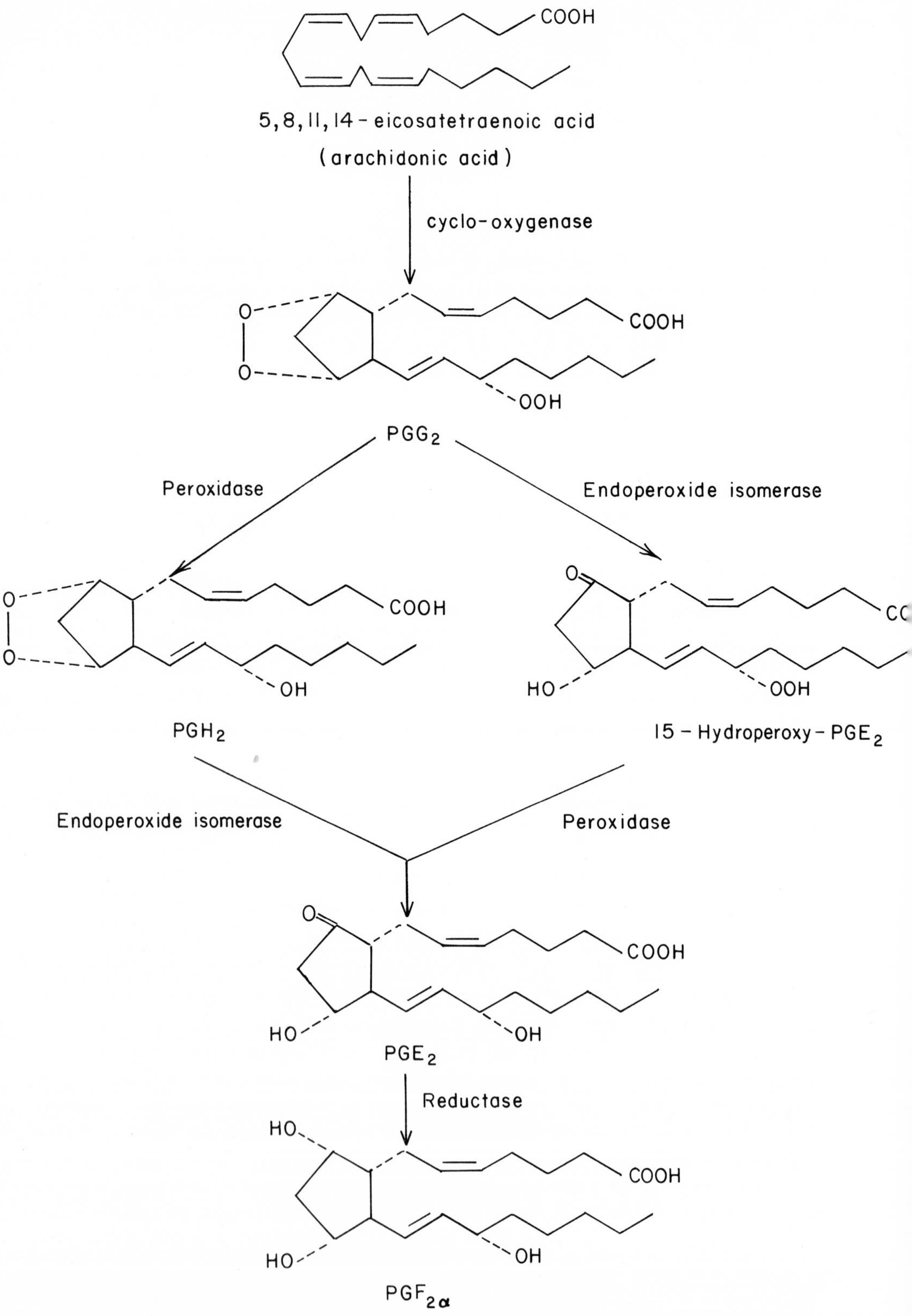

Fig. 13-14. Biosynthetic pathway for the conversion of arachidonic acid to prostaglandins.

The Greek letter applies only to the F series and refers to the configuration of the hydroxyl group at C_9. In the F_α series the hydroxyl group projects below the

OH   OH

- - C_7 - - C_7

C_8 C_8

PGF$_\alpha$ OH   OH   PGF$_\beta$

plane of the ring in the same orientation as the hydroxyl group at C_{11}. In the F_β series the hydroxyl group at C_9 is above the plane of the ring. Only the F_α series is naturally occurring.

PGG, one of the endoperoxide forms, is an intermediate in the synthesis of the other prostaglandins. Its endoperoxide structure is cleaved in further biosynthetic reactions, and the PGG intermediates are converted to PGA, PGB, PGE, or PGF products. There are fourteen naturally occurring products that have been discovered: PGA$_1$, PGA$_2$, 19α-hydroxy PGA$_1$, 19α-hydroxy PGA$_2$, PGB$_1$, PGB$_2$, 19α-hydroxy PGB$_2$, PGE$_1$, PGE$_2$, PGE$_3$, PGF$_{1\alpha}$, PGF$_{2\alpha}$, and PGF$_{3\alpha}$.

Although fourteen naturally occurring prostaglandins have been identified, only six of them are widely distributed in the body. These are known as the primary prostaglandins and are PGE$_1$, PGF$_{1\alpha}$, PGE$_2$, PGF$_{2\alpha}$, PGE$_3$, and PGF$_{3\alpha}$.

Biosynthesis. Prostaglandins are formed in the cells upon which they act. They differ from hormones in this respect, for a hormone is formed in a different cell and is then transported through the blood to its site of action. Prostaglandins are biosynthesized from three twenty–carbon fatty acids. *Eicosatrienoic* and *eicosatetraenoic* acids are derived from linoleic acid ($C_{18}\Delta_{9,12}$), the essential fatty acid that is required in the diet. Arachidonic acid is the common name for eicosatetraenoic acid. Eicosapentaenoic acid is derived from linolenic acid ($C_{18}\Delta_{9,12,15}$), which also is supplied entirely from the diet. Eicosatrienoic acid is the precursor of prostaglandins of the 1 series, for example, PGE$_1$, PGA$_1$, etc., eicosatetraenoic acid of the 2 series, and eicosapentaenoic acid of the 3 series. A lack of linoleic acid intake causes an illness known as essential fatty acid deficiency. This suggests that prostaglandins of the 1 and 2 series probably are necessary for optimal health. By contrast, no requirement for linolenic acid ever has been demonstrated in man, suggesting that prostaglandins of the 3 series probably are not essential for good health even though they normally are present in people who eat a well-balanced diet.

Prostaglandins are not stored in tissues; the precursor fatty acids are stored at position 2 of the phosphoglycerides contained in cellular membranes. When prostaglandin formation is required, the appropriate twenty–carbon atom polyenoic acids are hydrolyzed from the fatty acid-containing phosphoglycerides by phospholipase A$_2$. The latter is a hydrolase specific for position *sn*-2 of the phosphoglyceride.

A microsomal enzyme system known as prostaglandin synthetase converts the released polyenoic fatty acid to the prostaglandin. The first step is a cyclooxygenase reaction in which the twenty–carbon atom polyenoic fatty acid is converted to an endoperoxide having a PGG ring structure. In the reaction the double bonds present originally in positions 8 and 11 of the twenty–carbon atom poly-enoic fatty acid precursor are involved in the formation of the five-membered ring. The cyclo-oxygenase reaction is rate-limiting for prostaglandin synthesis. Three binding sites are present on the enzyme; they are specific for oxygen, a polyenoic fatty acid, and an activator, respectively. In endothelial cells *angioten-sin II* is a physiologic cyclo-oxygenase activator. The antiinflammatory drugs *aspirin, indomethacin,* and *tylenol* decrease prostaglandin synthesis by inhibit-ing the cyclooxygenase reaction.

Fig. 13-14 shows the biosynthetic pathway for the formation of PGE_2 and $PGF_{2\alpha}$, the prostaglandins that are derived from arachidonic acid. The peroxide intermediates in this pathway have very short half-lives. In some tissues, how-ever, endoperoxides such as PGG_2 have more potent prostaglandin effects than either PGE_2 or $PGF_{2\alpha}$. This has led to the view that the endoperoxide intermedi-ates actually may be the substances that mediate the effects that have been attributed to the prostaglandins themselves. Moreover, in platelets, arachidonic acid is converted to other metabolites in addition to the usual prostaglandins. These include *12-hydroxy-5,8,10,14-eicosatetraenoic acid* (HETE) and the *thromboxanes,* which contain a six-membered ring having an oxygen atom as a part of the ring structure. Thromboxane A_2, which is extremely labile, is a very potent platelet-aggregating agent and may be the prostaglandin-like substance that mediates the platelet's role in blood coagulation.

Inactivation. Two enzymatic reactions are involved in the inactivation of the prostaglandins. One is *15α-hydroxy prostaglandin dehydrogenase,* which catalyzes the conversion of the 15α-hydroxyl group to a keto group. The other is a *Δ13-prostaglandin reductase,* which catalyzes the reduction of the *trans* double bond in position 13,14.

Function. Prostaglandins have been reported to have numerous functions, some which are seemingly contradictory. However, certain of the conflicting actions are understandable in light of the fact that there are fourteen known naturally occurring prostaglandins. It is likely that there are several opposing pairs of prostaglandins, the members of which have opposite effects. Two general physiologic effects have been attributed to the prostaglandins—an effect on the contractile state of smooth muscle and a modulating effect on target tissues that respond to adenohypophyseal trophic hormones. The latter observation has given rise to the idea that prostaglandins may modulate the activity of the adenyl and guanyl cyclase systems, thereby regulating the response of the cell to external stimuli. Recent work suggests that PGE's regulate cellular cAMP levels, whereas PGF's regulate cellular cGMP content. In turn, the cellular response to certain stimuli is thought to depend on the relative cAMP and cGMP concentrations inside the cell. These observations imply that the prostaglandins are intimately involved in basic cellular response mechanisms.

Prostaglandins are among the most potent biologic substances discovered thus far; as little as 1 ng/ml causes contraction of animal smooth muscle preparations. Some of the main physiologic functions that are modulated by the prostaglandins are listed in Table 13-4. They have potential therapeutic use in such diverse areas as the treatment of hypertension, prevention of conception, induction of abortions, relief of bronchial asthma and nasal congestion, and healing of peptic ulcers. Because of the diversity of actions of the different types of prostaglandins, much confusion exists as to their effects on various diseases and bodily functions. For example, PGE's stimulate intestinal smooth muscle contraction and lower blood pressure, whereas PGF's exhibit the intestinal effect but are not hypotensive.

Table 13-4. Functions regulated by prostaglandins

FUNCTION	TISSUE	EFFECT
Muscular contraction	Heart	Contractility
	Arteries	Blood pressure
	Bronchi	Diameter of air passages
	Gastrointestinal tract	Motility
	Uterus	Contractility
Lipolysis	Adipocytes	Inhibition
Secretion	Stomach	HCl production
Permeability	Capillaries	Inflammation, swelling
Electrolytes	Kidney tubule	Na^+, H_2O retention
Coagulation	Platelets	Aggregation

Additional uncertainty is entailed in the many different actions of each prostaglandin. The use of a natural prostaglandin often produces unwanted side effects in addition to the desired effect; for example, diarrhea often accompanies the antihypertensive action. A major pharmacologic goal has been to synthesize prostaglandin analogues that have more limited and specific effects than do the naturally occurring substances. Among the most promising pharmacologic preparations are 16,16-dimethyl PGE_2, 15-methyl $PGF_{2\alpha}$, and 13-dehydro $PGF_{2\alpha}$.

Growth-regulating polypeptides

In addition to growth hormone there are a number of polypeptides and proteins that stimulate the proliferation of mammalian cells in culture. These substances have been isolated from serum or tissue extracts and range in molecular weight from 300 to 120,000. Although their role in human health is uncertain at present, they are potentially important for processes such as wound healing and nerve regeneration as well as for an understanding of certain types of cancer. One of the most widely studied of these growth factors is *somatomedin*, a protein with molecular weight of 8000 that increases sulfate incorporation into cartilage and stimulates bone growth. Recent evidence indicates that it acts by inhibiting adenyl cyclase, which decreases the intracellular cAMP content. Somatomedin has insulin-like effects, and it may be closely related to a substance with similar actions that is called *nonsuppressible insulin-like activity* (NISLA). In addition to the growth stimulating factors, evidence recently has been presented for the existence of factors that inhibit mitogenesis in cultured mammalian cells. These inhibitory factors, called *chalones*, are proteins with molecular weight of 30,000 to 50,000. They are synthesized within the cell and then bind to surface receptors, thereby blocking the access of growth-promoting polypeptides to these receptors. Although the role of mitogenic substances and chalones in human disease is not yet known, it is possible that these substances may become extremely important in the future.

REFERENCES

Agranoff, B. W.: Neurotransmitters and synaptic transmission, Fed. Proc. 34:1911, 1975. *A very brief overview containing many excellent references.*

Axelrod, J.: The pineal gland: a neurochemical transducer, Science 184:1341, 1974. *A brief summary written by the major contributor to this field.*

De Luca, H. F.: Recent advances in our understanding of the vitamin D endocrine system, J. Lab. Clin. Med. 87:7, 1976. *A superb short review containing many of the most recent ideas on this subject.*

Malarkey, W. B.: Recently discovered hypothalamic-pituitary hormones, Clin. Chem. 22:5, 1976. *A brief, up-to-date summary.*

O'Malley, B. W., and Means, A. R.: Female steroid hormones and target cell nuclei, Science **183:** 610, 1974. *This short review describes the current concepts concerning the transcriptional mechanism of steroid hormone action.*

Pastan, I. H., Johnson, G. S., and Anderson, W. B.: Role of cyclic nucleotides in growth control, Annu. Rev. Biochem. **44:**491, 1975. *A comprehensive presentation that discusses the general question of how cell growth is regulated.*

Robison, G. A., Butcher, R. W., and Sutherland, E. W.: Cyclic AMP, New York, 1971, Academic Press, Inc. *A comprehensive treatise on cyclic AMP and its role in hormonal action.*

Samuelsson, B., Granström, E., Green, K., Hamberg, M., and Hammarström, S.: Prostaglandins, Annu. Rev. Biochem. **44:**669, 1975. *A detailed review of recent advances in this rapidly changing field.*

Schally, A. V., Arimura, A., and Kastin, A. J.: Hypothalamic regulatory hormones, Science **179:** 341, 1973. *A synopsis of this rapidly growing area.*

Sutherland, E. W.: Studies on the mechanism of hormone action, Science **177:**401, 1972. *This article describes the discovery of cAMP.*

Tager, H. S., and Steiner, D. F.: Peptide hormones, Annu. Rev. Biochem. **43:**509, 1974. *A comprehensive review with numerous references.*

Walter, R., editor: Neurophysins: carriers of peptide hormones, Ann. N.Y. Acad. Sci. **248:**1-512, 1975. *This volume contains many of the recent research findings in this new and exciting area.*

Clinical examples

CASE 1: HYPERPARATHYROIDISM

A 53-year-old woman was admitted to the hospital with symptoms that had begun 6 months earlier. These included abdominal and back pain, myalgia, easy fatigability, and constipation. The physical findings were within normal limits. Laboratory studies revealed a serum calcium concentration of 11.8 mg/dl, a phosphorus concentration of 2.9 mg/dl, and an alkaline phosphatase activity of 6.8 Bodansky units (normal = 2.0 to 4.5 Bodansky units). Calcium oxalate crystals were present in the urinary sediment. While on a controlled diet with a daily calcium intake of 300 mg, the patient's 24-hr urine collections for calcium ranged from 380 to 450 mg. The patient underwent a surgical exploration of the neck, and a parathyroid adenoma was found and excised. Following complete recovery from surgery, the serum calcium was found to be 9.9 mg/dl and the phosphrous 4.0 mg/dl.

Biochemical questions

1. What is the relationship of parathyroid hormone to the plasma calcium and phosphorus concentrations?
2. How do calcitonin and vitamin D affect plasma calcium and phosphate?
3. What other information is required in order to adequately assess plasma calcium transport and the biologic effectiveness of a given plasma calcium concentration?
4. What is the most likely explanation for the elevated plasma alkaline phosphatase activity seen in the patient?

Case discussion

Parathyroid hormone raises the concentration of plasma calcium and lowers the concentration of plasma inorganic phosphate. One of its targets is bone, where it causes calcium to be mobilized from the hydroxyapatite crystals and released into the blood. A second target organ is the kidney, where it has at least two actions. One is to increase calcium reabsorption by the tubular cells. The other is to stimulate the conversion of 25-hydroxycholecalciferol to the hormonally active form of vitamin D, 1,25-dihydroxycholecalciferol.

Calcitonin has opposite effects from parathyroid hormone, lowering the calcium and raising the inorganic phosphate concentrations of the plasma. Calcitonin and parathyroid hormones are polypeptides, and they act by binding to cell-surface receptors, thereby regulating the cyclic nucleotide content of the target cells.

Vitamin D has actions similar to parathyroid hormone in that it raises the calcium concentration of the plasma. The active metabolite is 1,25-dihydroxycholecalciferol, which, as described above, is formed in the kidney. This vitamin D metabolite increases calcium absorption in the intestine, a process that probably is secondary to the synthesis of a calcium-binding protein in the intestinal mucosa. When the plasma calcium concentration is raised in this manner, bone calcification is facilitated. Although it may seem paradoxical, the hormonally active forms of vitamin D also can cause calcium mobilization from bone. Apparently, the primary action of vitamin D is to raise the blood calcium concentration. If dietary calcium is available, it is the source of the serum calcium, and bone calcification occurs. By contrast, calcium will be obtained from the bones if the need for additional blood calcium cannot be met from the dietary source. For calcium mobilization from bone to occur, elevated amounts of parathyroid hormone also must be present in the blood.

In order to assess properly the biologic effectiveness of a given plasma calcium concentration, one also must know the plasma albumin concentration. Two forms of calcium exist in the plasma. Some of the calcium is bound to albumin, and the remainder is in equilibrium with the bound material and is known as the free or ionized calcium. The biologic effectiveness of the plasma calcium is dependent on the free concentration.

Therefore, in order to evaluate the biologic effectiveness of any total calcium concentration, one must be certain that the distribution between free and bound forms is in the normal range. In ordinary clinical practice it is assumed that the extent and strength of binding between albumin and calcium is the same in all individuals. Based on this assumption, the plasma albumin concentration is taken as a measure of the fraction of total plasma calcium that is bound. If the albumin concentration in the plasma is within the normal range, the distribution between bound and free calcium is considered to be normal, and the total plasma calcium concentration is used as a measure of the biologically effective calcium content in the blood (see Table 2-7 for additional details).

When the alkaline phosphatase content of the plasma is elevated, the tissue or origin of this enzyme is usually either liver or bone. No evidence for obstructive liver disease is presented in this case, so there is no reason to suspect that the enzyme is derived from the liver. Conversely, alkaline phosphatase is released from bone when excessive bone destruction and calcium mobilization occur. This is expected in hyperparathyroidism. Therefore, it is logical to assume that the elevated plasma alkaline phosphatase is derived from bone. Based on this reasoning, the laboratory finding of alkaline phosphatase elevation supports the diagnosis of hyperparathyroidism.

REFERENCES

Avioli, L. V.: The diagnosis of primary hyperparathyroidism, Med. Clin. North Am. 52:451, 1968.
Goldsmith, R. S.: Hyperparathyroidism, N. Engl. J. Med. 281:367, 1969.
Kolata, G. B.: Vitamin D: investigations of a new steroid hormone, Science 187:635, 1975.

CASE 2: INSULINOMA

A 36-year-old woman was referred to a university hospital for evaluation of spells of dizziness and weakness. These spells typically lasted for 10 min and were occurring with increasing frequency. The spells usually came on after a large meal and could be terminated by eating candy or drinking fruit juice. After each episode the patient was hungry and tired, and her memory was blurred. The patient's physical examination was within normal limits except for mild obesity. She claimed to have gained 20 kg during the preceding 2 years. After a 13-hr fast her blood glucose concentration was 38 mg/dl. After a 5-hr glucose tolerance test her blood glucose was 46 mg/dl. Celiac angiography revealed an abnormality in the body and tail of the pancreas. The patient developed one of her spells while a medical student was in her room, and he was able to obtain a blood sample during the episode. This sample contained 19 mg/dl of glucose. The patient was transferred to the surgical service, and an insulin-secreting pancreatic adenoma (tumor) was removed, requiring resection of 90% of the pancreas.

Biochemical questions

1. An insulinoma is an insulin-secreting tumor. How did the presence of such a tumor explain the patient's symptoms?
2. Proinsulin was found in large quantities in this patient's plasma. What is the relationship of proinsulin to insulin?
3. What effects of increased insulin secretion might have predisposed this woman to obesity?
4. What digestive problems might result from excision of 90% of the pancreas?

Case discussion

The insulinoma was producing insulin. Because of the excessive amount of insulin-secreting tissue, too much insulin was released after dietary carbohydrate intake. This caused hypoglycemia during the 5-hr glucose tolerance test and after meals, producing the spells of weakness and dizziness. In addition to this normal insulin release when carbohydrate was ingested, the tumor also was secreting some insulin continuously. This inappropriate insulin release caused the low blood glucose concentration during prolonged fasting.

Proinsulin is the prohormone form of insulin that is made in the β-cells of the pancreatic islets. It has little or no insulin-like actions. After synthesis on the ribosomes, proinsulin is transported to the Golgi apparatus and stored in granules. Proinsulin is converted to insulin in these granules by proteolytic cleavage, but the conversion is incomplete. When insulin is discharged from the β-cell, some proinsulin that remains in the granule also is released. Likewise, C-peptide that is split out in the conversion of proinsulin to insulin is released during insulin secretion, but it too has no insulin-like activity.

Insulin acts on adipocytes, enhancing fatty acid storage as triglyceride. It facilitates glucose entry into the adipocyte, increasing the availability of the triose backbone glycerol 3-phosphate needed for triglyceride synthesis. It also provides glucose carbon atoms for fatty acid synthesis. In addition, it increases the content of lipoprotein lipase in the adipose tissue. This enzyme catalyzes the hydrolysis of chylomicron and VLDL triglycerides, a step that is required in order to transfer these fatty acids into the adipocyte for resynthesis into triglyceride. Much of the fatty acid stored in the adipose tissue is delivered to the adipocytes in the form of lipoprotein triglycerides, either VLDL from the liver or chylomicrons from the intestine. Therefore, the elevated lipoprotein lipase activity also favors triglyceride formation in the adipose tissue. Those adipose tissue effects that were mediated by the excessive insulin production could have contributed to the recent weight gain noted by this patient.

In addition to polypeptide hormones, the pancreas makes many digestive enzymes. These include amylase for dietary starches, lipase for triglycerides, and trypsin for proteins, as well as several others. Since 90% of the pancreas was excised, the remaining 10% may not produce sufficient amounts of these enzymes to adequately digest large meals. This might lead to malnutrition and weight loss in spite of an adequate diet. Because of this possibility, six or more smaller feedings rather than three regular meals each day might be recommended.

REFERENCES

Firestone, A. J., and Wohl, M. C.: Hypoglycemia—a complex problem, Med. Clin. North Am. 54: 531, 1970.

Levine, R., and Haft, D. E.: Carbohydrate homeostasis, N. Engl. J. Med. 283:175, 237, 1970.

Van Obberghen, E., Somers, G., Devis, G., Vaughan, G. D., Malaisse-Lagae, F., Orci, L., and Malaisse, W. J.: Dynamics of insulin release and microtubular-microfilamentous system, J. Clin. Invest. 52:1041, 1973.

CASE 3: HYPERTHYROIDISM

A 48-year-old woman was admitted to the hospital because of weight loss, palpitations, weakness, and exophthalmos. She stated that a goiter, which had been present for years, had recently begun to enlarge. She was extremely irritable, could not tolerate heat, and was short of breath. Physical examination revealed bilateral eyelid lag. The thyroid gland was diffusely enlarged, and a bruit was audible over the right lobe. Her heart was enlarged, a prominent apical thrust was noted, and there was a soft systolic heart murmur along the left sternal border. Laboratory examination revealed that the hemoglobin was 11.7 g/dl and that the hematocrit was 38%. The basal metabolic rate was 145% of normal. Plasma T_4 and T_3 were grossly elevated, and the ^{131}I uptake over the thyroid gland was very high (18% in 4 hr). A diagnosis of hyperthyroidism was made.

Biochemical questions

1. What are T_3 and T_4, and how are they related to the thyroid gland?
2. How are TRH and TSH involved in the regulation of thyroid hormone production and secretion?
3. What is the mechanism of increased ^{131}I uptake in hyperthyroidism?

Case discussion

T_3 and T_4 are the thyroid hormones, triiodothyronine and thyroxine, respectively. Both are tyrosine derivatives, and their chemical structures are shown on p. 632. Although lesser amounts of T_3 are released by the thyroid gland, it has a more potent effect than T_4 in producing the hypermetabolic effects of the thyroid hormones. T_4 is converted to T_3 in the target cells, and it is likely that T_3 actually is the metabolically active form of the thyroid hormone. In this sense, T_4 may be considered as a prohormone.

TSH is released from the anterior pituitary and stimulates T_3 and T_4 production and release. In turn, TSH release is stimulated by TRH, which is made in the hypothalamus. When the plasma T_3 and T_4 concentrations are elevated, TRH production and release is inhibited. This leads to decreased T_3 and T_4 release from the thyroid gland (Fig. 13-1).

T_3 and T_4 contain iodine atoms attached to their phenolic rings. Thyroglobulin, the protein precursor of these hormones that is contained in the thyroid cells, has many iodinated tyrosine residues. The iodine is obtained from iodide ions in the blood plasma, and the thyroid cells have the capacity to take up and concentrate iodide ions. In hyperthyroidism the thyroid gland is more active than normal. It synthesizes more thyroglobulin, T_3, and T_4 and takes up much larger amounts of iodine than in the euthyroid (normal) state. Therefore, when ^{131}I is administered to a hyperthyroid patient, a larger fraction of the dose is concentrated within the thyroid gland than in a euthyroid subject. This is useful clinically in two ways. First, small quantities of ^{131}I can be administered as a diagnostic test of thyroid function. After administration, the radiation emanating from the thyroid gland can be measured at various times by placing a scanning device over the neck. Greater than normal uptakes occur in hyperthyroidism, and less than normal uptakes occur if the thyroid gland is hypofunctioning (hypothyroidism). Second, the enhanced iodine uptake can be used to treat hyperthyroidism. If larger amounts of ^{131}I are administered, enough ^{131}I will concentrate in the thyroid to provide intense but localized radiation to the glandular cells. This will destroy many of the T_3- and T_4-producing cells, reducing the excessive function of the thyroid and correcting the hyperthyroidism. In some respects this is a safer form of treatment than surgical removal of a large portion of the hyperactive gland. It is not without some danger, however, for ^{131}I treatment can lead in some cases to either hypothyroidism or even thyroid cancer.

REFERENCES

Oppenheimer, J. H.: Initiation of thyroid hormone action, N. Engl. J. Med. 292:1063, 1975.

Werner, S. C., and Ingbar, S. H., editors: The thyroid gland, ed. 4, New York, 1976, Harper & Row, Publishers.

CASE 4: CUSHING'S SYNDROME

A 35-year-old male was admitted to the hospital because of irritability and emotional lability together with muscle weakness and easy fatigability. Physical examination revealed that his trunk was obese but his arms and legs were quite lean. He had a rounded facial appearance and a small, nontender hump at the junction of his neck and back. Fullness was noted in the supraclavicular regions, and purple striae were present in the subaxillary areas. Laboratory examination revealed that the plasma cortisol concentration was elevated (20 mg/dl), the fasting blood glucose was 150 mg/dl, and the urinary excretion of 11-hydroxyandrosterone and 11-hydroxyetiocholanolone were greatly increased. A diagnosis of Cushing's syndrome was made.

Biochemical questions

1. From what substance is cortisol synthesized?
2. How is cortisol synthesis regulated?
3. How is cortisol transported in the blood plasma?
4. What are the metabolic effects of cortisol in man?

5. How does cortisol function at the molecular level?
6. How are 11-hydroxyandrosterone and 11-hydroxyetiocholanolone related to cortisol?
7. Would you expect that excessive quantities of vanillylmandelic acid (VMA) also would be excreted in the urine?

Case discussion

Cortisol, the main glucocorticoid hormone in man, is synthesized in the zona fasciculata of the adrenal cortex. Cholesterol is the precursor of cortisol as well as of the approximately fifty other steroids that can be isolated from the adrenal cortex. Most of the cholesterol utilized for steroid hormone synthesis is taken up from plasma lipoproteins and stored in the adrenal cortical cells as cholesteryl esters. The pathway for steroid synthesis, which was described in detail in Chapter 10, is reviewed schematically in Fig. 13-15. Progesterone is the common intermediate in steroid hormone synthesis.

Cortisol synthesis is regulated by ACTH, a polypeptide hormone secreted by the anterior pituitary gland. ACTH release, in turn, is stimulated by CRH, produced in the hypothalamus. Cortisol production is decreased by high concentrations of plasma cortisol or ACTH, both of which inhibit CRH release (see Fig. 13-12). ACTH stimulates cortisol production in two ways. It activates the hydrolysis of cholesteryl esters through a cAMP-dependent mechanism. It also activates the desmolase reaction in which a six-carbon atom fragment is cleaved from the cholesterol side chain (see p. 500).

Like the other steroid hormones, cortisol is transported in the blood plasma by a specific carrier protein, transcortin. This protein is a plasma globulin that is synthesized in the liver. It also transports corticosterone, the other major gluco-

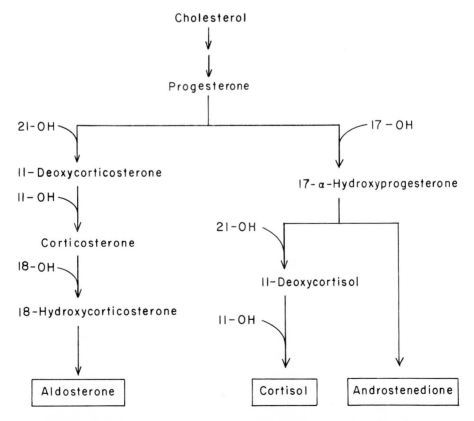

Fig. 13-15. Summary of steroid hormone biosynthetic pathway in adrenal cortex.

corticoid produced by the adrenal cortex.

One of the main metabolic effects of the glucocorticoid hormones is to raise the blood glucose concentration by stimulating gluconeogenesis. Body proteins are catabolized to provide the amino acid substrates for gluconeogenesis, leading to decreases in muscle mass and weakness. Although the molecular mechanism of this action is not known with certainty, there is some evidence that glucocorticoids operate through a transcriptional mechanism. In experimental studies with rat liver the production of tryptophan pyrrolase and tyrosine aminotransferase, a transaminase, is increased by administration of glucocorticoid hormones. This involves binding of the hormone to cytoplasmic protein receptors and transport of the steroid-receptor complex into the cell nucleus. Once inside the nucleus, the hormone-receptor complex binds either to acidic nuclear proteins or to DNA itself and activates the transcription of mRNA, which is specific for the synthesis of these amino acid catabolic enzymes. These enzymes then facilitate the conversion of amino acids into fragments that can enter the Krebs cycle and flow into the gluconeogenic pathway (see pp. 273 to 275). The glucocorticoids have additional actions, such as indirectly facilitating the stimulation of lipolysis, destruction of lymphocytes, and reduction in the number of circulating eosinophilic leukocytes. These effects, however, are not understood completely at the molecular level.

Cortisol is catabolized by reduction of its 3-keto group and Δ_4 double bond as shown in Fig. 10-20. In the final step the two-carbon atom side chain is removed, leading to the formation of a 17-keto group. The 17-keto metabolites of cortisol are 11-hydroxyandrosterone and 11-hydroxyetiocholanolone. Almost all of the cortisol metabolites are excreted in the urine.

VMA is the main urinary excretion product of the catecholamines. It is elevated in diseases associated with excessive catecholamine production, such as pheochromocytoma. If a tumor of the adrenal medulla or other chromaffin tissue is present, VMA excretion would be increased. Cushing's disease results from overproduction of glucocorticoids by the *adrenal cortex* and, of itself, does not involve the adrenal medulla or other catecholamine-producing tissues. Hence there is no reason to suspect overproduction of either epinephrine or norepinephrine in this case, and one would *not* expect VMA excretion to be affected if this diagnosis is correct.

REFERENCES

Baxter, J. D., and Forsham, P. H.: Tissue effects of glucocorticoids, Am. J. Med. **53:**573, 1972.

Liddle, G. W.: Pathogenesis of glucocorticoid disorders, Am. J. Med. **53:**639, 1972.

CASE 5: PSORIASIS

A 23-year-old male was referred to a dermatologist by his family physician because of an extensive eruption of the skin. The patient stated that he had several patches of dry, scaling skin since he was about 10 years old, but the problem was localized and went almost unnoticed. In the last year, however, the lesions increased markedly in number and severity. Physical examination revealed that his elbows, knees, scalp, and trunk were covered with silvery gray, scaling plaques that had sharp margins. A scaly area was removed by gentle scraping, revealing numerous minute bleeding points. Many of the lesions were in the form of a ring, with some clearing in the center and a more actively involved circumference. Although the patient was not aware of similar severe skin problems in his relatives, he did recall that several of them had chronic problems with scaling of the scalp that was attributed to seborrheic dermatitis (dandruff). Based on these considerations, a diagnosis of psoriasis was made. The patient agreed to enter the hospital for additional study of this problem. Biopsies were obtained of the lesions as well as uninvolved areas of skin. As compared with the normal skin, the lesions were found to contain larger amounts of cGMP and arachidonic acid and less cAMP.

Biochemical questions

1. How are cAMP and cGMP formed, and how are these cyclic nucleotides thought to be related to cell proliferation?
2. How is the cAMP content of a cell regulated?
3. What enzymes are involved in cAMP and cGMP synthesis and breakdown?
4. How does cAMP exert its metabolic effect on cells?
5. Discuss the role of polyunsaturated fatty acids in prostaglandin synthesis.
6. What prostaglandins might be produced in greater quantities because of an arachidonic acid excess?
7. What metabolic effects have been attributed to the prostaglandins, and is there any suspected link between cyclic nucleotides and the prostaglandins?
8. Suggest a mechanism whereby the arachidonic acid increase might account for the cyclic nucleotide changes and the skin lesions of psoriasis.

REFERENCE

Hammarström, S., Hambert, M., Samuelsson, B., Duell, E. A., Stawiski, M., and Voorhees, J. J.: Increased concentration of nonesterified arachidonic acid, 12L-hydroxy-5,8,10,14-eicosatetraenoic acid, prostaglandin E_2, and prostaglandin $F_{2\alpha}$ in epidermis of psoriasis, Proc. Natl. Acad. Sci. U.S.A. 72:5130, 1975.

CASE 6: BRONCHIAL ASTHMA

An 11-year-old boy with a history of allergies to dust, mold, and pollen was brought to the hospital emergency room at 8:30 PM on a Saturday. It had been a warm, clear day, and he had played outdoors for most of it. He also had helped his father mow the lawn and clean out the flower beds that afternoon. In the early evening he began to cough and complain of difficulty in breathing and nasal congestion. A wheeze soon became audible, and he had difficulty in breathing. Physical examination confirmed the suspected diagnosis of acute bronchial asthma, and the boy was treated by administration of epinephrine and aminophylline, a methyl xanthine drug. He improved rapidly and was able to go home within 2 hr.

Biochemical questions

1. What effects on plasma glucose and free fatty acid concentration would be expected from the administration of epinephrine?
2. How are β-adrenergic receptors, cyclic nucleotides, and protein phosphorylation related to the action of epinephrine?
3. In what way would a methyl xanthine drug modulate the metabolic effect of epinephrine at the molecular level?
4. What enzymes are involved in the metabolism and inactivation of the administered epinephrine?
5. Wheezing is caused by construction of the bronchioles, leading to labored and difficult expiration. CO_2 is retained. What kind of an acid-base imbalance would such a condition produce?
6. The commonly used treatment for acute asthma involves the administration of catecholamines and methyl xanthines as was done in this case. Based on this information, why might the administration of a β-adrenergic blocking drug such as propranolol be dangerous in this child? Explain in molecular terms.

REFERENCES

Franklin, W.: Treatment of severe asthma, N. Engl. J. Med. 290:1469, 1974.
McFadden, E.: Acute bronchial asthma, N. Engl. J. Med. 288:221, 1973.

CASE 7: ZOLLINGER-ELLISON SYNDROME

A 42-year-old man was admitted to the hospital with complaints of diarrhea, weight loss, vomiting, and abdominal pain. He had a past history of a duodenal ulcer. An x-ray examination after oral administration of barium revealed edematous mucosal folds in the stomach and upper small intestine. Gastroscopy demonstrated the presence of a superficial ulcer in the pyloric region. A histamine stimulation test revealed excessive fluid and H^+ secretion in the stomach and excessive levels of gastrin in the blood plasma. A diagnosis of a gastrin-secreting tumor of either the pancreas or duodenum, the Zollinger-Ellison syndrome, was made.

Biochemical questions

1. How do the known metabolic actions of gastrin explain the findings made in this case?

2. Gastrin is a polypeptide hormone. Based on this information, what is the most likely molecular mechanism of its action?

3. Gastrin contains seventeen amino acids, but a larger form also is present in the blood. Based on your knowledge of insulin, suggest a possible relationship between these two forms of gastrin.

4. What other hormones are produced by the gastrointestinal tract? What are their actions?

REFERENCES

Isenberg, J., Walsh, J., and Grossman, M. I.: Zollinger-Ellison syndrome, Gastroenterology **65**:140, 1973.

Walsh, J. H., and Grossman, M. I.: Gastrin, N. Engl. J. Med. **292**:1324, 1377, 1975.

CASE 8: PITUITARY INSUFFICIENCY

A 48-year-old female complained of weakness and lack of energy. Her basal metabolic rate was low, and she was diagnosed as suffering from hypothyroidism. She was treated with oral thyroid hormone replacement with some improvement. Recently, however, she began to complain of severe, constant headaches and decreased visual acuity. Ophthalmologic examination confirmed a decrease in the visual fields and early optic nerve atrophy. Skull x-rays indicated an irregularity in the sella turcica, and a pneumoencephalogram suggested the presence of a mass in the chiasmatic cystern. A frontal craniotomy was performed, and a 7-g tumor was removed from the region of the pituitary gland.

Biochemical questions

1. Why might a pituitary tumor have the clinical symptoms of hypothyroidism?
2. In addition to the thyroid, what other endocrine glands might be affected by the loss of anterior pituitary function? What hormones might be deficient as a result of this?
3. If the posterior pituitary also was damaged by this tumor, what hormonal deficiencies might result?
4. In many cases of hypothyroidism, the parenteral administration of TSH does not produce any improvement. In this case of pituitary insufficiency, however, administration of TSH over a period of time would eventually lead to normal thyroid function. Explain these differences.
5. Would the administration of TRH also be effective in restoring thyroid function in hypothyroidism resulting from anterior pituitary insufficiency?

REFERENCES

Nabarro, J. D.: Pituitary tumors and hypopituitarism, Br. Med. J. **1**:492, 1972.

ADDITIONAL QUESTIONS AND PROBLEMS

1. Why would a patient with hyperparathyroidism (excessive secretion of parathyroid hormone) be prone to bone fractures, even with minimal trauma, and the development of calcium-containing kidney stones?

2. What types of hormonal imbalance might occur if the pituitary gland of a 9-year-old girl were inadvertently destroyed during x-ray therapy for a brain tumor? If the patient survived, would she be expected to experience a normal puberty?

3. Why would having only a large quantity of black, sugarless coffee for breakfast produce an elevation in the plasma free fatty acid concentration?

4. Why do some people develop hypothyroidism, a disease resulting from a decrease in circulating thyroid hormone, as a result of an iodine-deficient diet?

5. Dexamethasone is a synthetic glucocorticoid that has a powerful cortisol-like action when taken orally. This drug was administered to a patient at bedtime. A blood sample taken the next morning contained a very low plasma cortisol concentration. How did dexamethasone produce a lowering of the plasma cortisol level?

6. Why can hormones such as cortisol and thyroxine be administered orally whereas glucagon, growth hormone, and ACTH must be given by injection?

7. Why would the prolonged administration of a fat-free diet eventually be expected to cause reduced prostaglandin levels in the body?

8. Explain why pregnancy can sometimes cause glucose intolerance and a hyperthyroid-like state.

9. The insulin content of a plasma specimen was inactivated using specific antibodies. Yet insulin-like metabolic effects were observed when samples of the plasma were added to human cells in culture. How can this finding be explained?

CHAPTER 14

COMPREHENSIVE CASE
ANALYSIS

OBJECTIVES

1. To illustrate how biochemical reactions are integrated into specialized physiologic processes
2. To demonstrate by selected problems how compartmentalized metabolic pathways interact
3. To analyze clinical problems in terms of their total known biochemical elements

In the preceding chapters the individual elements of biochemical information required to explain human physiologic and disease processes have been presented in molecular terms. In order to facilitate the presentation of this material it was necessary to separate the various biochemical phenomena into unified subsections such as acid-base balance, energetics, and lipid metabolism. The clinical cases selected to illustrate the relationship of a particular area of biochemistry to human health have been selected in a very limited manner, especially in the beginning chapters. This was done for two reasons. First, it emphasized the basic aspects of biochemistry covered in the language statements without undue digression into areas that would distract from the subject at hand. Second, the biochemical facts needed to understand certain aspects of the clinical cases were often not yet described at the point where the cases were introduced. For example, diabetes mellitus was discussed as an illustration of a problem in carbohydrate metabolism. Abnormalities in lipid metabolism also occur in diabetes, but these were either omitted or treated very briefly because the metabolism of lipids had not yet been discussed. In fact, neither carbohydrate nor lipid metabolism exists in a vacuum; both are part of the metabolic whole that maintains life but that becomes deranged when a disease develops.

At this point, since the basic biochemical concepts have all been introduced, it is possible to undertake a much broader analysis of clinical problems. Accordingly, this chapter contains no language statement, but consists entirely of a series of clinical problems. The cases presented here were carefully selected with several aims in mind. First, some problems that were examined earlier in a limited manner are set forth again, this time with sufficient depth of analysis that the metabolic relationships necessary to full understanding are apparent. Second, certain problems that could not readily be discussed in a limited manner can here be explored. Third, students of the health sciences should be aware that in some problem areas our knowledge of biochemistry is not sufficiently complete to provide as deep an understanding as might be wished. These areas will quickly become apparent to the reader.

BLOOD COAGULATION: AN INTEGRATED ENZYMATIC CASCADE MECHANISM

Blood coagulation is mediated by a series of plasma clotting factors, most of which are *proteolytic enzymes*. They are present in the plasma at all times in *proenzyme* or *zymogen* form. The coagulation sequence is an example of a cascade mechanism that is one of the most striking examples of biochemical or *enzymatic amplification*. In a typical sequence the substrate of an initial reaction is a proenzyme; the product is the active enzyme that catalyzes the conversion of another proenzyme to its active form. *Fibrin*, the matrix of the clot, is available in the plasma in concentrations of about 0.5 g/dl as an inactive precursor form, *fibrinogen*. Breakdown of the fibrin clot is catalyzed by another proteolytic enzyme, *plasmin*, that also exists in proenzyme form, *plasminogen*. These very complex enzyme systems are carefully regulated so that, except in relatively rare pathologic situations, they operate in concert to maintain homeostasis.

Case 1: Thrombophlebitis, anticoagulants, and hemorrhage

A 52-year-old woman consulted her physician because of a painful and swollen left lower limb. The symptoms had developed rapidly, and she was soon unable to walk. Her family physician correctly diagnosed the condition as acute thrombophlebitis, and the patient was hospitalized immediately. Treatment was begun with heparin, an anticoagulant drug that is administered by injection. After several days, oral administration of a coumarin type of anticoagulant was begun. The patient's prothrombin time was 12 sec, the control value, before the coumarin was started. This gradually increased to 36 sec as the dosage was raised, and a maintenance dose of coumarin that kept the prothrombin time between 30 and 40 sec was given daily. The patient gradually improved, and the heparin was discontinued. She was discharged after 3 weeks of hospitalization with a supply of coumarin, and she was advised to report weekly for laboratory determinations of her prothrombin time. She followed this advice for several months. Because of the inconvenience and expense, she discontinued both her visits to the physician and her weekly blood tests. However, she continued to take the coumarin anticoagulant. Six weeks later she again consulted her physician because she was passing large quantities of bright red blood in her urine. She was hospitalized immediately and a water-soluble vitamin K analogue, menadione, was administered parenterally. Prior to the administration of this drug, the patient's prothrombin time was 73 sec. The coumarin anticoagulant was discontinued, and additional menadione and vitamin K were given. Hematuria ceased completely within 24 hr, and a final determination indicated that the patient's prothrombin time was 13 sec.

Biochemical questions

1. What biochemical events are involved in the formation and dissolution of a thrombus?
2. What mechanism prevents the clotting factors from functioning continuously in the plasma?
3. How does the coagulation mechanism exhibit enzymatic amplification?
4. In what ways does thrombin resemble trypsin and chymotrypsin?
5. How do the coumarin anticoagulants decrease the tendency of blood to coagulate?
6. Why is vitamin K effective in overcoming the actions of the coumarin anti-

coagulants? Why was menadione given initially rather than natural vitamin K?

7. If a patient with obstructive jaundice resulting from inoperable carcinoma of the head of the pancreas were to develop a prolonged prothrombin time, would he respond better to vitamin K administered orally or parenterally?

Case discussion

This patient's initial problem, thrombophlebitis, was caused by the formation of a blood clot in one of the deep veins of the left leg. The reason for the formation of these clots, or thrombi, is not presently understood. We know that blood will clot if it is withdrawn from the circulatory system. This can be prevented by the addition of an *anticoagulant* to the blood that is drawn. Most of the anticoagulants that are employed for this purpose function through the chelation of calcium ions; calcium is essential for several reactions involved in clot formation. These anticoagulants, such as ethylenediaminetetraacetate (EDTA), cannot be used therapeutically, for ionic calcium is vital to many other physiologic processes and its chelation would produce death. In addition to clots formed in the test tube, clots will also form within the blood vessels if the proper stimulus, for example, injury to the vessel wall, is provided. The formation of a clot may be lifesaving, for it prevents hemorrhage after a laceration. Unfortunately the process may also cause illness, for example, myocardial infarction secondary to the formation of a thrombus in a coronary artery damaged by atherosclerosis or, as in this patient, obstruction to venous drainage from a limb. A serious complication that may develop from thrombosis of the leg veins, such as has occurred in this case, is *pulmonary embolism*. This can occur if a portion of the clot dislodges and is carried through the circulation into the pulmonary vessels, where it blocks blood flow through one or more segments of the lungs.

Mechanism of blood coagulation.   Although the mechanism causing thrombus formation in thrombophlebitis is unknown, the way in which vascular injury triggers clot formation has been elucidated to some extent. When vascular injury occurs, platelets adhere to collagen and subendothelial fibers in the damaged area of the vessel wall and aggregate with one another. In certain vessels this is associated with vasoconstriction in the injured area. A platelet plug containing trapped neutrophilic leukocytes forms at the point of damage. Platelet aggregation is associated with or causes activation of the plasma coagulation mechanism, leading to the formation of a fibrin clot that seals off the damaged area. The platelets provide phospholipids, probably in the form of a membrane, that activate the clotting mechanism. Activation occurs because the soluble enzymes are provided a surface on which to act. This has the effect of concentrating the enzymes in a localized area, leading to faster reaction rates. The phospholipid material is called *platelet factor III*.

The basic principle is that most of the necessary ingredients required to make fibrin are available in inactive form in the plasma. These circulating clotting factor precursors are proteins and, except for fibrinogen and factor V, are also proenzymes. The clotting factors can be divided into the *intrinsic* system, which exists in the plasma, and the *extrinsic* system, which is derived from the injured tissues. As shown in Fig. 14-1 the intrinsic and extrinsic clotting factors converge at factor X_a. Beyond that point, both systems operate through a common mechanism that involves the activation of prothrombin. The intrinsic system requires factors VIII, IX, XI, and XII. The extrinsic system requires factor VII and *tissue factor*, a material that is released into the plasma as a result of tissue damage.

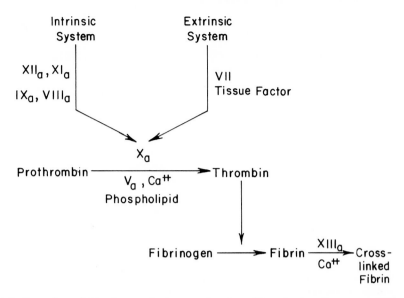

Fig. 14-1. Overview of blood coagulation mechanism. The clotting factors are abbreviated by a Roman numeral. A subscript "a" is placed after the clotting factor to indicate that it is in the activated form. Thus X_a indicates the activated form of factor X. The phospholipid required for the activation of prothrombin is provided by platelet membranes.

Table 14-1. Clotting factors

FACTOR	NAME	FUNCTION
I	Fibrinogen	Converted to fibrin
II	Prothrombin	Converted to thrombin
IV	Calcium	Cofactor for several reactions
V	Proaccelerin	Thrombin formation (not an enzyme)
VII	Proconvertin	Factor X activation
VIII	Antihemophilic globulin	Factor X activation
IX	Plasma thromboplastin component (PTC)	Formation of factor VIII
X	Stuart	Thrombin formation
XI	Plasma thromboplastin antecedent (PTA)	Factor IX activation
XII	Hageman	Factor XI activation
XIII	Fibrin-stabilizing factor (transglutaminase)	Cross-linking of fibrin

In both systems the end result is the formation of a fibrin plug from its circulating soluble precursor, *fibrinogen*. This conversion is catalyzed by the enzyme *thrombin*, which is formed from *prothrombin* through the actions of factors V and X as well as Ca^{++} and the phospholipids that are derived from platelets.

Clotting factors. The clotting factors are listed in Table 14-1. Most of them are referred to clinically by their Roman numeral designation. When the activated form of the clotting factor is being referred to, this is designated by placing a subscript "a" after the Roman numeral (see Fig. 14-1). The relationship of the various clotting factors in the intrinsic system is illustrated schematically in Fig. 14-2. Most if not all of the clotting factors are synthesized in the liver. They are released into the plasma in proenzyme form and must be enzymatically converted to the active form before they become functional. Thus the clotting factors are almost immediately available for action when they are needed. The intrinsic system is initiated by activation of factor XII. In a series of additional reactions the intrinsic system is primed for the activation of factor X. The latter step requires activated factor IX as well as the active form of factor VIII. The reaction also requires Ca^{++} and phospholipids. Factor VIII also is known as *antihemophilic globulin* because it is deficient in the disease hemophilia. Active factor X, together with Ca^{++}, phospholipids, and factor V, catalyzes the conversion of prothrombin to thrombin. The latter enzyme catalyzes the conversion of fibrinogen to fibrin. Factor XIII (also known as *transglutaminase*) and Ca^{++} catalyze the cross-linking of the fibrin clot. Fibrin is insoluble in plasma and acts as a plug to seal off the injured area. The phospholipid required for both the activation of factor X and the conversion of prothrombin to thrombin is derived from the aggregated platelets.

As noted in Fig. 14-1, factor X also is the key intermediate in the conversion of prothrombin to thrombin in the extrinsic clotting system. In this case, however, factor X is activated through the action of factor VII and tissue factor, so that the extrinsic system does *not* involve factors VIII, IX, XI, or XII.

These enzymatic sequences are known as a *cascade* or *enzymatic amplification mechanism*. The basic principle of enzymatic amplification is illustrated in

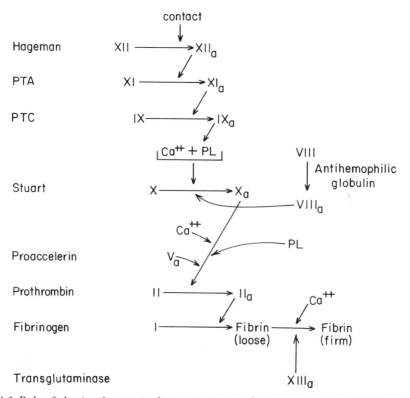

Fig. 14-2. Role of clotting factors in the intrinsic coagulation mechanism. PL, Phospholipid.

the following hypothetical example. There are six major steps that lead to fibrin formation in the intrinsic system. Assume that each step results in the formation of an active enzyme that can catalyze the conversion of ten molecules of its substrate. In other words, one stimulus leads to the formation of one active factor XII, and this activated factor XII in turn converts ten molecules of XI into factor XI_a. If this progresses, as shown in Fig. 14-2, the single contact stimulus will lead to the formation of 10^6 molecules of fibrin. In this way a relatively small stimulus can lead to an enormous biochemical response if it is amplified sufficiently through a series of enzymatic steps.

Diseases involving clotting factors. Diseases that are characterized by abnormal or excessive bleeding result from an inherited deficiency or absence of a clotting factor. *Hemophilia A* results from the lack of factor VIII. *Hemophilia B*, which is caused by lack of factor IX, is also known as *Christmas disease.* Factor XI deficiency produces *hemophilia C*, while factor V deficiency results in *congenital parahemophilia.* Deficiencies of the other clotting factors have also been observed clinically. Of the clinical cases resulting from clotting factor deficiencies, approximately 80% involve factor VIII and 10% to 20% involve factor IX. All of the other factors combined account for only 1%.

If a bleeding problem exists that is suspected of involving the clotting factors, two tests often are done on the patient's plasma in an attempt to localize the abnormality. One of these is called the *prothrombin time.* This measures factors V, VII, and X and prothrombin. If any one of these factors is missing, the clotting time of the plasma will be prolonged. The second test is the *partial thromboplastin time*, which measures factors VIII, IX, XI, and XII. Once the defect is localized to one of these two segments of the clotting mechanism, more complex tests can be applied to the patient's plasma in order to pinpoint the precise clotting factor that is deficient.

Conversion of fibrinogen to fibrin. The mechanism of fibrinogen conversion to fibrin will be considered in some detail in order to illustrate the type of enzymatic processes that occur in the coagulation cascade reactions. Fibrinogen is a glycoprotein that has a molecular weight of 3.3×10^5. It is composed of six polypeptide chains. The usual concentration of fibrinogen is 400 mg/dl, making it the third most abundant plasma protein. Fibrinogen is converted into the clot material fibrin by the enzyme thrombin.

The clotting factors can be thought of as triggering the conversion of the zymogen, prothrombin, to thrombin, the active enzyme. Prothrombin is made up of a single polypeptide chain with a molecular weight of 7.2×10^4; thrombin has a molecular weight of 3.7×10^4 and consists of two polypeptide chains joined by disulfide bonds. The β-chain of thrombin is similar to trypsin, and thrombin has trypsinlike activity. In other words, thrombin is a proteolytic enzyme that cleaves specifically at arginyl and lysyl bonds. Like chymotrypsin, its active center contains a free serine hydroxyl group and is inactivated by diisopropylfluorophosphate. It should be remembered that, like thrombin, trypsin and chymotrypsin also are biosynthesized and released in zymogen forms.

Thrombin catalyzes the cleavage of one peptide bond in each of four chains of fibrinogen. Four fibrinopeptides are derived, coming from four of the six N-terminal ends of fibrinogen. The fibrin monomer units so produced polymerize as a result of physical interactions between groups that are unmasked when the fibrinopeptides are removed. The fibrin polymer that is formed by this polymerization process then is transformed into a firm clot through cross-linking of γ-carboxyl groups from glutamine with ϵ-amino groups of lysine. This *transpeptidation* reac-

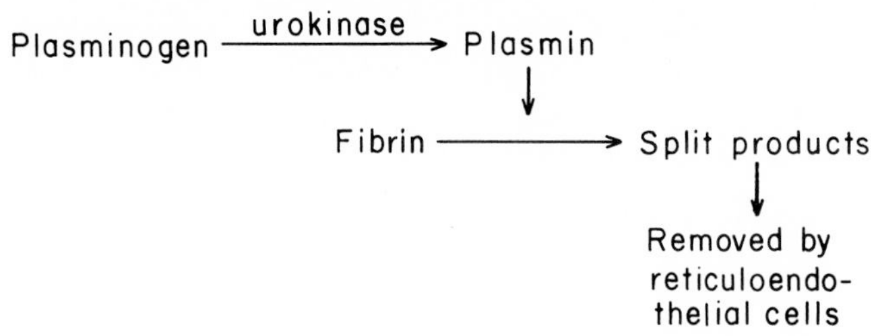

Fig. 14-3. Mechanism of fibrinolysis.

tion is catalyzed by factor XIII and Ca^{++}. Hence the bonds that stabilize the fibrin polymer in the form of a firm clot are covalent.

There are two types of clinical problems that involve fibrinogen. One is *afibrinogenemia,* a condition in which no fibrinogen is present in the plasma. This is a very serious disease in which there is a very high risk of hemorrhage. The other, *dysfibrinogenemia,* is caused by the production of an abnormal fibrinogen molecule. Although still serious, this is a less lethal condition than afibrinogenemia.

Fibrinolysis. Breakdown and removal of the fibrin clot is part of the healing process and is mediated by the plasminogen-plasmin system, as shown in Fig. 14-3. Plasminogen is a circulating proenzyme. When activated by a protease, *urokinase,* or an analogous enzyme, it is converted to the active form, *plasmin.* The latter catalyzes the dissolution of the fibrin clot. The split products that are formed as the result of fibrin dissolution are of different sizes. They are cleared rapidly from the circulation by the reticuloendothelial system.

Coumarin anticoagulants and vitamin K

One method of reducing the tendency of blood to clot intravascularly is to lower the concentration of one or more of the circulating clotting factors. This can be done by administration of a coumarin derivative, for example, dicumarol, which interferes with the biosynthesis of prothrombin as well as of factors VII, IX, and X in the liver.

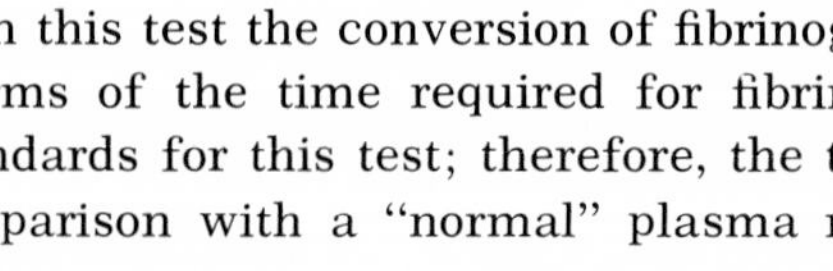

Dicumarol

The assay of the amounts of these factors in the blood plasma is usually the *prothrombin time.* In this test the conversion of fibrinogen is observed. Results are expressed in terms of the time required for fibrin formation. There are no true absolute standards for this test; therefore, the time of fibrin formation is estimated by comparison with a "normal" plasma run concomitantly as a control.

Important new insight into the role of vitamin K in prothrombin biosynthesis recently has been obtained. Vitamin K is necessary for the formation of the Ca^{++} binding sites of prothrombin. In order to be activated, prothrombin must be bound to phospholipids. Ca^{++} is required for this binding process. The Ca^{++} binding sites of prothrombin are formed by the introduction of a second carboxyl group into the glutamyl side chains located in the amino-terminal region of the protein. When dicumarol is administered, the prothrombin that is produced has a very low Ca^{++} binding capacity. In this inactive prothrombin, glutamate residues exist in place of γ-carboxyglutamate. Therefore, vitamin K somehow facilitates the carboxylation of glutamate residues in prothrombin and perhaps factors VII, IX, and X. When the action of vitamin K is blocked by dicumarol, Ca^{++} cannot bind to prothrombin because the protein lacks the added carboxyl groups. If it is artificially converted to thrombin, however, the defective molecule is active catalytically. In fact, the segment of prothrombin that is γ-carboxylated and binds calcium actually is removed during the activation process. Therefore, the activation reaction is blocked and clotting is abnormal when there is a vitamin K deficiency or pharmacologic interference with vitamin K action.

The fat-soluble vitamin K is required for normal function of the blood coagulation system. Naturally occurring substances with vitamin K activity are analogues of 2-methyl-1,4-naphthoquinone, and they contain a long aliphatic chain attached to position C_3 (Chapter 1). However, synthetic 2-methyl-1,4-naphthoquinone, or *menadione*, can act as vitamin K even though it lacks any alkyl substitution at position C_3. The natural vitamins K are lipid substances. Therefore, in order to be absorbed from the intestine, natural vitamin K requires adequate emulsification by bile as well as normal function of the fat-absorption mechanisms. Vitamin K deficiency in man can result from biliary tract obstruction, pancreatic failure, or an intestinal disease in which fat absorption is defective (for example, celiac disease). Humans normally obtain part of their vitamin K requirement from lipid-containing foods in the diet, and the remainder is synthesized by the intestinal bacterial flora. Vitamin K deficiency can lead to bleeding.

In certain clinical situations (such as in this patient with *thrombophlebitis*), pathologic clot formation occurs. Treatment of such conditions include partial inhibition of the normal clotting mechanism in an attempt to prevent additional thrombosis. This was the purpose in administering the coumarin drug. Examination of the structure of coumarin anticoagulants such as dicumarol reveals that they are analogues of vitamin K. Indeed, the coumarins act as competitive inhibitors of vitamin K and therefore inhibit the carboxylation of the vitamin K–dependent coagulation factors produced by the liver. This is manifested clinically by an increase in prothrombin time. The tendency toward clot formation in this test is reduced because there is less factor VII, X, and prothrombin. The other anticoagulant that is often used clinically is *heparin* (Chapter 7). Heparin must be given parenterally, for the complex mucopolysaccharide would be digested if it were administered orally. It begins to exert its anticoagulant effect immediately, and that is why this patient was given this drug as soon as she was admitted for treatment. In contrast, the coumarins require several days to exert their effect, for time must elapse for the body to clear the carboxylated prothrombin that remains in the plasma. Heparin exerts a direct effect on the circulating clotting system, but precisely how it acts is not known.

In the present case the prothrombin time was not determined for 6 weeks, and for unknown reasons the coumarin drug exerted an excessive effect that resulted in hemorrhage. The immediate treatment is the rapid neutralization of the effect of the coumarin. Vitamin K, which has lipid solubility properties, would be extremely difficult to inject directly into the plasma. Vitamin K administered orally would require hours to days to reach the liver in large quantities. Menadione, a substance with vitamin K activity, is quite water soluble and can be injected directly into the blood. It is taken up by the liver, and the coumarin effect is counteracted more quickly.

REFERENCES

Davie, E. W., and Fujikawa, K.: Basic mechanisms in blood coagulation, Annu. Rev. Biochem. **44:**799, 1975.

Weiss, H. J., editor: Platelets and their role in homeostasis, Ann. N.Y. Acad. Sci. **201:**1, 1972.

INTEGRATION OF METABOLIC PATHWAYS

There are three main classes of nutrients from which man derives calories; these are the carbohydrates, fats, and proteins. These nutrients undergo various metabolic transformations, depending on body requirements. Thus glucose may be either stored as glycogen, oxidized through the pentose phosphate and glucuronate pathways, or converted to pyruvate. Likewise, fatty acids may be oxidized, stored as triglycerides, or incorporated into structural lipids. In previous discussions the intermediary metabolism of each nutrient class was considered separately. However, these processes occur in a concerted manner, so that various pathways and systems in the body do not operate at cross purposes. The control mechanisms are hormonal in many instances; in others, concentrations of the various metabolites from one pathway regulate flux through a second pathway.

Case 2: Acute starvation—gluconeogenesis, lipid mobilization, ketosis, and acidosis

An 18-year-old woman who was a college freshman was brought to the emergency room of a university hospital by her roommates. She was pale, dizzy, and extremely weak. Her friends related the following story. The patient had been moderately obese when she entered school; she was 63 inches tall and weighed 167 pounds. During her first month on campus she did not do well socially, whereas her friends who were slimmer attracted attention from numerous men. Despondent, the patient presented herself to the student health office requesting diet medication. She was advised that no effective diet pills were available, and a reducing diet was prescribed. This consisted of 1500 kcal/day, and she was told to expect a weight loss of approximately 1 to 2 pounds/week. The patient realized that it would be well into the spring semester before she attained her desired weight. Determined to make herself attractive at a more rapid pace, the patient decided not to eat anything except fluids and vitamins. She had begun this starvation diet 8 days earlier and had taken only water, black coffee, and two multiple vitamin capsules daily. After several days, she became listless and did not wish to attend classes. She had collapsed while taking a shower on the eighth day of her fast and had been rushed to the hospital by her alarmed friends. Physical examination revealed the odor of acetone on the patient's breath. Urinalysis was positive for acetone but not for glucose, and the urine pH was 5.5. Her plasma glucose was 60 mg/dl and free fatty acids were 1.1 μeq/ml.

Biochemical questions

1. What is the status of carbohydrate and lipid metabolism in this patient?
2. What is the main source of the plasma glucose in starvation?
3. Discuss the nitrogen balance of this patient.
4. What is the mechanism of ketosis in this case?
5. What is the mechanism of acidosis in this case?
6. Would you expect the patient's respiratory rate to be slow or rapid?

Case discussion

The young woman is suffering from a self-imposed metabolic ketoacidosis. Because of starvation, her body metabolism is completely deranged in an attempt to maintain homeostasis. In the earlier consideration of obesity (case 2, Chapter 1) it was suggested that the *complete* withdrawal of food to lose weight was unwise. The biochemistry of the carbohydrates can be elucidated by considering the sequence of events as starvation progresses.

Under normal conditions the body tissues of a well-fed 70 kg person represent caloric storage to the extent of 14,000 kcal of fat, 24,000 kcal of muscle protein, 480 kcal of muscle glycogen, 280 kcal of liver glycogen, and 80 kcal of circulating glucose. These stores serve as the food for the starving person, but an adequate fluid intake must be maintained. The striated muscle, comprising some 40% of the body weight and consuming more than 50% of the total body oxygen supply, normally derives much of its basal energy from free fatty acids. After an overnight fast, glycolysis accounts for only about 20% of muscle oxygen consumption. The myocardium has similar demands for glucose, but it can readily accommodate by utilizing free fatty acids, acetoacetate, β-hydroxybutyrate, and lactate.

Blood cells, particularly erythrocytes, oxidize about 34 g of glucose/day. In mature erythrocytes, which have no mitochondria, this occurs by an anaerobic glycolysis and the lactate that is produced is released into the plasma. This contributes significantly to the resting blood lactate levels.

The brain consumes 110 to 145 g of glucose/day and accounts for 20% to 25% of the total body oxygen consumption. This obviously places a great demand on the blood glucose supply, and in the absence of adequate food, gluconeogenesis is required. This occurs in the liver and kidney. It is also possible that some gluconeogenesis can occur in the brain. It is now accepted that after prolonged starvation the brain can adapt to the oxidation of ketone bodies. Starvation does not decrease the oxygen demand of the brain.

The liver is a total metabolic factory. Like the brain, its oxygen consumption is about 20% of the total and also does not decrease on starvation. It produces but cannot metabolize ketone bodies. It metabolizes glucose, amino acids, fatty acids, and glycerol to meet its own energy needs.

The renal medulla is almost free of mitochondria and is therefore dependent on glycolysis. The cortex, however, favors fatty acid and ketone bodies for its energy supply. During starvation the kidney, unlike adipose tissue, liver, and muscle, does not lose much of its weight.

It would therefore appear that aerobic glycolysis provides significant energy for the major body tissues when dietary glucose supplies are adequate. However, even overnight starvation may result in a tendency toward fatty acid oxidation. This ability to adapt provides the body with a defense against starvation.

Withdrawal of food lowers the blood glucose to fasting levels. This initiates responses of lowered plasma insulin and secretion of glucagon and cortisol, a hormonal state that stimulates glycogenolysis and gluconeogenesis. Protein is hydrolyzed and amino acids are deaminated in the liver; the carbon skeletons of the amino acids enter the gluconeogenic pathway in order to produce glucose and maintain the blood glucose levels. Negative nitrogen balance results, with excretion of the nitrogen more in the form of NH_4^+ than urea; the former conserves Na^+ and K^+, which necessarily accompany the excreted keto acids in the urine. Only so much body nitrogen can be lost before irreversible changes take place.

Relationship between glucose and fatty acid utilization.   During the first day of fasting most of the patient's glycogen stores were depleted. In order to meet the demands for energy, fatty acid was mobilized from her adipose tissue stores for catabolism by the heart, muscles, liver, kidneys, and other organs. Under ordinary circumstances both glucose and fatty acids are catabolized to acetyl

CoA, which then enters the Krebs cycle (Chapter 6). However, in the present situation no glucose is flowing down the glycolytic pathway, and the respiratory energy is derived almost completely from oxidation of fat. If the *respiratory quotient* were measured in this patient, the value would be almost 0.75. The respiratory quotient is the ratio of carbon dioxide produced to oxygen consumed. If carbohydrate were the only oxidative substrate, the respiratory quotient would be 1.0. It would be 0.8 with protein and 0.7 with fat as the only oxidative substrate.

Regulatory mechanisms usually ensure that excessive amounts of both glucose and fatty acid do not continuously feed into the acetyl CoA pool, for this would be wasteful. When such a situation begins to develop, the concentration of acetyl CoA within the mitochondria rises. Pyruvate carboxylase, an allosteric enzyme that is acetyl CoA dependent, is activated. Pyruvate is converted to oxaloacetate instead of acetyl CoA, providing more acceptor for the acetyl CoA coming in from fatty acid β-oxidation, as shown in Fig. 14-4. This regulatory mechanism prevents an excessive buildup of acetyl CoA and, at the same time, provides increased capacity to catabolize acetyl CoA through the Krebs cycle. Regulation also can occur at the level of delivery of fatty acid to the mitochondrion. The fatty acid is derived either from circulating lipids, chiefly free fatty acid, or intracellular triglycerides. The degradation of triglycerides is controlled by lipases that in some tissues are activated hormonally through the cAMP mechanism. When glucose concentration is high, insulin levels would be high, intracellular cAMP low, and lipolysis minimal. Any fatty acid that would be produced would tend to be resynthesized into triglyceride because of the high concentration of L-glycerol 3-phosphate, as illustrated in Fig. 14-5. The glucose-directed resynthesis of triglycerides occurs in the adipocytes, decreasing free fatty acid release into the blood. As a result, less fatty acid would be available for β-oxidation.

Additional regulation can occur at the level of glucose metabolism. Since there is adequate substrate for oxidation and sufficient ATP is being produced, it would be wasteful to continue to convert large amounts of glucose to pyruvate. Under these conditions glycolysis is reduced through a negative allosteric effect at the phosphofructokinase step. Since glycolysis is inhibited, the excess available glucose would be stored as glycogen or oxidized through the pentose phosphate pathway to provide NADPH for biosynthetic reactions. This is illustrated in Fig. 14-6. Details about each of these pathways are given in Chapter 7.

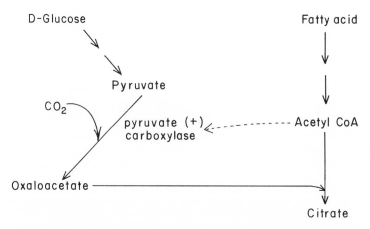

Fig. 14-4. Regulation of pyruvate metabolism by acetyl CoA.

In the present patient there is not an overabundance of circulating glucose. The insulin concentration is extremely low, and in its place there are high levels of circulating glucagon and epinephrine. Free fatty acid is being released from the adipose tissue and is elevated in the plasma, providing the fatty acid that is needed for oxidation by the tissues. Intracellular triglycerides also are providing fatty acid for this purpose, and essentially all of the acetyl CoA is being derived from these sources, not from glucose.

Mechanism of ketosis. Ketone bodies are made in the liver from acetyl CoA (Chapter 10). They are secreted in large quantities when excessive amounts of fatty acid are being oxidized and glucose availability is limited, as in the present case of starvation. Normally the amount of fatty acid that is converted to acetyl CoA is regulated by the availability of L-glycerol 3-phosphate, derived from glucose through the glycolytic pathway (Fig. 14-5). When acetyl CoA concentrations rise, the acyl CoA formed from fatty acid is resynthesized into triglycerides, with L-glycerol 3-phosphate serving as the acceptor to which the acyl groups are esterified. This regulatory mechanism does not operate in periods of starvation because there is insufficient glucose for glycerol phosphate formation. As a result, the input of acetyl CoA exceeds the ability of the Krebs cycle to oxidize it, and the excess is

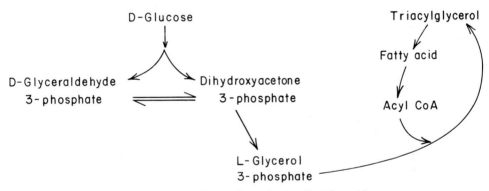

Fig. 14-5. Glucose-directed synthesis of triglycerides.

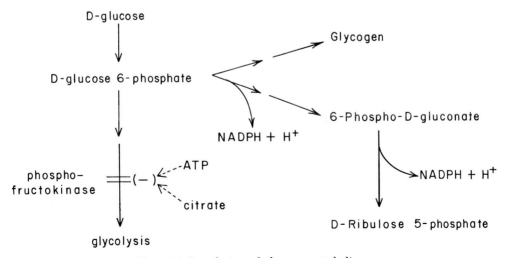

Fig. 14-6. Regulation of glucose metabolism.

shunted into ketone bodies. As described in Chapter 8, this usually is not wasteful. Ketone bodies released from the liver are taken up and serve as an oxidative substrate for many organs, including the brain.

Mechanism of acidosis. Acetoacetic and β-hydroxybutyric acids are strong acids relative to the bicarbonate buffer system of the plasma. The large influx of the acids into the plasma, which is buffered at pH 7.4, leads to an input of hydrogen ions. In an attempt to buffer this, bicarbonate is converted to carbonic acid and thence to CO_2. Eventually the total CO_2 content of the plasma becomes depleted, the buffering capacity of the plasma is exceeded, and the pH decreases, as was described in Chapter 4. In terms of the Henderson-Hasselbalch equation, the pH falls because the bicarbonate concentration is decreased.

$$pH = pK_a + \log \frac{[HCO_3^-]}{[\text{dissolved } CO_2]}$$

This condition is known as metabolic acidosis. Since appreciable changes in pH are detrimental and can be fatal, the body initially attempts to compensate for the acidosis and returns the pH toward 7.4. This is accomplished by the respiratory system. Inspection of the above equation indicates that pH is dependent on the $[HCO_3^-]/[CO_2]$ ratio. Since bicarbonate has decreased, it is necessary to decrease the dissolved CO_2 concentration, the denominator, in order to return the ratio toward its physiologic value of 20. The dissolved CO_2 is dependent on the $\bar{p}CO_2$, so it is necessary to reduce the CO_2 gas content in the plasma. This is accomplished by hyperventilation. Although this compensation relieves the immediate threat, it actually causes further derangement of the system by lowering the total CO_2 content of the body. This is the condition of our patient at the time of admission. Her ketone body production is so high that she is excreting ketones in her urine; this, of course, is producing electrolyte imbalances (Chapter 4). While her body has compensated for the metabolic acidosis and her pH probably is in the range of 7.2 to 7.3, she is suffering from severe acid-base abnormalities because the buffering capacity in terms of total CO_2 is probably quite low, perhaps in the range of 10 to 15 meq/liter.

Gluconeogenesis. Hypoglycemia may result in convulsions and even death. In extreme circumstances the brain can utilize ketone bodies as an oxidative substrate; however, it still requires glucose for optimal activity. When there is no input of glucose from the diet and glycogen stores are depleted, the body makes glucose from the glucogenic amino acids. Except for leucine and perhaps lysine, all amino acids have at least some glucogenic properties. The major amino acid inputs into glucose production, which are either transaminations or oxidative deaminations, are discussed in Chapters 6 and 9. Gluconeogenesis is stimulated by cortisol, by glucagon, and by high concentrations of fatty acid. It occurs predominantly in the liver. Only the liver and the kidney contain glucose 6-phosphatase, the enzyme necessary so that intracellular glucose 6-phosphate can be released into the blood as glucose.

The amino acids used for gluconeogenesis are derived from plasma and tissue proteins. In situations such as this the overall status of protein nutrition can be estimated from nitrogen intake and excretion. Because of the lack of food intake, the patient was not replacing the amino acid output. There was a net utilization of body protein that produced a negative nitrogen balance, as described in Chapter 1.

REFERENCES

Felig, P.: Amino acid metabolism in man, Annu. Rev. Biochem. 44:933, 1975.

Owen, O. E., and Richard, G. A., Jr.: Fuels consumed by man: the interplay between carbohydrates and fatty acids, Progr. Biochem. Pharmacol. 6:177, 1971.

PANCREATITIS: A MULTIENZYME, MULTISYSTEM DISEASE

In an earlier chapter (case 2, Chapter 3) a case of acute pancreatitis was presented in a highly simplified manner. Because the pancreas is both an exocrine and an endocrine organ, pancreatic disease is accompanied by biochemical findings related to both the endocrine and digestive systems. At this point pancreatitis can be reexamined on a broader basis.

Case 3: Acute pancreatitis

P. J., a 36-year-old man, was first seen in the emergency room at 6 PM. He stated that at approximately 4:15 PM that afternoon he had felt a sharp, severe pain that increased in intensity for about 45 min and that had since remained constant. He had never before experienced pain of this type. He was sitting quite still in a chair, and when he moved to an examining table he stated that the pain worsened. He looked pale, sweaty, and sick. His respirations were shallow, and he stated that deeper breathing was more painful. His blood pressure was 100/60 mm Hg, his pulse was 110 and regular, and his temperature was 37.5° C. Significant physical findings were limited to the abdomen, which was rigid to palpation. There was tenderness (pain) when the epigastrium was gently pressed; when the examiner's hand was suddenly removed, the pain was momentarily increased (rebound tenderness).

The patient was admitted with orders that he be given nothing by mouth. An intravenous catheter was inserted, and he was given a mixture of glucose and saline solution. A nasogastric tube was employed for continuous aspiration of stomach contents.

Laboratory data obtained during the acute phase of this disease are shown in Table 14-2. Because of the possibility of gallbladder disease, intravenous cholangiography was performed soon after admission: the results were equivocal and the common bile duct appeared normal. The marked elevation of the serum amylase, together with lack of any previous history of peptic ulcer, were grounds for a tentative diagnosis of acute pancreatitis. Thirteen days after the onset of symptoms, all laboratory findings but one had returned to normal levels, and the patient was asymptomatic. The fasting blood glucose remained elevated for 3 weeks, and the postprandial blood glucose determinations were high for an additional month. Three months after the first signs of disease, gallstones were visualized by radiographic procedures and the gallbladder was removed surgically.

Biochemical questions

1. Which digestive enzymes are produced in the pancreas?
2. How would digestion and absorption be affected if the enzymes produced by the pancreas were lacking?
3. Which hormones are produced in the pancreas? In what form are they stored in that organ?
4. What are the major ionic components of the pancreatic secretions? Do they occur in constant concentration?
5. What kinds of electrolyte abnormalities would be induced by lengthy employment of nasogastric suction?

Table 14-2. Laboratory data for patient P. J. (see text for details)

DATE AND TIME	SERUM AMYLASE*	SERUM LIPASE†	SERUM CALCIUM (mg/dl)	SERUM SODIUM (meq/liter)	SERUM POTASSIUM (meq/liter)	SERUM CHLORIDE (meq/liter)	SERUM BICARBONATE (meq/liter)	FASTING BLOOD SUGAR (mg/dl)
4/27 (1850)	2100	2.0	10.0	140	4.2	103	22	97
4/28 (0730)	2400	3.8	9.5	142	4.0	104	23	
4/29 (0800)	1800	4.0	9.7	141	4.1	103	21	←— Intravenous
4/30 (0830)	1200	3.7	8.2	144	3.7	101	24	glucose drip
5/1 (0800)	800	3.8	8.7	140	3.2	98	29	
5/2 (0830)	950	3.7	8.4	141	2.9	93	34	
5/3 (0900)	825	3.6	8.8	140	2.9	91	37	
5/4 (0830)	950	5.2	9.2	142	3.4	93	33	
5/5 (0830)	700	2.9	9.5	141	3.7	99	29	—→
5/7 (0700)	400	3.0						120
5/10 (0800)	137	1.9	10.1					125

*Amylase reported in Somogyi units/dl.
†Lipase reported in Cherry-Crandall units/ml.

6. Would treatment with glucose and saline completely correct the electrolyte abnormalities?
7. What controls regulate the production of pancreatic secretions?
8. Are there natural inhibitors of any component of the pancreatic secretions? Do these occur in man?

Case discussion

The endocrine elements of the pancreas produce proinsulin as the storage form of this blood sugar regulating hormone. Proinsulin is produced in the β-cells of the islets of Langerhans as a single large peptide that, on activation, undergoes cleavage into two smaller peptides joined by disulfide bonds. There is simultaneous elimination of a third peptide fragment and several amino acid residues (see Chapter 13). The major action of insulin is to lower the concentration of glucose in the blood by facilitating entry of glucose into skeletal muscle, adipose tissue, and so forth. Insulin also facilitates amino acid uptake and protein biosynthesis, although its effect on the latter mechanism is not clear. Release of insulin is regulated in some way by the amount of glucose in the blood. A second pancreatic hormone is glucagon, made in the α-cells of the islets of Langerhans. Glucagon is a much smaller peptide than insulin. Glucagon activity resides in a peptide only twenty-nine amino acid residues long, which has been sequenced. There is now evidence that glucagon is also made as a somewhat larger prohormone. There is also good evidence for a second source of glucagon, produced in endocrine cells of the upper gastrointestinal tract. This is the so-called enteroglucagon, which appears to be substantially larger than the pancreatic hormone, but which has not yet been highly purified. The two forms of the hormone cross-react in immunologic tests. The effects of glucagon are opposite to those of insulin and mimic the glycogenolytic effects of epinephrine. Glucagon not only increases the concentration of glucose in the blood, it also raises the concentrations of free fatty acids. It promotes lipolysis in adipose tissue by activation of adenyl cyclase, which increases concentrations of cAMP in adipose tissue.

The exocrine elements of the pancreas are responsible for production of amylase, phospholipase, nucleotidases, carboxypeptidases, and the major proteinases, trypsin and chymotrypsin. As far as is known, amylase and the nucleotidases are produced and stored in the active forms. Lipase is not produced as a proenzyme, but does require a cofactor, termed colipase (see Chapter 8), and bile acids for activation. Phospholipases appear to be formed as proenzymes, or zymogens. The carboxypeptidases and the proteinases are also formed and stored as proenzymes. Most of these pancreatic enzymes are very potent; if they were released in the pancreas in fully active form, the pancreas would probably undergo autolysis. Two mechanisms exist to prevent this. First, storage in the form of proenzymes drastically reduces or eliminates enzyme activity. Second, an antitryptic protein is also produced in the pancreas. This severely inhibits the small amount of proteolytic activity that could occur in the organ should some activation of the proteolytic proenzymes occur. An antitryptic globulin is also found in the blood plasma, but it is not clear if this is identical with the pancreatic protein. When the pancreatic proenzymes are released into the duodenum they are activated by the duodenal enzyme known as *enterokinase*. The first traces of free trypsin serve to promote the activation of more trypsinogen and chymotrypsinogen. In other words, the process becomes autocatalytic. The other enzymes are similarly activated.

All of the pancreatic enzymes are contained in a juice that is a mixture of ions derived from blood plasma. The juice is formed continuously, with maximum production during the digestive process that follows a meal. The concentrations of Na^+ and K^+ in pancreatic juice approximate those in the blood plasma, but the Cl^- and HCO_3^- concentrations are quite different; they are 40 and 110 meq/liter, respectively. The high concentration of HCO_3^- is required to bring the products of gastric digestion to an alkaline pH suitable for trypsin and chymotrypsin. *Secretin,* a hormone produced in the duodenum and released by the HCl that reaches the intestine from the stomach, stimulates output of inorganic ions by the pancreas and makes the pancreatic juice richer in inorganic ions and water. *Pancreozymin,* another hormone produced in the duodenum by influx of fat and protein or by the products of gastric digestion, promotes the release from the pancreas of the digestive proenzymes. Thus the exact composition of the pancreatic juice is not entirely constant but reflects the nature of material entering the duodenum. The daily production of pancreatic juice in a normal adult varies between 1500 and 3000 ml, with virtually all of the fluid being resorbed during passage through the lower portions of the small intestine and the large bowel.

Loss of the pancreatic enzymes, by surgical drainage or other means, would seriously hinder digestion. The incompletely digested material would not be readily absorbed. While some of it might be further broken down by bacteria resident in the intestinal tract, this generally would occur at too low a point to permit much useful uptake; as a consequence, the undigested part of the food intake would be wasted in the excreta. Loss of the pancreatic juice would also involve a significant loss of water, leading to dehydration, and a loss of important ionic components of the body fluids.

Pathology of acute pancreatitis. The underlying pathology of acute pancreatitis is probably the result of tissue autolysis following premature, intrapancreatic enzyme activation, especially of the proteinases. The most common causative factors are (1) obstruction of the pancreatic ductules with a continued stimulus for secretion, (2) reflux of bile into the pancreatic duct system, (3) certain viral infections, including mumps, hepatitis, and infectious mononucleosis, (4) a variety of nutritional or metabolic disturbances, and (5) severe alcoholism.

In acute pancreatitis caused by biliary reflux it has been proposed that the bile salts attack cell membranes, removing some of the essential phospholipids. As a result, proenzymes are prematurely activated, a process that leads to altered membrane and vascular permeability. The results may range from simple edema to hemorrhagic necrosis. This accounts for the symptoms of severe pain and mild systemic shock mentioned in the case description. Continued activation of proenzymes and extended permeability changes are both responsible for increased concentrations of amylase and lipase in the blood serum. Indeed, the concentration of amylase in the blood may become so great that it exceeds the renal threshold, and in some cases of acute pancreatitis it is possible to detect and measure amylase activity in the urine, especially during the most acute phase of the disease. In this patient the serum amylase was markedly elevated almost from the beginning, and the amylase concentration returned to normal values (50 to 150 Somogyi units/dl) rather slowly. The maximal increase of lipase activity was reached more slowly and took even longer to return to normal (0.05 to 1.0 Cherry-Crandall units/ml.

Biochemical basis of therapy. It is useful to reduce as much as possible any stimuli to further production of the pancreatic secretion. This was accomplished by the order to give nothing by mouth, which eliminated stimulation by gastric digestion products, and by nasogastric suction, which kept HCl from entering the duodenum. On the other hand, continued suction has a significant effect on electrolyte balance. Note from the laboratory data (Table 14-2) the fall in serum K^+ and Cl^- concentrations, since these are major ions of the gastric juice. A partial physiologic compensation for loss of serum Cl^- was accomplished by retention of serum HCO_3^-. We can conclude from these data that the therapeutic management introduced secondary problems of induced hypokalemia and a metabolic alkalosis (see Chapter 4).

Note also from the laboratory data that the serum Ca^{++} concentration dropped significantly during the acute stages of pancreatitis. Until recently it was felt that the decline in serum Ca^{++} concentration resulted from binding to fatty acids released by autolysis of lipid in the pancreas, forming insoluble soaps. It now seems clear that by itself this explanation is inadequate. In the first place, removal of serum Ca^{++} should be readily offset by mobilization from bone, which constitutes a large calcium reservoir. Secondly, replacement of Ca^{++} by intravenous administration often fails to restore the normal serum concentration, even though the amount given is equal to or greater than the estimated loss. It has recently been found that acute pancreatitis is often accompanied by hypercalcitoninemia, with the peak value of serum calcitonin preceding the greatest fall in serum Ca^{++}. The stimulation for release of calcitonin is the output of glucagon by the injured pancreas. Secondary stimuli may come from enteroglucagon and from *gastrin*, another product of the pancreatic islets, which is a very potent stimulator of calcitonin release.

There is frequently no synchrony between the rate of decline of the serum Ca^{++} concentration and the peak of the rise in serum calcitonin. This makes it appear probable that both of the causes cited above contribute to the problem. Saponification of fats should cause a drop in Mg^{++} concentration as well as that of Ca^{++}, and hypomagnesemia has been observed in pancreatitis as well as in hypercalcitoninemia. Consequent to changes in serum Ca^{++} and Mg^{++} concentration, there should be a significant decrease in serum P_i. While this was not noted in the case at hand, it can be predicted that a hyperphosphaturia could have been demonstrated, on the basis of calcitonin release.

Other biochemical considerations. The transient diabetic-like state and the hypocalcemia usually subside as the injured pancreas heals, except in cases where the serum Ca^{++} concentration drops to 7.0 mg/dl or less, when the clinical outlook becomes grave.

It is important to realize that the pancreas is the site of production of hormones in addition to glucagon and insulin, and that the physiologic function of these other hormones is only now becoming a matter of clinical concern. Similarly, the endocrine functions of the intestinal tract are not insignificant. As the present example illustrates, there is an essential interplay between the hormones and enzymes of the pancreas and the hormones of the intestine. In addition, these effects are further modulated by the thyroid-parathyroid system and the osteocytes of the bones, all of which can be assessed through changes in chemistry of the blood.

This patient was treated in a conservative manner. The major supportive therapy consisted of glucose and saline infusions and medication for relief of pain. Had the situation been more acute or had the attending physician been inclined to more vigorous therapy, several additional measures might have been taken. Calcium salts might have been added to the glucose infusion to offset the decreased serum Ca^{++} concentration. Similarly, potassium chloride or phosphate might have been added to correct electrolyte imbalance further.

As mentioned earlier, human blood contains an antitryptic globulin; however, in acute pancreatitis, with its attendant increased output of proteolytic enzymes, the effects of the antitryptic globulin may be insufficient. In this situation a naturally occurring antitrypsin derived from soybeans can be administered. This agent is a polypeptide with a molecular weight of approximately 6000. It is a heat-resistant, dialyzable material that acts by a mechanism that is not well understood but that strongly inhibits the activity of trypsin.

REFERENCES

Canale, D. D., and Donabedian, R. K.: Hypercalcitoninemia in acute pancreatitis, J. Clin. Endocrin. Metab. **40:**738, 1975.

Cantarow, A., and Trumper, M.: Clinical biochemistry, ed. 7 (ed. by A. L. Latner), Philadelphia, 1975, W. B. Saunders Co.

Dreiling, D. A., Janowitz, H. D., and Perrier, C. V.: Pancreatic inflammatory disease, New York, 1964, Harper & Row, Publishers.

Makhlouf, G. M.: The neuroendocrine design of the gut, the play of chemicals in a chemical playground, Gastroenterol. **67:**159, 1974.

Unger, R. H.: Alpha- and beta-cell interrelationships in health and disease, Metabol. **23:**581, 1974.

OSTEOGENESIS IMPERFECTA, A CONNECTIVE TISSUE DISEASE

Collagen is a major protein of connective tissue. It comprises approximately 5% of the total body weight. In certain instances defects in collagen biosynthesis occur, giving rise to serious disease.

Case 4: Osteogenesis imperfecta

A 7-year-old girl was referred to the university dental clinic because of a delayed pattern of dental eruption. Her family dentist was also concerned about malformations of the teeth that had erupted, as well as their unusually opalescent appearance. In the course of examination the staff dentist was informed that the patient had previously suffered two fractures of her left arm and observed also that the child was hard of hearing. He arranged a consultation with pediatric consultants, whose findings are summarized as follows.

The ligaments of the joints in this child's arms and legs were unusually lax. X-ray examination of the rib cage showed lack of symmetry on the right and left. Her skin had a waxy, transparent appearance, she sweated during the examination and complained of the heat in the room. A definite hearing loss was confirmed by audiometry, and her tympanae (and sclerae) appeared blue and thinned. Her resting heart rate was 105 beats per min (normal 75 to 90) and her respiratory rate was 60 per min (normal 25). Her basal metabolic rate was measured as 142% of normal. An intravenous tolerance test for proline showed she was intolerant of the test dose. Examination of her urine showed an excessive amount of pyrophosphate.

Biochemical questions

1. Name a protein especially rich in the amino acid proline.
2. Which vitamin is involved in the conversion of proline to hydroxyproline?
3. What is the structure of hydroxylysine?
4. Is there a metabolic pathway for removal of the ϵ-NH_2 group of lysine? How would you classify the enzyme that catalyzes the reaction?
5. How are the monomeric peptides of collagen cross-linked in the triple helix, given the fact that the monomers contain no cysteine?
6. In terms of total energy balance, what is the consequence of pyrophosphate loss in the urine?

7. Would you expect a normal P/O ratio in tissues from patients with osteogenesis imperfecta? Why?
8. Does the synthesis of collagen require exclusively intracellular enzymes?

Case discussion

Osteogenesis imperfecta is an inheritable disease long thought to be almost entirely a problem of the skeletal system. It is now known to involve virtually all tissues, since the cause of the disease is a defect in the synthesis of collagen, which is a component of most tissues. The expression of the disease varies widely; in the most severe cases infants are born dead with multiple fractures resulting from the stress of delivery. In the mildest forms it is not incompatible with normal life spans, and one patient has been known to reach the ninth decade of life. The disease is frequently first detected by virtue of the associated dental problems that can include deformed teeth or delayed eruption or by a history of repeated bone fractures. In more severe forms of the disease a hypermetabolic state exists, recognized by abnormally rapid heart and respiratory rates, increased oxygen consumption, excessive sweating, and heat intolerance.

The microscopic histology of tissues from patients with this disease reveals abnormalities of the collagen component. Since collagen is a major structural element of the sclerae, the skin, the tendons, and cartilage, it is not surprising that these tissues are, among others, affected by the disease. The major organic component of the sclerae is collagen, and the abnormality is reflected by a bluish, thinned appearance that is typical. Similarly, the skin appears transparent, as noted in the case description. Until recently, the biochemical lesion in collagen production was not clear. To understand the newest evidence requires a review of collagen structure and biosynthesis.

Biochemistry of collagen.   As noted earlier (see Chapter 2) collagen consists of three peptide chains wound together in the manner of a left-handed cable. It is very rich in glycine and proline and also contains numerous lysine residues. Part of the proline and lysine residues are hydroxylated after the individual peptides have been constructed. When the collagen molecules are first formed the three peptide chains are only slightly cross-linked, but as the collagen ages, cross-linking becomes more extensive. Cross-linking is possible because some of the ϵ-NH_2 groups of lysyl or hydroxylysyl residues are attacked by a copper-dependent oxidase that removes the amino group, leaving in its place an aldehyde group. The modified residue is known as an allysyl residue, as shown below.

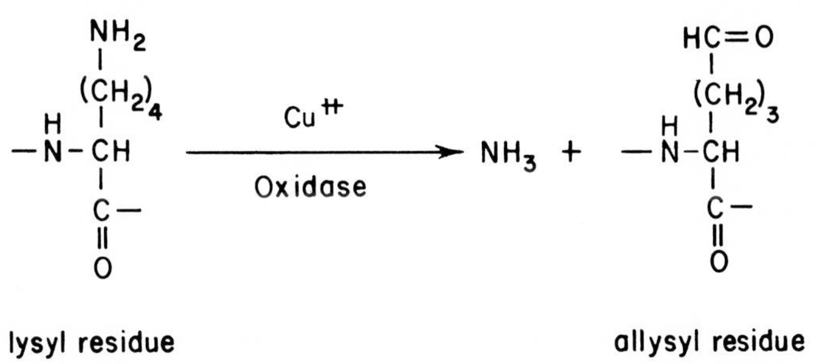

lysyl residue   allysyl residue

An allysyl residue in one peptide chain can undergo an aldol condensation with a hydroxylysyl residue in another, or a lysyl or hydroxylysyl residue in one chain can form an aldimine with an allysyl residue in another. These are the major means of cross-linking the peptide chains.

Table 14-3. Subunit structure and function of collagens

COLLAGEN TYPE	CHAINS CONTAINED	MOLECULAR FORM	MAJOR COLLAGEN COMPONENT
I	$\alpha1(I)$, $\alpha2$	$[\alpha1(I)]_2\alpha2$	Bone, mature skin, tendon, dentin, dental pulp, and liver
II	$\alpha1(II)$	$[\alpha1(II)]_3$	Cartilage
III	$\alpha1(III)$	$[\alpha1(III)]_3$	Young skin, cardiovascular system, some basement membranes
IV	$\alpha1(IV)$	$[\alpha1(IV)]_3$	Some basement membranes

Nature of the individual peptide chains. The collagens of many vertebrate tissues have been examined. When the triple helix is dissociated by weakly acidic media, at least five distinct peptide chains can be distinguished by electrophoresis or column chromatography. These are designated as $\alpha1(I)$, $\alpha1(II)$, $\alpha1(III)$, $\alpha1(IV)$, and $\alpha2$. These chains differ in their amino acid composition, but the exact sequence of each is not known at the present time. All collagen preparations examined thus far are trimers, composed of one or more of the chains described above. Some examples of the major collagenous components of mammalian tissues are listed in Table 14-3.

Biosynthesis of collagen. How genes are selected to produce a particular form of collagen required by a given cell is not known. Nor is it known if a cell can produce more than one collagen type at a time, or if the production of one collagen type automatically represses synthesis of another. From the data given in Table 14-3 it is clear that cells of a given tissue produce only selected collagen types.

The earliest ribosomal products in collagen synthesis bear at their N-terminus a sequence of amino acids, some 200 residues long, which includes cysteine. Recall that cysteine does not appear in collagen itself. This extra length of the initial peptide functions to align the three chains that form the triple helix. This alignment sequence is shown in Fig. 14-7 by a wavy line. The several peptide chains, that is, $\alpha1(I)$ and $\alpha2$, are produced simultaneously. Hydroxylation of prolyl and lysyl residues occurs while the nascent peptide chain is still attached to the ribosomes, but some recent evidence indicates it may continue after the peptide detaches from the ribosomes. The hydroxylases require Fe^{+2}, α-ketoglutarate, and a reducing agent such as a pteridine or ascorbate. The reaction employs molecular oxygen, and in the course of the reaction the α-ketoglutarate is converted to succinate. Glycosylation can occur, with addition of galactose or glucosylgalactose to the δ-carbon atom of selected hydroxylysyl residues, via glycosidic linkages. The number of carbohydrate groups may vary from one to thirty per peptide chain. Usually, only approximately half of the hydroxylysines are glycoslyated. The function of the attached carbohydrate is not known. The aligned peptides form a triple helix.

The product of these reactions is known as procollagen; how procollagen moves from within cells to the extracellular space is still a controversial matter. According to one theory, procollagen moves from the endoplasmic reticulum to the Golgi apparatus, which develops secretory granules that deliver the product from the odontoblasts or osteoblasts, depending on the tissue examined, to the site of collagen deposition. In the case of the rat incisor, label from ^{3}H-proline appeared in predentin within 30 min to 4 hr. The procollagen is converted to collagen by action

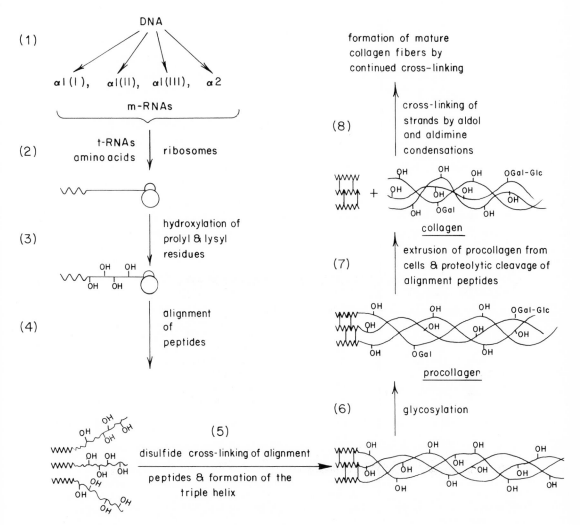

Fig. 14-7. The synthesis of collagen involves numerous steps. *1*, Genes coding for the appropriate monomer types are transcribed to corresponding mRNA's. How the particular genes are recognized is not understood. *2*, Ribosomal translation of the message produces a series of peptides, each of which carries at its N-terminus a sequence of amino acids, containing cysteine, which will later be removed by extracellular proteolysis. The purpose of this sequence is to group the collagen monomers in correct alignment. *3*, Hydroxylation of the nascent peptides begins while the strands are still attached to the ribosomes and may continue after they are detached. *4*, The peptides are aligned by formation of disulfide bonds between the alignment portions of the separate strands. *5*, Formation of the disulfide cross-links induces triple helix formation involving the monomers, which are not all of the same length. *6*, Glycosylation takes place, adding galactose or galactosylglucose units to the strands of the triple helix. *7*, The procollagen is extruded from the cells, then the alignment peptide portions are cleaved off by the extracellular enzyme, procollagen peptidase, to give a collagen molecule. *8*, The strands in collagen are further cross-linked by reactions involving allysine, hydroxylysine, lysine, etc. See text for further details.

of an extracellular protease, procollagen peptidase. Presumably, similar mechanisms exist in other tissues.

Collagen abnormality in osteogenesis imperfecta. Skin fibroblasts can be grown in culture. Fibroblasts from mature skin produce largely type I collagen, with only a small amount of type III. In fibroblasts from fetal skin, type III collagen may account for half or more of the total. When cells from a child who had died of severe osteogenesis imperfecta were compared with age-matched controls, it was found that the diseased cells produced a much higher proportion of type III collagen than the controls. Since skin was the only tissue studied in culture, it is not clear if the quality of collagen produced in other tissues was equally affected. There is evidence that the deposition of mineral in bony tissues is dependent on the collagen type, so it is presumed that substitution of type III collagen for type I could account for the symptoms of abnormal hard tissues associated with this disease.

Other metabolic abnormalities in osteogenesis imperfecta. There is a significant increase in oxygen consumption in patients with osteogenesis imperfecta, and there may be a concomitant increase in the pulse and respiratory rate. Together, these symptoms are indicative of a hypermetabolic state. Even isolated leukocytes demonstrate increased metabolism, so the situation is not limited to hard tissues. The exact cause of the hypermetabolism is not entirely clear, but it can be shown that patients with active, untreated disease excrete significantly greater than normal quantities of pyrophosphate in their urine. The blood serum may also contain increased amounts of pyrophosphate. How pyrophosphate escapes from cells without hydrolysis to orthophosphate is not known. In any event, the urinary loss of pyrophosphate represents a waste of energy that may account, at least in part, for the hypermetabolic state.

REFERENCES

Armstrong, D., and VanWormer, D.: Inorganic serum pyrophosphate in patients with osteogenesis imperfecta, Clin. Chem. 21:104, 1975.

Bornstein, P.: Biosynthesis of collagen, Annu. Rev. Biochem. 43:567, 1974.

Cropp, G. J. A., et al.: Physiological evidence of hypermetabolism in osteogenesis imperfecta, Pediatrics 49:375, 1972.

Gallup, P., Blumenfeld, O., and Seifter, S.: Structure and metabolism of connective tissue proteins, Annu. Rev. Biochem. 41:617, 1972.

Humbert, J. R., et al.: Increased oxidative metabolism by leukocytes of patients with osteogenesis imperfecta and of their relatives, J. Peds. 78:648, 1971.

Miller, E. J., and Matukas, V. J.: Biosynthesis of collagen, Fed. Proc. 33:1197, 1974.

Penttinen, R. P., et al.: Abnormal collagen metabolism in cultured cells in osteogenesis imperfecta, Proc. Natl. Acad. Sci. U.S.A. 72:586, 1975.

Summer, G. K., and Patton, W. C.: Intravenous proline tolerance in osteogenesis imperfecta, Metabolism 17:46, 1968.

Weinstock, M., and Leblond, C. P.: Formation of collagen, Fed. Proc. 33:1205, 1974.

CENTRAL NERVOUS SYSTEM DISEASES

The biochemistry of the nervous system is one of the most active areas of current biomedical investigation. Although large strides have been made in the last decade, the complexity of the brain is so great that we still know little of how it functions at the biochemical level. Therefore, only in a very few instances do we currently have any understanding of either neurologic or psychiatric diseases in terms of biochemical abnormalities. Several of these have already been mentioned; for example, the central nervous system degeneration that accompanies lipid storage diseases such as Niemann-Pick or Gaucher's disease (Chapter 8). A major recent advance in the area was the elucidation of the metabolic defect in *heredopathia atactica polyneuritiformis*, commonly known as *Refsum's disease*. This is a very rare, inherited abnormality that produces a number of neurologic defects. Since this disease is not widespread, elucidation of the biochemical defect that produces it does not of itself represent a major contribution in terms of human health. The finding is of great potential importance, however, because it offers a new metabolic approach for investigation of many other neurologic degenerative diseases whose etiology is presently unknown.

Case 5: Refsum's disease

A 34-year-old male was referred to the Clinical Center of the National Institutes of Health because of severe neurologic problems. He did not have a sense of smell, and he could not see in dim light or in the dark. He had tinnitus and progressive deafness, requiring the use of a hearing aid. Physical examination revealed cerebellar ataxia, motor weakness, and distal sensory loss. Muscle wasting was noted in his arms and legs, and a fast tremor affecting the fingers and wrists was observed. Ophthalmologic examination revealed that the visual fields were severely restricted, and retinitis pigmentosa was observed. A liver biopsy revealed the presence of excessive numbers of lipid droplets in the hepatocytes. Further analysis of these lipids by gas-liquid chromatography (Chapter 8) indicated that phytanic acid comprised about 30% of the tissue fatty acids. Moreover, similar analyses revealed that the plasma contained 45 mg/dl of phytanic acid, the normal value being about 0.2 mg/dl. A diagnosis of Refsum's disease was made.

Biochemical questions

1. What is the source of phytanic acid in humans?
2. What is the metabolic defect in Refsum's disease?
3. How might the observed metabolic abnormality produce the central and peripheral nervous system abnormalities seen in this patient?

Case discussion

The structure of phytanic acid is shown on p. 368. It contains twenty carbon atoms, sixteen of which are present in a straight chain. There are four methyl branches, occurring at C_3, C_7, C_{11}, and C_{15}, counting from the carboxyl group. Phytanic acid is synthesized from phytol, a component of many foods that are derived from plants. The structure of phytol corresponds to that of phytanic acid except that C_1 is an alcohol group rather than a carboxylate group. When phytol is ingested into the body, it is oxidized to form phytanic acid. This is part of the normal catabolic pathway for phytol, and the further degradation of phytanic acid ensures that neither phytol nor any of its derivatives accumulate in the body.

The first step in phytanic acid catabolism is an α-oxidation in which its carboxyl group is removed as CO_2. In this reaction the original C_2 of phytanic acid is converted to a carboxyl group, 2-hydroxyphytanic acid being the intermediate in this conversion.

$$\underset{\overset{|}{R-CH-CH_2-COOH}}{CH_3} \rightarrow \underset{\overset{|}{R-CH-CH-COOH}}{\overset{CH_3\ \ OH}{}} \rightarrow \underset{\overset{|}{R-CH-COOH}}{CH_3} + CO_2$$

Notice that phytanic acid cannot be oxidized by β-oxidation, for there is a methyl group on C_3. This prevents the formation of the β-keto intermediate, the final step in β-oxidation prior to thiolytic cleavage of the acetyl CoA group through the action of CoASH (p. 390). When the carboxyl group of phytanic acid is removed in the initial oxidation, however, the methyl branch moves to the α-carbon atom and the β-carbon atom becomes a methylene group.

$$\underset{\alpha\ \ \ \ \ \beta\ \ \ \ \ \alpha}{R-CH_2-\overset{\overset{\displaystyle CH_3}{|}}{CH}-CH_2-COOH} \rightarrow \underset{\beta\ \ \ \ \ \alpha}{R-CH_2-\overset{\overset{\displaystyle CH_3}{|}}{CH}-COOH} + CO_2$$

β-Carbon is blocked β-Carbon can be oxidized

Therefore, the new nineteen-carbon atom acid that is formed, *pristanic acid*, is able to undergo normal β-oxidation. When the three-carbon atom fragment is removed in the β-oxidation pathway, *propionyl CoA* rather than the usual acetyl CoA is released from this branched chain segment.

$$\underset{}{R-\overset{\overset{\displaystyle O}{||}}{C}-\overset{\overset{\displaystyle CH_3}{|}}{CH}-\overset{\overset{\displaystyle O}{||}}{C} \sim SCoA + CoASH} \rightarrow R-\overset{\overset{\displaystyle O}{||}}{C} \sim SCoA + CH_3-CH_2-\overset{\overset{\displaystyle O}{||}}{C} \sim SCoA$$

Propionyl CoA

Likewise, two other fragments of pristanic acid that contain the methyl branches also give rise to propionyl CoA, while the ω-terminal fragment gives rise to a four-carbon atom acyl CoA. In this way pristanic acid is completely degraded by β-oxidation. The key to the mechanism is the initial α-oxidation step that unblocks the branched chain, permitting it to enter the β-oxidation pathway.

Refsum's disease is caused by a defect in the α-oxidation step. Since there is no defect in the oxidation of phytol, it is readily converted to phytanic acid. The phytanic acid that is formed cannot be α-oxidized and therefore cannot undergo subsequent β-oxidation. Because of this, phytanic acid accumulates in the plasma and tissues of patients who have Refsum's disease. One of the main therapeutic procedures is to provide a diet in which phytol is low or absent in order to reduce the source of the phytanic acid.

It is not known at this time how this metabolic abnormality produces the neurologic damage, such as cerebellar degeneration or peripheral neuropathy. Two possibilities are foremost. One is that the phytanic acid accumulation itself produces the damage. The other is that a different nerve cell reaction that also happens to involve an α-oxidation is defective. In this regard, there are some fatty acids in the brain that contain an odd number of carbon atoms. These probably are formed through α-oxidation of ordinary fatty acids that contain an even number of carbon atoms. What role these odd-carbon number fatty acids play in the nervous system is as yet unknown, so one cannot speculate as to what defects might become apparent if their production was blocked.

The phytanic acid itself might cause a neurologic defect if its branched acyl chain were to interfere with the normal lipid bilayer structure when it is contained in the membrane phospholipid groups. Such a packing defect could cause altered membrane conductivity or permeability, leading to conduction defects or cell damage. There is some precedent for such a mechanism. Branched chain fatty acids accumulate in vitamin B_{12} deficiency. In certain severe forms of pernicious anemia, neuropathy develops, and this is thought to be associated with the incorporation of the branched chain fatty acids into the nerve cell membrane. To what extent the neurologic disorders in Refsum's disease are caused by a similar mechanism remains to be determined.

REFERENCES

Mize, C. E., Herndon, J. H., Jr., Blass, J. P., Milne, G. W. A., Follansbee, C., Laudat, P., and Steinberg, D.: Localization of the oxidative defect in phytanic acid degradation in patients with Refsum's disease, J. Clin. Invest. 48:1033, 1969.

Steinberg, D., Vroom, F. Q., Engel, W. K., Cammermeyer, J., Mize, C. E., and Avigan, J.: Refsum's disease—a recently characterized lipidosis involving the nervous system, Ann. Intern. Med. 66:365, 1967.

CYSTINOSIS: AN ERROR IN AN AMINO ACID TRANSPORT SYSTEM

Reference was made in earlier chapters to inheritable diseases associated with the intermediary metabolism of one or more amino acids. Problems may also arise from defects in the systems of facilitated diffusion responsible for moving amino acids into or out of certain cells. The disease, cystinosis, is just such a problem and is described in the next case.

Case 6: Cystinuria and cystine stones

A 14-year-old boy was seen in the emergency room with complaints of left flank pain of considerable severity and of sudden onset. He related that for the past few weeks his urine had been intermittently cloudy. His temperature was moderately elevated (38.8°C), and a urinalysis revealed the presence of both erythrocytes and leukocytes, some unorganized debris, and scattered, fine hexagonal crystals in the urinary sediment. The patient was referred to the urology department. X-ray studies revealed the presence of several abnormally dense areas that were interpreted as renal stones. A 24-hr urine collection was obtained and analyzed for cystine. The results indicated a cystine content of 0.45 g (normal, 0.05 to 0.15 g/24 hr, depending on age, sex, and diet). Further studies revealed higher than normal urinary excretion of ornithine, lysine, and arginine. When this patient was given an oral cystine tolerance test, his plasma cystine concentration showed a significant increase above the normal. The patient was treated with a low-protein diet and penicillamine.

Biochemical questions

1. What is the physicochemical basis for the formation of cystine stones in the urinary tract?
2. What is the chemical difference between cysteine and cystine? Is one convertible to the other?
3. Is cystine an essential amino acid?
4. Why are cystine and cysteine important amino acids in many proteins?
5. What important nonprotein metabolites require either cysteine or cystine for their production?
6. Can cysteine form dimers other than cystine?
7. What is the relation between the metabolism of cysteine and methionine?
8. How does the chemical structure of cysteine compare to that of penicillamine?
9. How is penicillamine useful in the treatment of cystinuria?
10. What is the connection between the excretion of cysteine and that of arginine, lysine, or ornithine?

Case discussion

Cystine is the least soluble of all the common amino acids. At pH 7 a saturated solution has a concentration of 1.66 mM, and, as shown in Fig. 14-8, the solubility decreases if the pH is lowered. Even at pH 8, which is approximately the upper limit of urinary pH, the solubility is less than 5 mM. If the concentration within

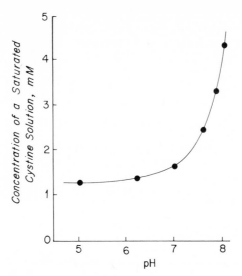

Fig. 14-8. Solubility of cystine as a function of pH.

a renal tubular cell exceeds the solubility, cystine will precipitate in the form of a crystalline mass that can grow until the cell is ruptured and dies. Continued growth of the mass may lead to obstruction of the urinary tract or to destruction of the kidney. Hematuria and pyuria are sometimes associated with formation of large kidney stones, as capillaries are damaged from the mechanical pressure, and the dying cells provide a focus for infection.

Cystine and cysteine are nonessential amino acids, but they occur in many enzymic and structural proteins and peptides. They are important because they provide the basis of disulfide bonds that serve to maintain specific conformations of the structures in which they occur (see Chapter 2). Cysteine is also important as a precursor of the 4-phosphopantetheine moiety of acyl carrier protein and CoASH. Cysteine is also required for synthesis of glutathione (see Chapters 2 and 9), as a source of taurine (see Chapter 10), and as a source of sulfate employed in conjugation of certain steroids. In this respect, cystine and cysteine are essentially equivalent, because of the ease with which one can be converted to the other, as shown below.

$$
\begin{array}{ccc}
\begin{array}{c}
S\!-\!\!-\!\!-\!\!-\!S \\
| \quad\quad | \\
CH_2 \quad CH_2 \\
| \quad\quad | \\
H_2NCH \quad H_2NCH \\
| \quad\quad | \\
COOH \quad COOH
\end{array}
&
\underset{-2H}{\overset{+2H}{\rightleftharpoons}}
&
\begin{array}{c}
SH \\
| \\
CH_2 \\
| \\
2\ H_2NCH \\
| \\
COOH
\end{array}
\\[2mm]
\text{Cystine} & & \text{Cysteine}
\end{array}
$$

The metabolism of cystine, cysteine, and methionine has some points in common, as is shown in Fig. 14-9. A major function of methionine is the activation of methyl groups by means of S-adenosylmethionine. When the methyl groups have been donated to suitable acceptors, one product is S-adenosylhomocysteine, which can be hydrolyzed to yield adenosine and homocysteine. Recall (see Chapter 9) that homocysteine can be methylated to regenerate methionine, but this pathway is normally not adequate to satisfy a continuing dietary need for methionine, for two reasons. First, homocysteine is like ornithine; it is not an amino acid

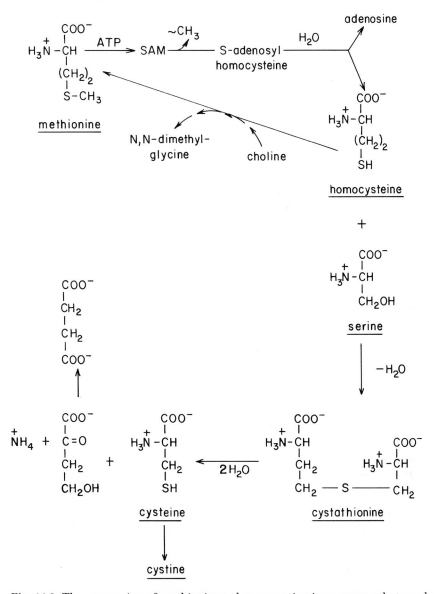

Fig. 14-9. The conversion of methionine to homocysteine in a manner that can be reversed and the irreversible conversion of homocysteine to cysteine and cystine.

commonly found in proteins so there is no good dietary source that could substitute for methionine. Second, homocysteine can also condense with serine to form cystathionine, a complex amino acid that is sometimes found in excessive amounts in the urine of cystinuric individuals. Cystathionine can be cleaved to generate cysteine and other fragments that ultimately enter the nitrogen pool or the Krebs cycle. It is important to remember that part of the excreted cystine and cysteine comes from the metabolism of methionine.

Cystinuria is essentially a disease of amino acid transport into the intestinal epithelium or out of the renal epithelium. The facilitated diffusion system (see Chapter 4), which regulates the movement of the dibasic amino acids, ornithine, lysine, and arginine, also regulates the transport of cystine. In the cystinuric kidney high loads of these dibasic amino acids inhibit the tubular uptake of cystine, with the result that it accumulates and may precipitate to form stones.

There is no known treatment that can replace or remedy the defective transport system, but it is possible to minimize the tendency to form stones by a purely chemical treatment. The treatment consists of providing a condensing partner for cysteine, other than itself, which makes a dimer more soluble than cystine. The chelating agent, penicillamine, is such a substance. It is sufficiently like cysteine to form a mixed disulfide dimer, shown below, which is significantly more soluble than cystine.

$$
\begin{array}{cc}
COO^- & COO^- \\
| & | \\
H_3\overset{+}{N}-CH & H_3\overset{+}{N}-CH \\
| & | \\
CH_2 & CH_2 \\
| & | \\
S\!-\!\!-\!\!-\!\!-\!\!-\!\!-\!\!-\!\!-\!\!S
\end{array}
\qquad
\begin{array}{cc}
COO^- & COO^- \\
| & | \\
H_3\overset{+}{N}-CH & H_3\overset{+}{N}-CH \\
| & | \\
CH_2 & H_3C-C-CH_3 \\
| & | \\
S\!-\!\!-\!\!-\!\!-\!\!-\!\!-\!\!-\!\!-\!\!S
\end{array}
$$

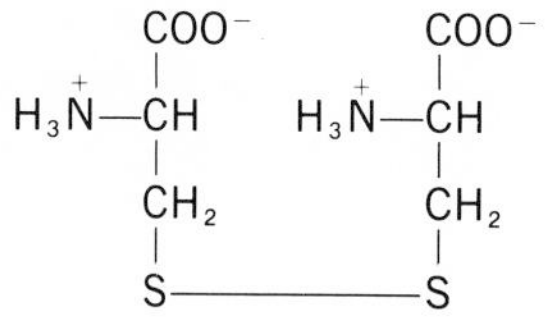

Cystine, a cysteine-cysteine dimer 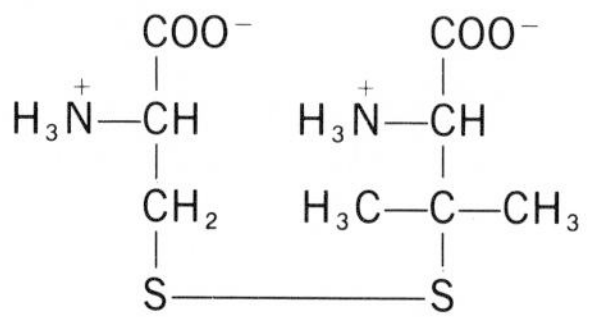   Cysteine-penicillamine dimer

Penicillamine treatment has its own drawbacks; as noted earlier (case 1, Chapter 3) it is a powerful chelator of metal ions. On the long-term basis required for treatment of an inheritable disease, penicillamine may give rise to symptoms of metal deficiencies or of deficiencies of the enzymes that require metals, such as the transaminases, and may induce a condition analogous to lathyrism (see Chapter 2). The chelating properties of penicillamine depend on the integrity of both the SH and the NH$_2$ groups, and it has recently been proposed that N-acetyl-penicillamine would be more useful in treatment of cystinuria, since the acetylated form has a very much lower chelating capacity while retaining a comparable capacity to form mixed disulfides with cysteine.

REFERENCES

Dent, C. E., et al.: Treatment of cystinuria, Br. Med. J. 1:403, 1965 (see also Br. Med. J. 1:284, 1968).

Frimpter, G. W.: Cystathioninuria in a patient with cystinuria, Am. J. Med. 46:832, 1969.

Schwartzman, L., Blair, A., and Segal, S.: Effect of transport inhibitors on dibasic amino acid exchange diffusion in rat-kidney cortex slices, Arch. Biochem. Biophys. 135:136, 1967.

Schwartzman, L., Blair, A., and Segal, S.: Exchange diffusion of dibasic amino acids, Arch. Biochem. Biophys. 135:120, 1967.

Segal, S., Schwartzman, L., Blair, A., and Bertoli, B.: Dibasic amino acid transport in rat-kidney cortex slices, Arch. Biochem. Biophys. 135:127, 1967.

Thier, S. O., and Segal, S.: Cystinuria. In Stanbury, J. B., Wyngaarden, J. B., and Fredrickson, D. S., editors: Biochemical basis of metabolic disease, New York, 1972, McGraw-Hill Book Co., pp. 1504-1519.

CHRONIC ALCOHOLISM: A METABOLIC DISEASE

The long-term, excessive intake of ethanol produces serious metabolic disturbances. Because the use of alcoholic beverages is culturally deeply rooted, health scientists must be aware of the problems posed by abuse of alcohol. Chronic alcoholism is responsible for a large number of costly accidents. Acute alcoholism is also a serious matter, especially in the operation of motor vehicles while intoxicated. This is still an unsolved medicolegal problem in much of the world.

The classic form in which chronic alcoholism usually presents itself is cirrhosis of the liver, but the liver is not the only organ in which the disease may be evident. The circulatory system and the central nervous system are also targets for the damaging effects of chronic alcoholism.

Case 7: Cirrhosis

E. O., a 54-year-old construction worker, was brought to the emergency room of a veteran's hospital from his job. In midmorning he had suffered a bout of hematemesis, apparently as a result of lifting some heavy beams. The following is an abstract of the intern's notes.

The patient appears to be about 10 kg underweight, a little dehydrated, and somewhat confused. . . . The patient has numerous carious teeth and a fetid breath. . . . The liver felt firm and enlarged, the abdomen was distended, and some pedal edema was noted. . . . The admission temperature was 42° C; the pulse was 100 and irregular. . . . The admission hemoglobin was 10.2 g/dl, with a hematocrit of 40%. . . . The initial urine specimen was tea-colored and strongly positive for protein. . . . A stool specimen collected the morning after admission gave a positive test for blood, supporting the results of an admission examination fecal smear. . . . The SGOT was elevated (220 King-Armstrong units). The total serum protein was 5.0 g/dl, the serum bilirubin was 5.0 mg/dl. . . . Serum electrolytes were $K^+ = 3.1$ meq/liter, $Na^+ = 152$ meq/liter, $Cl^- = 91$ meq/liter, and total $CO_2 = 29$ meq/liter. The BUN was 8.0 mg/dl. This patient had a previous history of abuse of alcohol.

Biochemical questions

1. How are the symptoms of hematemesis and edema related to the function of the liver?
2. What is the significance of the elevated SGOT? Would you expect that the SGPT would also be elevated?
3. What is the explanation of the increased serum bilirubin?
4. Would the increase be in the direct or indirect bilirubin fraction?
5. What metabolite was responsible for the unusual color of the urine?
6. How can you account for the low hemoglobin?
7. How does the damaged state of the liver relate to the clotting properties of the blood?
8. Alcohol is metabolized to acetyl CoA. Which coenzymes are involved in this process?
9. How is alcohol distributed in the tissues?
10. From what sort of electrolyte disturbance did this patient suffer?
11. Can you relate alcohol intake to mental performance and motor coordination?

Case discussion

Hemorrhage. Hematemesis (vomiting of blood) is obviously abnormal; in the present case it was apparently brought about by violent physical exertion. It presumably results from the rupture of dilated and weakened veins in the esophagus, which gave way when the blood pressure rose in response to exertion. Esophageal varices, as the dilated veins are known, are frequently associated with liver disease. Once initiated, the bleeding may continue for some time, since the diseased liver often is not able to produce normal amounts of prothrombin, an essential clotting factor. In this case the blood that escaped from the injured esophageal vessels was finally regurgitated, the event that brought the patient to the hospital.

Edema. The intern also noted abdominal distention, swelling of the feet, lowered hemoglobin, and moderate dehydration. All of these findings relate to inability of the diseased liver to produce normal amounts of important circulating proteins or the metabolites from which they are derived. Albumin, produced in the liver, is the most important contributor to the colloid osmotic pressure that pulls extracellular fluid from the interstitial spaces back into the capillary bed. When the albumin concentration is reduced, fluid tends to accumulate between the cells, where it produces swelling or distention. Reduced serum albumin also interferes with normal water reabsorption from the kidney, leading to dehydration. Note that the total serum protein was only 5.0 g/dl (normal, 6 to 8 g/dl). Although the serum proteins were not fractionated into albumin and globulins, it is quite safe to predict that the greatest depression would have been found in the albumin fraction, since tissues other than the liver contribute to globulin synthesis. The ratio of albumin to globulins (A/G ratio), which usually ranges from 1.8 to 2.3, might well have been reduced to 1 or even less.

Other effects on protein synthesis.   The low hemoglobin is caused in part by loss of blood from the vessels. It must be kept in mind, however, that the liver is important in hemoglobin synthesis. It provides the glycine that is an essential component of aminolevulinic acid, which ultimately becomes incorporated into the structure of heme (Chapter 3). Liver disease may therefore produce anemia even in the absence of actual blood loss.

Jaundice.   Bile pigment metabolism is an important function of the liver. The most important bile pigment is bilirubin, derived from the breakdown of the heme moiety of hemoglobin. Liver disease may result in one of two types of abnormal pigment metabolism. In the first type the damaged liver is unable to conjugate bilirubin with glucuronic acid (Chapter 3). In the second type the quantity of circulating pigment increases because of swelling or edema of the liver itself, so that the flow of the pigment-containing bile is mechanically impeded. When bilirubin diglucuronide is delivered to the gut, some of it will be reabsorbed as a slightly different pigment, urobilinogen, which is formed by the action of intestinal microbial flora on the parent compound, bilirubin. The presence of urobilinogen in urine leads to the formation of urobilin, especially when voided urine stands in the light and air. Urobilin was probably responsible for the peculiar color of the urine noted in the case description. Chronic alcoholism frequently produces liver damage, leading to both types of abnormal pigment metabolism as is evidenced in this case.

Electrolyte disturbances.   The diseased liver was contributing to electrolyte abnormalities. The serum K^+ was decreased and the serum Na^+ was increased. At the same time the serum Cl^- was decreased and the serum total CO_2 was increased. Part of the cation imbalance is a result of the edema; as the colloid osmotic pressure is reduced by the lowered serum albumin, fluid accumulates in the tissue spaces and the flow of blood through the kidney is reduced. This triggers the release of aldosterone, and Na^+ is conserved at the expense of K^+. In effect, the kidneys are reacting to a lowered *circulating* fluid volume, even though the body is retaining fluid in the abdominal cavity (ascites) and in the lower extremities (edema). The disparity between cations and anions, which may be estimated from the quoted electrolyte concentrations, is probably balanced by organic acid anions such as lactate and keto acids, since the damage to liver metabolism is by no means restricted to processes of protein biosynthesis. Cirrhotic livers are frequently infiltrated with fat that cannot be transported from the liver to the adipose tissue. Exactly why the fat cannot be transported is not entirely clear, but several possibilities have been suggested. It may be that synthesis of apolipoproteins is diminished, the assembly of lipoproteins may be defective, or perhaps the secretion of lipoproteins from the liver is abnormal. Whatever the precise reasons, fatty infiltration is by itself a sign of liver disease, and even before it becomes sufficiently severe to cause death of the hepatocytes, many of the normal metabolic functions of the liver have been damaged to a degree measurable by appropriate chemical tests. This patient has developed a metabolic acidosis because of an overproduction of lactate and keto acids; whether it would be compensated or uncompensated cannot be determined from the data given.

Serum enzymes. Both the serum alkaline phosphatase and the SGOT levels were elevated, a reflection of cellular damage with escape of the enzymes from the injured tissue. Although alkaline phosphatase is contained in several tissues, there is little doubt that most of the activity that appears in the serum of patients with cirrhosis comes from the liver, especially in situations in which the flow of bile from the liver is mechanically impeded.

Diseases of other tissues may also be associated with elevations of serum transaminases and alkaline phosphatase. However, the evidence of impaired protein synthesis, deranged bilirubin metabolism, hemorrhage, and electrolyte imbalance strongly point to the liver. Differential diagnosis of liver and heart disease can frequently be made by the assay of specific transaminases. In cardiac disease the SGPT shows less increase than the SGOT. The reverse is true in liver disease. Because the assay for SGPT is technically more complex, it is frequently omitted. Many other enzyme assays are available for the detection of liver disease and for differentiation of hepatic and cardiac disease as well. No single enzyme assay is entirely diagnostic of damage to a specific tissue, but by proper combination of individual enzyme assays and testing for specific isozymes it is frequently possible to narrow the list of alternatives.

Absorption and distribution of alcohol. Alcohol is readily absorbed when consumed in aqueous solution. Some absorption occurs in the stomach but most takes place in the intestine. Absorbed alcohol is distributed in all body fluids in proportion to the water content of the fluid. Consequently, alcohol may be detected and quantitatively measured in blood, urine, cerebrospinal fluid, and the water vapor borne by expired air. All of these materials have been employed in forensic or medicolegal practice, but the most common samples used are blood and expired air. Absorption and distribution of methanol and isopropanol, which are also sometimes encountered, appear to follow similar rules. Methanol and isopropanol are sometimes contaminants of ethanol, and sometimes they are ingested by themselves.

Determination of alcohol in biologic materials. Nearly all of the chemical methods that have been legally accepted for alcohol determination depend on its oxidation either to acetaldehyde, acetic acid, or CO_2 and water. The products formed depend on the circumstances of the analysis. The most widely used method converts alcohol to CO_2 and H_2O according to the following equation.

$$CH_3CH_2OH + 2K_2Cr_2O_7 + 8H_2SO_4 \longrightarrow 2CO_2 + 2K_2SO_4 + 2Cr_2(SO_4)_3 + 11H_2O$$

Excess dichromate can be measured by iodometric back-titration, so the dichromate reduced can be equated to alcohol in the sample. Variants of this chemical procedure have been accepted by legal authorities. More recently, analysis by means of gas-liquid chromatography has also gained acceptance. Similar methods have been developed for analysis of alcohol in samples of breath and urine. These have the advantage of being noninvasive analyses.

Clinical significance of alcohol intoxication. It is possible, by the means described above, to measure the amount of alcohol in the blood and other tissues. The largest single use of such determinations arises from the need to identify drivers intoxicated by alcohol. Since there is no precise definition of drunkenness, the relationship between alcohol concentration and the stage of intoxication can be summarized as follows:

1. Normal: No impairment of mental or physical skills; blood or urine alcohol from 5 to 15 mg/dl.

2. Social stage: Blood concentration up to 50 mg/dl, urine from 10 to 60 mg/dl; results from one or two bottles of beer or 1 to 2 ounces of whiskey; effects are slight, mild euphoria or a feeling of well-being with normal behavior tests. Virtually all individuals in this state would be described by an impartial bystander as exhilarated but sober.

3. Preintoxication stage: Blood from 50 to 150 mg/dl, urine from 60 to 200 mg/dl; results from two to six bottles of beer or 2 to 6 ounces of whiskey; release of social inhibitions with some signs of instability; diminished neuromuscular coordination; judgment and control required for quick response may be impaired. In this stage one or two out of three individuals would not be regarded as sober. In many jurisdictions blood concentrations of 100 mg/dl are accepted as legal evidence of intoxication in terms of motor vehicle operation. In this concentration range the chemical evidence of intoxication may be regarded as a legally rebuttable presumption.

4. Intoxicated (confused) stage: Blood concentration from 150 to 300 mg/dl, urine from 200 to 400 mg/dl; results from six or more bottles of beer or 6 or more ounces of whiskey; the symptoms and behavior are such that three out of three individuals in this stage would not be regarded as sober. Responses are typical for recognizable intoxication, and virtually all jurisdictions would accept concentrations of more than 200 mg/dl in the blood as the basis of a direct legal charge against a driver. Speech is impaired, and motor skills are uncoordinated.

5. Stuporous stage: Blood concentrations from 300 to 400 mg/dl, urine from 375 to 500 mg/dl; results from one pint or more of whiskey; subjects respond only to very strong stimuli.

6. Comatose stage: Blood concentrations from 400 to 600 mg/dl or more; subject cannot be easily aroused, shows depressed reflexes, hypothermia, and stertorous breathing; death may result.

Clearance of alcohol from the body. Not more than 2% of ingested alcohol is excreted via the lungs or the kidneys in unoxidized form. The remainder must be removed by biologic oxidation. Oxidation appears to proceed at a more or less constant rate, approximately 100 mg/kg/hr. This is equivalent to removing approximately 11 ml/hr of pure ethanol for a 70-kg individual. Since the decrease in alcohol concentration is a linear function of time, it requires about 90 min to clear the tissues of the alcohol load from 1 ounce of whiskey or one bottle of beer. No effective means of altering this clearance rate has been found.

Biologic oxidation of alcohol. The liver is the major site of alcohol oxidation. Two NAD^+-linked enzymes, *alcohol dehydrogenase* and *acetaldehyde dehydrogenase*, convert ethanol to acetaldehyde and acetate, respectively. Some of the acetate is converted to hepatic acetyl CoA, but most of the acetate is released to the blood and conveyed to the extrahepatic tissues. There is good evidence that little of the acetate enters the Krebs cycle, and the remainder must take other pathways. Presumably, some of it may be stored as long-chain fatty acids. This conclusion is suggested by experiments with human volunteers. The data demonstrate depression of gluconeogenesis and CO_2 formation during alcohol intoxication. Furthermore, any fatty acid biosynthesis that occurred from hepatic cytoplasmic acetyl CoA would contribute to the fatty content of the liver.

The $NADH/NAD^+$ ratio is significantly elevated (twofold to threefold) in the cytoplasm of hepatocytes during the oxidation of alcohol, with a resultant shift in the oxidation-reduction potential of the cytoplasm (see Chapter 5). Since neither form of the nucleotide can cross the mitochondrial membrane, reoxidation of NADH becomes a pressing problem. Part of the NADH could be oxidized by one of the several shuttle mechanisms that carry reducing equivalents from the cytoplasm to the mitochondria (see Chapter 6), and part could be oxidized by coupling with some strictly cytoplasmic reaction, for example, the oxidation of pyruvate to lactate. Note that this would affect the pH of body fluids (see Chapter 4). It is not possible to readily apportion precisely the task of NADH oxidation between these separate avenues, but it has been experimentally established that the ratio of lactate to pyruvate increases severalfold during alcohol oxidation, as does the ratio of L-α-glycerophosphate to dihydroxyacetone phosphate. Furthermore, the oxidation of alcohol actually impairs the oxidation of added pyruvate, perhaps by diverting it to lactate. The activity of glutamate dehydrogenase is inhibited by the excess of intramitochondrial NADH. Similarly, the conversion of α-ketoglutarate to succinate and of malate to oxaloacetate is depressed, evidence of an altered oxidation-reduction state even within the mitochondria. As a result, the synthesis of phosphoenolpyruvate, essential for gluconeogenesis (see Chapters 6 and 7), will also be depressed. These specific examples clearly indicate that the pathways of intermediary metabolism are distorted by direct effects on certain enzymes. It should be kept in mind that the mild to moderate acidosis resulting from increased concentrations of lactate and acetate will act as a generalized metabolic depressant, with measurable changes in electrolyte compositions of the extracellular and intracellular fluids.

Effects of alcohol on the endoplasmic reticulum. The endoplasmic reticulum, which in subcellular systems is isolated in fragments termed microsomes, is involved in the metabolism of many substances foreign to the cells. Many drugs are inactivated (detoxified) by the hepatic microsomal enzymes. When these drugs are ingested on a chronic basis, there is a proliferation of the smooth endoplasmic reticulum that can be demonstrated by microscopic examination. This can also be demonstrated by an increased activity of many microsomal enzymes. Barbiturates and tranquilizers, among other drugs, are frequently taken by some alcoholics. Alcohol alone also produces the increased microsomal enzyme activities, provided the tissue concentration of alcohol is typical of chronic alcoholism for a sufficient period of time. The heightened activity of microsomal enzymes results in a faster inactivation of drugs when the individual is not burdened with alcohol. Thus anesthesia of sober chronic alcoholic individuals requires larger doses of the drug than normals. On the other hand, when the alcoholic is burdened with alcohol, it appears that metabolism of the alcohol takes precedence over metabolism of sedative and tranquilizing drugs. For this reason, the ingestion of these drugs along with alcohol is dangerous, and may produce death by virtue of the slow inactivation of barbiturates, meprobamate, or similar agents.

Effects of alcohol on the central nervous system. The signs of alcohol intoxication have already been described, and it is obvious that many of the stigmata reflect involvement of the central nervous system. The injection of alcohol produces small decreases in brain norepinephrine and 5-hydroxytryptamine concentrations, but these are not usually significant. Injection of acetaldehyde at an equivalent concentration can cause significant decreases in brain norepinephrine. It is therefore possible that some of the effects of acute alcohol intoxication may be a result of selective alterations in brain catecholamine concentrations.

Broader nutritional aspects of chronic alcoholism. It is generally observed that chronic alcoholics have very poor dietary habits. Since alcohol has a caloric equivalent of 7 kcal/g, a "jigger" of whiskey furnishes approximately 120 kcal. It is therefore not surprising that chronic alcoholics may obtain 1200 to 1500 kcal/day from alcohol alone. If the beverages consumed include much beer or wine the figure may be even greater, since these drinks include calories not derived from alcohol. Nutritionists speak of such energy sources as "empty calories," insofar as these calories are not associated with vitamins, minerals, or essential amino acids necessary for growth and maintenance (see Chapter 1). The dietary inadequacies add to the metabolic disturbances described above. In dealing with the treatment of alcoholic patients it is important to employ a sound diet among the corrective measures.

REFERENCES

Duritz, G., and Truit, E. B., Jr.: Importance of acetaldehyde in the action of ethanol on brain norepinephrine and 5-hydroxytryptamine, Biochem. Pharmacol. 15:711, 1966.

Gonzales, T. A., et al.: Legal medicine, pathology and toxicology, New York, 1954, Appleton-Century-Crofts, Inc.

Hawkins, R. D., and Kalant, H.: Metabolism of ethanol and its metabolic effects, Pharmacol. Rev. 24:67, 1972.

Israel, Y., and Mardones, S., editors: The biology of alcohol, New York, 1971, Plenum Publishing Corp.

Ladd, M., and Gibson, R. B.: Legal-medical aspects of blood tests to determine intoxication, Va. Law Rev. 29:749, 1943 (see also Ann. Intern. Med. 18:564, 1943).

Lieber, C. S.: The metabolism of alcohol, Sci. Am. 234:25, 1976.

Clinical examples for further analysis

The following cases deal with the interrelationship of metabolic pathways, enzyme systems, and organs. By analyzing them the reader will be able to integrate the material presented in previous chapters and obtain more insight into the whole entity of human metabolism. These cases also afford the opportunity of analyzing clinical situations in terms of all of their biochemical ramifications rather than in artificially isolated subsections.

CASE 8: PHEOCHROMOCYTOMA

Mild hypertension was noted in a 54-year-old woman during a routine physical examination. She complained of intermittent episodes of dizziness, fatigue, and nausea associated with palpitations and headache. Diagnostic evaluation revealed a left suprarenal mass and an elevated 24-hr excretion of 3-methoxy-4-hydroxymandelic acid (VMA) on three separate occasions. Her fasting blood glucose concentration was elevated. After 9 days of treatment with an α-adrenergic blocking agent, she underwent abdominal exploratory surgery. A large left adrenal mass was removed; it was diagnosed histologically as a pheochromocytoma, an epinephrine-secreting tumor.

Biochemical questions

1. What is the relationship of the tumor to the elevated VMA excretion?
2. Discuss the probable mechanism of the elevated fasting blood glucose levels.
3. What plasma lipid abnormalities might be produced by this tumor?
4. What beneficial effects might be expected from the use of an α-adrenergic blocking drug?

REFERENCES

Kirkendall, W. M., Liechty, R. D., and Culp, D. A.: Diagnosis and treatment of patients with pheochromocytoma, Arch. Intern. Med. 115:529, 1965.
Packman, R. C., and O'Neal, L. W.: Pheochromocytoma, J.A.M.A. 212:780, 1970.

CASE 9: PRIMARY ALDOSTERONISM

A 44-year-old man who complained of headaches was discovered to be hypertensive. Hypokalemia was noted; the plasma potassium concentration remained between 2.5 and 3 meq/liter during 3 weeks of observation. Peripheral vein renin levels were consistently less than 1 ng/ml, and they remained low even when the patient was placed on a very low sodium diet. Urinary aldosterone excretion was consistently high, even when a very high sodium diet was given. A venogram showed a questionable lesion in the left adrenal gland. A left adrenalectomy was performed. On follow-up examination 2 months later, the patient was found to be normotensive and normokalemic.

Biochemical questions

1. Explain the hypertension produced by aldosteronism.
2. Explain the hypokalemia in this case. Would you expect any abnormality in total body sodium content?
3. Why was the plasma renin level low in this case?
4. Would you expect any trend in the angiotensin levels in this patient's plasma?
5. Would an 11-β-hydroxylase inhibitor be expected to lower the aldosterone output? What other effects might such a drug have on hormone production by the adrenal cortex?
6. Would dexamethasone, a synthetic glucocorticoid that inhibits ACTH release, be expected to decrease aldosterone production? Explain.

REFERENCES

Blaine, E., Davis, J., and Harris, P.: A steady state control analysis of the renin-angiotensin-aldosterone system, Circ. Res. 30:713, 1972.
Conn, J. W.: Primary aldosteronism, a new clinical syndrome, J. Lab. Clin. Med. 45:3, 1955.

CASE 10: ACROMEGALY

A 28-year-old woman noted that her tongue and feet were becoming larger. She complained of fatigue and lassitude. When she noted that her teeth were separating, she consulted her dentist. He noted that her mandible was enlarged and referred the patient to the university hospital. X-ray examination revealed increased thickness of the fingers and toes and a thickened heel pad. A diabetic-type glucose tolerance test was obtained, and the plasma free fatty acid was elevated. Skull x-ray films revealed enlargement of the sella turcica, a finding consistent with a pituitary tumor. Plasma growth hormone levels were grossly elevated. The patient was treated by proton beam irradiation of the pituitary, and the disease was arrested.

Biochemical questions

1. What types of hormonal replacement therapy should be considered for this patient after pituitary obliteration?
2. What are the differences in the relationship between the hypothalamus and the hormones of the anterior and posterior pituitary?
3. Explain the mechanism of the plasma free fatty acid elevation.
4. Explain the reason for the diabetic glucose tolerance test.

REFERENCE

Cryer, P. E., and Daughaday, W. H.: Regulation of growth hormone in acromegaly, J. Clin. Endocrinol. 29:386, 1969.

CASE 11: OBESITY

A 33-year-old man was admitted to the hospital because of extreme obesity. He was 63 inches tall and weighed 440 pounds. The patient had worked previously as a laborer, but because of a recent severe weight gain, he found it impossible to work. Prior to admission he had been sedentary and his weight had been constant for the last 2 months. All attempts to lose weight by reduction of his dietary intake had failed. A thorough physical, laboratory, and x-ray examination revealed no other abnormalities. The patient underwent gastric bypass surgery. In this procedure the stomach was reduced to less than half of its original size by closing off the distal portion, and the proximal segment was connected to the duodenum. Hypoventilation occurred postoperatively, and the patient required the use of a respirator for 12 hr. He did well subsequently and lost 60 pounds in the first year following surgery.

Biochemical questions

1. How many calories per day had this man been eating in order to maintain his weight at 440 pounds?
2. What difficulties in protein nutrition might the patient experience following surgery?
3. The gastric bypass operation may lead to a deficiency of one vitamin in particular. What is this vitamin, and what would be the mechanism through which the deficiency would occur?
4. What type of hormonal disturbance might result from bypassing the distal segment of the stomach?
5. Had the respirator not been employed postoperatively, what type of acid-base abnormality might have developed?
6. What types of clinical research studies could be done to determine whether the patient's adipose tissue was functioning normally?
7. Instead of gastric bypass, some surgeons recommend partial bypass of the small intestine in cases of severe, intractable obesity. Which of the two operative procedures might produce more severe nutritional complications?

REFERENCES

Mason, E. E., and Ito, C.: Gastric bypass, Ann. Surg. 170:329, 1969.
Nelson, R. A., et al.: Physiology and natural history of obesity, J.A.M.A. 223:627, 1973.

CASE 12: ANOREXIA NERVOSA

A psychotic patient was referred to a psychiatric hospital for diagnostic evaluation. Among the problems noted on admission was the patient's refusal to eat. The referring physician indicated that since the patient had first come to his attention she had lost 10 kg of body weight. The resident in charge of the patient ordered gavage feedings of a liquid formula that contained 100 g of glucose, 50 g of protein hydrolysate, and 10 g of corn oil/liter. The patient received 1200 ml of this formula in 24 hr. On the next day it was pointed out that the patient had a positive test for glucose in the urine, and that the blood sugar was abnormally elevated. The resident responded by an order to reduce the feeding volume to 400 ml that day, with the volume to be increased by 100 ml each day for the following week. The glucosuria promptly subsided.

Biochemical questions

1. What are the factors that regulate blood glucose concentration?

2. What enzymes intrinsic to carbohydrate metabolism would decline in activity during a fast sufficient to cause a 10-kg loss of body weight?
3. How would the energy charge of the adenylate nucleotides change during such a fast?
4. In what ways was the prescribed diet inadequate?
5. How would you have modified the diet?
6. How did the change in feeding pattern eliminate the observed symptoms?
7. Before treatment was started, would this patient's urine have contained a normal concentration of creatinine?
8. Would you expect a normal concentration of circulating thyroxine in this patient? Explain.

REFERENCES

Bliss, E. L., and Branch, C. H. H.: Anorexia nervosa, New York, 1960, Paul B. Hoeber, Inc.

Frazier, S.: Anorexia nervosa, Dis. Nerv. Syst. **26:** 155, 1965.

CASE 13: VITAMIN D-RESISTANT RICKETS

H. B., a 47-year-old man, was in good health until 5 years ago. While getting out of his automobile at that time, he suffered acute low back pain that has persisted. He also complains of intermittent pain in his right leg and foot. He has been seen in several medical centers without substantial relief, and the persistent pain and problems of locomotion have forced him to retire. Shortly before the present visit, he noted spontaneous onset of bilateral rib pain.

Examination revealed extreme bilateral tenderness over the rib cage and generalized myalgias. There was significant muscle wasting, especially in the lower extremities. H. B. could not squat, hop, or walk on his toes, nor could he get out of a chair unassisted. There was no evidence of joint deformities or of rheumatoid disease.

Malabsorption studies revealed a serum carotene of 140 μg/dl (normal, 50 to 300 μg/dl) and a 48 hr fat excretion of 3.76 g. X-ray examination revealed diffuse demineralization of the bones. Other laboratory data are shown in Table 14-4.

Biochemical questions

1. What is the biochemical relationship between serum Ca^{++} and serum P_i concentrations?
2. What governs P_i reabsorption in the renal tubules?
3. What is the active form of vitamin D?
4. In what organ is ergosterol converted to the active form of vitamin D?
5. What are the target organs of parathyroid hormone?
6. What is the function of alkaline phosphatase in bone metabolism?
7. How would the malabsorption syndrome affect Ca^{++} balance in man?

Table 14-4. Laboratory data for patient H. B.

TEST	DATE AND RESULT OF ANALYSIS						
	2/7	4/25	7/11	8/3	8/19	9/7	9/23
BUN (mg/dl)	22	21				19	
Creatinine (mg/dl)	1.0	1.1				0.8	
Serum Ca^{++} (mg/dl)	9.5	8.9	9.3	8.6	9.4	9.5	9.3
Serum P_i (mg/dl)	1.5	1.4	1.5	1.4	1.9	1.6	2.1
Urinary Ca^{++} (mg/24 hr)			206	124	104		63.0
Urinary P_i (mg/24 hr)			1450	1148	979		2486
Urine pH	5.0	5.5				5.5	6.0
Creatinine clearance (ml/min)		85	91	89		107	119
Alkaline phosphatase (Bodansky units)	5.9	6.1					7.0
Acid phosphatase (Bodansky units)	0.6						1.7
Tubular reabsorption of phosphate (%)		56		57			

REFERENCES

Aurbach, G. D., et al.: Structure, synthesis and mechanism of action of parathyroid hormone, Recent Prog. Horm. Res. 28:353, 1972.

DeLuca, H. F.: Recent advances in our understanding of the vitamin D endocrine system, J. Lab. Clin. Med. 87:7, 1976.

CASE 14: MULTIPLE MYELOMA

F. R. is a 66-year-old man who had fallen out of a tree 4 years ago without any apparent fractures. Since that time he has complained of pain over the ribs. The pain is worsened by coughing or straining and has become more severe in the past 2 months. F. R. has had a firm, marble-sized mass in his neck for the past 10 to 15 years; the size had remained constant until the past 6 months. At that point the mass enlarged to its present size. He denies pain in the mass, hoarseness, dysphagia, or respiratory obstruction as well as easy bruising or bleeding, anemia, shortness of breath on exertion, palpitation, or tachycardia.

In early December of last year an x-ray film obtained by his local physician revealed multiple rib fractures, and he was referred to a university hospital. Laboratory data obtained on admission are given below.

Hemoglobin	15.5 g/dl
Hematocrit	45%
White cell count	8700 cells/mm³
Urinalysis	pH = 7.0, sp. gr. = 1.022, 1 + protein

BUN	15 mg/dl
Creatinine	1.3 mg/dl
Serum Na^+	140 meq/liter
Serum K^+	4.1 meq/liter
Serum Cl^-	106 meq/liter
Serum CO_2	17 meq/liter
Serum Ca^{++}	8.7 mg/dl
Serum P_i	3.6 mg/dl
Alkaline phosphatase	4.7 Bodansky units
Acid phosphatase	0.4 Bodansky units
Uric acid	7.6 mg/dl

A chest x-ray film revealed healed or healing fractures of the right tenth and eleventh ribs. Rib x-ray films showed osteolytic lesions on the right and left sides. Cervical views showed degenerative changes, but a soft tissue view of the neck showed no calcification of the mass. The results of bone marrow aspiration and serum protein electrophoresis were judged to be normal. Serum and urine immunoelectrophoresis revealed monoclonal L-type light chains. No myeloma proteins were present.

Biochemical questions

1. What does the term "monoclonal" signify with respect to immunoglobulins?

2. How do light (L) and heavy (H) immunoglobulin chains differ?
3. In what way are L and H chains similar?
4. How are L and H chains held together in a typical IgG molecule?
5. Explain the occurrence of light chains in the urine.
6. What plasma protein should be measured in order to evaluate the clinical significance of the plasma calcium concentration?

REFERENCES

Edelman, G. M.: Antibody structure and molecular immunology, Science 180:830, 1973.

Third ICN-UCLA Symposium on Molecular Biology, New York, 1974, Academic Press, Inc.

Turino, G. M., editor: Plasma cell dyscrasias, Am. J. Med. 44:256, 1968.

Williams, R. C., et al.: Light chain disease, Ann. Intern. Med. 65:471, 1966.

CASE 15: DIABETIC KETOACIDOSIS

A 24-year-old man who has had diabetes for 5 years had been well until 2 days ago when he developed fever, nausea, and vomiting. The vomiting continued and on the day before admission to the hospital he developed diarrhea with ten loose stools during the 24 hr before admission. He had not taken his insulin for 24 hr because he was unable to eat. On admission he was semiconscious, with a respiratory rate of 35. Acetone could be smelled on his breath. He had the physical findings of moderate to severe dehydration. A catheterized urine specimen was obtained; this was 4+ for glucose and strongly positive for ketone bodies. A blood sample was drawn and sent to the laboratory for the determination of blood sugar, BUN, creatinine, sodium, potassium, chloride, and total CO_2. An arterial blood sample was sent for determination of blood pH and $\bar{p}CO_2$. The blood serum was strongly positive for ketones. He was immediately given 100 units of insulin. An intravenous drip of hypotonic saline solution with 1 ampule of sodium bicarbonate (50 ml containing 44 meq) was started. In approximately 45 min the laboratory results were available. The data were as follows: blood sugar, 960 mg/dl; BUN, 40 mg/dl; creatinine, 2.9 mg/dl; serum sodium, 138 meq/liter; serum potassium, 5.9 meq/liter; serum chloride, 94 meq/liter; and total CO_2, 3 meq/liter. The arterial pCO_2 was 18 mm Hg and the blood pH was 7.05. Serum osmolality was 390 mOsm (normal = 285 to 295 mOsm). Because of the low CO_2 and the high blood sugar, an additional 200 units of insulin was administered and the rapid intravenous drip of hypotonic saline solution was continued. One ampule of sodium bicarbonate was given by rapid intravenous injection. During the first 2 hr of therapy the patient received a total of $3\frac{1}{2}$ liters of hypotonic saline solution and 2 ampules of sodium bicarbonate. After 2 hr, blood was again drawn and sent to the laboratory. Results of this study were as follows: blood sugar, 880 mg/dl; total CO_2, 5 meq/liter; and pH, 7.1. Plasma ketones were strongly positive. The patient was still semiconscious but appeared better hydrated and had begun to excrete urine. Because of the continued low CO_2, the presence of severe ketonemia, and the continued elevation of blood sugar levels, another 300 units of insulin was given, half intravenously and half intramuscularly. Administration of hypotonic saline solution was continued intravenously but at a slower rate. Two hours after the second injection of insulin, additional blood studies were ordered. The results then showed the following: blood sugar, 400 mg/dl, and total CO_2, 10 meq/liter. The patient's condition was definitely improved, and he was no longer hyperventilating. He was now conscious and could give his name and the date. No further insulin was given at this time, and the intravenous infusion was changed to a solution containing hypotonic saline solution, 5% dextrose, and 50 meq/liter of potassium phosphate. This was run at a rate of about 250 ml/hr. Another blood sample was sent to the laboratory 2 hr later. The blood sugar was now 250 mg/dl and the serum potassium was 4.0 meq/liter. The total CO_2 was 16 meq/liter, and the pH was 7.35.

The plasma ketones were only trace positive in undiluted serum.

Biochemical questions

1. Why did severe hyperglycemia develop in this patient?
2. How did ketosis develop in the face of an elevated blood glucose concentration?
3. How did insulin function to help correct the metabolic abnormality? How was enzyme induction involved in the effect of insulin?
4. Describe the mechanism that caused the acidosis to develop, and trace the logic of the therapy with respect to fluid and electrolyte balance.

REFERENCE

Williams, R. H., and Ensinck, J. W.: Current studies regarding diabetes, Arch. Intern. Med. 128:820, 1971.

CASE 16: PROTEIN-LOSING GASTROENTEROPATHY

A 47-year-old woman was referred to a university hospital with a 10-year history of dependent edema, loose and foul-smelling stools, and intermittent carpopedal spasm. On physical examination she was found to be extremely thin, and she looked malnourished. There were bilateral pleural effusions, ascites, and anasarca. Her total plasma protein was 3.5 g/dl, the albumin being 0.8 g/dl. Serum calcium was 8 mg/dl, and the serum magnesium was 0.5 mg/dl. She also was anemic, with a hemoglobin of 9 g/dl and a hematocrit of 28%. A diagnosis of gastrointestinal protein loss secondary to intestinal lymphangiectasia was made. Treatment consisted of administration of vitamin D, calcium, magnesium, and a diet containing medium-chain triglycerides.

Biochemical questions

1. Which of the patient's abnormalities can be related to the low plasma albumin concentration?
2. Discuss the possible mechanisms for the anemia that was noted.
3. If a liver biopsy could be obtained from this patient, what kinds of biochemical studies might be done to evaluate the capacity of the patient's liver to synthesize protein?
4. How do the roles of calcium and magnesium differ in terms of physiologic functions?
5. Steatorrhea accompanied this abnormality. What vitamins may become deficient in the body because of chronic steatorrhea? What kinds of biochemical abnormalities might result from these vitamin deficiencies?
6. Why are many patients with steatorrhea able to readily utilize dietary medium-chain triglycerides? Discuss the metabolism of medium-chain fatty acids in man.
7. A severely malnourished patient sometimes must be fed for long periods of time entirely by the intravenous route. In ordinary foods the primary carbohydrates are starch, sucrose, and lactose. The main source of nitrogen is protein. Which of these substances can be included in intravenous feeding mixtures, and which of them would have to be modified?

REFERENCE

French, A. B.: Protein-losing gastroenteropathies, Gastroenterology 50:422, 1966.

ABBREVIATIONS

A	Adenine
A site	Site on ribosome that holds aminoacyl tRNA, *see also* P site
ACP	Acyl carrier protein
ACTH	Adrenocorticotropin
ADH	Vasopressin, an anti-diuretic hormone
ADP	Adenosine 5'-diphosphate
AICAR	5-Aminoimidazole-4-carboxamide ribotide
Ala	Alanine
ALA	Aminolevulinic acid
AMP	Adenosine 5'-monophosphate
Arg	Arginine
Asn	Asparagine
Asp	Aspartic acid
ATP	Adenosine 5'-triphosphate
B_6	*See* Vitamin B_6
B_{12}	*See* Vitamin B_{12}
BAL	British antilewisite, 2,3-dimercapto-propanol
BCCP	Biotin carboxyl-carrier protein
BMR	Basal metabolic rate
BUN	Blood urea nitrogen
BV	Biologic value of proteins
C	Cytosine
cAMP	Cyclic AMP, adenosine 3',5'-monophosphate
CAP	Catabolite activator protein

CDP	Cytidine 5'-diphosphate
Cer	Ceramide
cGMP	Cyclic GMP, guanosine 3',5'-monophosphate
Chln	Choline
CMP-NANA	Cytidine monophosphate N-acetyl neuraminic acid
CoA	Coenzyme A
COMT	Catecholamine O-methyltransferase
CPG III	Coproporphyrinogen III
CPK	Creatine phospho-kinase
CRH	Corticotropin-releasing hormone
CTP	Cytidine 5'-triphosphate
Cys	Cysteine
Cys-Gly	Cysteinylglycine
dAMP	Deoxyadenosine 5'-monophosphate
dATP	Deoxyadenosine 5'-triphosphate
dCTP	Deoxycytidine 5'-triphosphate
DFP	Diisopropyl fluoro-phosphate
dGTP	Deoxyguanosine 5'-triphosphate
DHF	Dihydrofolate
DIT	3,5-Diiodotyrosine
DNA	Deoxyribonucleic acid
DNase	Deoxyribonuclease
dopa	3,4-Dihydroxyphenyl-alanine

L-dopa	L-Dihydroxyphenyl-alanine	GHRH	Growth hormone–releasing hormone
dopamine	3,4-Dihydroxyphen-ethylamine	GHRIH	Somatostatin, or growth hormone release–inhibiting hormone
DPG	2,3-Diphospho-glycerate		
dTMP	Deoxythymidine 5'-monophosphate	Glc	Glucose (if no confusion results, glucose is also indicated by G)
dTTP	Deoxythymidine 5'-triphosphate		
EDTA	Ethylenediamine tetra-acetic acid	GLC	Gas-liquid chroma-tography
EF-1, EF-2	Elongation factors 1 and 2 in protein biosynthesis	GlcA	Glucuronic acid
		GlcNAc	N-Acetyl glucosamine
		Gln	Glutamine
EKG	Electrocardiogram	Glu	Glutamate
ES	Enzyme-substrate complex	Gly	Glycine
		GMP	Guanosine 5'-monophosphate
FAD	Flavin adenine dinucleotide	GOT	Glutamate oxalo-acetate transami-nase
FADH₂	Reduced form of flavin adenine dinu-cleotide		
		GPT	Glutamate-pyruvate transaminase
FAICAR	5-Formamidoimida-zole-4-carboxamide ribotide		
		GSH	Glutathione, γ-glutamylcysteinyl-glycine
fGAR	Formylglycinamide ribotide		
fMet-tRNAᶠ	Formylmethionyl tRNA	GSSG	Glutathione, oxidized
		GTP	Guanosine 5'-triphosphate
FMN	Flavin mononucleo-tide	HbA	Hemoglobin, major adult form
Fru	Fructose		
FSH	Follicle-stimulating hormone	HBDH	Hydroxybutyrate dehydrogenase
		HbF	Hemoglobin, fetal form
G	Guanine		
ΔG	Change in free energy of a given system	HbS	Hemoglobin, form found in sickle cell anemia
GA	Glucuronic acid		
GABA	γ-Aminobutyric acid		
Gal	Galactose	HDL	High-density lipopro-tein
GalNAc	N-Acetyl galactos-amine		
		HGPRT	Hypoxanthine-guanine phospho-ribosyl transferase
GAR	Glycinamide ribotide		
GDP	Guanosine 5'-diphosphate		
		His	Histidine
GDPMan	Guanidine diphos-phate mannose	HMGCoA	3-Hydroxyl-3-methyl-glutaryl coenzyme A
GFR	Glomerular filtration rate	HTC cells	Hepatoma tissue culture cells
GH	Growth hormone	Hyp	Hydroxyproline

"i" gene	*Escherichia coli* gene for *lac* operon repressor		MDH	Malate dehydrogenase
ICDH	Isocitrate dehydrogenase		MDR	Minimum daily requirement of nutrients
IDL	Intermediate-density lipoprotein		Met	Methionine
			MIT	3-Monoiodotyrosine
IdoA	Iduronic acid		mOsm	Milliosmolar
IF-1, IF-2, IF-3	Protein synthesis initiation factors 1, 2, and 3		MRH	Melanocyte-stimulating hormone–releasing hormone
IgA, IgG, IgM	Classes of immunoglobulins		MRIH	Melanocyte hormone–releasing-inhibiting hormone
Ile	Isoleucine		mRNA	Messenger ribonucleic acid
IMP	Inosine 5'-monophosphate		MSH	Melanocyte-stimulating hormone
IPU	2-Isopropyl-4-pentenoylurea		NAD^+	Nicotinamide adenine dinucleotide
IQ	Intelligence quotient		NADH	Reduced form of nicotinamide adenine dinucleotide
ITP	Inosine 5'-triphosphate			
IUPAC	International Union of Pure and Applied Chemistry		$NADP^+$	Nicotinamide adenine dinucleotide phosphate
K_m	Michaelis constant of an enzyme for a given substrate		NADPH	Reduced form of nicotinamide-adenine dinucleotide phosphate
αKG	α-Ketoglutarate			
lac operon	*Escherichia coli* operon for genes used in catabolism of lactose		$Na^+/K^+ATPase$	Sodium plus potassium activated adenosinetriphosphatase
LCAT	Lecithin-cholesterol acyl transferase			
LDH	Lactate dehydrogenase		NHI	Nonheme iron
LDL	Low-density lipoprotein		NSILA	Nonsuppressible insulin-like activity
Leu	Leucine			
LH	Luteinizing hormone		OAA	Oxaloacetate
LRH/FSHRH	Luteinizing and follicle-stimulating hormone-releasing hormone		Orn	Ornithine
			P site	Site on ribosome that holds peptidyl tRNA
			Ψ	Pseudouridine, 5-ribosyl uracil
LSD	Lysergic acid diethylamide		PBG	Porphyrobilinogen
Lys	Lysine		$\bar{p}CO_2$	Partial pressure of carbon dioxide
Man	Mannose			
MAO	Monoamine oxidase		PDH	Pyruvate dehydrogenase
Mb	Myoglobin		PEP	Phosphoenolypyruvate
MD	Muscular dystrophy		PFK	Phosphofructokinase

PG	Prostaglandin	SAM	S-Adenosylmethionine
Phe	Phenylalanine	SDA	Specific dynamic action of foods
P_i	Inorganic phosphate ion	SDH	Succinate dehydrogenase
P/O	Phosphorylation efficiency of oxidative phosphorylation	Ser	Serine
		SGOT	Serum glutamate oxaloacetate transaminase
$\bar{p}O_2$	Partial pressure of oxygen	SGPT	Serum glutamate pyruvate transaminase
polA	*Escherichia coli* DNA polymerase I		
PP_i	Inorganic pyrophosphate ion	SLE	Systemic lupus erythematosus
PPP	Pentose phosphate pathway	sRNA	Soluble RNA, more properly called tRNA
PRH	Prolactin-releasing hormone	T_3	Triiodothyronine
PRIH	Prolactin-release–inhibiting hormone	T_4	Thyroxine
		TBG	Thyroxin-binding globulin
Pro	Proline	THF	Tetrahydrofolate
PRPP	Phosphoribosylpyrophosphate	Thr	Threonine
		TLC	Thin-layer chromatography
PTA	Plasma thromboplastin antecedent	TPP	Thiamine pyrophosphate
PTC	Plasma thromboplastin component	TRH	Thyrotropin-releasing hormone
R17	An RNA bacteriophage of *Escherichia coli*	tRNA	Transfer ribonucleic acid
RBC	Erythrocytes, red blood cells	Trp	Tryptophan
		TSH	Thyroid-stimulating hormone
RDA	Recommended daily dietary allowance of nutrients	Tyr	Tyrosine
		U	Uridine
RNA	Ribonucleic acid	UDPG	Uridine diphosphate glucose
RNase	Ribonuclease		
rRNA	Ribosomal ribonucleic acid	UPG III	Uroporphyrinogen III
		UTP	Uridine 5'-triphosphate
30S	Small subunit of bacterial ribosome that sediments at force of 30 Svedberg units	UV	Ultraviolet
		V_{max}	Maximum initial velocity of an enzyme reaction
50S	Large subunit of bacterial ribosomes that sediments at force 50 Svedberg units	Val	Valine
		Vitamin B_1	Thiamine
		Vitamin B_2	Riboflavin
SAH	S-Adenosylhomocysteine	Vitamin B_6	Pyridoxine
		Vitamin B_{12}	Cobalamin

| VLDL | Very low–density lipoprotein | WBC | Leukocytes, white blood cells |
| VMA | 4-Hydroxy-3-methoxymandelic acid (vanillyl-mandelic acid) | XP | Xeroderma pigmentosum |

APPENDIXES

Appendix A. Recommended daily dietary allowances designed for maintenance of

	AGE (years)	WEIGHT kg	WEIGHT lbs	HEIGHT cm	HEIGHT in	ENERGY (kcal)[b]	PROTEIN (g)	FAT-SOLUBLE VITAMINS Vitamin A activity RE[c]	Vitamin A activity IU	Vitamin D (IU)	Vitamin E activity[e] (IU)
Infants	0.0-0.5	6	14	60	24	kg × 117	kg × 2.2	420[d]	1,400	400	4
	0.5-1.0	9	20	71	28	kg × 108	kg × 2.0	400	2,000	400	5
Children	1-3	13	28	86	34	1,300	23	400	2,000	400	7
	4-6	20	44	110	44	1,800	30	500	2,500	400	9
	7-10	30	66	135	54	2,400	36	700	3,300	400	10
Males	11-14	44	97	158	63	2,800	44	1,000	5,000	400	12
	15-18	61	134	172	69	3,000	54	1,000	5,000	400	15
	19-22	67	147	172	69	3,000	54	1,000	5,000	400	15
	23-50	70	154	172	69	2,700	56	1,000	5,000		15
	51+	70	154	172	69	2,400	56	1,000	5,000		15
Females	11-14	44	97	155	62	2,400	44	800	4,000	400	12
	15-18	54	119	162	65	2,100	48	800	4,000	400	12
	19-22	58	128	162	65	2,100	46	800	4,000	400	12
	23-50	58	128	162	65	2,000	46	800	4,000		12
	51+	58	128	162	65	1,800	46	800	4,000		12
Pregnant						+300	+30	1,000	5,000	400	15
Lactating						+500	+20	1,200	6,000	400	15

From Recommended dietary allowances, ed. 8, Publ. No. 1694, Washington, D.C., 1974, Food and Nutrition Board,
[a]The allowances are intended to provide for individual variations among most normal persons as they live in the
to provide other nutrients for which human requirements have been less well defined.
[b]Kilojoules (kJ) = 4.2 × kcal.
[c]Retinol equivalents.
[d]Assumed to be all as retinol in milk during the first six months of life. All subsequent intakes are assumed to be
[e]Total vitamin E activity, estimated to be 80 percent as α-tocopherol and 20 percent other tocopherols.
[f]The folacin allowances refer to dietary sources as determined by *Lactobacillus casei* assay. Pure forms of folacin
[g]Although allowances are expressed as niacin, it is recognized that on the average 1 mg of niacin is derived from
[h]This increased requirement cannot be met by ordinary diets; therefore, the use of supplemental iron is

good nutrition of practically all healthy people in U.S.A.

WATER-SOLUBLE VITAMINS							MINERALS					
Ascorbic acid (mg)	Folacin[f] (µg)	Niacin[g] (mg)	Riboflavin (mg)	Thiamin (mg)	Vitamin B$_6$ (mg)	Vitamin B$_{12}$ (µg)	Calcium (mg)	Phosphorus (mg)	Iodine (µg)	Iron (mg)	Magnesium (mg)	Zinc (mg)
35	50	5	0.4	0.3	0.3	0.3	360	240	35	10	60	3
35	50	8	0.6	0.5	0.4	0.3	540	400	45	15	70	5
40	100	9	0.8	0.7	0.6	1.0	800	800	60	15	150	10
40	200	12	1.1	0.9	0.9	1.5	800	800	80	10	200	10
40	300	16	1.2	1.2	1.2	2.0	800	800	110	10	250	10
45	400	18	1.5	1.4	1.6	3.0	1,200	1,200	130	18	350	15
45	400	20	1.8	1.5	2.0	3.0	1,200	1,200	150	18	400	15
45	400	20	1.8	1.5	2.0	3.0	800	800	140	10	350	15
45	400	18	1.6	1.4	2.0	3.0	800	800	130	10	350	15
45	400	16	1.5	1.2	2.0	3.0	800	800	110	10	350	15
45	400	16	1.3	1.2	1.6	3.0	1,200	1,200	115	18	300	15
45	400	14	1.4	1.1	2.0	3.0	1,200	1,200	115	18	300	15
45	400	14	1.4	1.1	2.0	3.0	800	800	100	18	300	15
45	400	13	1.2	1.0	2.0	3.0	800	800	100	18	300	15
45	400	12	1.1	1.0	2.0	3.0	800	800	80	10	300	15
60	800	+2	+0.3	+0.3	2.5	4.0	1,200	1,200	125	18+[h]	450	20
80	600	+4	+0.5	+0.3	2.5	4.0	1,200	1,200	150	18	450	25

National Academy of Sciences-National Research Council.
United States under usual environmental stresses. Diets should be based on a variety of common foods in order

half as retinol and half as β-carotene when calculated from international units.

may be effective in doses less than one fourth of the recommended dietary allowance.
each 60 mg of dietary tryptophan.
recommended.

Appendix B.   International classification of enzymes

MAIN CLASS AND SUBCLASS	PROSTHETIC GROUP OR COENZYME	EXAMPLE
1 Oxidoreductases		
1.1 Acting on $=$CHOH donors		
1.1.1 With NAD or NADP as acceptors	NAD, NADP	Lactate or malate dehydrogenases
1.1.3 With O_2 as acceptor	FAD	Glucose oxidase
1.2 Acting on $=$C$=$O donors		
1.2.1 With NAD or NADP as acceptors	NAD, NADP	Glyceraldehyde 3-phosphate dehydrogenase
1.2.3 With O_2 as acceptor	FAD	Xanthine oxidase
1.3 Acting on $=$CH$-$CH$=$ donors		
1.3.1 With NAD or NADP as acceptors	NAD, NADP	Dihydrouracil dehydrogenase
1.3.2 With FAD as acceptor	FAD	Acyl CoA dehydrogenase
1.4 Acting on $=$CH$-$NH$_2$ donors		
1.4.3 With O_2 as acceptor	FAD, pyridoxal phosphate	Amino acid oxidases
2 Transferases		
2.1 Transferring 1$-$C groups		
2.1.1 Methyltransferases	Tetrahydrofolate (THF)	Guanidomethyl transferase
2.1.2 Hydroxymethyl and formyl transferases	THF	Serine hydroxymethyl transferase
2.1.3 Carboxyl or carbamoyl transferases		Ornithine trans-carbamylase
2.3 Acyl transferases		Choline acetyl transferase, palmitoyl CoA-carnitine transferase
2.4 Glycosyl transferases	UDP	Galactosyl transferase
2.6 Transferring $-$NH$_2$ groups		
2.6.1 Aminotransferases	Pyridoxal phosphate	Transaminases
2.7 Transferring phosphorus-containing groups		
2.7.1.2 ATP-glucose 6-phospho-transferase		Glucokinase
3 Hydrolases		
3.1 Cleaving ester linkages		
3.1.1 Carboxylic ester hydrolases		Lipases
3.1.3 Phosphomonoester hydrolases		Phosphatases
3.1.4 Phosphodiester hydrolases		Phosphodiesterase
3.2 Cleaving glycosides		
3.2.1 Glycosidases		Amylases
3.2.2 *N*-Glycosidases		Nucleosidases
3.4 Cleaving peptide bonds		
3.4.1 α-Aminopeptide amino acid hydrolases		Leucine amino-peptidase
3.4.2 α-Carboxypeptide amino acid hydrolases		Carboxypeptidases
3.4.4 Peptidopeptide hydrolases		Pepsin, trypsin, etc.
4 Lyases		
4.1 $-$C$-$C$-$ lyases		

Appendix B. International classification of enzymes — cont'd

MAIN CLASS AND SUBCLASS	PROSTHETIC GROUP OR COENZYME	EXAMPLE
4.1.1 Carboxy lyases	Thiamine pyro-phosphate (TPP), NAD, CoASH, lipoic acid	Pyruvate decarboxylase, α-ketoglutarate dehydrogenase complex
4.1.2 Aldehyde lyases		Aldolase
4.2 —C—O— lyases		
4.2.1 Hydrolyases		Fumarase
5 Isomerases		
5.1 Racemases and epimerases		
5.1.3 Acting on carbohydrates		Ribulose 5-phosphate epimerase
5.2 *Cis-trans* isomerases		Maleylacetoacetate isomerase
5.3 Intramolecular oxidoreductases		
5.3.1 Interconverting aldoses and ketoses		Glucose phosphate isomerase
5.4 Intramolecular transferases	Vitamin B_{12}, cobamide	Methylmalonyl-CoA mutase
6 Ligases		
6.1 Forming C—O bonds		
6.1.1 Amino acid–RNA ligases		Activating enzymes
6.2 Forming C—S bonds		
6.2.1 Acyl CoA ligases	CoASH	Thiokinases
6.3 Forming C—N bonds		
6.3.4.1 Xanthosine 5'-phosphate: ammonia ligase		GMP synthetase
6.3.4.2 UTP:ammonia ligase (ADP)		CTP synthetase
6.3.4.3 Formate:tetrahydrofolate ligase (ADP)		Formyltetrahydro-folate synthetase
6.4 Forming C—C bonds		
6.4.1.1 Pyruvate:CO_2 ligase (ADP)	Biotin	Pyruvate carboxylase
6.4.1.2 Acetyl CoA:CO_2 ligase (ADP)	Biotin	Acetyl CoA carboxylase
6.4.1.3 Propionyl CoA:CO_2 ligase (ADP)	Biotin	Propionyl CoA carboxylase

Appendix C. Four-place logarithms

N	0	1	2	3	4	5	6	7	8	9	PROPORTIONAL PARTS								
											1	2	3	4	5	6	7	8	9
10	0000	0043	0086	0128	0170	0212	0253	0294	0334	0374	*4	8	12	17	21	25	29	33	37
11	0414	0453	0492	0531	0569	0607	0645	0682	0719	0755	4	8	11	15	19	23	26	30	34
12	0792	0828	0864	0899	0934	0969	1004	1038	1072	1106	3	7	10	14	17	21	24	28	31
13	1139	1173	1206	1239	1271	1303	1335	1367	1399	1430	3	6	10	13	16	19	23	26	29
14	1461	1492	1523	1553	1584	1614	1644	1673	1703	1732	3	6	9	12	15	18	21	24	27
15	1761	1790	1818	1847	1875	1903	1931	1959	1987	2014	*3	6	8	11	14	17	20	22	25
16	2041	2068	2095	2122	2148	2175	2201	2227	2253	2279	3	5	8	11	13	16	18	21	24
17	2304	2330	2355	2380	2405	2430	2455	2480	2504	2529	2	5	7	10	12	15	17	20	22
18	2553	2577	2601	2625	2648	2672	2695	2718	2742	2765	2	5	7	9	12	14	16	19	21
19	2788	2810	2833	2856	2878	2900	2923	2945	2967	2989	2	4	7	9	11	13	16	18	20
20	3010	3032	3054	3075	3096	3118	3139	3160	3181	3201	2	4	6	8	11	13	15	17	19
21	3222	3243	3263	3284	3304	3324	3345	3365	3385	3404	2	4	6	8	10	12	14	16	18
22	3424	3444	3464	3483	3502	3522	3541	3560	3579	3598	2	4	6	8	10	12	14	15	17
23	3617	3636	3655	3674	3692	3711	3729	3747	3766	3784	2	4	6	7	9	11	13	15	17
24	3802	3820	3838	3856	3874	3892	3909	3927	3945	3962	2	4	5	7	9	11	12	14	16
25	3979	3997	4014	4031	4048	4065	4082	4099	4116	4133	2	3	5	7	9	10	12	14	15
26	4150	4166	4183	4200	4216	4232	4249	4265	4281	4298	2	3	5	7	8	10	11	13	15
27	4314	4330	4346	4362	4378	4393	4409	4425	4440	4456	2	3	5	6	8	9	11	13	14
28	4472	4487	4502	4518	4533	4548	4564	4579	4594	4609	2	3	5	6	8	9	11	12	14
29	4624	4639	4654	4669	4683	4698	4713	4728	4742	4757	1	3	4	6	7	9	10	12	13
30	4771	4786	4800	4814	4829	4843	4857	4871	4886	4900	1	3	4	6	7	9	10	11	13
31	4914	4928	4942	4955	4969	4983	4997	5011	5024	5038	1	3	4	6	7	8	10	11	12
32	5051	5065	5079	5092	5105	5119	5132	5145	5159	5172	1	3	4	5	7	8	9	11	12
33	5185	5198	5211	5224	5237	5250	5263	5276	5289	5302	1	3	4	5	6	8	9	10	12
34	5315	5328	5340	5353	5366	5378	5391	5403	5416	5428	1	3	4	5	6	8	9	10	11
35	5441	5453	5465	5478	5490	5502	5514	5527	5539	5551	1	2	4	5	6	7	9	10	11
36	5563	5575	5587	5599	5611	5623	5635	5647	5658	5670	1	2	4	5	6	7	8	10	11
37	5682	5694	5705	5717	5729	5740	5752	5763	5775	5786	1	2	3	5	6	7	8	9	10
38	5798	5809	5821	5832	5843	5855	5866	5877	5888	5899	1	2	3	5	6	7	8	9	10
39	5911	5922	5933	5944	5955	5966	5977	5988	5999	6010	1	2	3	4	5	7	8	9	10
40	6021	6031	6042	6053	6064	6075	6085	6096	6107	6117	1	2	3	4	5	6	8	9	10
41	6128	6138	6149	6160	6170	6180	6191	6201	6212	6222	1	2	3	4	5	6	7	8	9
42	6232	6243	6253	6263	6274	6284	6294	6304	6314	6325	1	2	3	4	5	6	7	8	9
43	6335	6345	6355	6365	6375	6385	6395	6405	6415	6425	1	2	3	4	5	6	7	8	9
44	6435	6444	6454	6464	6474	6484	6493	6503	6513	6522	1	2	3	4	5	6	7	8	9
45	6532	6542	6551	6561	6571	6580	6590	6599	6609	6618	1	2	3	4	5	6	7	8	9
46	6628	6637	6646	6656	6665	6675	6684	6693	6702	6712	1	2	3	4	5	6	7	7	8
47	6721	6730	6739	6749	6758	6767	6776	6785	6794	6803	1	2	3	4	5	5	6	7	8
48	6812	6821	6830	6839	6848	6857	6866	6875	6884	6893	1	2	3	4	4	5	6	7	8
49	6902	6911	6920	6928	6937	6946	6955	6964	6972	6981	1	2	3	4	4	5	6	7	8
50	6990	6998	7007	7016	7024	7033	7042	7050	7059	7067	1	2	3	3	4	5	6	7	8
51	7076	7084	7093	7101	7110	7118	7126	7135	7143	7152	1	2	3	3	4	5	6	7	8
52	7160	7168	7177	7185	7193	7202	7210	7218	7226	7235	1	2	2	3	4	5	6	7	7
53	7243	7251	7259	7267	7275	7284	7292	7300	7308	7316	1	2	2	3	4	5	6	6	7
54	7324	7332	7340	7348	7356	7364	7372	7380	7388	7396	1	2	2	3	4	5	6	6	7
N	0	1	2	3	4	5	6	7	8	9	1	2	3	4	5	6	7	8	9

*Interpolation in this section of the table is inaccurate.

N	0	1	2	3	4	5	6	7	8	9	PROPORTIONAL PARTS								
											1	2	3	4	5	6	7	8	9
55	7404	7412	7419	7427	7435	7443	7451	7459	7466	7474	1	2	2	3	4	5	5	6	7
56	7482	7490	7497	7505	7513	7520	7528	7536	7543	7551	1	2	2	3	4	5	5	6	7
57	7559	7566	7574	7582	7589	7597	7604	7612	7619	7627	1	2	2	3	4	5	5	6	7
58	7634	7642	7649	7657	7664	7672	7679	7686	7694	7701	1	1	2	3	4	4	5	6	7
59	7709	7716	7723	7731	7738	7745	7752	7760	7767	7774	1	1	2	3	4	4	5	6	7
60	7782	7789	7796	7803	7810	7818	7825	7832	7839	7846	1	1	2	3	4	4	5	6	6
61	7853	7860	7868	7875	7882	7889	7896	7903	7910	7917	1	1	2	3	4	4	5	6	6
62	7924	7931	7938	7945	7952	7959	7966	7973	7980	7987	1	1	2	3	3	4	5	6	6
63	7993	8000	8007	8014	8021	8028	8035	8041	8048	8055	1	1	2	3	3	4	5	5	6
64	8062	8069	8075	8082	8089	8096	8102	8109	8116	8122	1	1	2	3	3	4	5	5	6
65	8129	8136	8142	8149	8156	8162	8169	8176	8182	8189	1	1	2	3	3	4	5	5	6
66	8195	8202	8209	8215	8222	8228	8235	8241	8248	8254	1	1	2	3	3	4	5	5	6
67	8261	8267	8274	8280	8287	8293	8299	8306	8312	8319	1	1	2	3	3	4	5	5	6
68	8325	8331	8338	8344	8351	8357	8363	8370	8376	8382	1	1	2	3	3	4	4	5	6
69	8388	8395	8401	8407	8414	8420	8426	8432	8439	8445	1	1	2	2	3	4	4	5	6
70	8451	8457	8463	8470	8476	8482	8488	8494	8500	8506	1	1	2	2	3	4	4	5	6
71	8513	8519	8525	8531	8537	8543	8549	8555	8561	8567	1	1	2	2	3	4	4	5	5
72	8573	8579	8585	8591	8597	8603	8609	8615	8621	8627	1	1	2	2	3	4	4	5	5
73	8633	8639	8645	8651	8657	8663	8669	8675	8681	8686	1	1	2	2	3	4	4	5	5
74	8692	8698	8704	8710	8716	8722	8727	8733	8739	8745	1	1	2	2	3	4	4	5	5
75	8751	8756	8762	8768	8774	8779	8785	8791	8797	8802	1	1	2	2	3	3	4	5	5
76	8808	8814	8820	8825	8831	8837	8842	8848	8854	8859	1	1	2	2	3	3	4	5	5
77	8865	8871	8876	8882	8887	8893	8899	8904	8910	8915	1	1	2	2	3	3	4	4	5
78	8921	8927	8932	8938	8943	8949	8954	8960	8965	8971	1	1	2	2	3	3	4	4	5
79	8976	8982	8987	8993	8998	9004	9009	9015	9020	9025	1	1	2	2	3	3	4	4	5
80	9031	9036	9042	9047	9053	9058	9063	9069	9074	9079	1	1	2	2	3	3	4	4	5
81	9085	9090	9096	9101	9106	9112	9117	9122	9128	9133	1	1	2	2	3	3	4	4	5
82	9138	9143	9149	9154	9159	9165	9170	9175	9180	9186	1	1	2	2	3	3	4	4	5
83	9191	9196	9201	9206	9212	9217	9222	9227	9232	9238	1	1	2	2	3	3	4	4	5
84	9243	9248	9253	9258	9263	9269	9274	9279	9284	9289	1	1	2	2	3	3	4	4	5
85	9294	9299	9304	9309	9315	9320	9325	9330	9335	9340	1	1	2	2	3	3	4	4	5
86	9345	9350	9355	9360	9365	9370	9375	9380	9385	9390	1	1	2	2	3	3	4	4	5
87	9395	9400	9405	9410	9415	9420	9425	9430	9435	9440	0	1	1	2	2	3	3	4	4
88	9445	9450	9455	9460	9465	9469	9474	9479	9484	9489	0	1	1	2	2	3	3	4	4
89	9494	9499	9504	9509	9513	9518	9523	9528	9533	9538	0	1	1	2	2	3	3	4	4
90	9542	9547	9552	9557	9562	9566	9571	9576	9581	9586	0	1	1	2	2	3	3	4	4
91	9590	9595	9600	9605	9609	9614	9619	9624	9628	9633	0	1	1	2	2	3	3	4	4
92	9638	9643	9647	9652	9657	9661	9666	9671	9675	9680	0	1	1	2	2	3	3	4	4
93	9685	9689	9694	9699	9703	9708	9713	9717	9722	9727	0	1	1	2	2	3	3	4	4
94	9731	9736	9741	9745	9750	9754	9759	9763	9768	9773	0	1	1	2	2	3	3	4	4
95	9777	9782	9786	9791	9795	9800	9805	9809	9814	9818	0	1	1	2	2	3	3	4	4
96	9823	9827	9832	9836	9841	9845	9850	9854	9859	9863	0	1	1	2	2	3	3	4	4
97	9868	9872	9877	9881	9886	9890	9894	9899	9903	9908	0	1	1	2	2	3	3	4	4
98	9912	9917	9921	9926	9930	9934	9939	9943	9948	9952	0	1	1	2	2	3	3	4	4
99	9956	9961	9965	9969	9974	9978	9983	9987	9991	9996	0	1	1	2	2	3	3	3	4
N	0	1	2	3	4	5	6	7	8	9	1	2	3	4	5	6	7	8	9

Appendix D.   Laboratory values

What is a normal laboratory value?

The normal laboratory value for a substance is that concentration that can be measured in tissue or body fluids from apparently healthy human beings. A "healthy human being" is not so easily defined. In the first place, if a series of analyses is done on a large number of individuals, the values obtained will cover some range of numbers. Further scrutiny will often reveal that the set can be divided into subsets with respect to age and sex. Sometimes there are additional influences that affect the distribution of values, for example, racial extraction, previous dietary history, patterns of physical activity, etc. Less frequently it is possible to show rhythmic changes produced by the menstrual cycle (certain blood electrolytes and hormones), by diurnal variations, and even by seasonal changes. It must also be accepted that humans exhibit a degree of individual variability in their biochemical makeup, just as they do in such obvious characteristics as weight, height, and personality. While many of these factors are poorly understood, they are nevertheless real.

For these and other reasons, it is customary to present normal ranges for body constituents, with the range chosen according to simple statistical principles. These are usually applied so that the cited range of values will include 95% of those values that can be predicted to occur in the entire population. It is sometimes necessary to establish *several* ranges. The concentration of hemoglobin in the blood, for example, depends on both age and sex. The concentration of creatine phosphokinase, a blood serum enzyme, appears to be affected significantly by sex but not by age. Some believe that the dependence could more properly be based on muscle mass, but the evidence is not conclusive. Still other constituents such as blood glucose seem to be independent of age and sex. Of the normal values cited on the inside of the front cover, those most obviously affected by age or sex have been indicated; where no such indication appears it may be assumed that the values apply generally.

What is an abnormal laboratory value?

An abnormal value may be strictly defined as any measurement of a body constituent that falls outside the normal range and in cases in which a disease has been diagnosed by other means. While correct, this is a very stringent definition. A more useful criterion of abnormality is that if any value falls more than two standard deviations from the mean of the normal range, it may be abnormal. Alternatively, such a value may be the result of a laboratory error.

By how much must a test result depart from the "normal" before it may be considered "abnormal"? From what has been said it should be clear that this question has no unique answer. It depends on the test in question and, to some extent, on the patient in question. Remember always that tests of some 5% of all perfectly normal individuals would not give "normal" results. This does not mean they are sick, only that they are different.

Compilations of laboratory values. A number of sources include compilations of normal and abnormal laboratory test values as well as information on the interpretation of diagnostic test results. Some of these sources are included in the following references.

REFERENCES

Cantarow, A., and Trumper, M.: Clinical biochemistry, ed. 7 (Latner, A., editor), Philadelphia, 1975, W. B. Saunders Co. *A good general source.*

Faulkner, W. R., King, J. W., and Damm, H., editors: Handbook of clinical laboratory data, ed. 2, Cleveland, 1968, The Chemical Rubber Co. *A larger, more expensive desk reference.*

Henry, R. J., et al.: Clinical chemistry, New York, 1974, Harper and Row, Inc. *An extensive and very comprehensive source, including data on methodology.*

Searcy, R. L.: Diagnostic biochemistry, New York, 1969, The McGraw-Hill Book Co. *Also extensive; it is very comprehensive in its coverage of clinical chemistry.*

Thompson, R. H. S., and Wootten, I. D. P.: Biochemical disorders in human disease, ed. 3, New York, 1970, Academic Press, Inc. *An elaborate review of biochemical aspects of disease; primarily for library reference.*

Wallach, J.: Interpretations of diagnostic tests, Boston, 1970, Little, Brown & Co. *A useful, paperback pocket-sized collection.*

INDEX*

---

*Boldface page numbers indicate
diagrams of chemical structure;
C following a page number indicates discussion in a clinical case.